T0181504

Lecture Notes in Artificial Intelligence 10024

Subseries of Lecture Notes in Computer Science

More information about this series at http://www.springer.com/series/1244

Eva Blomqvist · Paolo Ciancarini
Francesco Poggi · Fabio Vitali (Eds.)

Knowledge Engineering and Knowledge Management

20th International Conference, EKAW 2016 Bologna, Italy, November 19–23, 2016 Proceedings

Editors
Eva Blomqvist
Linköping University
Linköping
Sweden

Paolo Ciancarini
University of Bologna
Bologna
Italy

Francesco Poggi
University of Bologna
Bologna
Italy

Fabio Vitali
University of Bologna
Bologna
Italy

ISSN 0302-9743 ISSN 1611-3349 (electronic)
Lecture Notes in Artificial Intelligence
ISBN 978-3-319-49003-8 ISBN 978-3-319-49004-5 (eBook)
DOI 10.1007/978-3-319-49004-5

Library of Congress Control Number: 2016955511

LNCS Sublibrary: SL7 – Artificial Intelligence

Printed on acid-free paper

This Springer imprint is published by Springer Nature
The registered company is Springer International Publishing AG
The registered company address is: Gewerbestrasse 11, 6330 Cham, Switzerland

Preface

This volume contains the proceedings of the 20th International Conference on Knowledge Engineering and Knowledge Management (EKAW 2016), held in Bologna, Italy, November 19–23, 2016.

This edition of the conference was specifically concerned with the impact of space and time on knowledge representation, and what we chose to call "evolving knowledge." Knowledge engineering has classically been about creating static, universal representations. Yet the world is rarely static: Everything changes, including the models, and real-world systems need to evolve along with the surrounding world. Also, what makes some representations valid in some contexts may make them invalid elsewhere (e.g., jurisdiction for laws).

This special focus concerns all aspects of the management and acquisition of knowledge representations of evolving, contextual, and local models. This includes change management, trend detection, model evolution, streaming data and stream reasoning, event processing, time- and space-dependent models, contextual and local knowledge representations, etc. We also wanted to put a special emphasis on the evolvability and localization of knowledge and the correct usage of these limits.

In addition to this specific focus, EKAW as usual covered all aspects of eliciting, acquiring, modeling, and managing knowledge, the construction of knowledge-intensive systems and services for the Semantic Web, knowledge management, e-business, natural language processing, intelligent information integration, personal digital assistance systems, and a variety of other related topics.

For the main conference we invited submissions for research papers that present novel methods, techniques, or analysis with appropriate empirical or other types of evaluation, as well as in-use papers describing novel applications of knowledge management and engineering in real environments and experience reports. We also invited submissions of position papers describing novel and innovative ideas, or problem analyses, that are still in an early stage but may guide future research in the area.

In addition to the regular conference submission resulting in a Springer conference proceedings paper in this book, the authors of the best EKAW papers were invited to submit an extended version of the paper to a *Semantic Web Journal* (IOS Press) special issue to be published in 2017. The extended papers will go through a new review process and it should be noted that the journal follows an open review process, providing for a very transparent evaluation of the submissions.

Overall, we received 226 abstract submissions of which 171 were in the end accompanied by a valid paper submission and included in the review process. The reviewing was performed by a Program Committee of 127 researchers in the field and the two Program Chairs. Each paper received at least three reviews, and we specifically thank the reviewers for engaging in lively discussions, especially when there were conflicting opinions on papers. In total, 51 submissions were accepted by the Program

Committee (30 % overall acceptance rate), out of which four are in-use papers and one is a position paper. All papers are present in this volume as full-length papers. However, in order to fit this high number of papers into the single-session model of the EKAW conference, we had to select a number of papers for shorter presentations, which means that the conference program included 30 long presentations, and 21 shorter presentations.

To complement the program, we invited three distinguished keynote speakers:

- Luc Steels (Institut de Biologia Evolutiva, Barcelona, Spain) presented a talk entitled "How Much Are Our Representations of Knowledge Influenced by Our Languages?"
- Chris Welty (Sr. Research Scientist, Google, USA) gave a talk entitled "Towards an Embedded Theory of Truth"
- Francesca Rossi (IBM Research and University of Padova, Italy) gave a talk titled "From Data to Knowledge: Trust and Ethics in Symbiotic AI/Human Systems"

The program chairs of EKAW 2016 were Fabio Vitali from the University of Bologna, Italy, and Eva Blomqvist from Linköping University, Sweden. The EKAW 2016 program also included a Doctoral Consortium that gave PhD students an opportunity to present their research ideas and results in a stimulating environment, to get feedback from mentors who are experienced research scientists in the community, to explore issues related to academic and research careers, and to build relationships with other PhD students from around the world. The Doctoral Consortium was intended for students at each stage of their PhD. All accepted presenters had an opportunity to present their work to an international audience, to be paired with a mentor, and to discuss their work with experienced scientists from the research community. The Doctoral Consortium was organized by Mathieu d'Aquin from the Open University, UK, and Valentina Presutti from ISTC-CNR in Italy.

In addition to the main research tracks, EKAW 2016 hosted four satellite workshops and two tutorials.

Workshops:

- OWLED - ORE 2016 — the 13th OWL: Experiences and Directions Workshop and 5th OWL Reasoner Evaluation Workshop
- EKM — the Second International workshop on Educational Knowledge Management
- Drift-a-LOD — the First workshop on Detection, Representation and Management of Concept Drift in Linked Open Data
- LK&SW-2016 — the Third Workshop on Legal Knowledge and the Semantic Web

Tutorials:

- Mapping Management and Expressive Ontologies in Ontology-Based Data Access, by Diego Calvanese, Benjamin Cogrel, and Guohui Xiao
- Modeling, Generating, and Publishing Knowledge as Linked Data, by Anastasia Dimou, Pieter Heyvaert, and Ruben Verborgh

The workshop and tutorial program was chaired by Matthew Horridge, Stanford University, USA, as well as Jun Zhao, University of Lancaster, UK.

Finally, EKAW 2016 also featured a demo and poster session. We encouraged contributions that were likely to stimulate critical or controversial discussions about any of the areas of the EKAW conference series. We also invited developers to showcase their systems and the benefit they can bring to a particular application. The demo and poster program of EKAW 2016 was chaired by Tudor Groza from the Garvan Institute of Medical Research, Australia, and Mari Carmen Suárez-Figueroa of the Universidad Politécnica de Madrid, Spain.

The conference organization also included Silvio Peroni, University of Bologna, Italy, as the sponsorship chair, Paolo Ciancarini, Angelo Di Iorio, and Silvio Peroni all from the University of Bologna, Italy, took care of local arrangements, Andrea Giovanni Nuzzolese, ISTC-CNR, Italy, acted as Web presence chair, and Francesco Poggi, University of Bologna, Italy, acted as proceedings chair. Paolo Ciancarini, University of Bologna, Italy, was the general chair of EKAW 2016.

Thanks to everybody, including attendees at the conference, for making EKAW 2016 a successful event.

November 2016

Eva Blomqvist
Paolo Ciancarini
Francesco Poggi
Fabio Vitali

Organization

Steering Committee

Frank Van Harmelen	VU University Amsterdam, The Netherlands
Mark Musen	Stanford University, USA
Eero Hyvönen	Aalto University, Finland
Jerome Euzenat	Inria and University of Grenoble-Alpes, France
Patrick Lambrix	Linköping University, Sweden
Sofia Pinto	Instituto Superior Tecnico
Riichiro Mizoguchi	Japan Advanced Institute of Science and Technology, Japan
Asunción Gómez Pérez	Universidad Politécnica de Madrid, Spain
Johanna Vöelker	University of Mannheim, Germany
Paul Compton	The University of New South Wales, Australia
Nathalie Aussenac-Gilles	IRIT CNRS
Krzysztof Janowicz	University of California, Santa Barbara, USA
Siegfried Handschuh	University of Passau, Germany
Philipp Cimiano	Bielefeld University, Germany
Guus Schreiber	VU University Amsterdam, The Netherlands
Heiner Stuckenschmidt	University of Mannheim, Germany
Enrico Motta	Knowledge Media Institute, The Open University
Steffen Staab	University of Koblenz-Landau, Germany
Stefan Schlobach	VU University Amsterdam, The Netherlands
Aldo Gangemi	Université Paris 13 and CNR-ISTC, France
Vojtěch Svátek	University of Economics, Prague, Czech Republic

Program Committee

Alessandro Adamou	Knowledge Media Institute, The Open University
Benjamin Adams	The University of Auckland, New Zealand
Mehwish Alam	Université Paris 13, France
Luigi Asprino	University of Bologna and STLab (ISTC-CNR), Italy
Sören Auer	University of Bonn and Fraunhofer IAIS, Germany
Nathalie Aussenac-Gilles	IRIT CNRS
Wouter Beek	VU University Amsterdam, The Netherlands
Eva Blomqvist	Linköping University, Sweden
Christopher Brewster	Aston Business School, Aston University, UK
Amparo E. Cano	Knowledge Media Institute, The Open University
Luca Cervone	CIRSFID - University of Bologna
Vinay Chaudhri	SRI International

Michelle Cheatham	Wright State University, USA
Paolo Ciancarini	University of Bologna, Italy
Philipp Cimiano	Bielefeld University, Germany
Paul Compton	The University of New South Wales, Australia
Olivier Corby	Inria, France
Oscar Corcho	Universidad Politécnica de Madrid, Spain
Claudia D'Amato	University of Bari, Italy
Mathieu d'Aquin	Knowledge Media Institute, The Open University
Aba-Sah Dadzie	Knowledge Media Institute, The Open University
Enrico Daga	The Open University
Victor de Boer	VU University Amsterdam, The Netherlands
Daniele Dell'Aglio	IFI, University of Zurich, Switzerland
Emanuele Della Valle	DEIB, Politecnico di Milano, Italy
Angelo Di Iorio	University of Bologna, Italy
Rim Djedidi	LIMICS Laboratory - Paris Nord 13 University, France
Zlatan Dragisic	Linköping University, Sweden
Henrik Eriksson	Linköping University, Sweden
Dieter Fensel	University of Innsbruck, Austria
Jesualdo Tomás Fernández-Breis	Universidad de Murcia, Spain
Antske Fokkens	VU University Amsterdam, The Netherlands
Aldo Gangemi	Université Paris 13 and CNR-ISTC, France
Serge Garlatti	Telecom Bretagne, France
Anna Lisa Gentile	University of Mannheim, Germany
Chiara Ghidini	FBK-irst
Luca Gilardoni	KIE
Jose Manuel Gomez-Perez	Expert System
Guido Governatori	Data61
Jon Atle Gulla	NTNU
Christophe Guéret	BBC
Peter Haase	metaphacts
Harry Halpin	World Wide Web Consortium
Karl Hammar	Jönköping University, Sweden
Tom Heath	Open Data Institute
Martin Hepp	Universität der Bundeswehr München, Germany
Pascal Hitzler	Wright State University
Rinke Hoekstra	University of Amsterdam/VU University Amsterdam, The Netherlands
Aidan Hogan	DCC, Universidad de Chile
Matthew Horridge	Stanford University, USA
Andreas Hotho	University of Würzburg, Germany
Zhisheng Huang	VU University Amsterdam, The Netherlands
Eero Hyvönen	Aalto University, Finland
Antoine Isaac	Europeana and VU University Amsterdam, The Netherlands
Valentina Ivanova	Linköping University, Sweden

Krzysztof Janowicz	University of California, Santa Barbara, USA
C. Maria Keet	University of Cape Town, South Africa
Robin Keskisärkkä	Linköping University, Sweden
Carsten Keßler	Hunter College, City University of New York, USA
Sabrina Kirrane	Vienna University of Economics and Business, Austria
Adila A. Krisnadhi	Wright State University, USA and Universitas Indonesia
Patrick Lambrix	Linköping University, Sweden
Agnieszka Lawrynowicz	Poznan University of Technology, Poland
Steffen Lohmann	Fraunhofer IAIS, Germany
Wolfgang Maass	Saarland University, Germany
Suvodeep Mazumdar	University of Sheffield, UK
Albert Meroño-Peñuela	VU University Amsterdam, The Netherlands
Riichiro Mizoguchi	Japan Advanced Institute of Science and Technology
Dunja Mladenic	Jozef Stefan Institute, Slovenia
Andrea Moro	Sapienza, Università di Roma, Italy
Enrico Motta	Knowledge Media Institute, The Open University
Tom Narock	Marymount University, USA
Axel-Cyrille Ngonga Ngomo	University of Leipzig, Germany
Vit Novacek	DERI, National University of Ireland, Galway, Ireland
Andrea Giovanni Nuzzolese	STLab, ISTC-CNR
Francesco Osborne	KMi, the Open University
Monica Palmirani	CIRSFID
Matteo Palmonari	University of Milano-Bicocca, Italy
Sujan Perera	knoesis
Silvio Peroni	University of Bologna, Italy
Wim Peters	University of Sheffield, UK
Mohammad Taher Pilevar	University of Cambridge, UK
Sofia Pinto	Instituto Superior Tecnico
Francesco Poggi	Università di Bologna, Italy
Axel Polleres	Vienna University of Economics and Business, Austria
Valentina Presutti	STLab (ISTC-CNR)
Yannick Prié	LINA - University of Nantes, France
Guilin Qi	Southeast University
Diego Reforgiato	Consiglio Nazionale delle Ricerche, Italy
Ulrich Reimer	University of Applied Sciences St. Gallen, Switzerland
Chantal Reynaud	LRI, University of Paris-Sud, CNRS, Université Paris-Saclay, France
Francesco Ronzano	TALN Group, Universitat Pompeu Fabra, Barcelona
Marco Rospocher	Fondazione Bruno Kessler
Sebastian Rudolph	Technische Universität Dresden
Marta Sabou	Vienna University of Technology, Austria
Harald Sack	Hasso-Plattner-Institute for IT Systems Engineering, University of Potsdam, Germany
Francois Scharffe	3Top

Simon Scheider	ETH Zürich, Institut für Kartographie und Geoinformation, Switzerland
Stefan Schlobach	VU University Amsterdam, The Netherlands
Jodi Schneider	University of Pittsburgh, USA
Thomas Schneider	University of Bremen, Germany
Luciano Serafini	Fondazione Bruno Kessler, Italy
Elena Simperl	University of Southampton, UK
Derek Sleeman	University of Aberdeen, UK
Pavel Smrz	Brno University of Technology, Czech Republic
Monika Solanki	University of Oxford, UK
Steffen Staab	University of Koblenz-Landau, Germany
Heiner Stuckenschmidt	University of Mannheim, Germany
Rudi Studer	Karlsruher Institut für Technologie (KIT), Germany
Mari Carmen Suárez-Figueroa	Universidad Politécnica de Madrid, Spain
Vojtěch Svátek	University of Economics, Prague, Czech Republic
Valentina Tamma	University of Liverpool, UK
Vladimir Tarasov	Jönkoping University, Sweden
Annette Ten Teije	VU University Amsterdam, The Netherlands
Ilaria Tiddi	Knowledge Media Institute - The Open University
Ioan Toma	STI Innsbruck, Austria
Tania Tudorache	Stanford University, USA
Marieke Van Erp	VU University Amsterdam, The Netherlands
Frank Van Harmelen	VU University Amsterdam, The Netherlands
Charles Vardeman Ii	University of Notre Dame, USA
Iraklis Varlamis	Harokopio University of Athens, Greece
Ruben Verborgh	Ghent University - iMinds, Belgium
Fabio Vitali	University of Bologna, Italy
Hannes Werthner	Vienna University of Technology, Austria
Fouad Zablith	American University of Beirut, Lebanon
Ondřej Zamazal	University of Economics, Prague, Czech Republic
Ziqi Zhang	University of Sheffield, UK

Contents

Research Papers

In-Use Papers

Position Paper

Research Papers

Automatic Key Selection for Data Linking

Manel Achichi[1(✉)], Mohamed Ben Ellefi[1], Danai Symeonidou[2], and Konstantin Todorov[1]

[1] LIRMM/University of Montpellier, Montpellier, France
{achichi,benelle,todorov}@lirmm.fr
[2] INRA, MISTEA Joint Research Unit, UMR729, 34060 Montpellier, France
danai.symeonidou@supagro.inra.fr

Abstract. The paper proposes an RDF key ranking approach that attempts to close the gap between automatic key discovery and data linking approaches and thus reduce the user effort in linking configuration. Indeed, data linking tool configuration is a laborious process, where the user is often required to select manually the properties to compare, which supposes an in-depth expert knowledge of the data. Key discovery techniques attempt to facilitate this task, but in a number of cases do not fully succeed, due to the large number of keys produced, lacking a confidence indicator. Since keys are extracted from each dataset independently, their effectiveness for the matching task, involving two datasets, is undermined. The approach proposed in this work suggests to unlock the potential of both key discovery techniques and data linking tools by providing to the user a limited number of merged and ranked keys, well-suited to a particular matching task. In addition, the complementarity properties of a small number of top-ranked keys is explored, showing that their combined use improves significantly the recall. We report our experiments on data from the Ontology Alignment Evaluation Initiative, as well as on real-world benchmark data about music.

1 Introduction

In recent years, the Web of Data has been constantly growing both in terms of quantity of the RDF datasets published publicly on the web and in terms of diversity of the domains that they cover. One of the most important challenges in this setting is creating semantic links among these data [1]. Among all possible semantic links that could be declared between resources found in different datasets, identity links, defined by the `owl:sameAs` statement, are of great importance and the ones that most of the attention is given to. Indeed, `owl:sameAs` links allow to see currently isolated datasets as one global dataset of connected resources. Considering the small number of existing `owl:sameAs` links on the Web today, this task remains a major challenge [1].

Due to the large amount of data already available on the Web, defining manually `owl:sameAs` links would not be feasible. Therefore, many approaches try to answer to this challenge by providing different strategies to automate this process. Datasets conforming to different ontologies, data described using

E. Blomqvist et al. (Eds.): EKAW 2016, LNAI 10024, pp. 3–18, 2016.
DOI: 10.1007/978-3-319-49004-5_1

different vocabularies, datasets described in different languages are only several of the examples that make this problem hard to solve.

Many of the existing link discovery approaches are semi-automatic and require manual configuration. Some of these approaches use *keys*, declared by a domain expert, to link. A key represents a set of properties that uniquely identifies every instance of a given class. Keys can be used as logical rules to link data ensuring high precision results in the linking process. Additionally, they can be exploited to construct more complex rules. Nevertheless, keys are rarely known and are very hard to declare even for experts. Indeed, experts may not know all the specificities of a dataset leading to overlook certain keys or even introduce erroneous ones. For this reason, several automatic key discovery approaches have been already proposed in the context of the Semantic Web [2–6].

In spite of that fact, applying the output of these approaches directly is, in most of the cases, impossible due to the characteristics of the data. Ontology and data heterogeneity are not the only issues that can arise while trying to apply keys directly for data linking. Even if the datasets conform to the same ontology and the vocabulary of the properties is uniform, this does not ensure the success of the linking process. Very often, key discovery approaches discover a very large number of keys. The question that arises is whether all the keys are equally important among them, or there are some that are more significant than others. So far, no approach provides a strategy to rank the discovered keys, by taking in consideration their effectiveness for the matching task at hand.

Bridging the gap between key discovery and data linking approaches is critical in order to obtain successful data linking results. Therefore, in this paper we propose a new approach that, given two datasets to be linked, provides a set of ranked keys, valid for both of them. We introduce the notion of "effectiveness" of a discovered key. Intuitively, a key is considered as effective if it is able to provide many correct `owl:sameAs` links. In order to measure the effectiveness of keys, a support-based key quality criterion is provided. Unlike classic approaches using support for the discovered keys, in this work we introduce a new global support for keys valid for a set of (usually two) datasets.

The proposed approach can be summarized in the following main steps. *(1) Preprocessing:* in this step, given two datasets to be linked, only properties that are shared by both datasets are kept. This ensures that a key can be applied on both the source and the target datasets, and not only on each of them independently. At this point it is important to state that we consider that the datasets use either common vocabularies or that the explicit mapping between the respective vocabularies is known. *(2) Merge:* the key candidates discovered in each dataset are then merged by computing their cartesian product (recall that a key is a set of properties). *(3) Ranking:* we introduce a ranking criterion on the set of merged keys that is a function of the respective supports of each merged key in each dataset, normalized by the dataset sizes. *(4) Keys combination:* finally, the combined use of several top-ranked merged keys is evaluated, showing an improvement of the recall of a given link discovery tool.

The rest of the paper is structured as follows. Section 2 overviews data linking and automatic keys discovery and link specification approaches. Then, Sect. 3

presents our key ranking technique, evaluated in Sect. 4. Conclusions and future work are provided in Sect. 5.

2 Related Work

Let us look onto the process of data linking from a global perspective. The majority of the existing linking tools implement a process that consists of three steps: (1) configuration and pre-processing, (2) instance matching and (3) post-processing. Step (1) aims on the one hand to reduce the search space by identifying sets of linking candidates and key properties to compare, and on the other hand – to model instances by using a suitable representation that renders them comparable (one can think of indexing techniques, automatic translation, *etc.*). Step (2) aims at deciding on a pair of instances whether they are equivalent or not, mostly relying on similarity of property values, evaluated by similarity measures defined in step (1). The output of step (2) is a set of matched instances, also known as a *link set*. Finally, step (3) allows to filter out erroneous matches or infer new ones, based on the link set provided in step (2).

The configuration step of the linking workflow described above contains two important sub-steps: (a) the choice of properties (or keys) across the two datasets whose values need to be compared, and (b) the choice of similarity measures to apply and their tuning. Our approach is tightly related to these sub-steps, although it does not fit into either of these categories. Indeed, we are not aware of the existence of other approaches that address the problem of key quality evaluation with respect to data linking, therefore, the current section looks into approaches relevant to both (a) and (b), as well as to the data linking process as a whole.

2.1 Automatic Linking Tools Configuration

Key Discovery. In order to link, many data linking approaches require a set of linking rules. Some data linking approaches use keys to build such rules. A key is a set of properties that uniquely identifies every resource of a given class. Nevertheless, keys are rarely known and also very hard to define even for expert.

In the context of Semantic Web, different key discovery approaches have been already proposed. Both [2,5] propose a key discovery approach that follows the semantics of a key as defined by OWL. This definition states that two instances are referring to the same real world entity if at least one value per property appearing in a key is equal. Unlike [5], [2] proposes a method that scales on large datasets, taking also into account errors or duplicates in the data. In [4,6], the authors propose an alternative definition for the keys that is valid when the data are locally complete. In this case, to consider that two instances are equal, all the set of values per property appearing in a key should be the same. Finally, in [7], a key discovery approach for numerical data is proposed.

Atencia *et al.* [3] observe that key extraction is conducted by state-of-the-art tools in an independent manner for two input datasets without taking into consideration the linking task ahead. The authors introduce the concept of a *linkkey* – a set of properties that are a key for two classes *simultaneously*, implying equivalence between resources that have identical values for the set of these properties.

Automatic Link Specification Algorithms. We consider the work on automatic link specification as related in terms of motivation to our approach and complementary in terms of application. Link specification is understood as the process of automatically building a set of linking rules (restrictions on the instances of the two datasets), choosing similarity measures to apply on corresponding property values across datasets together with their respective thresholds [8]. Several approaches have been introduced so far, mostly based on machine learning techniques, such as FEBRL [9], an extension of SILK [10], RAVEN [11] or, more recently, EAGLE [8]. Contrarily to key discovery methods, these approaches mainly focus on the automatic selection, combination and tuning of similarity measures to apply on the values of comparable properties. The identification of properties to compare is done by matching algorithms and no key computation is implied in this process. The efficiency of these algorithms can be improved if the system knows on which properties and on what types of values the similarity measures will be applied.

2.2 Data Linking

Data linking has evolved as a major research topic in the semantic web community over the past years, resulting in a number of approaches and tools addressing this problem. Here, instead of making an inventory of these techniques, surveyed in [12,13], we scrutinize the main characteristics that unite or differentiate the most common approaches.

The majority of the off-the-shelf linking tools [14–19] produce an RDF *linkset* of `owl:sameAs` statements relating equivalent resources and the linking process is commonly semi-automatic. As discussed above, the user has to configure manually a number of input parameters, such as the types of the instances to compare (with certain exceptions like [18] where ontology matching techniques are applied to identify the equivalent classes automatically), the properties (or property chains) to follow, since most linking tools adopt a property-based link discovery philosophy, the similarity measure(s) and thresholds to apply on the literals and possibly an aggregation function for several measures. The bigger part of the existing approaches are conceived as general purpose linking methods and are designed to handle monolingual RDF data.

What differentiates these tools in the first place is the techniques of automatic preprocessing that are embedded in their architecture. Scalability and computational efficiency are major issues when dealing with data linking problems on the web scale. To reduce the search space, [19] cluster data items, based on their similarity with respect to their properties. Indexing techniques are used to reduce the number of instance comparisons by Rong *et al.* [20] using similarity of vectors as a proxy for instance relatedness. Similarly, Shao *et al.* [16] and Kejriwal *et al.* [21] apply a blocking technique, which consists in using inverted indexing to generate candidate linking sets. SILK [14] relies on indexing all target resources by the values of one or more properties used as a search term. LIMES [15] relies on the triangle inequality property of metric spaces to reduce the number of comparisons and thus the time complexity of the task.

The linking tools vary with respect to their abilities to handle different degrees and types of data heterogeneity. Indeed, most of the tools are able to cope with minor differences in spelling in the string literals by applying string matching techniques, but only a few are able to deal with more complex heterogeneities and just a couple of them try to resolve the problem of multilingualism (using different natural languages in data description), as Lesnikova *et al.* do, although in a very restricted scenario of only two languages [22].

2.3 Positioning

The approach that is proposed in this paper attempts to close the gap between automatic key discovery algorithms and the data linking process. As observed above, the majority of key discovery techniques do not effectively facilitate the task of selection of properties whose values to compare in the linking process, due the large number of keys produced and the lack of confidence indicator coupled with the keys. Our method suggests to unlock the potential of key-based techniques by providing to the user of a data linking tool a limited number of quality keys, well-suited to the particular matching task. The only key-based approach that looks into the usefulness of keys for two datasets simultaneously, and not independently from one another, is [3]. In contrast to our approach, the set of *linkkeys* produced in [3] is unordered which does not allow to effectively select a key or decide on the use of one key as opposed to another.

As compared to automatic link specification algorithms cited in Subsect. 2.1, our approach can be seen as complementary: we focus on the identification of a limited set of properties that can be used to effectively link datasets, while leaving the choice of the similarity measures, their combination and tuning to the user, or to the auto-configuring link specification methods given above. The automatic selection of keys can potentially improve the quality of link specification methods by restricting considerably the similarity space.

3 Automatic Key Ranking Approach

Given two RDF datasets, candidates to be linked, our approach aims at ranking the keys that are valid for *both* datasets. These keys can be used successfully as link specifications by link discovery frameworks. Before introducing the approach, recall the OWL definition of a key. A key is a set of properties, such that if two resources share at least one value for every property participating in this key, these resources are considered as equal, or formally:

$$\forall X, \forall Y, \forall Z_1, \ldots, Z_n, \wedge c(X) \wedge c(Y) \bigwedge_{i=1}^{n} (p_i(X, Z_i) \wedge p_i(Y, Z_i)) \Rightarrow X = Y, \quad (1)$$

where X and Y are instances of the class c and $p_i(X, Z_i) \wedge p_i(Y, Z_i)$ expresses that both X and Y share the same value Z_i for every property p_i in the key.

In next section, we describe how do we select keys that are valid for the two datasets. Afterwards, we describe our ranking approach on the set of these keys.

3.1 Selecting Mutual Keys for Two Datasets and Merging

We start by giving one of our initial hypothesis. The number of available vocabularies has been growing with the growth of the LOD cloud, resulting in datasets described by a mixture of reused vocabulary terms. It is therefore often the case that two different datasets to be linked are described by different vocabularies. To answer to that, ontology alignment methods [23] are used in order to create mappings between vocabulary terms. In this paper, we assume that equivalence mappings between classes and properties across the two input datasets are declared (either manually, or by the help of an ontology matching tool). These mappings will be used to obtain keys that are valid for *both* datasets.

Algorithm 1 gives an overview of the main steps of our approach, also depicted in Fig. 1. Overall, given two datasets to be linked, this algorithm returns a set

Algorithm 1. The merged keys ranking algorithm.

Input: D_S and D_T, a pair of datasets candidates to be linked.
Output: A set of merged and ranked keys: *rankedMergedKeys*

1 $M \leftarrow$ Mapping(D_S, D_T);
2 $KeysD_S \leftarrow$ keysDiscovery(D_S, M);
3 $KeysD_T \leftarrow$ keysDiscovery(D_T, M);
4 $MergedKeys \leftarrow$ keysMerging($KeysD_S$, $KeysD_T$);
5 $rankedMergedKeys \leftarrow$ mergedKeysRanking(D_S, D_T, $MergedKeys$);
6 **return** *rankedMergedKeys*;

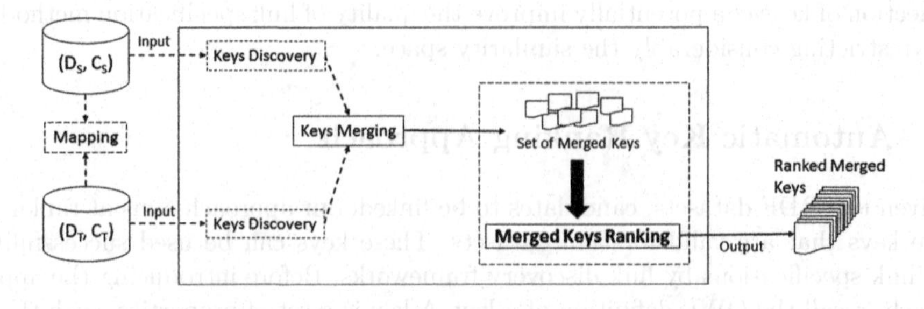

Fig. 1. The processing pipeline of Algorithm 1

of ranked keys valid for both datasets. In addition to that, every proposed key is given a score, allowing to rank keys according to their impact on the data linking process. This process is described step by step below.

First, given the datasets D_S and D_T containing instances of a class C, a set of property mappings M between the two datasets is computed. As described in [5], property mappings allow the identification of properties that belong to both datasets simultaneously.

A key discovery step is applied to both datasets independently allowing the discovery of valid keys in each dataset. Only mapped properties, appearing in M, will be contained in the discovered keys. For this step, existing key discovery tools such as SAKey [2] or ROCKER [4] can be used to obtain keys for a given class C.

However, even if keys consist of properties that belong to both datasets, nothing ensures that the discovered keys found in each dataset independently will be the same. Indeed, there can be cases where something found as a key in one dataset it is not true in the other. Since key discovery approaches learn keys from the data, the generality of each dataset affects the generality of the discovered keys. For example, if a dataset contains people working in a specific university, it is possible to discover that the last name is a key. Thus, to deal with this challenge a merging step is performed. Indeed, merging keys coming from different datasets allows to verify the validity of discovered keys and to obtain more meaningful keys since they are applicable to more than one datasets. Different strategies for key merging could be applied. In this work, we apply a merging strategy proposed in [5] providing minimal keys valid in both datasets.

The result is a set of merged keys considered as valid for *both* datasets. However, the number of merged keys produced by the algorithm can be significantly high, which makes manual selection difficult, particularly in the lack information of the keys suitability for the data linking task. Therefore, we introduce a novel ranking method for merged keys to identify the most suitable keys to be used in the link specification, introduced in next section.

3.2 Merged Keys Ranking

As described before, the merged keys are valid for both datasets. However, these keys may vary in terms of "effectiveness" in the linking process. Therefore, we propose to first to assign a score reflecting the "effectiveness" of a discovered key and second use this score to rank the discovered keys among them.

In general, it is very common that not all the properties are used to describe every instance of a given class. This happens often due to the nature of the property or the incompleteness of the data and may have significant impact on the quality of the discovered keys with respect to the linking task. While many properties apply to every instance of a class, there exist cases of properties that have values only for certain instances (the property "spouse" for a person applies only to people that are married). In addition, in the case when data are incomplete, an instance may not have a value for a specific property even if a value exists in reality. This can lead to the discovery of wrong keys since not all the possible scenarios are visible in the data. Since it is very hard to differentiate these two cases automatically and a manual identification would not be feasible due to the size of the existing datasets, we use the notion of support to measure the completeness of a key. The support measures the presence of a set of properties in a dataset. Intuitively, we tend to trust more keys that are valid for many instances in the data, i.e., keys with high support.

Basing ourselves on the support definition initially given by Atencia *et al.* in [6], we redefine this measure in order to provide a ranking score for properties with respect to a given dataset.

Let D be an RDF dataset described by an ontology O. For a given class $C \in O$, let I_C be the set of instances of type C and P the set of properties having an element of I_C as a subject and let G_C be the subgraph defined by the set of triples of I_C and P, $G_C = \{< i, p, . >: i \in I_C, p \in P\}$.

Definition 1 (Property Support Score). *The support of a property* $p \in P$ *with respect to the pair (D, C) is defined by:*

$$supportProp(p, D, C) = \left| \bigcup_{i \in I_C} < i, p, . > \right| \frac{1}{|I_C|}.$$

In other words, $supportProp(p, D, C) = N \frac{1}{|I_C|}$ means that N instances of type C in the dataset D have a value for the property p ($supportProp(p, D, C) \in [0, 1]$).

As keys for a given class can be composed of one or several properties, we introduce a ranking score for keys based on the supports of their properties, again with respect to their dataset.

Definition 2 (Key Support Score). *Let* $K = \{p_1, ..., p_n\}$ *be a key corresponding to the pair* (D, C)*, where* $p_j \in P, j \in [1, n]$*. We define the support of* K *with respect to* (D, C) *as*

$$supportKey(K, D, C) = \left| \bigcup_{i \in I_C} < i, K, . > \right| \frac{1}{|I_C|},$$

where $< i, K, . >$ *means that* $\forall p_j \in K, \exists < i, p_j, . > \in G_C$.

In other words, $supportKey(K, D, C)$ can be seen as a measure of the **co-occurrence** of $\{p_1, ..., p_n\}$ in G_C.

To illustrate, let us consider a source dataset D_S having 300 instances of type C_S. Respectively, let D_T be a target dataset having 100 instances of type C_T, where C_S and C_T are two mapped (equivalent) classes, potentially sharing instances. Let K_i and K_j be two merged keys, obtained as described in Algorithm 1, with the following supports for (D_S, C_S) and (D_T, C_T), respectively:

$$supportKey(K_i, D_S, C_S) = \frac{160}{300}; \quad supportKey(K_i, D_T, C_T) = \frac{40}{100};$$

$$supportKey(K_j, D_S, C_S) = \frac{110}{300}; \quad supportKey(K_j, D_T, C_T) = \frac{90}{100}.$$

Obviously, the challenge that arises here is how to rank the merged keys in order to ensure a maximum instance representativeness.

We note that key support score expresses the importance of a merged key with respect to each dataset, however, it is still necessary to provide a ranking function allowing to measure the importance of the merged keys for *both* datasets *simultaneously*.

An intuitive strategy to compute the final support of a merged key, given the supports computed locally in each dataset, would be to compute the average score of these supports. Nevertheless, this strategy would fail to capture all the different scenarios that could lead to a support value. For example, a key having supports 1 and 0.4 in datasets 1 and 2, would have the same merged support than a key having supports of 0.7 and 0.7 in datasets 1 and 2 respectively. Thus, we propose a multiplication function between already computed key supports which ensures better results in the context of data linking evaluation. Consequently, we adopt this ranking function as defined below.

Definition 3 (Merged Keys Rank Function). *We define the rank of a merged key K with respect to two datasets D_S and D_T and two classes C_S and C_T as:*

$$mergedKeysRank(K) = supportKey(K, D_S, C_S) \times supportKey(K, D_T, C_T).$$

Applying the ranking to our example, we obtain the following scores:

$$globalRank(K_j) = 0.33; \quad globalRank(K_k) = 0.22; \quad globalRank(K_i) = 0.21;$$

Therefore, in this example, the key K_j is more important than K_i which means that intuitively should lead to better data linking results.

4 Evaluation

In order to confirm the effectiveness of the proposed approach, we have conducted an experimental evaluation applying two state-of-the-art key discovery tools: SAKey and ROCKER. We have used two different datasets, a real-world dataset coming from the DOREMUS project[1] and a synthetic benchmark provided by the Instance Matching Track of the Ontology Alignment Evaluation Initiative (OAEI) 2010[2]. The current experiments were applied on links generated semi-automatically using the linking tool SILK. In this evaluation, we highlight a set of issues raised during these experiments. But first, let us define the criteria and the measures used for this evaluation. Two aspects are taken into account through the keys ranking performed using our approach, first the *correctness* that determines whether the discovered links are correct and second, the *completeness* that determines whether all the correct links are discovered. These criteria are evaluated by the help of three commonly used evaluation metrics:

[1] http://www.doremus.org.

[2] http://oaei.ontologymatching.org/2010/.

- ***Precision***: expresses the ratio between the cardinalities of the set of valid matchings and all matching pairs identified by the system.
- ***Recall***: expresses the ratio between the cardinalities of the set of valid matchings and the all matching pairs that belong in the reference alignment.
- ***F-Measure***: is computed by the following formula:

$$\text{F-Measure} = 2 * \frac{Precision * Recall}{Precision + Recall}$$

We note that all considered pairs of datasets are using the same ontology model, hence, the ontology mapping process is not considered in our experiments. We first execute SAKey or ROCKER on each dataset in order to identify the set of keys. However, we emphasize the fact that advanced key exceptions like pseudo-keys or almost keys are not the focus of this paper, therefore, only traditional keys are discovered. These keys are then merged and ranked according to their support score. We launch SILK iteratively as many times as the number of the retrieved keys and produce an F-measure at each run by the help of the reference alignment of our benchmark data. We expect to find a monotonic relation between the ranks of keys and the F-measure values produced by SILK by using these keys. Note that the purpose of these experiments is not to evaluate the performance of the linking tools, but to evaluate the quality of the automatically computed ranks of keys. In other words, we assess whether the generated links are increasingly correct in an ascending order of the ranked keys.

4.1 Experiments on the DOREMUS Benchmark

The data in our first experiment come from the DOREMUS project and consists of bibliographical records found in the music catalogs of two major French institutions – *La Bibliothque Nationale de France (BnF)* and *La Philharmonie de Paris (PP)*. These data describe music works and contain properties such as work titles ("Moonlight Sonata"), composer (Beethoven), genre (sonata), opus number, *etc.*. The benchmark datasets were built based on these data with the help of music librarian experts of both institutions, providing at each time sets of works that exist in both of their catalogs, together with a reference alignment. The data were converted from their original MARC format to RDF using the marc2rdf prototype[3] [24]. We consider two benchmark datasets[4], each manifesting a number data heterogeneities:

(1) **DS1** is a small benchmark dataset, consisting of a source and a target dataset form the BnF and the PP, respectively, each containing 17 music works. These data show recurrent heterogeneity problems such as letters and numbers in the property values, orthographic differences, missing catalog numbers and/or opus numbers, multilingualism in titles, presence of diacritical characters, different value distances, different properties describing the same information, missing

[3] https://github.com/DOREMUS-ANR/marc2rdf.

[4] Doremus datasets, together with their reference alignments, are available at http://lirmm.fr/benellefi/doremus-bench.

properties (lack of description) and missing titles. SAKey produced eight keys in this scenario. The three top-ranked merged keys using our approach are:

1. K1: $\{P3\_has\_note\}$
2. K2: $\{P102\_has\_title\}$
3. K3: $\{P131\_is\_identified\_by, P3\_has\_note\}$,

where $P3\_has\_note$, $P102\_has\_title$, $P131\_is\_identified\_by$ and $P3\_has\_note$ correspond to a *comment*, *title*, *composer* and *creation date* of a musical work, respectively.

As we can see in Fig. 2(a), our ranking function ensures a decrease of the F-measure with the decrease of the key-rank, in the prominent exception of the top-ranked key, which obtains a very low value of F-Measure. This is explained by the nature of the property $P3\_has\_note$. This property describes a comment in a free format text written by a cataloguer providing information on the works, creations or authors of such works. The values for this property for the same work are highly heterogeneous (most commonly they are completely different) across the two institutions, which introduces noise and considerably increases the alignment complexity between these resources.

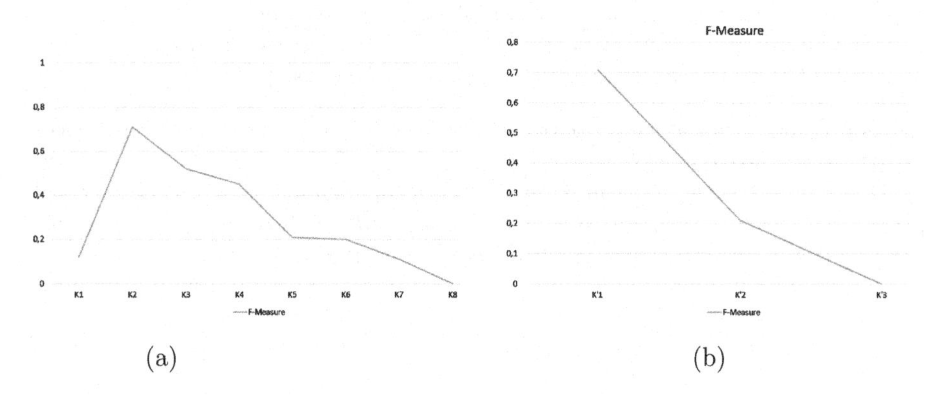

Fig. 2. Results by using SAKey on DS1: (a) by considering all properties, (b) without the property *has_note*

Thus, we decided to conduct a second experiment on the same data by removing the property *has_note* in order to confirm our observation. Figure 2 (b) reports the results of this experiment and shows a net decrease of the curve. Overall, the experiment showed that our ranking approach is efficient and the misplaced key is due to the heterogeneous nature of data.

The same experiment has been conducted using this time the key discovery approach ROCKER. The results are reported in Fig. 3 showing that the keys were well ranked. Note that, due to the different keys identification definition used by ROCKER, the problematic property *has_note* did not appear in the keys produced by the system.

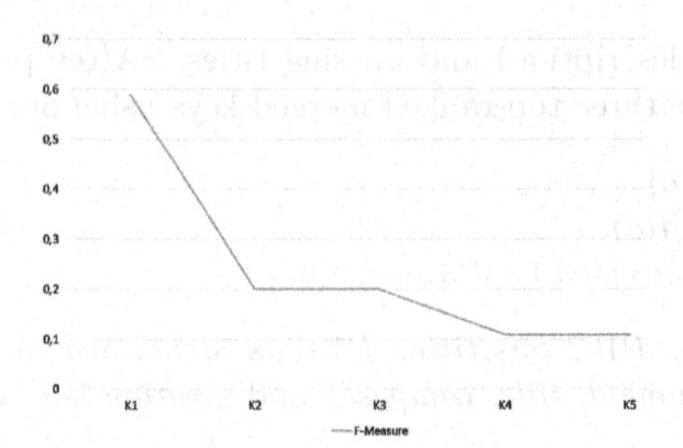

Fig. 3. Results on DS1 by using ROCKER

(2) DS2 is a benchmark dataset consisting of a source and a target dataset from the BnF and the PP, respectively, each composed of 32 music works. Contrarily to DS2, these datasets consist of blocks that are highly similar in their description works (i.e., works of the same composer and with same titles).

The results on this dataset by using SAKey are reported in Fig. 4(a). The three top-ranked merged keys are:

1. K1: $\{P3\_has\_note, P102\_has\_title, P131\_is\_identified\_by\}$
2. K2: $\{P3\_has\_note, P102\_has\_title, U35\_had\_function\_of\_type\}$
3. K3: $\{P3\_has\_note, P131\_is\_identified\_by, P3\_has\_note\}$

As their names suggest the properties *P3_has_note* (in *K1* and the first property in *K2*), *P102_has_title*, *P131_is_identified_by*, *U35_had_function_of_type* and *P3_has_note* (the third property in *K3*) correspond to a *creation date*, *title*, *composer*, *function of the composer* and *comment* on a musical work, respectively.

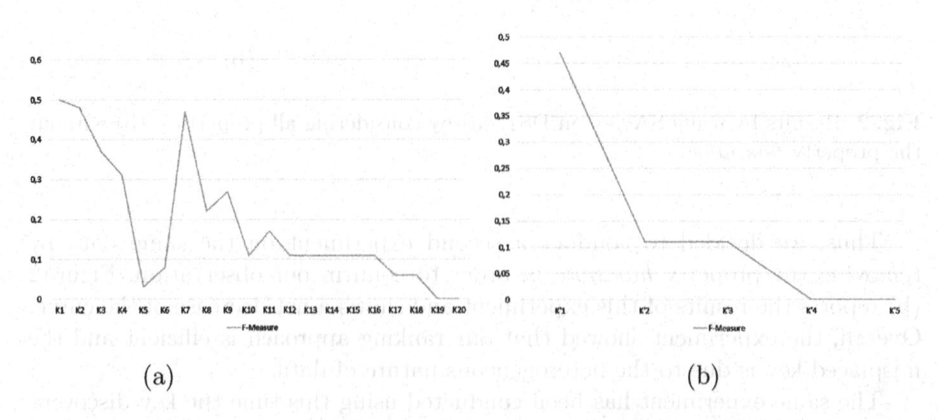

Fig. 4. Results by using SAKey on DS2: (a) by considering all properties, (b) without the property *has_note*

The results of this experiment are similar to the first one. Not considering the property *P3_has_note* improves considerably (see Fig. 4(a) and (b)) the keys ranking. Indeed, as shown in Fig. 4(a), the key $K5$ which is composed by the properties *P102_has_title*, *U35_had_function_of_type* and *P3_has_note* has significantly lowered the *f-measure* value; which is not the case of the keys in Fig. 4(b).

4.2 Experiments on the OAEI Benchmark Data

In the second series of experiments, we apply our ranking approach on keys identified in datasets proposed in the instance matching track of OAEI 2010. In this work, we report the obtained results on the dataset *Person1*. The results by using SAKey and ROCKER are shown in Fig. 5(a) and (b), respectively, where one can notice that there is an overall decrease in the F-Measure values in the two cases. Note that in Fig. 5(a), there are some problematic key-ranks, showing increase in F-measure while the ranks descend. We observed that SILK achieves better results comparing string characters than numeric characters. Indeed, this explains why we have had an increasing curve between the keys $K7$ and $K8$, knowing that they are composed of *street* and *house_number* properties (*street* and *surname* properties), respectively.

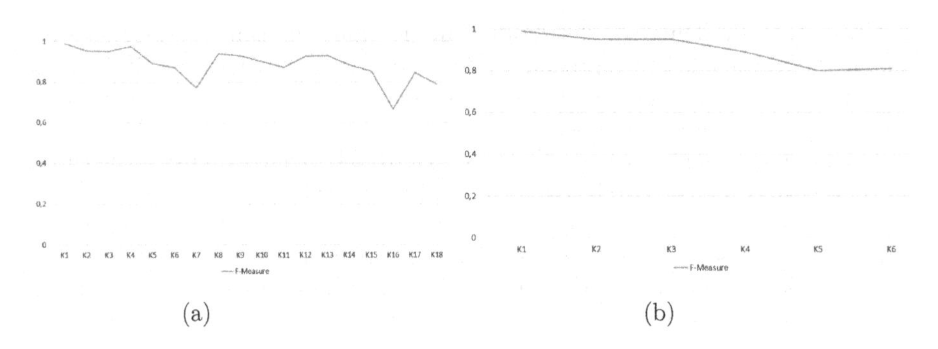

(a) (b)

Fig. 5. Results on the dataset Person1: (a) by using SAKey, (b) by using ROCKER

The three top ranked merged keys (in Fig. 5(a)) on the dataset Person1 using SAKey are:

1. K1: $\{soc\_sec\_id\}$,
2. K2: $\{given\_name, postcode\}$
3. K3: $\{surname, postcode\}$,

where the properties *soc_sec_id*, *given_name*, *surname* and *postcode* correspond to the *social security number*, *given name*, *surname* and *postal code address* of a person, respectively. In the same manner, we reiterated the experiment using ROCKER which gives better results as shown in Fig. 5(b).

4.3 Top Ranked Keys Complementarity

In this evaluation, we want to examine whether using the k (we have taken $k = 3$) top-ranked keys in combination can improve the linking scores as compared to using only one of the top-ranked keys (e.g., the first one) for linking. As discussed above, even if a key is discovered as a first-rank key, nothing ensures that the vocabulary used in both datasets to describe that key is homogeneous. To answer to that, combining a set of top ranked keys would lead to better linking results.

Table 1. Results of the combination of the three top-ranked keys on the DOREMUS datasets.

	SAKey						ROCKER			
	Dataset 1			Dataset 2			Dataset 1			Dataset 2
	F	P	R	F	P	R	F	P	R	
K1	*0.12*	*0.12*	*0.11*	*0.5*	**0.75**	*0.37*	*0.59*	0.8	*0.47*	No merged key has been identified.
K2	0.71	0.9	0.58	0.48	0.7	0.37	0.2	0.66	0.11	
K3	0.52	1	0.35	0.37	0.56	0.28	0.2	0.66	0.11	
K1+K2+K3	**0.54**	**0.44**	**0.7**	**0.51**	*0.63*	**0.43**	**0.62**	*0.75*	**0.52**	

Notice that by doing so, the *recall* value remains the same or increases as compared to the single key approach, while the *precision* may increase (if the proportion of the positive matching pairs becomes larger than the negative matching pairs) as it may as well decrease.

As shown in Table 1, the experiments on DOREMUS datasets using the three top ranked keys increased relatively (in bold in the table) the *F-Measure* with respect to the first-rank key (where the improved values are in italics) and significantly the *recall* scores (more positive matching pairs were recovered). Thus, it seems reasonable to conclude that merging the matching results retrieved from the top ranked keys allows to improve significantly the results in terms of *recall*, while this cannot guarantee an improvement in *precision*.

5 Conclusion and Future Work

This paper presents an approach that allows to select automatically a number of merged keys, relevant for a given pair of input datasets, and rank them with respect to their "effectiveness" for the task of discovering `owl:sameAs` links between them. The effectiveness of a merged key is defined as a function of the combination of its respective supports on each of the two input datasets. The proposed method allows to reduce significantly the user effort in the selection of keys used as a parameter of a data linking tool, such as SILK or LIMES. In this way, we attempt to bridge the gap between configuration-oriented approaches, such as automatic key discovery and automatic link specification, and the actual process of data linking. We also look into the complementarity properties of a small set of top-ranked keys and show that their combined use improves significantly the recall. To demonstrate our concepts, we have conducted a series of experiments on data coming from the OAEI campaign, as well as on real-world data from the field of classical music cataloguing.

In near future, we plan to improve our ranking criterion by defining it as a function of the estimated intersection of the sets of instances covered by a given key across two datasets.

Acknowledgements. This work has been partially supported by the French National Research Agency(ANR) within the DOREMUS Project, under grant number ANR-14-CE24-0020.

References

1. Bizer, C., Heath, T., Berners-Lee, T.: Linked data-the story so far. In: Semantic Services, Interoperability and Web Applications, pp. 205–227 (2009)
2. Symeonidou, D., Armant, V., Pernelle, N., Saïs, F.: SAKey: scalable almost key discovery in RDF data. In: Mika, P., et al. (eds.) ISWC 2014, Part I. LNCS, vol. 8796, pp. 33–49. Springer, Heidelberg (2014)
3. Atencia, M., David, J., Euzenat, J.: Data interlinking through robust linkkey extraction. In: ECAI, pp. 15–20 (2014)
4. Soru, T., Marx, E., Ngomo, A.N.: ROCKER: a refinement operator for key discovery. WWW **2015**, 1025–1033 (2015)
5. Pernelle, N., Saïs, F., Symeonidou, D.: An automatic key discovery approach for data linking. J. Web Semant. **23**, 16–30 (2013)
6. Atencia, M., David, J., Scharffe, F.: Keys and pseudo-keys detection for web datasets cleansing and interlinking. In: ten Teije, A., Völker, J., Handschuh, S., Stuckenschmidt, H., d'Acquin, M., Nikolov, A., Aussenac-Gilles, N., Hernandez, N. (eds.) EKAW 2012. LNCS, vol. 7603, pp. 144–153. Springer, Heidelberg (2012)
7. Symeonidou, D., Sanchez, I., Croitoru, M., Neveu, P., Pernelle, N., Saïs, F., Roland-Vialaret, A., Buche, P., Muljarto, A., Schneider, R.: ICCS, pp. 222–236 (2016)
8. Ngonga Ngomo, A.-C., Lyko, K.: EAGLE: efficient active learning of link specifications using genetic programming. In: Simperl, E., Cimiano, P., Polleres, A., Corcho, O., Presutti, V. (eds.) ESWC 2012. LNCS, vol. 7295, pp. 149–163. Springer, Heidelberg (2012)
9. Christen, P.: Febrl: an open source data cleaning, deduplication and record linkage system with a graphical user interface. In: SIGKDD, pp. 1065–1068. ACM (2008)
10. Isele, R., Jentzsch, A., Bizer, C.: Efficient multidimensional blocking for link discovery without losing recall. In: WebDB (2011)
11. Ngomo, A.-C.N., Lehmann, J., Auer, S., Höffner, K.: Raven-active learning of link specifications. In: International Conference on Ontology Matching, pp. 25–36 (2011). CEUR-WS.org
12. Ferrara, A., Nikolov, A., Scharffe, F.: Data linking for the semantic web. Semantic Web: Ontology and Knowledge Base Enabled Tools, Services, and Applications, vol. 169 (2013)
13. Nentwig, M., Hartung, M., Ngomo, A.-C.N., Rahm, E.: A survey of current link discovery frameworks. Semantic Web, pp. 1–18 (2015, preprint)

14. Jentzsch, A., Isele, R., Bizer, C.: Silk-generating RDF links while publishing or consuming linked data. In: ISWC, Citeseer (2010)
15. Ngomo, A.N., Auer, S.: LIMES - a time-efficient approach for large-scale link discovery on the web of data. In: IJCAI, pp. 2312–2317 (2011)
16. Shao, C., Hu, L., Li, J., Wang, Z., Chung, T.L., Xia, J.: RiMOM-IM: a novel iterative framework for instance matching. J. Comput. Sci. Technol. **31**(1), 185–197 (2016)
17. Jiménez-Ruiz, E., Cuenca Grau, B.: LogMap: logic-based and scalable ontology matching. In: Aroyo, L., Welty, C., Alani, H., Taylor, J., Bernstein, A., Kagal, L., Noy, N., Blomqvist, E. (eds.) ISWC 2011, Part I. LNCS, vol. 7031, pp. 273–288. Springer, Heidelberg (2011)
18. Nikolov, A., Uren, V.S., Motta, E., De Roeck, A.: Integration of semantically annotated data by the KnoFuss architecture. In: Gangemi, A., Euzenat, J. (eds.) EKAW 2008. LNCS (LNAI), vol. 5268, pp. 265–274. Springer, Heidelberg (2008)
19. Araujo, S., Hidders, J., Schwabe, D., De Vries, A.P.: Serimi-resource description similarity, RDF instance matching, interlinking. arXiv preprint arXiv:1107.1104 (2011)
20. Rong, S., Niu, X., Xiang, E.W., Wang, H., Yang, Q., Yu, Y.: A machine learning approach for instance matching based on similarity metrics. In: Cudré-Mauroux, P., et al. (eds.) ISWC 2012, Part I. LNCS, vol. 7649, pp. 460–475. Springer, Heidelberg (2012)
21. Kejriwal, M., Miranker, D.P.: Semi-supervised instance matching using boosted classifiers. In: Gandon, F., Sabou, M., Sack, H., d'Amato, C., Cudré-Mauroux, P., Zimmermann, A. (eds.) ESWC 2015. LNCS, vol. 9088, pp. 388–402. Springer, Heidelberg (2015)
22. Lesnikova, T., David, J., Euzenat, J.: Interlinking english, Chinese RDF data using babelnet. In: Proceedings of the 2015 ACM Symposium on Document Engineering, pp. 39–42. ACM (2015)
23. Shvaiko, P., Euzenat, J.: Ontology matching: state of the art and future challenges. IEEE Transactions on knowledge and data engineering **25**(1), 158–176 (2013)
24. Achichi, M., Bailly, R., Cecconi, C., Destandau, M., Todorov, K., Troncy, R.: Doremus: doing reusable musical data. In: ISWC PD (2015)

Selection and Combination of Heterogeneous Mappings to Enhance Biomedical Ontology Matching

Amina Annane[1,2](✉), Zohra Bellahsene[1], Faiçal Azouaou[2], and Clement Jonquet[1,3]

[1] Université de Montpellier, Laboratoire d'Informatique, de Robotique et de Microélectronique (LIRMM), Montpellier, France
amina.annane@lirmm.fr

[2] Ecole Nationale Supérieure en Informatique (ESI), Algiers, Algeria

[3] Center for Biomedical Informatics Research, Stanford University, Stanford, USA

Abstract. This paper presents a novel background knowledge approach which selects and combines existing mappings from a given biomedical ontology repository to improve ontology alignment. Current background knowledge approaches usually select either manually or automatically a limited number of different ontologies and use them as a whole for background knowledge. Whereas in our approach, we propose to pick up only relevant concepts and relevant existing mappings linking these concepts all together in a specific and customized background knowledge graph. Paths within this graph will help to discover new mappings. We have implemented and evaluated our approach using the content of the NCBO BioPortal repository and the Anatomy benchmark from the Ontology Alignment Evaluation Initiative. We used the mapping gain measure to assess how much our final background knowledge graph improves results of state-of-the-art alignment systems. Furthermore, the evaluation shows that our approach produces a high quality alignment and discovers mappings that have not been found by state-of-the-art systems.

Keywords: Ontology matching · Background knowledge · Repository of ontologies · Biomedical ontologies · BioPortal

1 Introduction

Ontology alignment is recognized by the scientific community as an important area of research because of its multiple applications in different domains [7]: ontology engineering, data integration, information sharing, etc. Especially in the biomedical domain that generates and manipulates a big volume of data. Ontology matching plays a key role in the development of biomedical research by facilitating the development of data warehouses articulated around common ontologies. Many works have been made to extract mappings automatically, mainly using lexical and structural matchers, but these matchers often fail when

E. Blomqvist et al. (Eds.): EKAW 2016, LNAI 10024, pp. 19–33, 2016.
DOI: 10.1007/978-3-319-49004-5_2

the ontologies to align have different structures and do not use the same vocabulary (different terms to describe the same concepts) [21]. In the recent years, the community has started to consider an alternative solution for automatic approaches in the use of *background knowledge* as a semantic mediator to discover mappings between ontologies. These background knowledge resources span from thesaurus, lexical resources, linked open data, one or several ontologies or a full repository of ontologies [18–20] and in our case, already existing mappings. The use of background knowledge has raised the following challenges: (1) selection: How to select the most useful background to align ontologies? (2) usage: How to use such knowledge in order to enhance alignment results? In all proposed approaches, the use of background knowledge was a complementary solution to traditional automatic approaches. In this paper, we propose a novel approach to align ontologies using only a background knowledge built from heterogeneous mappings, the main idea is to combine the knowledge formalized in mappings produced manually by human experts, to mappings produced automatically by simple lexical matching to discover new mappings between the ontologies to be aligned. The main contributions of this paper are:

- A novel approach to align ontologies using a background knowledge graph automatically built from existing mappings
- A novel measure called *Path Confidence Measure* to select the most accurate from several candidates mappings derived from the previously built background knowledge graph.

We have implemented and evaluated our approach using the content of the NCBO BioPortal[1] repository and the Anatomy benchmark[2] from the Ontology Alignment Evaluation Initiative. The obtained results show that our approach produces a high quality alignment, and discovers mappings not found by state-of-the-art alignment systems.

The rest of this paper is organized as follows. Section 2 defines ontology matching and common biomedical ontology mappings. Section 3 describes our novel approach exploiting existing mappings extracted from a given repository to align biomedical ontologies. Section 4 presents the proposed Path Confidence Measure. Section 5 describes the implementation of our approach. Section 6 provides the evaluation results of our approach. Section 7 discusses related work. Finally, Sect. 8 concludes our paper and points out future work.

2 Preliminaries

2.1 Ontology Matching

Ontology matching is the process of finding correspondences between two given ontologies O_1 and O_2. Each correspondence can be formalized by a quadruplet

[1] http://bioportal.bioontology.org/.
[2] http://oaei.ontologymatching.org/2015/anatomy/index.html.

$\prec e_1, e_2, r, n \succ$ with $e_1 \in O_1$ and $e_2 \in O_2$, r is a relationship between two given entities e_1 and e_2, and n is the confidence value of this relationship (generally, a value between 0 and 1) [7]. In this paper, we deal only with equivalence relationship between entities.

We distinguish the direct matching which has only the two ontologies to be aligned as an input, from the indirect matching which uses external resources, that we call Background Knowledge (BK), to enhance the quality of direct matching. These resources may be one mediator ontology, a set of ontologies, an existing alignment. The common schema to perform an alignment using a BK is composed of two steps: anchoring and deriving relations [19,20]. Anchoring consists in finding for source and target entities their equivalent entities in the BK. This step is generally done by using a lexical matcher. The second step consists in deriving relations between the entities of ontologies to align according to the relations between the anchored entities in the BK.

2.2 Biomedical Ontologies Mapping

The number of biomedical ontologies is too big to allow manual alignment of all of them (the repository NCBO BioPortal stores more than 500 biomedical ontologies). In addition, their size is also very large (e.g., SNOMEDCT, Gene Ontology). Therefore, interconnecting manually all biomedical ontologies is not feasible. However, we can find some reliable manually produced mappings in several resources such as UMLS[3] [3], the OBO Foundry [6] and the NCBO BioPortal[4] [11]. For instance, the OBO Foundry ontology developers produce Xref relations between the concepts of their ontologies(more than 141 ontologies) that can be considered mappings (latter called OBO mappings). As another example CUI (Concept Unique Identifier) mappings that are produced by the US National Library of Medicine team. When an ontology or a terminology is integrated in the UMLS Meta-Thesaurus, a CUI is manually assigned to each concept, grouping concepts together. These manually produced mappings are the formalization of human experts knowledge that we aim to exploit to enhance biomedical ontology matching.

3 Overview of Our Approach

Our approach aims to reuse mappings that can be extracted from a repository of ontologies to discover new ones, especially by combining manually and automatically produced mappings. Indeed, we hypothesis that manual mappings may be the bridge that overcomes the limitations of automatic matchers. As we can see in Fig. 1, our approach involves five steps: (1) Extraction of different kinds of mappings between all ontologies stored in the repository to construct the *Global Mapping Graph*, (2) Anchoring the concepts of the source ontology on

[3] Unified Medical Language System.

[4] Not all mappings in BioPortal are manually produced, see Sect. 5.1 for more information about NCBO BioPortal mappings.

the resulted graph, (3) Selection of mappings that may help to discover new ones using resulted anchors. The selected mappings are organized in the form of a graph called the *Specific Mapping Graph*, (4) Anchoring the concepts of the target ontology on the *Specific Mapping Graph* and extract all paths between the source and target anchors (candidate mappings. Finally (5) Filtering discovered candidates mappings to keep only the most reliable ones according to a given aggregation strategy.

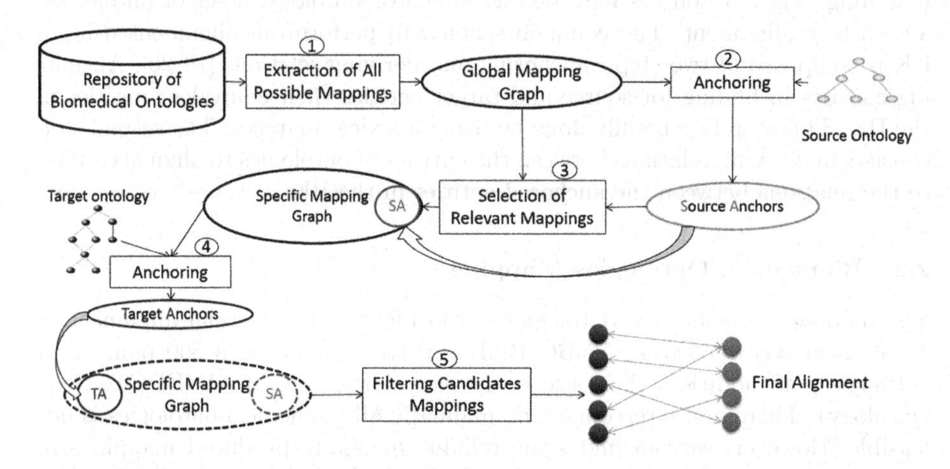

Fig. 1. Overview of the proposed approach

3.1 Building the Global Mapping Graph

In the biomedical domain the most known resources of manually produced mappings are: (i) ontologies produced by the OBO Foundry, (ii) ontologies integrated in UMLS. For a given repository of ontologies, to build the *Global Mapping Graph* we start by checking for each ontology if it is an OBO ontology, or if it is integrated in UMLS. Then, we extract from each one its manually produced mappings (OBO from the first category and CUI mappings from the second one). After that, we use a lexical matcher or any other efficient matcher to match each ontology with all others ontologies in the repository. We add these mappings produced automatically to those produced manually. For each extracted mapping we keep the source and the target concepts, the ontology of each concept, the set of labels of each concept and the provenance of this mapping (OBO, CUI, etc.). We can add any other sets of relevant mappings to enrich the final set of extracted mappings. At the end of the mappings extraction step we obtain a large set of mappings. We merge these mappings to obtain the *Global Mapping Graph* (naturally some mappings have common concepts). We note that this step is done just for once; the *Global Mapping Graph* is an independent resource that can be exploited to match any couple of ontologies. In case of enriching the repository with a new ontology, we will only extract its related mappings with other ontologies, and adding them to the resulted graph.

3.2 Anchoring Source Concepts

The second step consists in anchoring source concepts on the *Global Mapping Graph*. If the source ontology is stored in the repository, the anchors are the source concepts themselves. Otherwise, the anchors can be found using a lexical matcher on the concept labels between the source ontology and all concepts of the *Global Mapping Graph*. In this case, the mappings returned by the lexical matcher will be the first selected mappings in the *Specific Mapping Graph*. The use of a lexical matcher offers the advantages of being fast (anchoring is a preprocessing stage) and effective in aligning biomedical ontologies [10]. For a given source concept we can get wrong anchors, for that we can imagine to use more sophisticated matchers but this choice could entail higher costs in terms of resources (time and memory). In our approach we propose to let the filter at the end (see Sect. 3.5).

3.3 Selection of the Specific Mapping Graph

This step allows selecting the appropriate fragment from the *Global Mapping Graph* for a given input ontology (Algorithm 1). For each concept in the list of source anchors, we select its direct mappings in the *Global Mapping Graph* (mappings of different provenance). For each new concept in the *Specific Mapping Graph*, we search for their direct mappings and so on, until no new concept is found. Indeed if a concept A is mapped directly to B, the concept B may be automatically or manually mapped to another concept C that has no mapping with A. Finally, we obtain the *Specific Mapping Graph* which is composed of all concepts related to the source ontology interconnected via selected mappings. It is interesting to note that this *Specific Mapping Graph* is not limited in number of used ontologies, our units are concepts, not ontologies.

3.4 Anchoring Target Concepts

This step is necessary only if the target ontology is not in the initial repository. Otherwise, the anchors are the target concepts themselves. Indeed, if a target concept belongs to a mapping related to the source ontology, this target concept should be already in the resulted *Specific Mapping Graph*. In the same manner (see Sect. 3.2), we can use any efficient lexical matcher to anchor target concepts on *Specific Mapping graph* concepts and add the returned alignment in it.

3.5 Filtering Candidates Mappings

To derive mappings between the source and the target ontologies, we search for all paths between the source anchors and the target anchors in the *Specific Mapping Graph*. In Fig. 2 we can find an example of paths between the concept (MA:1012) and the concept (NCIT:C32337). One source concept may have several target concepts (several mapping candidates). Indeed, mappings composing the *Specific Mapping Graph*, in particular automatically produced ones, may be

Algorithm 1. Specific Mapping Graph Selection

Input: *GlobalMappingGraph*, *sourceAnchors*, *MappingsResultedFromAnchoring*
Output: *SpecificMappingGraph*
  if *sourceOntology* $\notin$ *BiomedicalOntologyRepository* **then**
    SpecificMappingGraph=*MappingsResultedFromAnchoring*
  end if
  for each $c \in$ *sourceAnchors* **do**
    listConcepts.add(c)
  end for
  next $\leftarrow 0$
  while *next* $<$ *listConcepts.size*() **do**
    $x \leftarrow$ *listConcepts.get*(*next*)
    Extract S from *GlobalMappingGraph*: all direct mappings of x
    for each $m \in S$ **do**
      if $m \notin$ *SpecificMappingGraph* **then**
        SpecificMappingGraph.add(m)
      end if
      if *m.targetConcept* $\notin$ *listConcepts* **then**
        listConcepts.add(*m.targetConcept*)
      end if
    end for
    next $++$
  end while
  return *SpecificMappingGraph*

not precise (or wrong) which lead to derive wrong mappings. The challenge is to select the most accurate candidate target concept, especially if we deal with 1:1 mappings (searching only for equivalence relationship). In our case, a candidate mapping corresponds to one or several paths linking the same source concept to the same target concept. Paths in Fig. 2 represents a candidate mapping between the concept (MA:1012) and the concept (NCIT:C32337). We have experimented different aggregation strategies (see Sect. 6.2) to select one mapping from several candidates for a given source concept, but these strategies produced a low recall. To improve the quality of the final alignment, we propose a novel measure to select for a given source concept the best mapping from several candidates. This measure is described in the next Section.

4 Path Confidence Measure

We define the type of a given path as a distinct sequence of provenances that forms this path, independently from intermediate concepts. For example, the type of path linking the concept (MA:1012) to the concept (MeSH:D17626) in Fig. 2 is OO (OBO_OBO). The types of path linking the concept (MA:1012) to the concept (NCIT:C32337) are: OO, OSO, OLLL, etc.

To enhance the selection of the final mappings, we propose the novel *Path Confidence Measure*(PCM) that takes the confidence value of given path type

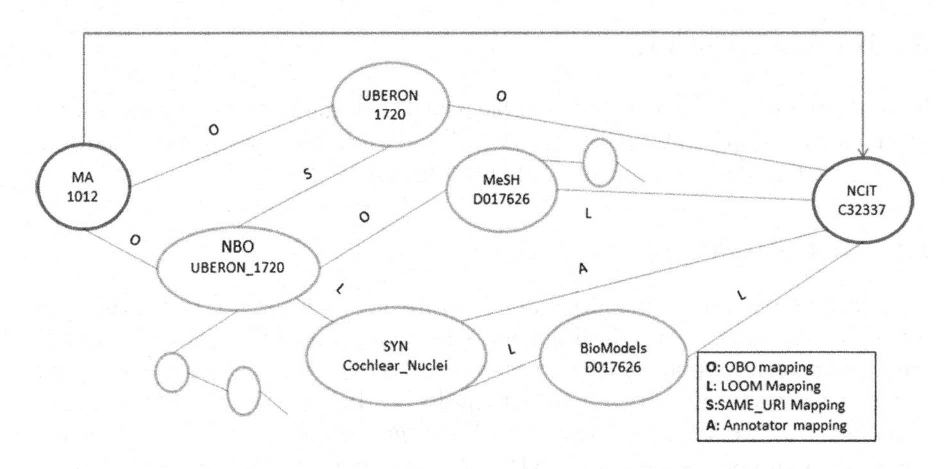

Fig. 2. Extracted mappings from BioPortal for the mouse anatomy concept 1012 (each concept is represented by the acronym of its ontology and its code within BioPortal)

into account. The confidence value is a score assigned to each path type according to its ability to discover correct mappings. This measure is inspired from the most frequent aggregation strategy (also called popularity in [16]) based on the hypothesis: for a given source concept, the most accurate target concept is the concept that has the highest number of paths linking it to this source concept. In this hypothesis we assume that all path types has the same confidence value. However, the quality of discovered mappings is different from one path type to another. Indeed, some types give better results than the others (see Sect. 5). For this purpose, we introduce the confidence value of a given path type as a coefficient to be multiplied by the number of paths of this type. The Path Confidence Measure for a given candidate mapping (C_s, C_t) is defined as the sum of the number of each path type linking C_s to C_t multiplied by its confidence value. We use the log function to avoid the over-estimation of a given candidate mapping due to a large number of a given path type. We add 1 to avoid $log(0)$ and we divide by the max sum to normalize values between 0 and 1. For a given candidate mapping (C_s, C_t), we compute the PCM value of the target concept C_t as follows:

$$PCM(C_s, C_t) = \frac{\sum_{i=1}^{n} log(1 + NP_i * CV_i)}{\max_{j=1}^{m} \sum_{i=1}^{n} log(1 + NP_{ji} * CV_i)}$$

where n is the number of different types of paths that lead to the target concept C_t from the source concept C_s; NP_i is the number of paths of type i linking C_s to C_t; CV_i is the confidence value of the path of type i; m is the number of concepts of the source ontology. This measure is proposed only to select for a given source concept, one target concept from several candidates.

5 Implementation

To evaluate our approach, we have implemented it using the reference repository of biomedical ontologies NCBO BioPortal and the ontologies of the Anatomy track from Ontology Alignment Evaluation Initiative 2015[5].

5.1 NCBO BioPortal

NCBO BioPortal is a community based repository. Currently, it is one of the richest repository in the biomedical domain with more than 500 biomedical ontologies. The repository offers a REST web services API.[6] In particular, mappings of different provenances[7] between stored ontologies. In addition of OBO and CUI mappings that we have previously explained, the repository generates automatically other mappings such as LOOM [10], SAME_URI and REST mappings. LOOM mappings are based on close lexical match between preferred names of concepts or a preferred name and a synonym. The lexical match involves removing white-space and punctuation from labels. SAME_URI mappings are based on exact match between the URI of concepts. Finally, REST mappings that are mappings uploaded manually by users of the portal, they represent the minority. In addition, the portal integrates an efficient Annotator [15] which can be used as a lexical matcher. For a given concept label, the Annotator returns a list of concepts that have the same label.

5.2 Anatomy Track

The Anatomy track consists in finding an alignment of 1516 mappings between the Adult Mouse Anatomy ontology (2738 concepts)and a part of the NCI Thesaurus (describing the human anatomy 3298 concepts). The task has a good share of non-trivial mappings.

Instead of creating a local repository of biomedical ontologies, we have chosen to use the NCBO BioPortal. Another factor that motivates our choice is the mappings of different provenances that are stored and accessible through its REST API. Consequently, BioPortal can be considered as a huge graph where nodes are concepts and edges are mappings with different provenances. With this vision, BioPortal can play the role of the *Global Mapping Graph* in our approach. Also, the source and the target ontologies of the Anatomy track are already stored in BioPortal, we do not need to anchor concepts (see Sects. 3.2 and 3.4), we can access directly to them using their URI. Consequently, to run our approach, we need just to execute the steps 3 and 5 of the proposed approach to produce the final alignment.

[5] http://oaei.ontologymatching.org/2015/.
[6] http://data.bioontology.org/documentation.
[7] http://www.bioontology.org/wiki/index.php/BioPortal_Mappings.

6 Evaluation

The selection of the *Specific Mapping Graph* step with the mouse anatomy (MA) as a source ontology and the NCBO BioPortal as *Global Mapping Graph* has produced a graph[8] combining 85192 concepts and 368371 mappings of different provenance (see Fig. 2). We have extracted the preferred label of each concept and annotate it using the BioPortal Annotator, because it works with a richest synonym dictionary which allows to discover mappings that the LOOM algorithm does not discover. Indeed, the LOOM algorithm is based only on close lexical match without using any complementary resources. Mappings are extracted in JSON format as we can see in [2], we note that no score is assigned to these mappings, we have just the information about their provenance. It is important to keep this information to be able to explain the provenance of a given derived mapping by the end. The distribution of extracted mappings per provenance is presented in Table 1. As we can see, the number of the annotator mappings is greater than the number of LOOM mappings, this can be explained by the fact that the annotator works only with exact string match whereas LOOM involves some pretreatment such as removing white-space and punctuation from labels.

Table 1. Number of extracted mappings per provenance

Provenance of mappings	Number of mappings
LOOM	196225
Annotator	78446
OBO	65305
CUI	17551
SAME_URI	10488
REST	356

6.1 Evaluation of Paths Types Quality

From the resulted *Specific Mapping Graph*, we have extracted all possible paths between the concepts of the source ontology *MA* and the concepts of the target ontology *NCIt*. Each path represents a candidate mapping that may be true or false according to the reference alignment provided by OAEI2015. We have computed the true positive mappings (mappings present in the reference alignment) and the false positive mappings (mappings absent in the reference alignment) for each type of path. Using these parameters, we have computed the precision, recall and F-Score for each type of path. Figure 3 represents the top 50 path types ranked according to the F-Score measure. Based on the obtained results, we can

[8] We have created the graph using the graph database Neo4J (https://neo4j.com/).

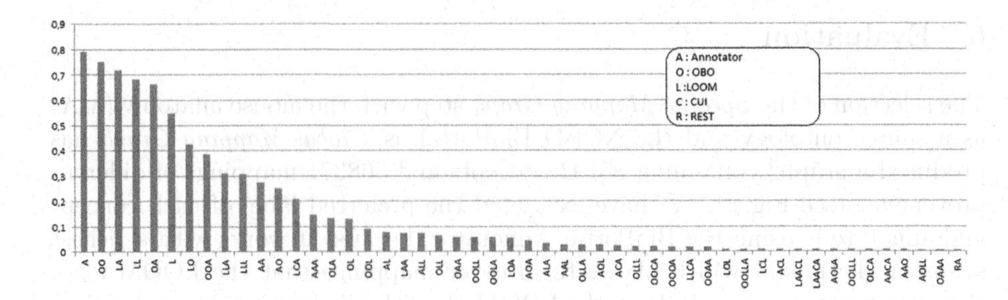

Fig. 3. Fscore per type of path

conclude that the best paths are the shortest ones: direct matching (paths of type A and L), paths with two steps; one mediator concept (OO,LA,LL,OA,LA) and paths with three steps; two mediator concepts (OOA,LLA,LLL). We note that the combination of manually and automatically produced mappings provides a good results(e.g., LA,OOA,LA). The longest paths return a few mappings candidates, and generally wrong ones (see Fig. 4).

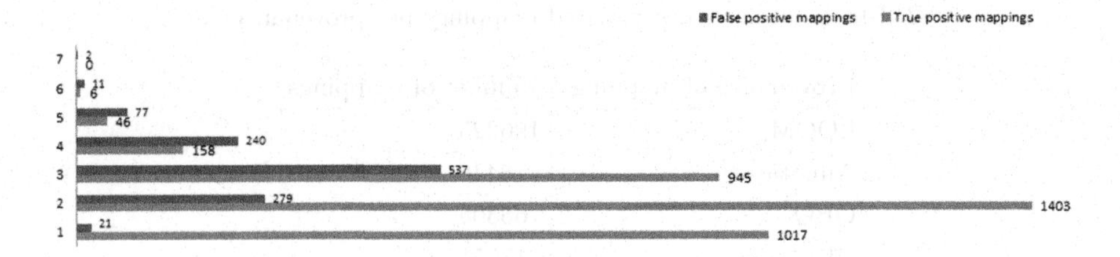

Fig. 4. True positive/False positive mappings per length of paths (number of steps)

According to this study, we have chosen to use the F-Score of each path's type as its confidence value to asses its ability to discover true positive mappings.

6.2 Evaluation of Final Alignment Quality

In order to evaluate the quality of the *Specific Mapping Graph*, we have compared mappings derived from it (mappings linking MA concepts to NCIt concepts) to the reference mappings of the Anatomy track. First of all, we have evaluated all mappings derived from the *Specific Mapping Graph* without any aggregation strategy. Then, we have experimented three strategies to select only one target concept for each source concept: (i) the first found; i.e. the final node of the shortest path leading to the target ontology (ii) the most visited target concept; it is the concept of the target ontology that has the highest number of paths from a given source concept and (iii) the target concept that has the greatest PCM score (path's type F-Score as confidence value). Then, we have compared the alignment produced by our approach to the final alignments of the four top systems in OAEI 2015 [4] for the Anatomy track. The results presented in Table 2 show that our final alignment is competitive with top alignment systems. Without any strategy of aggregation, our final alignment has the best precision but relatively a low recall, what gives it the worst F-Score. However, the use of any aggregation strategy improve the recall, and lets our final alignment having the second position after AML system. We note that AML and LogMapBio [14] systems use already biomedical ontologies as BK. Also, AML implement several features that help improving the final alignment. The best F-Score is obtained using the PCM measure for the selection of final mappings. Indeed, the proposed measure promotes paths with high confidence.

Table 2. Quality evaluation of the discovered mappings

Systems		Mappings	Correct	Incorrect	Precision	Recall	F-Score
Resulted BK	All mappings	2247	1416	831	**0,934**	0,630	0,753
	First found	1504	1366	138	0,901	0,909	0,905
	Most frequent	1504	1372	132	0,905	0,912	0,909
	PCM	1503	1395	108	0,920	**0,928**	**0,924**
AML		1477	1412	66	**0,931**	**0,956**	**0,944**
LogMapBio		1549	1366	183	0,901	0,882	0,891
LogMap		1397	1282	115	0.846	0,918	0,88
XMAP		1414	1312	102	0,865	0,928	0,896

6.3 Specific Mapping Graph: Usefulness Evaluation

The mapping gain [8] is a measure proposed to asses the usefulness of a BK for a given task of alignment. It measures how many new mappings have been found in an alignment A thanks to a given BK comparing to another alignment B. For clarity, we recall here the formula of this measure. Given two alignments A and B between ontologies S and T, the mapping gain between A and B is defined as the fraction of mappings in A that are not in B.

$$MG(A, B) = Min(\frac{C_s(A \cap \neg B)}{C_s(B)}, \frac{C_t(A \cap \neg B)}{C_t(B)})$$

where C_s and C_t denote respectively the sets of concepts in the alignments (A and B) and belong respectively to the source and the target ontologies.

To evaluate the usefulness of the *Specific Mapping Graph* as a BK, we have computed the mapping gain using the previous formula replacing A by our final derived alignment (with PCM) and B by one of alignments produced by the four top systems in the OAEI 2015[9] (see Table 3).

[9] http://oaei.ontologymatching.org/2015/results/anatomy/index.html.

Table 3. Mapping gain using resulted BK

Systems	# Absent concepts of MA	# Absent concepts of NCIT	Mapping gain
AML	77	195	5 %
LogMapBio	134	247	9 %
XMAP	188	302	13 %
LogMap	218	337	16 %

Based on analysis done in [8], the authors conclude that if the use of a BK provides a mapping gain greater than 2 %, the BK could be considered as useful. According to that, the *Specific Mapping Graph* is useful for all these systems (state-of-the-art alignment systems). We can observe that the resulted BK is significantly useful for XMAP and LogMap because they do not use any biomedical ontologies as a BK. The other systems already use biomedical ontologies as a BK. AML uses three ontologies (Uberon, DOID and Mesh) which represents 292 591 concepts. LogMap uses top ten ontologies returned by the algorithm presented in [5]. The first ontology returned by this algorithm is SNOMEDCT which contains 324129 concepts. In the last both cases we observe the large number of concepts comparing to the *Specific Mapping Graph*'s concepts number (85192 concepts). We observe also that even if AML and LogMap use a biomedical BK, the *Specific Mapping Graph* allows to enhance their results. Table 4 presents the number of reference mappings found by our approach, missed by the other systems.

Table 4. Mappings found by our approach, missed by top alignment systems

AML	LogMapBio	XMAP	LogMap
20	87	161	133

7 Related Work

The selection of the appropriate BK to enhance biomedical ontology matching is an active research issue. Several approaches have been proposed to address it. To avoid the complexity of an automatic selection, many approaches usually manually select the relevant BK. For examples, WordNet is used in [20], DOLCE in [17]. The manual selection does not guarantee the enhancement of a given task of alignment, and requires a wide range of knowledge. For this purpose, several automatic approaches have been defined to select the appropriate BK as those described in [18,19]. The most similar work to this paper is done in [12]. Their approach consists in aligning the source and the target ontologies with each ontology in a set of intermediate ontologies. Then, compose the different produced alignments to derive mappings between source and target ontologies.

The authors do not extract manually produced mappings and they do not extract mappings between intermediates ontologies. Using their approach, one can derive only mappings with one mediator concept (paths of two steps only). In the same manner [5] propose to compose mappings after selecting dynamically five ontologies from BioPortal. However, and as we can see in Fig. 4, paths of length three (two mediator concepts) and four (three mediator concepts) return many reference mappings. For example, 945 reference mappings are returned by three-step-paths. This can be the explanation of the high F-Score obtained by our approach (0.928) comparing to the F-Score obtained in their experimentation (0.847 and 0.913 respectively).

Recently, other measures have been proposed to select the most appropriate set of ontologies (which represents the BK) as the effectiveness [13] and the mapping gain [8] measures. The drawback of the proposed measures resides in the fact that they select the whole ontologies (many thousands of concepts) even if we need just for a fragment from these ones. Furthermore, dealing with whole ontologies makes it necessary to limit the number of selected ontologies. In our approach, there is no limitation of the number of selected ontologies, our units are concepts. We select only concepts that may help us to discover new mappings without considering the number of used ontologies. In [8] the selection is based on the mapping gain score. The ontologies with a low mapping gain (less then the defined threshold) are eliminated even if they contain some concepts that may help to discover reliable mappings. In our case, we do not select specific ontologies but we work with all ontologies in the repository at the same time. We propose to follow mappings of different provenances, and select progressively potential useful concepts. Therefore, we combine the lexical overlapping with the human knowledge from mappings produced manually without eliminating any candidate mediator concept.

Furthermore, in all other approaches, the selection and the combination of different ontologies is based only on mappings produced automatically, they do not distinguish different types of mappings (different provenances). They are based mainly on the lexical overlapping between the BK and ontologies to be aligned. This criteria does not guarantee the selection of the best BK. For instance, the huge biomedical ontology SNOMED-CT with its rich lexical content may always be ranked first to match biomedical ontologies, even if more appropriate BK are available as Uberon for Anatomy in [5]. The use of SNOMED-CT needs more resources, memory to manage the whole ontology and time to anchor concepts on it.

Moreover, the *Specific Mapping Graph* could be reused as a resource to map the source ontology with any other ontology. If a new ontology is added to the initial repository, we just need to extract its related mappings with the concepts in the *Specific Mapping Graph* and integrate them. In the previous approaches, one will need to restart the selection process from scratch. The probability of not finding an anchor for a given concept in a rich repository of biomedical ontologies as NCBO BioPrtal (8150126 concepts) is very low. In this case, we can search on the web for ontologies that may contain this concept as proposed in [1,18].

8 Conclusion and Future Work

This paper deals with the selection and the combination of heterogeneous existing mappings, produced manually and automatically, stored in a biomedical repository, to discover new ones. Our approach is based on building the *Specific Mapping Graph* as a BK. Such graph allows to get an alignment of high quality between ontologies to be aligned without using complex lexical and structural measures. One source concept may have several candidates target concepts. To select the most accurate one, we have proposed the *Path Confidence Measure* that takes the confidence of a given path type into account.

The presented evaluation shows that our approach provides good results, competitive to those of state-of-the-art systems. Also, that the reuse of existing mappings allows discovering mappings missed by the previous approaches.

The explanation of final mappings is one of challenges of ontology matching [21]. Indeed, it is very important to be able to justify the provenance of a given mapping instead of a simple score. In our approach, each found mapping is deducted from one or several paths. The edges of paths are tagged with their provenance. Consequently, all found mappings are explained.

Moreover, we have evaluated our approach using one benchmark (Anatomy benchmark). For a better evaluation, we will evaluate it on other OAEI biomedical benchmarks. Also to improve the quality of the final alignment, we plan to study the impact of the variation of the PMC threshold on the F-Score, currently no threshold is applied. Also, the coherence of automatically produced BioPortal mappings has been critiqued in [9]. For this purpose, we plan to integrate a semantic verification into our approach to improve the quality of produced alignment. Currently our approach is used to derive only 1:1 mappings. We will experiment the usefulness of our method to derive n:m mappings. This will be possible if we extract not only mappings but also fragments of ontologies (sequence of concepts linked with is_a relationship) that connect two concepts in the *Specific Mapping Graph* if they belong to the same ontology.

Acknowledgment. This work was achieved during a LIRMM-ESI collaboration within the SIFR project funded in part by the French National Research Agency (grant ANR-12-JS02-01001), as well as by University of Montpellier, the CNRS and the EU H2020 MSCA program.

References

1. Aleksovski, Z., Kate, W.T., Van Harmelen, F.: Exploiting the structure of background knowledge used. In: 1st International Conference on Ontology Matching, vol. 225, pp. 13–24 (2006)
2. Annane, A., Emonet, V., Azouaou, F., Jonquet, C.: Multilingual mapping reconciliation between english-french biomedical ontologies. In: 6th International Conference on Web Intelligence, Mining, Semantics, WIMS, pp. 13:1–13:12 (2016)
3. Bodenreider, O.: The Unified Medical Language System (UMLS): integrating biomedical terminology. Nucleic Acids Res. **32**, 267–270 (2004)

4. Cheatham, M., et al.: Results of the ontology alignment evaluation initiative 2015. In: 10th ISWC Workshop on Ontology Matching, pp. 60–115 (2015)
5. Chen, X., Xia, W., Jiménez-Ruiz, E., Cross, V.V.: Extending an ontology alignment system with bioportal: a preliminary analysis. In: ISWC, pp. 313–316 (2014)
6. Smith, B., et al.: The OBO foundry: coordinated evolution of ontologies to support biomedical data integration. Nat. Biotechnol. **25**(11), 1251–1255 (2007)
7. Euzenat, J., Shvaiko, P.: Ontology Matching. Springer, Heidelberg (2013)
8. Faria, D., et al.: Automatic background knowledge selection for matching biomedical ontologies. PloS one **9**(11), e111226 (2014)
9. Faria, D., Jiménez-Ruiz, E., Pesquita, C., Santos, E., Couto, F.M.: Towards annotating potential incoherences in bioportal mappings. In: Mika, P., et al. (eds.) ISWC 2014, Part II. LNCS, vol. 8797, pp. 17–32. Springer, Heidelberg (2014)
10. Ghazvinian, A., Noy, N.F., Musen, M.A.,et al.: Creating mappings for ontologies in biomedicine: simple methods work. In: AMIA, pp. 198–202 (2009)
11. Ghazvinian, A., Noy, N.F., Jonquet, C., Shah, N., Musen, M.A.: What four million mappings can tell you about two hundred ontologies. In: Bernstein, A., Karger, D.R., Heath, T., Feigenbaum, L., Maynard, D., Motta, E., Thirunarayan, K. (eds.) ISWC 2009. LNCS, vol. 5823, pp. 229–242. Springer, Heidelberg (2009)
12. Gross, A., Hartung, M., Kirsten, T., Rahm, E.: Mapping composition for matching large life science ontologies. In: ICBO, pp. 109–116 (2011)
13. Hartung, M., Gross, A., Kirsten, T., Rahm, E.: Effective mapping composition for biomedical ontologies. In: ESWC, pp. 176–190 (2012)
14. Jiménez-Ruiz, E., Grau, B.C., Solimando, A., Cross, V.V.: Logmap family results for OAEI 2015. In: 10th International Workshop on Ontology Matching, pp. 171–175 (2015)
15. Jonquet, C., Shah, N., Musen, M.: The open biomedical annotator. In: AMIA Summit on Translational Bioinformatics, pp. 56–60 (2009)
16. Locoro, A., David, J., Euzenat, J.: Context-based matching: design of a flexible framework and experiment. J. Data Semant. **3**(1), 25–46 (2014)
17. Mascardi, V., Locoro, A., Rosso, P.: Automatic ontology matching via upper ontologies: a systematic evaluation. IEEE Trans. Knowl. Data Eng. **22**(5), 609–623 (2010)
18. Quix, C., Roy, P., Kensche, D.: Automatic selection of background knowledge for ontology matching. In: International Workshop on Semantic Web Information Management, p. 5. ACM (2011)
19. Sabou, M., d'Aquin, M., Motta, E.: Exploring the semantic web as background knowledge for ontology matching. In: Spaccapietra, S., Pan, J.Z., Thiran, P., Halpin, T., Staab, S., Svatek, V., Shvaiko, P., Roddick, J. (eds.) Journal on Data Semantics XI. LNCS, vol. 5383, pp. 156–190. Springer, Heidelberg (2008)
20. Safar, B., Reynaud, C., Calvier, F.: Techniques d'alignement d'ontologies basées sur la structure dune ressource complémentaire. 1ères Journées Francophones sur les Ontologies, pp. 21–35 (2007)
21. Shvaiko, P., Euzenat, J.: Ontology matching: state of the art and future challenges. IEEE Trans. Knowl. Data Eng **25**(1), 158–176 (2013)

Populating a Knowledge Base with Object-Location Relations Using Distributional Semantics

Valerio Basile[1(✉)], Soufian Jebbara[2], Elena Cabrio[1], and Philipp Cimiano[2]

[1] Université Côte d'Azur, Inria, CNRS, I3S, Sophia Antipolis, France
valerio.basile@inria.fr, elena.cabrio@unice.fr
[2] Bielefeld University, Bielefeld, Germany
{sjebbara,cimiano}@cit-ec.uni-bielefeld.de

Abstract. The paper presents an approach to extract knowledge from large text corpora, in particular knowledge that facilitates object manipulation by embodied intelligent systems that need to act in the world. As a first step, our goal is to extract the prototypical location of given objects from text corpora. We approach this task by calculating relatedness scores for objects and locations using techniques from distributional semantics. We empirically compare different methods for representing locations and objects as vectors in some geometric space, and we evaluate them with respect to a crowd-sourced gold standard in which human subjects had to rate the prototypicality of a location given an object. By applying the proposed framework on DBpedia, we are able to build a knowledge base of 931 high confidence object-locations relations in a fully automatic fashion (The work in this paper is partially funded by the ALOOF project (CHIST-ERA program)).

1 Introduction

Embodied intelligent systems such as robots require world knowledge to be able to perceive the world appropriately and perform appropriate actions on the basis of their understanding of the world. Take the example of a domestic robot that has the task of tidying up an apartment. A robot needs, e.g., to categorize different objects in the apartment, know where to put or store them, know where and how to grasp them, and so on. Encoding such knowledge by hand is a tedious, time-consuming task and is inherently prone to yield incomplete knowledge. It would be desirable to develop approaches that can extract such knowledge automatically from data.

To this aim, in this paper we present an approach to extract object knowledge from large text corpora. Our work is related to the machine reading and open information extraction paradigms aiming at learning generic knowledge from text corpora. In contrast, in our research we are interested in particular in extracting knowledge that facilitates object manipulation by embodied intelligent systems that need to act in the world. Specifically, our work focuses on the problem of

E. Blomqvist et al. (Eds.): EKAW 2016, LNAI 10024, pp. 34–50, 2016.
DOI: 10.1007/978-3-319-49004-5_3

relation extraction between entities mentioned in the text[1]. A *relation* is defined in the form of a tuple $t = (e_1; e_2; ...; e_n)$ where the e_i are entities in a predefined relation r within document D [1]. We develop a framework with foundations in *distributional semantics*, the area of Natural Language Processing that deals with the representation of the meaning of words in terms of their distributional properties, i.e., the context in which they are observed. It has been shown in the literature that distributional semantic techniques give a good estimation of the *relatedness* of concepts expressed in natural language (see Sect. 3 for a brief overview of distributional semantics principles). Semantic relatedness is useful for a number of tasks, from query expansion to word association, but it is arguably too general to build a general knowledge base, i.e., a triple like <`entity1, relatedTo, entity2`> might not be informative enough for many purposes.

Distributional Relation Hypothesis. We postulate that the relatedness relation encoded in distributional vector representations can be made more precise based on the type of the entities involved in the relation, i.e., if two entities are distributionally related, the natural relation that comes from their respective types is highly likely to occur. For example, the *location* relation that holds between an *object* and a *room* is represented in a distributional space if the entities representing the object and the room are highly associated according to the distributional space's metric.

Based on this assumption, as a first step of our work, we extract the prototypical location of given objects from text corpora. We frame this problem as a ranking task in which, given an object, our method computes a ranking of locations according how protoypical a location they are for this object. We build on the principle of distributional similarity and map each location and object to a vector representation computed on the basis of words these objects or locations co-occur with in a corpus. For each object, the locations are then ranked by the cosine similarity of their vector representations.

The paper is structured as follows. Section 2 discusses relevant literature, while Sect. 3 provides a background on word and entity vector spaces. Section 4 describes the proposed framework to extract relations from text. Section 5 reports on the creation of the goldstandard, and on the experimental results. Section 6 describes the obtained knowledge base of object locations, while conclusions end the paper.

2 Related Work

Our work relates to the three research lines discussed below, i.e.: *(i)* machine reading, *(ii)* supervised relation extraction, and *(iii)* encoding common sense knowledge in domain-independent ontologies and knowledge bases.

[1] In the rest of the paper, the labels of the entities are identifiers from DBpedia URIs, stripped of the namespace http://dbpedia.org/resource/ for readability.

The Machine Reading Paradigm. In the field of knowledge acquisition from the Web, there has been substantial work on extracting taxonomic (e.g. hypernym), part-of relations [15] and complete qualia structures describing an object [8]. Quite recently, there has been a focus on the development of systems that can extract knowledge from any text on any domain (the open information extraction paradigm [13]). The DARPA Machine Reading Program [2] aims at endowing machines with capabilities for lifelong learning by automatically reading and understanding texts (e.g. [12]). While such approaches are able to quite robustly acquire knowledge from texts, these models are not sufficient to meet our objectives since: *(i)* they lack visual and sensor-motor grounding, *(ii)* they do not contain extensive object knowledge. Thus, we need to develop additional approaches that can harvest the Web to learn about usages, appearance and functionality of common objects. While there has been some work on grounding symbolic knowledge in language [29], so far there has been no serious effort to compile a large and grounded object knowledge base that can support cognitive systems in understanding objects.

Supervised Relation Extraction. While machine reading attempts to acquire general knowledge by reading texts, other works attempt to extract specific relations applying supervised techniques to train classifiers. A training corpus in which the relation of interest is annotated is typically assumed (e.g. [6]). Another possibility is to rely on the so called *distant supervision* assumption and use an existing knowledge base to bootstrap the process by relying on triples or facts in the knowledge base to label examples in a corpus (e.g. [17,18,36,38]). Other researchers have attempted to extract relations by reading the Web, e.g. [4]. Our work differs from these approaches in that, while we are extracting a specific relation, we do not rely on supervised techniques to train a classification model, but rather rely on semantic relatedness and distributional similarity techniques to populate a knowledge base with the relation in question.

Ontologies and KB of Common Sense Knowledge. DBpedia[2] is a large-scale knowledge base automatically extracted from semi-structured parts of Wikipedia. Besides its sheer size, it is attractive for the purpose of collecting general knowledge given the one-to-one mapping with Wikipedia (allowing us to exploit the textual and structural information contained in there) and its position as the central hub of the Linked Open Data cloud.

YAGO [34] is an ontology automatically created by mapping relations between WordNet synsets such as hypernymy and relations between Wikipedia pages such as links and redirects to semantic relations between concepts. Despite its high coverage, for our goals YAGO suffers from the same drawbacks of DBpedia, i.e. a lack of general relations between entities that are not instance of the DBpedia ontology, such as common objects. While a great deal of relations and properties of named entities are present, knowledge about, e.g. the location or the functionality of entities is missing.

[2] http://dbpedia.org.

ConceptNet[3] [23] is a semantic network containing lots of things computers should know about the world. While it shares the same goals of the knowledge base we aim at building, ConceptNet is not a Linked Open Data resource. In fairness, the resource is in a graph-like structure, thus RDF triples could be extracted from it, and the building process provides a way of linking the nodes to DBpedia entities, among other LOD resources. However, we cannot integrate ConceptNet directly in our pipeline because of the low coverage of the mapping with DBpedia—of the 120 DBpedia entities in our gold standard (see Sect. 5) only 23 have a correspondent node in ConceptNet.

OpenCyC[4] attempts to assemble a comprehensive ontology and knowledge base of everyday common sense knowledge, with the goal of enabling AI applications to perform human-like reasoning. While for the moment in our work we focus on specific concepts and relations relevant to our scenario, we will consider linking them to real-world concepts in OpenCyc.

3 Background: Word and Entity Vector Spaces

Word space models (or distributional space models, or word vector spaces) are abstract representations of the meaning of words, encoded as vectors in a high-dimensional space. A word vector space is constructed by counting *cooccurrences* of pairs of words in a text corpus, building a large square n-by-n matrix where n is the size of the vocabulary and the cell i, j contains the number of times the word i has been observed in cooccurrence with the word j. The i-th row in a cooccurrence matrix is a n-dimensional vector that acts as a *distributional* representation of the i-th word in the vocabulary. Words that appear in similar contexts often have similar representations in the vector space; this similarity is geometrically measurable with a distance metric such as cosine similarity, defined as the cosine of the angle between two vectors. This is the key point to linking the vector representation to the idea of semantic relatedness, as the *distributional hypothesis* states that "words that occur in the same contexts tend to have similar meaning" [16]. Several techniques can be applied to reduce the dimensionality of the cooccurrence matrix. Latent Semantic Analysis [21], for instance, uses Singular Value Decomposition to prune the less informative elements while preserving most of the topology of the vector space, and reducing the number of dimensions to 100–500.

In parallel, neural network-based models have recently began to rise to prominence. To compute word embeddings, several models rely on huge amounts of natural language texts from which a vector representation for each word is learned by a neural network. Their representations of the words are based on *prediction* as opposed to *counting* [3].

Vector spaces created on word distributional representations have been successfully proven to encode word similarity and relatedness relations [9,31,32],

[3] http://conceptnet5.media.mit.edu/.

[4] http://www.opencyc.org/; as RDF representations: http://sw.opencyc.org/.

while word embeddings have proven to be a useful feature in many natural language processing tasks [10,22,33] in that they often encode semantically meaningful information of a word.

4 Word Embeddings for Relation Extraction

This section presents our framework to extract relations from natural language text. The methods are based on distributional semantics, but present different approaches to compute vector representations of entities: one is based on a word embedding approach (Sect. 4.1), the other on a LSA-based representation of DBpedia entities (Sect. 4.2). We present one framework for which we test different ways of calculating the vector embeddings, each one having its own specificities and strengths.

4.1 A Word Space Model of Entity Lexicalizations

In this section, we propose a neural network-based word embedding method for the automatic population of a knowledge base of object-location relations. As outlined in Sect. 1, we frame this task as a ranking problem and score the vector representation for object-location pairs with respect to how prototypical the location is for the given object. Many word embedding methods encode useful semantic and syntactic properties [20,26,28] that we leverage for the extraction of object-location relations. In this work, we restrict our experiments to the skip-gram method [25]. The objective of the skip-gram method is to learn word representations that are useful for predicting context words. As a result, the learned embeddings often display a desirable linear structure [26,28]. In particular, word representations of the skip-gram model often produce meaningful results using simple vector addition [26]. For this work, we trained the skip-gram model on a corpus of roughly 83 million Amazon reviews [24].

Motivated by the compositionality of word vectors, we derive vector representations for the entities as follows: considering a DBpedia entity such as `Public_Toilet` (we call this label the *lexicalization*), we clean it by removing parts in parenthesis, convert it to lower case, and split it into its individual words. We retrieve the respective word vectors from our pretrained word embeddings and sum them to obtain a single vector, namely, the vector representation of the entity: $vector(public\_toilet) = vector(public) + vector(toilet)$. The generation of entity vectors is trivial for "single-word" entities, such as `Cutlery` or `Kitchen`, that are already contained in our word vector vocabulary. In this case, the entity vector is simply the corresponding word vector. With this derived set of entity vector representations, we compute cosine vector similarity score for object-location pairs. This score is an indicator of how typical the location for the object is. Given an object, we can create a ranking of locations with the most likely location candidates at the top of the list (see Table 1).

Table 1. Locations for a sample object, extracted by computing cosine similarity on skip-gram-based vectors.

Object	Location	Cosine similarity
Dishwasher	Kitchen	.636
	Laundry_room	.531
	Pantry	.525
	Wine_cellar	.519

4.2 Distributional Representations of Entities

Vector representations of words (Sect. 4.1) are attractive since they only require a sufficiently large text corpus with no manual annotation. However, the drawback of focusing on words is that a series of linguistic phenomena may affect the vector representation. For instance, a polysemous word as *rock* (stone, musical genre, metaphorically strong person, etc.) is represented by a single vector where all the senses are conflated.

NASARI [7], a resource containing vector representations of most of DBpedia entities, solves this problem by building a vector space of concepts. The NASARI vectors are actually distributional representations of the entities in BabelNet [30], a large multilingual lexical resource linked to Wordnet, DBpedia, Wiktionary and other resources. The NASARI approach collects cooccurrence information of concepts from Wikipedia and then applies a LSA-like procedure for dimensionality reduction. The context of a concept is based on the set of Wikipedia pages where a mention of it is found. As shown in [7], the vector representations of entities encode some form of semantic relatedness, with tests on a sense clustering task showing positive results. Table 2 shows a sample of pairs of NASARI vectors together with their pairwise cosine similarity ranging from −1 (totally unrelated) to 1 (identical vectors).

Table 2. Examples of cosine similarity computed on NASARI vectors.

	Cherry	Microsoft
Apple	.917	.325
Apple_Inc	.475	.778

Following the hypothesis put forward in the introduction, we focus on the extraction of object-location relations by computing the cosine similarities of object and location entities. We exploit the alignment of BabelNet with DBpedia, thus generating a similarity score for pairs of DBpedia entities. For example, the DBpedia entity `Dishwasher` has a cosine similarity of .803 to the entity `Kitchen`, but only .279 with `Classroom`, suggesting that the appropriate location for a generic dishwasher is the kitchen rather than a classroom. Since cosine similarity is a graded value on a scale from −1 to 1, we can generate, for a given object, a ranking of candidate locations, e.g., the rooms of a house. Table 3 shows a sample of object-location pairs of DBpedia labels, ordered by the cosine similarity of their respective vectors in NASARI. Prototypical locations for the objects show up at the top of the list as expected, indicating a relationship between the semantic relatedness expressed by the cosine similarity of vector representations and the actual locative relation of entities.

Table 3. Locations for a sample object, extracted by computing cosine similarity on NASARI vectors.

Object	Location	Cos. similarity
Dishwasher	Kitchen	.803
	Air_shower_(room)	.788
	Utility_room	.763
	Bathroom	.758

5 Evaluation

This section presents the evaluation of the proposed framework for relation extraction (Sect. 4). We collected a set of relations rated by human subjects to provide a common benchmark, and we test several methods with varying values for their parameters. We then adopt the best performing method to automatically build a knowledge base and test its quality against the manually created gold standard dataset.

5.1 Gold Standard

To test our hypothesis, we collected a set of human judgments about the likelihood of objects to be found in certain locations. To select the objects and locations for this experiment, every DBpedia entity that falls under the category `Domestic_implements`, or under one of the `narrower` categories than `Domestic_implements` according to SKOS[5], is considered an object; every DBpedia entity that falls under the category `Rooms` is considered a location. This step results in 336 objects and 199 locations.

To select suitable object-location pairs for the creation of the gold standard, we need to filter out odd or uncommon examples of objects or locations like `Ghodiyu` or `Fainting_room`. For example, the rankings produced by the cosine similarity of NASARI vectors (Table 3) are cluttered with results that are less prototypical because of their uncommonness. An empirical measure of

[5] Simple Knowledge Organization System: https://www.w3.org/2004/02/skos/.

commonness of entities could be used to rerank or filter the result to improve its generality. To this extent, we use the URI counts extracted from the parsing of Wikipedia with the DBpedia Spotlight tool for entity linking [11]. These counts are derived, for each DBpedia entity, from the number of incoming links to its correspondent Wikipedia page. We use it as an approximation of the notion of commonness of locations, e.g., a `Kitchen` (URI count: 742) is a more common location than a `Billiard_room` (URI count: 82). Table 4 shows an example of using such counts to filter out irrelevant entries from the ranked list of candidate locations for the entity `Paper_towel` according to NASARI-based similarity.

Table 4. Locations for `Paper_towel`, extracted by computing cosine similarity on NASARI vectors with URI count. Locations with frequency <100 are in gray.

Location	URI count	Cosine similarity
Air_shower_(room)	0	.671
Public_toilet	373	.634
Mizuya	11	.597
Kitchen	742	.589

We rank the 66,864 pairs of `Domestic_implements` and `Rooms` using the aforementioned entity frequency measure and select the 100 most frequent objects and the 20 most frequent locations (2,000 object-location pairs in total). Examples of pairs: (`Toothbrush,Hall`), (`Wallet, Ballroom`) and (`Nail_file, Kitchen`).

In order to collect the judgments, we set up a crowdsourcing experiment on the Crowdflower platform[6]. For each of the 2,000 object-location pairs, contributors were asked to rate the likelihood of the object to be in the location out of four possible values:

- **−2** (**unexpected**): finding the object in the room would cause surprise, e.g., it is unexpected to find a bathtub in a cafeteria.
- **−1** (**unusual**): finding the object in the room would be odd, the object feels out of place, e.g., it is unusual to find a mug in a garage.
- **1** (**plausible**): finding the object in the room would not cause any surprise, it is seen as a normal occurrence, e.g., it is plausible to find a funnel in a dining room.
- **2** (**usual**): the room is the place where the object is typically found, e.g., the kitchen is the usual place to find a spoon.

Contributors are shown ten examples per page, instructions, a short description of the entities (the first sentence from the Wikipedia abstract), a picture (from Wikimedia Commons, when available), and the list of possible answers as labeled radio buttons.

[6] http://www.crowdflower.com/.

After running the crowdsourcing experiment for a few hours, we collected 12,767 valid judgments (455 were deemed "untrusted" by Crowdflower's quality filtering system based on a number of test questions we provided). Most of the pairs have received at least 5 separate judgments, with some outliers collecting more than one hundred judgments each. The average agreement, i.e. percentage of contributors that answered the most common answer for a given question, is 64.74 %. The judgments are skewed towards the negative end of the spectrum, as expected, with 37 % pairs rated unexpected, 30 % unusual, 24 % plausible and 9 % usual. The cost of the experiment was 86 USD.

5.2 Ranking Evaluation

The proposed methods produce a ranking on top of a list of locations, given an input object. To test the validity of our methods we need to compare their output against a gold standard ranking. The latter is extracted from the dataset described in Sect. 5.1 by assigning to each object-location pair the average of the numeric values of the judgments received. For instance, if the pair (`Wallet`, `Ballroom`) has been rated −2 (unexpected) six times, −1 (unusual) three times, and never 1 (plausible) or 2 (usual), its score will be about −1.6, indicating that a `Wallet` is not very likely to be found in a `Ballroom`. The pairs are then ranked by this averaged score on a per-object basis.

As a baseline, we apply two simple methods based on entity frequency. In the *location frequency* baseline, the object-location pairs are ranked according to the frequency of the location. The ranking is thus the same for each object, since the score of a pair is only computed based on the location. This method makes sense in absence of any further information on the object: e,g, a robot tasked to find an unknown object should inspect "common" rooms such as a kitchen or a studio first, rather than "uncommon" rooms such as a pantry. The second baseline (*link frequency*) is based on counting how often every object is mentioned on the Wikipedia page of every location and vice versa. A ranking is produced based on these counts. An issue is that they could be sparse, i.e., most object-location pairs have a count of 0, thus sometimes producing no value for the ranking for an object. This is the case for rather "unusual" objects and locations.

For each object in the dataset, we compare the location ranking produced by our algorithms to the gold standard ranking and compute the Normalized Discounted Cumulative Gain (NDCG), a measure of rank correlation used in information retrieval that gives more weight to the results at the top of the list than at its bottom. This choice of evaluation metric follows from the idea that it is more important to guess the position in the ranking of most likely locations for a given object than to the least likely locations. Table 5 shows the average NDCG across all objects: methods *NASARI-sim* (Sect. 4.2) and *SkipGram-sim* (Sect. 4.1), plus the two baselines introduced above. Both our methods outperform the baselines with respect to the gold standard rankings.

Table 5. Average NDCG of the produced rankings against the gold standard rankings.

Method	NDCG
Location frequency baseline	.851
Link frequency baseline	.875
NASARI-sim	.903
SkipGram-sim	.912

5.3 Precision Evaluation

The NDCG measure gives a complete account of the quality of the produced rankings, but it is not easy to interpret apart from comparisons of different outputs. To gain a better insight into our results, we provide an alternative evaluation based on the "precision at k" measure. This Information Retrieval measure is the number of retrieved items that are ranked in the top-k part of the retrieved list and of the relevance ranking. In our experiments, for a given object, precision at k is the number of locations among the first k of the produced rankings that are also among the top-k locations in the gold standard ranking. It follows that, with $k = 1$, precision at 1 is 1 if the top returned location is the top location in the gold standard, and 0 otherwise. We compute the average of precision at k for $k = 1$ and $k = 3$ across all the objects. The results are shown in Table 6.

Table 6. Average precision at k for $k = 1$ and $k = 3$.

Method	Precision at 1	Precision at 3
Location frequency baseline	.000	.008
Link frequency baseline	.280	.260
NASARI-sim	.390	.380
SkipGram-sim	.350	.400

As for the rank correlation evaluation, our methods outperform the baselines. The location frequency baseline performs very poorly, due to an idiosyncrasy in the frequency data, that is, the most "frequent" location in the dataset is *Aisle*. This behavior reflects the difficulty in evaluating this task using only automatic metrics, since automatically extracted scores and rankings may not correspond to common sense judgment.

The NASARI-based similarities outperform the SkipGram-based method when it comes to guessing the most likely location for an object, as opposed to the better performance of SkipGram-sim in terms of precision at 3 and rank correlation (Sect. 5.2).

We explored the results and found that for 19 objects out of 100, NASARI-sim correctly guesses the top ranking location but SkipGram-sim fails, while the opposite happens 15 out of 100 times. We also found that the NASARI-based method has a lower coverage than the other method., due to the coverage of the original resource (NASARI), where not every entity in DBpedia is assigned a vector (objects like `Back- pack` and `Comb`, and locations like `Loft` are all missing). The SkipGram-based method also suffer from this problem, however, only for very rare or uncommon objects and locations (as `Triclinium` or `Jamonera`). These findings suggest that the two methods could have different strengths and weaknesses. In the following section we show two strategies to combine them.

5.4 Hybrid Methods: Fallback Pipeline and Linear Combination

The results from the previous sections highlight that the performance of our two main methods may differ qualitatively. In an effort to overcome the coverage issue of NASARI-sim, and at the same time experiment with hybrid methods to extract location relations, we devised two simple ways of combining the SkipGram-sim and NASARI-sim methods. The first method is based on a fallback strategy: given an object, we consider the pair similarity of the object to the top ranking location according to NASARI-sim as a measure of confidence. If the top ranked location among the NASARI-sim ranking is exceeding a certain threshold, we consider the ranking returned by NASARI-sim as reliable. Otherwise, if the similarity is below the threshold, we deem the result unreliable and we adopt the ranking returned by SkipGram-sim instead. The second method produces an object-location similarity scores by linear combination of the NASARI and SkipGram similarities. The similarity score for the generic pair o, l is thus given by $sim(o,l) = \alpha sim_{NASARI}(o,l) + (1-\alpha) sim_{SkipGram}(o,l)$, where parameter α controls the weight of one method w.r.t. the other.

Table 7. Rank correlation and precision at k for the method based on fallback strategy.

Method	NDCG	precision at 1	precision at 3
Fallback strategy (threshold=.4)	.907	.410	.393
Fallback strategy (threshold=.5)	.906	.400	.393
Fallback strategy (threshold=.6)	.908	.410	.406
Fallback strategy (threshold=.7)	.909	.370	.396
Fallback strategy (threshold=.8)	.911	.360	.403
Linear combination (α=.0)	.912	.350	.400
Linear combination (α=.2)	.911	.380	.407
Linear combination (α=.4)	.913	.400	.423
Linear combination (α=.6)	.911	.390	.417
Linear combination (α=.8)	.910	.390	.410
Linear combination (α=1.0)	.903	.390	.380
Max	.911	.410	.413

Table 7 shows the obtained results, with varying values of the parameters *threshold* and α. The line labeled *Max* shows the result obtained by choosing the highest similarity between NASARI-sim and SkipGram-sim, for comparison. While the NDCG is basically not affected, both precision at 1 and precision at 3 show an increase in performance with respect to any of the previous methods.

6 Building a Knowledge Base of Object Locations

In the previous section, we tested how the proposed methods succeed in determining the relation between given objects and locations on a closed set of entities (for the purpose of evaluation). In this section we return to the original motivation of this work, that is, to collect location information about objects in an automatic fashion.

All the methods introduced in this work are based on some measure of relatedness between entities, expressed as a real number in the range [−1,1] interpretable as a sort of confidence score relative to the target relation. Therefore, by imposing a threshold on the similarity scores and selecting only the object-location pairs that score above said threshold, we can extract a high-confidence

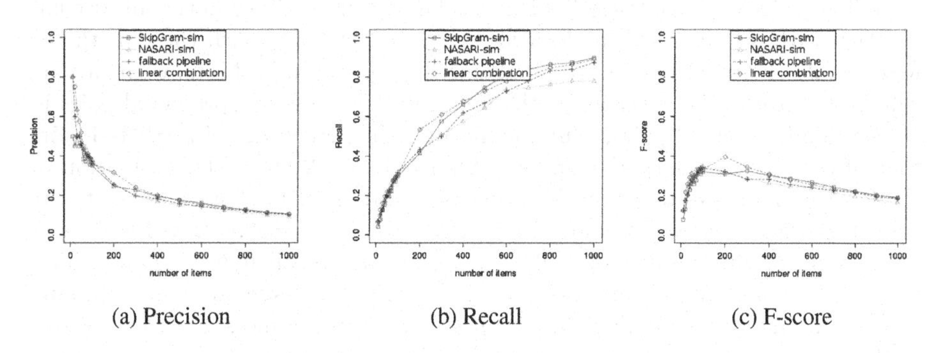

(a) Precision (b) Recall (c) F-score

Fig. 1. Evaluation on automatically created knowledge bases ("usual" locations).

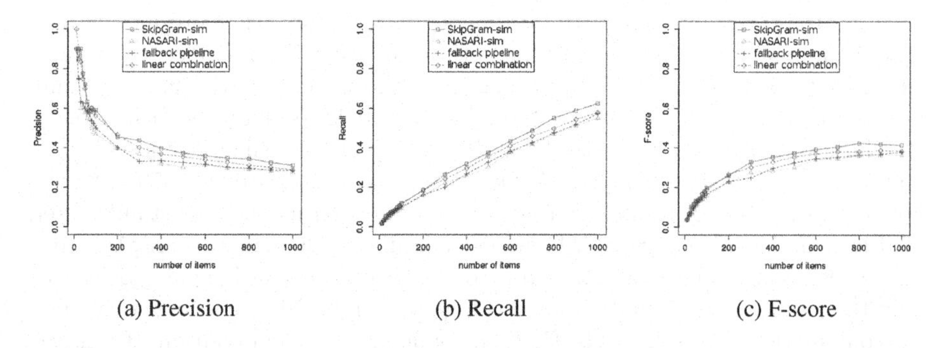

(a) Precision (b) Recall (c) F-score

Fig. 2. Evaluation on automatically created knowledge bases ("plausible" and "usual" locations).

set of object-location relations to build a new knowledge base from scratch. Moreover, by using different values for the threshold, we are able to control the quality and the coverage of the produced relations.

We test this approach on the gold standard dataset introduced in Sect. 5, using the version with data aggregated by Crowdflower: the constributors' answers are aggregated using relative majority, that is, each object-location pair has exactly one judgment assigned to it, corresponding to the most popular judgment among all the contributors that answered that question. We extract two lists of relations from this dataset to be used as a gold standard for experimental tests: one list of the 156 pairs rated 2 (*usual*) by the majority of contributors, and a larger list of the 496 pairs rated either 1 (*plausible*) or 2 (*usual*). The aggregated judgments in the gold standard have a confidence score assigned to them by Crowdflower, based on a measure of inter-rater agreement. Pairs that score low on this confidence measure ($\leqslant 0.5$) were filtered out, leaving respectively 118 pairs in the "usual" set 496 pairs in the "plausible or usual" set.

We order the object-location pairs produced by our two main methods by similarity score, and select the first n from the list, with n being a parameter. We also add to the comparison the results of the two hybrid methods from Sect. 5.4, with the best performing parameters in terms of precision at 1, namely the fallback strategy with threshold on similarity equal to 0.6 and the linear combination with $\alpha = 0.4$. For the location relations extracted with these methods, we compute the precision and recall against the gold standard sets, with varying values of n. Here, the precision is the percentage of correctly predicted pairs in the set of all predicted pairs, while the recall is the percentage of predicted pairs that also occur in the gold standard. Figures 1 and 2 show the evaluation of the four methods evaluated against the two aggregated gold standard datasets described above. Figures 1c and 2c, in particular, show F-score plots for a direct comparison of the performance. The precision and recall figures show similar performances for all the methods, with the SkipGram-sim method obtaining a generally higher recall. The SkipGram-sim method produces generally better-quality sets of relations. However, if the goal is high precision, the other methods may be preferable.

Given these results, we can aim for a high-confidence knowledge base by selecting the threshold on object-location similarity scores that produces a reasonably high precision knowledge base in the evaluation. For instance, the knowledge base made by the top 50 object-location pairs extracted with the linear combination method ($\alpha = 0.4$) has 0.52 precision and 0.22 recall on the "usual" gold standard (0.70 and 0.07 respectively on the "usual" or "plausible" set, see Figs. 1a and 2a). The similarity scores in this knowledge base range from 0.570 to 0.866. Following the same methodology that we used to construct the gold standard set of objects and locations (Sect. 5.1), we extract all the 336 `Domestic_implements` and 199 `Rooms` from DBpedia, for a total of 66,864 object-location pairs. Selecting only the pairs whose similarity score is higher than 0.570, according to the linear combination method, yields 931 high confidence location relations. Of these, only 52 were in the gold standard set of pairs (45 were rated "usual" or "plausible" locations), while the remaining 879 are new, such as (`Trivet`, `Kitchen`),

(`Flight_bag`, `Airport_lounge`) or (`Soap_dispenser`, `Unisex_public_toilet`). The distribution of objects across locations has an arithmetic mean of 8.9 objects per location and standard deviation 11.0.`Kitchen` is the most represented location with 89 relations, while 15 out of 107 locations are associated with one single object.[7]

7 Conclusion and Future Work

This paper presents novel methods to extract object relations, focusing on the typical locations of common objects. The proposed approaches are based on distributional semantics, where vector spaces are built that represent words or concepts in a high-dimensional space. We then map vector distance to semantic relatedness, and instantiate a specific relation that depends on the type of the entities involved (e.g., an object highly related to a room indicates that the room is a typical location for the object)[8].

The NASARI-based scoring method is a concept-level vectors space model derived from BabelNet. The skip-gram model (Sect. 4.1) is trained on Amazon review data and offers a word-level vector space which we exploit for scoring object-location pairs. Experiments on a crowdsourced dataset of human judgments show that they offer different advantages. To combine their strengths, we test two combination strategies, and show an improvement on their performances. Finally, we select the best parameters to extract a new, high-precision knowledge base of object locations.

As future work, we would like to employ *retrofitting* [14] to enrich our pre-trained word embeddings with concept knowledge from a semantic network such as ConceptNet or WordNet [27] in a post-processing step. With this technique, we might be able to combine the benefits of the concept-level and word-level semantics in a more sophisticated way to bootstrap the creation of an object-location knowledge base. We believe that this method is a more appropriate tool than the simple linear combination of scores. By specializing our skip-gram embeddings for relatedness instead of similarity [19] even better results could be achieved. Apart from that, we would like to investigate knowledge base embeddings and graph embeddings [5,35,37] that model entities and relations in a vector space in more detail. By defining an appropriate training objective, we might be able to compute embeddings that encode directly object-location relations and thus are tailored more precisely to our task at hand. Finally, we used the frequency of entity mentions in Wikipedia as a measure of commonality to drive the creation of a gold standard set for evaluation. This information, or equivalent measures, could be integrated directly into our relation extraction framework, for example in the form of a weighting scheme, to improve its predictions accuracy.

---

[7] The full automatically created knowledge base is available at http://project.inria.fr/aloof/files/2016/04/objectlocations.nt_.gz.

[8] All the datasets resulting from this work are available at https://project.inria.fr/aloof/data/.

As main limitation of our current work, it needs to be stressed that the relation in question (here *isLocatedAt*) is predicted in all cases where the semantic relatedness is over a certain threshold. Thus, the method described is not specific for the particular relation given. In fact, the relation we predict is a relation of general (semantic) association. In our particular case, the method works due to the specifiy of the types invovled (room and object), which seem to be specific enough to restict the space of possible relations. It is not clear, however, to which other relations our method would generalize. This is left for future investigation. In particular, we intend to extend our method so that a model can be trained to predict a particular relation rather than a generic associative relationship.

References

1. Bach, N., Badaskar, S.: A Review of Relation Extraction (2007)
2. Barker, K., Agashe, B., Chaw, S.Y., Fan, J., Friedland, N., Glass, M., Hobbs, J., Hovy, E., Israel, D., Kim, D.S., Mulkar-Mehta, R., Patwardhan, S., Porter, B., Tecuci, D., Yeh, P.: Learning by reading: a prototype system, performance baseline and lessons learned. In: Proceedings of the 22nd National Conference on Artificial Intelligence, vol. 1, pp. 280–286. AAAI 2007 (2007)
3. Baroni, M., Dinu, G., Kruszewski, G.: Don't count, predict! a systematic comparison of context-counting vs. context-predicting semantic vectors. In: Proceedings of ACL 2014 (vol. 1: Long Papers), June 2014
4. Blohm, S., Cimiano, P., Stemle, E.: Harvesting relations from the web - quantifiying the impact of filtering functions. In: Proceedings of the Twenty-Second AAAI Conference on Artificial Intelligence, pp. 1316–1321 (2007)
5. Bordes, A., Weston, J., Collobert, R., Bengio, Y.: Learning structured embeddings of knowledge bases. Artif. Intell. (Bengio), 301–306 (2011)
6. Bunescu, R.C., Mooney, R.J.: A shortest path dependency kernel for relation extraction. In: HLT/EMNLP. http://acl.ldc.upenn.edu/H/H05/H05-1091.pdf
7. Camacho-Collados, J., Pilehvar, M.T., Navigli, R.: HLT-NAACL (2015)
8. Cimiano, P., Wenderoth, J.: Automatically learning qualia structures from the web. In: Proceedings of the ACL-SIGLEX Workshop on Deep Lexical Acquisition. DeepLA 2005, pp. 28–37 (2005)
9. Ciobanu, A.M., Dinu, A.: Alternative measures of word relatedness in distributional semantics. In: Joint Symposium on Semantic Processing, p. 80 (2013)
10. Collobert, R., Weston, J., Bottou, L., Karlen, M., Kavukcuoglu, K., Kuksa, P.: NLP (almost) from scratch. J. Mach. Learn. Res. **12**, 2493–2537 (2011)
11. Daiber, J., Jakob, M., Hokamp, C., Mendes, P.N.: Improving efficiency and accuracy in multilingual entity extraction. In: Proceedings of I-Semantics (2013)
12. Etzioni, O.: Machine reading at web scale. In: Proceedings of the 2008 International Conference on Web Search and Data Mining, WSDM 2008, p. 2 (2008)
13. Etzioni, O., Fader, A., Christensen, J., Soderland, S., Mausam, M.: Open information extraction: the second generation. In: Proceedings of IJCAI, IJCAI 2011, vol. 1 (2011)
14. Faruqui, M., Dodge, J., Jauhar, S.K., Dyer, C., Hovy, E., Smith, N.A.: Retrofitting word vectors to semantic lexicons. In: Proceedings of NAACL (2015)
15. Girju, R., Badulescu, A., Moldovan, D.: Learning semantic constraints for the automatic discovery of part-whole relations. In: Proceedings of the NAACL 2003, vol. 1 (2003)

16. Harris, Z.: Distributional structure. Word **10**(23), 146–162 (1954)
17. Hoffmann, R., Zhang, C., Ling, X., Zettlemoyer, L.S., Weld, D.S.: Knowledge-based weak supervision for information extraction of overlapping relations. In: Proceedings of ACL 2011, pp. 541–550 (2011)
18. Hoffmann, R., Zhang, C., Weld, D.S.: Learning 5000 relational extractors. In: Proceedings of ACL 2010, pp. 286–295 (2010)
19. Kiela, D., Hill, F., Clark, S.: Specializing word embeddings for similarity or relatedness. In: Proceedings of EMNLP 2015 (September), pp. 2044–2048 (2015)
20. Köhn, A.: What's in an embedding? Analyzing word embeddings through multilingual evaluation. Proc. EMNLP **2015**(2014), 2067–2073 (2015)
21. Landauer, T.K., Dutnais, S.T.: A solution to platos problem: the latent semantic analysis theory of acquisition, induction, and representation of knowledge. Psychol. Rev. **104**(2), 211–240 (1997)
22. Le, Q., Mikolov, T.: Distributed representations of sentences and documents. In: ICML, pp. 1188–1196 (2014)
23. Liu, H., Singh, P.: Conceptnet — a practical commonsense reasoning toolkit. BT Technol. J. **22**(4), 211–226 (2004)
24. McAuley, J.J., Pandey, R., Leskovec, J.: Inferring networks of substitutable and complementary products. In: KDD (2015)
25. Mikolov, T., Corrado, G., Chen, K., Dean, J.: Efficient estimation of word representations in vector space. In: Proceedings of ICLR 2013 (2013)
26. Mikolov, T., Sutskever, I., Chen, K., Corrado, G.S., Dean, J.: Distributed representations of words and phrases and their compositionality. In: Advances in Neural Information Processing Systems, pp. 3111–3119 (2013)
27. Miller, G.A.: Wordnet: a lexical database for English. Commun. ACM **38**(11), 39–41 (1995)
28. Mitchell, J., Lapata, M.: Vector-based models of semantic composition. In: Computational Linguistics (June), pp. 236–244
29. Mooney, R.J.: Learning to connect language and perception. In: Proceedings of the 23rd National Conference on Artificial Intelligence, AAAI 2008, vol. 3, pp. 1598–1601 (2008)
30. Navigli, R., Ponzetto, S.P.: BabelNet: the automatic construction, evaluation and application of a wide-coverage multilingual semantic network. Artif. Intell. **193**, 217–250 (2012)
31. Radinsky, K., Agichtein, E., Gabrilovich, E., Markovitch, S.: A word at a time: computing word relatedness using temporal semantic analysis. In: Proceedings of WWW 2011, pp. 337–346. ACM (2011)
32. Reisinger, J., Mooney, R.J.: Multi-prototype vector-space models of word meaning. In: Proceedings of ACL 2010, pp. 109–117. Association for Computational Linguistics (2010)
33. Santos, C.D., Zadrozny, B.: Learning character-level representations for part-of-speech tagging. In: Proceedings of the 31st ICML, pp. 1818–1826 (2014)
34. Suchanek, F.M., Kasneci, G., Weikum, G.: Yago: a core of semantic knowledge. In: Proceedings of WWW 2007, pp. 697–706. ACM, New York (2007)
35. Sun, Y., Lin, L., Tang, D., Yang, N., Ji, Z., Wang, X.: Modeling mention, context and entity with neural networks for entity disambiguation. In: IJCAI International Joint Conference on Artificial Intelligence, pp. 1333–1339 (2015)
36. Surdeanu, M., Tibshirani, J., Nallapati, R., Manning, C.D.: Multi-instance multi-label learning for relation extraction. In: Proceedings of EMNLP-CoNLL 2012, pp. 455–465 (2012)

37. Weston, J., Bordes, A., Yakhnenko, O., Usunier, N.: Connecting language and knowledge bases with embedding models for relation extraction. In: EMNLP, pp. 1366–1371
38. Xu, W., Hoffmann, R., Zhao, L., Grishman, R.: Filling knowledge base gaps for distant supervision of relation extraction. In: Proceedings of ACL 2013, vol. 2: Short Papers, pp. 665–670 (2013)

Ontology Forecasting in Scientific Literature: Semantic Concepts Prediction Based on Innovation-Adoption Priors

Amparo Elizabeth Cano-Basave[1], Francesco Osborne[2(✉)], and Angelo Antonio Salatino[2]

[1] Aston Business School, Aston University, Birmingham, UK
a.cano-basave@aston.ac.uk

[2] Knowledge Media Institute, Open University, Milton Keynes, UK
{francesco.osborne,angelo.salatino}@open.ac.uk

Abstract. The ontology engineering research community has focused for many years on supporting the creation, development and evolution of ontologies. Ontology forecasting, which aims at predicting semantic changes in an ontology, represents instead a new challenge. In this paper, we want to give a contribution to this novel endeavour by focusing on the task of forecasting semantic concepts in the research domain. Indeed, ontologies representing scientific disciplines contain only research topics that are already popular enough to be selected by human experts or automatic algorithms. They are thus unfit to support tasks which require the ability of describing and exploring the forefront of research, such as trend detection and horizon scanning. We address this issue by introducing the Semantic Innovation Forecast (SIF) model, which predicts new concepts of an ontology at time $t + 1$, using only data available at time t. Our approach relies on lexical innovation and adoption information extracted from historical data. We evaluated the SIF model on a very large dataset consisting of over one million scientific papers belonging to the Computer Science domain: the outcomes show that the proposed approach offers a competitive boost in mean average precision-at-ten compared to the baselines when forecasting over 5 years.

Keywords: Topic evolution · Ontology forecasting · Ontology evolution · Latent semantics · LDA · Innovation priors · Adoption priors · Scholarly data

1 Introduction

The mass of research data on the web is growing steadily, and its analysis is becoming increasingly important for understanding, supporting and predicting the research landscape. Today most digital libraries (e.g., ACM Digital Library, PubMed) and many academic search engines (e.g., Microsoft Academic Search[1],

[1] http://academic.research.microsoft.com/.

E. Blomqvist et al. (Eds.): EKAW 2016, LNAI 10024, pp. 51–67, 2016.
DOI: 10.1007/978-3-319-49004-5_4

Rexplore [21], Saffron [18]) have adopted taxonomies and ontologies for representing the domain of research areas. For example, researchers and publishers in the field of Computer Science are now well familiar with the ACM classification and use it regularly to annotate publications.

However, these semantic classifications are usually hand-crafted and thus are costly to produce. Furthermore, they grow obsolete very quickly, especially in rapidly changing fields such as Computer Science. To alleviate this task is possible to use approaches for ontology evolution and ontology learning. The first task aims to extend, refine and enrich an ontology based on current domain knowledge [23,26]. For example, an ontology of research areas should be updated regularly by including topics which emerged after the last version of the ontology was published. Ontology learning aims instead to automatically generate ontologies by analysing relevant sources, such as relevant scientific literature [20]. Nonetheless, these ontologies still reflect the past, and can only contain concepts that are already popular enough to be selected by human experts or automatic algorithms. Hence, while they are very useful to produce analytics and examine historical data, they hardly support tasks which involve the ability to describe and explore the forefront of research, such as trend detection and horizon scanning. It is thus crucial to develop new methods to allow also the identification of emerging topics in these semantic classifications.

Nonetheless, *predicting the emergence of semantic concepts*, is still a challenge. To the best of our knowledge, predicting the future iteration of a ontology and the relevant concepts that will extend it, which we refer to as *ontology forecasting*, is a novel open question.

For the particular case of scholarly data, being able to predict new research areas can be beneficial for researchers, who are often interested in emerging research areas; for academic publishers, which need to offer the most up-to-date contents; and for institutional funding bodies and companies, which have to make early decisions about critical investments.

In this paper, we address this challenge by presenting a novel framework for the prediction of new semantic concepts in the research domain, which relies on the incorporation of lexical innovation and adoption priors derived from historical data. The main contributions of this work can be summarised as follows:

1. We approach the novel task of ontology forecasting by predicting semantic concepts in the research domain;
2. We introduce two metrics to analyse the linguistic and semantic progressiveness in scholarly data;
3. We propose a novel weakly-supervised approach for the forecasting of innovative semantic concepts in scientific literature;
4. We evaluate our approach in a dataset of over one million documents belonging to the Computer Science domain;
5. Our findings demonstrate that the proposed framework offers competitive boosts in mean average precision at ten for forecasts over 5 years.

2 Related Work

The state of the art presents several approaches for identifying topics in a collection of documents and determining their evolution in time. The most adopted technique for extracting topics from a corpus is Latent Dirichlet Allocation (LDA) [4], which is a generative statistical model that models topics as a multinomial distribution over words. LDA has been extended in a variety of ways for incorporating research entities. For example, the Author-Topic model (ATM) [24] included authorship information in the generative model. Bolelli et al. [6] extended it even further by introducing the Segmented Author-Topic model, which also takes in consideration the temporal ordering of documents to address the problem of topic evolution. In scenarios where it already exists a taxonomy of research areas [21], it is also possible to use entity linking techniques [7] for mapping documents to related concepts. For example, the Smart Topic Miner [22], an application used by Springer Nature for annotating proceedings books, maps keywords extracted from papers to the automatically generated Klink-2 Computer Science Ontology [20] with the aim of selecting a comprehensive set of structured keywords.

The approaches for topic evolution can be distinguished in discriminative and generative [13]. The first ones consider topics as a distribution over words or a mixture over documents and analyse how these change in time using a variety of indexes and techniques [25]. For example, Morinaga and Yamanishi [19] employed a Finite Mixture Model to represent the structure of topics and analyse diachronically the extracted component and Mei and Zhai [16] correlated term clusters via a temporal graph model. However, these methods do not take advantage of the identification of lexical innovations and their adoption across years, but rather focus only on tracking changes in distributions of words.

The second class of approaches for topic evolution employ instead generative topic models [5] on document streams. For example, Gohr et al. [11] used Probabilistic Latent Semantic Analysis and proposed a folding-in techniques for a topic adaptation under an evolving vocabulary. He et al. [13] characterised the analysis of the evolution of topics into the independent topic evolution (ITE) and accumulative topic evolution (ATE) approaches. However, these models do not cater for the identification of novel topics, but rather caters for tracking change of existing ones.

In addition, some approaches aim at supporting ontology evolution by predicting extensions of an ontology. For example, Pesquita and Couto [23] introduced a method for suggesting areas of biomedical ontologies that will likely be extended in the future. Similarly Wang et al. [26] proposed an approach for forecasting patterns in ontology development, with the aim of suggesting which part of an ontology will be next edited by users. Another relevant approach is iDTM (infinite dynamic topic model) [1], which studies the birth, death and evolution of topics in a text stream. iDTM can identify the birth of topics appearing on a given epoch, such topics are considered new when compared to previous epochs. In contrast to their work, our proposed model addresses the prediction of new topics in *future* epochs based on past data rather than identifying topics on the current epoch. In addition, our work is different from all previous approaches because we aim at predicting new classes (concepts) that will appear in the future representations of an ontology.

3 Language and Semantic Progressiveness in Scientific Literature

Previous work has studied the role of language evolution and adoption in online communities showing that users' conformity to innovation can impact the churn or grow of a community [9]. Inspired by this fact, we follow the intuition that language innovation and adoption could impact the generation and expiration of semantic concepts modelling a shared conceptualisation of a domain.

This section presents a motivation for predicting semantic concepts in scientific literature based on the study of the use of language in scholarly data. The following Subsect. 3.1 introduces the dataset used in this paper and presents an analysis of the evolution of language in the field of Computer Science during the course of 14 years in Subsects. 3.2 and 3.3.

3.1 Dataset Description

Our dataset comprises of a collection of research articles relevant to the Computer Science field extracted from Scopus[2], one of the largest databases of peer-reviewed literature. The full 14 years collection ranges from 1995–2008 with a total of 1,074,820 papers. Each year consists of a set of papers categorised within a semantic representation of the Computer Science domain. Such ontological representation is generated per two year-corpus starting from 1998 using the Klink-2 algorithm [20].

The Klink-2 algorithm combines semantic technologies, machine learning and knowledge from external sources (e.g., the LOD cloud, web pages, calls for papers) to automatically generate large-scale ontologies of research areas. It was built to support the Rexplore system [21] a system that integrates statistical analysis, semantic technologies and visual analytics to provide support for exploring and making sense of scholarly data. In particular, the ontology generated by Klink-2 enhances semantically a variety of data mining and information extraction techniques, and improves search and visual analytics.

The classical way to address the problem of classifying research topics has been to adopt human-crafted taxonomies, such as the ACM Computing Classification System and the Springer Nature Classification. However, the ontology created by Klink-2 presents two main advantages over these solutions. Firstly, human-crafted classifications tend to grow obsolete in few years, especially in fields such as Computer Science, where the most interesting topics are the emerging ones. Conversely, Kink-2 can quickly create a new ontology by running on recent data. Secondly, Klink-2 is able to create huge ontologies which includes very large number of concepts which do not appear in current manually created classifications.

[2] Scopus, https://www.elsevier.com/solutions/scopus.

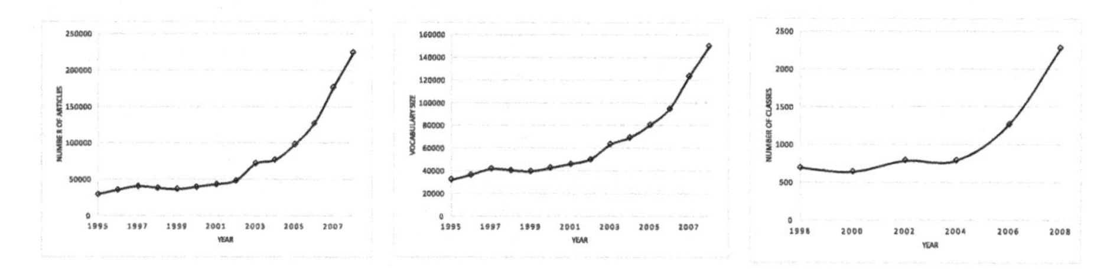

Fig. 1. From left to right, (a) number of articles per year, (b) vocabulary size per year, (c) number of classes per year.

For example, the current version of the full Klink-2 Computer Science ontology includes 17 000 concepts and about 70 000 semantic relationships.

The data model of the Klink-2 ontology is an extension of the BIBO ontology which in turn builds on SKOS. It includes three semantic relations: *skos:broaderGeneric*, which indicates that a topic is a sub-area of another one (e.g., Linked Data is considered a sub-area of Semantic Web); *relatedEquivalent*, which indicates that two topics can be treated as equivalent for the purpose of exploring research data (e.g., Ontology Matching, Ontology Mapping); and *contributesTo*, which indicates that the research outputs of one topic significantly contribute to research into another (e.g., Ontology Engineering contributes to Semantic Web, but arguably it is not its sub-area).

The ontologies associated to different years were computed by feeding to Klink-2 all publications up to that year, to simulate the normal situation in which Klink-2 regularly updates the Computer Science ontology according to most recent data. Figure 1 presents general statistics of the dataset including number of articles, size of the vocabularies and number of semantic concepts per year ontology. Each paper is represented by its title and abstract. Vocabulary sizes where computed after removing punctuation, stopwords and computing Porter stemming [27]. The data presented in Fig. 1 indicates that as years go by the production of scholarly articles for the Computer Science increases. Moreover, it shows that as more articles are introduced each year, novel words – not mentioned in previous years– are also appearing. When analysing the number of semantic concept over time we see that every year there is also an augmentation of the ontological concepts describing the Computer Science field. The following subsections analyse language and ontology evolution on this dataset.

3.2 Linguistic Progressiveness

Language innovation in a corpus refers to the introduction of novel patterns of language which do not conform to previously existing patterns [9]. Changes in time on the use of lexical features within a corpus characterise the language evolution of such corpus. To characterise such changes, here we first generate a language model – probability distribution over sequences of words [15]– per year. For this analysis we use the Katz back-off smoothing language model [14].

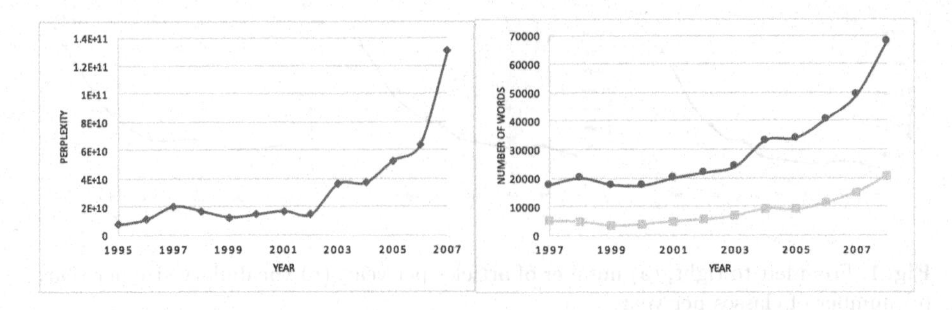

Fig. 2. From left to right, (a) Language models's perplexity per year; (b) Number of new words per year (•), number of adopted words per year (■).

This model estimates the conditional probability of a word given the number of times such word has been seen in the past.

To analyse differences in language models between consecutive years we use the perplexity metric. Perplexity is commonly used in Natural Language Processing to evaluate how well a language model predicts an unseen test set [8]. To analyse changes in language patterns for consecutive years we: (1) obtained the language model for year $t(lm_t)$ then; (2) we computed perplexity comparing lm_t to the unseen corpus at $t+1$.

Perplexity predicts word-error rate well when only in-domain training data is used, but poorly when out-of-domain text is added [8]. Figure 2, left, shows that for the Computer Science domain perplexity increases as time goes by. Therefore, language models representing language patterns trained in previous years provide poor predictions when tested on future datasets, indicating that language models can become outdated.

To analyse the impact of lexical innovation in language model changes, we perform a progressive analysis based on *lexical innovation* and *lexical adoption*. Let D_t be the collection of papers from corpus at year t. Let V_t be the vocabulary of D_t; we define a *lexical innovation* in D_t, LI_t, as the set of terms appearing in V_t, which were not mentioned in V_{t-1}[3]. We also define a *lexical adoption* in D_t, LA_t, as the set of terms appearing in LI_t which also appear in V_{t+1}. Figure 2, right, shows that while the number of novel words in Computer Science is high in consecutive years, only few of these words are adopted.

Based on these two metrics we introduce the ***linguistic progressiveness metric***, LP_t as the ratio of *lexical adoption* and *lexical innovation*, i.e., $LP_t = \frac{|LA_t|}{|LI_t|}$. The higher the adoption of innovative terms the more progressive the language used in a domain. In Fig. 3, left, the data indicates that the Computer Science domain has had a tendency towards being linguistically progressive. The following subsection studies the impact of innovation and adoption on semantic concepts in temporally consecutive ontologies of a domain.

[3] Notice that we are following a one step memory approach, further historical data could be used in future research.

3.3 Semantic Progressiveness

Ontology evolution refers to the maintenance of an ontological structure by adapting such structure with new data from a domain [28]. Such adaptation can result in both the generation or expiration of an ontology's concepts and properties. Hence the introduction of new classes that better describe the conceptualisation of a domain can be considered to be a semantic innovation. In this subsection we analyse the introduction of new concepts to an ontological per consecutive year.

Let (D_t, O_t) represent a tuple where D_t is a collection of articles belonging to year t and O_t is the corresponding ontology representation computed with Klink-2 over the D_t collection. Let CI_t be the conceptual innovation in D_t, which we define as the set of concepts appearing in O_t, which were not mentioned in O_{t-1}. Also let CA_t be the conceptual adoption in D_t, which consists on the set of concepts in CI_t that also appear in O_{t+1}. Based on these definitions we introduce the ***semantic progressiveness metric***, CP_t, as the ratio of conceptual adoption and conceptual innovation, i.e., $CP_t = \frac{|CA_t|}{|CI_t|}$.

Figure 3, right, shows that the ontologies extracted for the Computer Science domain indicate a tendency to be less semantically progressive. A tendency towards a lower semantic progressiveness can be understood as a tendency towards having a more stable representation of the domain. Notice that the semantic progressiveness metric do not account for churn of semantic concepts but focuses only of innovation and adoption.

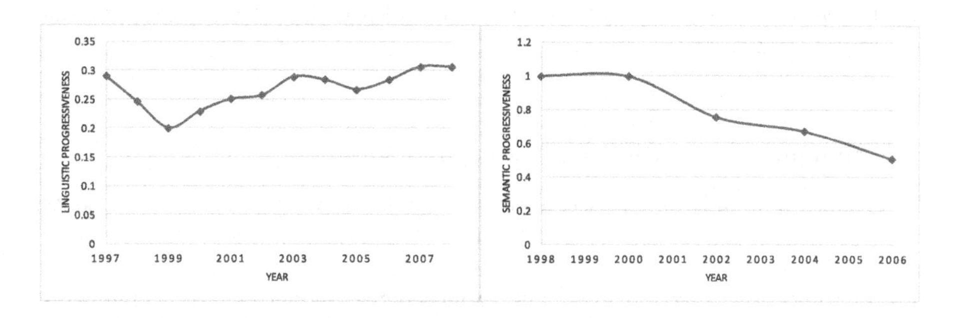

Fig. 3. From left to right, (a) linguistic progressiveness per year, (b) semantic progressiveness per year

Both linguistic and semantic progressiveness characterise the rate of change on the language and semantic conceptualisations used in a research field over the years. This constant evolution of a scientific area motivates us to study the prediction of semantic concepts that will likely enhance the current semantic representation of a research domain. The following section introduces our proposed model for forecasting concepts appearing on an ontology based on historical data.

4 Framework for Forecasting Semantic Concepts Based on Innovation-Adoption Priors

The proposed framework relies on the representation of an ontology's class as a topic word distribution. Learning topic models from text-rich structured data has been successfully used in the past [2,3,10]. Our proposed framework focuses on the task defined as follows: *Given a set of documents at year t and a set of historical priors, forecast topic word distributions representing new concepts in the ontology* O_{t+1}.

The proposed framework breaks down into the following phases: (1) Predicting new semantic concepts with the Semantic Innovation Forecast (SIF) model; (2) Incorporating innovation priors; Inferring topics with SIF; (3) Matching predicted topics to the forecast year's semantic concepts' gold standard
The overall pipeline is depicted in Fig. 4.

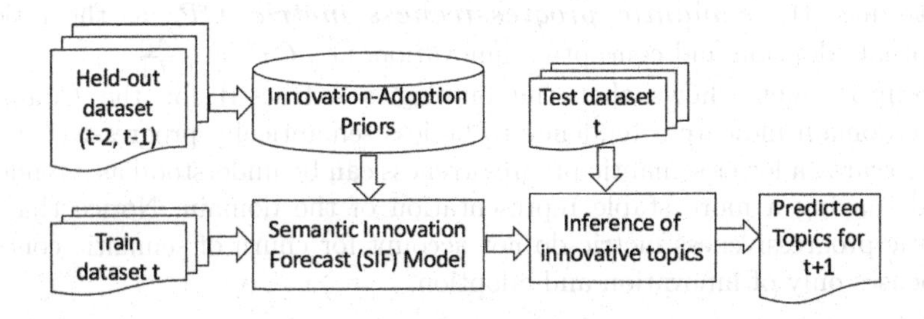

Fig. 4. Pipeline of the proposed framework for predicting semantic concepts using innovation/adoption priors.

4.1 Semantic Innovation Forecast (SIF) Model

We propose a weakly-supervised approach for forecasting innovative concepts based on lexical innovation-adoption priors. We introduce the Semantic Innovation Forecast (SIF) model which forecasts future semantic concepts in the form of topic-word distributions. The proposed SIF model favours the generation of innovative topics by considering distributions that enclose innovative and adopted lexicons based on word priors computed from historical data.

Assume a corpora consisting of a collection of documents grouped by consecutive years. Let a corpus of documents written at year t be denoted as $D_t = \{d_1, d_2, \dots_{D_d}\}$. Let each document be represented as a sequence of N_d words denoted by $(w_1, w_2, \dots, w_{N_d})$; where each word in a document is an element from a vocabulary index of V_t.

We assume that when an author writes an article, she first decides whether the paper will be innovative or will conform to existing work. In the proposed generative model we consider that if a paper is innovative then a topic is drawn from an innovation specific topic distribution θ. In such case each word in the article is generated from either the background word distribution ϕ_0 or the multinomial word distribution for the innovation-related topics ϕ_z.

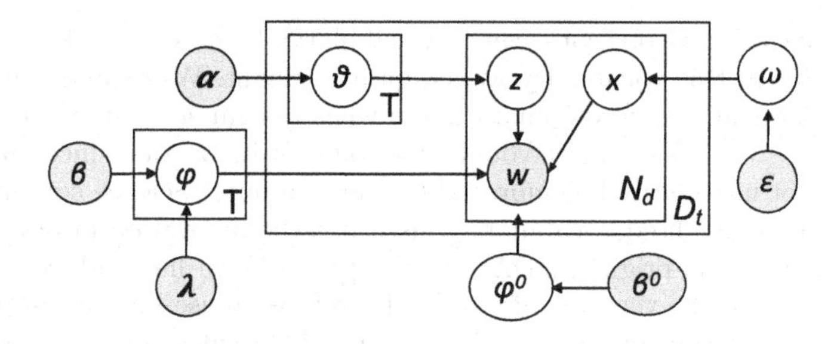

Fig. 5. Semantic innovation forecasting model

The generative process for SIF is as follows:

- Draw $\omega \sim \text{Beta}(\epsilon), \varphi^0 \sim \text{Dirichlet}(\beta^0), \varphi \sim \text{Dirichlet}(\beta)$.
- For each topic z draw $\phi_z \sim \text{Dirichlet}(\lambda \times \beta_z^T)$.
- For each document $m \in \{1 \dots D\}$,
  - Choose $\theta_m \sim \text{Dirichlet}(\alpha)$
  - For each word $n \in \{1 \dots N_d\}$ in document m,
    - ∗ draw $x_{m,n} \sim \text{Bernoulli}(\omega)$;
    - ∗ if $x_{m,n} = 0$,
      - · draw a word $w_{m,n} \sim \text{Multinomial}(\varphi^0)$;
    - ∗ if $x_{m,n} = 1$,
      - · draw a topic $z_{m,n} \sim \text{Multinomial}(\theta)$,
      - · draw a word $w_{m,n} \sim \text{Multinomial}(\varphi_{z_{m,n}})$.

The SIF model can be considered as an adaptation of a smoothed LDA [4], where we have added a per token latent random variable x which acts as a switch. If $x = 0$, words are generated from a background distribution, which accumulates words common to conformer articles. While if $x = 1$, words are sampled from the topic-specific multinomial ϕ_z. Moreover, SIF encodes word priors generated from historical data, such priors encapsulate innovation and adoption polarity in the matrix λ and are explained in more detail in the following Subsection.

4.2 Incorporating Innovation-Adoption Priors

Word priors enable us to have a preliminary or prior model of the language related to a topic of interest in the absence of any other information about this topic. A word prior is a probability distribution that expresses one's belief about a word's relevance to, in this case, being characteristic of innovative topics, when no other information about it is provided. Since the aim is to discover new semantic concepts, we propose to use lexical innovation and lexical adoption as indicators of lexicons characterising innovative word distributions.

The procedure to generate such ***innovation-adoption priors*** is as follows; to compute priors for a SIF model at time t we make use of two vocabularies, the one at year $t-1$ and $t-2$. From these vocabularies we identify innovative (at $t-2$) and adopted (at $t-1$) lexicons as described in Subsect. 3.2.

The union of these lexicons constitute a vocabulary of size K. Then for each term $w \in \{1, \ldots K\}$ in this vocabulary we assign it a weight. We experimented with different weights and we found an optimum when assigning 0.7 if $w \in LI_{t-2}$ and 0.9 if $w \in LA_{t-1}$. This setting favours adoption over innovation since innovative words may not necessarily be embraced by the Computer Science community in the future. This weighted vocabulary constitutes the innovation priors λ.

Compared to the original LDA model [4] in SIF we have added a dependency link of ϕ on the vector λ of size K. Therefore we use innovation priors as supervised information and modify the topic-word Dirichlet priors for innovation classification.

4.3 SIF Inference

We use Collapsed Gibbs Sampling [12] to infer the model parameters and topic assignments for a corpus at year $t+1$ given observed documents at year t. Such sampling estimates empirically the target distribution. Let the index $t = (m, n)$ denote the n_{th} word in document m and let the subscript $-t$ denote a quantity which excludes data from the n_{th} word position in document m, the conditional posterior of x_t is:

$$P(x_t = 0 | \mathbf{x}_{-t}, \mathbf{z}, \mathbf{w}, \beta^0, \epsilon) \propto \frac{\{N_m^0\}_{-t} + \epsilon}{\{N_m\}_{-t} + 2\epsilon} \times \frac{\{N_{w_t}^0\}_{-t} + \beta^0}{\sum_{w'} \{N_{w'}\}_{-t} + V\beta^0}, \quad (1)$$

where N_m^0 denotes the number of words in document m assigned to the background component, N_m is the total number of words in document m, $N_{w_t}^0$ is the number of times word w_t is sampled from the background distribution.

$$P(x_t = 1 | \mathbf{x}_{-t}, \mathbf{z}, \mathbf{w}, \beta, \epsilon) \propto \frac{\{N_m^s\}_{-t} + \epsilon}{\{N_m\}_{-t} + 2\epsilon} \times \frac{\{N_{w_t}^s\}_{-t} + \beta}{\sum_{w'} \{N_{w'}\}_{-t} + V\beta}, \quad (2)$$

where N_m^s denotes the number of words in document m sampled from the topic distribution, $N_{w_t}^s$ is the number of times word w_t is sampled from the topic specific distributions.

The conditional posterior for z_t is:

$$(z_t = j | \mathbf{z}_{-t}, \mathbf{w}, \alpha, \beta) \propto \frac{N_{d,j}^{-t} + \alpha_j}{N_d^{-t} + \sum_j \alpha_j} \cdot \frac{N_{j,w_t}^{-t} + \beta}{N_j^{-t} + V\beta}, \quad (3)$$

where N_d is the total number of words in document d, $N_{d,j}$ is the number of times a word from document d has been associated with topic j, N_{j,w_t} is the number of times word w_t appeared in topic j, and N_j is the number of words assigned to topic j.

When the assignments have been computed for all latent variables, then we can estimate the model parameters $\{\boldsymbol{\theta}, \boldsymbol{\varphi}, \boldsymbol{\varphi}^{\mathbf{0}}, \boldsymbol{\omega}\}$. For our experiments we set the symmetric prior $\epsilon = 0.5$, $\beta_0 = \beta = 0.01$. We learn the asymmetric prior $\boldsymbol{\alpha}$ directly from the data using maximum-likelihood estimation [17] and updating this value every 40 iterations during the Gibbs sampling. In our experiments we run the sampler for 1000 iterations, stopping once the log-likelihood of the learning data has converged under the learning model.

5 Experimental Setup

Here we present the experimental set up used to assess the SIF framework. We evaluate the accuracy of SIF in a semantic-concept forecasting task.

We perform this task by applying our framework on the dataset described in Sect. 3.1. Each collection of documents per year is randomly partitioned into three independent subsets contains respectively 20 %, 40 % and 40 % of the documents. For a given document collection at year t, the 20 % partition represents a held-out dataset used to derive innovation priors (Dp_t); while the other two partitions represent the training ($Dtrain_t$) and testing sets($Dtest_t$).

5.1 Forecasting with SIF

To forecast semantic concepts for a corpus at year $t+1$, we assume no information from $t+1$ is known at the time of the forecast. We train a SIF model on year t with $Dtrain_t$ using innovative priors computed on the held-out datasets for the two previous years: Dp_{t-1} and Dp_{t-2}. Then using the trained model on year t we perform inference over $Dtest_t$ and consider this output to be the forecast for concepts aiming to match those in CI_{t+1} (concept innovation at $t+1$, see Subsect. 3.3). The output of this last step is a set of topics that are effectively sets of word distributions, which we use to compare against our gold standard.

5.2 Gold Standard

We build our gold standard by generating a one-topic model per semantic-concept appearing in CI_{t+1}. This is performed by applying the standard LDA model [4] over the test dataset for documents belonging to each concept at year $t+1$.

Table 1 shows some examples of the gold standard computed for each innovative semantic concept of each year. The one-topic model representation of a semantic-concept provides a word distribution, which can be compared against the ones generated with SIF.

5.3 Baselines

We compare SIF against four baselines. For a year t forecasting for year $t+1$:

Table 1. Examples of semantic concepts' gold-standard. For a given year, we present a semantic concept and an extract of the word distribution representing such concept. Each distribution is derived from a one-topic standard LDA model computed from documents belonging to such concept. Words are presented stemmed, weights assigned to each word are omitted in this example.

Year	Semantic concept	Top 10 LDA words
2000	Anthropomorph robot	Robot, control, humanoid, human, anthropomorph, mechan, system, design, skill, method
2002	Context-free-grammar	Languag, grammar, model, context-fre, system, algorithm, gener, method, show, paper
2004	Video-stream	Video, stream, network, rate, system, applic, adapt, bandwidth, packet, internet
2006	3D-reconstruct	Reconstruct, imag, model, algorithm, structur, camera, point, surfac, data, base
2008	Open-access	Access, open, research, journal, repositori, publish, articl, develop, data, institut

1. ***LDA Topics*** (LDA); referring to word distributions weighted by latent topics extracted from the training $Dtrain_t$. This setting makes no assumption over innovative/adopted lexicons. It outputs a collection of n topics per training set, which are compared against the gold standard.
2. ***LDA Innovative Topics*** (LDA-I); computes topics based on documents containing at least one word appearing in LI_t.
3. ***LDA Adopted Topics*** (LDA-A); computes topics based only on documents containing at least one word appearing in LA_t.
4. ***LDA Innovation/Adoption Topics*** (LDA-IA): this baseline filters documents based on words appearing λ_t.

Baselines 2–4 represent three strong baselines, which consider innovative and adopted lexicons.

5.4 Estimating the Effectiveness of SIF

To estimate the effectiveness of SIF we consider how similar the predicted semantic concepts for $t+1$ are from the reference gold standard concepts for that year. To this end we based the similarity scores using the cosine similarity metric [15]. This metric ranges from 0 (no similarity exists between compared vectors) to 1 (the compared vectors are identical), therefore scoring a similarity higher than 0.5 indicates that the compared vectors are similar.

To compute this similarity metric we used the word vector representation of a predicted topic and of the topics generated for that year's gold standard. Therefore when forecasting for $t+1$ we computed the cosine similarity between the predicted candidate topic x and each of the topic y in CI_{t+1}, keeping as matches the similar ones.

We evaluated the semantic concept forecast task as a ranked retrieval task, where the appropriate set of forecast concepts are given by the top retrieved topic distributions. To measure the effectiveness on this task we used the Mean Average Precision (MAP) metric [15], a standard metric for evaluating rank retrieval results. For our experiments we computed MAP@10 to measure the mean of precision scores obtained from the top 10 predicted topics ranked based on topic-word distributions. The higher the word weights assigned on a topic the higher in the rank the topic is within the set of predicted topics.

6 Experimental Results and Evaluation

In this section we report the experimental results obtained for the semantic concept forecasting task. SIF and LDA require defining the number of topics to extract before applying on the data[4]. For our experiments we considered a fixed number of 100 topics, making no assumption on the expected number of new concepts appearing on the forecast year. These 100 topics are ranked based on topic-word distributions. The evaluation is done over the top 10 forecast topics using MAP@10.

Results in all experiments are computed using 2-fold cross validation over 5 runs of different random splits of the data to evaluate results' significance. Statistical significance is done using the T-test. The evaluation consists in assessing the following:

(1) Measure and compare SIF against the proposed baselines introduced in Subsect. 5.3.
(2) Investigate whether the proposed SIF approach effectively forecasts future semantic concepts.

6.1 Semantic Concept Forecast Results

Table 2 presents MAP results for SIF and the four baselines. The first three columns of Table 2 shows: (i) the year in which the model was trained; (ii) the year from where the innovative priors were derived for that setting; (iii) the year for which semantic concepts are forecast.

All baselines except LDA offer competitive results. LDA achieves a poor average result of 16 % over the 5 forecast years. For the predictions of 2002 and 2004, LDA fails to generate concepts matching those from the gold standard. This is expected since LDA alone do not make assumptions over linguistic innovation and adoption, therefore it's unlikely that the LDA-based generated topic based on past data will predict future concepts. However, pre-filtering documents containing either innovative lexicons, adopted lexicons or both appear instead to have a positive effect in the forecasting task.

[4] The data generated in the evaluation are available on request at http://technologies.kmi.open.ac.uk/rexplore/ekaw2016/OF/.

Table 2. MAP@10 for SIF and baselines. The number of topics is set to 100 for all five models. The value highlighted in bold corresponds to the best results obtained in MAP@10. A $\star$ denotes that the MAP@10 of SIF significantly outperforms the baselines. Significance levels: $p - value < 0.01$.

Year forecast	Year trained	Year prior	SIF	LDA	LDA-A	LDA-I	LDA-IA
2000	1999	1997–1999	0.7031	0.125	0.4761	0	0.408
2002	2001	1999–2001	0.8750	0	0.8227	0.6428	0.7486
2004	2003	2001–2003	0.9060	0	0.5822	0.5726	0.6347
2006	2005	2003–2005	0.8755	0.3069	0.7853	0.8385	0.6893
2008	2007	2005–2006	0.988	0.398	0.681	0.5661	0.7035
AVG			**0.8695**$\star$	0.1659	0.6694	0.524	0.6368

In particular, the use of LDA-A over LDA-I gives a boost on MAP of 14.54 %, indicating that *adopted words features are better predictors of innovative semantic concepts.* LDA-A also improves in average upon the LDA-IA baseline with a boost of 3 %. The proposed SIF model however outperforms significantly all four baselines with an average boost: over LDA of 70 %; over LDA-A of 20 %; over LDA-I of 34 %; over LDA-IA of 23 % (significant at $p < 0.01$). We could have expected LDA-IA to achieve closer results to SIF, since it is computed on documents filtered using both innovative and adopted lexicons. However, LDA-IA do not assign any preference over distributions of words containing either of such lexicons. In contrast, SIF takes innovation priors as a weighting strategy to build a prior model of language which is potentially used in future semantic concepts. The model is learnt over the full training set allowing to make use of both documents containing innovative and adopted lexicons and otherwise. The above results show the effectiveness of SIF for semantic concept forecasting over the baselines.

Table 3 presents examples of SIF's predicted topics that obtained a match in the forecast year's gold-standard (GS). While SIF do not forecast a specific name for the new semantic concept, the information provided by the word distribution gives context to the predicted concept. Table 3 presents top 10 words for the forecast SIF and GS representation however similarity computations where made using the whole topic-word representations. When comparing the SIF prediction vs the GSs we observe very close matches in 2000–2006 while for 2008 it is interesting to observe the appearance of words such as islam, victim, terror which don't match the top 10 of the corresponding GS (notice however they may appear in the further topic-word representation of the GS), however the word hate within the GS gives a insight of the use of mechatronics in violence-related scenarios.

Table 3. Examples of semantic concepts forecast with SIF for each year. The second row describes the semantic concept matching the predicted topic obtained with SIF. SIF columns presents top 10 words extracted from the word distribution of the SIF topic prediction. GS columns present top 10 words extracted from the one-topic LDA distribution.

2000		2002		2004		2006		2008	
Wireless network		Asynchronous transfer mode		Image threedimension		Cryptography		Mechatronics	
SIF	GS	SIF	GS	SIF	GS	SIF	GS	SIF	GS
Control	Control	Network	Network	Activ	Model	Method	Model	Robot	Robot
System	System	Service	Servic	Function	Algorithm	Structur	Method	model	model
Propos	Propos	System	Applic	Show	Function	Data	Algorithm	Base	Propos
Network	Applic	Mobil	System	Result	Data	Protocol	System	Perform	Simul
Servic	Network	Protocol	Mobil	Image	Result	Secur	Data	Simul	Process
Data	Servic	Wireless	Protocol	Respons	Image	Inform	Process	Islam	Mechan
Time	Commun	Rout	Base	Effect	Measure	Signatur	Scheme	Time	Control
Perform	Compu	Perform	Perform	Patient	Cell	Authenti	User	Control	Applic
Distribut	Manag	Packet	Algorithm	Clinic	Structure	Detec	Protocol	Applic	Dynam
Traffic	Schem	Control	Packet	Visual	Patient	Attack	Secur	Victim	Hate
Protocol	Mobil	Scheme	Control	Brain	Surfac	Sequenc	Inform	Terror	Best

7 Conclusions and Future Work

This work focused on the task of *semantic concept forecasting*, which aims at predicting classes which will be added to an ontology at time $t+1$ when only information up to time t is available. To approach this task we proposed the concepts of linguistic and semantic progressiveness, and introduced a strategy to encode lexical innovation and adoption as innovation priors. Based on these concepts we introduced the Semantic Innovation Forecast Model (SIF), which is a generative approach relying on historical innovation priors for the prediction of word distributions characterising a semantic concept.

In SIF each semantic concept is represented as a distribution of words obtained from the one-topic model of the collection of documents belonging to such concept. To this end we applied the proposed approach on a very large dataset belonging to the Computer Science domain, consisting of over one million papers on the course of 14 years. Our data analysis included the introduction of two novel metrics namely the linguistic and semantic progressiveness; which gave insights on the semantic trends in the Computer Science domain. Our experiments indicate that adopted lexicon are better predictors for semantic classes. Our experimental results also proof that the proposed approach is useful for the innovative semantic concept forecasting task. The SIF model outperforms the best baseline LDA-A showing an average significant boost of 23 %.

To the best of our knowledge this is the first approach to address the ontology forecasting task in general and in particular the first one in addressing the prediction of new semantic concepts. We believe that research on the prediction of semantic concepts in particular and in general the forecast of changes in an ontology can be beneficial to different areas of research not limited to the study of scholarly data. For the future, we plan to keep working on the integration between explicit and latent semantics, improve further the performance of our approach and introduce graph-structure information into the model. We also intend to use this approach for detecting innovative authors and forecast topic trends.

Acknowledgements. We would like to thank Elsevier BV and Springer DE for providing us with access to their large repositories of scholarly data.

References

1. Ahmed, A., Xing, E., Timeline.: A dynamic hierarchical Dirichlet process model for recovering birth/death and evolution of topics in text stream. Uncert. Artif. Intell. (2010)
2. Andrzejewski, D., Zhu, X., Craven, M., Recht, B.: A framework for incorporating general domain knowledge into latent Dirichlet allocation using first-order logic. In: Proceedings of 22nd International Joint Conference on Artificial Intelligence, IJCAI 2011, vol. 2, pp. 1171–1177. AAAI Press (2011)
3. Bicer, V., Tran, T., Ma, Y., Studer, R.: TRM – learning dependencies between text and structure with topical relational models. In: Alani, H., et al. (eds.) ISWC 2013. LNCS, vol. 8218, pp. 1–16. Springer, Heidelberg (2013)
4. Ng, A.Y., Blei, D.M., Jordan, M.I.: Latent Dirichlet allocation. In. J. Mach. Learn. Res. **3**, 993–1022 (2003)
5. Bolelli, L., Ertekin, Ş., Giles, C.L.: Topic and trend detection in text collections using latent Dirichlet allocation. In: Boughanem, M., Berrut, C., Mothe, J., Soule-Dupuy, C. (eds.) ECIR 2009. LNCS, vol. 5478, pp. 776–780. Springer, Heidelberg (2009)
6. Bolelli, L., Ertekin, S., Zhou, D., Giles, C. L.: Finding topic trends in digital libraries. In: Proceedings of 9th ACM/IEEE-CS Joint Conference on Digital Libraries, JCDL 2009, pp. 69–72. ACM, New York (2009)
7. Bunescu, R.C., Pasca, M.: Using encyclopedic knowledge for named entity disambiguation. In: EACL, vol. 6, pp. 9–16 (2006)
8. Chen, S., Beeferman, D., Rosenfeld, R.: Evaluation metrics for language models (1998)
9. Danescu-Niculescu-Mizil, C., West, R., Jurafsky, D., Leskovec, J., Potts, C.: No country for old members: user lifecycle and linguistic change in online communities. In: Proceedings of 22nd International Conference on World Wide Web, WWW 2013, pp. 307–318 (2013)
10. Deng, H., Han, J., Zhao, B., Yu, Y., Lin, C. X.: Probabilistic topic models with biased propagation on heterogeneous information networks. In: Proceedings of 17th ACM SIGKDD International Conference on Knowledge Discovery and Data Mining, KDD 2011, pp. 1271–1279. ACM, New York (2011)
11. Gohr, A., Hinneburg, A., Schult, R., Spiliopoulou, M.: Topic evolution in a stream of documents. In: SDM, pp. 859–872 (2009)
12. Griffiths, T., Steyvers, M.: Finding scientific topics. Proc. Natl. Acad. Sci. U.S.A. **101**(Suppl. 1), 52285235 (2004)
13. He, Q., Chen, B., Pei, J., Qiu, B., Mitra, P., Giles, L.: Detecting topic evolution in scientific literature: how can citations help? In: Proceedings of 18th ACM Conference on Information and Knowledge Management, CIKM 2009, pp. 957–966. ACM, New York (2009)

14. Katz, S.M.: Estimation of probabilities from sparse data for the language model component of a speech recognizer. IEEE Trans. Acoust. Speech Sig. Process. **35**, 400–401 (1987)
15. Manning, C.D., Raghavan, P., Schütze, H.: Introduction to Information Retrieval. Cambridge University Press, New York (2008)
16. Mei, Q., Zhai, C.: Discovering evolutionary theme patterns from text: an exploration of temporal text mining. In: Proceedings of 11th ACM SIGKDD International Conference on Knowledge Discovery in Data Mining, pp. 198–207. ACM (2005)
17. Minka, T.: Estimating a Dirichlet distribution. Technical report (2003)
18. Monaghan, F., Bordea, G., Samp, K., Buitelaar, P.: Exploring your research: sprinkling some saffron on semantic web dog food. In: Semantic Web Challenge at the International Semantic Web Conference, vol. 117, pp. 420–435. Citeseer (2010)
19. Morinaga, S., Yamanishi, K.: Tracking dynamics of topic trends using a finite mixture model. In: 10th ACM SIGKDD International Conference on Knowledge Discovery and Data Mining (2004)
20. Osborne, F., Motta, E.: Klink-2: integrating multiple web sources to generate semantic topic networks. In: 14th International Semantic Web Conference (2015)
21. Osborne, F., Motta, E., Mulholland, P.: Exploring scholarly data with rexplore. In: Alani, H., et al. (eds.) ISWC 2013, Part I. LNCS, vol. 8218, pp. 460–477. Springer, Heidelberg (2013)
22. Osborne, F., Salatino, A., Birukou, A., Mottam, E.: Automatic classification of springer nature proceedings with smart topic miner. In: Groth, P., Simperl, E., Gray, A., Sabou, M., Krötzsch, M., Lecue, F., Flöck, F., Gil, Y. (eds.) ISWC 2016. LNCS, vol. 9982, pp. 383–399. Springer, Heidelberg (2016)
23. Pesquita, C., Couto, F.M.: Predicting the extension of biomedical ontologies. PLoS Comput. Biol. **8**(9), e1002630 (2012)
24. Rosen-Zvi, M., Griffiths, T., Steyvers, M., Smyth, P.: The author-topic model for authors and documents. In: Proceedings of 20th Conference on Uncertainty in Artificial Intelligence, pp. 487–494. AUAI Press (2004)
25. Tseng, Y.-H., Lin, Y.-I., Lee, Y.-Y., Hung, W.-C., Lee, C.-H.: A comparison of methods for detecting hot topics. Scientometrics **81**(1), 73–90 (2009)
26. Wang, H., Tudorache, T., Dou, D., Noy, N.F., Musen, M.A.: Analysis and prediction of user editing patterns in ontology development projects. J. Data Semant. **4**(2), 117–132 (2015)
27. Willett, P.: The porter stemming algorithm: then and now. Program **40**(3), 219–223 (2006)
28. Zablith, F., Antoniou, G., d'Aquin, M., Flouris, G., Kondylakis, H., Motta, E., Plexousakis, D., Sabou, M.: Ontology evolution: a process-centric survey. Knowl. Eng. Rev. **30**(01), 45–75 (2015)

Leveraging the Impact of Ontology Evolution on Semantic Annotations

Silvio Domingos Cardoso[1,2](✉), Cédric Pruski[1], Marcos Da Silveira[1], Ying-Chi Lin[3], Anika Groß[3], Erhard Rahm[3], and Chantal Reynaud-Delaître[2]

[1] LIST, Luxembourg Institute of Science and Technology, 5, Avenue des Hauts-Fourneaux, 4362 Esch-sur-alzette, Luxembourg
{silvio.cardoso,cedric.pruski,marcos.dasilveira}@list.lu

[2] LRI, University of Paris-Sud XI, Orsay, France
chantal.reynaud@lri.fr

[3] Institute of Computer Science, Universität Leipzig, P.O. Box 100920, 04009 Leipzig, Germany
{lin,gross,rahm}@informatik.uni-leipzig.de

Abstract. This paper deals with the problem of maintenance of semantic annotations produced based on domain ontologies. Many annotated texts have been produced and made available to end-users. If not reviewed regularly, the quality of these annotations tends to decrease over time due to the evolution of the domain ontologies. The quality of these annotations is critical for tools that exploit them (e.g., search engines and decision support systems) and need to ensure an acceptable level of performance. Although the recent advances for ontology-based annotation systems to annotate new documents, the maintenance of existing annotations remains under studied. In this work we present an analysis of the impact of ontology evolution on existing annotations. To do so, we used two well-known annotators to generate more than 66 million annotations from a pre-selected set of 5000 biomedical journal articles and standard ontologies covering a period ranging from 2004 to 2016. We highlight the correlation between changes in the ontologies and changes in the annotations and we discuss the necessity to improve existing annotation formalisms in order to include elements required to support (semi-) automatic annotation maintenance mechanisms.

Keywords: Ontology evolution · Semantic annotations · Life sciences

1 Introduction

The use of ontologies, or more generally speaking Knowledge Organization Systems (KOS) [1] (which includes classification schemes, thesauri or ontologies), to annotate documents, is a current practice in order to make their semantic explicit for computers. This is for instance the case in the biomedical domain where main interests for healthcare professionals to annotate documents are twofold: (1) to transfer these documents to other institutions/people (e.g., to

E. Blomqvist et al. (Eds.): EKAW 2016, LNAI 10024, pp. 68–82, 2016.
DOI: 10.1007/978-3-319-49004-5_5

accelerate the reimbursement process, to request second opinion, etc.); (2) to easily retrieve patient information. Secondary uses of these annotations are often foreseen for decision support systems, public health analysis, patient recruitment for clinical trials, etc. In the biomedical field the entities annotated include diseases, parts of the body, genes, etc. [2]. There are many structured forms to represent annotations, basically the inputs and outputs from clinical documents when it is processed by software as text processors (e.g. GATE, NCBO Annotator, MetaMap) can be expressed as annotations [2]. This is usually done by associating concept code or label of a given KOS to an element of the document (see Fig. 1). Through this link, human and computers can have an unambiguous understanding of the content of the document.

However, the dynamic nature of KOS may affect the annotations each time a new version is released. Actually, new KOS concepts can be added, obsolete ones can be removed and existing concepts may have their definition refined through the modification of their attribute values [3]. In consequence, changes in concepts can alter their semantics and therefore create a mismatch between the versions of the same concept (e.g. version 1 can be more abstract or more specific than version 2) impacting the validity of the semantic annotation. Following this observation, it is important to constantly evaluate and adapt the annotations to insure an optimal use of the annotated data. Nevertheless, the revision can hardly be done manually by virtue of the huge amount of existing annotations. Therefore, there is an urgent need for intelligent tools to support domain experts in this task.

In this paper our objectives are twofold. First, we aim at quantifying the impact of KOS evolution on the associated annotations to justify the need of automatic tools for maintaining the validity of annotations over time. This is done through systematic analyses of 66 millions of annotations obtained using biomedical journal articles and 13 successive versions of two standard medical KOS: ICD-9-CM and MeSH which will complement existing reviews that usually focus on one specific ontology [4]. Second, we discuss the capabilities of existing annotation models that deal with KOS evolution and propose new key features to cope with this problem.

The remainder of the paper is structured as follow: in Sect. 2 we review related work of the field semantic annotation evolution. Section 3 describes the experiments we have conducted to obtain the results presented in Sect. 4. Section 5 discusses the results and introduces our model to deal with annotation maintenance. Section 6 concludes the paper and outlines future work.

2 Related Work

Semantic annotation is the central notion of this work. However, many definitions can be found in the literature. According to Oren et al. [5], the term annotation can denote the process of annotating as well as the result of this process. Moreover, they distinguish three families of annotations. *Informal annotations* that are not machine-readable, (e.g. a handwritten margin annotation in a book).

Formal annotations that are machine-understandable but are not defined using ontological terms, (e.g. highlights in a html document). Last, and the kind of annotation we are referring to in this paper, *ontological annotations* that are machine-understandable and are taken from an ontology (see Fig. 1).

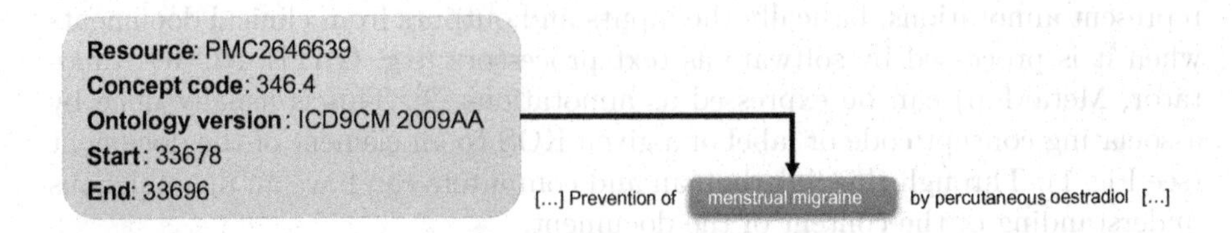

Fig. 1. Example of annotation using the concept recognition process for a PubMed document. The term *menstrual migraine* is annotated with the KOS concept 346.4 that belongs to ICD-9-CM version 2009AA (UMLS)

2.1 Existing Annotation Models

To represent annotations in the biomedical field, Luong and Dieng-Kuntz [6] defined the following annotation model:

$$\mathcal{SA} = (R_a, C_a, P_a, L, T_a) \quad (1)$$

Where:

R_a: set of resources, for instance, an RDF resource.
C_a: set of concept names defined in ontology ($C_a \subset R_a$)
P_a: set of property, for instance, an rdf:type ($P_a \subset R_a$)
L: set of literal values, for example, "Fever", "Malaria Fever", etc.
T_a: set of triples (s,p,v) where $s \in R_a$, $p \in P_a$ and $v \in (R_a \cup L)$

Gross et al. [7] and Hartung et al. [8] gave a more complete definition of an annotation, taking evolution aspect into account which was missing in Luong et al. model. In their work an annotation is defined as:

$$\mathcal{AM} = (I_u, ON_v, Q, A) \quad (2)$$

Where:

$I_u = (I, t)$: is an instance source. It consists of a set of instances $I = \{i_j, ..., i_n\}$, e.g., molecular biological objects such as genes or proteins, at timestamp t. Instances are described by an accession ID.

ON_v: is an ontology in the version v that contains (C, R, t), it comprises a set of concepts $C = \{c_1, ..., c_n\}$ and relationships $R = \{r_1, ..., r_m\}$ released at time t.

Q: is a set of quality indicators (ratings) of annotations. The quality indicators may be numerical values or come from predefined quality taxonomies, e.g., the evidence codes for provenance information or stability indicators.

A: is a set of annotations. A single annotation $a \in A$ is denoted by $a = (i, c, \{q\})$, i.e. an instance item $i \in I_u$ is annotated with an ontology concept $c \in ON_v$ and a set of quality indicators (ratings) $\{q\} \in Q$

Recently, the W3C has published a new candidate recommendation for expressing annotation[1]. An annotation includes a body and a target and the relation between these two entities that may vary according to the intention of the annotation. This model is the foundation of a more general framework for sharing and reusing annotated information across different hardware and software platforms. However, this model is still not sufficient to deal with evolution issues as we will show in the following sections.

2.2 Annotation Evolution Techniques

As mentioned, the dynamic of knowledge leads to frequent revisions of KOS content which, sometimes, impacts the definition of the semantic annotations associated with documents (as illustrated in Fig. 2) [9]. The most recent approaches to analyse the evolution of the annotations is focused on biological domain, in particular on GO annotated documents. Traverso-Ribón et al. [10] developed the AnnEvol framework to compare two versions of a dataset (for instance, UnitProt-GOA and Swiss-Prot) and to verify the entities in the $dataset_{(i)}$ and $dataset_{(i+1)}$ that are similar and those which are different, using evolution criteria (e.g. obsoleted, removed and added annotations).

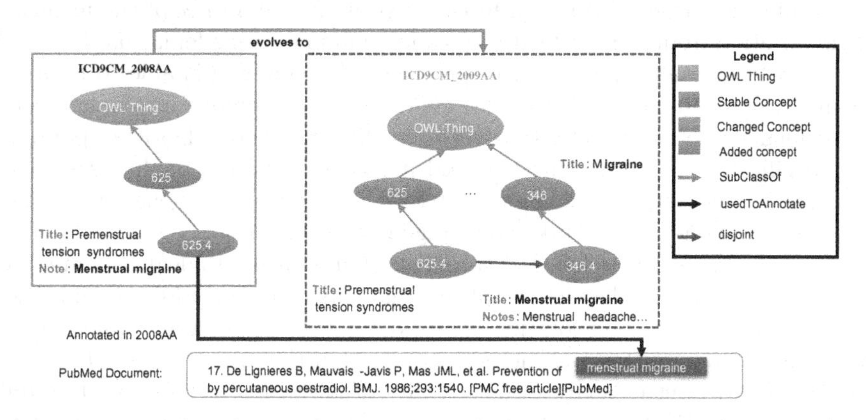

Fig. 2. Annotation evolution case study. A subset of a document is annotated with *Menstrual migraine*, an attribute of the concept 625.4 of ICD-9-CM version 2008AA. In the next version the attribute of 625.4 is removed and added as a new concept 346.4. This change has caused a mismatch between the annotation created with the older version and the concept of the new KOS version

Groß et al. [11] provide a method to test to what degree changes of GO and GO annotations (GOAs) may affect functional enrichment analyses, analyzing two real-world experimental datasets as well as 50 generated datasets.

[1] http://www.w3.org/TR/annotation-model/.

They proposed two types of stability measures to assess the impact of ontology and annotation changes. Differently from AnnEvol, Groß et al. deal with other change types, besides *add* and *delete*, such as, *merge* (merge of two or more categories into one category). They also verified strong structural changes as *addR* (insertion of a new relationship r), *delR* (deletion of an existing relationship r). However, these changes do not significantly impact on GOAs. As result they concluded that term-enrichment results are significantly affected by ontology and annotation evolution.

Luong and Dieng-Kuntz [6] developed the CoSWEM framework to investigated annotation evolution and explored a rule-based approach to detect and correct basic annotation inconsistencies, such as deletion. This approach converts ontologies to RDF(S) files and detects annotations affected by their evolution, as well as potentially inconsistent annotations using CORESE. Afterwards, inconsistent annotations are detected and corrected. This work focuses on expressive and small-sized ontologies and can hardly be applied to large biomedical ones, because the implemented reasoning techniques require the power of description logics (not always used in biomedical controlled terminologies) to decide on the validity of the annotations.

Frost and Moore [12] proposes a novel algorithm for optimizing gene set annotations to best match the structure of specific empirical data sources. The proposed method uses entropy minimization over variable clusters (EMVC). It filters the annotations for each gene set to remove inconsistent annotations. The results show that EMVC can filter between 92 % and 67 % of the inconsistent annotation from MSigDB C4 v4.0 cancer modules using leukemia data and MSigDB C2 v1.0 using p53 data, respectively. This method is able to improve the annotations but does not produce good results to improve incomplete gene sets or identify new gene sets. It is very sensitive to several algorithm parameters, specifically, the cluster method and it can be computationally expensive. Furthermore, the author's highlight that EMVC only works in gene set domain, thus other domains can not take advantage of this approach.

In summary, we concluded that the existing approaches to deal with annotation evolution just handle with simple changes (like concept addition and deletion), and only study the evolution of GO ontology. Furthermore, almost all of the works do not propose any method to maintain the annotations. Therefore, it is necessary to better analyze the stability of KOS annotations based on different KOS like ICD-9-CM and verify possible features to take into account to properly maintain semantic annotations in biomedical and clinical use cases.

3 Experimental Assessment of the Impact of KOS Evolution on Semantic Annotation

To bridge the gaps underlined in the previous section, we decided to conduct an empirical analysis regarding the evolution of the KOS and annotations. The lessons we learn through these experiments will allow us to come up with new proposal to deal with semantic annotation evolution issues. The used material and the adopted assessment methodology are detailed in this section.

3.1 Material

As our objective aims at analysing the evolution of semantic annotation, we have to work on several versions of an annotated corpus. Since no gold standard containing successive sets of annotated documents, we had to build our own environment. To this end, we used two annotation tools (based on distinct annotation methods), two different medical standard KOS and their associated successive versions, an ontology *Diff* tool to be able to identify the evolution of the concepts used to produce the annotations and a collection of biomedical documents. The documents were collected from the 2014 Clinical Decision Support Track (TREC 2014) campaign. It contains 733,138 biomedical articles about generic medical records. All documents from this database are open access documents from PubMed Central PMC. For our analyses we selected 5000 documents randomly.

The set of KOS is composed of several versions of medical KOS, represented in OWL format and used as "reference ontology" for text annotation. In order to annotate the documents, we selected two KOS: International Classification of Diseases, Ninth Revision, Clinical Modification (ICD-9-CM); and Medical Subject Headings (MeSH). We collected 13 official versions of each KOS released between 2004 and 2016 in UMLS and we transformed them into OWL files.

Regarding the annotation tools, the selection criteria were: be open source, allow selecting the reference ontology, provide APIs, have good documentation, and have been extensively used for research and/or commercial purposes. We first selected General Architecture for Text Engineering (GATE) [13]. It provides support for Ontology-Aware NLP, allowing loading any ontology as RDF file and then uses a gazetteer to obtain lookup annotations that have the text offset (offset is a pair {start, end} that indicates the distance, in terms of characters, from the beginning of the document. {start} indicates the position of the first character of the text while {end} indicates position of the last character), instance and class URI. The second selected tool is the NCBO Annotator. It is part of the NCBO Annotator framework and uses a dictionary built by extracting from KOS all concepts' label and/or other associated attributes (e.g., synonyms) that syntactically identify concepts [14]. Both annotators utilize different algorithms to produce the annotations. In this case, GATE uses Ontology-Aware NLP and NCBO Annotator uses MGrep. Moreover, NCBO Annotator also allows using other KOS to annotate the term, if a mapping exists between the concepts of both KOS. For instance, *melanoma* could also be annotated with the concept C0025202 (from NCI Thesaurus), or C0025202 (from SNOMED CT).

We used COnto-Diff [15] to determine an expressive and invertible *diff* evolution mapping between two versions of an ontology. It calculates basic change operations (insert/update/delete) from two KOS versions expressed in either OWL or OBO based on predefined set of rules defining basic and complex transformations (e.g., concept merging, concept splitting, move of concept, etc.)

3.2 Method

To identify and quantify the impact of changes affecting KOS concepts involved in annotations (as illustrated in Fig. 2), we proposed the methodology depicted in Fig. 3.

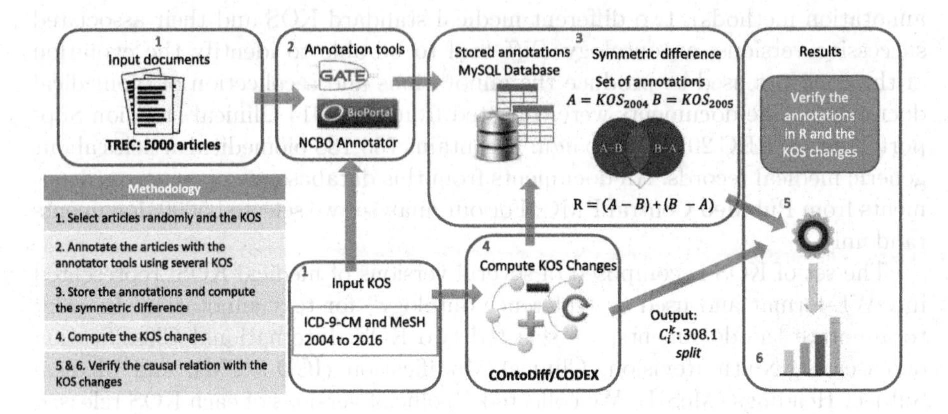

Fig. 3. The experimental protocol. The numbers in red correspond to the six steps explained in the text.

The six steps of the methodology are the following:

1. We randomly selected 5000 documents from the TREC corpus and collected the 13 KOS versions of ICD-9-CM and MeSH (from 2004 to 2016).
2. We used GATE and NCBO Annotator to annotate these documents. We configured GATE and NCBO Annotator to use one specific KOS version and repeated the annotation process for each version. We filtered the annotations produced by both annotators according to [16] (e.g., keep the longest match concept for an annotation).
3. We regrouped all annotations in one database. We then computed the symmetric difference $A_{m,n} \Delta A_{m,n+1}$ between the two annotation sets ($A_{m,n}$ and $A_{m,n+1}$) generated for a document R_m using two successive KOS versions (K_n and K_{n+1}) as the following:

$$A_{m,n} \Delta A_{m,n+1} := \{a \mid a \in A_{m,n} \wedge a \notin A_{m,n+1}\} \cup \{a \mid a \in A_{m,n+1} \wedge a \notin A_{m,n}\} \tag{3}$$

   a is an annotation that can be described as $\{i, Offset, c\}$ where i is an instance at position $Offset$ annotated with a KOS concept c. The symmetric difference allows us to identify annotations that have been removed, added and modified.
4. To identify KOS changes, each pair of two KOS successive versions was input into COnto-Diff to compute the KOS difference. The difference was stored into another MySQL database and has been reused to explain the changes.

5. We compared the 13 annotation sets of each document by pairs [2004–2005, 2005–2006 ...] to identify what changed in the annotations and to find correlations with the KOS changes identified by COnto-Diff. An annotation a is considered as evolved to a' if the $Offset$ or/and the c of a are different from those of a' and there is an overlap of both $Offsets$.
6. Finally, we analysed the generated subset of annotations/KOS changes in order to understand the impact of KOS changes on the annotations.

4 Results

The methodology described in the previous section has allowed us to produce more than 66 millions of annotations. The amount of annotations varies according to the used annotation tools (GATE or NCBO Annotator) as depicted in Figs. 4 and 5. The difference between the two sets of annotations results from the method used to annotate the documents (they are not using only exact match). A general observation can be made based on Figs. 4 and 5.

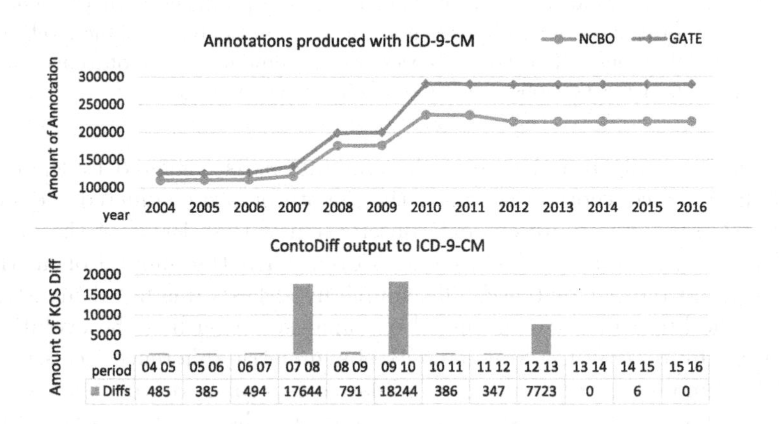

Fig. 4. Amount of annotation and KOS changes (green) produced with 13 versions of ICD-9-CM. The annotations from NCBO Annotator are represented in (blue circles) and GATE (orange diamond). The y-axis represents the amount of annotations/changes and the x-axis the KOS versions over time. (Color figure online)

We observe a huge increase in the amount of produced annotations in the periods 2007/2008 and 2009/2010 using ICD-9-CM (Fig. 4). This increase is accompanied by the changes that occurred in the KOS during these periods according to COnto-Diff output. On the other hand, the amount of annotations in the period 2012–2013 is not increased even though there were many KOS changes. We observe an average of words/label of 8,746 during this period and thus the annotators are not able to produce annotations for these changed labels. Hence, we can conclude that the change of the number of annotations does not necessarily correspond to the amount of KOS changes. In the future work, we will analyse what kinds of KOS changes trigger which types of annotation changes since not all kind of changes in the KOS has the same impact on the annotations (e.g., some KOS changes do not change the annotations).

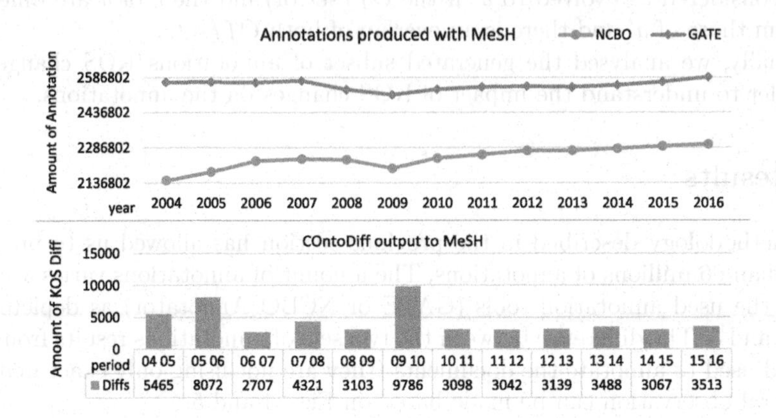

Fig. 5. Amount of annotation and KOS changes (green) produced with 13 versions of MeSH. The annotations from NCBO Annotator are represented in (blue circles) and GATE (orange diamond). The y-axis represents the amount of annotations/changes and the x-axis the KOS versions over time. (Color figure online)

In order to verify if a change in the annotations is triggered by the evolution of the KOS concepts or a gap in the annotator, we conducted the step 3 in Sect. 3.2. The first (quite evident) observation is that 100 % of the annotation changes are caused by KOS changes even when the annotation methods not only produce exact matches. This simple hypothesis was not demonstrated before in the literature. We continued our analyses regarding the evolution of annotations by refining the previous sets of symmetric difference (see step 5 in Sect. 3.2). If more than one concept candidate exists to annotate a text, we used selection criteria: (1) the most recent concept and the one with largest offset, as proposed by [16]. For instance, a text with the words *chronic kidney disease* can be annotated as *kidney disease* or *chronic kidney disease*, we select only the later concept. This decision can generate changes in the annotation from one KOS version to another (change operations). One of these changes is a shift of the offsets before and after the evolution while part of these offsets overlaps. For instance, in 2007 we have the annotation "*personality disorders*". After a KOS change in 2008 the new annotation is "*schizoid personality*" (of which "*personality*" is overlapped with the previous offset). For such case, we compute a (2) *chgOffset* operation. We formally define these conditions in Eq. (4):

$$Evolution(a_i, a_{i+1}) \longrightarrow \begin{cases} recentCp(a_i, a_{i+1}) \wedge bigOffset(a_i, a_{i+1}), & \text{if } 1 \\ chgOffset(a_i, a_{i+1}), & \text{if } 2 \end{cases} \quad (4)$$

As result we observe that the new KOS versions do not necessarily produce more annotations despite the increasing size of the KOS over time [9] (cf. Figs. 6

and 7). Analysing the amount of annotations and the types of changes occurring in the KOS, we observed that some minor changes which do not affect the semantics of the concepts still might impact the annotations. For instance, the concept 780.39 in ICD-9-CM version 2007AA (Seizures) evolves to (Seizure) in ICD-9-CM version 2008AA. However, both annotators did not recognize that the concepts have the same meaning and therefore the associated annotations are different from one version to the next.

We also observed that there are some periods in the KOS evolution history which are more stable and this stability is also reflected in the evolution of the annotations (e.g. the two periods 2010/2011 and 2013/2014 in ICD-9-CM on Figs. 4 and 6).

Changes in the KOS have also different impact depending on the amount of annotations a concept is associated with. This is for instance the case for the concept 084.4 of ICD-9-CM period 2007/2008 which is associated with 3143 annotations distributed in 162 documents in our corpus while concept V15.03 of ICD-9-CM period 2012/2013 is associated with only one annotation. If a single KOS change affects many annotations, it may require a huge amount of time if the maintenance of the annotation is done manually by domain experts.

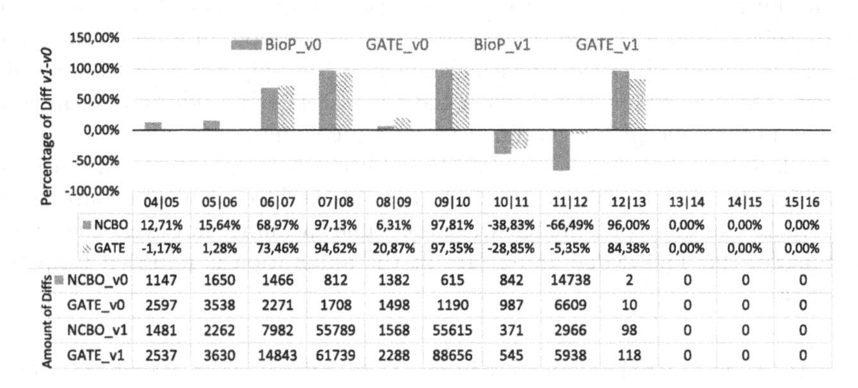

	04|05	05|06	06|07	07|08	08|09	09|10	10|11	11|12	12|13	13|14	14|15	15|16
NCBO	12,71%	15,64%	68,97%	97,13%	6,31%	97,81%	-38,83%	-66,49%	96,00%	0,00%	0,00%	0,00%
GATE	-1,17%	1,28%	73,46%	94,62%	20,87%	97,35%	-28,85%	-5,35%	84,38%	0,00%	0,00%	0,00%
NCBO_v0	1147	1650	1466	812	1382	615	842	14738	2	0	0	0
GATE_v0	2597	3538	2271	1708	1498	1190	987	6609	10	0	0	0
NCBO_v1	1481	2262	7982	55789	1568	55615	371	2966	98	0	0	0
GATE_v1	2537	3630	14843	61739	2288	88656	545	5938	118	0	0	0

Fig. 6. Differences in two successive annotation sets produced with ICD-9-CM. The blue (solid) colour represents the annotations that belong to NCBO Annotator, and the orange (hashed) colour to GATE. (Color figure online)

We then analyse how these annotations evolve. In Table 1, we present 5 use cases showing how the annotations evolve over time and their relation with the evolution of KOS. A concept is stable if no change occurred from one KOS version to the next (second column in Table 1). In the first use case (in 2008), *hepatitis* is associated to the concept 573.3 which did not change between 2008 and 2009 (i.e. a stable concept). In 2009, another concept (571.42) was also used to annotate the term *hepatitis*. Our selection criteria define that we will select the concept with the longest title (*autoimmune hepatitis*). We also observed that this concept (571.42) changed in 2009 (a split was detected).

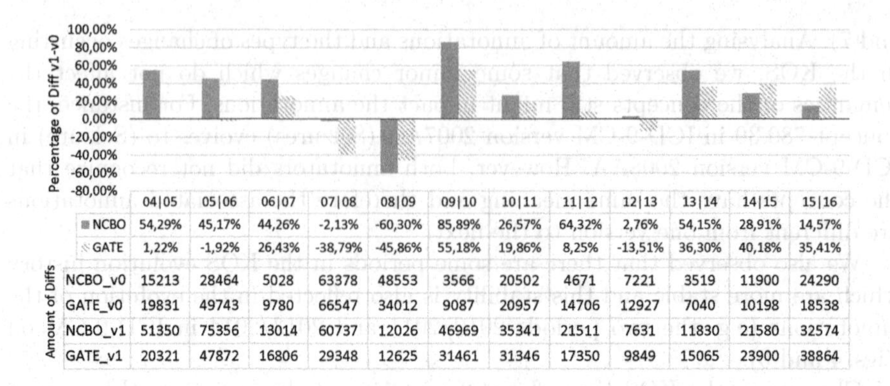

Fig. 7. Differences in two successive annotation sets produced with MeSH. The blue (solid) colour represents the annotations that belong to NCBO Annotator, and the orange (hashed) colour to GATE. (Color figure online)

The second use case illustrates a situation where both concepts changed (i.e., 625.4 had an attribute deleted, and 346.4 is a new concept).

The third use case presents the inverse situation of use case 1, i.e., an annotation evolves from a change concept to a stable concept. In a depth analysis, this case is mainly observed when more general concepts are used to annotate the text. This behaviour occurs when the annotator is not able to determine if a change in the concept has modified its meaning or not.

The last two use cases describe the addition or removal of annotations. Regarding the removal of annotations, we also verified that there are some cases where the concept remains with the same meaning, however, the annotator misses this knowledge and as result the annotation is removed from the document.

Table 1. Use cases for annotation evolution. These different cases are referred in the paper as: case 1: *stable_to_change*; case 2: *change_to_change*; case 3: *change_to_stable*; case 4: *addition*; case 5: *removal*.

Use case	KOS version	Annotation		Concept	KOS change
1	2008	Hepatitis	Change	573.3	Stable concept
	2009	Autoimmune hepatitis		571.42	Split
2	2008	Menstrual migraine	Change	625.4	delAtt
	2009	Menstrual migraine		346.4	addC
3	2009	Acute renal failure	Change	584.9	ChgAttValue
	2010	Renal failure		586	Stable concept
4	2008	Abdominal tomography	Addition	88.02	AddA
5	2004	Bulimia	Removal	307.51	ChgAttValue

Figures 8 and 9 show how often these use cases are observed in the corpus annotated with ICD-9-CM and MeSH using GATE and NCBO Annotator, respectively. In general, we observe that changes in ICD-9-CM have less impact on the annotations than those in MeSH. The low expressiveness of ICD-9-CM can be justified as the annotators tend to apply exact match techniques for these kinds of KOS. Semantic-based techniques are more used for KOS with high expressiveness. These differences are better observed by comparing Figs. 8 and 9 to see how the annotation technique influences the final annotation results regarding to the expressiveness of the KOS. The use case 2 and 5 (*change_to_change* and *removal*, respectively) are more frequent in the MeSH based annotations. Thus, annotations based on ICD-9-CM evolve quite similarly for GATE and NCBO Annotator, while the annotations based on MeSH evolve differently, depending on the used annotator.

Taking into account the annotators techniques only, we observe that GATE also tends to preserve existing annotations while the rates of new annotations over deleted ones are quite similar for both annotators. More precisely, the rates of use cases 1 and 2 over the deleted ones (GATE has more than double of NCBO) explain the results presented in Fig. 4 (number of annotations increases faster for GATE).

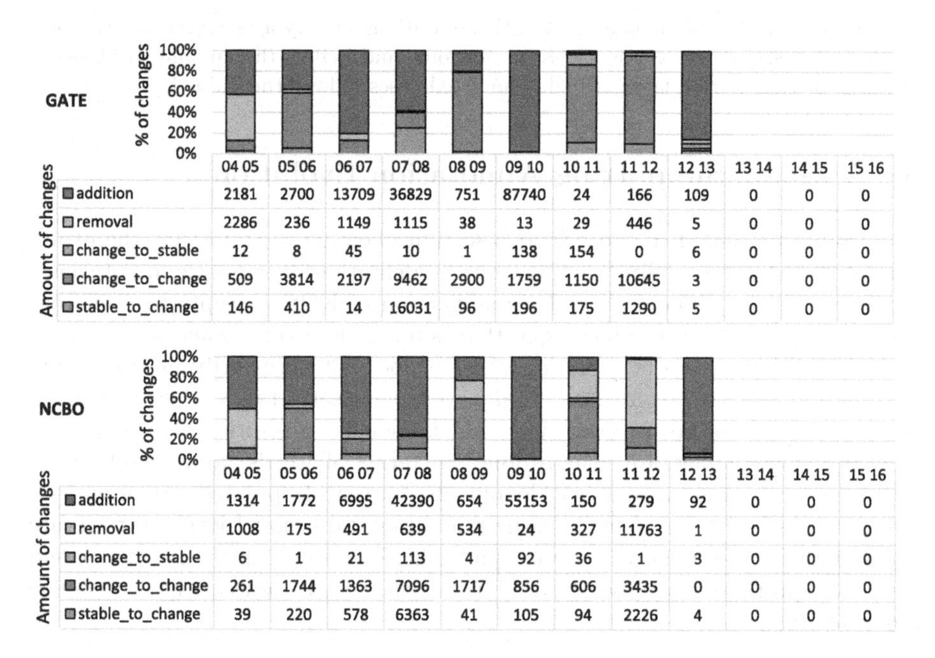

GATE

	04 05	05 06	06 07	07 08	08 09	09 10	10 11	11 12	12 13	13 14	14 15	15 16
addition	2181	2700	13709	36829	751	87740	24	166	109	0	0	0
removal	2286	236	1149	1115	38	13	29	446	5	0	0	0
change_to_stable	12	8	45	10	1	138	154	0	6	0	0	0
change_to_change	509	3814	2197	9462	2900	1759	1150	10645	3	0	0	0
stable_to_change	146	410	14	16031	96	196	175	1290	5	0	0	0

NCBO

	04 05	05 06	06 07	07 08	08 09	09 10	10 11	11 12	12 13	13 14	14 15	15 16
addition	1314	1772	6995	42390	654	55153	150	279	92	0	0	0
removal	1008	175	491	639	534	24	327	11763	1	0	0	0
change_to_stable	6	1	21	113	4	92	36	1	3	0	0	0
change_to_change	261	1744	1363	7096	1717	856	606	3435	0	0	0	0
stable_to_change	39	220	578	6363	41	105	94	2226	4	0	0	0

Fig. 8. Distribution of changes of ICD-9-CM annotations. The y-axis represents the percentage of changes, the x-axis the KOS versions, and bellow the amount of observed changes for each period is described. The listed cases follows the Table 1

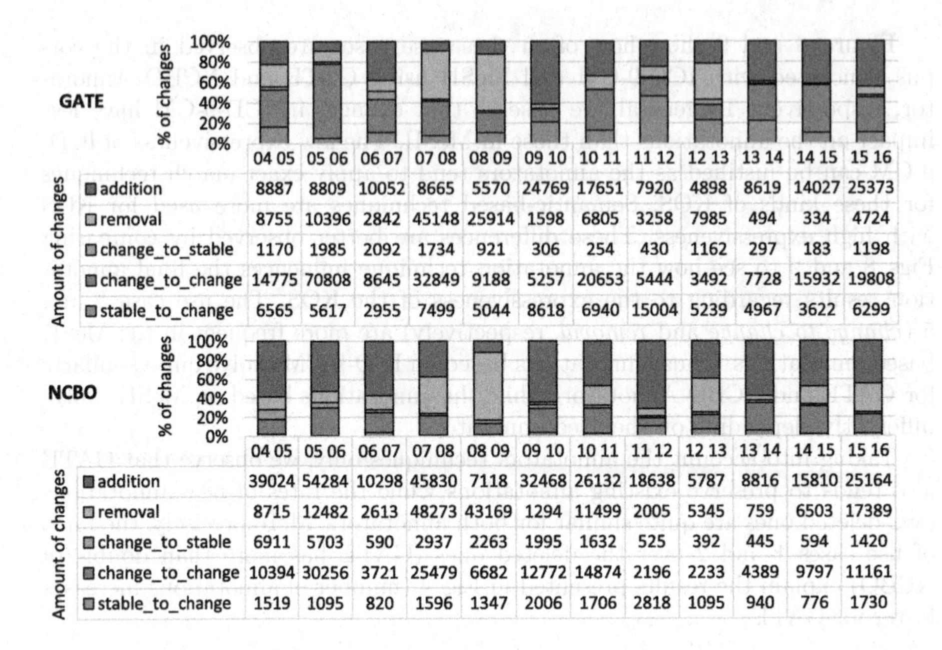

GATE	04 05	05 06	06 07	07 08	08 09	09 10	10 11	11 12	12 13	13 14	14 15	15 16
addition	8887	8809	10052	8665	5570	24769	17651	7920	4898	8619	14027	25373
removal	8755	10396	2842	45148	25914	1598	6805	3258	7985	494	334	4724
change_to_stable	1170	1985	2092	1734	921	306	254	430	1162	297	183	1198
change_to_change	14775	70808	8645	32849	9188	5257	20653	5444	3492	7728	15932	19808
stable_to_change	6565	5617	2955	7499	5044	8618	6940	15004	5239	4967	3622	6299

NCBO	04 05	05 06	06 07	07 08	08 09	09 10	10 11	11 12	12 13	13 14	14 15	15 16
addition	39024	54284	10298	45830	7118	32468	26132	18638	5787	8816	15810	25164
removal	8715	12482	2613	48273	43169	1294	11499	2005	5345	759	6503	17389
change_to_stable	6911	5703	590	2937	2263	1995	1632	525	392	445	594	1420
change_to_change	10394	30256	3721	25479	6682	12772	14874	2196	2233	4389	9797	11161
stable_to_change	1519	1095	820	1596	1347	2006	1706	2818	1095	940	776	1730

Fig. 9. Distribution of changes of MeSH annotations. The y-axis represents the percentage of changes, the x-axis the KOS versions, and bellow the amount of observed changes for each period is described. The listed cases follows the Table 1

5 A Model Supporting Annotation Evolution

The results presented in the previous section allow us to state that the evolution of the KOS has a direct impact on the definition of semantic annotations. However, we also showed that the modification of KOS concepts has different impacts depending on the technique that is implemented to generate the annotations. Furthermore, the evolution of KOS does not necessarily produce more information (see Figs. 6 and 7). Actually, we have observed that KOS are becoming more and more precise over time, which means the addition of new specific concepts whose labels are usually long (in terms of words) and therefore are contained very rarely in medical documents. Our study pointed out important features to take into account, at semantic annotation model level, to facilitate the maintenance of annotation over time. These features can be used to extend the model proposed by Gross et al. [7] (see Sect. 2.1). In consequence, we define our model as:

$$SAM = (I_u, ON_v, R_a, Offset, Q, H, A, SemRel, U_f)$$

Where:

- *Offset* is an element to describe the location of the element to be annotated in a given resource. From an evolution perspective, this is important for linking annotations of different versions and also for distinguishing annotations related to the same element but are annotated differently.

- H is an element to describe which attribute of the concept (e.g., title, synonym, preferred terms, etc.) was used to produce the annotation. This element is really important since the annotation is usually defined based on the value of one concept attribute. If the corresponding concept has one of its attribute changed but not the one used to annotate, it is maybe not needed to modify the annotation.
- *SemRel* is an element to describe the semantic relationship between the KOS concept and the annotated part of the resource. For instance, one sentence can be annotated as equivalent to a concept, more/less specific, partial match, etc. Thus, in the case of removal of a concept, the annotated sentence can be linked to the super-class of the concept and have the relation changed to "less specific".
- U_f is an element to point to the previous version of the annotation. This element is used to keep an evolution chain of annotations.

Our proposal, allowing to link annotation versions, can also be used to improve the W3C proposal by creating an additional property called "evolved to" that links the element "annotation" to itself allowing then to create a chain of annotation version.

6 Conclusion

In this paper we made an empirical analysis of the evolution of biomedical annotations and its relation to the KOS changes. We used for that a set of documents annotated with GATE and NCBO Annotator using 13 different versions of two well-known biomedical KOS (ICD-9-CM and MeSH). We observed that there is a correlation between KOS and annotation changes. Then we regrouped the annotation changes according to the type of information that was modified and the way it was done. We obtained five different cases of changes (see Sect. 4) and verified how the annotations evolve during the KOS evolution. In a second step we analysed different annotation models in order to verify if they can represent (or if we can infer from their elements) all criteria required to classify the annotation changes. As a result of this step, we propose an extended annotation model designed to support evaluations and maintenance of annotations. However, we are still working on the maintenance methods that will use this model and other external information (e.g., KOS changes, background knowledge, etc.) to select the most adapted maintenance strategy for the annotations. We plan to continue our empirical analysis to refine the types of changes in the annotations and to determine fine grained correlations between types of changes in the KOS and types of changes in the annotations.

Acknowledgment. This work is supported by the National Research Fund (FNR) of Luxembourg and Deutsche Forschungsgemeinschaft (DFG) under the ELISA research project.

References

1. Hodge, G.: Systems of Knowledge Organization for Digital Libraries: Beyond Traditional Authority Files. ERIC, Washington (2000)
2. Comeau, D., Doan, R., Ciccarese, P., Cohen, K., Krallinger, M., Leitner, F., Lu, Z., Peng, Y., Rinaldi, F., Torii, M., Valencia, A., Verspoor, K., Wiegers, T., Wu, C., Wilbur, W.: BioC: a minimalist approach to interoperability for biomedical text processing. Database: J. Biol. Databases Curation **2013**, bat064 (2013)
3. Dos Reis, J.C., Pruski, C., Da Silveira, M., Reynaud-Delaître, C.: Understanding semantic mapping evolution by observing changes in biomedical ontologies. J. Biomed. Inf. **47**, 71–82 (2014)
4. Hartung, M., Kirsten, T., Gross, A., Rahm, E.: OnEX: exploring changes in life science ontologies. BMC Bioinform. **10**(1), 1 (2009)
5. Oren, E., Möller, K., Scerri, S., Handschuh, S., Sintek, M.: What are semantic annotations? Relatório técnico. DERI Galway **9**, 62 (2006)
6. Luong, P.H., Dieng-Kuntz, R.: A rule-based approach for semantic annotation evolution in the coswem system. In: Kon, M., Lemire, D. (eds.) Canadian Semantic Web. Semantic Web and Beyond, vol. 2, pp. 103–120. Springer US, New York (2006)
7. Gross, A., Hartung, M., Kirsten, T., Rahm, E.: Estimating the quality of ontology-based annotations by considering evolutionary changes. In: Paton, N.W., Missier, P., Hedeler, C. (eds.) DILS 2009. LNCS, vol. 5647, pp. 71–87. Springer, Heidelberg (2009)
8. Hartung, M., Kirsten, T., Rahm, E.: Analyzing the evolution of life science ontologies and mappings. In: Bairoch, A., Cohen-Boulakia, S., Froidevaux, C. (eds.) DILS 2008. LNCS (LNBI), vol. 5109, pp. 11–27. Springer, Heidelberg (2008)
9. Da Silveira, M., Dos Reis, J.C., Pruski, C.: Management of dynamic biomedical terminologies: current status and future challenges. Yearb. Med. Inf. **10**(1), 125–133 (2015)
10. Traverso-Ribón, I., Vidal, M.-E., Palma, G.: AnnEvol: an evolutionary framework to description ontology-based annotations. In: Ashish, N., Ambite, J.-L. (eds.) DILS 2015. LNCS, vol. 9162, pp. 87–103. Springer, Heidelberg (2015)
11. Groß, A., Hartung, M., Prfer, K., Kelso, J., Rahm, E.: Impact of ontology evolution on functional analyses. Bioinformatics **28**(20), 2671–2677 (2012)
12. Frost, H.R., Moore, J.H.: Optimization of gene set annotations via entropy minimization over variable clusters (EMVC). Bioinformatics (Oxf. Engl.) **30**(12), 1698–1706 (2014)
13. Cunningham, H.: GATE, a general architecture for text engineering. Comput. Humanit. **36**(2), 223–254 (2002)
14. Whetzel, P.L., Noy, N.F., Shah, N.H., Alexander, P.R., Nyulas, C., Tudorache, T., Musen, M.A.: BioPortal: enhanced functionality via new web services from the national center for biomedical ontology to access and use ontologies in software applications. Nucleic Acids Res. **39**(suppl 2), W541–W545 (2011)
15. Hartung, M., Groß, A., Rahm, E.: COnto-Diff: generation of complex evolution mappings for life science ontologies. J. Biomed. Inf. **46**(1), 15–32 (2013)
16. Doğan, R.I., Leaman, R., Lu, Z.: NCBI disease corpus: a resource for disease name recognition and concept normalization. J. Biomed. Inf. **47**, 1–10 (2014)

Capturing the Ineffable: Collecting, Analysing, and Automating Web Document Quality Assessments

Davide Ceolin[1(✉)], Julia Noordegraaf[2], and Lora Aroyo[1]

[1] VU University Amsterdam, Amsterdam, The Netherlands
{d.ceolin,lora.aroyo}@vu.nl
[2] University of Amsterdam, Amsterdam, The Netherlands
j.j.noordegraaf@uva.nl

Abstract. Automatic estimation of the quality of Web documents is a challenging task, especially because the definition of quality heavily depends on the individuals who define it, on the context where it applies, and on the nature of the tasks at hand. Our long-term goal is to allow automatic assessment of Web document quality tailored to specific user requirements and context. This process relies on the possibility to identify document characteristics that indicate their quality. In this paper, we investigate these characteristics as follows: (1) we define features of Web documents that may be indicators of quality; (2) we design a procedure for automatically extracting those features; (3) develop a Web application to present these results to niche users to check the relevance of these features as quality indicators and collect quality assessments; (4) we analyse user's qualitative assessment of Web documents to refine our definition of the features that determine quality, and establish their relevant weight in the overall quality, i.e., in the summarizing score users attribute to a document, determining whether it meets their standards or not. Hence, our contribution is threefold: a Web application for nichesourcing quality assessments; a curated dataset of Web document assessments; and a thorough analysis of the quality assessments collected by means of two case studies involving experts (journalists and media scholars). The dataset obtained is limited in size but highly valuable because of the quality of the experts that provided it. Our analyses show that: (1) it is possible to automate the process of Web document quality estimation to a level of high accuracy; (2) document features shown in isolation are poorly informative to users; and (3) related to the tasks we propose (i.e., choosing Web documents to use as a source for writing an article on the vaccination debate), the most important quality dimensions are accuracy, trustworthiness, and precision.

1 Introduction

Automatically estimating the quality of Web documents is a compelling, yet intricate issue. It is compelling because the huge amount of Web documents

E. Blomqvist et al. (Eds.): EKAW 2016, LNAI 10024, pp. 83–97, 2016.
DOI: 10.1007/978-3-319-49004-5_6

we can access makes their manual evaluation a costly operation. So, to guarantee we access the best documents available on the Web on a given matter, an automated assessment is needed. However, quality is a rather inflated term, that assumes different meanings in different contexts and with different subjects. Quality assessments vary depending on their context (what is the document used for), author (who is judging the document), time (e.g., users may change their assessments about documents as soon they acquire new knowledge), etc. Quality assessments are hard to capture, hence we call them "ineffable".

This paper investigates strategies for capturing such ineffable judgments and assessing their characteristics. In particular, our focus is on the quality assessment of Web documents to be used for professional use (i.e., by journalists and media scholars). Our ultimate goal is to automate the process of document quality assessment, and the contribution of this paper in this direction is threefold. Firstly, we introduce a nichesourcing application for collecting Web document quality assessments (WebQ[1]). Secondly, we present a curated dataset of Web documents (on the topic of vaccinations) enriched with a set of features we extracted, and a set of quality assessments we nichesourced[2]. Thirdly, we describe a thorough set of analyses we performed on these assessments, from which we derive that: (1) given an explicit task at hand, subjects with similar background will provide coherent assessments (i.e., assessments agree with document similarity, measured in terms of shared entities, sentiment, emotions, trustworthiness); (2) users find it difficult to judge document quality based on quantitative features (entities, sentiment, emotions, trustworthiness) extracted from them; however (3) such features are useful to automate the process of quality assessment. The user studies analyzed are based on limited – but highly specialized – judgments, so these findings provide useful insights on how to progress this research.

The rest of the paper is structured as follows. Section 2 introduces related work. Section 3 describes the application we developed for collecting quality assessments, WebQ. Section 4 describes the two case studies we performed, along with the results collected, that are discussed in Sect. 5. Section 6 concludes.

2 Related Work

The problem of assessing the quality of Web documents and, in general, (Web) data and information, is compelling and has been tackled in many contexts.

The ISO 25010 Model [9] is a standard model for data quality. From this model, we select those data quality dimensions that apply also to Web documents (e.g., precision, accuracy) and ask the users of WebQ to rate Web documents on them. This set of quality dimensions has been extended to include other measures tailored to Web documents, like neutrality and readability.

The problem of identifying the documents of higher quality for a given purpose is common in information retrieval. Bharat et al. [2] copyrighted a method

[1] The tool is running at http://webq3.herokuapp.com, the code is available at https://github.com/davideceolin/webq.

[2] The dataset is available at https://github.com/davideceolin/WebQ-Analyses.

for clustering online news content based on freshness and quality of content. Clearly, their approach differs from ours as they focus on news, and they aim at clustering documents. However, one of the key features for determining the quality of documents is the (estimated) authoritativeness of the source, both in their and in our approach. Kang and Kim [10] find links between specific quality requirements and user queries. We do not make use of queries: we preselect documents (to guarantee that documents get an even number of assessments) and we predefine the task the users are asked to perform (to allow controlling the definition of quality adopted by users). We still analyze user assessments to derive their specific definition of quality, and might consider analyzing user queries in the future, when we will expand the dataset and tasks at hand.

Following up on the use of specific metadata as markers for quality, Amento et al. [1] use link-based metrics to make quality predictions, showing that these perform as good as content-based ones. In our case, we focus on features we can automatically extract from the documents using AlchemyAPI and WOT. We will consider other features (including link-based ones) in the future.

Regarding the use of niche- or crowdsourcing for collecting information and, in particular, quality assessments, Lee et al. [11] provide a framework tailored to organizations. Zhu et al. [14] propose a method for collaboratively assessing the quality of Web documents that shows some similarity with ours (e.g., we both collect collaborative quality assessments), but the assessments we collect are based on specific tasks, while they rely on contributions via browser plugins. Currently, we focus on niches for collecting quality assessments because the definition of 'quality' is different for different types of users; so, for us, it is necessary to have a controlled user study. In the future, we plan to make use of crowdsourcing, adopting methods for extracting ground truth like CrowdTruth [8].

While this paper proposes a framework that aims at generically identifying markers for quality of Web documents, we evaluate such framework with an emphasis on Digital Humanities applications. Digital Humanities scholars are professionals that are used to critically evaluate the sources they deal with, hence we target this specific class of users to investigate how to extend source criticism practices to cover Web documents as well. Source criticism is the process of evaluating traditional information sources that is common in the (Digital) Humanities. De Jong and Schellers [5] provide an overview of source criticism methods, evaluated in terms of predictive and congruent validity. We will advance such evaluations to identify which Web document features determine their quality. This paper extends the work we presented at the Web Science conference, where we began the exploration of how it is possible to assess the quality of Web documents, especially for the Digital Humanities [4]. In that, we outlined a pipeline for assessing document quality and we provide a preliminary evaluation based on a manual assessment. Here we develop an application for nichesourcing such assessments and we deeply analyze them and their predictability.

Lastly, one aspect that we consider when estimating the quality of Web documents is their provenance. Provenance analysis is used to assess the quality of humanities sources, as Howell and Prevenier mention [7]. In Computer Science, Hartig and Zhao [6] use temporal qualities of provenance traces to assess the quality of Web data. More extensively, Zaveri et al. [13] provide a review on quality assessment for Linked Data. We also investigated the assessment of crowdsourced annotations using provenance analysis [3,12].

3 Nichesourcing Web Document Quality Assessments

To collect and analyze judgments about Web documents, we developed the tool WebQ, that aims at understanding three main aspects of Web document quality:

- whether (professional) users are able to estimate the quality of Web documents based on limited sets of features of these documents (e.g., the sentiment of these documents, or the list of entities extracted from them);
- whether assessments are coherent enough over multiple documents and among diverse assessors (i.e., whether assessors assess similar documents in a similar manner; similarity is measured in terms of shared entities, sentiment, emotions, trustworthiness), to allow their automated learning;
- how the overall quality assessments can be explained in terms of specific quality dimensions (precision, accuracy, etc.) when focusing on specific tasks.

3.1 Document Features and Document Quality Dimensions

We characterize documents by means of features we automatically extract about them. In Sect. 4 we analyze the existence of correlations between these automatically extracted features and the nichesourced features of quality.

Document Features. These are a series of attributes we automatically extract by means of Web APIs. These features aim at identifying commonalities among documents, opening up for the possibility of predicting their qualities (provided that features and qualities correlate). These features are:

Entities, Sentiment, Emotions. We use AlchemyAPI[3] to extract all the entities mentioned in the documents, along with an assessment of their relevance to the document. Also, AlchemyAPI provides us with a quantification of the sentiment expressed by the document (positive or negative, and its strength), and its emotions (joy, fear, sadness, disgust and anger, and their strength).
Trustworthiness. In this case, we use the Web Of Trust API[4] to obtain crowdsourced trustworthiness assessments about the source publishing the article.

Document Quality Dimensions. These are a series of abstractions of the documents qualifying the information therein contained. We ask the users to assess the documents based on each quality dimension reported as follows:

Overall Quality provides an overall indication of the quality of a document. It summarizes the other quality dimensions in a single value representing the suitability of the document for a given task, in a given context.

[3] http://www.alchemyapi.com.
[4] http://www.mywot.com.

Accuracy quantifies the level of the truthfulness of the document information.
Precision determines whether the document information is precise or vague.
Completeness determines whether the document information is complete.
Neutrality determines whether a particular stance (e.g., pro or anti a given topic) is represented in the document.
Readability quantifies whether the document reads well.
Trustworthiness quantifies the perceived level of trustworthiness of the information in the document. Note that the Web Of Trust score refers to the source, while this quality refers to the specific document evaluated.

3.2 Structure of WebQ

Below we describe the structure of WebQ, illustrated in Fig. 1.

Architecture. The application is developed based on the Flask Python library[5]. As backend storage for Web document assessments, we use MongoDB[6].

Annotations. We use AnnotatorJs[7] to allow users to indicate which specific parts of a document mark particular qualities of the whole document. AnnotatorJs is a javascript library run on the client side that records the document annotations by sending HTTP messages to a storage server. We adapted to this purpose the Annotation Store[8], which relies on ElasticSearch[9].

HTTP Proxy. We developed an HTTP proxy to provide the users with the Web documents to be annotated within WebQ. This proxy allows the system to present the documents within our application and allows users to annotate them by enabling AnnotatorJs. In this manner, the users see the exact same document they would see on the Web, but they are able to annotate it, remaining in the context of our application. This proxy is tailored to the documents in our dataset and renders them at their best. In particular, it addresses the following issues:

- **replace relative paths with absolute ones** in image, CSS and link addresses, so the page can refer to the absolute addresses of the accessory files;
- **correctly detect and utilize charsets** to properly render the documents;
- **forward the browser headers** because some websites allow being accessed only via (some) browsers, and not being scraped. The proxy accesses them programmatically on behalf of a browser.

In the future, we will extend our dataset, so we will extend further this proxy.

---

[5] http://flask.pocoo.org/.
[6] http://mongodb.com.
[7] http://annotatorjs.org.
[8] https://github.com/openannotation/annotator-store.
[9] https://www.elastic.co/.

Randomizer. WebQ is designed for collecting Web document quality assessments via one or more user studies. In such a scenario, users access the application more or less simultaneously. We assign to each user a random sequence of documents to assess (we set the length of such sequence to six), but we also guarantee that the dataset is uniformly assessed: documents should get approximatively

$$n_{ass} = |dataset| \text{div} |users|$$

assessments, where $|dataset|$ is the cardinality of the document dataset (50), div is the integer division and $|users|$ is the cardinality of the set of users. Offline, we generate n_{ass} random permutations of documents. We split them in consecutive sequences of six documents, uniquely assigned to users when they register.

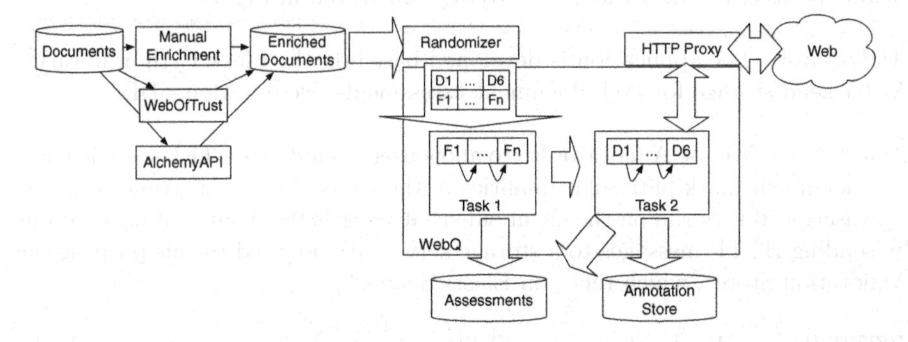

Fig. 1. Overview of the WebQ application. The document set is enriched by using AlchemyApi, Web of Trust, and manually. A random selection of six documents is presented to the users for the first task: identifying the highest quality documents on the basis of the value of one feature. After all the features (sentiment, etc.) have been evaluated, users assess each of the six documents assigned (task 2). Documents are rendered through an HTTP proxy, to allow annotating them within the app.

3.3 Tasks Description

In WebQ we ask the users to perform two tasks. The first task aims at exploring whether single document features could be used as quality indicators. The second task aims at collecting assessments about the documents presented. The two tasks are described as follows, first in general terms, and then, in Sect. 4, as adapted according to a specific scenario for the two case studies.

Task 1. Task 1 is structured as follows:

1. We assign to each user a set of six documents from our overall dataset.
2. We identify six classes of potentially useful features about the documents, namely: the document's sentiment and emotions, its trustworthiness, its title, its source and the list of entities we extract from it.

3. We show the values for each of these features to the user. First, we present the user with the lists of entities extracted from the six documents, then we present the user with the sentiment and the emotions detected in each document, and so on, each feature at a time. Users do not know the documents, they only know the values of the features we present. Every time we present features we shuffle the document order, and we change document identifiers.
4. We ask the user to select which documents among these six she will use as a source for her article, based on the information displayed.
5. Lastly, we ask the user to make the selection again, on the basis of all the features presented together.

Task 2. We ask users to assess the quality of each article in depth. Based on the same selection of six articles the user was assigned to in task 1, she:

1. Reads the article
2. Assesses the overall quality of the article, as well as the following quality dimensions: accuracy, precision, completeness, readability, neutrality, trustworthiness. Assessments are indicated in a 1 to 5 Likert scale.
3. Highlights in the article the words or sentences that motivate her assessments, tagging each selection with the name of the corresponding quality dimension and indicating if it represents a positive or negative observation.
4. Revises their quality assessments (step 2.) if she wishes so.

4 Case Studies

In this section, we describe the two case studies we ran. Both case studies are based on the same set of documents, which we describe as follows.

4.1 Dataset and Scenario

The dataset we base our experiments on is composed of Web documents about the vaccination debate triggered by the measles outbreak that happened at Disneyland, California, in 2015[10]. This dataset contains 50 documents, diversified in terms of **stance** (some are pro vaccinations, some anti, some neutral) and **type of source** (e.g., we include: official reports, editorial articles, blog posts).

The scenario we hypothesize is that users have to write an article about the vaccination debate triggered by such measles outbreak. We propose diverse types of Web documents to the users, and we ask to select those they would use as a source for their article (i.e., those they consider of a higher quality). Thus, we consider selection a marker of relatively high quality.

4.2 Case Study 1 - Journalism Students

Experimental Setup. The first case study involved a class of 20 last-year journalism students from the University of Amsterdam. The students performed both tasks of WebQ in a time frame that lasted between 45 and 60 min.

[10] The dataset is available at https://goo.gl/cLDTtS.

Results. We present here a series of analyses on the results collected.

Document Assessments Collected. We collected 104 complete assessments about the diverse quality dimensions of the documents and 238 annotations.

Comparison of the Two Document Assessments in Task 2. We asked users to assess the documents twice: when they first read the documents, and after having highlighted the motivations for their assessments. These two assessments show no significant difference using a Wilcoxon Signed-rank test at 95 % confidence.

Document Assessments Predictability. The first analysis we perform regards the predictability of Web documents assessments. Only two or three assessments are provided per document, but if users assess the documents coherently enough (i.e., following similar policies), and if the features we extracted (entities, sentiment, emotions, trustworthiness) are considered by the users' policies, then we might be able to automatically learn such predictions. Table 1 shows the results of such predictions using the Support Vector Classification algorithm.

Table 1. Accuracy of 10-fold cross-validation using Support Vector Classification with different combinations of features, and predicting either 5 classes (as in the 1–5 Likert scale used in WebQ) or 2 classes (i.e., high- and low-quality documents). We calculated the performance for all possible combinations of the four classes of features. For each cardinality of such combination (1, 2, 3, 4) we show the best-performing combination.

Features used	SVC 5 classes	SVC 2 classes
Trustworthiness	48 %	75 %
Sentiment, trustworthiness	46 %	78 %
Sentiment, emotions, trustworthiness	38 %	72 %
Sentiment, emotions, trustworthiness, entities	39 %	72 %

Correlation Between Quality Dimensions and Overall Quality. Table 2 shows the results for each quality dimension.

Correlation Between Document Selection (Task 1) and Document Assessments (Task2). In task 1 we ask the users to select documents they think are of high quality based on diverse document features. If many users select a document, we derive that it has a high probability to be of high quality. Since each document has been proposed to only either two or three users, we compute such probability using a smoothing factor that allows accounting for the uncertainty due to the small samples observed (see Eq. (1)). Smoothing allows treating differently documents that have been proposed two or three times: if a document has never been selected when it has been proposed two times, its probability to be of high quality is 0.25; if it has been proposed three times, 0.2. This allows us

Table 2. Correlation between each quality dimension and the overall quality score attributed to the documents.

Quality dimension	Correlation with overall quality
Accuracy	0.89
Completeness	0.69
Neutrality	0.46
Relevance	0.63
Trustworthiness	0.80
Readability	0.67
Precision	0.77

to compare probabilities based on a different amount of evidence in an unbiased manner. The resulting probability is equivalent to the expected value of a Beta probability distribution with a non-informative prior: we add 1 and 2 to the numerator and denominator exactly because we do not know a priori if a given document is of high or low quality (hence its probability of being of high quality is 50 %).

$$P = \frac{\#selection + 1}{\#samples + 2} \quad (1)$$

In task 2, users assess these same documents. Table 3 shows the correlation between the probability from task 1 and the overall quality score from task 2. Entities, sentiment, and title show a poor correlation, close to zero: probabilities from these features (task 1) are not correlated with assessments from task 2. Trustworthiness, sources and all show a slightly higher but still weak correlation: between 20 % and 30 % of the times, their probabilities agree with assessments.

Table 3. Correlation (spearman) between the probability of documents to be selected in task 1 and their overall quality assessment from task 2.

Feature shown (task 1)	Correlation with overall quality (task 2)
Entities	−0.07
Sentiment	0.09
Trustworthiness	0.20
Sources	0.29
Title	−0.07
All	0.20

User Evaluation. We asked the users to complete a questionnaire about their experience[11]. The quantitative results of the 13 respondents (52 % of the total) are reported in Table 4, which shows the percentage of users that indicated a

[11] The questionnaire is available at http://goo.gl/forms/2pIjjpIp0PtyPxd72.

feature or quality as important. Moreover, the majority (~70 %) of users gave a low score (1 or 2 on a 1–5 Likert scale) to the whole experience, to its easiness, and to the fact that the experiment resembles their process when writing an article. Users agree on the importance of most of the features and qualities we identified, but they negatively assess the experience they had. We use such information to improve the experiment design in the next case study, as we explain below.

Table 4. Results of the user evaluation questionnaire.

Feature	Users choosing it	Quality	Users choosing it
Sentiment	0 %	Accuracy	30.8 %
Entities	23.1 %	Completeness	23.1 %
Emotions	0 %	Neutrality	15.4 %
Source	76.9 %	Precision	30.8 %
Title	46.2 %	Trustworthiness	**69.2 %**
Trustworthiness	**100 %**	Relevance	38.5 %

Quality Definition and Qualitative Analysis of Annotations and Remarks. Lastly, from a qualitative evaluation of the annotations and of the remarks collected, we derive that users assume that *the documents of higher quality are those showing the following qualities: high trustworthiness, high accuracy, and high precision.*

4.3 Case Study 2 - Media Scholars

Experimental Setup. This case study involves 20 media scholars (RMA and Ph.D. students as well as senior scholars) attending the Research School for Media Studies (RMeS) summer school in Utrecht (27 May 2016). Based on the user evaluation of case study 1, we add a walk-through session to guide the users in the application, and we improve the task descriptions and the user experience (e.g., landing pages). The users had about 45 min at their disposal.

Results. We present the results obtained and their analyses.

Document Assessments Collected. In this experiment, we collected 47 complete assessments about the documents in our dataset and 89 annotations.

Comparison of the Two Document Assessments in Task 2. We observe no significance difference between the two series of assessments, for any quality dimension.

Document Assessments Predictability. Like with the previous case study, we use 10-fold cross-validation to test the predictability performance of Support Vector Classifier on the overall quality assessment. Results are reported in Table 5.

Correlation Between Quality Dimensions and Overall Quality. Table 6 shows the results for each quality dimension.

Table 5. Accuracy of the prediction of the overall quality assessments case study 2. We show the best-performing combination of features per set cardinality (1, 2, 3, 4).

Features used	SVC 5 classes	SVC 2 classes
Trustworthiness	63 %	89 %
Sentiment, trustworthiness	53 %	86 %
Sentiment, entities, trustworthiness	34 %	85 %
Sentiment, entities, trustworthiness, emotions	34 %	85 %

Table 6. Correlation between each quality dimension and the overall quality score.

Quality dimension	Correlation with overall quality
Accuracy	0.89
Completeness	0.69
Neutrality	0.45
Relevance	0.64
Trustworthiness	0.78
Readability	0.66
Precision	0.76

Correlation Between Document Selection (Task 1) and Document Assessments (Task2). We computed the probability of documents to be of high quality based on the number of selections collected in task 1 (see Eq. (1)). Table 7 shows the correlation between such probability and the overall quality from task 2. Again, the probabilities show a weak correlation with the quality assessments.

Table 7. Correlation (spearman) between the probability of documents to be selected in task 1 and their overall quality assessment from task 2.

Feature shown (task 1)	Correlation with overall quality (task 2)
Entities	0.38
Sentiment	0.19
Trustworthiness	0.21
Sources	0.25
Title	0.15
All	0.24

User Evaluation. The results of the user evaluation questionnaire[12] are reported in Table 8. To these quantitative results, we add the fact that users indicate accuracy and also indicators from social media (e.g., discussion on the

[12] The questionnaire is available at http://goo.gl/forms/ZwvaqDidGeC8FCXm1.

topic, likes) as possible quality markers and that the majority of the users (75 %–100 %) rate the experience and its easiness fairly (2–3 in a 1–5 scale). Users disagree on whether or not this resembles the process of writing an article. Only four participants responded to the questionnaire.

Table 8. Results of the user evaluation questionnaire

Feature	Users choosing it	Quality	Users choosing it
Sentiment	0 %	Accuracy	25 %
Entities	0 %	Completeness	0 %
Emotions	0 %	Neutrality	25 %
Source	100 %	Precision	0 %
Title	50 %	Trustworthiness	50 %
Trustworthiness	100 %	Relevance	25 %

Quality Definition and Qualitative Analysis of Annotations and Remarks. From a qualitative evaluation of the annotations and of the remarks collected, we can derive that users assume that *the documents of higher quality are those showing the following qualities: high trustworthiness, high accuracy, and high precision.*

4.4 Comparison Between Case Study 1 and 2

We compare the results obtained in case study 1 and 2. We use a Wilcoxon signed-rank test to compare the performance obtained by support vector machines (Tables 1 and 5). We observe no significant difference neither with 2 nor with 5 classes. Also comparing the correlations between the quality dimensions and the overall quality (Tables 2 and 6), we observe no significant difference. Neither the results of Tables 3 and 7, i.e., the correlation between probabilities of a document to be selected and its quality, show any significant difference between task 1 and 2. The second user questionnaire has been completed only by a very limited number of users. A Wilcoxon signed-rank test and a χ^2 test agree that the results from the two case studies are not significantly different but, in this case, the sample sizes are so small that we can hardly rely on these results.

5 Discussion

Our long-term goal is to allow automatic assessment of Web document quality tailored to specific user requirements and context. Such a process relies on the possibility to identify document features that indicate quality (if these exist). In this paper, we perform two case studies that shed a light on how professionals evaluated Web documents. Here we discuss the results presented in Sect. 4 by means of a series of statements that emerge from the analysis of the results. Even though the sets of assessments are small, they are large enough to support the statistical test run in Sect. 4. Only the tests run to compare the evaluation test are based on a very small dataset, and thus are less conclusive.

User Assessments are Stable and Coherent. In both case studies, we observe that the first and the second document assessments are not significantly different. Moreover, in both cases, we can use Support Vector Classifier to automatically learn and predict the quality of documents. This means that, even if users assess different documents (the same document has been assessed by three users at most), assessments are coherent enough to be learned. The features we identified (entities, sentiment, emotions, trustworthiness) correlate with these judgments enough to allow using them as features for prediction, at least in this case.

User Assessments are Highly Related to the Task at Hand. The extremely high similarity between the results in Tables 2 and 6 shows that, when assessing the quality of documents, the task at hand is the most important factor. Here the users were asked to pretend they were writing an article about the vaccination debate. So, they focused on identifying the most accurate and trustworthy documents. Neutrality is the least significant quality of these documents because, to represent the whole spectrum of the debate, users have to consider also the least neutral documents, provided that they are accurate enough. Different tasks can imply different quality requirements. This facilitates the definition of future user studies that will provide assessments that are mergeable to the existing dataset (provided, for instance, that they show no statistically significant difference between the existing ones, or that this difference is manageable). So, we will scale up our current approach: even though different case studies will have to be based on limited groups of (diverse) users, their contributions will be used to incrementally build a larger set of document assessments. To guarantee that assessments are handled and merged properly, keeping track of their provenance will be crucial. In this light, although in some cases we observe that by considering only a subset of features we obtain a better performance (up to +6 % in some cases), we still prefer to consider all the features we collected so far. In fact, we do not know if, by extending the set of documents considered (or by diversifying the tasks at hand), some of the features could gain or loose importance, and it may be extremely difficult (if not impossible) to know when this would happen.

Features in Isolation are Hardly Meaningful (but the User Experience Plays a Role Here). Showing entities, sentiment, and emotions, trustworthiness, title and source (especially in isolation) is hardly useful to users to decide if a document is of high quality or not (see Tables 3 and 7). The fact that these features are profitably used to learn the quality assessments of the documents using SVC means that they are good markers of quality (e.g., the fact that a given document expresses an extremely positive sentiment or show specific entities is correlated with its quality). Nevertheless, users are hardly able to determine the document quality on the basis of a quantification of such features. What is true is that in the second case study, although the performance is still pretty low, the results are slightly better than those of the first use case. This might be due to the different user background (more senior level scholars in case study 2), as to the fact that we improved the setup of the WebQ application and explained the logic behind it better in the introduction and walk through.

The Application Setup Should Take the User Experience into Consideration. We aim at collecting annotations from users, so we need to balance a couple of trade-offs between the application requirements and user-based constraints. First, our target users have a professional background that is not necessarily Information or Computer Science. So, even if the application is able to capture all the necessary information, the way its functionality is presented to the user and the way she is guided plays an important role. In fact, after having better explained the logic of the setup of the application we observed (both via the questionnaire and via a post-study discussion) an improvement in the perception of the experience from case study 1 to 2. Second, our goal is to collect as many assessments as possible, but we must take into account that the user attention decreases over time. So, in a situation like case study 2, we need to either extend the duration of the experiment or to reduce the number of documents assessed by each user (e.g., to preserve a uniform number of assessments per document).

6 Conclusion

Automatically assessing the quality of Web documents is crucial to benefit from the vast amount of online information. In this paper, we present WebQ, a Web application to nichesource quality assessments. We also describe two datasets of Web documents, enriched with assessments resulting from two case studies involving journalists and media scholars. WebQ provides the necessary functionalities (i.e., rating and annotating documents) to collect such assessments, and the user evaluations collected allowed fine tuning it. Our last contribution is a set of thorough analyses on the resulting dataset. Through such analyses, we showed that if we assign a clearly defined task to users with a similar background we can obtain uniform document quality assessments. These can be automatically estimated (in our case, using SVC) but, given their tight relation to the context, their provenance needs to be precisely tracked to allow their future reuse. Also, by decomposing overall quality assessments into quality dimensions, we can identify which quality definition (expressed in terms of quality dimensions) is adopted by users. For the task performed (selecting documents to be used as a source for an article on the vaccination debate), the most important dimensions are accuracy, precision, and trustworthiness. We show that the results collected in the two case studies are assimilable: this allows creating a uniform collection of document assessments. Lastly, the user experience in such application matters, and while it is a delicate balance, small changes lead to improvements.

We plan to extend our application in several directions. We will consider other typologies of users and extend the tasks evaluated. Clearly, we intend to extend also the dataset of documents considered, and to incorporate additional features in our models, including link- and network-based features (e.g., based on document interlinking) and social media-based features (e.g., the number of likes a given article received on social media sites, or the number of followers a given blog has). Besides nichesourcing, we will also make use of crowdsourcing, to reach out more contributors. However, such step will require particular attention to assimilate expert and laymen assessments. Lastly, as a consequence of such extension, we will have to consider methods for scaling up our prediction models.

Acknowledgements. This work was supported by the Amsterdam Academic Alliance Data Science (AAA-DS) Program Award to the UvA and VU Universities. We thank the students of the UvA journalism course and the RMeS summer school participants for participating our user studies.

References

1. Amento, B., Terveen, L., Hill, W.: Does authority mean quality? Predicting expert quality ratings of web documents. In: SIGIR, pp. 296–303. ACM (2000)
2. Bharat, K., Curtiss, M., Schmitt, M.: Method and apparatus for clustering news online content based on content freshness and quality of content source. US Patent 9,361,369 (2016). https://www.google.com/patents/US9361369
3. Ceolin, D., Groth, P., Maccatrozzo, V., Fokkink, W., van Hage, W.R., Nottamkandath, A.: Combining user reputation and provenance analysis for trust assessment. J. Data Inf. Qual. **7**(1–2), 6:1–6:28 (2016)
4. Ceolin, D., Noordegraaf, J., Aroyo, L., van Son, C.: Towards web documents quality assessment for digital humanities scholars. In: WebSci, pp. 315–317. ACM (2016)
5. De Jong, M., Schellens, P.: Toward a document evaluation methodology: what does research tell us about the validity and reliability of evaluation methods? (2000)
6. Hartig, O., Zhao, J.: Using web data provenance for quality assessment. In: SWPM (2009)
7. Howell, M., Prevenier, W.: From Reliable Sources: An Introduction to Historical Methods. Cornell University Press, Ithaca (2001)
8. Inel, O., et al.: CrowdTruth: machine-human computation framework for harnessing disagreement in gathering annotated data. In: Mika, P., et al. (eds.) ISWC 2014, Part II. LNCS, vol. 8797, pp. 486–504. Springer, Heidelberg (2014)
9. International Organization for Standardization: ISO/IEC 25012: 2008 software engineering - software product quality requirements and evaluation (SQuaRE) - data quality model. Technical report, ISO (2008)
10. Kang, I.H., Kim, G.: Query type classification for web document retrieval. In: SIGIR, pp. 64–71. ACM (2003)
11. Lee, Y.W., Strong, D.M., Kahn, B.K., Wang, R.Y.: AIMQ: a methodology for information quality assessment. Inf. Manage. **40**(2), 133–146 (2002)
12. Nottamkandath, A., Oosterman, J., Ceolin, D., de Vries, G.K.D., Fokkink, W.: Predicting quality of crowdsourced annotations using graph kernels. In: Jensen, C.D., Marsh, S., Dimitrakos, T., Murayama, Y. (eds.) IFIPTM 2015. IFIPAICT, vol. 454, pp. 134–148. Springer International Publishing, New York (2015)
13. Zaveri, A., Rula, A., Maurino, A., Pietrobon, R., Lehmann, J., Auer, S.: Quality assessment for linked data: a survey. Seman. Web J. **7**(1), 63–93 (2015). http://www.semantic-web-journal.net/content/quality-assessment-linked-data-survey
14. Zhu, H., Ma, Y., Su, G.: Collaboratively assessing information quality on the web. In: ICIS sigIQ Workshop (2011)

Active Integrity Constraints for Multi-context Systems

Luís Cruz-Filipe[1(✉)], Graça Gaspar[2], Isabel Nunes[2], and Peter Schneider-Kamp[1]

[1] Department of Mathematics and Computer Science, University of Southern Denmark, Odense, Denmark
{lcf,petersk}@imada.sdu.dk

[2] BioISI—Biosystems & Integrative Sciences Institute, Faculty of Sciences, University of Lisbon, Lisbon, Portugal
{gg,in}@di.fc.ul.pt

Abstract. We introduce a formalism to couple integrity constraints over general-purpose knowledge bases with actions that can be executed to restore consistency. This formalism generalizes active integrity constraints over databases. In the more general setting of multi-context systems, adding repair suggestions to integrity constraints allows defining simple iterative algorithms to find all possible grounded repairs – repairs for the global system that follow the suggestions given by the actions in the individual rules. We apply our methodology to ontologies, and show that it can express most relevant types of integrity constraints in this domain.

1 Introduction

Integrity constraints (ICs) for databases have been an important topic of research since the 1980s [1]. An early survey [26] already identified over 90 relevant types of integrity constraints. Since then, significant effort has been focused not only on identifying inconsistencies, but also on repairing inconsistent databases.

The same problem has been studied in other domains of knowledge representation. Integrity constraints for deductive databases [2] were also considered in the 1980s. More recently, interest for integrity constraints has arisen in the ontology domain, with several approaches on how to define them and how to check their satisfaction [14,19,21]. Given its challenges, the more complex problem of repairing inconsistent knowledge bases has not received as much attention.

In this paper, we address the problem of computing repairs by combining two ideas: clausal-form integrity constraints for multi-context systems (MCSs) [11] and active integrity constraints (AICs) for relational databases [16]. We demonstrate the expressiveness of our formalism and show how it can be used to compute repairs for inconsistent MCSs in general, and for ontologies, in particular.

This work was supported by the Danish Council for Independent Research, Natural Sciences, grant DFF-1323-00247, and by FCT/MCTES/PIDDAC under centre grant to BioISI (Centre Reference: UID/MULTI/04046/2013).

E. Blomqvist et al. (Eds.): EKAW 2016, LNAI 10024, pp. 98–112, 2016.
DOI: 10.1007/978-3-319-49004-5_7

Contribution. The main contribution of this paper is a notion of AIC for MCSs, which enables us to compute repairs for inconsistent MCSs automatically, requiring only decidability of entailment in the individual contexts. Particularized to ontologies, our framework is expressive enough to capture all types of integrity constraints identified as relevant in [14], as we exemplify in the text.

The step from ICs for MCSs to AICs for MCSs is inspired by the similar step in the database case [16]. However, we draw more significant benefits in this more general setting. AICs are ICs that also specify possible repair actions in their head. In the database case, every clausal IC can be transformed into an AIC automatically. The goal, though, is to restrict in order to establish preferences among different possible repairs. In the general case, such a transformation would require solving complex abduction problems [17].

Using AICs, we can automatically compute repairs for inconsistent MCSs, bypassing the need to solve such reasoning problems. The price to pay is the need to prove that an AIC is valid (Definition 4). The key observation here is that AICs should be written with a very clear semantic idea in mind, typically by an engineer with a deep knowledge of the underlying system, who should be able to show their validity formally. Thus, in practice, the complexity involved in computing each repair is moved to a one-time verification of validity of AICs.

Structure. We review previous work in Sect. 2, summarizing the key notions from [5,8,11]. Section 3 introduces AICs for MCSs, showing that they generalize the corresponding notion for relational databases, and studies their properties in general. Section 4 focuses on the case of ontologies and evaluates our formalism against the classes of integrity constraints identified in [14]. Section 5 discusses how algorithms to compute repairs in the database setting can be adapted to the general case of MCSs. We conclude in Sect. 6.

1.1 Related Work

Database Repairs. ICs for databases have been extensively studied throughout the last decades, and we restrict ourselves to works most directly related to ours.

Integrity constraints are typically grouped in different syntactic categories [26]. Many important classes can be expressed as first-order formulas, and can also be written in denial (clausal) form – the fragment expressable in our formalism.

Whenever an integrity constraint is violated, the database must be *repaired* to regain consistency. The problem of database repair is to determine whether such a transformation is possible, and many authors have invested in algorithms for computing database repairs efficiently. Typically, there are several possible ways of repairing an inconsistent database, and several criteria have been proposed to evaluate them. Minimality of change [13,27] demands that the database be changed as little as possible, while the common-sense law of inertia [23] states that every change should have an underlying reason. While these criteria narrow down the possible database repairs, it is commonly accepted that human interaction is ultimately required to choose the "best" possible repair [25].

Active Integrity Constraints (AICs). The formalism of AICs, introduced in [16], addresses the issue of choosing among several possible repairs. An AIC specifies not only an integrity constraint, but it also gives indications on how inconsistent databases can be repaired through the inclusion of *update actions*, which can be addition and removal of tuples from the database – a minimal set that can implement the three main operations of database updates [1].

The original, declarative, semantics of AICs defined *founded* repairs [5], in which every action is *supported*: it occurs in the head of a constraint that is violated if that action is not included. Despite this characterization, there are unnatural founded repairs where two actions mutually support each other, but do not have support from other actions. The same authors then proposed *justified* repairs [7], which however are not intuitive and pose further problems [9]. Furthermore, justified repairs are intrinsically linked to the syntactic structure of databases, and cannot be adapted to other knowledge representation formalisms.

Grounded repairs [8] form a middle ground between both semantics, requiring support for arbitrary subsets of the repair. They are grounded fixed points of the intuitive operation of "applying one action from the head of each AIC that is not satisfied", which is in line with the intuitive motivation for studying AICs.

Founded and justified repairs can be computed via revision programming [7]. Alternatively, an operational semantics for AICs [9] was implemented for SQL databases [10]. There, repairs are leaves of particular trees, yielding a semantics equivalent to the declarative one when existence of a repair is an NP-complete problem. For grounded and justified repairs, where this existence problem is Σ_2^P-complete, the trees still contain all repairs, but may also include spurious leaves – requiring a post test that brings the overall complexity to the theoretical limit.

Multi-context Systems (MCSs). MCSs, as defined in [3], can be informally described as collections of logic knowledge bases – the *contexts* – connected by Datalog-style *bridge rules*. Since their introduction, several variants of MCSs have been proposed that add to their potential fields of application. Relational MCSs [15] were proposed as a way to allow a formal first-order syntax, introducing variables and aggregate expressions in bridge rules, and extending the semantics of MCSs accordingly. Managed MCSs [4], which we describe in Sect. 2, further generalize MCSs by abstracting from the possible actions that change individual knowledge bases. Other variants, which are not directly relevant for this work, are discussed in [11]. A different line of research deals with repairing *logical* inconsistency of an MCS (non-existence of a model) [12].

ICs in Ontologies. Integrating ICs with ontology-based systems poses several challenges, mainly due to the open-world assumption and the absence of the unique name assumption [14,20,22,24]. In this context, ICs are conventionally modelled as T-Box axioms [19], but variants based on hybrid knowledge bases, auto-epistemic logic, modal logic, and grounded circumscription have recently been proposed. For an overview of these proposals see Sect. 2 in [21]. For details on how some of these can be expressed by ICs over MCSs, using a systematic interpretation of ontologies as MCSs, see Sect. 4.5 in [11]. The interpretation we use in Sect. 4 is a variant of the one presented therein.

2 Background

AICs for Databases. Let Σ be a first-order signature without function symbols. A database is a set of ground atoms over Σ, and an update action is an expression of the form $+a$ or $-a$, where a is a ground atom over Σ. An *active integrity constraint* (AIC) over a database DB is a rule r of the form

$$p_1, \ldots, p_m, \mathsf{not}\ (p_{m+1}), \ldots, \mathsf{not}\ (p_\ell) \Longrightarrow \alpha_1 \mid \cdots \mid \alpha_k \tag{1}$$

where each p_i is an atom over the database's signature, every variable free in $p_{m+1}, \ldots, p_\ell$ occurs in $p_1, \ldots, p_m$, and each update action α_i is either $-p_j$ for some $1 \leq j \leq m$ or $+p_j$ for $m < j \leq \ell$.[1] The *body* of r is $\mathsf{body}(r) = p_1, \ldots, p_m, \mathsf{not}\ (p_{m+1}), \ldots, \mathsf{not}\ (p_\ell)$, and the *head* of r is $\mathsf{head}(r) = \alpha_1 \mid \ldots \mid \alpha_k$.

If r is ground, then DB *satisfies* r, denoted $\mathsf{DB} \models r$, if $\mathsf{DB} \not\models p_i$ for some $1 \leq i \leq m$ or $\mathsf{DB} \models p_i$ with $m < i \leq \ell$. In general, $\mathsf{DB} \models r$ if DB satisfies all ground instances of r. Otherwise, r is *applicable* in DB [16]. If η is a set of AICs, then $\mathsf{DB} \models \eta$ if $\mathsf{DB} \models r$ for every $r \in \eta$.

A set of update actions $\mathcal{U}$ is *consistent* if it does not contain both $+a$ and $-a$ for any ground atom a. Given a consistent $\mathcal{U}$, we write $\mathcal{U}(\mathsf{DB})$ for the result of applying all actions in $\mathcal{U}$ to DB, and say that $\mathcal{U}$ is a *weak repair* for $\langle \mathsf{DB}, \eta \rangle$ if: (i) every action in $\mathcal{U}$ changes DB and (ii) $\mathcal{U}(\mathsf{DB}) \models \eta$. $\mathcal{U}$ is a *repair* if $\mathcal{V}(\mathsf{DB}) \not\models \eta$ for every $\mathcal{V} \subsetneq \mathcal{U}$ [5], and $\mathcal{U}$ is *grounded* if, for every $\mathcal{V} \subsetneq \mathcal{U}$, there exists a ground instance r of a rule in η such that $\mathcal{V}(\mathsf{DB}) \not\models r$ and $\mathsf{head}(r) \cap (\mathcal{U} \setminus \mathcal{V}) \neq \emptyset$ [8].

Multi-context Systems. We now describe the variant of multi-context systems we use: *managed multi-context systems* (also abbreviated to MCSs) [4].

A *relational logic* L is a tuple $\langle \mathsf{KB}, \mathsf{BS}, \mathsf{ACC}, \Sigma \rangle$, where KB is the set of well-formed knowledge bases of L (sets of well-formed formulas), BS is a set of possible belief sets (candidate models), $\mathsf{ACC} : \mathsf{KB} \to 2^{\mathsf{BS}}$ is a function assigning to each knowledge base a set of acceptable belief sets (its models), and Σ is a signature generating first-order sublanguages of $\bigcup \mathsf{KB}$ and $\bigcup \mathsf{BS}$.

A *managed multi-context system* is a collection of managed contexts $\{C_i\}_{i=1}^n$, with each $C_i = \langle L_i, \mathsf{kb}_i, \mathsf{br}_i, D_i, \mathsf{OP}_i, \mathsf{mng}_i \rangle$ where: $L_i = \langle \mathsf{KB}_i, \mathsf{BS}_i, \mathsf{ACC}_i, \Sigma_i \rangle$ is a relational logic; $\mathsf{kb}_i \in \mathsf{KB}_i$; D_i (the *import domain*) is a set of constants from Σ_i; OP_i is a set of operation names; $\mathsf{mng}_i : \wp(\mathsf{OP}_i \times \bigcup \mathsf{KB}_i) \times \mathsf{KB}_i \to \mathsf{KB}_i$ is a *management function*; and br_i is a set of *managed bridge rules*, with the form

$$(i : o(p)) \leftarrow (i_1 : p_1), \ldots, (i_q : p_q), \mathsf{not}\ (i_{q+1} : p_{q+1}), \ldots, \mathsf{not}\ (i_m : p_m) \tag{2}$$

such that $o \in \mathsf{OP}_i$, $p \in \bigcup \mathsf{KB}_i$, $1 \leq i, i_j \leq n$, and each p_j is a belief[2] of L_{c_j}.

Intuitively, kb_i is the knowledge base of context C_i and OP_i are the names of the operations that can be applied to change it. The management function

[1] In [16], existentially quantified variables can also occur in negative literals. This was not discussed in subsequent work, and we ignore it for simplicity of presentation.

[2] Technically, P_p is a *relational element* of C_{i_p}: it can include variables, which when instantiated yield elements of $\bigcup \mathsf{BS}_{i_p}$ – see [4] for details.

defines the semantics of these operations: $\mathsf{mng}_i(O, \mathsf{kb})$ is the result of applying the operations in O to kb. Bridge rules govern the interaction between contexts.[3]

A *belief state* for an MCS $M = \{C_i\}_{i=1}^n$ is a set $S = \{S_i\}_{i=1}^n$ such that each $S_i \in \mathsf{BS}_i$. A ground instance of bridge rule (2) is *applicable* in S if $p_i \in S_i$ for $1 \leq i \leq q$ and $p_i \notin S_i$ for $q < i \leq m$; the variables in the rule can only be instantiated by elements of the import domain D_i. A belief state is an *equilibrium* for M if it is stable under application of all bridge rules, i.e.:

$$S_i \in \mathsf{ACC}_i(\mathsf{mng}_i(\{\mathsf{head}(r) \mid r \in \mathsf{br}_i \text{ applicable in } S\}, \mathsf{kb}_i))$$

In general, M can have zero, one or several equilibria; if at least one exists, then M is *logically consistent.* We present examples of MCSs in the next sections.

Integrity Constraints for General-Purpose Knowledge Bases. ICs for MCSs [11] generalize clausal ICs to a generic framework for reasoning systems – covering not only relational databases, but also deductive databases, peer-to-peer systems and ontologies, among others. Syntactically, ICs are bridge rules with empty head, forming an added layer on top of an MCS that does not affect its semantics.

As MCSs may have several equilibria, satisfaction of a set of ICs η can be *weak* – there is an equilibrium satisfying all rules in η – or *strong* – all equilibria satisfy all rules in η. In order to avoid vacuous quantifications, strong satisfaction only holds for logically consistent MCSs. In general these properties are undecidable [11], but if entailment in every context is decidable then satisfaction of a set of ICs is in most cases as hard as the hardest entailment decision problem.

In this paper, we do not explicitly mention the set of ICs when clear from the context. Moreover, our development applies both to weak and strong satisfaction, and we simply say that an MCS is *consistent* if it satisfies the given set of ICs. We explicitly write "logical consistency" for existence of an equilibrium.

3 Active Integrity Constraints

We begin by defining active integrity constraints over multi-context systems.

Definition 1. *An AIC over an MCS $M = \{C_i\}_{i=1}^n$ is a rule r of the form*

$$(i_1 : P_1), \ldots, (i_m : P_m), \mathsf{not}\ (i_{m+1} : P_{m+1}),\ \ldots, \mathsf{not}\ (i_\ell : P_\ell) \\ \Longrightarrow (j_1 : \alpha_1) \mid \cdots \mid (j_k : \alpha_k) \quad (3)$$

where $1 \leq i_p, j_q \leq n$, each P_p is a belief in C_{i_p}, each update action $\alpha_q \in \mathsf{OP}_{j_q} \times \bigcup \mathsf{KB}_{j_q}$*, and all variables in $P_{m+1}, \ldots, P_\ell$ occur in $P_1, \ldots, P_m$.*

This definition follows the one for databases (1), and we define *body* and *head* of r similarly. Equation (3) also generalizes ICs for MCSs: each AIC corresponds

[3] For the sake of presentation, we simplified the management function, which in the original work is allowed to return several possible effects for each action.

to an IC by ignoring its head, immediately yielding notions of weak and strong satisfaction for an AIC. We also say that r is *applicable* to an MCS M if $M \not\models r$. Intuitively, in this case M should be repaired by applying actions in $\mathsf{head}(r)$.

The reasoning capabilities of MCSs dictate that we cannot restrict the actions in the head of an AIC syntactically (as in the database world, see Sect. 2). We thus relax this requirement by only demanding that the actions are capable of solving the inconsistency. It is also not reasonable to require that every action in $\mathsf{head}(r)$ be able to solve every inconsistency detected by $\mathsf{body}(r)$: since inconsistencies may be triggered by derived information, they may have different origins, and the different actions may be solutions for those different causes.

We are interested in sets of update actions that are applied simultaneously, i.e. the order in which actions are executed should be irrelevant. This corresponds to the consistency requirement usually considered in databases.

Definition 2. *Let $M = \{C_i\}_{i=1}^n$ be an MCS, $\mathcal{U}$ be a finite set of update actions, and $\mathcal{U}_i$ be the set of actions in $\mathcal{U}$ affecting C_i.*

$\mathcal{U}_i$ is consistent *w.r.t. kb_i if, for every permutation $\alpha_1, \ldots, \alpha_k$ of the elements of $\mathcal{U}_i$, $\mathsf{mng}_i(\mathcal{U}_i, \mathsf{kb}_i) = \mathsf{mng}_i(\alpha_1, \mathsf{mng}_i(\ldots, \mathsf{mng}_i(\alpha_k, \mathsf{kb}_i)\ldots))$. $\mathcal{U}$ is consistent w.r.t. M if each $\mathcal{U}_i$ is consistent w.r.t. kb_i, and in this case we write $\mathcal{U}(M)$ for the result of applying each $\mathcal{U}_i$ to each kb_i.*

Example 1. We consider a concrete toy example of a deductive database with two unary base relations p and q, a view consisting of a relation r such that $r(x) \leftrightarrow p(x) \vee q(x)$, and the integrity constraint $\neg r(a)$.

We formalize this as an MCS $M = \langle C_E, C_I \rangle$ where C_E is an extensional database including predicates p and q (but not r), C_I is the view context including predicate r (but not p or q), and they are connected by the bridge rules

$$(I : r(X)) \leftarrow (E : p(X)) \qquad (I : r(X)) \leftarrow (E : q(X)).$$

Furthermore, mng_E allows addition and removal of any tuples to C_E, using operations add and del, while mng_I does not allow any changes. (See [11] for details of this construction.)

From the structure of M, we know that $r(a)$ can only arise as a deduction from $p(a)$ or $q(a)$ (or both), so it makes sense to write an AIC

$$(I : r(a)) \Longrightarrow (E : \mathsf{del}(p(a))) \mid (E : \mathsf{del}(q(a))).$$

The actions on the head of this AIC solve the problem in all future states of M, since C_I cannot change. However, restoring consistency may require performing both actions (if the database contains both $p(a)$ and $q(a)$).

This example also illustrates an important point: repair actions are written with a particular structure of the MCS in mind.

Definition 3. *The set of* variants *to an MCS M, denoted $\mathsf{vrt}(M)$, is*

$$\mathsf{vrt}(M) = \{\mathcal{U}(M) \mid \mathcal{U} \text{ is a finite set of update actions over } M\}.$$

Restrictions on the actions in the head of AICs only range over $\mathsf{vrt}(M)$, which contains all possible future evolutions of M.

Definition 4. *An AIC r of the form* (3) *is* valid *w.r.t. an MCS M if:*
- *for every logically consistent* $M' \in \mathsf{vrt}(M)$ *such that* $M' \not\models r$, *there is* $\mathcal{U} \subseteq \mathsf{head}(r)$ *with* $\mathcal{U}(M') \models r$;
- *for every* $\alpha \in \mathsf{head}(r)$, *there is* $M' \in \mathsf{vrt}(M)$ *with* $M' \not\models r$ *and* $\alpha(M') \models r$.

These conditions require that the set of suggested actions be complete (it can solve all inconsistencies) and that it does not contain useless actions.

Example 2. The AIC in Example 1 is valid: the only possible changes to M are in kb_E, which only contains information about p and q, thus, in any element of $\mathsf{vrt}(M)$ the only way to derive $r(a)$ is still from either $p(a)$ or $q(a)$. The second condition follows by considering M' with $\mathsf{kb}_E = \{p(a)\}$ and $\mathsf{kb}_E = \{q(a)\}$.

Proposition 1. *Deciding whether an AIC is valid is in general undecidable.*

Proof (sketch). Let L be a logic with an undecidable entailment problem, C be a context over L with $\mathsf{add} \in \mathsf{OP}_C$ such that $\mathsf{mng}_C(\mathsf{add}(\varphi), \Gamma) = \Gamma \cup \{\varphi\}$, and $M = \{C\}$. Assume also that $\mathsf{vrt}(M)$ includes all knowledge bases over L. Then $(C : \neg B) \Longrightarrow (C : \mathsf{add}(A))$ is valid iff $A \models_L B$. □

In practice, proving validity of AICs should not pose a problem: AICs are written by humans with a very precise semantic motivation in mind, and this means that the conditions in Definition 4 should be simple for a human to prove.

We now show that the framework we propose generalizes the database case. A database DB can be seen as an MCS $M(\mathsf{DB})$, defined as having a single context over first-order logic, whose knowledge base is DB, with management function allowing addition (+) or removal (−) of facts, and where the only set of beliefs admissible w.r.t. a given database is the set of literals that are true in that database (see [11] for a detailed definition).

Proposition 2. *Every AIC over a database* DB *yields a valid AIC over* $M(\mathsf{DB})$.

Proof (sketch). We write a generic AIC over a database (1) as the AIC

$$(1 : p_1), \ldots, (1 : p_m), \mathsf{not}\ (1 : p_{m+1}), \ldots, \mathsf{not}\ (1 : p_\ell) \Longrightarrow (1 : \alpha_1) \mid \cdots \mid (1 : \alpha_k)$$

over $M(\mathsf{DB})$. If DB does not satisfy the body of (1), then it can always be repaired by performing exactly one of the actions in its head [6], establishing both conditions for validity. □

Definition 5. *Let* $M = \{C_i\}_{i=1}^n$ *be an MCS,* η *be a set of AICs over M and* $\mathcal{U}$ *be a finite set of update actions.* $\mathcal{U}$ *is a* weak repair *for* $\langle M, \eta \rangle$ *if* $\mathcal{U}$ *is consistent w.r.t. M and* $\mathcal{U}(M) \models \eta$. *Furthermore,* $\mathcal{U}$ *is* grounded *if: for every* $\mathcal{V} \subsetneq \mathcal{U}$, *there is an AIC* $r \in \eta$ *such that* $\mathcal{V}(M) \not\models r$ *and* $\mathsf{head}(r) \cap (\mathcal{U} \setminus \mathcal{V}) \neq \emptyset$.

The definitions of weak and grounded repair directly correspond to those for the database case (Sect. 2). The notion of grounded repair implies, in particular, minimality under inclusion [8].

4 Application: The Case of Ontologies

This section is devoted to examples illustrating how our framework can be applied to the particular case of integrity constraints over ontologies.

Previous work [3,11] shows how to view an ontology as a context of an MCS. In the present work, we refine this interpretation by representing an ontology as *two* contexts: one for the A-Box, one for the T-Box, connected by bridge rules that port every instance from the former into the latter. (This is reminescent of how deductive databases are encoded in MCSs, see [11]). This finer encoding allows us, in particular, to reason about asserted instances (which are given in the A-Box) and those that are derived using the axioms (see Example 5).

We further assume that the A-Box only contains instances of atomic concepts or roles ($C(t)$ or $R(t,t')$). This option does not restrict the expressive power of the ontology, but it helps structure AICs: to include instance axioms about e.g. $C \sqcup D$, one instead defines a new concept $E = C \sqcup D$ in the T-Box and includes instance axioms about E in the A-Box (see also Example 7).

Definition 6. *A description logic $\mathcal{L}$ is represented as the relational logic $L_{\mathcal{L}} = \langle \mathsf{KB}_{\mathcal{L}}, \mathsf{BS}_{\mathcal{L}}, \mathsf{ACC}_{\mathcal{L}}, \Sigma_L \rangle$, where:*

- $\mathsf{KB}_{\mathcal{L}}$ *contains all well-formed knowledge bases of $\mathcal{L}$;*
- $\mathsf{BS}_{\mathcal{L}}$ *contains all sets of queries in the language of $\mathcal{L}$;*
- $\mathsf{ACC}_{\mathcal{L}}(\mathsf{kb})$ *is the singleton set containing the set of queries to which* kb *answers "Yes".*
- $\Sigma_{\mathcal{L}}$ *is the first-order signature underlying $\mathcal{L}$.*

An ontology $\mathcal{O} = \langle T, A \rangle$ based on $\mathcal{L}$ induces the multi-context system $M(\mathcal{O}) = \langle \mathsf{Ctx}(T), \mathsf{Ctx}(A) \rangle$ where $\mathsf{Ctx}(T) = \langle L_{\mathcal{L}}, T, \mathsf{br}_T, \Sigma_0, \emptyset, \emptyset \rangle$ with

- br_T *contains all rules of the form $(T : C)(X) \leftarrow (A : C)(X)$ where C is a concept, and $(T : R)(X,Y) \leftarrow (A : R)(X,Y)$ where R is a role;*
- Σ_0 *is the set of constants in $\Sigma_{\mathcal{L}}$;*

and $\mathsf{Ctx}(A) = \langle L_{\mathcal{L}}, A, \emptyset, \Sigma_0, OP, \mathsf{mng} \rangle$ where OP and mng *are the set of allowed update operation names and their definition.*

The management function does not allow changes to the T-Box; the particular operations in the A-Box depend on the concrete ontology. This is in line with our motivation that writing AICs requires knowledge of the system's deductive abilities (expressed by the T-Box), which should not change.

We now evaluate the expressivity of our development by showing how to formalize several types of ICs over ontologies. We follow the classification in Sect. 4.5 of [14], which describes families of ICs determined by OWL engineers and ontologists as the most interesting, as well as other types of ICs considered in the scientific literature. Several classes of ICs are syntactically similar, so we do not include examples for all categories in [14], but explain in the text how the missing ones can be treated.

Most of our examples are adapted from [14], which frames them in a variant of the Lehigh University Benchmark [18], an ontology designed with the goal of providing a realistic scenario for testing. This ontology considers concepts student, gradStudent, class and email, and roles hasEmail, enrolled and webEnrolled. Our semantics is: class is a concept including all classes of a common course; enrolled(c,s) holds if student s is enrolled in course s; and webEnrolled holds if the student is furthermore to be contacted only electronically.[4] The actual contents of the A-Box are immaterial for our presentation, and we restrict ourselves to the fragment of the T-Box containing the following axioms.

$$\begin{array}{ll} \mathsf{gradStudent} \sqsubseteq \mathsf{student} & \exists \mathsf{enrolled.student} \sqsubseteq \mathsf{class} \\ \mathsf{webEnrolled} \sqsubseteq \mathsf{enrolled} & \exists \mathsf{hasEmail.email} \sqsubseteq \mathsf{student} \\ & \exists \mathsf{webEnrolled}^R.\mathsf{class} \sqsubseteq \exists \mathsf{hasEmail} \end{array}$$

4.1 Functional Dependencies

Functional dependencies are one of the most frequently occurring families of ICs: requirements that certain relations be functional on one argument. In our example, this applies to hasEmail: two distinct students cannot have the same e-mail.

Since ontologies do not have the Unique Name Assumption, we cannot distinguish individuals by checking name equality (as in databases), but must query the ontology instead. Furthermore, while in the database world such violations can only be repaired by removing one of the offending instances, in ontologies, we can also add the information that two individuals are the same.

Example 3. Suppose that the management function includes operations add and del to add or remove a particular instance from the A-Box, as well as assertEqual, establishing equality of two individuals. Under these assumptions, we can express funcionality of e-mail as the following AIC.

$$\begin{aligned} &(A : \mathsf{hasEmail}(X, Z)),\ (A : \mathsf{hasEmail}(Y, Z)), \mathsf{not}\ (T : (X = Y)) \\ &\qquad \Longrightarrow (A : \mathsf{del}(\mathsf{hasEmail}(X, Z))) \mid (A : \mathsf{assert}(X = Y)) \end{aligned} \tag{4}$$

Observe that, if T explicitly proves that $X \neq Y$, then only the first action can be used, as asserting equality between X and Y would lead to an inconsistency. However, if this is not the case then the second action is also a repair possibility, and hence this AIC is valid. There are several possibilities for the implementation of assert: it can add the equality $X = Y$ to the A-Box, but it can also syntactically replace every occurrence of one of them for the other.

Several other types of dependencies (e.g. key constraints, uniqueness constraints, functionality constraints) are expressed by similar formulas. Likewise, max-cardinality constraints can be represented as AICs with similar types of actions in the head (deleting some instances or unifying some individuals).

[4] This semantics is slightly changed from that of [14], in order to make some aspects of our example more realistic.

4.2 Property Domain Constraints

This family of ICs specifies that the domain of a role should be a subset of a particular concept. In case such a constraint is violated, the offending element has to be added as an instance of that concept. The treatment of these ICs is thus very similar to the database case.

Example 4. To model that only students can be enrolled in courses, we write the following AIC.

$$(T : \mathsf{enrolled}(X, Y), \mathsf{not}\ (T : \mathsf{student}(Y)) \Longrightarrow (A : \mathsf{add}(\mathsf{student}(Y))) \qquad (5)$$

We could also add the action $(A : \mathsf{del}(\mathsf{enrolled}(X, Y)))$ to the head of this AIC; note that it would only restore consistency in the case where this fact is explicitly stated in the A-Box and not otherwise derivable. Property range constraints (restricting the range of a role) can be similarly treated.

4.3 Specific Type Constraints

In many applications, it is interesting to minimize redundancy in the A-Box. In particular, in the presence of inclusion axioms, it is often desirable only to include instances pertaining to the most specific type class of each individual.

Example 5. Since $\mathsf{gradStudent} \sqsubseteq \mathsf{student}$, we guarantee that the A-Box only contains instances of the most specific class a student belongs to by writing:

$$(A : \mathsf{gradStudent}(X)), (A : \mathsf{student}(X)) \Longrightarrow (A : \mathsf{del}(\mathsf{student}(X))) \qquad (6)$$

Thus, if the A-Box contains e.g. student(john) and gradStudent(john), then the axiom student(john) will be removed. The system will still be able to derive student(john), but only in context C_T (using the information in the T-Box). The separation of the A-Box and T-Box in different contexts is essential to express this integrity constraint in our formalism. Constraints that distinguish between assertions explicitly stated in the A-Box and derived ones have been considered e.g. in [22].

4.4 Min-Cardinality Constraints

We now consider a more interesting type of ICs: min-cardinality constraints. Inconsistencies arising from the violation of such constraints are hard to repare automatically, as such a repair requires "guessing" which instances to add. Using AICs and adequate management functions, we can even specify the construction of "default" values that may depend on the actual ontology.

Example 6. We want to express that each class must have a minimum of 10 students. Classes with less enrolled students should be closed, and those students moved to the smallest remaining class using an operation redistribute.

$$(T : (\leq 10.\mathsf{enrolled})(X)) \Longrightarrow (A : \mathsf{redistribute}(\neg\mathsf{class}(X))) \qquad (7)$$

For this AIC to be valid, redistribute must check whether students are enrolled or webEnrolled and change the appropriate instance in the A-Box. This also uses the knowledge that instances of enrolled cannot be derived in other ways.

A similar kind of constraints are totality constraints, which require that a role be total on one of its arguments. In our example, we could require every student to be enrolled in some class, and use an adequate management function to add non-enrolled students to e.g. the smallest class.

4.5 Missing Property Value Constraints

We now turn our attention to a kind of ICs that is also very common in ontologies: disallowing unnamed individuals for particular properties [22].

Example 7. Our ontology specifies that all students that are webEnrolled in a class must have an e-mail address. However, for the purpose of contacting these individuals, this e-mail address must be explicitly provided. We address this issue with the following AIC.

$$(T : (\exists \mathsf{hasEmail})(X)),\ \mathsf{not}\ (T : \mathsf{hasEmail}(X,Y)) \\ \Longrightarrow (A : \mathsf{unregister}(\neg\exists \mathsf{webEnrolled}^R(X))) \qquad (8)$$

Here, unregister replaces the axiom $\mathsf{webEnrolled}(X)$ with $\mathsf{enrolled}(X)$, as it makes sense to keep the student enrolled in the course. Validity of this AIC follows from observing that the only possible ways to derive $\exists\mathsf{hasEmail}(X)$ are either from an explicit assertion $\mathsf{hasEmail}(X,Y)$ or indirectly from $\mathsf{webEnrolled}(Z,X)$.

This example also justifies our requirement that the A-Box can only contain instances of atomic concepts or roles. If the A-Box were allowed to contain e.g. ∃hasEmail(john), then AIC (8) would no longer be valid. By restricting to atomic concepts, the only way to perform a similar change would be by defining a new concept as equivalent to ∃hasEmail – and this information would be present in the T-Box, making it clear that AICs should consider it.

4.6 Managing Unnamed Individuals

Finally, we illustrate how we can write AICs in different ways to control whether they range over all individuals of a certain class, or only over named ones.

Example 8. For ecological reasons, we want all students with an e-mail address to be enrolled in the web version of courses. We can write this as follows.

$$(T : (\mathsf{hasEmail})(Y,Z)),\ (T : \mathsf{enrolled}(X,Y)), \mathsf{not}\ (T : \mathsf{webEnrolled}(X,Y)) \\ \Longrightarrow (A : \mathsf{webEnroll}(\mathsf{webEnrolled}(X,Y))) \qquad (9)$$

Operation webEnroll will replace $\mathsf{enrolled}(X,Y)$ with $\mathsf{webEnrolled}(X,Y)$, dually to unregister in the previous example.

Alternatively, we could consider writing

$$(T : (\exists \mathsf{hasEmail})(Y)),\ (T : \mathsf{enrolled}(X,Y)), \mathsf{not}\ (T : \mathsf{webEnrolled}(X,Y)) \\ \Longrightarrow (A : \mathsf{webEnroll}(\mathsf{webEnrolled}(X,Y))) \quad (10)$$

In this particular context, this formulation is undesirable, as it will also affect individuals who do not have a known e-mail address. By writing an explicit variable in the first query of the body, as in (9), we guarantee that we only affect those individuals whose e-mail address is known.

Similar considerations about the two possible ways to formulate this type of ICs can be found in [22].

5 Computing Repairs

In [9], we showed how to use active integrity constraints to compute repairs for inconsistent databases, by using the actions in the head of unsatisfied AICs to build a *repair tree* whose leaves were the repairs. We showed how the construction of the tree could be adapted to the different types of repairs considered originally in [7]; in particular, for the case of grounded repairs (which is the one we are interested in this work), it is enough to expand each node with the actions in the heads of the AICs that are not satisfied in that node.

We adapt this construction to the framework of AICs over MCSs. As we will see, the algorithms have to be adapted to this more general scenario, but we can still construct all grounded repairs for a given (inconsistent) MCS automatically, as long as entailment in all contexts is decidable.

Definition 7. *Let M be an MCS and η be a set of integrity contraints over M. The* repair tree *for $\langle M, \eta\rangle$, $\mathcal{T}_{\langle M,\eta\rangle}$, is defined as follows.*

- *Each node is a set of update actions.*
- *A node n is* consistent *if: (i) $n(M)$ is logically consistent and (ii) if n' is the parent of n, then n is a consistent set of update actions w.r.t. $n'(M)$.*
- *Each edge is labeled with a closed instance of a rule.*
- *The root of the tree is the empty set $\emptyset$.*
- *For each consistent node n and rule r, if $n(M) \not\models r$ then $n' = n \cup \mathcal{U}$ is a child of n if (i) $\mathcal{U} \subseteq \mathsf{head}(r)$, (ii) $n'(M) \models r$ and (iii) if $\mathcal{U}' \subseteq \mathcal{U}$ then $(n \cup \mathcal{U}')(M) \not\models r$.*

In the database case [9], it is straightforward to show that repair trees are finite, since the syntactic restrictions on database AICs guarantee that each rule can only be applied at most once in every branch. In the general MCS case, this is not true, as the following example shows.

Example 9. Consider an ontology (represented as an MCS as in Sect. 3) with four concepts B_1, B_2, B_3 and D. The T-Box contains axioms

$$B_1 \sqsubseteq D \qquad \text{and} \qquad B_2 \sqcap B_3 \sqsubseteq D$$

and the A-Box is $\{B_1(a), B_3(a)\}$. Furthermore, we have integrity constraints

$$(T:D)(a) \Longrightarrow (A:\mathsf{del}(B_1)(a)) \mid (A:\mathsf{del}(B_3)(a)) \quad (r_1)$$

$$\mathsf{not}\ (T:B_1)(a), \mathsf{not}\ (T:B_2)(a) \Longrightarrow (A:\mathsf{add}(B_2)(a)) \quad (r_2)$$

Following this construction, we obtain the tree on the right, and its leaf is a grounded repair.

$$\emptyset \xrightarrow{r_1} \{\mathsf{del}(B_1)(a)\} \xrightarrow{r_2} \{\mathsf{del}(B_1)(a), \mathsf{add}(B_2)(a)\} \xrightarrow{r_1} \{\mathsf{del}(B_1)(a), \mathsf{add}(B_2)(a), \mathsf{del}(B_3)(a)\}$$

Lemma 1. *$\mathcal{T}_{\langle M,\eta\rangle}$ is finite.*

Proof. By definition, every node of $\mathcal{T}_{\langle M,\eta\rangle}$ has a finite number of descendants, since there are only finitely many ground instances of AICs with a finite number of actions in each one's head. By construction, in every branch the labels of the nodes form an increasing sequence (w.r.t. set inclusion), and each node is again a subset of the (finite) set of all actions in the heads of all rules. Therefore, $\mathcal{T}_{\langle M,\eta\rangle}$ has finite depth and finite degree, hence it is finite. □

Lemma 2. *Every grounded repair for $\langle M,\eta\rangle$ is a leaf of $\mathcal{T}_{\langle M,\eta\rangle}$.*

Proof. Let $\mathcal{U}$ be a grounded repair for M and η. By definition of grounded repair, if $\mathcal{U}' \subseteq \mathcal{U}$ then there is a ground instance r of an AIC such that: there exists $\mathcal{V} \subseteq \mathsf{head}(r) \cap \mathcal{U}$ such that $(\mathcal{U}' \cup \mathcal{V})(M) \models r$. This directly yields a branch of the repair tree ending at $\mathcal{U}$. □

(This is essentially the same argument for showing that, in the database case, grounded repairs are well-founded, see [8].)

$\mathcal{T}_{\langle M,\eta\rangle}$ is constructed as the well-founded repair tree in the database case [9].[5] In both cases, this tree may, in general, contain leaves that are not grounded repairs [8]. Under the assumption that P $\neq$ NP, this cannot be avoided, since existence of grounded repairs for databases is a Σ_2^P-complete problem [8].

Complexity. The proof of Lemma 1 shows that the depth of $\mathcal{T}_{\langle M,\eta\rangle}$ is polynomial in the size of the grounded instances of η. Therefore, given an oracle that decides whether an MCS satisfies a set of AICs, the problem of existence of a grounded repair for $\langle M,\eta\rangle$ is Σ_2^P-complete: $\mathcal{T}_{\langle M,\eta\rangle}$ can be built in non-deterministic polynomial time (guessing which rule to apply at each node and using the oracle to decide whether the descendant is a leaf), and the validation step can be done in co-NP time (if the leaf is not a grounded repair, then we guess the subset that violates the definition and use the oracle to confirm this).

[5] There is also a notion of repair tree for databases in [9], but it relies on the ability of inferring heads of AICs automatically, which does not exist in the MCS setting.

6 Discussion and Conclusions

Validity. At the end of Example 1, we pointed out that restoring consistent w.r.t. an AIC r may require applying several actions in $\mathsf{head}(r)$. This suggests allowing sets of actions (rather than actions) in the heads of AICs. Besides increasing the complexity of our development, it is not clear that this change would bring significant benefits. In terms of computing repairs, we already cover those cases, since we add sets of actions when going from a node to its descendents. Also, it is not clear that there exists a situation when *every* possible inconsistent MCS requires a set of actions to repair.

One could also remove the second condition of validity of an AIC, i.e. allow the actions in the head to be insufficient to restore consistency of some MCSs. This would remove some burden from the programmer who has to specify the AICs, and would not affect the performance of the algorithms in Sect. 5. However, it would contradict the original motivation for AICs [16]: that the actions in the head of a rule should provide the means for restoring consistency.

Variants of AICs. The authors of [16] also considered *conditioned active integrity constraints*, where the actions on the head of AICs are guarded by additional conditions that have to be satisfied. In their setting, conditioned AICs do not add expressive power to the formalism, as they can be split into several unconditioned AICs (with more specific bodies) preserving the notions of consistency and repairs. In our setting, this transformation is not possible, and it would thus be interesting to study conditioned active integrity constraints over multi-context systems. However, we point out that the management function *can* use information about the actual knowledge bases in its implementation, so some conditions can actually be expressed in our setting (see Example 6).

Conclusion. We proposed active integrity constraints for multi-context systems and showed that, using them, we can compute grounded repairs for inconsistent MCSs automatically. Although validity of AICs is in general undecidable, we showed that we can cover the most common types of ICs in our framework.

References

1. Abiteboul, S.: Updates a new frontier. In: Gyssens, M., Paredaens, J., Van Gucht, D. (eds.) ICDT 1888. LNCS, vol. 326, pp. 1–18. Springer, Heidelberg (1988)
2. Asirelli, P., de Santis, M., Martelli, M.: Integrity constraints for logic databases. J. Log. Program. **2**(3), 221–232 (1985)
3. Brewka, G., Eiter T.: Equilibria in heterogeneous nonmonotonic multi-context systems. In: AAAI, pp. 385–390. AAAI Press (2007)
4. Brewka, G., Eiter, T., Fink, M., Weinzierl, A.: Managed multi-context systems. In: IJCAI, pp. 786–791. IJCAI/AAAI (2011)
5. Caroprese, L., Greco, S., Sirangelo, C., Zumpano, E.: Declarative semantics of production rules for integrity maintenance. In: Etalle, S., Truszczyński, M. (eds.) ICLP 2006. LNCS, vol. 4079, pp. 26–40. Springer, Heidelberg (2006)

6. Caroprese, L., Truszczyński, M.: Declarative semantics for active integrity constraints. In: Garcia de la Banda, M., Pontelli, E. (eds.) ICLP 2008. LNCS, vol. 5366, pp. 269–283. Springer, Heidelberg (2008)
7. Caroprese, L., Truszczyński, M.: Active integrity constraints and revision programming. Theor. Pract. Log. Program. **11**(6), 905–952 (2011)
8. Cruz-Filipe, L.: Grounded fixpoints and active integrity constraints. In: ICLP, OASICS. Dagstuhl (2016, accepted)
9. Cruz-Filipe, L., Engrácia, P., Gaspar, G., Nunes, I.: Computing repairs from active integrity constraints. In: TASE, pp. 183–190. IEEE (2013)
10. Cruz-Filipe, L., Franz, M., Hakhverdyan, A., Ludovico, M., Nunes, I., Schneider-Kamp, P., repAIrC: a tool for ensuring data consistency by means of active integrity constraints. In: KMIS, pp. 17–26. SciTePress (2015)
11. Cruz-Filipe, L., Nunes, I., Schneider-Kamp, P.: Integrity constraints for general-purpose knowledge bases. In: Gyssens, M., et al. (eds.) FoIKS 2016. LNCS, vol. 9616, pp. 235–254. Springer, Heidelberg (2016). doi:10.1007/978-3-319-30024-5_13
12. Eiter, T., Fink, M., Ianni, G., Schüller, P.: Towards a policy language for managing inconsistency in multi-context systems. In: Workshop on Logic-Based Interpretation of Context: Modelling and Applications, pp. 23–35 (2011)
13. Eiter, T., Gottlob, G.: On the complexity of propositional knowledge base revision, updates, and counterfactuals. Artif. Intell. **57**(2–3), 227–270 (1992)
14. Fang, M.: Maintaining integrity constraints in semantic web. Ph.D. thesis, Georgia State University (2013)
15. Fink, M., Ghionna, L., Weinzierl, A.: Relational information exchange and aggregation in multi-context systems. In: Delgrande, J.P., Faber, W. (eds.) LPNMR 2011. LNCS, vol. 6645, pp. 120–133. Springer, Heidelberg (2011)
16. Flesca, S., Greco, S., Zumpano, E.: Active integrity constraints. In: PPDP, pp. 98–107. ACM (2004)
17. Guessoum, A.: Abductive knowledge base updates for contextual reasoning. J. Intell. Inf. Syst. **11**(1), 41–67 (1998)
18. Guo, Y., Pan, Z., Heflin, J.: LUBM: a benchmark for OWL knowledge base systems. J. Web Sem. **3**(2–3), 158–182 (2005)
19. Motik, B., Horrocks, I., Sattler, U.: Bridging the gap between OWL, relational databases. Web Semant.: Sci. Serv. Agents World Wide Web **7**(2), 74–89 (2011)
20. Motik, B., Rosati, R.: Reconciling description logics and rules. J. ACM, **57** (2010). Article Nr 30
21. Ouyang, D., Cui, X., Ye, Y.: Integrity constraints in OWL ontologies based on grounded circumscription. Front. Comput. Sci. **7**(6), 812–821 (2013)
22. Patel-Schneider, P.F., Franconi, E.: Ontology constraints in incomplete and complete data. In: Cudré-Mauroux, P., et al. (eds.) ISWC 2012, Part I. LNCS, vol. 7649, pp. 444–459. Springer, Heidelberg (2012)
23. Przymusinski, T.C., Turner, H.: Update by means of inference rules. J. Log. Program. **30**(2), 125–143 (1997)
24. Tao, J., Sirin, E., Bao, J., McGuinness, D.L.: Integrity constraints in OWL. In: AAAI. AAAI Press (2010)
25. Teniente, E., Olivé, A.: Updating knowledge bases while maintaining their consistency. VLDB J. **4**(2), 193–241 (1995)
26. Thalheim, B.: Dependencies in Relational Databases. Teubner-Texte zur Mathematik. B.G. Teubner, Leipzig (1991)
27. Winslett, M.: Updating Logical Databases. Cambridge Tracts in Theoretical Computer Science. Cambridge University Press, Cambridge (1990)

Evolutionary Discovery of Multi-relational Association Rules from Ontological Knowledge Bases

Claudia d'Amato[1](✉), Andrea G.B. Tettamanzi[2], and Tran Duc Minh[2]

[1] University of Bari, Bari, Italy
claudia.damato@uniba.it
[2] Université Côte d'Azur, Inria, CNRS, I3S, Nice, France
andrea.tettamanzi@unice.fr, tdminh2110@yahoo.com

Abstract. In the Semantic Web, OWL ontologies play the key role of domain conceptualizations, while the corresponding assertional knowledge is given by the heterogeneous Web resources referring to them. However, being strongly decoupled, ontologies and assertional knowledge can be out of sync. In particular, an ontology may be incomplete, noisy, and sometimes inconsistent with the actual usage of its conceptual vocabulary in the assertions. Despite of such problematic situations, we aim at discovering hidden knowledge patterns from ontological knowledge bases, in the form of multi-relational association rules, by exploiting the evidence coming from the (evolving) assertional data. The final goal is to make use of such patterns for (semi-)automatically enriching/completing existing ontologies. An evolutionary search method applied to populated ontological knowledge bases is proposed for the purpose. The method is able to mine intensional and assertional knowledge by exploiting problem-aware genetic operators, echoing the refinement operators of inductive logic programming, and by taking intensional knowledge into account, which allows to restrict the search space and direct the evolutionary process. The discovered rules are represented in SWRL, so that they can be straightforwardly integrated within the ontology, thus enriching its expressive power and augmenting the assertional knowledge that can be derived from it. Discovered rules may also suggest new (schema) axioms to be added to the ontology. We performed experiments on publicly available ontologies, validating the performances of our approach and comparing them with the main state-of-the-art systems.

Keywords: Description logics · Pattern discovery · Evolutionary algorithms

1 Introduction

The Semantic Web [3] is the new vision of the Web aiming at making Web contents machine readable besides of human readable. For the purpose, Web resources are semantically annotated with metadata referring to ontologies that

E. Blomqvist et al. (Eds.): EKAW 2016, LNAI 10024, pp. 113–128, 2016.
DOI: 10.1007/978-3-319-49004-5_8

are formal conceptualizations of domains of interest acting as shared vocabularies where the meaning of the annotations is formally defined. As such, annotated web resources represent the assertional knowledge, given the intensional definitions provided with ontologies. Assertional and intensional ontological knowledge will be referred to as ontological knowledge base. In the SW view, data, information, and knowledge are connected following best practices and exploiting standard Web technologies, e.g. HTTP, RDF and URIs. This allows to share and link information that can be read automatically by computers meanwhile creating a global space of resources semantically described. The description of data/resources in terms of ontologies represents a key aspect in the SW. Interestingly, ontologies are also equipped with powerful deductive reasoning capabilities. However, due to the heterogeneous and distributed nature of the SW, ontological knowledge bases (KBs)[1] may turn out to be incomplete and noisy w.r.t. the domain of interest. Specifically, an ontology is incomplete when it is logically consistent (i.e., it contains no contradiction) but it lacks of information (e.g., assertions, disjointness axioms, etc.) w.r.t. the domain of reference; an ontology is noisy when it is logically consistent but it contains invalid information w.r.t. the reference domain. These situations may prevent the inference of relevant information or cause incorrect information to be derived.

By exploiting the evidence coming from the (assertional) knowledge, data mining techniques could be fruitfully exploited for discovering hidden knowledge patterns from ontological KBs, to be used for enriching an ontology both at terminological (schema) and assertional (facts) level, even in presence of incompleteness and/or noise. We present a method, based on evolutionary algorithms, for discovering hidden knowledge patterns in the form of multi-relational association rules (ARs) coded in SWRL [14], which can be added to the ontology thus enriching its expressive power and increasing the assertional knowledge that can be derived from it. Additionally, discovered rules may suggest new axioms to be added to the ontology, such as transitivity and symmetry of a role, and/or concept/role inclusion axioms. Even if related works focussing on a similar goal can be found in the SW community (see [11,12,15,16,24]) and in the ILP community (see [7,19,21]), to the best of our knowledge, our work represents the first proposal that is able to discover hidden knowledge patterns from ontological knowledge bases while: (i) taking into account the background/ontological knowledge; (ii) exploiting the efficiency of genetic algorithms jointly with reasoning capabilities. Evolutionary algorithms (EAs) [5,9] are bio-inspired stochastic optimization algorithms, which exploit two principles that allow populations of organisms to adapt to their surrounding environment: genetic inheritance and survival of the fittest. Each individual of the population represents a point in the space of the potential solutions for the considered problem. The evolution is obtained by iteratively applying a small set of stochastic operators, known as *mutation*, *recombination*, and *selection*. Mutation randomly perturbs a candidate solution; recombination decomposes two distinct solutions and then randomly

[1] By *ontological knowledge base*, we refer to a populated ontology, namely an ontology where both the schema and instance level are specified. The expression will be interchangeably used with the term ontology.

mixes their parts to form novel solutions; selection replicates the most successful solutions found in a population at a rate proportional to their relative quality. Given enough time, the resulting process tends to find globally optimal solutions to the problem in the same way as in nature populations of organisms tend to adapt to their surrounding environment. We build on these ideas and we combine them with recent works on relational ARs discovery from populated KBs in the SW [4,11], with the final goal of proposing an EA for discovering mulit-relational ARs. The rationale for using EAs as a meta-heuristic is to mitigate the combinatorial explosion usually characterizing purely ILP-based methods when applied to rich representations, such as Description Logics [4], while maintaining the quality of the results. Our solution is experimentally evaluated and comparisons with the main state-of-the art systems are provided.

The rest of the paper is organized as follows. In the next section, the problem definition and basics are introduced. The EA-based method for discovering multi-relational ARs from ontological KBs is presented in Sect. 3; its experimental evaluation is illustrated in Sect. 5. The main characteristics and value added of our proposal with respect to the state of the art are analyzed in Sect. 4. Conclusions are drawn in Sect. 6.

2 Basics

We refer to ontological KBs described in Description Logics (DLs) [2] (representing the theoretical foundation of OWL), and we do not fix any specific DL. As usual in DLs, we refer to a KB $\mathcal{K} = \langle \mathcal{T}, \mathcal{A} \rangle$ defined by the set $\mathcal{T}$ of the terminological axioms, named the TBox, and the set $\mathcal{A}$ of assertional axioms, named the ABox The formal meaning of the axioms is given in terms of model-theoretic semantics. As for reasoning services, *instance checking*, which assesses if an individual is instance of a given concept, and *concept subsumption*, which consists in checking whether a concept (role) is subsumed by another concept (role), are exploited. DLs adopt the *open-world assumption* (OWA) which has consequences on answering class-membership queries. Specifically, it may happen that an individual, that cannot be proved to be instance of a certain concept, is not necessarily a counterexample for it, rather it would be only interpreted as a case of insufficient (incomplete) knowledge for proving the assertion (for details see [2]).

In the following, the general definition of relational AR for an ontological KB $\mathcal{K}$ is given. Hence, the problem we want to address is formally defined.

Definition 1 (Relational Association Rule). *Let* $\mathcal{K}= \langle \mathcal{T}, \mathcal{A} \rangle$ *be a populated ontological KB. A* relational association rule *r for* $\mathcal{K}$ *is a Horn-like clause of the form: body* $\rightarrow$ *head, where: (a) body is a generalization of a set of assertions in* $\mathcal{K}$ *co-occurring together; (b) head is a consequent that is induced from* $\mathcal{K}$ *and body*

Definition 2 (Problem Definition).
Given:
- *a populated ontological knowledge base* $\mathcal{K} = \langle \mathcal{T}, \mathcal{A} \rangle$*;*
- *a minimum "frequency threshold",* θ_f*;*
- *a minimum "head coverage threshold",* θ_{hc}*;*
- *a minimum "confidence improvement threshold",* θ_{ic}*;*

Discover: *all frequent hidden patterns w.r.t* θ_f*, in the form of multi-relational ARs, that may induce new assertions for* $\mathcal{K}$*.*

Intuitively, a *frequent hidden pattern* is a generalization of a set of concept/role assertions co-occurring reasonably often (w.r.t. a fixed frequency threshold) together, thus showing an underlying form of correlation that is exploited for obtaining new assertions.

For representing the rules to be discovered (following Definition 2), the Semantic Web Rule Language (SWRL) [14] is adopted which straightforwardly extends the set of OWL axioms of a given ontology with Horn-like rules.[2]

Definition 3 (SWRL Rule). *Given a KB* $\mathcal{K}$*, a SWRL rule is an implication of the form:* $B_1 \wedge B_2 \wedge \dots B_n \rightarrow H_1 \wedge \cdots \wedge H_m$*, namely between an antecedent* $B_1 \wedge \cdots \wedge B_n$*, called rule body, and a consequent* $H_1 \wedge \cdots \wedge H_m$ *called rule head. Each* $B_1, \dots, B_n, H_1, \dots H_m$ *is called* atom.

An atom *is a unary or binary predicate of the form* $P_c(s)$*,* $P_r(s_1, s_2)$*,* $\mathsf{sameAs}(s_1, s_2)$ *or* $\mathsf{differentFrom}(s_1, s_2)$*, where the predicate symbol* P_c *is a concept name in* $\mathcal{K}$*,* P_r *is a role name in* $\mathcal{K}$*,* s, s_1, s_2 *are* terms*. A* term *is either a variable (denoted by* x, y, z*) or a constant (denoted by* a, b, c*) standing for an individual name or data value.*

The discovered rules can be generally called *multi-relational* rules since multiple binary predicates $P_r(s_1, s_2)$ with different role names of $\mathcal{K}$ could appear in a rule.

The intended meaning of a rule is: whenever the conditions in the antecedent hold, the conditions in the consequent must also hold. Due to the *safety condition* (see Definition 4), a rule having more than one atom in the head can be equivalently transformed into multiple rules, each one having the same body and a single atom in the head. We will consider, w.l.o.g., only SWRL rules (hereafter just "rules") with one atom in the head.

2.1 Language Bias

In this section, the adopted *language bias* is specified. It consists of a set of constraints giving a tight specification of the patterns worth considering, thus allowing to reduce the search space. We manage rules having only atomic concepts and/or role names of $\mathcal{K}$ as predicate symbols, and individual names as constants. Only *connected* [11] and non-redundant [15] rules satisfying the *safety condition* [13] are considered. Additionally, to guarantee decidability, only *DL-safe*

[2] The results is a KB with an enriched expressive power. More complex relationships than subsumption can be expressed. For details see [13].

rules are managed [17], that is rules interpreted under the DL-safety condition consisting in binding all variables in a rule only to explicitly named individuals in $\mathcal{K}$.[3] In the following, the formal definitions for the properties listed above are reported.

Given an atom A, let $T(A)$ denote the set of all the terms occurring in A and let $V(A) \subseteq T(A)$ denote the set of all the variables occurring in A, e.g. $V(C(x)) = \{x\}$ and $V(R(x,y)) = \{x,y\}$. Such notation may be extended to rules straightforwardly.

Definition 4 (Safety Condition). *Given a KB $\mathcal{K}$ and a rule $r = B_1 \wedge B_2 \wedge \dots B_n \rightarrow H$, r satisfies the* safety condition *if all variables appearing in the rule head also appear in the rule body; formally if: $V(H) \subseteq \bigcup_{i=1}^{n} V(B_i)$,*

Definition 5 (Connected Rule). *Given a KB $\mathcal{K}$ and a rule $r = B_1 \wedge B_2 \wedge \dots B_n \rightarrow H$, r is* connected *iff every atom in r is transitively connected to every other atom in r.*

Two atoms B_i and B_j in r, with $i \neq j$, are connected *if they share at least a variable or a constant i.e. if $T(B_i) \cap T(B_j) \neq \emptyset$.*

Two atoms B_1 and B_k in r are transitively connected *if there exist in r atoms $B_2, \dots, B_{k-1}$, with $k \leq n$, s.t. for all $i, j \in \{1, \dots, k\}$, $i \neq j$, $T(B_i) \cap T(B_j) \neq \emptyset$.*

Definition 6 (Non-redundant Rule). *Given a KB $\mathcal{K}$ and a rule $r = B_1 \wedge B_2 \wedge \dots B_n \rightarrow H$, r is a non-redundant rule if no atom in r is entailed by other atoms in r wrt $\mathcal{K}$, i.e., if $\forall i \in \{0, 1, \dots, n\}$, with $B_0 = H$, results: $\bigwedge_{j \neq i} B_j \not\models_{\mathcal{K}} B_i$,*

Example 1 (Redundant Rule). Given $\mathcal{K}$ with $\mathcal{T} = \{\mathsf{Father} \sqsubseteq \mathsf{Parent}\}$ and the rule $r = \mathsf{Father}(x) \wedge \mathsf{Parent}(x) \rightarrow \mathsf{Human}(x)$ where Human is a primitive concept, r is redundant since the atom $\mathsf{Parent}(x)$ is entailed by the atom $\mathsf{Father}(x)$ wrt $\mathcal{K}$.

2.2 Metrics for Rule Evaluation

Given a set of discovered rules, metrics for assessing the quality of a rule and for assessing if it is actually of interest for the goal of Definition 2, are necessary. In the following, we first summarize standard metrics adopted for the purpose. Successively, we present additional metrics to be adopted, jointly with the motivation for introducing them.

Given a rule $r = B_1 \wedge \dots \wedge B_n \rightarrow H$, let us denote:

- $\Sigma_H(r)$ the set of distinct bindings of the variables occurring in the head of r, formally: $\Sigma_H(r) = \{\mathit{binding}\ V(H)\}$
- $E_H(r)$ the set of distinct bindings of the variables occurring in the head of r provided that the body and the head of r are satisfied, formally: $E_H(r) = \{\mathit{binding}\ V(H) \mid \exists\ \mathit{binding}\ V(B_1 \wedge \dots \wedge B_n) : B_1 \wedge \dots \wedge B_n \wedge H\}$. Since rules are connected, $V(H) \subseteq V(B_1 \wedge \dots \wedge B_n)$

[3] When added to an ontology, DL-safe rules are decidable and generate sound results but not necessarily complete.

– $M_H(r)$ the set of distinct bindings of the variables occurring in the head of r also appearing as binding for the variables occurring in the body of r, formally: $M_H(r)= \{binding\ V(H) \mid \exists\ binding\ V(B_1 \wedge \cdots \wedge B_n) : B_1 \wedge \cdots \wedge B_n\}$

Standard metrics (as given e.g. in [1]) modified for copying with rich representations and ensuring monotonicity when atoms are added to a rule body (as argued in [4,11]) are reported below.

Definition 7 (Rule Support). *Given a rule $r = B_1 \wedge \ldots \wedge B_n \rightarrow H$, its support is given by the number of distinct bindings of the variables in the head, formally:*

$$\text{supp}(r) = |E_H(r)|. \tag{1}$$

Definition 8 (Head Coverage for a Rule). *Given a rule $r = B_1 \wedge \ldots \wedge B_n \rightarrow H$, its head coverage is given by the proportion of the distinct variable bindings from the head of the rule that are covered by the predictions of the rule:*

$$\text{headCoverage}(r) = |E_H(r)|/|\Sigma_H(r)|. \tag{2}$$

Definition 9 (Rule Confidence). *Given a rule $r = B_1 \wedge \ldots \wedge B_n \rightarrow H$, its confidence is defined as the ratio of the number of distinct bindings of the predicting variables in the rule head and the number of their bindings in the rule body:*

$$\text{conf}(r) = |E_H(r)|/|M_H(r)|. \tag{3}$$

An issue with these definitions, and particularly Definition 9, is that an implicit closed-world assumption is made, since no distinction between *false* predictions, i.e., bindings σ matching r such that $\mathcal{K} \models \neg H\sigma$, and *unknown* predictions, i.e., bindings σ matching r such that both $\mathcal{K} \models H\sigma$ and $\mathcal{K} \models \neg H\sigma$, is made. On the contrary, reasoning on ontological KBs is grounded on the OWA. Additionally, our goal is to maximize correct predictions, not just describing the available data. To circumvent this limitation the following metric, generalizing the *PCA Confidence* [11], is introduced.

Definition 10 (Rule Precision). *Given a rule $r = B_1 \wedge \ldots \wedge B_n \rightarrow H$, its precision is given by the ratio of the number of correct predictions made by r and the total number of correct and incorrect predictions (predictions logically contradicting $\mathcal{K}$), leaving out the predictions with unknown truth value.*

This metric expresses the ability of a rule to perform correct predictions, but it is not able to take into account the induced knowledge, that is the *unknown* predictions. In order to evaluate/quantify the induced predictions, the metrics proposed for this purpose in [10] are also considered. They are briefly recalled in the following:

– *match rate*: number of predicted assertions in agreement with facts in the complete ontology, out of all predictions;
– *commission error rate*: number of predicted assertions contradicting facts in the full ontology, out of all predictions;
– *induction rate*: number of predicted assertions that are not known (i.e., for which there is no information) in the complete ontology, out of all predictions.

3 Evolutionary Discovery of Relational Association Rules

Given a populated ontological KB, our goal is to discover frequent hidden patterns in the form of multi-relational ARs to be exploited for making predictions of new assertions in the KB. The discovered rules are DL-Safe and represented in SWRL (see Sect. 2), hence, they can be straightforwardly integrated with the existing ontology, thus resulting in a KB with an enriched expressive power [13,14]. To achieve this goal, we propose to search the space of the SWRL rules that respect the language bias (as defined in Sect. 2.1) using an EA. The algorithm maintains a population of patterns (the individuals) and makes it evolve by iteratively applying a number of genetic operators. A pattern is the genotype of an individual and the corresponding rule is its phenotype. Since, like [11], our goal is to discover rules capable of making a large number of predictions, the fitness of a pattern is the head coverage (see Definition 8) of the rule constructed using the first atom of the pattern as the head and the remaining atoms as the body.

The approach we propose may be regarded as alternative and complementary to level-wise generate-and-test algorithms for discovering relational ARs from RDF datasets [11] and recent proposals that take into account terminological axioms and deductive reasoning capabilities [4].

3.1 Representation

As in [4,11], a pattern is represented as a list of atoms of the form $C(x)$ or $R(x, y)$, respecting the language bias, to be interpreted in conjunctive form. For each discovered frequent pattern, a multi-relational AR is constructed by considering the first atom in the list as the head of the rule and the remaining atoms as the rule body.

The genetic operators of initialization, recombination, and mutation, described in the following sections, are designed to enforce the language bias. An important consequence of the fact that patterns are intended to be transformed into rules for evaluation is that the order of atoms counts only insofar as one atom is in the head position (and, therefore, the head of the rule) or it is not (and, therefore, in the body of the rule). The relative position of atoms that are not in the head position is irrelevant.

3.2 Initialization

The initial population is seeded by n random patterns, randomly generated according to Algorithm 1. This CREATENEWPATTERN() initialization operator requires a list A_f of frequent atoms, which is computed once and for all before launching the evolutionary process, and returns a new random pattern. A frequent atom is a pattern r consisting of a single atom of the form $C(x)$ or $R(x, y)$, such that $\text{supp}(r) \geq \theta_f$ (cf. Definition 7). A new pattern is seeded with a frequent pattern picked at random from A_f and a random target length between 2 and MAX_RULE_LENGTH is chosen; the specialization operator (detailed in

Algorithm 1. The CREATENEWPATTERN() Operator.

Input: a global variable A_f: a list of frequent atoms;
Output: r: a new, random pattern.
1: $length \sim \lceil \mathcal{U}(2, \text{MAX\_RULE\_LENGTH}) \rceil$
2: pick an atom $a \in A_f$ at random
3: $r \leftarrow a$
4: **while** r.SIZE() $< length_{p'}$ **do**
5: $r \leftarrow$ SPECIALIZE(r)
6: **return** r

Algorithm 2. The Recombination Operator RECOMBINE(p, r).

Input: p, r: the two patterns to be recombined;
Output: p', r': two patterns that are a recombination of the input patterns.
1: $L \leftarrow p \cup r$
2: $length_{p'} \sim \lceil \mathcal{U}(2, \text{MAX\_RULE\_LENGTH}) \rceil$
3: $length_{r'} \sim \lceil \mathcal{U}(2, \text{MAX\_RULE\_LENGTH}) \rceil$
4: $p' \leftarrow \top$
5: **while** p'.SIZE() $< length_{p'}$ **do**
6: pick an atom $a \in L$ at random
7: fix a so that $p' \wedge a$ respects the language bias
8: $p' \leftarrow p' \wedge a$
9: $r' \leftarrow \top$
10: **while** r'.SIZE() $< length_{r'}$ **do**
11: pick an atom $a \in L$ at random
12: fix a so that $r' \wedge a$ respects the language bias
13: $r' \leftarrow r' \wedge a$
14: **return** p', r'

Algorithm 4), which adds a random atom to an existing pattern while respecting the language bias, is then called repeatedly, until the target length is attained.

3.3 Recombination

The recombination (or crossover) operator produces two offspring patterns from two parent patterns, by randomly exchanging their body atoms and fixing, if necessary, their variables so that they respect the language bias.

The operator, detailed in Algorithm 2, proceeds by creating a set L including all the atoms in the two input patterns and choosing a target length for the two offspring; then, atoms are picked from L at random and added to either pattern until the target length is attained, possibly changing their variables to ensure the language bias is respected.

Recombination is performed with probability p_{cross}.

3.4 Mutation

The mutation operator is based on the idea of specialization and generalization operators in inductive logic programming. Roughly speaking, a specialization operator appends a new atom to a pattern while preserving the language bias, whereas a generalization operator removes a body atom from a pattern while preserving the language bias.

Mutation is applied to every child pattern (resulting from recombination or not) with a small probability $p_{\text{mut}} \ll 1$.

Algorithm 3. The Mutation Operator MUTATE(r).

Input: r: the pattern to be mutated;
Output: r': the mutated pattern.
1: **if** r.GETHEADCOVERAGE() $> \theta_{\text{mut}}$ **then**
2:     **if** r.SIZE() $<$ MAX_RULE_LENGTH **then**
3:         $r' \leftarrow$ SPECIALIZE(r)
4:     **else**
5:         $r' \leftarrow$ CREATENEWPATTERN()
6: **else**
7:     **if** r.SIZE() > 2 **then**
8:         $r' \leftarrow$ GENERALIZE(r)
9:     **else**
10:         $r' \leftarrow$ CREATENEWPATTERN()
11: **return** r'

Algorithm 4. The Specialization Operator SPECIALIZE().

Input: r: the pattern to be specialized;
Output: r': the specialized pattern.
1: $X \sim \mathcal{U}(0,1)$ {Extract a uniform random number from $[0,1)$}
2: **if** $X < \frac{1}{2}$ **then**
3:     pick a concept name $C \in \mathcal{N}_C^{\text{freq}}$ at random
4:     $r' \leftarrow$ ADDCONCEPTATOM(r, C)
5: **else**
6:     pick a role name $R \in \mathcal{N}_R^{\text{freq}}$ at random
7:     **if** $X < \frac{3}{4}$ **then**
8:         $r' \leftarrow$ ADDROLEATOMWITHFRESHVAR(r, R)
9:     **else**
10:         $r' \leftarrow$ ADDROLEATOMWITHWITHALLVARSBOUND(r, R)
11: **return** r'

Mutation, summarized in Algorithm 3, applies the specialization operator, if the head coverage of the rule corresponding to the pattern is above a given threshold θ_{mut}, or the generalization operator, if its head coverage is below θ_{mut}, to the pattern undergoing it.

The specialization operator is detailed in Algorithm 4. A specialization for a given pattern may be generated by applying one of the operators, defined in [4]:

- ADDCONCEPTATOM, which adds an atom whose predicate symbol is a concept name in the ontology and its variable argument already appears in the pattern to be specialized. The predicate symbol can already appear in the pattern, in that case, a different variable name has to be used;
- ADDROLEATOMWITHFRESHVAR or WITHWITHALLVARSBOUND, which add an atom whose predicate symbol is a role name in the ontology and at least one of its variable arguments is shared with one or more atoms in the pattern while the other could be a shared or a new variable. The predicate symbol could be already existing in the pattern.

The operators are applied so that, at each step of the specialization process, rules in agreement with the language bias (see Sect. 2) are obtained. We refer the reader to [4] for a detailed description of these operators.

The generalization operator simply removes the last atom from a pattern. Given the way patterns are created and specialized, this guarantees that the resulting pattern respects the language bias.

3.5 Fitness and Selection

A pattern is evaluated by first constructing a rule from it, using the first atom of the pattern as its head and the remaining atoms as its body. Fitness is defined as the head coverage of the rule: $f(r) = \text{headCoverage}(r)$.

Selection is performed as in the breeder algorithm [20] by truncation with parameter τ: the n patterns in the population are sorted by decreasing fitness and the $\lfloor \tau n \rfloor$ fittest individuals are selected for reproduction. The remaining individuals are replaced by the offspring of the selected individuals.

3.6 Consistency Check

Inconsistent rules, i.e., rules that are unsatisfiable when considered jointly with the ontology, are of no use for knowledge base enrichment and have thus to be discarded.[4] Notice that this case should never occur if the ontological KB is consistent and noise-free. Nevertheless, since the proposed method can be also applied to noisy ontologies, it may happen that an unsatisfiable rule/pattern (when considered jointly with the ontology) is extracted, particularly if low *frequency* and *Head Coverage* thresholds (see Sect. 2.2 for details about the adopted metrics and related discussions) are considered.

Since checking rules for consistency may be very computationally expensive, we have decided not to check patterns for consistency during evolution. Instead, we defer this check and we apply it to the final population.

The satisfiability check is performed by calling an off-the-shelf OWL reasoner. Our current implementation is able to use two state-of-the-art OWL reasoners, namely Pellet [22] and Hermit [18]. However, we have observed that both reasoners fail to give an answer within a reasonable time for some patterns. This happens relatively seldom and not necessarily with the same patterns for either reasoner; however, given the large number of pattern our algorithm generates, these cases have a high chance of occurring in every run. As a workaround, we have introduced a time-out, which is an additional parameter of the algorithm, after which the reasoner is interrupted. When this happens, we discard the problematic pattern, since we have observed that, in general, patterns that take too long to be checked are either inconsistent or uninteresting.

The overall flow of the EA may be summarized as in Algorithm 5. The parameters of the algorithm are summarized in Table 1.

The rules corresponding to the patterns returned by the EA are straightforwardly obtained and coded in SWRL by considering, for each pattern, the first atom as the head of the rule and the remaining as the rule body.

4 Related Works

The exploitation of data mining methods for discovering hidden knowledge patterns is not new in the SW context. First proposals have been formalized

[4] As remarked in [15], the satisfiability check is useful only if disjointness axioms occur in the ontology. This check can be omitted (thus saving computational cost) if no disjointness axioms occur.

Algorithm 5. Evolutionary algorithm for the discovery of multi-relational ARs from a populated ontological KB.

Input: $\mathcal{K}$: ontological KB; θ_f: frequency threshold; θ_{hc}: *head coverage* threshold;
Output: *pop*: set of frequent patterns discovered from $\mathcal{K}$
1: Compute A_f, a list of frequent atoms in $\mathcal{K}$.
2: Initialize population *pop* of size n.
3: $g \leftarrow 0$
4: **while** $g <$ MAX_GENERATIONS **do**
5:     **for** $i = 0, 1, \ldots, n-1$ **do**
6:         compute fitness for $pop[i]$
7:     sort *pop* by decreasing fitness
8:     **for** $i = \lfloor \tau n \rfloor, \lfloor \tau n \rfloor + 2, \ldots, n-2$ **do**
9:         $pop[i] \leftarrow pop[i \textbf{ mod } \lfloor \tau n \rfloor]$
10:         $pop[i+1] \leftarrow pop[i+1 \textbf{ mod } \lfloor \tau n \rfloor]$
11:         **with probability** p_{cross} **do** RECOMBINE($pop[i]$, $pop[i+1]$)
12:         **with probability** p_{mut} **do** MUTATE($pop[i]$)
13:         **with probability** p_{mut} **do** MUTATE($pop[i+1]$)
14:     $g \leftarrow g + 1$
15: Remove redundant and inconsistent rules from the final population *pop*
16: **return** *pop*

Table 1. Parameters of the evolutionary algorithm.

Parameter	Description
n	Population size
MAX_GENERATIONS	Maximum number of generations
MAX_RULE_LENGTH	Maximum pattern length
p_{cross}	Crossover rate
p_{mut}	Mutation rate
θ_{mut}	Head coverage threshold for mutation
τ	Truncation proportion
T/O	Reasoner time-out

in [15,16], where solutions for discovering frequent patterns in the form of, respectively, DATALOG clauses and conjunctive queries from hybrid sources of knowledge (i.e. a rule set and an ontology) have been presented. These methods are grounded on a notion of *key*, standing for the basic entity/attribute to be used for counting elements for building the frequent patterns. Unlike these methods, our solution focuses on an ontological KB and does not require any notion of *key* and as such it is able to discover any kind of frequently hidden knowledge patterns in the ontology. A method for learning ARs from RDF datasets, with the goal of inducing a schema ontology has been proposed in [24], while a method for inducing new assertional knowledge from RDF datasets has been presented in [11] and further optimized in [12]. Differently from our approach, these two methods do not take into account any background/ontological knowledge and do not exploit any reasoning capabilities. Furthermore, our solution allows to discover rules that can be directly added to the ontology, which is not the case for the existing methods.

As regards exploiting EAs in combinations with ILP, several started to appear in the literature at the beginning of the new millennium. An EA has been exploited as a wrapper around a population of ILP algorithms in [21]; alternatively, a hybrid approach combining an EA and ILP operators has been proposed in [6–8]. A similar idea is also followed by [19,23], in which a genetic algorithm is used to evolve and recombine clauses generated by a stochastic bottom-up local search heuristic. The rationale for using evolutionary algorithm as a meta-heuristic for ILP is to mitigate the combinatorial explosion generated by the inductive learning of rich representations, such as those used in description logics [4], while maintaining the quality of the results.

5 Experiments and Results

We tested our method on the same publicly available ontologies used in [4]: Financial,[5] describing the banking domain; Biological Pathways Exchange (BioPAX) Level 2 Ontology,[6] describing biological pathway data; and New Testament Names Ontology (NTN),[7] describing named things (people, places, and other classes) in the New Testament, as well as their attributes and relationships. Details are reported in Table 2.

Table 2. Key facts about the ontological KBs used.

Ontology	# Concepts	# Roles	# Indiv.	# Declared assertions	# Decl.+derived assertions	# Disjoint. axioms
Financial	59	16	1000	3359	3814	15
BioPAX	40	33	323	904	1671	15
NTMerged	47	27	695	4161	6863	5

The first goal of our experiments consisted in assessing the ability of the discovered rules to predict new assertional knowledge for a considered ontological KB. For that purpose, different samples of each ontology have been built for learning multi-relational ARs (as presented in Sect. 3) while the full ontology versions have been used as a testbed. Specifically, for each ontology three samples have been built by randomly removing, respectively, 20 %, 30 %, and 40 % of the concept assertions, according to a stratified sampling procedure. We ran the EA-based algorithm by repeating for each run the sampling procedure. For the purpose a Dell Laptop with Ubuntu Operating System, CPU Core I5 and 4GB RAM has been used. We performed 10 runs for each ontology and parameter setting, finally using the following parameters setting which resulted the best setting over the several runs: $n = 1000$, MAX_GENERATIONS $= 1000$, MAX_RULE_LENGTH $= 10$, $p_{\text{cross}} = 0.6$, $p_{\text{mut}} = 0.4$, $\theta_{\text{mut}} = 0.2$, $\tau = \frac{1}{5}$, $\theta_f = 1$, $\theta_{hc} = 0.01$, $\theta_{ic} = 0.001$. As for the reasoner, Pellet reasoner has been used and as for the reasoner time-out (T/O), after some preliminary tests, we concluded that 10 seconds were enough to reduce the number of discarded patterns to a

[5] http://www.cs.put.poznan.pl/alawrynowicz/financial.owl.
[6] http://www.biopax.org/release/biopax-level2.owl.
[7] http://www.semanticbible.com/ntn/ntn-view.html.

minimum; nevertheless, in the experiments we have also considered time-outs of 20 and 30 seconds to be on the safe side. As a results, three sets of 10 runs were performed for each ontology, one for each combination of sample and time-out, and the final population of each run were filtered using three time-outs, yielding a total of nine sets of 10 results. As in [11], we applied the discovered rules to the full ontology versions and collected all predictions, that is the head atoms of the instantiated rules. All predictions already contained in the reduced ontology versions were discarded while the remaining predicted facts were considered for the evaluation. Specifically, a prediction is assessed as *correct* if it is contained/entailed by the full ontology version and as *incorrect* if it is inconsistent with the full ontology version. Results (see Table 3) have been averaged over the different runs for each parameter setting and have been measured in terms of: *precision* (see Definition 10), *match rate*, *commission error rate*, and *induction rate* (see Sect. 2.2).

These results fully confirm the capability of the proposed approach to discover accurate rules (precision = 1 on all samples of all ontologies considered)

Table 3. Average (± standard deviation) performance metrics on each ontology.

Ontology	Sample	T/O	Match rate	Comm. rate	Ind. rate	Precision	Number of # predictions
Financial	20 %	10s	0.983 ± 0.017	0	0.017 ± 0.17	1.0	32,607 ± 39,099
		20s	0.983 ± 0.017	0	0.017 ± 0.17	1.0	32,607 ± 39,099
		30s	0.983 ± 0.017	0	0.017 ± 0.17	1.0	32,607 ± 39,099
	30 %	10s	0.970 ± 0.034	0	0.030 ± 0.034	1.0	64,875 ± 60,514
		20s	0.970 ± 0.034	0	0.030 ± 0.034	1.0	64,875 ± 60,514
		30s	0.970 ± 0.034	0	0.030 ± 0.034	1.0	64,875 ± 60,514
	40 %	10s	0.933 ± 0.105	0	0.067 ± 0.105	1.0	47,264 ± 49,700
		20s	0.933 ± 0.105	0	0.067 ± 0.105	1.0	47,264 ± 49,700
		30s	0.933 ± 0.105	0	0.067 ± 0.105	1.0	47,264 ± 49,700
BioPAX	20 %	10s	0.808 ± 0.087	0	0.192 ± 0.087	1.0	21,065 ± 8,914
		20s	0.807 ± 0.085	0	0.193 ± 0.085	1.0	22,397 ± 8,737
		30s	0.807 ± 0.085	0	0.193 ± 0.085	1.0	22,397 ± 8,737
	30 %	10s	0.877 ± 0.056	0	0.123 ± 0.056	1.0	19,697 ± 8,846
		20s	0.877 ± 0.056	0	0.123 ± 0.056	1.0	19,697 ± 8,847
		30s	0.877 ± 0.056	0	0.123 ± 0.056	1.0	19,697 ± 8,847
	40 %	10s	0.877 ± 0.056	0	0.113 ± 0.056	1.0	19,621 ± 12,811
		20s	0.877 ± 0.056	0	0.113 ± 0.056	1.0	19,621 ± 12,811
		30s	0.877 ± 0.056	0	0.113 ± 0.056	1.0	19,621 ± 12,811
NTMerged	20 %	10s	0.578 ± 0.118	0	0.422 ± 0.118	1.0	3,324,264 ± 891,161
		20s	0.572 ± 0.119	0	0.428 ± 0.119	1.0	3,702,706 ± 826,273
		30s	0.571 ± 0.119	0	0.429 ± 0.119	1.0	3,748,387 ± 827,350
	30 %	10s	0.707 ± 0.080	0	0.293 ± 0.080	1.0	3,489,818 ± 1,089,094
		20s	0.705 ± 0.081	0	0.295 ± 0.081	1.0	3,781,877 ± 1,415,805
		30s	0.705 ± 0.081	0	0.295 ± 0.081	1.0	3,790,930 ± 1,408,588
	40 %	10s	0.665 ± 0.131	0	0.335 ± 0.131	1.0	3,564,421 ± 1,290,532
		20s	0.664 ± 0.131	0	0.336 ± 0.131	1.0	3,643,770 ± 1,320,093
		30s	0.662 ± 0.131	0	0.338 ± 0.131	1.0	3,708,683 ± 1,363,246

Table 4. Comparison of EA vs. RARD and AMIE w.r.t. the number of rules discovered.

Ontology	Samp.	T/O	# Rules			Top	# Predictions		
			EA	RARD	AMIE	m	EA	RARD	AMIE
Financial	20 %	10s	94.1 ± 33.7	177	2	2	14,442 ± 17,280	29	208
		20s	94.1 ± 33.7				14,442 ± 17,280		
		30s	94.1 ± 33.7				14,442 ± 17,280		
	30 %	10s	86 ± 32	181	2	2	29,890 ± 29,576	57	197
		20s	86 ± 32				29,890 ± 29,576		
		30s	86 ± 32				29,890 ± 29,576		
	40 %	10s	78 ± 50	180	2	2	18,958 ± 21,954	85	184
		20s	78 ± 50				18,958 ± 21,954		
		30s	78 ± 50				18,958 ± 21,954		
BioPax	20 %	10s	144.1 ± 46.2	298	8	8	1,902.3 ± 755.7	25	2
		20s	144.4 ± 46.7				2,045.6 ± 740.9		
		30s	144.4 ± 46.7				2,045.6 ± 740.9		
	30 %	10s	188.2 ± 25.5	283	8	8	1,653.1 ± 779.1	34	2
		20s	188.2 ± 25.5				1,653.1 ± 779.1		
		30s	188.2 ± 25.5				1,653.1 ± 779.1		
	40 %	10s	159.3 ± 37.7	272	0	8	1,704.4 ± 1,437	50	0
		20s	159.3 ± 37.7				1,704.4 ± 1,437		
		30s	159.3 ± 37.7				1,704.4 ± 1,437		
NTMerged	20 %	10s	1,035.4 ± 588.7	243	1,129	10	85,457 ± 25,754	620	420
		20s	1,044.4 ± 592.8				97,622 ± 24,878		
		30s	1,045.9 ± 592.6				98,470 ± 25,261		
	30 %	10s	942.4 ± 217.1	225	1,022	10	103,962 ± 32,449	623	281
		20s	945.6 ± 218				114,940 ± 41,960		
		30s	946.1 ± 218.4				11,940 ± 41,960		
	40 %	10s	893.7 ± 473.5	239	1,063	10	101,102 ± 38,777	625	332
		20s	895.6 ± 473.9				102,569 ± 38,828		
		30s	897 ± 473.2				103,100.4 ± 38,903		

and, which is even more relevant, to come up with rules that induce previously unknown facts (induction rate > 0), with a very large absolute number of predictions by the standards of alternative state-of-the art approaches.

The second goal of our experiments consisted in comparing the performance of the proposed evolutionary method to those of the two state-of-the-art level-wise generate-and-test algorithms which are closest to it in purpose, namely the multi-relational association rule discovery (RARD) method proposed by d'Amato *et al.* [4] and AMIE [11]. The comparison has been performed by considering the top m rules, wrt. their match rate, with m equal to: i) the number of rules discovered by AMIE, when few rules were discovered; ii) to 10 for the other cases. Averaged results are reported in Table 4, further corroborating the claim that the proposed evolutionary algorithm can substantially boost the performance of multi-relational AR discovery. The large number of predictions made, on average, by the rules discovered by the evolutionary algorithm, depends on our language bias, which allows open rules (such that $V(B) \setminus V(H) \neq \emptyset$): open rules may generate substantially larger number of predictions than closed rules.

6 Conclusions

We presented an evolutionary method for discovering multi-relational ARs, coded in SWRL, from ontological KBs, to be used primarily for enriching assertional knowledge. The proposed approach has been experimentally evaluated through its application to publicly available ontologies and compared to the two most relevant state-of-the-art algorithms having the same goal.

For the future, we intend to focus on two main aspects: (1) scalability, by considering experimenting our method on datasets from the Linked Data Cloud; (2) reducing the search space for discovering ARs by further exploiting the expressive power of the representation language by considering the presence of hierarchy of roles.

References

1. Agrawal, R., Imielinski, T., Swami, A.N.: Mining association rules between sets of items in large databases. In: Proceedings of the International Conference on Management of Data, pp. 207–216. ACM Press (1993)
2. Baader, F., Calvanese, D., McGuinness, D.L., Nardi, D., Patel-Schneider, P.F. (eds.): The Description Logic Handbook: Theory, Implementation, and Applications. Cambridge Univ. Press, Cambridge (2003)
3. Berners-Lee, T., Hendler, J., Lassila, O.: The semantic web. Sci. Am. 28–37 (2001)
4. d'Amato, C., Staab, S., Tettamanzi, A., Tran, M., Gandon, F.: Ontology enrichment by discovering multi-relational association rules from ontological knowledge bases. In: Proceedings of SAC 2016. ACM (2016)
5. DeJong, K.: Evolutionary Computation: A Unified Approach. MIT Press, Cambridge (2002)
6. Divina, F.: Hybrid genetic relational search for inductive learning. Ph.D. thesis, Vrije Universiteit Amsterdam (2004)
7. Divina, F.: Genetic Relational Search for Inductive Concept Learning: A Memetic. LAP LAMBERT Academic Publishing, Saarbrücken (2010)
8. Divina, F., Marchiori, E.: Evolutionary concept learning. In: Langdon, W.B. et al. (eds.) GECCO 2002: Proceedings of the Genetic and Evolutionary Computation Conference, New York, USA, 9–13 July 2002, pp. 343–350. Morgan Kaufmann (2002)
9. Eiben, A., Smith, J.: Introduction to Evolutionary Computing. Springer, Berlin (2003)
10. Fanizzi, N., d'Amato, C., Esposito, F.: Learning with kernels in description logics. In: Železný, F., Lavrač, N. (eds.) ILP 2008. LNCS (LNAI), vol. 5194, pp. 210–225. Springer, Heidelberg (2008)

11. Galárraga, L., Teflioudi, C., Hose, K., Suchanek, F.: AMIE: association rule mining under incomplete evidence in ontological knowledge bases. In: Proceedings of the 22th International Conference on World Wide Web (WWW 2013), pp. 413–422. ACM (2013)
12. Galárraga, L., Teflioudi, C., Hose, K., Suchanek, F.: Fast rule mining in ontological knowledge bases with AMIE+. VLDB J. **24**(6), 707–730 (2015)
13. Horrocks, I., Patel-Schneider, P.F.: A proposal for an OWL rules language. In: Proceedings of the International Conference on World Wide Web, pp. 723–731. ACM (2004)
14. Horrocks, I., Patel-Schneider, P.F., Boley, H., Tabet, S., Grosof, B., Dean, M.: SWRL: a semantic web rule language combining OWL and RuleML (2004). https://www.w3.org/Submission/SWRL/
15. Józefowska, J., Lawrynowicz, A., Lukaszewski, T.: The role of semantics in mining frequent patterns from knowledge bases in description logics with rules. Theor. Pract. Logic Program. **10**(3), 251–289 (2010)
16. Lisi, F.A.: AL-QuIn: an onto-relational learning system for semantic web mining. Int. J. Semant. Web Inf. Syst. **7**(3), 1–22 (2011)
17. Motik, B., Sattler, U., Studer, R.: Query answering for OWL-DL with rules. Web Semant. **3**(1), 41–60 (2005)
18. Motik, B., Shearer, R., Horrocks, I.: Hypertableau reasoning for description logics. J. Artif. Intell. Res. **36**, 165–228 (2009)
19. Muggleton, S., Tamaddoni-Nezhad, A.: QG/GA: a stochastic search for progol. Mach. Learn. **70**(2–3), 121–133 (2008)
20. Mühlenbein, H., Schlierkamp-Voosen, D.: The science of breeding, its application to the breeder genetic algorithm (BGA). Evol. Comput. **1**(4), 335–360 (1993). Winter
21. Reiser, P.G.K., Riddle, P.J.: Scaling up inductive logic programming: an evolutionary wrapper approach. Appl. Intell. **15**(3), 181–197 (2001)
22. Sirin, E., Parsia, B., Grau, B.C., Kalyanpur, A., Katz, Y.: Pellet: a practical OWL-DL reasoner. Web Semant. **5**(2), 51–53 (2007)
23. Tamaddoni-Nezhad, A., Muggleton, S.H.: A genetic algorithms approach to ILP. In: Matwin, S., Sammut, C. (eds.) ILP 2002. LNCS (LNAI), vol. 2583, pp. 285–300. Springer, Heidelberg (2003)
24. Völker, J., Niepert, M.: Statistical schema induction. In: Antoniou, G., Grobelnik, M., Simperl, E., Parsia, B., Plexousakis, D., De Leenheer, P., Pan, J. (eds.) ESWC 2011, Part I. LNCS, vol. 6643, pp. 124–138. Springer, Heidelberg (2011)

An Incremental Learning Method to Support the Annotation of Workflows with Data-to-Data Relations

Enrico Daga[1(✉)], Mathieu d'Aquin[1], Aldo Gangemi[2,3], and Enrico Motta[1]

[1] Knowledge Media Institute (KMI) - The Open University, Walton Hall, Milton Keynes MK76AA, UK
{enrico.daga,mathieu.daquin,enrico.motta}@open.ac.uk
[2] National Research Council (CNR), Via Gaifami 18, 95126 Catania, Italy
aldo.gangemi@cnr.it
[3] Paris Nord University, Sorbonne Cite CNRS UMR7030, Paris, France

Abstract. Workflow formalisations are often focused on the representation of a process with the primary objective to support execution. However, there are scenarios where what needs to be represented is the effect of the process on the data artefacts involved, for example when reasoning over the corresponding data policies. This can be achieved by annotating the workflow with the semantic relations that occur between these data artefacts. However, manually producing such annotations is difficult and time consuming. In this paper we introduce a method based on recommendations to support users in this task. Our approach is centred on an incremental rule association mining technique that allows to compensate the cold start problem due to the lack of a training set of annotated workflows. We discuss the implementation of a tool relying on this approach and how its application on an existing repository of workflows effectively enable the generation of such annotations.

1 Introduction

Research in workflows has been characterized on a variety of aspects, spanning from representation and management to preservation, reproducibility, and analysis of process executions [11,13,14,16,17]. Recently, a data-centric approach for the representation of data relying systems has been proposed with the aim to simulate the impact of process executions on the data involved, particularly to perform reasoning on the propagation of data policies [4,6]. This approach puts the data objects as first class citizens, aiming to represent the possible semantic relations among the data involved. Annotating data intensive workflows is problematic for various reasons: (a) annotation is time consuming and it is of primary importance to support the users in such activity, and (b) workflow descriptions are centred on the processes performed and not on the data, meaning that some form of remodelling of the workflow is required. In this paper we introduce a method based on recommendations to support users in producing data-centric

E. Blomqvist et al. (Eds.): EKAW 2016, LNAI 10024, pp. 129–144, 2016.
DOI: 10.1007/978-3-319-49004-5_9

annotations of workflows. Our approach is centred on an incremental rule association mining technique that allows to compensate the cold start problem due to the lack of a training set of annotated workflows. We discuss the implementation of a tool relying on this approach and how its application on an existing repository of workflows (the "My experiment"[1] repository) effectively enables the generation of such annotations. In the next Section we introduce the related work. Section 3 describes the approach and Sect. 4 how it has been implemented in a tool that allows to annotate workflows as data-centric descriptions. In Sect. 5 we present the results of an experiment performed with real users where we measured how this method impacts the sustainability of the task. Finally, we discuss some open challenges and derive some conclusions in the final Sect. 6.

2 Related Work

In this paper we introduce a novel approach to recommend (semantic, data-centric) annotations for workflows. Research on process formalization and description covers a variety of aspects, from the problem of reproducibility to the ones of validation, preservation, tracing and decay [3,7,11,20,22]. Several models have been proposed for describing workflow executions, like the W3C PROV Model[2], the Provenance Model for Workflows (OPMW)[3] and more recently the Publishing Workflow Ontology (PWO)[4] introduced in [9]. A recent line of research is focused on understanding the activities behind processes in workflows, with the primary objective to support preservation and reusability of workflow components, particularly in the context of scientific workflows [2,10]. We place our work in the area of semantic annotation of workflows. Semantic technologies have been used in the past to analyze the components of workflows, for example to extract common structural patterns [8]. Recently more attention has been given to the elicitation of the activity of workflows in a knowledge principled way, for example searching for common motifs in scientific workflows [10] or labelling data artifacts to produce high level execution traces (provenance) [1]. This research highlighted the need for adding semantics to the representation of workflows and the challenges associated with the problem of producing such annotations [1]. Recently a number of repositories of scientific workflows have been published - Wings[5], My experiments[6], SHIWA[7] are the prominent examples. We selected the My experiments repository as data source for our study. For this reason, we will use the terminology of the SCUFL2 model[8] when discussing how our approach deals with the workflow formalization.

[1] My experiment: http://www.myexperiment.org/.
[2] W3C PROV: https://www.w3.org/TR/prov-overview/.
[3] OPMW: http://www.opmw.org/.
[4] PWO: http://purl.org/spar/pwo.
[5] Wings: http://www.wings-workflows.org/.
[6] My experiments: http://www.myexperiment.org/.
[7] SHIWA: http://www.shiwa-workflow.eu/wiki/-/wiki/Main/SHIWA+Repository.
[8] SCUFL2: https://taverna.incubator.apache.org/documentation/scufl2/.

There are several approaches to recommendation using clustering techniques (Support Vector Machines (SVM), Latent Semantic Aanalysis (LSA), to name a few). Formal Concept Analysis (FCA) [21] found a large variety of applications [19], and the literature reports several approaches to incremental lattice construction [15], including the Godin [12] algorithm, used in the present work. FCA found application in knowledge discovery as a valuable approach to association rule mining (ARM) [19]. In the context of FCA, association rules are generated from closed item sets, where the association rule to be produced relates attributes appearing in the intent of the same concept. A large number of studies focused on how to reduce the number of item sets to explore in order to obtain a complete set of minimal rules [19]. In the scenario of the present study, where the lattice changes incrementally, generating all the possible association rules would be a waste of resources. The algorithm proposed in the present work is *on demand*, as it only extracts the rules that are relevant for the item to annotate. Our algorithm receives as input an item set, and *retrieves* from the lattice the association rules associated with a relevance score. In other words, we follow an approach unusual with respect to the literature, attacking the ARM problem as an Information Retrieval (IR) one.

The approach presented in this paper uses the Datanode ontology [5], a hierarchy of possible relations between data objects. The ontology defines a unique type - *Datanode* - and 114 relations, starting from a single top property: *relatedWith*, having the class *Datanode* as `rdfs:domain` and `rdfs:range`. Datanode relations can express meta-level aspects (e.g. *describes/describedBy*, *hasAnnotation/isAnnotationOf*), containment (e.g. *hasPart/isPartOf*, *hasSection/isSectionOf*) as well as a properties like derivation (e.g. *hasCopy/isCopyOf*, *processedInto/processedFrom*), among others. Relations are organised by the means of the `rdfs:subPropertyOf` property. For example, *processedInto* is a subproperty of *hasDerivation*, as it is possible to derive a new data object from another also in other ways, for example generating an unprocessed copy - *hasCopy*. In the present work, a *datanode* is any data object that can be the input or output of a workflow processor. Instead on characterizing the activities of a workflow (like in [10]), Datanode can be applied to describe it in terms of relations between the input and the output of processors[9]. The resulting network of data objects can be used to reason upon the propagation of policies, for example in the context of a Smart City data hub [4,6].

3 Recommendations for Data-Centric Workflow Annotations

Our approach to the problem is an iterative supervised annotation process supported by incremental recommendations. Figure 1 provides an overview of the approach by listing the elements and their dependency, organised in four phases.

Phase 1. The starting point is an encoded artefact representing the workflow structure and its metadata (like the ones available through My experiments).

[9] Datanode: http://purl.org/datanode/ns/.

The workflow code is first translated into a data centric graph, where nodes are data objects manipulated by processors and arcs the relations among them. The result of this transformation is a directed graph with anonymous arcs (named IO port pairs in the Figure), being these arcs the items to be annotated by the user.

Phase 2. Each IO port pair is then associated with a set of features automatically extracted from the workflow metadata.

Phase 3. Extracted features constitute the input of the recommendation engine, designed using the Formal Concept Analysis (FCA) framework. This method is an incremental association rules mining technique that exploits incoming annotations to incrementally produce better recommendations.

Phase 4. Features of the IO port pair, alongside the workflow documentation and the recommendations, are the input of the user that is requested to select a set of annotations from a fixed vocabulary (the Datanode ontology).

In this section we focus on the first three phases of the approach: the workflow to data graph transformation (Sect. 3.1); the features extraction method (Sect. 3.2); and the recommendation engine (Sect. 3.3), leaving the last one to Sect. 4.

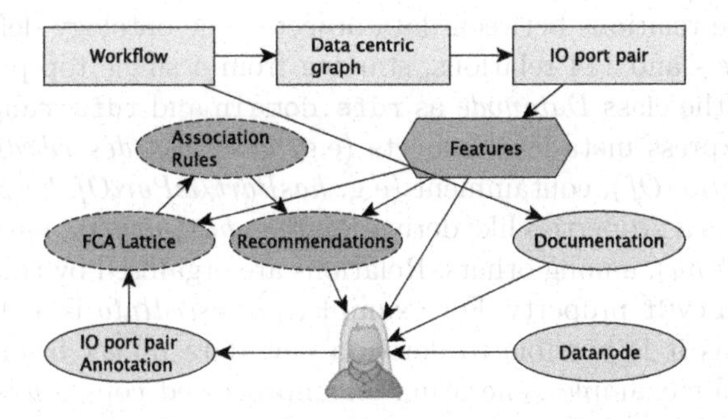

Fig. 1. Description of the approach and dependencies. Elements of phase 1 are represented in blue rectangles on top. Phase 2 includes the features generation (the only stretched exagon). Elements of Phase 3 are depicted as pink ovals with dashed borders and phase 4 ones as light yellow ovals. (Color figure online)

3.1 Workflows as Data-Centric Graphs

Workflows are built on the concept of *processor* as unit of operation[10]. A *processor* constitutes of one or more input and output *ports*, and a specification of the operation to be performed. Processors are then linked to each other through

[10] In this paper we use the terminology of the SCUFL2 specification. However, the basic structure is a common one. In the W3C PROV-O model this concept maps to the class *Activity*, in PWO with *Step*, and in OPMW to *WorkflowExecutionProcess*, just to mention few examples.

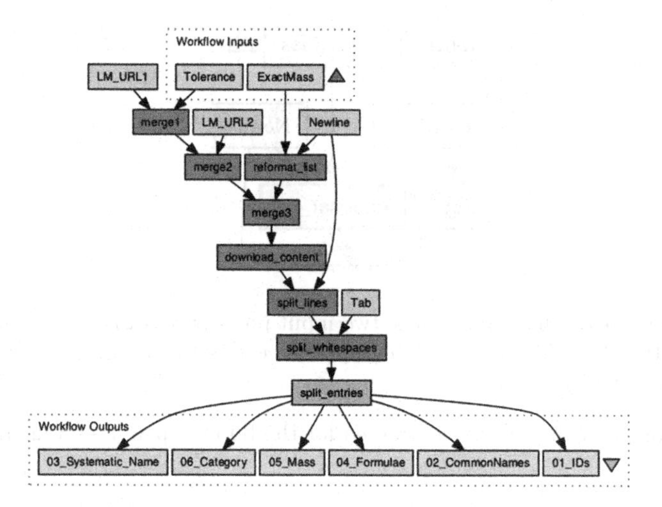

Fig. 2. A workflow from the My Experiment repository: "LipidMaps Query".

a set of data links connecting an output port to the input of another processor resulting in a composite tree-like structure. Figure 2 shows an example of a workflow taken from the "My Experiment" repository [11].

The objective of our work is to describe what happens inside the processors by expressing the relation between *input* and *output*. For example, the processor depicted in Fig. 3 has two input ports (1 and 2) and one output (3). For this processor, we generate two links connecting the input data objects to the output one, through two anonymous arcs: $1 \rightarrow 3$ and $2 \rightarrow 3$. We name these arcs "IO port pairs" (input-output port pairs), and these are the items we want to be annotated. In this example, the IO port pair $1 \rightarrow 3$ could be annotated with the Datanode relation $refactoredInto$, while the IO port pair $2 \rightarrow 3$ would not be annotated as only referring to a configuration parameter of the processor and not to an actual data input. For the present work we translated 1234 Workflows from the My Experiments repository, resulting in 30612 IO port pairs (although we will use a subset of them in the user evaluation).

3.2 Extracting Features from Workflow Descriptions

As described in the previous Section, the workflow description is translated in a graph of IO port pairs connected by unlabelled links. In order to characterize the IO port pair we exploit the metadata associated with the components of the workflow involved: the input and output port and the processor that includes them. For each of these elements we extract the related metadata as key/value pairs, which we use as core features of the IO port pair. Applying this approach to the My Experiments corpus we obtained 26900 features. Table 1 shows an example of features extracted for the IO port pairs described in Fig. 3.

[11] "LipidMaps Query" workflow from My experiment: http://www.myexperiment.org/workflows/1052.html.

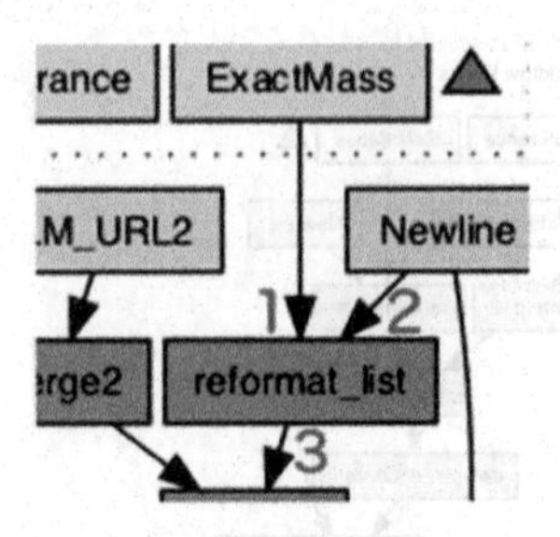

Fig. 3. This processor has three ports: two input ports (1 and 2) and one output port (3). We can translate this model into a graph connecting the data objects of the inputs to the one of the output.

Table 1. Sample of the features extracted for the IO port pair 1 → 3 in the example of Fig. 3.

Type	Value
From/FromPortName	string
To/ToPortName	split
Activity/ActivityConfField	script
Activity/ActivityType	http://ns.taverna.org.uk/2010/activity/beanshell
Activity/ActivityName	reformat_list
Activity/ConfField/derivedFrom	http://ns.taverna.org.uk/2010/activity/ localworker/org.embl.ebi.escience.scuflworkers. java.SplitByRegex
Activity/ConfField/script	List split = new ArrayList();if (!string.equals("")) { String regexString = ","; if (regex != void) ...
Processor/ProcessorType	Processor
Processor/ProcessorName	reformat_list

However, the objective of these feature sets is to support the clustering of the annotated IO port pair through finding similarities with IO port pairs to be annotated. At this stage of the study we performed a preliminary evaluation of the distribution of the features extracted. We discovered that very few of them were shared between a significant number of port pairs (see Fig. 4). In order to increase the number of shared features we generated a set of *derived features* by extracting bags of words from lexical feature values and by performing Named Entity Recognition on the features that constituted textual annotations (labels and comments), when present. Moreover, from the extracted entities we also added the related DBPedia categories and types as additional features. As example, Table 2 shows a sample of the bag of words and entities extracted from the features listed in the previous Table 1.

Table 2. Example of derived features (bag of words and DBPedia entities) generated for the IO port pair $1 \rightarrow 3$.

Type	Value
From/FromPortName-word	string
To/ToPortName-word	split
From/FromLinkedPortDescription-word	single
From/FromLinkedPortDescription-word	possibilities
From/FromLinkedPortDescription-word	orb
From/FromLinkedPortDescription-word	mass
FromToPorts/DbPediaType	wgs84:SpatialThing
FromToPorts/DbPediaType	resource:Text_file
FromToPorts/DbPediaType	resource:Mass
FromToPorts/DbPediaType	Category:State_functions
FromToPorts/DbPediaType	Category:Physical_quantities
FromToPorts/DbPediaType	Category:Mathematical_notation

The generation of derived features increased the number of total features significantly (up to 59217), while making the distribution of features less sparse, as reported in Fig. 5.

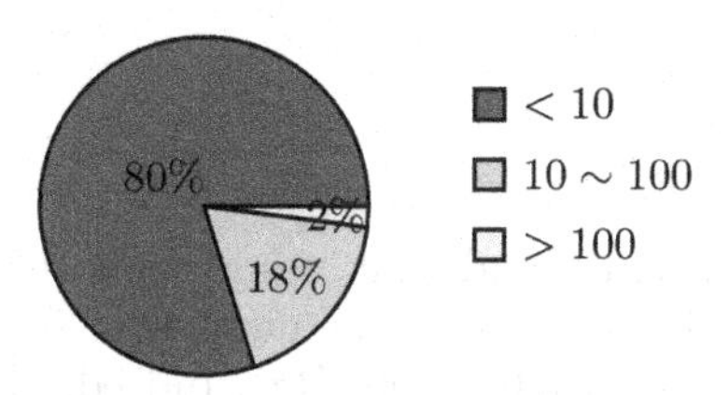

Fig. 4. Distribution of features extracted from the workflow descriptions.

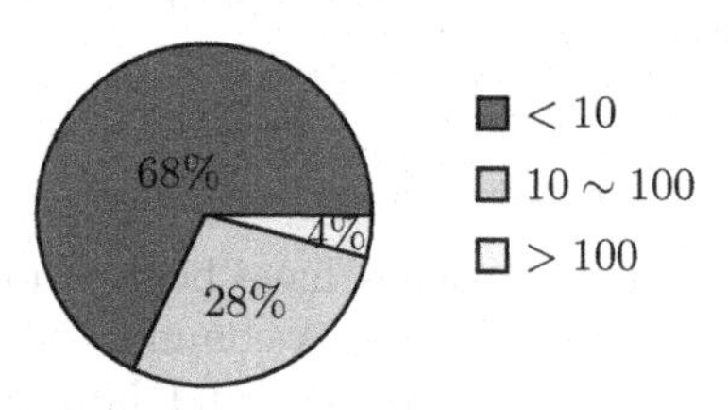

Fig. 5. Distribution of features (including derived features).

3.3 Retrieval of Association Rules and Generation of Recommendations

Generating recommendations usually requires an annotated corpus to be available as training set. While repositories of workflows (especially scientific workflows) exist, they are not annotated with data-to-data relations. In order to overcome this problem we opted for an incremental approach, where the recommendations are produced according to the available annotated items *on demand*. The rules needed are of the following form:

$$(f^1, f^2, ..., f^n) \rightarrow (a^1, a^2, ..., a^n)$$

where $f1, \ldots, fn$ are the features of the IO port pairs and $a1, \ldots, an$ are the data-to-data relations used to annotate them. Our approach relies on extracting association rules from a concept lattice built through FCA incrementally. Such a lattice is built on a formal context of items and attributes. In FCA terms, the items are the IO port pairs and the attributes their features as well as the chosen annotations. Each node of the FCA lattice is a closed concept, mapping a set of items all having a given set of attributes. A FCA concept would then be a collection of IO port pairs all having a given set of features and/or annotations. In a FCA lattice, concepts are ordered from the top concept (supremum), including all items and (usually) no shared features, to the bottom concept (infimum), including all the available features and a (possibly) empty set of items. The lattice is built incrementally using the Godin algorithm [12]. The algorithm (re)constructs the lattice integrating at each iteration a new item - the IO port pair, with its set of attributes (the features and annotations altogether). Association rules are extracted from the FCA lattice, where the key point is the co-occurrence of features f and annotations a in the various FCA concepts.

The following Listing 1.1 gives a sample of an association rule we want to mine from the lattice:

Listing 1.1. Example of association rule mined from the FCA lattice.

```
(ProcessorName−word: base,
 FromPortName: base64,
 ActivityName−word: decode,
 ActivityType: http://ns.taverna.org.uk/2010/activity/beanshell,
 ProcessorName−word: array,
 FromPortName−word: base64,
 ToPortName: bytes,
 ActivityName−word: 64,
 ActivityConfField: mavenDependency,
 ActivityName−word: array,
 ActivityConfField: derivedFrom,
 ProcessorName−word: decode,
 ActivityName−word: byte)
   → (dn:hasDerivation, dn:refactoredInto)
```

Several approaches have been studied to generate and rank association rules from a FCA lattice. A common problem in this scenario is the number of rules that can be extracted, and how to reduce them effectively [18]. Indeed, the number of rules can increase significantly with the number of concepts of the FCA lattice. Generating all of them is time consuming as the lattice becomes larger. Precomputing the rules is not a valid solution, as the lattice will change for any new item inserted. In this scenario, we are forced to compute the rules *live* for each new item to be annotated.

The above considerations motivate a set of new requirements for implementing a rule mining algorithm that is effective in this scenario:

1. generate only rules that have annotations in the body
2. generate only rules that are applicable to the candidate item to be annotated
3. only use one rule for each recommendation (head of the rule), to avoid redundancies
4. rank the rules to show the most relevant first

In order to satisfy the requirements above we propose an algorithm to mine association rules *on-demand*, by considering two sets of attributes as constraints for the head and body of the rules.

The algorithm we propose has three inputs: (1) a FCA Lattice; (2) the set of attributes of the item for which we need recommendations (the set of attributes that needs to be in the body of the rules); and (3) the set of attributes we want to be part of the recommendations (the set of attributes that can be in the rule head). Listing 1.2 illustrates the algorithm for extracting rules *on-demand.* Input is a lattice L, a set of attributes as possible recommendations (target rule head: H) and a set of attributes for which we need recommendations (target rule body: B). The algorithm assumes the two sets to be disjoint. The algorithm traverses the lattice starting from the bottom, adding the *infimum* to a FIFO queue - lines 3–5. For each concept in the queue, first assess whether its attributes contains items from both the target head and body. If it doesn't, the concept (and related paths in the lattice) can be skipped - lines 7–11. Otherwise, the parent concepts are added to the queue, and the concept considered to rule extraction - line 13. The non empty intersections of attributes with the target head and body form a candidate rule $b \rightarrow h$.

Listing 1.2. Algorithm to mine association rules from a lattice *on demand*:

```
1   // L: the lattice; H: attributes in the rule head; B: attributes in the rule body
2   mineRules(L,H,B):
3     C ← [] // an empty FIFO list of concepts
4     R ← [] // an empty set of Rules (indexed by their head).
5     add(inf(L), C) // add the infimum of L to C
6     while !empty(C):
7        c ← first(C) // remove one concept from the top of the queue
8        h=retain(attributes(c),H) // attributes in c in the head of rule
9        if empty(h): continue // move to another concept
10       b ← retain(attributes(c),B) // attributes in c allowed in the body of the rule
11       if empty(b): continue // move to another concept
12       // Add the concept parents to the queue.
13       addAll(parents(L,c),C)
14       // Examine b → h measures (s: support, k: confidence, r: relevance)
15       // support (s): items satisfying the rule divided by all items
16       s ← count(objects(c)) / count(objects(supremum(L)))
17       if s = 0: continue // A supremum rule includes this one
18       // confidence (k): support divided by the items only satisfying b
19       I ← [] // items only satisfying the body
20       for p in parents(c):
21        if (attributes(p) ∩ h) = ∅:
22          if attributes(p) = b: add(objects(p), I)
23        end
24       end
25       if count(I) = 0: k ← 1
26       else:
27         k ← count(objects(c)) / count(I)
28       end
29       // relevance (r): intersection of B with b, divided by B
30       r ← count(B ∩ b) / count(B)
31       // check this rule is the best so far with this head
32       if hasRuleWithHead(R,h):
33          rule ← getRuleWithHead(R,h)
34          if relevance(rule) > r: continue
35          if relevance(rule) = r:
36              if confidence(rule) > k: continue
37              if confidence(rule) = k:
38                  if support(rule) >= s: continue
39              end
40          end
41       end
42       rule ← (h,b,s,k,r) // the new rule, or the best so far for head
43       add(rule, R)
44    end
45  return R
```

The association rule derived is scored by support (s), confidence (k) and a third measure inspired from information retrieval and called *relevance* (r) - lines 15–30. The definitions of these measures, considering a rule $b \rightarrow h$, is as follows:

- Support s ($b \rightarrow h$): the ratio of items satisfying $b \cup h$ to all the items in the lattice - line 16;
- Confidence k ($b \rightarrow h$): the ratio of items satisfying $b \cup h$ to the items satisfying b - lines 19–28;
- Relevance r ($b \rightarrow h$): the degree of overlap between the body of the rule b and the set of features of the candidate item B. It is calculated as the size of the body divided by the size of the intersection between the body of the rule and the features of the candidate item - line 30.

Only the rule with best score for a given head is kept in the list of rules - lines 31–43. Our ranking algorithm will privilege relevance over confidence and support, in order to boost the rules (recommendations) that are more likely to be relevant for the candidate item.

Since this is an iterative process, at the very beginning there will be no recommendation. New annotations will feed the reference corpus (the FCA lattice) and the system will start to generate association rules. Our hypothesis is that the quality of the rules and therefore their usefulness in supporting annotations, increase with the size of the annotated items (this will be part of the evaluation in Sect. 5).

4 Implementation of the Approach

The approach described in the Sect. 3 has been implemented in the Dinowolf (Datanode in workflows) tool[12] based on the SCUFL2 worfklow specification[13] and the taxonomy of data-to-data relations represented by the Datanode ontology. While Dinowolf has been implemented leveraging the Apache Taverna[14] library, it can work with any input following the SCUFL2 specification. When a workflow is loaded, the system performs a preliminary operation to extract the IO port pairs and to precompute the related set of features following the methods described in Sects. 3.1 and 3.2. In order to expand the feature set with derived features - bag of words and entities from DBPedia - the system relies on Apache Lucene[15] for sentence tokenization (considering english stopwords), DBPedia Spotlight[16] for named entity recognition, and the DBPedia[17] SPARQL endpoint for feature expansion with categories and entity types. The tool includes

[12] Dinowolf: http://github.com/enridaga/dinowolf.

[13] SCUFL2 Specification: https://taverna.incubator.apache.org/documentation/scufl2/.

[14] Apache Taverna: https://taverna.incubator.apache.org/.

[15] Apache Lucene: https://lucene.apache.org/core/.

[16] DBPedia Spotlight: http://spotlight.dbpedia.org/.

[17] DBPedia: http://dbpedia.org/.

three views: (1) a Workflows view, listing the workflows to be annotated; (2) a Workflow details view, including basic information and a link to the external documentation at My Experiments; and a (3) Annotation view, focused on providing details of the features of the IO port pair to annotate. The task presented to the users is the following:

1. Choose an item from the list of available workflows;
2. Select an IO port pair to access the Annotation view;
3. The annotation view shows the features associated with the selected IO port pair alongside a list of data node relationships exploiting a set of rules extracted from the FCA lattice as recommendations, and the full Datanode hierarchy as last option;
4. The user can select one or more relations by picking from the recommended ones or by exploring the full hierarchy. Recommended relations, ranked following the approach described in Sect. 3.3, are offered with the possibility to expand the related branch and select one of the possible subrelations as well;
5. Alternatively, the user can skip the item, if the IO port pair does not include two data objects (it is the case of a configuration parameter set as input for the processor);
6. Finally, the user can postpone the task if she feels unsure about what to choose and wants first explore other IO port pairs of the same workflow;
7. The user iteratively annotate all the port pairs of a workflow. At each iteration, the system makes use of the previous annotations to recommend the possible relations for the next selected IO port pair.

This system has been used to perform the user based experiments that constitute the source of our evaluation.

5 Experimental Evaluation

Our main hypothesis is that the approach presented can boost the task of annotating workflows as data-to-data annotated graphs. In particular, we want to demonstrate that the quality of the recommendations improves while the annotated cases grow in number. In order to evaluate our approach we performed a user based evaluation. We loaded twenty workflows from "My Experiments"[18] in Dinowolf and asked six users to annotate the resulting 260 IO port pairs. The users, all members of the research team of the authors, have skills that we consider similar to the ones of a data manager, for example in the context of a large data processing infrastructure like the one of [4]. In this experiment, users were asked to annotate each one of the IO port pairs with a semantic relation from a fixed vocabulary (the Datanode ontology), by exploiting the workflow documentation, the associated feature set and the recommendations provided. The workflows were selected randomly and were the same for all the participants, who were requested also (a) to follow the exact order proposed by the tool, (b)

[18] My Experiments: http://www.myexperiments.org.

to complete all portpairs of a workflow before moving to the next; (c) to only perform an action when confident of the decision, otherwise to postpone the choice (using the"Later" action); (d) to select the most specific relation available - for example, to privilege *processedInto* over *hasDerivation*, when possible. Each user worked on an independent instance of the tool (and hence lattice) and performed the annotations without interacting with other participants. During the experiment the system monitored a set of measures:

- the time required to annotate an IO port pair;
- how many annotations were selected from recommendations;
- the average rank of the recommendations selected, calculated as a percentage of the overall size of the recommendation list; and
- the average of the relevance score of the recommendations selected.

Figures 6, 7, 8 and 9 illustrate the results of our experiments with respect of the above measures. In all diagrams, the horizontal axis represents the actions performed in chronological order, placing on the left the initial phase of the experiment going towards the right until all 260 IO port pairs were annotated. The diagrams ignore the actions marked as "Later", resulting on few jumps in users' lines, as we represented in order all actions including at least one annotation from at least a single user. Figure 6 shows the evolution of the time spent by each user on a given annotation page of the tool before a decision was made. The diagram represents the time (vertical axis) in logarithmic scale, showing how, as more annotations are made and therefore more recommendations are generated, the effort (time) required to perform a decision is reduced. Figure 7 illustrates the progress of the ratio of annotations selected from recommendations. This includes cases where a subrelation of a recommended relation has been selected by the user. While it shows how recommendations have an impact from the very beginning of the activity, it confirms our hypothesis that the cold-start problem is tackled through our incremental approach. Figure 8 depicts the average rank of selected recommendations. The vertical axis represents the score placing at the top the first position. This confirms our hypothesis that the quality of recommendations increases, stabilizing within the upper region after a critical mass of annotated items is produced, reflecting the same behavior observed in Fig. 7.

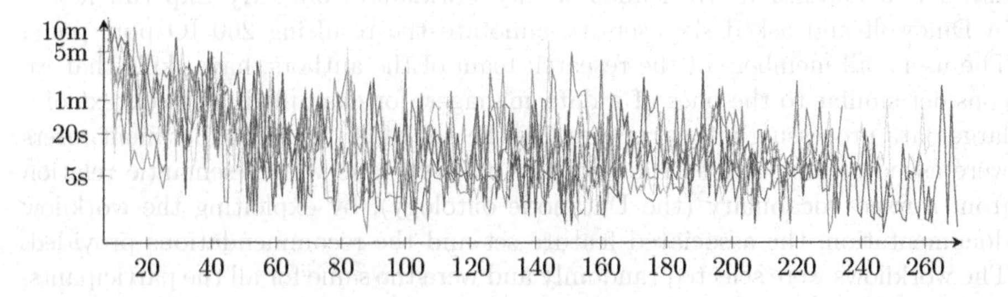

Fig. 6. Evolution of the time spent by each user on a given annotation page of the tool before a decision was made.

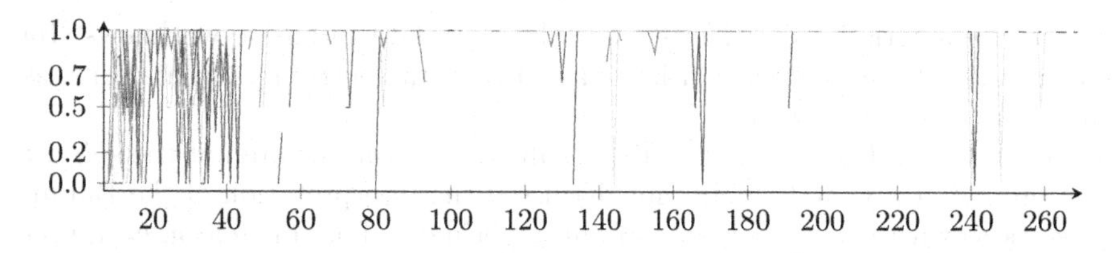

Fig. 7. Progress of the ratio of annotations selected from recommendations.

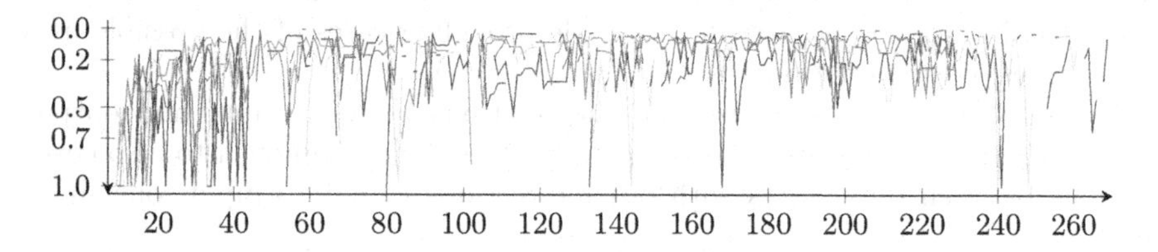

Fig. 8. Average rank of selected recommendations. The vertical axis represents the score placing at the top the first position.

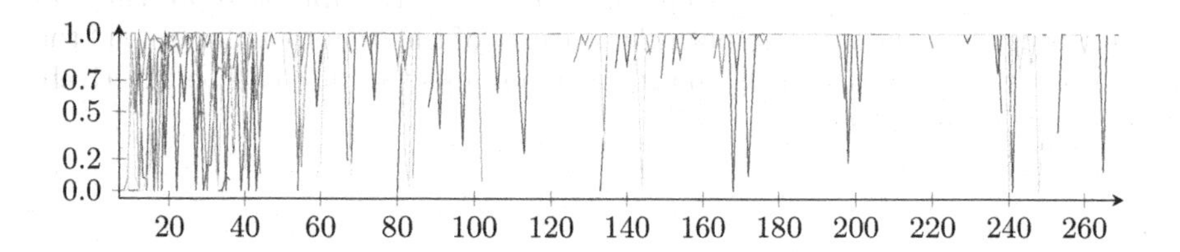

Fig. 9. Progress of the average relevance score of picked recommendations.

Finally, we illustrate in Fig. 9 how the average relevance score of picked recommendations changes in time. The relevance score, computed as the portion of features matching a given recommendation that overlaps with the features of the item to be annotated, increases partly because the rules become more abstract (contain less features), partly reflecting the behavior of the ranking algorithm and matching the result of Fig. 8.

6 Conclusions

In this article we proposed a novel approach to support the semantic annotation of workflows with data centric relations. We showed through applying this approach on a set of workflows from the My Experiments repository that it can effectively reduce the effort required to achieve this task for data managers and workflow publishers. We plan to integrate the presented approach with the methodology described in [4] in order to support Data Hub managers in the annotation of the data manipulation processes required to compute the propagation of policies associated with the data involved. We have enough confidence to believe that the characteristics of scientific workflows as data intensive workflows [16] are equivalent, because they can be reduced to data centric representations, as demonstrated in Sect. 3.1.

The quality and consistency of the resulting annotations are not the subject of the present study, and we did not discussed the interpretation of the Datanode relations with the participants of our experiment. For this reason each user operated on a separate instance of the tool, to reduce the possibility that inconsistent usage of relations would negatively impact the quality of the association rules generated. However, we received feedback that encourages to better document the Datanode ontology, for example providing cases of the possible uses and misuses of each relation.

In this work we only focused on the relations between input and output within workflow processors. It is possible to extend this approach to also cover relations between data items with other directions (input to input, output to input, etc.).

The FCA component of the Dinowolf Tool is based on an incremental lattice construction algorithm. We plan to integrate a lattice *update* algorithm in order to support modifications to the annotations.

However, the incremental learning of association rules approach presented in this paper is independent from both the features of the item to annotate and the nature of the annotations. This opens the hypothesis that it could be effectively reused in other scenarios.

References

1. Alper, P., Belhajjame, K., Goble, C.A., Karagoz, P.: *Label*Flow: exploiting workflow provenance to surface scientific data provenance. In: Ludäscher, B., Plale, B. (eds.) IPAW 2014. LNCS, vol. 8628, pp. 84–96. Springer, Heidelberg (2015). doi:10.1007/978-3-319-16462-5_7
2. Belhajjame, K., Corcho, O., Garijo, D., Zhao, J., Missier, P., Newman, D., Bechhofer, S., Garc a Cuesta, E., Soiland-Reyes, S., Verdes-Montenegro, L., et al.: Workflow-centric research objects: first class citizens in scholarly discourse. In: Proceedings of Workshop on the Semantic Publishing (SePublica 2012) 9th Extended Semantic Web Conference Hersonissos, Crete, Greece, 28 May 2012 (2012)
3. Belhajjame, K., Zhao, J., Garijo, D., Garrido, A., Soiland-Reyes, S., Alper, P., Corcho, O.: A workflow prov-corpus based on taverna and wings. In: Proceedings of the Joint EDBT/ICDT 2013 Workshops, pp. 331–332. ACM (2013)
4. Daga, E., d'Aquin, M., Adamou, A., Motta, E.: Addressing exploitability of smart city data. In: 2016 IEEE Second International Smart Cities Conference (ISC2). IEEE (2016)
5. Daga, E., d'Aquin, M., Gangemi, A., Motta, E.: Describing semantic web applications through relations between data nodes. Technical report kmi-14-05, Knowledge Media Institute, The Open University, Walton Hall, Milton Keynes (2014). http://kmi.open.ac.uk/publications/techreport/kmi-14-05

6. Daga, E., d'Aquin, M., Gangemi, A., Motta, E.: Propagation of policies in rich data flows. In: Proceedings of the 8th International Conference on Knowledge Capture, K-CAP 2015, New York, NY, USA, pp. 5:1–5:8 (2015). http://doi.acm.org/10.1145/2815833.2815839
7. Di Francescomarino, C., Ghidini, C., Rospocher, M., Serafini, L., Tonella, P.: Semantically-aided business process modeling. In: Bernstein, A., Karger, D.R., Heath, T., Feigenbaum, L., Maynard, D., Motta, E., Thirunarayan, K. (eds.) ISWC 2009. LNCS, vol. 5823, pp. 114–129. Springer, Heidelberg (2009)
8. Ferreira, D.R., Alves, S., Thom, L.H.: Ontology-based discovery of workflow activity patterns. In: Daniel, F., Barkaoui, K., Dustdar, S. (eds.) BPM 2011. LNBIP, vol. 100, pp. 314–325. Springer, Heidelberg (2012). doi:10.1007/978-3-642-28115-0_30
9. Gangemi, A., Peroni, S., Shotton, D., Vitali, F.: A pattern-based ontology for describing publishing workflows. In: Proceedings of the 5th International Conference on Ontology and Semantic Web Patterns, WOP 2014, vol. 1302, Aachen, Germany, pp. 2–13. CEUR-WS.org (2014). http://dl.acm.org/citation.cfm?id=2878937.2878939
10. Garijo, D., Alper, P., Belhajjame, K., Corcho, O., Gil, Y., Goble, C.: Common motifs in scientific workflows: an empirical analysis. Future Gener. Comput. Syst. **36**, 338–351 (2014)
11. Garijo, D., Gil, Y.: A new approach for publishing workflows: abstractions, standards, and linked data. In: Proceedings of the 6th Workshop on Workflows in Support of Large-scale Science, WORKS 2011, NY, USA, pp. 47–56 (2011). http://doi.acm.org/10.1145/2110497.2110504
12. Godin, R., Missaoui, R., Alaoui, H.: Incremental concept formation algorithms based on galois (concept) lattices. Comput. Intell. **11**(2), 246–267 (1995)
13. Gómez-Pérez, J.M., Corcho, O.: Problem-solving methods for understanding process executions. Comput. Sci. Eng. **10**(3), 47–52 (2008)
14. Hettne, K., Soiland-Reyes, S., Klyne, G., Belhajjame, K., Gamble, M., Bechhofer, S., Roos, M., Corcho, O.: Workflow forever: Semantic web semantic models and tools for preserving and digitally publishing computational experiments. In: Proceedings of the 4th International Workshop on Semantic Web Applications and Tools for the Life Sciences, SWAT4LS 2011, NY, USA, pp. 36–37 (2012). http://doi.acm.org/10.1145/2166896.2166909
15. Kuznetsov, S.O., Obiedkov, S.A.: Comparing performance of algorithms for generating concept lattices. J. Exp. Theor. Artif. Intell. **14**(2–3), 189–216 (2002)
16. Liu, J., Pacitti, E., Valduriez, P., Mattoso, M.: A survey of data-intensive scientific workflow management. J. Grid Comput. **13**(4), 457–493 (2015)
17. Palma, R., Corcho, O., Hotubowicz, P., Pérez, S., Page, K., Mazurek, C.: Digital libraries for the preservation of research methods and associated artifacts. In: Proceedings of the 1st International Workshop on Digital Preservation of Research Methods and Artefacts, DPRMA 2013, NY, USA, pp. 8–15 (2013). http://doi.acm.org/10.1145/2499583.2499589
18. Poelmans, J., Elzinga, P., Viaene, S., Dedene, G.: Formal concept analysis in knowledge discovery: a survey. In: Croitoru, M., Ferré, S., Lukose, D. (eds.) ICCS 2010. LNCS (LNAI), vol. 6208, pp. 139–153. Springer, Heidelberg (2010). doi:10.1007/978-3-642-14197-3_15
19. Poelmans, J., Kuznetsov, S.O., Ignatov, D.I., Dedene, G.: Formal concept analysis in knowledge processing: a survey on models and techniques. Expert Syst. Appl. **40**(16), 6601–6623 (2013)

20. Weber, I., Hoffmann, J., Mendling, J.: Semantic business process validation. In: Proceedings of the 3rd International Workshop on Semantic Business Process Management (SBPM 2008). CEUR-WS Proceedings, vol. 472 (2008)
21. Wille, R.: Formal concept analysis as mathematical theory of concepts and concept hierarchies. In: Ganter, B., Stumme, G., Wille, R. (eds.) Formal Concept Analysis. LNCS (LNAI), vol. 3626, pp. 1–33. Springer, Heidelberg (2005)
22. Wolstencroft, K., Haines, R., Fellows, D., Williams, A., Withers, D., Owen, S., Soiland-Reyes, S., Dunlop, I., Nenadic, A., Fisher, P., et al.: The taverna workflow suite: designing and executing workflows of web services on the desktop, web or in the cloud. Nucleic Acids Res. **41**, W557–W561 (2013)

A Query Model to Capture Event Pattern Matching in RDF Stream Processing Query Languages

Daniele Dell'Aglio[1,2(✉)], Minh Dao-Tran[3], Jean-Paul Calbimonte[4], Danh Le Phuoc[5], and Emanuele Della Valle[2]

[1] Department of Informatics, University of Zurich, Zurich, Switzerland
dellaglio@ifi.uzh.ch
[2] Dipartimento di Elettronica, Informatica e Bioingegneria, Politecnico of Milano, Milano, Italy
{daniele.dellaglio,emanuele.dellavalle}@polimi.it
[3] Institute of Information Systems, Vienna University of Technology, Vienna, Austria
dao@kr.tuwien.ac.at
[4] Institute of Information Systems, HES-SO Valais-Wallis and LSIR, EPFL, Lausanne, Switzerland
jean-paul.calbimonte@hevs.ch
[5] Technical University of Berlin, Berlin, Germany
danh.lephuoc@tu-berlin.de

Abstract. The current state of the art in RDF Stream Processing (RSP) proposes several models and implementations to combine Semantic Web technologies with Data Stream Management System (DSMS) operators like windows. Meanwhile, only a few solutions combine Semantic Web and Complex Event Processing (CEP), which includes relevant features, such as identifying sequences of events in streams. Current RSP query languages that support CEP features have several limitations: EP-SPARQL can identify sequences, but its selection and consumption policies are not all formally defined, while C-SPARQL offers only a naive support to pattern detection through a timestamp function. In this work, we introduce an RSP query language, called RSEP-QL, which supports both DSMS and CEP operators, with a special interest in formalizing CEP selection and consumption policies. We show that RSEP-QL captures EP-SPARQL and C-SPARQL, and offers features going beyond the ones provided by current RSP query languages.

1 Introduction

Processing heterogeneous and dynamic data is a challenging research topic and has a wide range of applications in real-world scenarios. Different models,

This research has been supported by the Austrian Science Fund (FWF) project P26471, the Nano-tera.ch D1NAMO project, and the Marie Skłodowska-Curie Programme H2020-MSCA-IF-2014 under Grant No. 661180.

E. Blomqvist et al. (Eds.): EKAW 2016, LNAI 10024, pp. 145–162, 2016.
DOI: 10.1007/978-3-319-49004-5_10

languages, and systems have been proposed in the last years to handle streams on the Web, combining Semantic Web technologies with Complex Event Processing (CEP) [18] and Data Stream Management Systems (DSMS) [5] features. These languages and systems, commonly labeled under the RDF Stream Processing (RSP) name, are solutions that extend SPARQL with stream processing features, based on either the CEP or DSMS paradigm.

A problem that recently emerged is the heterogeneity of those solutions [11,13]. Every RSP engine has unique features that are not replicable by others; moreover, even when the same feature is supported by two or more engines, the behavior and the produced output can be different and hardly comparable. In our previous work, namely RSP-QL [14] and LARS [7], we developed models to capture the RSP features inspired by the DSMS paradigm, e.g., time-based sliding windows and aggregations over streams.

In this paper, we study the integration of the currently available CEP features in RSP engines into RSP-QL, by investigating the research question: *"Is it possible to extend RSP-QL to enable the detection of expressive event patterns over RDF streams?"* We give an answer with RSEP-QL, an RSP query model that incorporates CEP at its core.

RSEP-QL is a reference model[1] and has several possible uses: (a) to provide a common framework to explain the behavior of existing RSP solutions, enabling their comparison; (b) to support software architects to design new RSP implementations; testers in designing benchmarks and evaluations; and researchers to have a general model to develop new research; (c) to act as a formal model to define a standardized language that embraces the most prominent features of existing RSP languages.

Combining CEP and DSMS features in a unique model is a step towards filling the gap between RSP and stream processing engines available on the non-semantically-aware systems on the market (e.g., Oracle Event Processor, ESPER, IBM InfoSphere Streams) [10]. There are indeed several motivations behind combining DSMS and CEP. It is clearly possible to mix different DSMS and CEP languages to achieve the desired tasks, but there are drawbacks, e.g., the need to learn multiple languages, the limited possibility for query optimizations, the potential higher amount of resources.

Our contributions are: (1) We elicit a set of requirements to design an RSP query model that supports both DSMS and CEP features. (2) We adapt our model to process RDF graphs as stream elements, following the current guidelines of the W3C RSP Community Group (RSP-CG).[2] (3) We introduce event patterns to capture CEP features of existing RSP engines, most notably the sequencing operator, and provide syntax and semantics as extensions of SPARQL. (4) We formally define selection and consumption policies, to capture the operational semantics of the CEP-inspired RSP engines, contrary to current approaches that consider policies at the implementation level.

[1] Cf. https://www.oasis-open.org/committees/soa-rm/faq.php.

[2] Cf. https://www.w3.org/community/rsp/.

2 Related Work and Requirements

RSP engines emerged in recent years, with the goal of extending RDF and SPARQL to process RDF streams. They can be broadly divided into two groups. RSPs influenced by CEP reactively process the input streams to identify relevant events and sequences of them. EP-SPARQL [3] is one of the first RSP that adopts some of these complex pattern operators. Other such recent approaches include Sparkwave [17] and Instans [20]. On the other hand, approaches inspired by DSMS exploit sliding window mechanisms to capture a recent and finite portion of the input data, enabling their processing through SPARQL operators [15] in an atemporal fashion. C-SPARQL [6], CQELS [19], and SPARQL$_{stream}$ [9] are representative examples of this group.

Currently, there is so far no RSP language that can combine both paradigms under a clearly defined semantics, leaving a gap for those use cases that require this query expressivity. However, some initial attempts exist. In C-SPARQL, one can access the timestamp of a statement and specify limited forms of temporal conditions. CQELS recently proposed to integrate sequencing and path navigation [12], although it does not include typical selection mechanisms of CEP [10]. In the following, we present a set of requirements to lead the design of RSEP-QL, based on an analysis of the state of the art in RSP, with a particular focus on the CEP features of EP-SPARQL, and C-SPARQL.

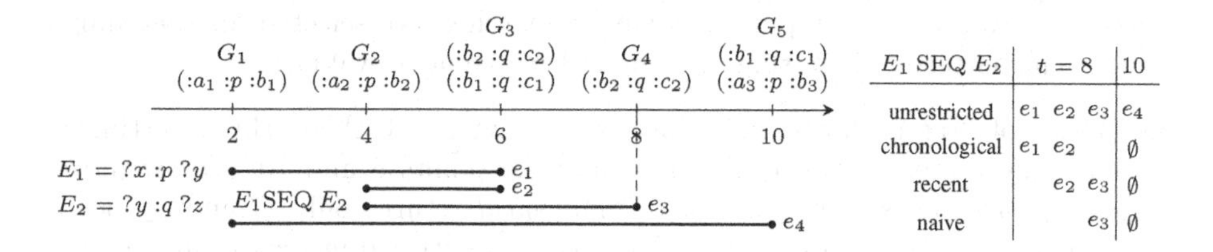

Fig. 1. Illustration of the running example. The stream, on the top left, composes of five items $(G_1, 2) \dots (G_5, 10)$. Events matched the pattern E_1 SEQ E_2 are depicted below the timeline. The bold lines denote the intervals that justify the events. The table on the right shows the results produced with regards to different policies.

[R1] RSEP-QL should process RDF graph-based streams. While in early RSP data models the stream data items are represented by single RDF statements, the recent standardization effort from W3C RSP-CG proposes to adopt RDF graphs as items[3]. The latter model generalizes of the former, as a stream of time-annotated RDF statements can be modeled as a stream of time-annotated RDF graphs, each containing one statement. In this sense, addressing [R1] is important to realize a generic RDF stream query model.

[3] Cf. http://goo.gl/pqUSri (last access: July 7, 2016).

[R2] RSEP-QL must preserve the DSMS features captured by RSP-QL. The introduction of CEP features in the model should not lead to incompatibilities with the RSP models we already captured in RSP-QL [14]. This requirement is important to guarantee that RSEP-QL is generic enough to model the operational semantics of different systems.

[R3] RSEP-QL should capture the CEP features of existing RSP engines. In this work, we focus on the SEQ operator: the most basic building block in CEP. Intuitively, E_1 SEQ E_2 identifies events matching pattern E_1 followed by those matching E_2. Even if it may seem straightforward to formalize this operator, its execution in different engines produces different and hardly comparable results. We, therefore, refine [R3] into two sub-requirements, associated with the two engines we aim at capturing, EP-SPARQL and C-SPARQL. To illustrate our idea, we use the RDF stream depicted in Fig. 1.

[R3.1] RSEP-QL should capture the EP-SPARQL SEQ behavior. To the best of our knowledge, EP-SPARQL is the RSP language with the largest support for CEP features, with a wide range of operators to define complex events, e.g., SEQ, OPTIONALSEQ, EQUALS and EQUALSOPTIONAL. EP-SPARQL supports three different policies [2]:

- *unrestricted*: all input elements are selected for matching the event patterns.
- *chronological*: only the earliest input that can be matched are selected for matching the event patterns; then, they are ignored in the next evaluations.
- *recent*: only the latest input that can be matched are selected for matching the event patterns; then, they are ignored in the next evaluations.

The table of Fig. 1 shows the different behaviors of these three settings. Assume that there are two evaluations at time points 8 and 10. *Unrestricted* returns e_1, e_2, e_3 at 8 and e_4 at 10. *Chronological* returns only e_1 and e_2 at 8. *Recent* returns only e_2 and e_3 at 8. Furthermore, both *chronological* and *recent* do not return any event at 10 because ($:a_1$ $:p$ $:b_1$) were already consumed by the previous evaluation.

Notably, the EP-SPARQL query does not change in the three cases, as the setting is a configuration parameter set at the startup of the engine. Moreover, independently on the setting, all the system outputs happen as soon as they are available.

[R3.2] RSEP-QL should capture the C-SPARQL SEQ behavior. C-SPARQL is based on DSMS techniques, but it has a naive support to some CEP features. C-SPARQL implements a function, named *timestamp* that takes as input a triple pattern and returns the time instant associated to the *most recent* matched triple. This function can be used inside a FILTER clause to express time constraints among events.

The evaluation in C-SPARQL strictly relies on the notion of time-based sliding window, which selects a portion of the stream to be used as input and the time instants on which evaluations occur. Wrt. the above example, with a sliding window with a length of 7 and that slides of 1 at each step, C-SPARQL outputs

e_3 at time 8 and has no output at 10, not because the input triples were consumed, but because it considers only the two triples (:b_1 :q :c_1) and (:a_3 :p :b_3) which do not match the sequencing pattern.

Remarks. While EP-SPARQL is an engine for performing CEP, C-SPARQL is a DSMS-inspired RSP engine that offers a naive support to event pattern matching. As shown above, even with simple event patterns, the two systems behave in completely different ways, and none of them is able to capture the other. It is out of the scope of this paper to determine which system is the most suitable to be used given a use case and the relative set of requirements. Our goal is to build a model able to capture the behavior of both engines. In this sense, satisfying both [R3.1] and [R3.2] is minimal to assess that RSEP-QL is a common framework to describe the semantics of RSP engines.

3 Anatomy of RSEP-QL Queries

A SPARQL query is defined by a signature of the form (E, DS, QF), that indicates the evaluation of an algebraic expression E over a set of data DS to produce an answer formatted according to a query form QF [16]. This section proposes RSEP-QL queries that extend SPARQL's queries with the following features: (1) the capability to take as input not only RDF graphs but also RDF streams; (2) a set of operators to access/process streams; and (3) an evaluation paradigm moving from one-time to continuous semantics.

3.1 Data Model

There are two main kinds of input data in the context of stream processing. The first are streams, defined as sequences of highly dynamic and time-annotated data such as sensor data and micro-posts. The second type is contextual (or background) data, which is usually static or quasi-static and is used to enrich the streams and solve more sophisticated tasks, e.g., sensor locations, user profiles. etc. In RSP, contextual data may be captured by RDF graphs, while streams are captured with RDF streams.

RDF Streams. To fulfill [R1], we adopt the notion of time-annotated RDF graphs as elements of RDF streams, following the data model under design by RSP-CG. We define a *timeline* T as an infinite, discrete, ordered sequence of time instants $(t_1, t_2, \ldots)$, where $t_i \in \mathbb{N}$ and for all $i > 0$, it holds that $t_{i+1} - t_i$ is a constant, called the *time unit* of T.

We now extend the definition of RDF graphs with time annotations and then define RDF streams as sequences of them.

Definition 1 (RDF Stream). *A* timestamped RDF graph *is a pair* (G, t), *where* G *is an RDF graph and* $t \in T$ *is a time instant. An* RDF stream S *is a (potentially) unbounded sequence of timestamped RDF graphs in a non-decreasing time order:*

$$S = (G_1, t_1), (G_2, t_2), (G_3, t_3), (G_4, t_4), \ldots$$

where, for every $i > 0$, (G_i, t_i) is a timestamped RDF graph and $t_i \leq t_{i+1}$.

Other streaming data model profiles exist and are currently under study by the RSP-CG. In this work, we focus on the model where the time annotation is represented by one time instant, as it is a usual case that appears in several scenarios.

Example 1. Figure 1 illustrates a stream $S = (G_1, 2), (G_2, 4), (G_3, 6), (G_4, 8), (G_5, 10), \ldots$, where each G_i contains the depicted RDF triples. □

Time-Varying Graphs. Statements in RDF graphs are atemporal and capture a given situation in a snapshot. We introduce the notion of time-varying graphs to capture the evolution of the graph over time (similar to time-varying relations in [4]).

Definition 2 (Time-Varying Graph). *A* time-varying graph $\overline{G}$ *is a function that relates time instants $t \in T$ to RDF graphs:*

$$\overline{G}: T \rightarrow \{G \mid G\, is\, an\, RDF\, graph\}.$$

An instantaneous RDF graph $\overline{G}(t)$ *is the RDF graph identified by the time-varying graph $\overline{G}$ at a given time instant t.*

RDF streams and time-varying graphs differ on the time information: while in the former time annotations are accessible and processable by the stream processing engine, in the latter there is no explicit time annotation. In this sense, t in Definition 2 can be viewed as a timestamp denoting the access time of the engine to the graph content.

3.2 RSEP-QL Dataset

A SPARQL dataset is a set of pairs (u, G), where $u \in I \cup \{def\}$[4] is an identifier for an RDF graph G. This section proposes the notion of dataset for RSEP-QL. It differs from SPARQL datasets in the presence of streams, and that RSEP-QL dataset elements may vary over time. Streams are potentially infinite, and the usage of windows allows to have a finite (and usually recent) view of portions of the streams for practical processing. We now introduce a generic notion of window functions, inspired by LARS [7].

Definition 3 (Window Function). *A* window function W *with a vector of* window parameters $\boldsymbol{p}$, *denoted as* $W[\boldsymbol{p}]$, *takes as input a stream S, a time instant $t \in T$ and produces a* substream *(aka.* window*) S' of S, i.e., a finite subsequence of S.*

---

[4] $def \notin I \cup L \cup B$ denoting the default graph. See [16] for the definitions of I, L, B.

This generic notion can be instantiated with specific parameters $\boldsymbol{p}$ to realize window functions used in practice. In the following, we present a set of window functions that constitute the basis of the operators defined in the next sections.

Time-Based (sliding) Windows. A *time-based window* function W^τ is defined through $\boldsymbol{p} = (\alpha, \beta)$, where α is the width and β is the sliding step. It slides every β time units and filters input graphs of the last α time units. Let $t' = \lfloor \frac{t}{\beta} \rfloor \cdot \beta$, we have that:

$$W^\tau[\boldsymbol{p}](S, t) = (G_j, t_j), \ldots, (G_k, t_k),$$

where $[j, k]$ is the maximal interval st. $\forall i \in [j, k]$: $(G_i, t_i) \in S \wedge t' - \alpha < t_i \leq t'$.

Landmark Windows. A *landmark window* function W^λ defined through $\boldsymbol{p} = (t_0)$ returns the content of the input stream from t_0:

$$W^\lambda[\boldsymbol{p}](S, t) = (G_j, t_j), \ldots, (G_k, t_k)$$

where $[j, k]$ is the maximal interval st. $\forall i \in [j, k]$: $(G_i, t_i) \in S \wedge t_0 \leq t_i \leq t$.

As we show below, landmark windows are useful to capture the behaviour of event pattern systems like EP-SPARQL. In fact, they offer views over large portions of the stream, without the eviction mechanism typical of sliding windows.

Identity Window. The *identity window* function W^{id} is introduced to give a uniform definition of event patterns evaluation later. It simply returns the input stream, that is:

$$W^{id}[\boldsymbol{p}](S, t) = S, \text{ and } \boldsymbol{p} \text{ is an empty vector.}$$

Interval Windows. The *interval-based* (or fixed) window function $W^\sqcup$ is defined through $\boldsymbol{p} = (t', t'')$ and returns the part of the input stream bounded by $[t', t'']$:

$$W^\sqcup[\boldsymbol{p}](S, t) = (G_j, t_j), \ldots, (G_k, t_k) \text{ where } \forall i \in [j, k]\colon (G_i, t_i) \in S \wedge t_i \in [t', t''].$$

For simplicity, we often omit the parameters $\boldsymbol{p}$ when it is clear from the context and write $W(S, t)$. Notably, window functions can be nested, for example, we can have $W^\sqcup(W^\tau(S, t), t)$. We denote the nesting by the $\bullet$ operator. Formally:

$$W_1 \bullet W_2(S, t) = W_1(W_2(S, t), t).$$

Example 2. Consider S from Example 1. Here are some results of applying the time-based, landmark, and interval window functions W^τ, W^λ, and $W^\sqcup$ on this stream:

$$\begin{aligned} W^\lambda[(1)](S, 8) &= (G_1, 2), (G_2, 4), (G_3, 6), (G_4, 8) \\ W^\tau[(5, 1)](S, 8) &= (G_2, 4)(G_3, 6), (G_4, 8) \\ W^\sqcup[(0, 5)] \bullet W^\lambda[(1)](S, 8) &= (G_1, 2), (G_2, 4). \end{aligned}$$

Dataset. We now formally define RSEP-QL datasets, as sets of pairs of an identifier $u \in I \cup \{def\}$ and either a window function applied to a stream or a time-varying graph.

Definition 4 (RSEP-QL Dataset). *An* RDF streaming dataset *SDS is a set consisting of an (optional) default time-varying graph* $\overline{G}_0$*,* $n \geq 0$ *named time-varying graphs, and* $m \geq 0$ *named window functions applied to a set of streams* $\mathbf{S} = \{S_1, \ldots, S_k\}$:

$$SDS = \{(def, \overline{G}_0)\} \cup \{(g_i, \overline{G}_i) \mid i \in [1, n]\} \cup \{(w_j, W_j(S_\ell)) \mid j \in [1, m],\ \ell \in [1, k]\}, \text{ where}$$

- $\overline{G}_0$ *is the default time-varying graph,*
- $g_i \in I$ *is the identifier of the time-varying graph* $\overline{G}_i$*,*
- $w_j \in I$ *is the identifier of the named window function* W_j *over the RDF stream* $S_\ell \in \mathbf{S}$.

We denote by $ids(SDS) = \{def\} \cup \{g_1, \ldots, g_n\} \cup \{w_1, \ldots, w_m\}$ *the set of symbols identifying the time-varying graphs and windows in SDS.*

An important difference that emerges comparing the SPARQL and the RSEP-QL dataset is that the former contains RDF graphs and is fixed in the sense that SPARQL datasets are composed according to the query (e.g. FROM clauses), and the set of elements included in a dataset does not vary over time. On the other hand, RSEP-QL datasets contain RDF streams and time-varying graphs that are updated as time proceeds.

Example 3. Let W_1^{λ} and $W_2^{\mathcal{T}}$ be a landmark and a time-based window functions with respective parameters $\boldsymbol{p}_1 = (1)$ and $\boldsymbol{p}_2 = (5, 1)$. Then, $SDS = \{(w_1, W_1^{\lambda}(S)), (w_2, W_2^{\mathcal{T}}(S))\}$ is an RDF streaming dataset, where S is from Example 1. □

3.3 RSEP-QL Patterns

To fulfill [R2] and [R3], we introduce RSEP-QL operators to enable DSMS and CEP features. We then extend SPARQL graph patterns to support these operators on streams.

In SPARQL, the construction of the query relies on graph patterns. The elementary building block for building graph patterns is *Basic Graph Patterns* (BGP), i.e. sets of triple patterns $(t_s, t_p, t_o) \in (I \cup B \cup L \cup V) \times (I \cup V) \times (I \cup B \cup L \cup V)$. More complex patterns are recursively defined on top of BGP using operators such as join and union[5].

Concerning DSMS operations, we introduce the *window graph pattern*, defined as an expression (WINDOW w_j P), where P is a SPARQL graph pattern and $w_j \in I$ is an IRI. Intuitively, WINDOW indicates that P should be evaluated over the content of the window identified by w_j in the dataset (similarly to the SPARQL GRAPH operator).

To support CEP features, we introduce *event patterns* as follows.

[5] Cf. https://www.w3.org/TR/sparql11-query for the whole list.

(1) If P is a Basic Graph Pattern, $w \in I$, then the expressions (EVENT w P) is an event pattern, named *Basic Event Pattern* (BEP)[6];
(2) If E_1 and E_2 are event patterns, then the expressions (FIRST E_1), (LAST E_1), (E_1 SEQ E_2) are event patterns;

To relate graph and event patterns, we define the *event graph pattern* as (MATCH E) where E is an event pattern.

3.4 Query Definition

Having all building blocks, it is now possible to define RSEP-QL queries.

Definition 5. *An* RSEP-QL query *Q is defined as (SE, SDS, ET, QF), where SE is an RSEP-QL algebraic expression, SDS is an RDF streaming dataset, ET is the sequence of time instants on which the evaluation occurs, and QF is the Query Form.*

The continuous evaluation paradigm is captured in the query signature through the set ET of execution times. Intuitively, this set represents the time instants on which the algebraic expression evaluation may occur. Note that this set is not explicitly defined by the query and in general it may be unknown at query registration time (as it can depend on the streaming content). In practice, ET can be expressed through report policies [8], which define rules to trigger the query evaluation. For example, C-SPARQL can be captured by a window close report policy, i.e., evaluations are periodically and determined by the window definition. EP-SPARQL and CQELS are regulated by content change report policy, i.e., evaluations occur every time a new item appears on the stream.

Example 4. This example presents an RSEP-QL query with CEP features. The MATCH clause describes an event pattern (E_1 SEQ E_2), where the BEPs E_1 and E_2 are defined on the respective landmark and time-based windows from Example 3. Their patterns are: E_1 = EVENT w_1 (?x :p ?y) and E_2 = EVENT w_2 (?y :q ?z).

```
SELECT ?x ?z
FROM NAMED :S  WIN [LND    9] AS :w1
FROM NAMED :S  WIN [RANGE 5] AS :w2
EVENT ON :w1 { ?x :p ?y. } AS E1
EVENT ON :w2 { ?y :q ?z. } AS E2
WHERE { MATCH { E1 SEQ E2 } }
```

4 RSEP-QL Semantics

We now proceed to define the evaluation semantics of the operators introduced in Sect. 3.3. Sections 4.1 and 4.2 present the semantics of the graph pattern and event pattern operators, respectively. Sections 4.3 and 4.4 address CEP selection and consumption policies to completely capture settings such as *chronological recent* of EP-SPARQL, or the naive sequencing of triples based on last their appearances like in C-SPARQL.

[6] We do not tackle here the case where $w \in I \cup V$, which is one of our future works.

4.1 Graph Pattern Evaluation Semantics

To cope with graph-based RDF streams, we adapt the graph pattern evaluation semantics from [14]. There, the evaluation semantics of a SPARQL operator is defined as a function that takes as input a graph pattern P and a SPARQL dataset DS having a default RDF graph G, and produces bags of solution mappings: partial functions that map variables to RDF terms. It is usually denoted as $[\![P]\!]_{DS(G)}$.

The RSEP-QL evaluation semantics of graph patterns considers the evaluation time instants and redefines the active graph notion. Given an RSEP-QL dataset SDS and an identifier $\iota \in ids(SDS)$ of one of its elements, we name *temporal sub-dataset*, denoted by SDS_ι, the active element of the dataset. The active element is $SDS_\iota = \overline{G}_i$ if $(\iota = g_i, \overline{G}_i) \in SDS$, or $SDS_\iota = W_j(S_\ell)$ if $(\iota = w_j, W_j(S_\ell)) \in SDS$.

Definition 6 (Graph Pattern Evaluation Semantics). *Given an RSEP-QL pattern P, an active time-varying graph or window identified by $\iota \in ids(SDS)$ of a streaming dataset SDS, and an evaluation time instant t, we define*

$$[\![P]\!]^t_{SDS_\iota}$$

as the evaluation *of P at t over the active element ι in SDS.*

We now briefly summarize the evaluation semantics of the graph patterns available in SPARQL, with a special focus on BGP and window graph patterns from Sect. 3.3.

Basic Graph Pattern. BGP evaluation in SPARQL is one of the few cases in which there is an actual access to the data stored in the active RDF graph. The idea behind the evaluation of BGPs in RSEP-QL is to exploit the SPARQL evaluation semantics. To make it possible, it is necessary to move from the active element ι of SDS and the evaluation time instant t to an RDF graph over which the BGP can be evaluated. We name this RDF graph the *snapshot of a temporal sub-dataset* at t, and it is defined as:

$$SDS_{g_i}(t) = \overline{G}_i(t) \quad \text{and} \quad SDS_{w_j}(t) = \bigcup_{(G_k,t_k)\in W_j(S_\ell,t)} G_k$$

By exploiting the snapshot of the temporal sub-dataset, it is possible to obtain an RDF graph given a streaming dataset and an active element. This RDF graph is the one over which the BGP has to be evaluated, following the SPARQL semantics.

Example 5. Take SDS from Example 3. We have

$$SDS_{w_2}(12) = \bigcup_{(G_k,t_k)\in W_2^{\tau}[(5,1)](S,12)} G_k = G_4 \cup G_5 = \{:a_3\ :p\ :b_3,\ \ :b_1\ :q\ :c_1,\ \ :b_2\ :q\ :c_2\}.$$

We can now define the evaluation of a basic graph pattern P as:

$$[\![P]\!]^t_{SDS_\iota} = [\![P]\!]_{SDS_\iota(t)} = \{\mu \mid dom(\mu) = var(P) \text{ and } \mu(P) \in SDS_\iota(t)\}. \quad (1)$$

Other SPARQL Graph Patterns. For other graph patterns, we maintain the idea of SPARQL of defining them recursively [16]. For example, the graph pattern P_1 *Join* P_2:

$$[\![P_1 \, Join \, P_2]\!]^t_{SDS_\iota} = [\![P_1]\!]^t_{SDS_\iota} \bowtie [\![P_2]\!]^t_{SDS_\iota} \tag{2}$$

where SDS_ι indicates the active time-varying graph or window in the RSEP-QL dataset SDS and P_1, P_2 are graph patterns. The evaluation of $P_1 \, Join \, P_2$ consists of joining the two multisets of solution mappings computed by evaluating P_1 and P_2 at time t with regards to the active part SDS_ι of the RDF streaming dataset SDS.

Window Graph Pattern. Finally, we define the evaluation semantics of the window graph patterns. Given a window identifier w_j and a graph pattern P, we have that:

$$[\![\text{WINDOW } w_j \ P]\!]^t_{SDS_\iota} = [\![P]\!]^t_{SDS_{w_j}} \tag{3}$$

The following example shows the application of Eqs. (1) and (3).

Example 6. Take SDS from Example 3 and its sub-temporal-dataset $SDS_{w_2}(12)$ from Example 5, let $P = \{?x \ {:}p \ ?y\}$. We have that:

$$[\![\text{WINDOW } w_2 \ P]\!]^{12}_{SDS_{def}} = [\![P]\!]^{12}_{SDS_{w_2}} = [\![?x \ {:}p \ ?y]\!]_{SDS_{w_2}(12)} = \{\{?x \mapsto {:}a_3, ?y \mapsto {:}b_3\}\}.$$

4.2 Event Pattern Evaluation Semantics

Similarly to Sect. 4.1, we define the evaluation semantics of event pattern operators by decomposing complex patterns into simple ones. The main difference is that this decomposition process should take into account the temporal aspects related to event matching, i.e., the evaluation should (i) produce time-annotated solution mappings, and (ii) control the time range in which a subpattern is processed. We address (i) by defining the notion of *event mapping* as a triple (μ, t_1, t_2) composed by a solution mapping and two time instants t_1 and t_2, representing the initial and final time instants that justify the matching, respectively. We assume that a partial order $\prec$ to compare timestamps is given. Depending on particular applications, specific ordering can be chosen. Regarding (ii), we associate the evaluation with an active window function that sets the boundaries of the valid ranges for evaluating event patterns.

Definition 7 (Event Pattern Evaluation Semantics). *Given an event pattern E, a window function W (*active window*), and an evaluation time instant $t \in ET$, we define*

$$\langle\!\langle E \rangle\!\rangle^t_W$$

as the evaluation *of E in the scope defined by W at t.*

Different from graph pattern evaluation semantics, in this case there is no explicit reference to data. This information is carried in the basic event patterns defined below.

Basic Event Patterns. Similar to BGPs, Basic Event Patterns (BEP) are the simplest building block. The idea behind their semantics is to produce a set of SPARQL BGP evaluations over the stream items from a snapshot of a temporal sub-dataset (identified by w_j), restricted by the active window function W:

$$\langle\langle \text{EVENT } w_j \; P \rangle\rangle_W^t = \{(\mu, t_k, t_k) \mid \mu \in [\![P]\!]_{G_k} \wedge (G_k, t_k) \in W \bullet W_j(S_\ell, t)\} \quad (4)$$

Example 7. We show how to evaluate $\langle\langle E_2 \rangle\rangle_{W^{id}}^8$ for $E_2 = (\text{EVENT } w_2 \; (?y \; {:}q \; ?z))$ from Example 4. First of all, from Example 2, we have that

$$W^{id} \bullet SDS_{w_2}(8) = W^{id} \bullet W_2^{\tau}(S, 8) = W_2^{\tau}(S, 8) = (G_1, 2), (G_2, 4)(G_3, 6), (G_4, 8).$$

Now we evaluate $[\![?y \; {:}q \; ?z]\!]_{G_k}$ for $1 \leq k \leq 4$. Only G_3 and G_4 have matches, which are $\mu_2 = \{?y \mapsto {:}b_1, ?z \mapsto {:}c_1\}$ and $\mu_2' = \{?y \mapsto {:}b_2, ?z \mapsto {:}c_2\}$. Combining with the timestamps 6 and 8 when G_3 and G_4 respectively appear in S, we have:

$$\langle\langle E_2 \rangle\rangle_{W^{id}}^8 = \{(\mu_2, 6, 6), (\mu_2', 6, 6), (\mu_2', 8, 8)\}.$$

It is worth comparing the evaluation semantics of a BEP with the one of a BGP as defined in Sect. 4.1. They both exploit the SPARQL BGP evaluation, but while the former defines an evaluation for each stream item (i.e., an RDF graph), the latter is a unique evaluation over the merge of the stream items in one RDF graph.

Other Event Patterns. Next is the semantics of other event patterns, starting with those that identify the *first* and *last* event matching a pattern, based on the ordering $\prec$.

$$\langle\langle \text{FIRST } E \rangle\rangle_W^t = \{(\mu, t_1, t_2) \in \langle\langle E \rangle\rangle_W^t \mid \nexists (\mu', t_3, t_4) \in \langle\langle E \rangle\rangle_W^t : (t_3, t_4) \prec (t_1, t_2)\} \quad (5)$$

$$\langle\langle \text{LAST } E \rangle\rangle_W^t = \{(\mu, t_1, t_2) \in \langle\langle E \rangle\rangle_W^t \mid \nexists (\mu', t_3, t_4) \in \langle\langle E \rangle\rangle_W^t : (t_1, t_2) \prec (t_3, t_4)\} \quad (6)$$

Let us now consider the SEQ operator. The evaluation of E_1 SEQ E_2 is defined as:

$$\langle\langle E_1 \text{ SEQ } E_2 \rangle\rangle_W^t = \{(\mu_1 \cup \mu_2, t_1, t_4) \mid (\mu_2, t_3, t_4) \in \langle\langle E_2 \rangle\rangle_W^t \wedge (\mu_1, t_1, t_2) \in \langle\langle \mu_2(E_1) \rangle\rangle_{W^{\sqcup}[0, t_3 - 1] \bullet W}^t\} \quad (7)$$

Intuitively, for each event mapping (μ_2, t_3, t_4) that matches E_2, Eq. (7) seeks for (a) *compatible* and (b) *preceding* event mappings matching E_1. The two demands are guaranteed by introducing constraints on the evaluation of E_1:

- (a) is imposed by, in E_1, substituting the shared variables with E_2 for their values from μ_2, denoted by $\mu_2(E_1)$.
- (b) is ensured by restricting the time range on which input graphs are used to match $\mu_2(E_1)$: we only consider graphs appearing before t_3, thus $W^{\sqcup}[0, t_3 - 1] \bullet W$.

Example 8 (cont'd). We show how $\langle\langle E_1 \text{ SEQ } E_2\rangle\rangle^8_{W^{id}}$ is evaluated. For $(\mu_2, t_3, t_4) = (\{?y \mapsto :b_1, ?z \mapsto :c_1\}, 6, 6) \in \langle\langle E_2\rangle\rangle^8_{W^{id}}$, we then evaluate:

$$\langle\langle \mu_2(E_1)\rangle\rangle^8_{W^{id}} = \langle\langle \text{EVENT } w_1\ (?x\ :p\ :b_1)\rangle\rangle^8_{W^{\sqcup}[0,5]\bullet W^{id}} = \langle\langle \text{EVENT } w_1\ (?x\ :p\ :b_1)\rangle\rangle^8_{W^{\sqcup}[0,5]}.$$

Similar to Example 7, we first see that $W^{\sqcup}[0, 5] \bullet W_1^{\lambda}(S, 8) = (G_1, 2), (G_2, 4)$. Then, evaluating $[\![?x\ :p\ :b_1]\!]_{G_k}$ for $k = 1, 2$ matches in only G_1. Therefore, the mapping satisfying conditions (a) and (b) is $(\mu_1, t_1, t_2) = (\{?x \mapsto :a_1, ?y \mapsto :b_1\}, 2, 2)$. Finally, Eq. (7) gives us $(\{?x \mapsto :a_1, ?y \mapsto :b_1, ?z \mapsto :c_1\}, 2, 6)$.

Similarly, with $(\mu_2', 6, 6)$ and $(\mu_2', 8, 8)$ from Example 7, we find a compatible and preceding match $(\{?x{\mapsto}:a_2, ?y{\mapsto}:b_2\}, 4, 4)$ for E_1. This gives us two more results: $(\{?x{\mapsto}:a_2, ?y{\mapsto}:b_2, ?z{\mapsto}:c_2\}, 4, 8)$ and $(\{?x{\mapsto}:a_2, ?y{\mapsto}:b_2, ?z{\mapsto}:c_2\}, 6, 8)$. □

Event Graph Pattern. Finally, we define the semantics of the MATCH operator. Being a graph pattern, its evaluation semantics is defined through the function in Definition 6. Intuitively, the function acts to remove the time annotations from event mappings and to produce a bag of solution mappings. Thus, the result of this operator can be combined with results of other graph pattern evaluations (i.e., other bags of solution mappings).

$$[\![\text{MATCH } E]\!]^t_{SDS_\iota} = \{\mu \mid (\mu, t_1, t_2) \in \langle\langle E\rangle\rangle^t_{W^{id}}\} \tag{8}$$

The initial active window function to E is W^{id}, which imposes no time restriction. Such restrictions can appear later by CEP operators like in Eq. (7).

Example 9 (cont'd). Applying MATCH on (E_1 SEQ E_2) from Example 8 returns:

$$[\![\text{MATCH } (E_1 \text{ SEQ } E_2)]\!]^8_{SDS_{def}} = \{\{?x \mapsto :a_i, ?y \mapsto :b_i, ?z \mapsto :c_i\} \mid i = 1, 2\}.$$

4.3 Event Selection Policies

Evaluating the SEQ operator as in Eq. (7) takes into account all possible matches from the two sub-patterns. This kind of evaluation captures only the *unrestricted* behavior of EP-SPARQL and C-SPARQL. With the purpose of formally capturing the CEP semantics of C-SPARQL and EP-SPARQL, we introduce in this section different versions of the sequencing operator that allows different ways of selecting stream items to perform matching, known as *selection policies*.

Firstly, for C-SPARQL's *naive* CEP behavior, Eq. (9) simply picks the two latest event mappings that match the two sub-patterns and compare their associated timestamps.

$$\begin{aligned}\langle\langle E_1 \text{ SEQ}^n\ E_2\rangle\rangle^t_W = \{(\mu_1 \cup \mu_2, t_1, t_4) \mid (t_1, t_2) \prec (t_3, t_4) \wedge \\ (\mu_1, t_1, t_2) \in \langle\langle \text{LAST } E_1\rangle\rangle^t_W \wedge (\mu_2, t_3, t_4) \in \langle\langle \text{LAST } E_2\rangle\rangle^t_W\}\end{aligned} \tag{9}$$

For the *chronological* and *recent* settings from EP-SPARQL, we need more involved operators SEQ^c and SEQ^r. In the sequel, let $W^\star = W^\sqcup[0, t_3 - 1] \bullet W$.

$$\begin{aligned}\langle\langle E_1 \text{ SEQ}^c\ E_2\rangle\rangle_W^t = \{(\mu_1 \cup \mu_2, t_1, t_4) \mid (\mu_2, t_3, t_4) \in \langle\langle E_2\rangle\rangle_W^t \wedge \\ \langle\langle \mu_2(E_1)\rangle\rangle_{W^\star}^t \neq \emptyset \wedge (\mu_1, t_1, t_2) \in \langle\langle \text{FIRST } \mu_2(E_1)\rangle\rangle_{W^\star}^t \wedge \\ (\nexists(\mu_2', t_3', t_4') \in \langle\langle E_2\rangle\rangle_W^t : \langle\langle \mu_2'(E_1)\rangle\rangle_{W^\star}^t \neq \emptyset \wedge (t_3', t_4') \prec (t_3, t_4))\}.\end{aligned} \tag{10}$$

Compared to (7), Eq. (10) selects an event mapping (μ_2, t_3, t_4) of E_2 that:

- has a compatible event mappings in E_1 which appeared before μ_2. This is guaranteed by the condition $\langle\langle \mu_1(E_2)\rangle\rangle_{W^\star}^t \neq \emptyset$ and the window function $W^\star = W^\sqcup[0, t_3 - 1] \bullet W$;
- is the first of such event mappings. This is ensured by stating that no such (μ_2', t_3', t_4') exists, where $(t_3', t_4') \prec (t_3, t_4)$.

Once (μ_2, t_3, t_4) is found, (μ_1, t_1, t_2) is taken from $\langle\langle \text{FIRST } \mu_2(E_1)\rangle\rangle_{W^\star}^t$, which makes sure that it is the first compatible event that appeared before (μ_2, t_3, t_4). Finally, the output event matching $E_1 \text{ SEQ}^c\ E_2$ is $(\mu_1 \cup \mu_2, t_1, t_4)$.

Equation (11) follows the same principle as Eq. (10), except that it selects the last instead of the first event mappings.

$$\begin{aligned}\langle\langle E_1 \text{ SEQ}^r\ E_2\rangle\rangle_W^t = \{(\mu_1 \cup \mu_2, t_1, t_4) \mid (\mu_2, t_3, t_4) \in \langle\langle E_2\rangle\rangle_W^t \wedge \\ \langle\langle \mu_2(E_1)\rangle\rangle_{W^\star}^t \neq \emptyset \wedge (\mu_1, t_1, t_2) \in \langle\langle \text{LAST } \mu_2(E_1)\rangle\rangle_{W^\star}^t \wedge \\ (\nexists(\mu_2', t_3', t_4') \in \langle\langle E_2\rangle\rangle_W^t : \langle\langle \mu_2'(E_1)\rangle\rangle_{W^\star}^t \neq \emptyset \wedge (t_3, t_4) \prec (t_3', t_4'))\}.\end{aligned} \tag{11}$$

Example 10 (cont'd). Continue with the setting in Example 8, one can check that:

$$\begin{aligned}\langle\langle E_1 \text{ SEQ}^n\ E_2\rangle\rangle_{W^{id}}^8 &= \{(\{?x \mapsto :a_2, ?y \mapsto :b_2, ?z \mapsto :c_2\}, 4, 8)\}; \\ \langle\langle E_1 \text{ SEQ}^c\ E_2\rangle\rangle_{W^{id}}^8 &= \left\{\begin{matrix}(\{?x \mapsto :a_1, ?y \mapsto :b_1, ?z \mapsto :c_1\}, 2, 6) \\ (\{?x \mapsto :a_2, ?y \mapsto :b_2, ?z \mapsto :c_2\}, 4, 6)\end{matrix}\right\}; \\ \langle\langle E_1 \text{ SEQ}^r\ E_2\rangle\rangle_{W^{id}}^8 &= \left\{\begin{matrix}(\{?x \mapsto :a_1, ?y \mapsto :b_1, ?z \mapsto :c_1\}, 2, 6) \\ (\{?x \mapsto :a_2, ?y \mapsto :b_2, ?z \mapsto :c_2\}, 4, 8)\end{matrix}\right\}.\end{aligned}$$

4.4 Event Consumption Policies

Selection policies are not sufficient to capture the behavior of EP-SPARQL in the chronological and recent settings. As described in Sect. 2, under these settings, stream items that contribute to an answer are not considered in the following evaluation iterations. We complete the model by formalizing this feature, known as *consumption policies*.

Let $ET = t_1, t_2, \ldots, t_n, \ldots$ be the set of evaluation instants. Abusing notation, we say that a window function w_j *appears* in an event pattern E, denoted by $w_j \hat{\in} E$, if E contains a basic event pattern of the form (EVENT w_j P).

Consumption policies which determine input for the evaluation will be covered next. Definition 8 is about a possible input for the evaluation while Definition 9 talks about the new incoming input. We first define such notions for a window in an RDF streaming dataset, and then lift them to the level of structures that refer to all windows in an event pattern.

Definition 8 (Potential Input and Input Structure). *Given an RDF streaming dataset SDS, we denote by* $I_i(w_j) \subseteq SDS_{w_j}(t_i)$ *a* potential input *at time* t_i *of the window identified by* w_j. *For initialization purposes, we let* $I_0(w_j) = \emptyset$.

Given an event pattern E, an input structure I_i *of E at time* t_i *is a set of potential inputs at* t_i *of all windows appearing in E, i.e.,* $I_i = \{I_i(w_j) \mid w_j \hat{\in} E\}$.

Definition 9 (Delta Input Structure). *Given an RDF streaming dataset SDS and two consecutive evaluation times* t_{i-1} *and* t_i, *where* $i > 1$, *the new triples arriving at a window* w_j *are called a* delta input, *denoted by* $\Delta_i(w_j) = SDS_{w_j}(t_i) \setminus SDS_{w_j}(t_{i-1})$. *For initialization purposes, let* $\Delta_1(w_j) = SDS_{w_j}(t_1)$.

Given an event pattern E, a delta input structure *at time* t_i *is a set of delta inputs at* t_i *of all windows appearing in E, i.e.,* $\Delta_i = \{\Delta_i(w_j) \mid w_j \hat{\in} E\}$.

We can now define consumption policies in a generic sense.

Definition 10 (Consumption Policy and Valid Input Structure). *A* consumption policy function $\mathcal{P}$ *takes an event pattern E, a time instance* $t_i \in ET$, *and a vector of additional parameters* $\mathbf{p}$ *depending on the specific policy, and produces an input structure for E.*

The resulted input structure is called valid *if it is returned by applying* $\mathcal{P}$ *on a set valid parameters* $\mathbf{p}$, *where the validity of* $\mathbf{p}$ *is defined based on each specific policy.*

This generic notion can be instantiated to realize specific policies in practice. For example, the policy $\mathcal{P}^u$ that captures the EP-SPARQL's unrestricted setting requires no further parameters, thus $\boldsymbol{p} = \emptyset$ and returns full input at evaluation time. To be more concrete:

$$\mathcal{P}^u(E, t_i) = \{I_i(w_j) = SDS_{w_j}(t_i) \mid w_j \hat{\in} E\}$$

For the chronological and recent settings, we describe here only informally the two respective functions $\mathcal{P}^c$ and $\mathcal{P}^r$. Their additional parameters include I_{i-1} (the input structure at t_{i-1}) and Δ_i (the delta input structure at t_i), and they return an input structure I_i such that its elements $I_i(w_j)$ contain $\Delta_i(w_j)$ and the triples in $I_{i-1}(w_j)$ that are not used to match E at t_{i-1}. The validity of input can be guaranteed by starting the evaluation with $I_1(w_j) = SDS_{w_j}(t_1)$ which is valid by definition. For the formal description of $\mathcal{P}^c$ and $\mathcal{P}^r$, we refer the reader to the extended version of the paper.[7]

[7] http://tinyurl.com/ekaw2016-195-ext (Hosted by Google Drive).

We now proceed to incorporate consumption policies into event patterns evaluation. The idea is to execute the evaluation function $\langle\langle . \rangle\rangle$ with a policy function $\mathcal{P}$, i.e., to evaluate an event pattern E with $\langle\langle E \rangle\rangle^{t}_{W,\mathcal{P}}$. Then, when the evaluation process reaches a BEP at leafs of the operator tree, $\mathcal{P}$ is used to filter out already consumed input. Formally:

$$\langle\langle \text{EVENT } w_j \ P \rangle\rangle^{t_i}_{W,\mathcal{P}} = [\![P]\!]_{\mathcal{I}},$$

where $\mathcal{I} = I_i(w_j) \cap (\bigcup_{(G_k,t_k)\in W\bullet W_j(S_\ell,t_i)} G_k)$ and $I_i(w_j) \in I_i = \mathcal{P}(E, t_i, I_{i-1}, \Delta_i)$.

Example 11. Similar to Example 10, one has

$$\langle\langle E_1 \ \text{SEQ}^c \ E_2 \rangle\rangle^{8}_{W^{id},\mathcal{P}^c} = \left\{ \begin{array}{l} (\{?x \mapsto :a_1, ?y \mapsto :b_1, ?z \mapsto :c_1\}, 2, 6) \\ (\{?x \mapsto :a_2, ?y \mapsto :b_2, ?z \mapsto :c_2\}, 4, 6) \end{array} \right\}.$$

Furthermore, carrying out the evaluation under the chronological policy ($\mathcal{P}^c$) will consume G_1, G_2, and G_3. Then, at time $t = 10$, there is no $(:a_1 \ :p \ :b_1)$ available to match the new coming triple $(:b_1 \ :q \ :c_1)$, and no event of the pattern E_1 SEQ^c E_2 is produced.

5 Conclusions and Outlook

The evaluation semantics of graph and event patterns presented in this paper constitutes a milestone towards defining a holistic query model for RSP that combines features from DSMS and CEP. We showed in [14] that RSP-QL, the model underlying RSEP-QL, covers the DSMS features of major RSP languages, and in this work, we introduced the CEP features. Moreover, RSEP-QL models both event patterns and their evaluation semantics taking into account the presence of selection and consumption policies. These policies are key to determine the answer that a query should produce for a given input stream. Thus, it is not possible to consider them as only technical/implementation related.

We have also shown that RSEP-QL complies with the set of requirements described in Sect. 2. First, it processes RDF graph-based streams [R1]. It is also capable of capturing the DSMS features of representative RSP languages [R2], as an inheritance from the expressivity of RSP-QL. Moreover, RSEP-QL captures the behavior of the sequential event pattern matching features of EP-SPARQL and C-SPARQL [R3], including the different selection and consumption policies that they provide. Table 1 shows the

Table 1. Coverage of DSMS/CEP features of RSEP-QL compared to EP-SPARQL and C-SPARQL

RSEP-QL	EP-SPARQL/C-SPARQL
W^λ + SEQ	EP-SPARQL unrestricted
W^λ + SEQ^c + $\mathcal{P}^c$	EP-SPARQL chronological
W^λ + SEQ^r + $\mathcal{P}^r$	EP-SPARQL recent
W^τ + SEQ^n	C-SPARQL SEQ (timestamp)
W^τ	C-SPARQL time-window

equivalence of the main features in RSEP-QL with their counterparts in EP-SPARQL and C-SPARQL. For instance, one can observe that an EP-SPARQL sequence pattern (with recent policy) can be captured by the SEQ^r operator and the $\mathcal{P}^r$ function on a landmark window in RSEP-QL.

Our formalization is able to capture a rich set of operators including time-based sliding windows and event patterns such as sequencing, and combines them. As a result, RSEP-QL offers expressivity beyond the capabilities of current RSPs. For example, RSEP-QL allows to define event patterns over more than one streams, e.g., given E_1 SEQ E_2, E_1 and E_2 can match over different streams. It is not possible to express this with an EP-SPARQL or C-SPARQL query, as the first operates on a unique stream, while the latter merges different input streams in a unique one.

Furthermore, the expressivity of RSEP-QL allows defining complex queries that combine both windows and event patterns. For instance, consider that in a social network we want to find the post made by a user that is then followed by a popular user, defined as someone that gets a lot of mentions in the last hour and has a lot of followers. In this case, a time window is needed to keep track of the number of mentions in the last hour. Then the sequence pattern is required to capture the fact that someone is followed after he made a post. The contextual information is used to look for the number of followers of a person, to determine if he is popular. Another example consists in enriching the event pattern matching with information from contextual streaming data and other streams.

Future works include enriching RSEP-QL with more CEP operators, e.g., DURING and NOT, and realizing other selection and consumption policies in CEP, e.g., *strict/partition contiguity*, *skip till next match*, and *skip till any match* [1] in RSEP-QL.

Another important aspect of this work is its compatibility with alternative data models. Even though we chose a particular model based on timestamped graphs, one can see that it can be converted, or in some case, extended if necessary, to other similar models. For example, data streams with interval timestamps can be easily incorporated into the event pattern evaluation semantics. Finally, the RSEP-QL model can also be helpful for the RSP community, as it provides the most comprehensive query processing model for RDF streams so far. We plan to align our model to the latest proposals of the W3C RSP group, as well as study how it can be adapted for the different profiles proposed in the RSP abstract model.

References

1. Agrawal, J., Diao, Y., Gyllstrom, D., Immerman, N.: Efficient pattern matching over event streams. In: SIGMOD, pp. 147–160 (2008)
2. Anicic, D.: Event processing and stream reasoning with ETALIS. Ph.D. thesis, Karlsruhe Institute of Technology (2011)
3. Anicic, D., Fodor, P., Rudolph, S., Stojanovic, N.: EP-SPARQL: a unified language for event processing and stream reasoning. In: WWW, pp. 635–644 (2011)

4. Arasu, A., Babu, S., Widom, J.: The CQL continuous query language: semantic foundations and query execution. VLDB J. **15**(2), 121–142 (2006)
5. Babcock, B., Babu, S., Datar, M., Motwani, R., Widom, J.: Models and issues in data stream systems. In: PODS, pp. 1–16. ACM (2002)
6. Barbieri, D.F., Braga, D., Ceri, S., Della Valle, E., Grossniklaus, M.: C-SPARQL: a continuous query language for RDF data streams. Int. J. Semant. Comput. **4**(1), 3–25 (2010)
7. Beck, H., Dao-Tran, M., Eiter, T., Fink, M.: LARS: a logic-based framework for analyzing reasoning over streams. In: AAAI, pp. 1431–1438 (2015)
8. Botan, I., Derakhshan, R., Dindar, N., Haas, L.M., Miller, R.J., Tatbul, N.: SECRET: a model for analysis of the execution semantics of stream processing systems. PVLDB **3**(1), 232–243 (2010)
9. Calbimonte, J.P., Jeung, H., Corcho, Ó., Aberer, K.: Enabling query technologies for the semantic sensor web. Int. J. Semant. Web Inf. Syst. **8**(1), 43–63 (2012)
10. Cugola, G., Margara, A.: Processing flows of information: from data stream to complex event processing. ACM Comput. Surv. **44**(3), 15:1–15:62 (2011)
11. Dao-Tran, M., Beck, H., Eiter, T.: Contrasting RDF stream processing semantics. In: Qi, G., Kozaki, K., Pan, J.Z., Yu, S. (eds.) JIST 2015. LNCS, vol. 9544, pp. 289–298. Springer, Heidelberg (2016). doi:10.1007/978-3-319-31676-5_21
12. Dao-Tran, M., Le-Phuoc, D.: Towards enriching CQELS with complex event processing and path navigation. In: HiDeSt, pp. 2–14 (2015)
13. Dell'Aglio, D., Balduini, M., Della Valle, E.: On the need to include functional testing in RDF stream engine benchmarks. In: 10th ESWC 2013 Conference Workshops: BeRSys 2013, AImWD 2013 and USEWOD 2013 (2013)
14. Dell'Aglio, D., Valle, E.D., Calbimonte, J., Corcho, O.: RSP-QL semantics: a unifying query model to explain heterogeneity of RDF stream processing systems. Int. J. Semant. Web Inf. Syst. **10**(4), 17–44 (2014)
15. Gutierrez, C., Hurtado, C., Vaisman, A.: Introducing time into RDF. IEEE Trans. Knowl. Data Eng. **19**(2), 207–218 (2007)
16. Harris, S., Seaborne, A.: SPARQL 1.1 Query Language (2013). http://www.w3.org/TR/sparql11-query/
17. Komazec, S., Cerri, D., Fensel, D.: Sparkwave: continuous schema-enhanced pattern matching over RDF data streams. In: DEBS, pp. 58–68 (2012)
18. Luckham, D.C.: The power of events - an introduction to complex event processing in distributed enterprise systems. ACM (2005)
19. Le-Phuoc, D., Dao-Tran, M., Xavier Parreira, J., Hauswirth, M.: A native and adaptive approach for unified processing of linked streams and linked data. In: Aroyo, L., Welty, C., Alani, H., Taylor, J., Bernstein, A., Kagal, L., Noy, N., Blomqvist, E. (eds.) ISWC 2011, Part I. LNCS, vol. 7031, pp. 370–388. Springer, Heidelberg (2011)
20. Rinne, M., Törmä, S., Nuutila, E.: SPARQL-based applications for RDF-encoded sensor data. In: SSN, vol. 904, pp. 81–96 (2012)

TAIPAN: Automatic Property Mapping for Tabular Data

Ivan Ermilov[1,2(✉)] and Axel-Cyrille Ngonga Ngomo[1,2]

[1] University of Leipzig, Institute of Computer Science, Leipzig, Germany
{iermilov,ngonga}@informatik.uni-leipzig.de
[2] AKSW Research Group, Leipzig, Germany
http://aksw.org/

Abstract. The Web encompasses a significant amount of knowledge hidden in entity-attributes tables. Bridging the gap between these tables and the Web of Data thus has the potential to facilitate a large number of applications, including the augmentation of knowledge bases from tables, the search for related tables and the completion of tables using knowledge bases. Computing such bridges is impeded by the poor accuracy of automatic property mapping, the lack of approaches for the discovery of subject columns and the mere size of table corpora. We propose TAIPAN, a novel approach for recovering the semantics of tables. Our approach begins by identifying subject columns using a combination of structural and semantic features. It then maps binary relations inside a table to predicates from a given knowledge base. Therewith, our solution supports both the tasks of table expansion and knowledge base augmentation. We evaluate our approach on a table dataset generated from real RDF data and a manually curated version of the T2D gold standard. Our results suggest that we outperform the state of the art by up to 85 % F-measure.

Keywords: Web tables · Knowledge base augmentation · Table expansion

1 Introduction

The Linked Data Web has developed from a mere idea to a set of more than 85 billion facts distributed across more than 10,000 knowledge bases[1] over less than 10 years. However, the Document Web is also growing exponentially, with a large proportion of the information contained therein not being available on the Data Web. Consequently, the gap between the Data Web and the Document Web keeps on growing with the addition of novel knowledge in either portion of the Web. Devising ways to bridge between the Document Web and the Linked Data Web has been the purpose of a number of works from different domains. The unstructured data on the Web is being transformed to RDF by means of

[1] http://lodstats.aksw.org.

E. Blomqvist et al. (Eds.): EKAW 2016, LNAI 10024, pp. 163–179, 2016.
DOI: 10.1007/978-3-319-49004-5_11

a combination of named entity recognition (see, e.g., [5,14,19]), entity linking (see, e.g., [2,22]) and relation extraction (see, e.g., [6,15]) approaches. However, such approaches can only deal with well-formed sentences and do not address other structures that are commonly found on the Document Web, in particular, tables. While a few approaches for disambiguating entities in tables have been developed in the past [1,23–25,27], porting the content of tables to RDF has been the subject of a limited number of approaches [11,13,16]. These approaches are however limited in the structure of the tables they can handle. For example, they partly rely on heuristics such as using the first non-numeric column of a table as subject for the triples that are to be extracted [10].

We present TAIPAN, a generic approach towards extracting RDF triples from tables. Given a table and a reference knowledge base, TAIPAN aims to (1) identify the column that contains the subject of the triples to extract, i.e., the *subject column*. To this end, our approach relies on maximizing the likelihood that the elements of a column (a) all belong to the same class and, (b) once disambiguated, will actually have property values that correspond to the properties found in the table; (2) detect properties that correspond to the columns of the tables. Here, TAIPAN maximizes the probability that the columns of the table will yield property values for the same property given the assumed assignment of the subject column; (3) facilitate the disambiguation and extraction of RDF from tables. Hence, the results of TAIPAN can be used to feed any entity disambiguation system for tables.

The rest of this paper is structured as follows: in Sect. 2 we describe our conceptual framework. Then, we employ this framework to define the problem tackled by TAIPAN formally (see Sect. 3). Thereafter, we use the same notation to explain our approach (see Sect. 4). We clarify implementation details in Sect. 5. In Sect. 6, we evaluate our approach on a manually curated portion of the T2D benchmark (which we dub T2D*) against the approaches proposed in [24] and [16,17]. In particular, we measure the accuracy of our subject column identification approach as well as the F-measure achieved by our property mapping approach. Section 7 gives an overview of related work and Sect. 8 concludes the paper.

2 Preliminary Definitions

In this section, we introduce the notation and definitions required to formalize the subject column identification and property mapping problems.

2.1 Tabular Data Model

For modeling tabular data we extend the canonical table model described in [4]. Essentially, the canonical table model distinguishes between the *header* of a table and the *data* of the same table (see Fig. 1). A table is represented as a tuple, where the header is a vector and the data is a matrix.

Definition 1. *A* table $T = (H, D)$ *is a tuple consisting of a* header H *and* data D, *where:*

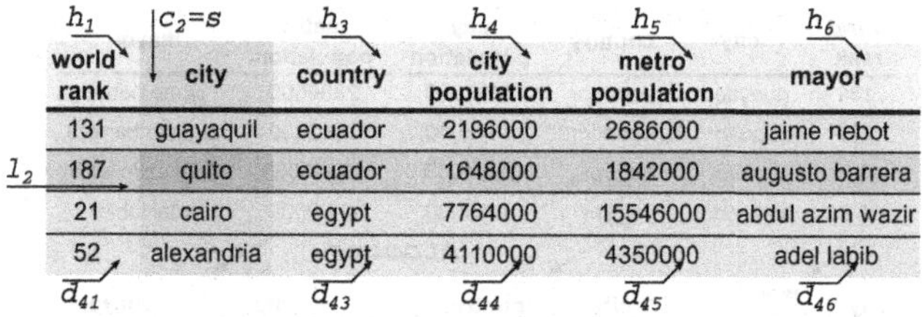

world rank	city	country	city population	metro population	mayor
131	guayaquil	ecuador	2196000	2686000	jaime nebot
187	quito	ecuador	1648000	1842000	augusto barrera
21	cairo	egypt	7764000	15546000	abdul azim wazir
52	alexandria	egypt	4110000	4350000	adel labib

Fig. 1. An example of a table from T2D gold standard with semantics from our table model

- *the header* $H = (h_1, h_2, \ldots, h_n)$ *is a vector of size* n *which contains* header elements h_i.
- *the data* $D = \begin{pmatrix} d_{1,1} & d_{1,2} & \cdots & d_{1,n} \\ d_{2,1} & d_{2,2} & \cdots & d_{2,n} \\ \vdots & \vdots & \ddots & \vdots \end{pmatrix}$ *is a* (m, n)*-matrix consisting of* n *columns and* m *rows.*

Consequently, we introduce the concept of table projections, where the data of a table is represented as a one-dimensional vector of value vectors.

Definition 2. *The* column projection *of a* table $T = (H, D)$ *is a table* $col(T) = (H, col(D))$, *where* $col(D) = (c_1, c_2, \ldots, c_n)$, *with* $c_n = (d_{1,n}, d_{2,n}, \ldots, d_{m,n})$. *Similarly, the row projection of a table* $row(T) = (H, row(D))$ *where* $row(D) = (l_1, l_2, \ldots, l_m)$, *with* $l_m = (d_{m,1}, d_{m,2}, \ldots, d_{m,n})$.

Hereafter, we will commonly work with the row projections of tables.

Informally, the *subject column* of a table T is a column that contains labels of resources that instantiate the main subject of a table. For instance, in a table taken from the T2D reference dataset [16] with the header H = (world rank, city, country, city population, metro population, mayor) (see Fig. 1), the main subject is city. Consequently, the second column is the subject column. In general, we assume that the subject column is to be connected to every other column in the reference table via a binary relation. Hence, we adopt the following functional definition:

Definition 3. *The subject column* s *is a column which divides table* T *into* $(n-1)$ *two-column tables (which we dub* ***atomic tables****), where the binary relation* ρ_i *between* s *and each of the other columns* c_i *in* T *corresponds to a property in a reference knowledge base* K *(e.g., see Fig. 2).*

Following the Definition 3, we define an atomic table as follows:

Definition 4. *An atomic table is a table* $T_i' = (H_i', D_i')$ *such as* $H_n' = (h_s, h_i)$ *and* $col(D_i') = (s, c_i)$, *where* h_s *is a header item of the subject column and* s *is a subject column.*

For example, in Figure 2, for the left-most atomic table $T_1' = (H_1', D_1')$, the header is $H_1' = $ (*city, world rank*). The column projection consists of subject column and the first column of the source table: $col(D_1') = (s, c_1)$, where $s = $ (guayaquil, quito, cairo, alexandria) and $c_1 = (131, 187, 21, 51)$.

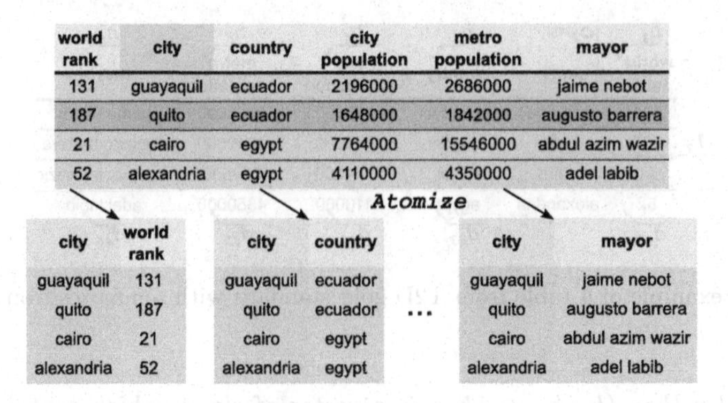

Fig. 2. Example of a table atomization

2.2 Knowledge Base Model

We now introduce the knowledge base model (derived from [4]) underlying our work. Let $\mathcal{U}$ be the set of all URIs, $\mathcal{B}$ be the set of all blank nodes, $\mathcal{L}$ be the set of all literals and Γ be the set of all *RDF terms* with $\Gamma = \mathcal{U} \cup \mathcal{B} \cup \mathcal{L}$. Furthermore, we make use of the following notions:

- $\mathcal{S}$ is the set of RDF subjects with $\mathcal{S} \subseteq \mathcal{U} \cup \mathcal{B}$,
- $\mathcal{R}$ is the set of RDF properties (relations) with $\mathcal{R} \subseteq \mathcal{U}$,
- $\mathcal{O}$ is the set of RDF objects, with $\mathcal{O} \subseteq \Gamma$,
- Π is the set of all *triples*, defined as $\Pi \subseteq \mathcal{S} \times \mathcal{R} \times \mathcal{O}$,
- $\mathbb{E}$ is the set of all *entities*, and
- $\mathcal{C}$ is the set of all classes, that is the subset of $\mathcal{U}$, which describes the classes of the entities $\mathbb{E}$ in Π.

Our basic assumption is that a binary relation between columns of a table can correspond to a property inside a knowledge base.

3 Problem Statement

TAIPAN aims to expose the semantics of tabular data. To this end, we address the following two subproblems.

3.1 Problem 1: Subject Column Identification

The problem of subject column identification can be formalized using previously introduced concepts as follows.

Problem 1. Given a table $col(T) = (H, col(D))$, where $col(D) = (c_1, c_2, \ldots, c_n)$, find a column c_i such that c_i satisfies Definition 3, i.e., such that $col(T)$ can be split into atomic tables which express the extension of a property $r \in \mathcal{R}$ or the inverse r^{-1} of such a property.

The subject column identification is an important preprocessing step, which has to be performed with the highest precision possible. Failing to identify subject column will lead to erroneous atomic tables and thus to less information being ported from T to the reference knowledge K. For example, for a correctly identified subject column $c_i = s$ dubbed `city` (see Fig. 1), the binary relation ρ_i between "cairo" and "abdul azim wazir" (i.e. ρ_i(*"cairo"*, *"abdul azim wazir"*)) can be mapped to a knowledge base such as DBpedia, where ρ_i corresponds to `dbo:mayor` property. Another important consequence of subject column identification is the possibility to decompose table into atomic tables.

3.2 Problem 2: Property Mapping

The property mapping of a table can be defined as a function λ, such as for each binary relation $\rho_i : s \rightarrow c_i$ between the subject column s and every other column of a table, it assigns a property inside a knowledge base. Therefore, for each ρ_i we have to find a mapping to a particular $r \in \mathcal{R}$. We denoted this mapping by λ and write $\lambda(\rho_i) = r$.

As table semantics are ambiguous, we cannot determine the definite correspondence between a binary relation in a table and a property inside a knowledge base. Moreover, a single binary relation can be mapped to several properties. However, relational tables are likely to have functional binary dependencies, which are mapped to particular functional properties inside a knowledge base. Therefore, given a single binary relation between columns and for each property $r \in \mathcal{R}$, we can define the probability of r being the correct binary relation ρ_i. We denote this probability $P(\lambda(\rho_i) = r)$. The problem at hand can now be reduce to finding the best mapping function λ, i.e., the λ that maximizes $P(\lambda(\rho_i) = r)$ for all ρ_i.

Problem 2. Given a table $col(T) = (H, col(D))$, where $col(D) = (c_1, c_2, \ldots, c_n)$ and $c_k = s$, find a mapping function λ, which maximizes the probability of having mapped each $\rho_i : s \rightarrow c_i$ to the correct $r_j \in \mathcal{R}$.

Note that by these means, we reduce the two tasks to the same core problem formulation. In the following, we will use this formulation to derive approaches for addressing the two problems at hand.

4 Approach

In this section we describe our solutions to the subject column identification and property mapping problems.

4.1 Subject Column Identification

To support column identification we extend an idea from distant supervision learning [12,18]. Essentially, we boil down the column identification to finding the column c_i in a table that has the most relations to other columns inside the same table. To find such a column, we begin by selecting m' rows of the given table T. Then, for each row, we disambiguate cell values against entities from a given reference knowledge base. Finally, we apply three triple patterns to find potential relations between each combination of columns. The approach derives two important features for each column: *support* and *connectivity*.

Definition 5. *The* support St_i *of a column* c_i *in a table* T *is the ratio between cells with disambiguated entities inside and total number of cells for a column.* $St_i = \frac{\sum_{j=1}^{|row(D)|} e_j}{|row(D)|}$, *where*

$$e_j = \begin{cases} 1, & \text{if } d_{ij} \text{ could be disambiguated to some } e \in \mathbb{E} \\ 0, & \text{otherwise} \end{cases} \quad (1)$$

Definition 6. *The* connectivity C_i *of a column* c_i *in a table* T *is the ratio between number of connections (i.e., properties) of the column to other columns inside the same table to the total number of columns.*

In our implementation, we evaluated the *support* of a particular column by using *AGDISTIS* [21] to disambiguate the entries d_{ij} (disambiguateEntities on line 4 in Algorithm 1) and used DBpedia as reference knowledge base. For example, given the table in Fig. 1, the entry d_{22} = `quito` was disambiguated as http://dbpedia.org/resource/Quito. All entities in the columns c_2, c_3 and c_6 of the example table could be disambiguated. Hence, their support is $\frac{4}{4} = 1$. In contrast, all numerical columns have support of 0. Our approach towards computing the support of all columns in a table is shown in Algorithm 1.

Algorithm 1. TAIPAN Column Support Evaluation. Runs in $\mathcal{O}(m'n)$ time.

```
   Data: Table T of size (m, n), m'
   Result: St - support vector for the table columns, Et - entity matrix
 1 Instantiate St, Et;
 2 for row = 1 to m' do
 3 |   for col = 1 to n do
 4 |   |   Et[row][col] ⟵ disambiguateEntities(T[row][col]);
 5 |   |   if |Et[row][col]| > 0 then
 6 |   |   |   St[col] ← St[col] + 1
 7 |   |   end
 8 |   end
 9 end
10 for col = 1 to n do
11 |   St[col] = St[col]/m' · 100 %
12 end
13 return St, Et
```

After the disambiguation, we now employ a set of triple patterns to find potential properties in a knowledge base as follows.

```
<%value> ?property <%value>
```

Listing 1.1. Entity-Entity Triple Pattern (1)

```
<%value> ?property "%value"@en
```

Listing 1.2. Entity-Literal Triple Pattern (2a)

```
<%value> ?property ?o .
FILTER regex(?o, ".*%value.*", "i")
```

Listing 1.3. Regex Entity-Literal Triple Pattern (2b)

These patterns are a heuristic mean to determine the set of potential properties between pairs of columns. To this end, we combine the results of the disambiguation step with the original cell values (for entries that could not be disambiguated). Correspondingly, `%value` is instantiated by using either the disambiguated entity from a column value (patterns 1 and 2a-b) or a column value itself (patterns 2a-b). For instance, to find relations between *city* and *city population* in our example, given that *quito* was disambiguated and 1648000 not, the triple patterns (2a-b) are used. In this case triple pattern (2b) will be instantianed as follows.

```
PREFIX dbpedia: <http://dbpedia.org/resource/>
dbpedia:Quito ?property ?o .
FILTER regex(?o, ".*1648000.*", "i")
```

Listing 1.4. Example of TP (2b) with instantiated variables

The retrieved properties from triple patterns are stored in a *connectivity tensor* of order 3 and of dimensions $m' \times n \times n$ (m' is the sample size for rows and stands for the number of rows used in the Algorithm 1 as disambiguated entities are used in the triple patterns). Each entry Cn_{ijk} contains the set of properties that were detected by the approach above for the pair of column entries d_{ij} and d_{ik}. The connectivity C_j of a column c_j can be inferred from Cn as follows:

$$C_j = \frac{\sum_{i=1}^{|row(D)|} \frac{\sum_{k=1}^{|col(D)|} |Cn_{ijk}|}{|col(D)|}}{|row(D)|}. \quad (2)$$

The evaluation of *connectivity tensor* is shown in Algorithm 2.

Algorithm 2. TAIPAN Column Connectivity Tensor Evaluation. Runs in $\mathcal{O}(mn^2)$ time.

Data: Table T of size (m, n), entity matrix Et, m'
Result: Cn, connectivity matrix for table T
1 Instantiate Cn;
2 **for** $row = 1$ *to* m' **do**
3 | **for** $col = 1$ *to* n **do**
4 | | **for** $otherCol = col + 1$ *to* n **do**
5 | | | Cn[row][col][otherCol], Cn[row][otherCol][col] ⟵ findRelation($T[row][col]$, $T[row][otherCol]$, Et)
6 | | **end**
7 | **end**
8 **end**
9 **return** Cn

For example, the connectivity of column `country` of our running example (see Fig. 1) can be evaluated as: $C_3 = \frac{\sum_{i=1}^{4} \frac{\sum_{k=1}^{6} |Cn_{i3k}|}{6}}{4}$.

$$Cn_{i3k} = \begin{pmatrix} \emptyset & country & \emptyset & populationTotal & \emptyset & citizen, official \\ \emptyset & country & \emptyset & \emptyset & \emptyset & \emptyset \\ \emptyset & \emptyset & \emptyset & \emptyset & populationTotal & citizen, official \\ \emptyset & country & \emptyset & \emptyset & \emptyset & \emptyset \end{pmatrix} \quad (3)$$

Given Cn_{i3k} as in Eq. 3, the connectivity evaluates to $C_3 = 0.375$.

After characterizing columns by means of their support and connectivity scores, we can use binary classifiers to classify columns of a table as being either subject columns or not. Binary classifiers used in the experiments as well as discussion on their performance are described in Sect. 6.2.

4.2 Property Mapping

In this section we describe our approach to find an adequate mapping function λ. Our approach assumes that a subject column has already been identified. As a first step, we take the header $H = (h_1, h_2, \ldots, h_n)$ of the input table T and for each element h_i retrieve seed properties from a reference set of potential properties. Then, the set of seed properties is ranked according to the property frequency inside the reference knowledge base K.

Given an identified subject column, a table of size (m, n) is atomized into $(n-1)$ two-column tables $T'_i = (H'_i, D'_i)$. Each atomic table represents exactly one binary relation ρ_i, which should have a correspondence to a property $r_j \in \mathcal{R}$ inside a knowledge base. For example, table shown in Fig. 1 is decomposed as shown in Fig. 2.

While connectivity performs well to identify subject column of a table, the connectivity tensor (i.e. properties found by triple patterns) does not contain the target properties from a knowledge base. Therefore, for each element h_i we retrieve seed properties in addition to properties extracted via triple patterns. To retrieve seed properties from a knowledge base, we perform a look up on an index created from the values of `rdfs:label` and `rdfs:comment`. This index is queried with the values of the table header such as $h_3 = country$.

To rank the properties, we employ a probabilistic model. The probability of a relation ρ_i for an atomic table $T_i' = (H_i', D_i')$ to map to a property r_j is defined as follows:

Definition 7. *A probability of relation ρ_i to correspond to property r_j equals to a number of pairs (s_m, d_{mi}) corresponding to property r_j divided by size of a table:* $P(\lambda(\rho_i) = r_j) = \frac{\sum_{m=1}^{|row(D)|} |(s_m, d_{mi}) \in r_j|}{|row(D)|}$.

For example, for the atomic table shown in Fig. 2 we would retrieve two properties from DBpedia knowledge base such as: *dbo:country* and *dbo:largestCity*. Let us assume the following knowledge base for the sake of simplicity:

City	dbo:country	dbo:largestCity
Guayaquil	Equador	Equador
London	UK	UK
Cairo	Egypt	Egypt
Alexandria	Egypt	N/A

We can calculate probabilities for the properties as: $P(h_3 = dbo : country) = \frac{3}{4}$, $P(h_3 = dbo : largestCity) = \frac{2}{4}$.

The property with the highest probability as defined in Definition 7 would be selected, i.e. `dbo:country`.

5 Implementation Details

In the implementation, we use DBpedia as a reference knowledge base. The properties are retrieved from DBpedia with triple patterns as well as from LOV.[2] LOV maintains a reverse index of classes and properties from different ontologies based on `rdfs:label` and `rdfs:comment`. The property ranking is performed as described in Sect. 4.2. For the property lookups LOV returns a score which quantify the relevance of each result. The score is based on TF/IDF and field norms.[3] To improve the precision of TAIPAN, we introduce a score threshold (i.e., we only accept properties which have a score higher than the specified threshold as candidates). As we can see in Fig. 3, the best performance is achieved when the threshold is set to 0.8, which the value we use throughout our experiments.

[2] http://lov.okfn.org/.

[3] https://www.elastic.co/guide/en/elasticsearch/guide/current/scoring-theory.html.

6 Experiments and Results

The goal of our experiments was to measure how well our column identification and our property mapping approaches perform. Hence, we compared the recall and precision achieved by our approach with that of the approaches presented in [24] (subject column identification) and [16] (property mapping). To the best of our knowledge, these are the best performing approaches on these tasks at the moment. The data used in our experiments and the source code of TAIPAN and the annotation interfaces used to curate T2D are available on Github.[4]

6.1 Experimental Setup

Hardware. The T2K algorithm [16] requires at least 100 GB RAM to run. Therefore, the experiments for T2K algorithm were performed on a virtual machine running Ubuntu 14.04 with 128 GB RAM and 4 CPU cores. All experiments with TAIPAN were evaluated on an Ubuntu 14.04 machine with 4 cores i7-2720QM CPU and 16 GB RAM.

Gold Standard. We aimed to use T2D entity-level Gold Standard (T2D), a reference dataset which consists of 1 748 tables and reflects the actual distribution of the data in the Common Crawl,[5] to evaluate our algorithms. However, the analysis of T2D showed a substantial amount of annotation mistakes such as[6]:

- Tables containing data about `dbo:Plant`, `dbo:Hospital` instances are annotated with the class `owl:Thing`.
- `rdfs:label` is used in an inflationary manner. For example, both first and last name of persons are marked as `rdfs:label`.
- Columns with country names is annotated with `dbo:collectionSize`.
- Columns with active drug ingredients is annotated with `dbo:commonName`.

It is noticeable, that T2D contains 978 tables annotated with owl:Thing class. An analysis of a random sample (50) of the tables from these 978 showed that all of them contain annotation mistakes.

To address T2D annotation problems, we asked expert users to annotate both subject columns and DBpedia properties. For the subject column identification annotation task, we had 15 expert users annotate 322 randomly picked tables from T2D with 2 annotators per table. We discarded the tables where the experts did not agree. As a result, the 116 tables that (1) had no subject column at all (4 tables) and (2) which possessed a subject column upon which the experts agreed (112 tables) were included into our manually curated dataset, which we dub T2D*. To assess the quality of T2D*, we calculated the F-measure achieved

[4] https://github.com/aksw/taipan.
[5] http://webdatacommons.org/webtables/goldstandard.html.
[6] For a complete analysis, see https://github.com/AKSW/TAIPAN-Datasets/tree/master/T2D.

by each annotator as proposed in [7]: $F = \frac{2 \cdot 116}{2 \cdot 116 + (322 - 116)} = 0.53$. According to [9], the interval (0.41, 0.60) represents moderate agreement strength. This hints at how difficult the problem at hand really is.

For the property annotation, we involved 12 Semantic Web experts. All experts were experienced DBpedia users or contributors. Each user annotated 20 tables (2 annotators per table). However, to reduce the time per annotation, we also displayed property suggestions from the LOV search engine. On average, each user spent approximately 30 min to complete the task. Out of 116 annotated tables, 90 (77.5 %) tables had properties upon which the experts agreed. Moreover, the experts agreed on 236 (53.5 %) properties for the 441 columns we considered in T2D* (subject columns excluded). Out of 236 annotated properties, the experts identified 104 (44 %) properties from DBpedia. The F-measure for the property annotation task is defined as $F = \frac{2 \cdot 236}{2 \cdot 236 + (441 - 236)} = 0.70$. According to [9] (0.61, 0.80) interval represents substantial agreement strength. Note that we shuffled the positions of the columns in the T2D* dataset randomly as in real-life scenarios the subject column can be at any position in a table (in contrast to most tables in T2D). The same holds for the subsequent dataset.

DBpedia Table Dataset (DBD). We also evaluated TAIPAN using a dataset generated directly from DBpedia concise bounded descriptions[7] (CBDs) dubbed **DBD**. We selected 200 random classes with at least 100 CBDs in each class. For each class, we generated 5 tables with 20 rows each (i.e. using 20 CBDs). Inside a table, each row corresponds to a CBD. The subject column was assigned the header `label` and contained the `rdfs:label` of the resource whose CBD was described by the row at hand. The headers of all other columns were values of `rdfs:label` of corresponding properties. The values of the columns are the values of corresponding properties. We selected only direct property/value pairs for CBDs, ignoring blank nodes. The resulting dataset contains 1 000 tables. The implementation of the data generator[8] as well as the DBD[9] are available on Github.

Training and Testing. Given that one usually only has a small number of annotated tables to train an extraction approach, we opted to use an inverse 10-fold cross-validation to evaluate TAIPAN. This means that each dataset was subdivided into 10 folds of the same size. 10 experiments were then ran, within which one fold was used for training and the 9 other folds for testing.

[7] https://www.w3.org/Submission/CBD/.
[8] https://github.com/aksw/TAIPAN-DBD-Datagen.
[9] https://github.com/AKSW/TAIPAN-Synth-Datagen/tree/master/DBpediaTableDataset/tables.

Table 1. Accuracy for subject column identification. Evaluation of support and connectivity features

	Rule-based	Support	Connectivity	Support-connectivity
T2D*	51.72 %	54.31 %	36.00 %	56.89 %
DBD	52.20 %	90.80 %	80.00 %	84.40 %

6.2 Subject Column Identification

According to [24], a simple rule-based approach (pick the left-most column which is not a number or date) for subject column identification achieves 83 % accuracy[10], while an SVM with an RBF kernel with the following 5 features increases accuracy up to 94 %: (1) fraction of cells with unique content, (2) fraction of cells with numeric content, (3) variance in the number of date tokens in each cell, (4) average number of words in each cell, and (5) column index from the left.

We recreated the experiment on T2D* and DBD. Our experiments (see Table 1) show that for T2D*, the rule-based approach (the baseline) achieves only 51.72 % accuracy, while the SVM proposed in [24] achieves 49.52 % accuracy in an inverse ten-fold cross-validation. Note that this performance is different from stipulated by the authors on their corpus.[11] On the other hand, selecting the column that achieves the highest support (see Table 1) already performs by 5.17 % better than the rule-based baseline. While selecting a column based on connectivity alone performs much worse than baseline, a linear combination of the support and connectivity features $\alpha \cdot St_i + (1 - \alpha) \cdot C_i$ with $\alpha = 0.3$ achieves further gain over the baseline (6.04 %).

In an effort to check whether more complex models would lead to even better results, we evaluated TAIPAN feature set with 7 different classifiers (see Table 2).[12] TAIPAN feature set includes all the features proposed by [24] with addition of connectivity and support. For T2D*, the best performing method for TAIPAN was based on SVM. This method achieves 80.74 % accuracy in an inverse tenfold cross validation and thus achieves 29.02 % gain over the baseline. The further experiments for DBD dataset showed that decision tree classifier performs the best on average for both T2D* and DBD. As a result, we selected decision tree classifier to be default setting for TAIPAN.

[10] Accuracy is defined as a ratio of correctly guessed subject columns to a number of overall guessed subject columns.

[11] We contacted the authors to obtain their corpus but were not provided access to it. Still, we followed the specification of the SVM in their paper exactly.

[12] We used the classifier implementations from scikit-learn python library at http://scikit-learn.org/. For more information on the implementation, please refer to the TAIPAN Github repository at https://github.com/AKSW/TAIPAN.

Table 2. Accuracy for subject column identification. TAIPAN

	T2D*	DBD
SVM	$(80.74 \pm 9.17)\%$	$(69.64 \pm 19.91)\%$
KNeighbors	$(36.94 \pm 15.17)\%$	$(87.36 \pm 3.37)\%$
SGD	$(34.29 \pm 30.69)\%$	$(39.69 \pm 22.46)\%$
Decision tree	$(\mathbf{72.59} \pm \mathbf{15.04})\%$	$(\mathbf{79.50} \pm \mathbf{5.76})\%$
Gradient boosting	$(75.77 \pm 11.93)\%$	$(67.35 \pm 2.29)\%$
Nearest centroid	$(51.11 \pm 9.84)\%$	$(59.19 \pm 4.09)\%$
SGD (perceptron loss function)	$(37.25 \pm 27.84)\%$	$(29.63 \pm 19.88)\%$

6.3 Property Mapping

We evaluated TAIPAN using our T2D* and DBD by comparing it with the state-of-the-art solution for table to knowledge base mapping T2K described in [16,17]. T2K is open-source and available online.[13] We do not compare T2K to TAIPAN on T2D due to substantial amount of annotation mistakes in T2D (see Sect. 6.1).

Table 3. Recall, precision and F-measure of TAIPAN and T2K algorithm

	T2D*			DBD		
	Recall	Precision	F-measure	Recall	Precision	F-measure
TAIPAN	72.12 %	39.27 %	50.85 %	84.31 %	86.01 %	85.15 %
T2K	36.54 %	48.72 %	41.76 %	0.002 %	0.002 %	0.002 %

We calculated the recall achieved by the approaches as the number of correctly mapped properties divided by the number of properties in a gold standard. The precision was computed as the number of correctly mapped properties divided by total number of mapped properties.

The results achieved by both approaches are shown in Table 3. For T2D*, T2K has a 9.5 % better precision than TAIPAN. However, TAIPAN achieves a 36 % better recall, hence outperforming T2K by 9 % F-measure. An error analysis of TAIPAN suggests that the 39 % precision it achieves can be improved significantly by enhancing the ranking of properties with heuristics from the whole table corpus and not only using the information available in a single table. For example, given the frequency of the header `Anglican Church` inside the data corpus `Frequency(''Anglican Church'') = 1`, it is possible that this property is not available in the reference knowledge base.

[13] http://dws.informatik.uni-mannheim.de/en/research/T2K.

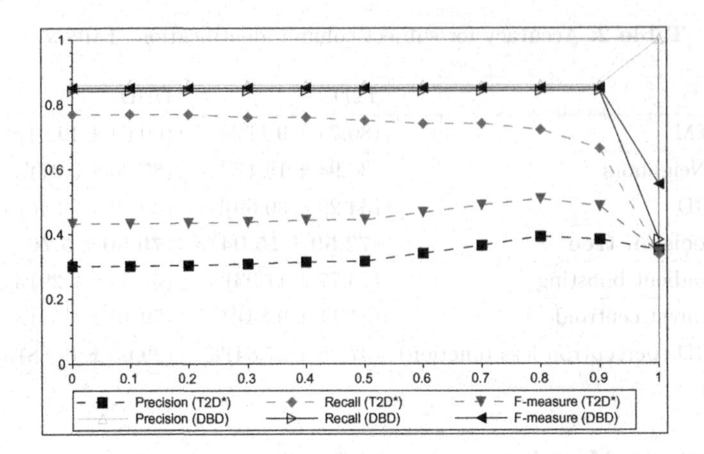

Fig. 3. Recall, precision and F-measure of TAIPAN as a function of a score threshold

For DBD, T2K could only match 6 columns correctly, resulting in under 1 % F-measure. TAIPAN achieved 85.15 % F-measure, significantly outperforming T2K. TAIPAN does not achieve a perfect property mapping because the DBD dataset contains columns homonymous columns from two different namespace, i.e., the ontology and the property namespace (for example, http://dbpedia.org/property/birthDate and http://dbpedia.org/ontology/birthDate). Overall, our results suggest that TAIPAN outperforms the state of the art significantly in both subject column identification and property mapping.

7 Related Work

In this paper, we focus on the problem of automatic mapping of web tables to ontologies. Semi-automatic and manual approaches, which rely on user input (e.g. [3,8]) as well as ontology alignment (e.g. [20]) are out of scope of this paper. Research on the topic of web tables is mostly carried out by two communities: Researchers from major search engines and researchers involved in open projects such as Common Crawl[14] and Web Data Commons[15]. A significant portion of the related work on web tables is enlisted on the Web Data Commons web site.[16] In general, WDC identified four different applications in the field of web tables: (1) data search, (2) table extension, (3) knowledge base construction, and (4) table matching. Approaches supporting data search are represented, for instance, by [1,23,24]. The authors describe creation of a `isA` database from webpages via Herst patterns and using it to identify column classes and relations between columns. In a table extension application, a local table is extended with additional columns based on the corpus of tables that are published on the Web.

[14] https://commoncrawl.org/.
[15] http://webdatacommons.org/.
[16] http://webdatacommons.org/webtables/.

In the table matching applications [11,13,16,17], most approaches perform three basic steps: (1) column class identification, (2) entity disambiguation and (3) relation extraction. Only recent work by Ritze et al. [16,17] made the T2D gold standard available.

Subject column identification is addressed to a larger extent by [24,26]. Wang et al. [26] propose a naive approach, where the subject column is simply the the first column from the left that satisfies a fixed set of rules. Venetis et al. [24] identify subject column using a SVM with an RBF kernel. However, they do not open-source their code or their data. To the best of our knowledge, we outperform both state of the art approaches w.r.t. the F-measure that we achieve.

8 Conclusions and Future Work

In this paper, we described novel approach for subject column identification and property mapping for web tables. We improved the T2D gold standard by curating it manually with the help of 20 Semantic Web experts and used this T2D* to evaluate our approach against the state-of-the-art. While we were able to achieve a recall and an F-measure that were considerably higher than the state-of-the-art, our evaluation also revealed that the precision of TAIPAN can still be improved. The improvements can be achieved by supplementing our property ranking with additional heuristics over the whole table corpus. Moreover, we noticed that a large portion of the columns (56 %) in our benchmark contained meaningful information that can be potentially mapped to other knowledge bases. We will thus extend our extraction approach to cover such cases in future work.

Acknowledgments. This work has been supported by Eurostars projects DIESEL (project no. 01QE1512C), the BMWI Project GEISER (project no. 01MD16014) as well as the European Union's H2020 research and innovation action HOBBIT (GA no. 688227).

References

1. Balakrishnan, S., Halevy, A., Harb, B., Lee, H., Madhavan, J., Rostamizadeh, A., Shen, W., Wilder, K., Wu, F., Yu, C.: Applying webtables in practice
2. Carmel, D., Chang, M.-W., Gabrilovich, E., Hsu, B.-J.P., Wang, K.: ERD'14: entity recognition and disambiguation challenge. In: ACM SIGIR Forum, vol. 48, pp. 63–77. ACM (2014)
3. Ermilov, I., Auer, S., Stadler, C.: CSV2RDF: user-driven CSV to RDF mass conversion framework. In: Proceedings of the ISEM 2013, Graz, Austria, 04–06 September 2013
4. Ermilov, I., Auer, S., Stadler, C.: User-driven semantic mapping of tabular data. In: Proceedings of 9th International Conference on Semantic Systems, I-SEMANTICS 2013, pp. 105–112. ACM, New York (2013)
5. Etzioni, O., Cafarella, M., Downey, D., Popescu, A.-M., Shaked, T., Soderland, S., Weld, D.S., Yates, A.: Unsupervised named-entity extraction from the web: an experimental study. Artif. Intell. **165**(1), 91–134 (2005)

6. Gerber, D., Ngomo, A.-C.N.: Extracting multilingual natural-language patterns for RDF predicates. In: ten Teije, A., Völker, J., Handschuh, S., Stuckenschmidt, H., d'Acquin, M., Nikolov, A., Aussenac-Gilles, N., Hernandez, N. (eds.) EKAW 2012. LNCS, vol. 7603, pp. 87–96. Springer, Heidelberg (2012)
7. Hripcsak, G., Rothschild, A.S.: Agreement, the F-measure, and reliability in information retrieval. J. Am. Med. Inform. Assoc. **12**(3), 296–298 (2005)
8. Knoblock, C.A., et al.: Semi-automatically mapping structured sources into the semantic web. In: Simperl, E., Cimiano, P., Polleres, A., Corcho, O., Presutti, V. (eds.) ESWC 2012. LNCS, vol. 7295, pp. 375–390. Springer, Heidelberg (2012). doi:10.1007/978-3-642-30284-8_32
9. Landis, J.R., Koch, G.G.: The measurement of observer agreement for categorical data. Biometrics **33**, 159–174 (1977)
10. Lehmberg, O., Ritze, D., Ristoski, P., Meusel, R., Paulheim, H., Bizer, C.: The Mannheim search join engine. Web Semant.: Sci. Serv. Agents World Wide Web **35**, 159–166 (2015)
11. Limaye, G., Sarawagi, S., Chakrabarti, S.: Annotating and searching web tables using entities, types and relationships. Proc. VLDB Endow. **3**(1–2), 1338–1347 (2010)
12. Mintz, M., Bills, S., Snow, R., Jurafsky, D.: Distant supervision for relation extraction without labeled data. In: Proceedings of Joint Conference of the 47th Annual Meeting of the ACL and the 4th International Joint Conference on Natural Language Processing of the AFNLP, vol. 2, pp. 1003–1011. Association for Computational Linguistics (2009)
13. Mulwad, V., Finin, T., Syed, Z., Joshi, A.: Using linked data to interpret tables. In: COLD, vol. 665 (2010)
14. Nadeau, D., Sekine, S.: A survey of named entity recognition and classification. Lingvisticae Investigationes **30**(1), 3–26 (2007)
15. Nakashole, N., Weikum, G., Suchanek, F.: Patty: a taxonomy of relational patterns with semantic types. In: Proceedings of 2012 Joint Conference on Empirical Methods in Natural Language Processing and Computational Natural Language Learning, pp. 1135–1145. Association for Computational Linguistics (2012)
16. Ritze, D., Lehmberg, O., Bizer, C.: Matching HTML tables to DBpedia. In: Proceedings of 5th International Conference on Web Intelligence, Mining and Semantics, p. 10. ACM (2015)
17. Ritze, D., Lehmberg, O., Oulabi, Y., Bizer, C.: Profiling the potential of web tables for augmenting cross-domain knowledge bases. In: Proceedings of 25th International Conference on World Wide Web, pp. 251–261. International World Wide Web Conferences Steering Committee (2016)
18. Snow, R., Jurafsky, D., Ng, A.Y.: Learning syntactic patterns for automatic hypernym discovery. In: Advances in Neural Information Processing Systems, vol. 17 (2004)
19. Speck, R., Ngonga Ngomo, A.-C.: Ensemble learning for named entity recognition. In: Mika, P., et al. (eds.) ISWC 2014, Part I. LNCS, vol. 8796, pp. 519–534. Springer, Heidelberg (2014)
20. Suchanek, F.M., Abiteboul, S., Senellart, P.: Paris: probabilistic alignment of relations, instances, and schema. Proc. VLDB Endow. **5**(3), 157–168 (2011)
21. Usbeck, R., Ngonga Ngomo, A.-C., Röder, M., Gerber, D., Coelho, S.A., Auer, S., Both, A.: AGDISTIS - graph-based disambiguation of named entities using linked data. In: Mika, P., et al. (eds.) ISWC 2014, Part I. LNCS, vol. 8796, pp. 457–471. Springer, Heidelberg (2014)

22. Usbeck, R., Röder, M., Ngonga Ngomo, A.-C., Baron, C., Both, A., Brümmer, M., Ceccarelli, D., Cornolti, M., Cherix, D., Eickmann, B., et al.: Gerbil: general entity annotator benchmarking framework. In: Proceedings of 24th International Conference on World Wide Web, pp. 1133–1143. International World Wide Web Conferences Steering Committee (2015)
23. Venetis, P., Halevy, A., Madhavan, J., Pasca, M., Shen, W., Wu, F., Miao, G., Wu, C.: Table search using recovered semantics (2010)
24. Venetis, P., Halevy, A., Madhavan, J., Paşca, M., Shen, W., Wu, F., Miao, G., Wu, C.: Recovering semantics of tables on the web. Proc. VLDB Endow. **4**(9), 528–538 (2011)
25. Wang, C., Chakrabarti, K., He, Y., Ganjam, K., Chen, Z., Bernstein, P.A.: Concept expansion using web tables. In: Proceedings of 24th International Conference on World Wide Web, pp. 1198–1208. International World Wide Web Conferences Steering Committee (2015)
26. Wang, J., Wang, H., Wang, Z., Zhu, K.Q.: Understanding tables on the web. In: Atzeni, P., Cheung, D., Ram, S. (eds.) ER 2012 Main Conference 2012. LNCS, vol. 7532, pp. 141–155. Springer, Heidelberg (2012)
27. Zhang, Z.: Towards efficient and effective semantic table interpretation. In: Mika, P., et al. (eds.) ISWC 2014, Part I. LNCS, vol. 8796, pp. 487–502. Springer, Heidelberg (2014)

Semantic Authoring of Ontologies by Exploration and Elimination of Possible Worlds

Sébastien Ferré(✉)

IRISA, Université de Rennes 1, Campus de Beaulieu, 35042 Rennes, France
ferre@irisa.fr

Abstract. We propose a novel approach to ontology authoring that is centered on semantics rather than on syntax. Instead of writing axioms formalizing a domain, the expert is invited to explore the possible worlds of her ontology, and to eliminate those that do not conform to her knowledge. Each elimination generates an axiom that is automatically derived from the explored situation. We have implemented the approach in prototype PEW (Possible World Explorer), and conducted a user study comparing it to Protégé. The results show that more axioms are produced with PEW, without making more errors. More importantly, the produced ontologies are more complete, and hence more deductively powerful, because more negative constraints are expressed.

1 Introduction

Ontology authoring is generally an essential step in the application of knowledge engineering, and Semantic Web technologies. Existing methodologies generally distinguish two phases: (a) *conceptualization*, and (b) *formalization* in a formal language, typically the Web Ontology Language (OWL) [7]. We are here concerned with the formalization phase, which presents a number of difficulties, in particular for beginners but not only. Some difficulties are related to the manipulation of a formal language. Tools like Protégé [10] have precisely been introduced to facilitate such manipulation. Other difficulties are related to the discrepancies that can arise between the original intention of the ontology author, and what the formal ontology really express [3,12]. For example, "only eats vegetables" does not imply "eats some vegetables"; or to know that "X is a woman" does not allow to infer that "X is not a man" unless it has been explicitely stated that "men and women form disjoint classes". Indeed, negative constraints, like class disjointness or inequalities between individuals, are often overlooked because they seem so obvious. Their omission is difficult to detect because they do not manifest themselves by erroneous inferences, but by missing inferences. In a previous paper [6], we have shown errors and important omissions in the Pizza ontology[1], albeit it is used as a model and pedagogical support

This research is supported by ANR project IDFRAud (ANR-14-CE28-0012-02).

[1] http://protege.stanford.edu/ontologies/pizza/pizza.owl.

E. Blomqvist et al. (Eds.): EKAW 2016, LNAI 10024, pp. 180–195, 2016.
DOI: 10.1007/978-3-319-49004-5_12

for years. For example, classes Food and Country are not disjoint, and it appears that a vegetarian pizza can actually contain meat and/or fish as ingredient.

We introduce a new approach to ontology authoring that is centered on semantics rather than on syntax. Rather than seeing an ontology as a set of axioms, we propose to see an ontology through its set of models, i.e. as the set of interpretations that satisfy the ontology. We informally call those models "possible worlds". In the same spirit, rather than seeing ontology authoring as the successive addition of axioms, we propose to see it as the successive elimination of "possible worlds". Each elimination of a subset of "possible worlds" generates an axiom so that we still obtain a set of axioms in the end. However, the generated axioms are only the result of the authoring process, not the means. The main advantage of this approach is to enable the ontology author to work at the level of instances – possible worlds – like for ontology population (particular knowledge), while actually defining the terminological level of the ontology (general knowledge). From a previous paper [6], we reuse *possible world exploration*, and the contribution of this paper is to support the *creation* of an ontology from scratch rather than the mere *completion* of an existing ontology. Another contribution is a user study comparing our prototype PEW to Protégé.

Section 2 presents related work on ontology authoring. Section 3 recalls the basics of description logics, and Sect. 4 recalls previous results about possible world exploration, and prototype PEW. Section 5 presents the extension of PEW for ontology authoring, and Sect. 6 sketches an example scenario for the formalization of hand anatomy. Section 7 details the methodology and results of our user study. Section 8 concludes with a few perspectives.

2 Related Work

Ontology editors such as Protégé [10] tend to favor the expression of positive constraints, i.e. axioms supporting the inference of positive facts: e.g., class hierarchy, domain and range of properties. Their users have a mostly syntactic view of their ontology, and are hardly exposed to their semantics. By semantics, we here mean which situations the ontology makes possible or not. Semantic feedback can be obtained by calling a reasoner to check the consistency of the ontology, or the satisfiability of a class expression. However, those calls are tedious and left to the users. OntoTrack [9] offers a graphical view of the ontology, and returns immediate semantic feedback when the ontology is modified. However, it only covers a fragment of OWL Lite, and the view is limited to class hierarchies. The use of *competency questions* has also been proposed [13] to specify what the ontology is expected to answer, and then to automatically test the ontology during authoring. To some extent, the exploratory approach of our work enables to generate and validate at the same time such competency questions through interaction. Another reasoner-assisted ontology authoring approach [8] adapts test-driven development from softwares to ontologies by defining for each type of axiom a test to be run before and after the insertion of each axiom.

Table 1. Syntax and semantics of some DL class and property constructors, followed by TBox and ABox axioms. Thereby C, D denote class expressions, R, S property expressions, a, b individual names, and r a property name.

Name	Syntax	Semantics
top	$\top$	$\Delta^{\mathcal{I}}$
bottom	$\bot$	$\emptyset$
negation	$\neg C$	$\Delta^{\mathcal{I}} \setminus C^{\mathcal{I}}$
conjunction	$C \sqcap D$	$C^{\mathcal{I}} \cap D^{\mathcal{I}}$
disjunction	$C \sqcup D$	$C^{\mathcal{I}} \cup D^{\mathcal{I}}$
nominal	$\{a\}$	$\{a^{\mathcal{I}}\}$
exist. restriction	$\exists R.C$	$\{x \in \Delta^{\mathcal{I}} \mid \text{for some } y \in \Delta^{\mathcal{I}}, (x, y) \in R^{\mathcal{I}} \text{ and } y \in C^{\mathcal{I}}\}$
univ. restriction	$\forall R.C$	$\{x \in \Delta^{\mathcal{I}} \mid \text{for all } y \in \Delta^{\mathcal{I}}, (x, y) \in R^{\mathcal{I}} \text{ implies } y \in C^{\mathcal{I}}\}$
inverse property	r^-	$\{(y, x) \in \Delta^{\mathcal{I}} \times \Delta^{\mathcal{I}} \mid (x, y) \in r^{\mathcal{I}}\}$
subclass	$C \sqsubseteq D$	$C^{\mathcal{I}} \subseteq D^{\mathcal{I}}$ TBox axioms
subproperty	$R \sqsubseteq S$	$R^{\mathcal{I}} \subseteq S^{\mathcal{I}}$
instance	$C(a)$	$a^{\mathcal{I}} \in C^{\mathcal{I}}$ ABox axioms
relation	$R(a, b)$	$(a^{\mathcal{I}}, b^{\mathcal{I}}) \in R^{\mathcal{I}}$
same	$a \doteq b$	$a^{\mathcal{I}} = b^{\mathcal{I}}$
different	$a \not\doteq b$	$a^{\mathcal{I}} \neq b^{\mathcal{I}}$

A number of controlled natural languages, such as CLOnE [4] or Rabbit [5] have been proposed to produce OWL axioms from sentences in natural language, and to verbalize OWL ontologies in natural language. They address the issue about the syntax of formal languages, not the issue about semantic feedback. Their contribution is therefore orthogonal to ours, and could complement it.

There also exists a number of (semi-)automated techniques to produce OWL axioms. Some tools detect common errors, and complete ontologies in a systematic way [3,11]. However, those approaches are not constructive but corrective. Moreover, they are often limited to disjointness axioms, the simplest form of negative constraints. Formal Concept Analysis [1,14] has been used in interactive ontology authoring. Experts are presented with a sequence of statements, and for each of them, they have to either confirm the statement, or produce a counter-example. It guarantees complete formalization for some DL fragments but it is rather expensive in terms of user interaction, and tends to patronize the expert.

3 Preliminaries

We here recall basic definitions of Description Logics (DL), which are the basis of OWL ontologies [7]. We briefly recap here the syntax and semantics of the sublanguage of OWL DL that is necessary for this work. We assume finite and disjoint sets N_I, N_C, and N_R, respectively called *individual names*, *class names* and *property names*. Table 1 shows how complex classes, complex properties, and axioms can be formed from these atomic entities. An *ontology* $\mathcal{O}$ is a set of axioms, which are often partioned in two subsets: a TBox containing general axioms about classes and roles, and an ABox containing particular axioms about individuals. The semantics of description logics is defined via interpretations $\mathcal{I} = (\Delta^\mathcal{I}, \cdot^\mathcal{I})$ composed of a non-empty set $\Delta^\mathcal{I}$ called the *domain of* $\mathcal{I}$ and a function $\cdot^\mathcal{I}$ mapping individuals to elements of $\Delta^\mathcal{I}$, classes to subsets of $\Delta^\mathcal{I}$ and properties to subsets of $\Delta^\mathcal{I} \times \Delta^\mathcal{I}$ (i.e., binary relations). This mapping is extended to complex classes and properties, and finally used to evaluate axioms (see Semantics in Table 1). We say $\mathcal{I}$ satisfies an ontology $\mathcal{O}$ (or $\mathcal{I}$ is a model of $\mathcal{O}$, written: $\mathcal{I} \models \mathcal{O}$) if it satisfies all its axioms. We say that an ontology $\mathcal{O}$ *entails* an axiom α (written $\mathcal{O} \models \alpha$) if all models of $\mathcal{O}$ are models of α. Finally, an ontology is *consistent* if it has a model and a class C is called *satisfiable* w.r.t. an ontology $\mathcal{O}$ if there is a model $\mathcal{I}$ of $\mathcal{O}$ with $C^\mathcal{I} \neq \emptyset$.

4 Possible World Exploration

We here recall from a previous paper [6] an approach for a safe and complete exploration of the possible worlds of an OWL DL ontology. No assumption is made on the ontology. It may contain instances or not. It may be limited to taxonomies or contain complex DL axioms. The exploration is based on navigation, where navigation places are situations made possible by the ontology (formally, satisfiable class expressions): e.g., "pizzas without topping" in the Pizza ontology. Navigation links enable to move from one place to another, i.e. from one situation to another (formally, transformations of class expressions). A key aspect is that those navigation links are automatically computed for each situation so as to never lead users to impossible situations. If we see the models $\mathcal{I}$ of an ontology $\mathcal{O}$ as "possible worlds", then each navigation place offers a view on a subset of possible worlds. If a navigation place is a situation described by the class expression C, then the subset of possible worlds is made of models $\mathcal{I}$ of the ontology that make the class expression satisfiable ($C^\mathcal{I} \neq \emptyset$). Therefore, navigation across situations supports in effect the exploration of possible worlds.

A prototype, PEW (Possible World Explorer), is available as open source[2]. It relies on HermiT [15] for all reasoning tasks. For the sake of concision, we here use the DL notation (see Sect. 3) for axioms and class expressions. In practice, PEW gives users the choice between the DL notation and the Manchester notation. In the following, examples are based on the Pizza ontology.

4.1 Views over Possible Worlds

At every navigation step, a view over the possible worlds selected by the current situation is presented to the ontology author. In order to formally define those views, we first define two sublanguages of class expressions. *Simple class expressions* serve as elementary situation descriptions, and *cognitively intuitive class expressions* combine them to describe complex situations. We called the latter

[2] http://www.irisa.fr/LIS/softwares/pew.

"cognitively intuitive" because they restrict negation to simple class expressions, which make situations easier to grasp for humans. Indeed, "humans would normally have no problems with handling the class of non-smokers or childless persons, while classes such as non-(persons having a big dog and a small cat) occur unnatural, contrived and are harder to cognitively deal with" [6].

Definition 1. *Given sets N_C, N_R, N_I of class names, property names, and individual names, the set $\mathcal{S}$ of* simple class expressions *is the set of class expressions of one of the forms (with $A \in N_C$, $r \in N_R$, $a \in N_I$): A, $\{a\}$, $\exists r.\top$, $\exists r.A$, $\exists r.\{a\}$, $\exists r^-.\top$, $\exists r^-.A$, $\exists r^-.\{a\}$, and their negations $\neg A$, $\neg\{a\}$, $\neg\exists r.\top$, etc.*

Definition 2. *The set $\mathcal{CI}$ of* cognitively intuitive class expressions *is inductively defined as follows:*

1. *every simple class expression is in $\mathcal{CI}$,*
2. *for $C_1, C_2 \in \mathcal{CI}$, $C_1 \sqcap C_2$ and $C_1 \sqcup C_2$ are in $\mathcal{CI}$,*
3. *for any property r and $C \in \mathcal{CI}$, $\exists r.C$ and $\exists r^-.C$ are in $\mathcal{CI}$.*
   The set $\mathcal{CI}[X]$ of pointed $\mathcal{CI}$ class expressions *denotes $\mathcal{CI}$ class expressions with symbol X occurring exactly once in the place of a subexpression.*

The situations that can be described by $\mathcal{CI}$ class expressions involve existing objects, their identity, their membership to atomic classes, and their interrelationships. The syntax and semantics of description logics entail that only tree-shape relationships can be expressed. Additionally, disjunction enables to express alternatives for some parts of the situation.

A pointed class expression $C(X) \in \mathcal{CI}[X]$ is used to put a *focus* on a subexpression of a class expression. It is a class expression with a hole in place of a subexpression, materialized by the meta-variable X: e.g., $C(X) = \mathit{Pizza} \sqcap \exists \mathit{ingredient}.X$ represents a pizza with an unspecified ingredient. Given a class expression D, the expression $C(D)$ denotes the expression $C(X)$ where D has been substituted for X, filling the hole. For example, given $D = \mathit{Meat} \sqcup \mathit{Fish}$, we have $C(D) = \mathit{Pizza} \sqcap \exists \mathit{ingredient}.(\mathit{Meat} \sqcup \mathit{Fish})$.

Definition 3. *Given an ontology $\mathcal{O}$, a* possible world view *$C(X)/D$ is specified by the combination of a pointed class expression $C(X)$ (the* context*), and a class expression D (the* focus*), such that $C(D)$ is a satisfiable class in $\mathcal{O}$. The focus is said* necessary *if $C(\neg D)$ is not satisfiable. A view is also composed of* instances *and* adjuncts*, derived from context and focus:*

- *$\mathit{Inst}(C(X)/D) = \{a \in N_I \mid \mathcal{O} \models C(D)(a)\}$ is the set of instances of $C(D)$;*
- *$\mathit{Adj}(C(X)/D) = \{E \in \mathcal{S} \mid C(D \sqcap E)$ is satisfiable in $\mathcal{O}\}$ is the set of* possible adjuncts *at focus. A positive simple expression E is called a* necessary adjunct *if $E \in \mathit{Adj}$ and $\neg E \notin \mathit{Adj}$*[3].

[3] When disjunctions are used, the definitions of satisfiable class and adjuncts are slighty more complex to ensure that all alternatives remain satisfiable. See [6] for details.

In a view, the expression $C(D)$ represents the situation, and its decomposition into context and focus enables the representation of the focus. That focus is essential for possible world exploration as it enables to look at the different objects of the situation. Only situations described by satisfiable classes, aka. "possible worlds", are considered. When the expression under focus cannot be negated in the context, the focus is said necessary. In the Pizza ontology, the view $Pizza \sqcap X/\exists hasIngredient.\top$ has a necessary focus because every pizza has necessarily an ingredient ($\mathcal{O} \models Pizza \sqsubseteq \exists hasIngredient.\top$). Therefore, PEW allows to check any class inclusion $C \sqsubseteq D$ by exploration when $C, D \in \mathcal{CI}$.

The instances of a view are the individuals that are a member of the class $C(D)$ in all models of the ontology. The adjuncts of a view are the simple class expressions E that are satisfied by the object at the focus of the situation in at least one model of the ontology. This is checked by evaluating the satisfiability of the situation after inserting E at the focus: $C(D \sqcap E)$. There are three cases for each positive simple class expression E. If E and $\neg E$ are possible adjuncts, we call E "ambivalent". If only E is possible, we call it "necessary". If only $\neg E$ is possible, we call it "impossible". The case where both E and $\neg E$ are not possible adjuncts is excluded because it would imply that $C(D)$ is not satisfiable.

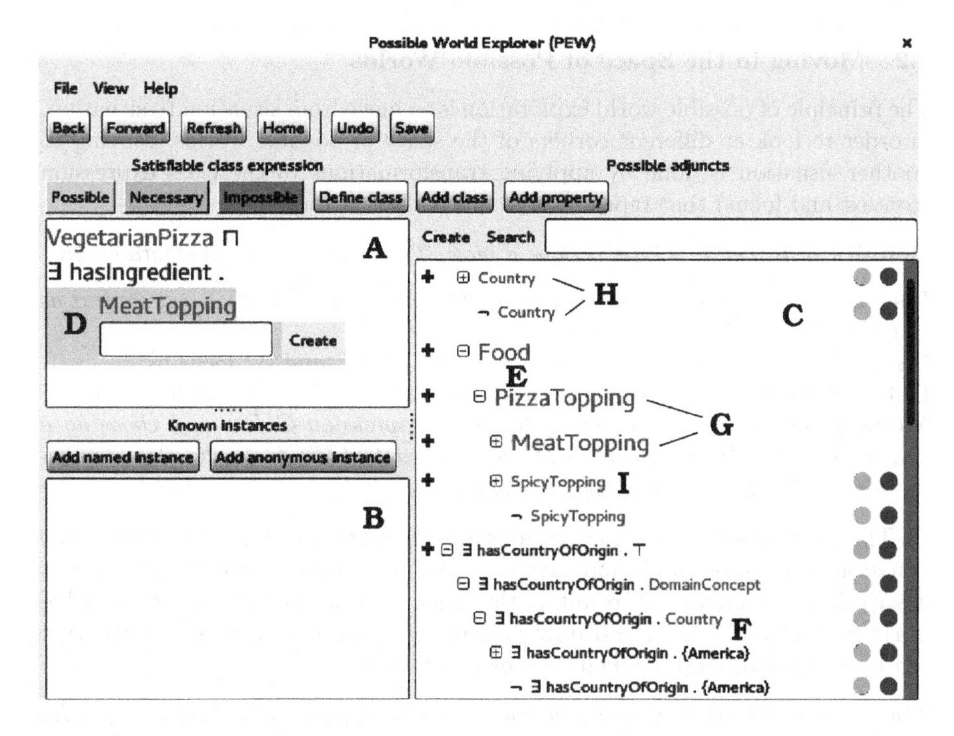

Fig. 1. Screenshot of PEW: looking at vegetarian pizzas with meat.

Figure 1 shows a screenshot of PEW. The user interface reflects the definition of views with the class expression at the top left (**A**), the instances at the bottom left (**B**), and the possible adjuncts at the right (**C**). The focus subexpression is highlighted by a background color (**D**) that is normally yellow, and becomes green when the focus is necessary. The described situation is here a "vegetarian pizza that has meat as an ingredient", with focus on the "meat". In the example, there are no known instance of the situation. The possible adjuncts are organized in a tree to reflect the class hierarchy, the property hierarchy, and the membership of individuals to classes. For example, we see that *PizzaTopping* is a subclass of *Food* (**E**), and that *America* is a member of *Country* (**F**). Necessary adjuncts are shown in a larger font (**G**), and impossible adjuncts are not displayed to avoid cluttering the interface. Ambivalent adjuncts appear as pairs $E, \neg E$, e.g. *Country*, ¬ *Country* (**H**). The unexpected facts that the screenshot tells us about the possible worlds of the Pizza ontology are that: (1) a vegetarian pizza can have meat as ingredient (**A, D**), and (2) a meat topping can be a country or not (**H**). Fact (1) is possible because vegetarian pizzas are defined to exclude meat as topping, but not as ingredient. Fact (2) is possible because a disjointness axiom between *Food* and *Country* is missing. The screenshot also shows reasonable facts: a meat topping can be spicy (**I**), and can have a country of origin, for example America (**F**).

4.2 Moving in the Space of Possible Worlds

The principle of possible world exploration is to move from situation to situation, in order to look at different corners of the space of possible worlds. Moving to another situation is done by applying transformations to the class expression (context and focus) that represents the current situation.

Definition 4. *Let $V = C(X)/D$ be a view. The available transformations are:*

Inserting an adjunct. *Choosing a possible adjunct $E \in Adj(V)$, set context as $C(D \sqcap X)$, and set focus as E.*
Inserting a disjunction. *Set context as $C(D \sqcup X)$, and set focus as $\top$.*
Deleting focus subexpression. *Keep context as $C(X)$, and set focus as $\top$.*
Moving the focus. *Considering the class expression $C(D)$, and choosing a subexpression D', set focus as D', and set context $C'(X)$ as the class expression $C(D)$ where D' has been replaced by X.*

The most important transformations are "inserting an adjunct" to extend the situation description, and "moving the focus" to choose where to extend that description. An important result is the *safeness* and *completeness* of possible world exploration, i.e. all and only possible and cognitively intuitive situation can be reached through navigation (see proofs in [6]).

Theorem 1. *Starting from the initial view $X/\top$ defined by context $C(X) = X$, and focus $D = \top$ ($C(D) = \top$), every reachable view has a satisfiable class expression (*safeness*), and every view defined by a satisfiable class expression in $\mathcal{CI}$ is reachable in a finite number of navigation steps (*completeness*).*

For example, the situation of Fig. 1 is reached by the successive insertion of two possible adjuncts: *VegetarianPizza*, $\exists hasIngredient.MeatTopping$. In PEW, inserting an adjunct is done by double-clicking it, and moving the focus is done by clicking on the desired subexpression. Inserting a disjunction or deleting the focus are done through the contextual menu.

5 Possible World Elimination for Ontology Authoring

Compared to our previous work, PEW now offers a wide range of *commands* to update an ontology. Previously, the only command was to make the current situation impossible, generating the axiom $C(D) \sqsubseteq \bot$. Although many axioms can be rewritten in that form, there were important limitations. First, the signature was fixed, limiting its application to the completion of existing ontologies, and forbidding the creation of new ontologies. Second, the restriction to atomic negations limited the expressivity of positive constraints such as class inclusion $C \sqsubseteq D$. Indeed, the latter axiom is equivalent to $C \sqcap \neg D \sqsubseteq \bot$, and therefore the right class expression D was limited to simple class expressions. Third, a number of simple axioms, while expressible, required a contrived formulation, and many navigation steps. For example, to state that class A is a subclass of B, it was necessary to select adjunct A, then select adjunct $\neg B$, and finally declare the resulting situation impossible. The following update commands are designed to minimize the use of explicit negation from the user point of view.

Update Commands. Let $C(X)/D$ be a possible world view. The current version of PEW supports the following commands to update the ontology.

New class. Extending the ontology signature with a new class name A.

New property. Extending the ontology signature with a new property name r.

Add instance. Extending the ontology signature with a new individual a, and making it an instance of the class expression with axiom $(C(D))(a)$.

Add subclass of class adjunct B. Creating a new class name A, and making it a subclass of B with axiom $A \sqsubseteq B$.

Add subproperty from relational adjunct $\exists s.\top$. Creating a new property r, and making it a subproperty of s with property axiom $r \sqsubseteq s$.

Add inverse property from relation adjunct $\exists s.\top$. Creating a new property r, and making it the inverse of s with axiom $r \equiv s^-$.

Add property constraint from relational adjunct $\exists s.\top$. Constrain property s to be any of functional, inverse functional, reflexive, irreflexive, symmetric, asymmetric, and/or transitive.

Impossible situation. Add axiom $C(D) \sqsubseteq \bot$.

Necessary focus. Add axiom $C(\neg D) \sqsubseteq \bot$. It has no effect if the focus is already necessary.

Impossible adjunct E. Add axiom $C(D \sqcap E) \sqsubseteq \bot$. This is equivalent to first inserting adjunct E at focus, and then triggering command "Impossible situation". It is forbidden on necessary adjuncts.

Necessary adjunct E. Add axiom $C(D \sqcap \neg E) \sqsubseteq \bot$. This is equivalent to first inserting adjunct E at focus, moving focus on E, and then triggering command "Necessary situation". It is useless on necessary adjuncts.

Disjoint adjuncts $E_1, \ldots E_n$. Add axiom $E_i \sqcap E_j \sqsubseteq \bot$ for each pair $\{E_i, E_j\}$. This command is equivalent to $\frac{n(n-1)}{2}$ "impossible situation" commands.

The main controls that trigger update commands are visible in Fig. 1. Above the class expression (**A**), there are buttons to make the current situation "impossible" or to make the current focus "necessary". Button "possible" has the effect to declare new classes, properties and individuals that may have been inserted directly in the class expression through the entry field in (**D**). Button "define class" allows to declare a new class name A, and to make it equivalent to the class expression ($A \equiv C(D)$). Above the instances (**B**), the first button allows to "add an instance to the class expression". The second button does the same with an anonymous individual. Above the tree of adjuncts (**C**), the two buttons allow to extend the signatures with "new classes and properties". Each positive adjunct has a blue cross on its left to "add subclasses" (when the adjunct is a class name), and "add subproperties" (when the adjunct is an existential restriction). Each possible adjunct has on its right a green dot to make it "a necessary adjunct", and a red dot to make it "an impossible adjunct". Other commands are available through the contextual menu of the tree of adjuncts. A general principle of the user interface is to provide as much immediate feedback as possible. For instance, when the focus is made necessary, its color switches from yellow to green; when an ambivalent adjunct is made necessary, the negative adjunct disappears, and the positive adjunct switches to a larger font.

Table 2. Translation of main DL axioms into PEW views and commands. Axioms are restricted to cognitively intuitive class expressions ($C, D \in \mathcal{CI}$).

DL axiom		PEW view	PEW update command
TBox	$C \sqsubseteq D$	$(C \sqcap X)/D$	Necessary focus
	$R \sqsubseteq S$	$X/\top$	Add subproperty R from relation adjunct $\exists S.\top$
ABox	$C(a)$	X/C	Add instance a
	$r(a, b)$	$X/\{a\}$	Necessary adjunct $\exists r.\{b\}$
	$a \doteq b$	$X/\{a\}$	Necessary adjunct $\{b\}$
	$a \neq b$	$X/\{a\}$	Impossible adjunct $\{b\}$

To assess the expressivity of PEW, Table 2 explains for each kind of axiom how to express it by specifying which view to reach by navigation, and which update command to trigger. Note that both TBox and ABox axioms are covered. The only restriction is that built class expressions must be cognitively intuitive, i.e. must have only negation on simple class expressions. Note that, compared to the previous version, command "necessary focus" enables class inclusions with complex class expressions on both sides. PEW is therefore *complete* w.r.t. the chosen sublanguage of OWL. The features of OWL2 that are not covered are literals and related features, cardinality restrictions, local reflexivity (*self*), property chains, and keys.

Similarly to possible adjuncts that are computed so as to avoid the construction of unsatisfiable class expressions, update commands are designed to avoid the production of inconsistencies in the ontology. If a command would produce an inconsistency, it is blocked and an error message is shown to the user. Commands "impossible adjunct" and "necessary adjunct" are only available on ambivalent adjuncts because they are either useless or inconsistent on necessary and impossible adjuncts.

6 Example Scenario: Ontology of Hand Anatomy

We here sketch a possible strategy to formalize the basics of the anatomy of a normal hand. (We strongly recommend the reader to watch the 8 min screencast at https://youtu.be/u4X0hq6et0Q.) First, new classes are created for the different types of elements of the hand: *Hand*, *Palm*, *Finger*, *Phalanx*, and *Nail*. Those 5 classes are made disjoint. Then, 5 subclasses are added to *Finger* for each finger (thumb, index, middle, ring, and little), and 3 subclasses to *Phalanx* for each phalanx (proximal, middle, and distal). Those two sets of subclasses are also made disjoint. The next step is to create property *hasDirectPart*, and its inverse *isDirectPartOf*. From there, the ontology signature is complete, and completing the ontology is a matter of exploring the possible worlds, and triggering "impossible" and "necessary" commands. For example, looking at hands with view $X/Hand$, the possible adjuncts show that at this stage a hand can contain any element, and be part of any element. Here, most possible adjuncts are either made necessary (e.g., $\exists HasDirectPart.IndexFinger$) or impossible (e.g., $\exists isDirectPartOf.Hand$). Some possible adjuncts may remain ambivalent, e.g. $\exists isDirectPartOf.\top$, to let open the possibility to make hands part of larger elements (e.g., arms). The following steps are to visit in turn each element type, similarly to hands, to apply the relevant constraints: e.g., distal phalanges have nails, proximal phalanges have no nail, thumbs have no middle phalanx. The domain of property r can be constrained by reaching view $X/\exists r.\top$, and its range by reaching view $(\exists r.X)/\top$, which is only one focus move from the former view.

7 User Study: Comparison with Protégé

We conducted a user study to compare PEW and Protégé in the task of designing an ontology from scratch. The ontology to be designed is the same as in the previous section, on hand anatomy. The advantages of this topic are that the related knowledge is well known to everybody, and that it is rich with many positive and negative constraints.

7.1 Methodology

The subjects of the user study are 30 postgraduate students in bioinformatics (Master of Bio-Informatics and Genomics at Université Rennes 1). Before the user study, subjects had been exposed to a short course on Semantic Web technologies (RDF, RDFS, OWL); but they had been exposed neither to Protégé nor to PEW. The user study was organized as a practical about OWL ontologies for the Semantic Web course. Students were simply told that their work would be used for a research experiment. Subjects were cast in two groups, one working with PEW, and the other one working with Protégé. Like in all practicals, students worked either alone or in a pair. In the following, we refer to the singleton or pair of students working together as a *team*. In total, 9 teams worked with PEW, and 8 teams worked with Protégé.

The user study lasted about 2 h, starting with 20 min of presentation and tutorial about the system of the group, ending with 10 min to fill a SUS questionnaire [2], and having a maximum of 1 h 30 min to design the requested ontology on hand anatomy. A document was distributed to each team with instructions, requirements, and a row of questions to guide the design of the ontology in a progressive and modular way. At the end of the sessions, we collected the OWL file of each team, and the anonymous SUS questionnaire filled by each student. The distributed document, the collected data, and analysis spreadsheets are available at http://www.irisa.fr/LIS/ferre/pub/ekaw2016.zip.

7.2 Results

Quantity and Correction. For each team, we counted the produced OWL declarations and axioms, and among them the number of erroneous axioms. We did not count uninformative or imprecise axioms as errors, only axioms that are inconsistent with the anatomy of a normal hand. The table below reports for each system the minimum, average, and maximum value of those measures across teams. It also reports the resulting precision of produced axioms: the individual precision is the average of precisions team-wise, while the collective precision is computed on the collection of all axioms produced with a system. The result is that more axioms were produced with PEW: 74 % increase on average, and nearly three-fold at maximum. The next paragraph shows that the increase is mostly due to one kind of axioms. Despite that increase, we do not observe more errors produced with PEW. This entails a higher precision for PEW, with a significant difference for collective precision. The fact that collective precision is higher than individual precision for PEW, and the inverse for Protégé says that the PEW teams that produced the more axioms made less errors than the average, and that, on the contrary, the Protégé teams that produced the more axioms made more errors.

Axiom Types. We analyzed the type of produced axioms, and counted for each team the number of axioms of each type. In addition to declarations and property axioms (e.g., inverse property), we distinguish between positive axioms

System	nb. declarations/axioms	nb. errors	Individual precision	Collective precision
Protégé	43 (68) 88	0 (5) 17	77 % (93 %) 100 %	92 %
PEW	58 (118) 241	0 (4) 18	69 % (95 %) 100 %	97 %

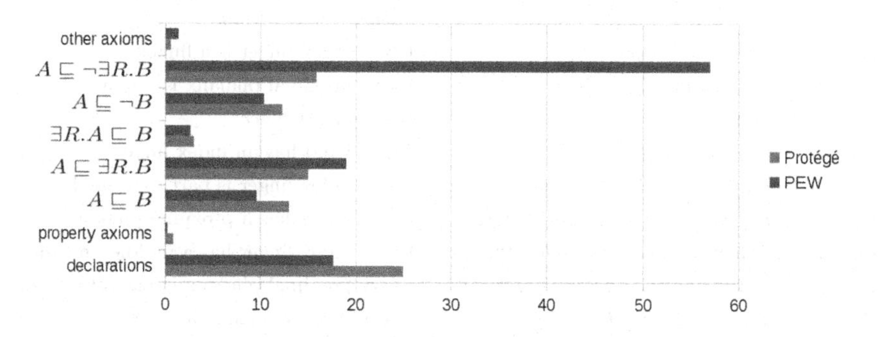

Fig. 2. Comparison of the average number of declarations and axioms per type.

in the form $C \sqsubseteq D$, and negative axioms in the form $C \sqsubseteq \neg D$ or equivalently $C \sqcap D \sqsubseteq \bot$. Then, we consider all combinations between atomic classes (A, B) and simple qualified existential restrictions ($\exists r.A$, $\exists r.B$), except combinations with two restrictions because they are rare and mostly errors. We decomposed the produced axioms to make them fit the previous types, when possible. For example, axiom $A \equiv B$ splits into $A \sqsubseteq B$ and $B \sqsubseteq A$; and an axiom declaring A as the domain of R translates to $\exists R.\top \sqsubseteq A$. For cardinality restrictions produced with Protégé, we counted minimum cardinality as an existential restriction, and maximum cardinality as a negated existential restriction. Figure 2 compares the average number of axioms produced by teams for each system, and for each type of axiom. The striking one difference is about complex negative axioms in the form $A \sqsubseteq \neg\exists R.B$, which were produced 3.5 times more often with PEW. A typical example is $\mathit{Thumb} \sqsubseteq \neg\exists \mathit{hasPart}.\mathit{MiddlePhalanx}$ stating that the thumb has no middle phalanx. An interesting example of complex axiom that was produced with PEW is $\exists \mathit{orientation}^{-}.\exists \mathit{isPartOf}.\exists \mathit{orientation}.\{\mathit{RIGHT}\} \sqsubseteq \{\mathit{RIGHT}\}$, stating that every part of a right element is a right element.

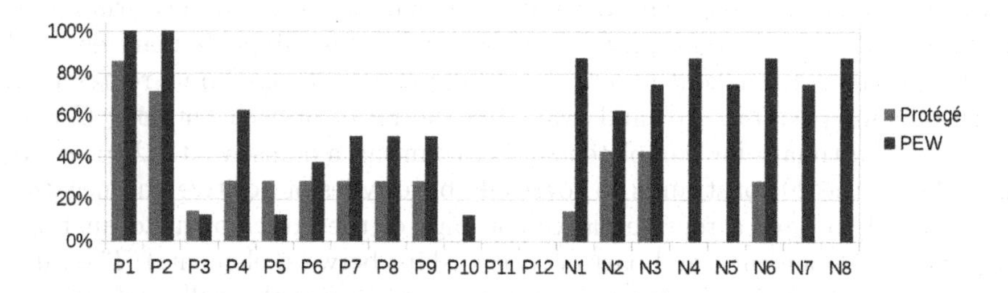

Fig. 3. Comparison of recall on 20 constraints: 12 pos. (P1–P12), 8 neg. (N1–N8).

Table 3. Sample of constraints (12 positive, 8 negative) that must be satisfied by a normal hand. They are organized in 5 types of DL axioms.

id	DL axiom	Informal description
	$A \sqsubseteq B$	*Subclass relationship*
P1	$IndexFinger \sqsubseteq Finger$	Every index finger is a finger
P2	$ProximalPhalanx \sqsubseteq Phalanx$	Every proximal phalanx is a phalanx
	$A \sqsubseteq \exists R.B$	*Relation existence*
P3	$Hand \sqsubseteq \exists hasPart.IndexFinger$	Every hand has an index finger
P4	$IndexFinger \sqsubseteq \exists isPartOf.Hand$	Every index finger is part of a hand
P5	$Finger \sqsubseteq \exists hasPart.ProximalPhalanx$	Every finger has a proximal phalanx
P6	$IndexFinger \sqsubseteq \exists hasPart.MiddlePhalanx$	Every index finger has a middle phalanx
P7	$ProximalPhalanx \sqsubseteq \exists isPartOf.Finger$	Every proximal phalanx is part of a finger
P8	$DistalPhalanx \sqsubseteq \exists hasPart.Nail$	Every distal phalanx has a nail
P9	$Nail \sqsubseteq \exists isPartOf.DistalPhalanx$	Every nail is part of a distal phalanx
	$\exists R.A \sqsubseteq B$	*Qualified property domain/range*
P10	$\exists hasPart.Palm \sqsubseteq Hand$	Only hands have a palm as a direct part
P11	$\exists isPartOf.Finger \sqsubseteq Phalanx$	Only phalanges are parts of fingers
P12	$\exists hasPart.Nail \sqsubseteq DistalPhalanx$	Only distal phalanges have a nail
	$A \sqsubseteq \neg B$	*Class disjointness*
N1	$Hand \sqsubseteq \neg Finger$	No hand is a finger
N2	$Thumb \sqsubseteq \neg IndexFinger$	No thumb is an index finger
N3	$ProximalPhalanx \sqsubseteq \neg DistalPhalanx$	No proximal phalanx is a distal phalanx
	$A \sqsubseteq \neg \exists R.B$	*Relation non-existence*
N4	$Hand \sqsubseteq \neg \exists hasPart.Hand$	No hand is made of a hand
N5	$Hand \sqsubseteq \neg \exists isPartOf.Finger$	No hand is part of a finger
N6	$Thumb \sqsubseteq \neg \exists hasPart.MiddlePhalanx$	No thumb has a middle phalanx
N7	$ProximalPhalanx \sqsubseteq \neg \exists isPartOf.Phalanx$	No proximal phalanx is part of a phalanx
N8	$ProximalPhalanx \sqsubseteq \neg \exists hasPart.Nail$	No proximal phalanx has a nail

Recall Estimate. Another question we wanted to answer is about the completeness of the produced ontologies. Indeed, producing more axioms does not imply a more complete formalization of the domain. Intuitively, measuring completeness amounts to counting the proportion of constraints that are entailed by the ontology. To make it practical, we have listed 20 constraints represented as DL axioms (see Table 3), and evaluated the recall of each ontology $\mathcal{O}$ over them as an estimate for completeness, i.e. counting axioms α s.t. $\mathcal{O} \models \alpha$. We have chosen the 20 constraints to cover all above types of positive and negative axioms, and to have a representative coverage of the basics of hand anatomy (types of hand elements, and part-of relationships between elements). The table below gives the minimum, average, and maximal values of recall for both systems, and for three sets of constraints: all of them, only positive ones, only negative ones. Globally, ontologies produced with PEW were more than twice as complete on average, and even 75 % complete in the best case. In fact, the least complete PEW ontology is still more complete than the average Protégé ontology. That difference is even stronger for negative constraints, and remains to a lesser degree for positive constraints. For negative constraints, PEW average recall reaches 80 %, 5 times higher than Protégé average recall (16 %). On negative constraints, the least complete PEW ontology (50 %) is still neatly more complete than the most complete Protégé ontology (38 %).

System	All constraints	Pos. constraints	Neg. constraints
Protégé	0 % (24 %) 45 %	0 % (29 %) 50 %	0 % (16 %) 38 %
PEW	35 % (56 %) 75 %	17 % (41 %) 58 %	50 % (80 %) 100 %

Figure 3 compares the average recall for each constraint. All negative constraints have a recall above 63 % with PEW, and below 43 % with Protégé.

SUS Questionnaire. The subjective perception of students about the systems was collected through the classic SUS questionnaire [2]. We got 13 answers for each system. The results show only small differences between the two systems. The score for PEW is 52 on average, ranging from 25 to 85, and is slightly better for Protégé, 58 on average, ranging from 18 to 80. Students tend to find PEW more complicated (*"I think that I would need the support of a technical person to be able to use this system"* 3.8 vs 3.1) but with less inconsistencies (*"I thought there was too much inconsistency in this system"* 2.2 vs 2.5) than Protégé, although the differences are small. Looking at extreme votes (1 and 5 on a 1-5 scale), it appears that PEW triggers more contrasted opinions with some students finding it easy to use, and others finding it very complicated. An interesting comment by a PEW user says that the difficulty was about the syntax of class expressions (in Manchester syntax), rather than about the tool itself.

7.3 Interpretation and Discussion

PEW is More Productive. We think that this is because in Protégé, there is a high step between the axioms that are easy to express in the interface, such as class hierarchy, class disjointness, domains and ranges, and other axioms that require to actually write class expressions. In PEW, there is also a step between simple class expressions that are readily available as possible adjuncts, and more complex class expressions that require several navigation steps. However, simple class expressions offer more expressivity than the easy axioms of Protégé, and navigating to complex class expressions is arguably simpler than writing them.

PEW Achieves a Better Precision. Errors in OWL axioms often come from cognitively unintuitive constructors, in particular universal restriction and general negation. For example, a Protégé team produced axiom $Hand \sqsubseteq \forall hasPart.RingFinger$, using $\forall$ instead of $\exists$. The restriction in PEW to cognitively intuitive class expressions implies that users are only asked to evaluate factual (i.e., existential and positive) situations as impossible or necessary.

PEW Achieves a Better Recall, Especially When There Are Many Negative Constraints. PEW presents a symmetry between positive and negative constraints because they are produced in the same way, just using command "necessary" for the former, and "impossible" for the later. On the contrary, Protégé has a strong bias towards positive constraints, apart from class disjointness. However, negative constraints are essential to the deductive potential of expressive ontologies.

PEW Usability is Encouraging. PEW is a prototype that is much less mature than Protégé but it got a SUS score not far behind Protégé. Its bad reception by a few proves that the design of the user interface must be improved. The good reception by a few others shows that there is ample room for such improvement. The main issue is the readability of class expressions. Another issue is the fact that commands in contextual menus (e.g., inverse property) are often overlooked compared to commands as buttons.

8 Conclusion

We have presented a semantic approach to ontology authoring based on possible world exploration and elimination. It has been implemented as prototype PEW, and a user study has demonstrated promising results in terms of quantity, precision, and recall of the produced axioms, and in terms of usability. The most notable result is the increase in recall, from 24 % with Protégé to 56 % with PEW, where 100 % would mean a complete OWL formalization of the domain knowledge for the selected OWL fragment. Future work will investigate long-run guidance for the systematic exploration of possible worlds in order to further improve recall; and the verbalization in natural language of class expressions to improve the readability of explored situations so as to further improve precision. We also plan to re-implement PEW as a Protégé plugin in order to combine their strengths, and favor its adoption.

Acknowledgement. We wish to thank Olivier Dameron for his precious support in the user study, as well as the students of master BIG for their kind participation.

References

1. Baader, F., Ganter, B., Sertkaya, B., Sattler, U.: Completing description logic knowledge bases using formal concept analysis. In: International Joint Conference Artificial Intelligence, pp. 230–235 (2007)
2. Brooke, J.: SUS: a quick and dirty usability scale. In: Jordan, P., Thomas, B., Weerdmeester, B., McClelland, A. (eds.) Usability Evaluation in Industry, pp. 189–194. Taylor and Francis, London (1996)
3. Corman, J.: Explorer les théorèmes d'une TBox. In: Journées francophones d'Ingénierie des Connaissances (2013)
4. Davis, B., Iqbal, A.A., Funk, A., Tablan, V., Bontcheva, K., Cunningham, H., Handschuh, S.: RoundTrip ontology authoring. In: Sheth, A.P., Staab, S., Dean, M., Paolucci, M., Maynard, D., Finin, T., Thirunarayan, K. (eds.) ISWC 2008. LNCS, vol. 5318, pp. 50–65. Springer, Heidelberg (2008)
5. Denaux, R., Dimitrova, V., Cohn, A.G., Dolbear, C., Hart, G.: Rabbit to OWL: ontology authoring with a CNL-based tool. In: Fuchs, N.E. (ed.) CNL 2009. LNCS, vol. 5972, pp. 246–264. Springer, Heidelberg (2010)
6. Ferré, S., Rudolph, S.: Advocatus diaboli – exploratory enrichment of ontologies with negative constraints. In: ten Teije, A., Völker, J., Handschuh, S., Stuckenschmidt, H., d'Acquin, M., Nikolov, A., Aussenac-Gilles, N., Hernandez, N. (eds.) EKAW 2012. LNCS, vol. 7603, pp. 42–56. Springer, Heidelberg (2012)
7. Hitzler, P., Krötzsch, M., Rudolph, S.: Foundations of Semantic Web Technologies. Chapman & Hall/CRC, Boca Raton (2009)
8. Keet, C.M., Lawrynowicz, A.: Test-driven development of ontologies. In: Sack, H., Blomqvist, E., d'Aquin, M., Ghidini, C., Ponzetto, S.P., Lange, C. (eds.) ESWC 2016. LNCS, vol. 9678, pp. 642–657. Springer, Heidelberg (2016). doi:10.1007/978-3-319-34129-3_39
9. Liebig, T., Noppens, O.: OntoTrack: a semantic approach for ontology authoring. Web Semant. **3**(2), 116–131 (2005)
10. Noy, N., Sintek, M., Decker, S., Crubezy, M., Fergerson, R., Musen, M.: Creating semantic web contents with Protege-2000. IEEE Intell. Syst. **16**(2), 60–71 (2001)
11. Poveda-Villalón, M., Suárez-Figueroa, M.C., Gómez-Pérez, A.: Validating ontologies with OOPS!. In: ten Teije, A., Völker, J., Handschuh, S., Stuckenschmidt, H., d'Acquin, M., Nikolov, A., Aussenac-Gilles, N., Hernandez, N. (eds.) EKAW 2012. LNCS, vol. 7603, pp. 267–281. Springer, Heidelberg (2012)
12. Rector, A., Drummond, N., Horridge, M., Rogers, J., Knublauch, H., Stevens, R., Wang, H., Wroe, C.: OWL pizzas: practical experience of teaching OWL-DL: common errors & common patterns. In: Motta, E., Shadbolt, N.R., Stutt, A., Gibbins, N. (eds.) EKAW 2004. LNCS, vol. 3257, pp. 63–81. Springer, Heidelberg (2004)
13. Ren, Y., Parvizi, A., Mellish, C., Pan, J.Z., van Deemter, K., Stevens, R.: Towards competency question-driven ontology authoring. In: Presutti, V., d'Amato, C., Gandon, F., d'Aquin, M., Staab, S., Tordai, A. (eds.) ESWC 2014. LNCS, vol. 8465, pp. 752–767. Springer, Heidelberg (2014)
14. Rudolph, S.: Acquiring generalized domain-range restrictions. In: Medina, R., Obiedkov, S. (eds.) ICFCA 2008. LNCS (LNAI), vol. 4933, pp. 32–45. Springer, Heidelberg (2008)
15. Shearer, R., Motik, B., Horrocks, I.: Hermit: a highly-efficient OWL reasoner. In: OWLED, vol. 432 (2008)

An RDF Design Pattern for the Structural Representation and Querying of Expressions

Sébastien Ferré(✉)

IRISA, Université de Rennes 1, Campus de Beaulieu, 35042 Rennes Cedex, France
ferre@irisa.fr

Abstract. Expressions, such as mathematical formulae, logical axioms, or structured queries, account for a large part of human knowledge. It is therefore desirable to allow for their representation and querying with Semantic Web technologies. We propose an RDF design pattern that fulfills three objectives. The first objective is the structural representation of expressions in standard RDF, so that expressive structural search is made possible. We propose simple Turtle and SPARQL abbreviations for the concise notation of such RDF expressions. The second objective is the automated generation of expression labels that are close to usual notations. The third objective is the compatibility with existing practice and legacy data in the Semantic Web (e.g., SPIN, OWL/RDF). We show the benefits for RDF tools to support this design pattern with the extension of SEWELIS, a tool for guided exploration and edition, and its application to mathematical search.

1 Introduction

Complex expressions account for a large part of human knowledge. Common examples are mathematical equations, logical formulae, regular expressions, or parse trees of natural language sentences. In the Semantic Web (SW) [4], they can be OWL axioms, SWRL rules, or SPARQL queries. It is therefore desirable to allow for their representation in RDF so that they can be mixed with other kinds of knowledge. For example, it should be possible to describe a theorem by its author, its discovery date, its informal description as a text, and its formal description as a mathematical expression, all in RDF. An expected advantage of the formal representation of expressions is the ability to search those expressions by their content, which is a problem that has been studied in *mathematical search* [1,3,6,12,16]. For example, we may wish to retrieve all expressions that are an integral in some variable x and whose body contains x^2 as a sub-expression. Correct answers are $\int x^2 + 1\,dx$ and $\int y^2 - y\,dy$. This example exhibits two difficulties in expression search. The first difficulty is to take into account the nested structure of expressions, e.g., the fact that the sub-expression x^2 must be in the scope of the integral. The second difficulty is to abstract over the name of bound variables, e.g., the variable x bound by the integral $\int\ dx$ can be renamed as y without changing the meaning of the expression. Another difficulty that we do not consider here is the need for logical

E. Blomqvist et al. (Eds.): EKAW 2016, LNAI 10024, pp. 196–211, 2016.
DOI: 10.1007/978-3-319-49004-5_13

and mathematical reasoning to recognize, for instance, that $x(x+1)$ or $\Sigma_{i=1}^{x} i$ implicitly contains x^2. The two difficulties above also apply to other kinds of expressions, and this paper addresses the more general problem of the structural representation and querying of expressions, only using mathematical expressions as a representative illustration.

The need for representing expressions in RDF has already been recognized, but to our knowledge, no generic solution has been proposed. In the domain of mathematical search, the survey by Lange [7] shows that few complete RDF representations have been proposed [10,14], and that none of them have been implemented and adopted because of awkward representations, and lack of support by RDF tools. More conservative approaches are easier to adopt but miss the objective of a tight integration of mathematical knowledge to the Semantic Web. For example, XML literals can be used to represent mathematical objects in RDF graphs but the content of literals is mostly opaque to RDF tools. In the Semantic Web, a well-known use case is the representation of complex classes in OWL ontologies. To this purpose, the OWL vocabulary defines a number of *structural* classes (e.g., `owl:Restriction`) and properties (e.g., `owl:onProperty`), which have no ontological meaning in themselves. Another example of a vocabulary that defines structural classes and properties is SPIN SPARQL Syntax [15], for the representation of SPARQL queries. OWL/RDF and SPIN use similar RDF patterns in their representations, and therefore offer a good basis for generalization.

In this paper, we propose an RDF design pattern for expressions that fulfills three objectives. The first objective (Sect. 2) is the structural representation of expressions in standard RDF, so that expressive structural querying (e.g., mathematical search) is made possible. We propose to re-use the standard RDF containers, and we introduce a small extension of Turtle and SPARQL notations for a more concise notation of descriptions and queries. The second objective (Sect. 3) is the generation of expression labels that are close to usual notations (e.g., using infix operators, symbols). This is important for RDF tools because expressions are generally represented by blank nodes, and because it would be tedious to manually attach a label to every expression and sub-expression. The third objective (Sect. 4) is the backward compatibility of systems using our expression design pattern with existing practice, and legacy data, in the Semantic Web (e.g., OWL/RDF, SPIN). This implies that legacy data need not be changed in order to benefit from the advantages of our design pattern, in particular the generation of labels. We claim that RDF tools can be adapted to support our design pattern, and that such support can enable users to describe and query complex expressions of any kind. We illustrate this (Sect. 5) by adapting SEWELIS [2], a tool for the guided exploration and edition of RDF graphs, and by applying it to mathematical search. We show (Sect. 6) that this approach is competitive with state-of-the-art in mathematical search w.r.t. expressiveness and readability of queries.

2 Representation of Expressions in RDF

In order to represent expressions in RDF, we rely on the fact that every expression can be represented as a syntax tree, and hence as a graph. A side-advantage of graphs compared to syntax trees is the possibility to share sub-expressions. Syntax tree leaves, i.e. *atomic expressions*, can be symbols, values, and variables; and syntax tree nodes, i.e. *compound expressions*, are labeled by symbols (e.g., operators, functions, quantifiers). Symbols are constants with a universal scope, and therefore, they are naturally represented by URIs. Note that those URIs could be derived from existing vocabularies, such as OpenMath[1]. A benefit of URIs is that, when a symbol (e.g., e) has different meanings (e.g., the base of the natural logarithm or the elementary charge), different URIs can be used to avoid ambiguity. A symbol can be linked to a URI as a label to specify its surface form, and nothing forbids different URIs to have the same label. Values such as numbers and strings are naturally represented by RDF literals. To account for the three facts that a variable can have several occurences, that distinct variables may have the same name, and that variables are not accessible out of their scope, we choose to represent variables as blank nodes, and to represent variable names as labels of those blank nodes. The same choice has been made in SPIN.

It remains to define the representation of compound expressions, i.e. tree nodes. A compound expression is completely defined by the node label, which we here call *constructor*, and the sequence of sub-expressions, which we here call *arguments*. We propose to represent a compound expression by RDF containers. An RDF container is generally a blank node (it can also be a URI), has a type (e.g., `rdf:Seq` for sequences), and is linked to its elements through the predefined predicates `rdf:_1` (1st element), `rdf:_2` (2nd element), etc. The idea is here to use the constructors of compound expressions (i.e., node labels) as container types, i.e. as subclasses of `rdfs:Container`. In each compound expression, the arity of the constructor determines the number of arguments. That representation is close to what is done in OWL/RDF or SPIN, except that membership properties `rdf:_`n are used for the arguments instead of *ad-hoc* properties (e.g., `owl:onProperty`). Section 4 explains how to reconcile the two approaches in practice.

The advantages of our RDF representation are its simplicity, and its conformance to existing standard. Indeed, no special vocabulary needs to be defined apart from the URIs that are used for symbols. It is true that there is little support for containers in existing tools. However, we show in this paper that notations and tools can be extended at a moderate cost, and it will be arguably easier to convince their developers to do so for a standard RDF notion, containers, than for a domain-specific vocabulary. Moreover, support for containers could be useful in other contexts than the management of expressions.

Figure 1 shows the graphical and Turtle forms of the RDF representation of the expression $\int x^2 + 1\,dx$. This expression contains mathematical operators

[1] http://www.openmath.org/cdindex.html.

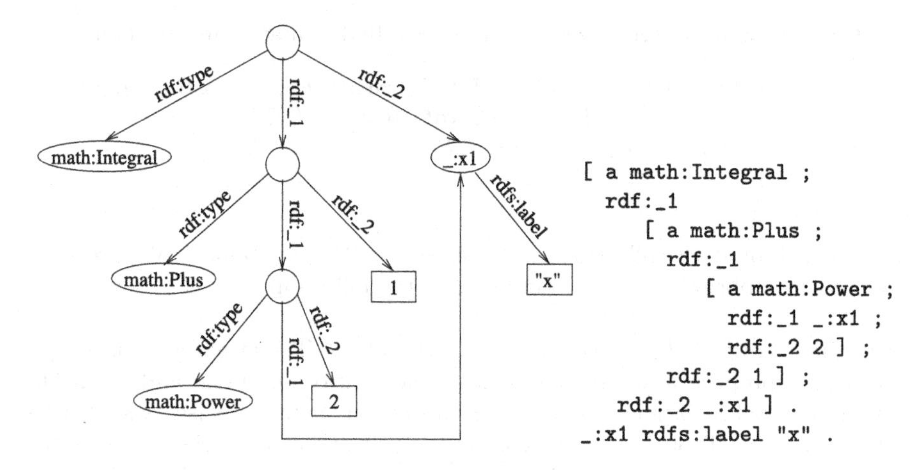

Fig. 1. RDF representation of the expression $\int x^2 + 1\,dx$: graphical (a), Turtle (b).

`math:Integral`[2] (integral), `math:Plus` (addition) and `math:Power` (power), which are all used as binary constructors. In the case of `math:Integral`, the second argument plays the role of the binding variable of the integral. The expression also contains the integer literals `1` and `2`, as well as a blank node `_:x1`, labelled `"x"`, to represent the variable x. The standard property `rdf:type` is used to link compound expressions to their constructor. This implies that a constructor is also an RDF class that contains all compound expressions based on that constructor. The standard properties `rdf:_`n are used to link compound expressions to their arguments. The standard property `rdfs:label` is used to link variable identifiers to variable names. In the Turtle notation, every compound expression appears as delimited by square brackets, and the standard property `rdf:type` is abbreviated by `a` (meaning "is a").

This representation correctly handles the full content of the expression while abstracting over possible variations in the presentation (e.g., brackets, various notations of integrals). This representation also makes it possible to distinguish different variables that have the same name (e.g., in $x + \int x\,dx$) by using distinct blank nodes. Invariance to the renaming of bound variables is addressed by separating the identity of the variable (as a blank node) and its concrete name (as a label). Indeed, invariance to renaming also applies to blank nodes: `_:x1` can be replaced by `_:x2` without changing the meaning of the RDF graph.

About RDF Collections. An alternative representation of compound expressions would be RDF collections (aka. RDF lists) whose elements would be the constructor followed by arguments. However it would require twice as many triples, and would make querying more costly. Its notation in Turtle and SPARQL would be more concise for expression $\int x^2 + 1\,dx$: `(math:Integral (math:Plus (math:Power _:x1 2) 1) _:x1)`. However, it is only because abbreviations have

[2] We here assume a namespace `math:` for mathematical constructors.

Table 1. Expansion rules for new Turtle/SPARQL syntactic abbreviations.

$$C(E_1,\ldots,E_n) \longrightarrow \texttt{[a } C\texttt{; rdf:\_1 } E_1\texttt{; ...; rdf:\_}n\ E_n\texttt{]}$$
$$\ldots E \ldots \longrightarrow \texttt{[rdfs:member* } E\texttt{]}$$
$$S\texttt{ is [}P_1\ O_1\texttt{; ...; }P_n\ O_n\texttt{]} \longrightarrow S\ P_1\ O_1\texttt{; ...; }P_n\ O_n$$

been defined for RDF collections and not yet for RDF containers. We precisely propose such abbreviations for containers in the following.

Turtle Abbreviations for Expressions. As the above Turtle notation is rather verbose, we propose a small extension of the syntax of Turtle and SPARQL with a few abbreviations in order to allow for more concise descriptions and queries (see Table 1). Those abbreviations work exactly the same as the Turtle/SPARQL notation for RDF collections (a.k.a. lists), where the form $(E_1 \ldots E_n)$ is expanded into the form `[rdf:first` E_1 `; rdf:rest` ... `[rdf:first` E_n `; rdf:rest rdf:nil]` ...]. The first abbreviation is a functional notation for RDF containers (hence for expressions), where container types (hence constructors) play the role of functions, and container elements play the role of arguments. The second and third abbreviations are ellipsis notations to reach sub-expressions in queries, and rely on transitive closures of the property `rdfs:member`, which is a super-property of the properties `rdf:_`n. Those notations can be used everywhere blank nodes can be used, including as a whole statement. The last abbreviation allows blank nodes, and hence the functional and ellipsis notations, to be used as a predicate-object list by prefixing it with the keyword `is`. With this extension, which we name Turtle+/SPARQL+ in the scope of this paper, the example in Fig. 1 can be written as follows, using the first abbreviation.

```
math:Integral(math:Plus(math:Power(_:x1,2),1),_:x1) .
_:x1 rdfs:label"x" .
```

Other abbreviations are illustrated below in queries. Note that it is easy to extend Turtle and SPARQL parsers to accept those abbreviations, and that no change at all is required on the RDF data model or the SPARQL query language.

SPARQL Expression Patterns. In order to validate the adequacy of our representation for structural search in SPARQL, we discuss the formulation of SPARQL queries for a few representative examples. SPARQL graph patterns are here used to constrain the shape of searched expressions. SPARQL variables are here used to match arbitrary sub-expressions, and to state equality constraints between several sub-expressions. This provides a way to solve the difficulty related to the renaming of bound variables: it suffices to introduce a SPARQL variable for each bound variable. For example, if we look for expressions like $\int x^2\,dx$ or $\int y^2\,dy$, we can use the following SPARQL+ query, where SPARQL variables stand for expressions (not values!).

```
SELECT ?e WHERE { ?e is math:Integral(math:Power(?x,2),?x) . }
```

The query in the introduction that retrieves the integrals in x whose body contains the term x^2 can then be expressed as follows, using the notation ...E... (see Table 1) to express the relation from an expression to its sub-expressions.

```
SELECT ?e WHERE { ?e is math:Integral(...math:Power(?x,2)..., ?x).}
```

This query returns the expressions $\int x^2\,dx$, $\int x^2 + 1\,dx$, $\int y^2 - y\,dy$, but not the expressions $\int x^2 + y\,dy$ and $\int 2x\,dx = x^2 + c$. Now, starting from the same example, assume that we want to retrieve the bodies of the integrals instead of the integrals themselves. After a reformulation to introduce a variable for the body of the integral, we obtain:

```
SELECT ?e WHERE { ?e is ...math:Power(?x,2)... .
                  math:Integral(?e,?x) . }
```

This query looks for an expression that contains x^2 as a sub-expression, and that appears as the body (1st argument) of an integral whose binding variable is x. As a conclusion, SPARQL provides enough expressivity to cover the needs of the structural querying of expressions. A comparison with existing query languages for mathematical search is given in Sect. 6.2.

3 Labelling of Expressions in RDF Tools

The main reason why RDF structural representations have not been widely adopted is the lack of support by RDF tools. This lack of support concerns in particular n-ary structures such as RDF containers and RDF collections (chained lists), which are necessary for the representation of expressions [7]. More essentially, this lack of support concerns blank nodes, which are often associated to n-ary structures. The problem with blank nodes is that they are notoriously difficult to present in query results, and in RDF tools in general [8]. Indeed, blank nodes have no names (anonymous resources), and their identifiers are contingent, and only relevant for internal use.

Our proposal is therefore to display blank nodes by the RDF structure they represent, rather than by their internal identifier as this is generally done. Turtle+ can be used to render expression contents. For example, the expression $\int x^2 + 1\,dx$ can be rendered by the string "`math:Integral(math:Plus(math:Power([rdfs:label "x"],2),1),[rdfs:label "x"])`". This is more concise than using standard Turtle, but this is still far from the usual mathematical notation.

We here propose to express annotations on constructors that can be used by RDF tools to generate usual representations of expressions. A similar approach has been followed in XML for the rendering of mathematical expressions from OpenMath and MathML-Content to LaTeX or MathML-Presentation [5]. The principle is that, when an expression has not been annotated explicitly with a label, a label will automatically be generated for it as an aggregation of the labels of its parts. The generated label need not be added to the store, but may simply be generated dynamically by the tool, as needed. By default, the functional

Table 2. Notation specifications (template and priorities) for a few constructors.

Constructor	Template	Priorities
math:Plus	`_ + _`	plus(plus,plus)
math:Minus	`_ - _`	plus(plus,times)
math:Times	`_ _`	times(times,times)
math:Power	`_^_`	power(atom,power)
math:Sin	`sin _`	plus(times)
math:Fact	`_!`	power(atom)
math:Integral	`∫ _ d_`	atom(plus,atom)
math:Set	`{_, _}`	atom(plus)

notation is used, like in Turtle+, but replacing the constructor and the arguments by their label. Of course, the label of an argument can itself be generated in the case it is a compound expression. Because those labels are only for display, all Unicode characters can be used, including mathematical symbols (e.g., ∫ for `math:Integral`, + for `math:Plus`, ^ for `math:Power`). Applying this to the above example generates the label "`∫(+(^(x,2),1),x)`" for the expression $\int x^2 + 1\,dx$.

Many mathematical operators use different notations than the functional notation: e.g., infix notation in $x + 1$, prefix notation in $sin\,x$, postfix notations in $n!$, and mixfix notations in $\int x^2\,dx$. With appropriate annotations of constructors, we could generate the label "`∫ x^2 + 1 dx`" for the above example, which is very close to the mathematical notation. However, those notations could lead to ambiguities, and brackets must be inserted according to the priority level of operators. For example, in the expression `math:Div(math:Plus(1,math:Power(_:x,2)),_:x)`, brackets are necessary around the addition, but not around the power, according to usual priorities. Therefore, the minimal bracketing of the expression leads to the label "`(1 + x^2)/x`". Without the brackets, the expression would be misinterpreted by human users, and adding superfluous brackets would make the expression label less readable.

We propose to use classical pretty-printing techniques to generate expression labels, based on the fixity and priority of operators. In order to allow RDF tools to perform this pretty-printing in a generic way, it is necessary to annotate constructors with all the necessary information. The necessary information comprises the *template* and the *priority signature* of the constructor. Table 2 gives the template and priority signature for a few constructors. A template is a string where the markers `_1, ..., _n` are placeholders for the (generated) labels of arguments. For example, a template for the addition is "`_1 + _2`". When arguments are placed in order, the generic marker `_` can be used instead: e.g., "`_ + _`". Alternately to strings, templates could be defined as XML literals, e.g., using the MathML presentation language, which would allow for a nice rendering in a browser [5]. A priority signature uses the functional notation with priority levels in place of the constructor and each argument. Priority levels are here named after their most representative operator, and are ordered in the usual way (e.g., 'times' represents higher priority than 'plus')[3]. For example, the priority signature of math:Minus, `plus(plus,times)`, says that substraction has the same priority level as addition, and is left associative. More precisely, it says that additive expressions can be used without brackets as left argument, but must be used with brackets as right argument. The general rule is that an argument expression must be bracketed if its priority level is lower than the priority level of the argument. By default, round brackets are used, but a template with one place holder (e.g., "{_}" for curly brackets) can be associated to each priority level that uses a different bracketing. Finally, some constructors expect a variable number of arguments. For example, the set $\{1, 2, 3\}$ can be represented by `math:Set(1,2,3)`, where the constructor `math:Set` has an arbitrary arity. In this case, we use two placeholders in the template (here "{_, _}") in order to specify the separator (here the comma) to be used between elements; and the priority signature uses only one argument, assuming that all arguments have the same priority level. Examples of generated labels from definitions in Table 2 are: "∫ `x^2 + 1 dx`", "`(a + b)^2 = a^2 + 2 a b + b^2`", "`{1, x, x^2, x^3}`".

4 Compatibility with Legacy RDF Structures

Blank nodes and structural classes/properties have been used in a number of circumstances for representing structures in RDF. For example, in OWL/RDF an existential restriction $\exists r.C$ is represented by a combination of the class `owl:Restriction`, and the properties `owl:onProperty` and `owl:someValuesFrom`, i.e. by the blank node `[ a owl:Restriction ; owl:onProperty` r `; owl:someValuesFrom` C `].` . The same representation principles are used for RDF collections with class `rdf:List` and properties `rdf:first` and `rdf:rest`, for SPIN SPARQL syntax, and in other circumstances [8].

OWL restrictions could equally well be represented as expressions, using our approach. Assuming the constructor `owl:Some` in the OWL namespace, an existential restriction $\exists r.C$ would be represented as `owl:Some` (r,C). That representation is close to OWL functional syntax[4], and are valid notations in Turtle+. A first advantage of our approach is that each construct is defined by a single constructor URI instead of a combination of classes and properties. A second advantage is a better separation between ontological classes and properties (e.g., `owl:equivalentClass`) and structural classes and properties (e.g., `owl:onProperty`). In our approach, the latter are only the constructors and the container membership properties `rdf:_`n. A third advantage is that natural labels for expressions can be generated in a more systematic way, as explained in Sect. 3. Indeed, a system only needs to read the annotations of constructors, and needs not be hard-coded w.r.t. an *ad-hoc* vocabulary. An advantage of OWL/RDF and similar approaches is that arguments have an explicit name rather than a position.

---

[3] The precise representation of priority levels is not detailed here. URIs could be used, possibly from a standard vocabulary, and compared with a standard transitive property, e.g., prio:isHigherThan.

[4] http://www.w3.org/TR/owl2-syntax/#Functional-Style_Syntax.

Table 3. Definitions of a few implicit constructors for OWL/RDF and SPIN.

Implicit constructor	Constructor class	Argument properties
owl:Some	owl:Restriction	(owl:onProperty,owl:someValuesFrom)
owl:All	owl:Restriction	(owl:onProperty,owl:allValuesFrom)
owl:And	owl:Class	(owl:intersectionOf)
sp:Select	sp:Select	(sp:resultVariables,sp:where)
sp:TriplePattern	sp:TriplePattern	(sp:subject,sp:predicate,sp:object)
sp:Filter	sp:Filter	(sp:expression)
sp:lt	sp:lt	(sp:arg1,sp:arg2)

Table 4. Different notations of a complex OWL class: Manchester (1), Turtle (2), Turtle+ (3), generated label (4).

1	*Pizza and hasTopping some MeatTopping and hasTopping some FishTopping*

2	`[ a owl:Class ;`
	`  owl:intersectionOf`
	`    ( ex:Pizza`
	`      [ a owl:Restriction ;`
	`        owl:onProperty ex:hasTopping ;`
	`        owl:someValuesFrom ex:MeatTopping ]`
	`      [ a owl:Restriction ;`
	`        owl:onProperty ex:hasTopping ;`
	`        owl:someValuesFrom ex:FishTopping ] ) ]`

3	`owl:And( ( ex:Pizza`
	`  owl:Some(ex:hasTopping,ex:MeatTopping)`
	`  owl:Some(ex:hasTopping,ex:FishTopping) ) )`

4	`"Pizza and has topping some Meat and has topping some Fish"`

In order to reconcile legacy data and the naming of arguments with functional notations and the systematic labelling of expressions, we introduce *implicit constructors*. An implicit constructor does not occur explicitly in the RDF graph of expressions, but it is mapped to a combination of structural classes and properties. It serves to translate from functional notation to standard RDF, and as a handle for the annotations about the generation of labels. Table 3 defines a few implicit constructors for OWL/RDF and SPIN expressions, annotating each implicit constructor *Cons* by a class C, and a sequence of properties $(P_1, \dots, P_n)$. Given such annotations, a Turtle+ expression $Cons(E_1, \dots, E_n)$ is translated to the blank node `[ a` C `;` P_1 E_1 `; ... ;` P_n E_n `]`. For example, `owl:Some` is defined as an implicit constructor for existential restrictions in OWL/RDF. From there, it is easy to get any of the three main syntaxes for OWL restrictions as generated labels, e.g. for restriction $\exists child.Doctor$. For the functional syntax, it is enough to define the label "`some`" on constructor `owl:Some` to produce label "`some(child,Doctor)`". For the Manchester syntax, `owl:Some` has to be defined as

Table 5. Different notations of a complex SPARQL query: SPARQL (1), Turtle (2), Turtle+ (3), generated label (4).

1	`SELECT ?x WHERE { ?x ex:age ?age . FILTER (?age < 18) }`

2	`[ a sp:Select ;`
	`  sp:resultVariables (_:x) ;`
	`  sp:where ([ a sp:TriplePattern ;`
	`              sp:subject _:x ;`
	`              sp:predicate ex:age ;`
	`              sp:object _:age ]`
	`            [ a sp:Filter ;`
	`              sp:expression [ a sp:lt ;`
	`                              sp:arg1 _:age ;`
	`                              sp:arg2 18 ] ]) ] .`
	`_:x a sp:Variable ; sp:varName "x" .`
	`_:age a sp:Variable ; sp:varName "age" .`

3	`sp:Select((_:x),`
	`          (sp:TriplePattern(_:x,ex:age,_:age)`
	`           sp:Filter(sp:lt(_:age,18)))) .`
	`_:x is sp:Variable("x") .`
	`_:age is sp:Variable("age") .`

4	`"SELECT ?x WHERE { ?x has age ?age . FILTER (?age < 18) }"`

a right-associative infix operator, like the power operator, but with template "_ some _": this produces label "child some Doctor". For the DL syntax, the constructor has instead to be defined as a mixfix operator with template "∃ _._", producing label "∃child.Doctor".

Tables 4, and 5 compare different notations of two complex expressions in two different languages: OWL, and SPARQL. In each table, the first line is the native syntax of the language (Manchester syntax for OWL). The second line is the Turtle notation of the RDF representation (OWL/RDF for OWL, SPIN for SPARQL). The third line is the Turtle+ functional notation based on implicit constructors. The fourth line is the generated label assuming that implicit constructors have been defined, and that appropriate labelling notations have been associated to them. Note how the generated label can be made very similar to the native syntax. For the OWL Manchester syntax, `owl:And` has a collection argument whose separator is " and ", and `owl:Some` is defined as a right associative infix operator. `ow:Some` is given higher priority than `owl:And`. For SPIN SPARQL queries, `sp:Select` has two collection arguments whose separators are the space character, `sp:TriplePattern` simply uses the template "_ _ _ .", `sp:lt` is defined as an infix operator, and `sp:Variable` is defined as a prefix operator with template "?_". The priority level for atomic graph patterns uses the template "{ _ }" as brackets instead of the default round brackets.

5 Implementation in SEWELIS and Application to Mathematical Search

We have evaluated our RDF design pattern by implementing it in an RDF tool, and by applying it to mathematical search. We chose to add expression support to SEWELIS, an RDF tool for the exploration and authoring of RDF graphs [2]. SEWELIS enables users to interactively build complex queries without the need to write anything, and hence without the risk of syntax errors. At each interaction step, query elements are computed from the dataset, and suggested to users for insertion in the query under construction. Therefore, the query and the results can be pretty-printed without the risk to introduce ambiguities. The same can be done in principle with other syntax-based editors, but SEWELIS has the additional advantage to be also semantic-based in that it provides only suggestions that match actual data, and it prevents users to fall on empty results. It also has the benefits to support exploratory search [9], when users do not have a precise idea of what they are looking for (Table 6).

Table 6. Different notations of an ingredient description: English (1), Turtle (2), Turtle+ (3), generated label (4).

1	*1 lb of green mango*

2	`[ a ex:Ingredient ;`
	`ex:ingredient ex:GreenMango ;`
	`ex:amount [ a ex:Measure ;`
	`ex:value 1 ;`
	`ex:unit ex:lb ] ]`

3	`ex:Ingredient(ex:GreenMango,ex:Measure(1,ex:lb))`
4	`"1 lb of green mango"`

We have extended SEWELIS with the RDF representation of expressions, and the automated generation of labels, as presented in previous sections. The pretty-printing of queries is based on the same principles as for the generation of labels (see Sect. 3), extended to expression patterns. Variables are noted like in SPARQL (e.g., `?X`), and a bare question mark `?` is an anonymous variable. The following list shows how the queries from Sect. 2 are displayed in SEWELIS, along with their meaning for recall. The underlined part represents the *focus* that indicates which part of the query answers are to be displayed.

- ∫ `?X^2 d?X`: the integrals in x of x^2;
- ∫ `...?X^2... d?X`: the integrals in x whose body contains x^2;
- ∫ `...?X^2... d?X`: the bodies of the integrals in x that contain x^2.

We have then applied SEWELIS to the exploration of a collection of 70 formulas taken from an official document[5] used for the French high-school final exam.

[5] http://www.lyc-monod-clamart.ac-versailles.fr/IMG/pdf/FormulaireBac2003.pdf.

```
(u / v) ' = (u ' v - u v ') / v^2
(1 / u) ' = - u ' / u^2
a^3 - b^3 = (a - b) (a^2 + a b + b^2)
a^3 + b^3 = (a + b) (a^2 - a b + b^2)
(a - b)^2 = a^2 - 2 a b + b^2
(a - b)^3 = a^3 - 3 a^2 b + 2 a b^2 - b^3
(a + b)^2 = a^2 + 2 a b + b^2
(a + b)^3 = a^3 + 3 a^2 b + 2 a b^2 + b^3
```

Fig. 2. The list of formulas containing a square, as displayed in SEWELIS.

This is a small dataset but our purpose is to demonstrate the applicability and benefits of our RDF design pattern, and not the scalability of SEWELIS and RDF representations. Note that our RDF expressions are made of standard RDF for which efficient stores and query engines exist. Figure 2 shows the subset of those formulas that contain a square, as displayed in SEWELIS. This list has been obtained as the answers to the query `...?^2...`, which can be reached in three navigation steps from the empty query: `...?...` (*contains...*), `...?^?...` (*a power...*), `...?^2...` (*of 2*). Figure 3 shows a complete screenshot of SEWELIS during the construction of a query. The query is at the left, and states that *the limit at positive infinity of something is equal to something*. The textfield marks the position of the focus, here on the body of the `limit` constructor. The middle column suggests the possible constructors at the focus, and the right column lists the possible expressions at the focus. The latter therefore contains the answers to the current query with respect to the current focus. Note that no blank node identifier is visible thanks to generated labels, even though all expressions *are* blank nodes. Suggestions can also be found and selected by auto-completion from the focus textfield. The next step for the user could be to select one of those expressions, and then to move the focus after the equal sign in order to discover the value of the limit for the chosen expression.

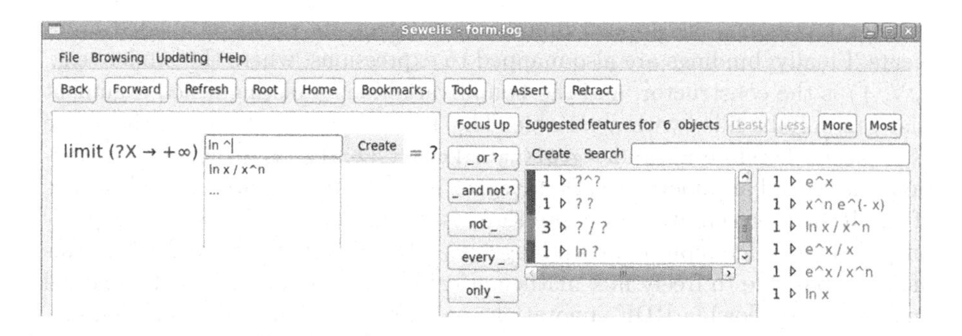

Fig. 3. A screenshot of SEWELIS where the current query retrieves the expressions whose limit at positive infinity is equal to something.

6 Related Work

We compare our approach first with other languages for representing expressions, and second with query languages for structural search among expressions. The latter are used for instance in proof assistants, such as Coq [3].

6.1 Representation Languages

The reference language for the representation of mathematical expressions is MathML [11], an XML dialect. In fact, MathML defines two languages: a *presentation* language, and a *content* language. Only the latter interests us because it represents the logical structure of expressions, and avoids ambiguity problems (e.g., the letter e that denotes either the Neperian constant or a variable) as well as synonymy problems (e.g., x/y and $\frac{x}{y}$ for division). The LaTeX language plays the same role as the presentation language of MathML, and therefore exhibits the same problems. However, presentation languages can be used for the labelling templates of expression constructors.

The (strict) content language of MathML is based on a small number of XML tags that encapsulate different types of expressions: `<cn>` (numbers), `<ci>` (identifiers), `<csymbol>` (symbols), `<cs>` (strings), `<apply>` (applications of functions and operators), `<bind>` and `<bvar>` (bindings). For example, the expression $\int x^2\,dx$ has the following MathML representation:

```
<bind><csymbol>integral</csymbol>
  <bvar><ci>x</ci></bvar>
  <apply><csymbol>power</csymbol> <ci>x</ci> <cn>2</cn> </apply>
</bind>
```

RDF expressions are expressive enough to represent all MathML contents. Numbers and strings are naturally mapped to RDF literals of different datatypes (e.g., `xsd:integer` for integers, `xsd:string` for strings). Symbols (e.g., functions, operators, constants) are naturally mapped to URIs, ideally defined in standard vocabularies. Identifiers are mapped to variables, i.e. blank nodes. Applications (of a function to arguments) are mapped to RDF expressions, where the constructor represents the applied function, and elements represent passed arguments. Finally, bindings are also mapped to expressions, where the binder (e.g., $\exists$, $\forall$, $\int$) is the constructor, and the bound variable is a distinguished argument that can only be filled with a variable. The Turtle+ representation of the above example is therefore `math:Integral(math:Power(_:x,2),_:x)`, where by convention, the second argument of `math:Integral` is the bound variable. An advantage of the RDF representation of expressions is its interoperability with a general-purpose knowledge representation language, RDF. Compared to MathML, this makes it possible to freely mix mathematical knowledge and non-mathematical knowledge by allowing RDF annotations on expressions and sub-expressions.

6.2 Query Languages

The approaches that consist in applying textual search methods by linearizing expressions [12] cannot correctly account for the nested structure of expressions, and for the issue of bound variables [16]. When looking for integrals in x whose body contains x^2, a textual search would have false positives such as $\int 2x\,dx = x^2 + c$ (x^2 is not in the scope of $\int$), and would have false negatives such as $\int y^2 - y\,dy$ (y instead of x). On the contrary, the approaches based on a structured query language [1,3,6,13] correctly account for nested structures and bound variables by reasoning directly on the structure of expressions, and by using *wild cards* as place-holders for variables and sub-expressions. MathWebSearch [6] defines an XML query language that extends MathML. Its XML syntax makes it difficult to use, and its expressiveness is limited. For example, it cannot express the relation between an expression and its sub-expressions (e.g., `...x^2...` in SPARQL+). The query language QMT by Rabe [13] has much in common with RDF and SPARQL, despite having a different style. However, it is applied at the theory level of mathematical knowledge, and provides no special support for expression patterns even though it inherits expression representations from OpenMath objects. For example, the above query can be expressed in QMT as follows (XQuery-style): **for** e_1 **in** obj(integral) **and** e_2 **in** subobj(arg$_1$(e_1), power) **where** arg$_1$(e_2)= var$_1$(e_1) **and** arg$_2$(e_2) =2 **return** e_1. The query language that is most similar to our approach is from Altamimi and Youssef [1]. They use an ASCII notation that is similar to the LaTeX notation, and a set of 6 wild cards that can be used in place of: one or several characters, one or several atoms, one or several expressions. The above example is expressed by the query `\int ...$1^2... d$1`. Our approach has a higher expressiveness, and a number of additional advantages for users. SPARQL and SEWELIS have a higher expressiveness by offering disjunction, negation, and the possibility to search for sub-expressions appearing in some context. For example, it is possible to look for the bodies of integrals in x that contain y^2 or y^3, where y is not x: $\int$ `...(not ?X)^(2 or 3)... d?X`, as displayed in SEWELIS. The 6 kinds of wild cards are covered by the combination of: the empty pattern `?` (a universal wild card), SPARQL variables for co-occurences of a same sub-expression, ellipsis ... for reaching sub-expressions, and classical queries for constraining the name and type of the atoms of expressions. An additional advantage when using SEWELIS is that users do not need to master the syntax of the query language because they are guided step after step in the construction of queries. This comes with the guarantee of non-empty results. Another advantage is that pretty-printing (UTF-8 characters, mixfix notations, etc.) can be used for expressions and queries because query elements are selected, not written.

7 Conclusion

We have proposed an RDF design pattern for the representation of expressions as containers that is compatible with existing practice such as in OWL/RDF and SPIN, and that allows for the expressive structural querying of expressions based on their contents and context of occurence. With a simple syntactic extension of Turtle and SPARQL, those expressions can be noted in a concise and familiar way, i.e. using the functional notation. With labelling annotations on expression constructors, human-readable labels can be automatically generated for each expression. We have adapted an existing RDF tool, SEWELIS, to support the presentation and manipulation of such expressions. As SEWELIS is a rather complex system, we are confident that similar adaptations can be applied to other RDF tools. We have also shown the benefits of using SEWELIS on the important task of mathematical search. We believe that the scope of our design pattern goes beyond mathematical expressions, as we have shown in this paper with OWL axioms and SPARQL queries, and that it is relevant to the representation of all kinds of structures and symbolic data in RDF.

References

1. Altamimi, M.E., Youssef, A.S.: A math query language with an expanded set of wildcards. Math. Comput. Sci. **2**(2), 305–331 (2008)
2. Ferré, S., Hermann, A.: Reconciling faceted search and query languages for the Semantic Web. Int. J. Metadata Semant. Ontol. **7**(1), 37–54 (2012)
3. Guidi, F., Schena, I.: A query language for a metadata framework about mathematical resources. In: Asperti, A., Buchberger, B., Davenport, J.H. (eds.) MKM 2003. LNCS, vol. 2594, pp. 105–118. Springer, Heidelberg (2003)
4. Hitzler, P., Krötzsch, M., Rudolph, S.: Foundations of Semantic Web Technologies. Chapman & Hall/CRC, London (2009)
5. Kohlhase, M., Müller, C., Rabe, F.: Notations for living mathematical documents. In: Autexier, S., Campbell, J., Rubio, J., Sorge, V., Suzuki, M., Wiedijk, F. (eds.) AISC 2008, Calculemus 2008, and MKM 2008. LNCS (LNAI), vol. 5144, pp. 504–519. Springer, Heidelberg (2008)
6. Kohlhase, M., Sucan, I.: A search engine for mathematical formulae. In: Calmet, J., Ida, T., Wang, D. (eds.) AISC 2006. LNCS (LNAI), vol. 4120, pp. 241–253. Springer, Heidelberg (2006)
7. Lange, C.: Ontologies and languages for representing mathematical knowledge on the Semantic Web. Semant. Web **4**(2), 119–158 (2013)
8. Mallea, A., Arenas, M., Hogan, A., Polleres, A.: On blank nodes. In: Aroyo, L., Welty, C., Alani, H., Taylor, J., Bernstein, A., Kagal, L., Noy, N., Blomqvist, E. (eds.) ISWC 2011, Part I. LNCS, vol. 7031, pp. 421–437. Springer, Heidelberg (2011)
9. Marchionini, G.: Exploratory search: from finding to understanding. Commun. ACM **49**(4), 41–46 (2006)
10. Marchiori, M.: The mathematical Semantic Web. In: Asperti, A., Buchberger, B., Davenport, J.H. (eds.) MKM 2003. LNCS, vol. 2594, pp. 216–224. Springer, Heidelberg (2003)
11. MathML: Mathematical markup language 3.0, W3C Recommendation (2010). http://www.w3.org/TR/MathML3/
12. Miner, R., Munavalli, R.: An approach to mathematical search through query formulation and data normalization. In: Kauers, M., Kerber, M., Miner, R., Windsteiger, W. (eds.) MKM/CALCULEMUS 2007. LNCS (LNAI), vol. 4573, pp. 342–355. Springer, Heidelberg (2007)

13. Rabe, F.: A query language for formal mathematical libraries. In: Jeuring, J., Campbell, J.A., Carette, J., Dos Reis, G., Sojka, P., Wenzel, M., Sorge, V. (eds.) CICM 2012. LNCS, vol. 7362, pp. 143–158. Springer, Heidelberg (2012)
14. Robbins, A.: Semantic MathML (2009). http://straymindcough.blogspot.fr/2009/06/
15. SPIN - SPARQL syntax, W3C Member Submission (2011). http://www.w3.org/Submission/2011/SUBM-spin-sparql-20110222/
16. Youssef, A.M.: Roles of math search in mathematics. In: Borwein, J.M., Farmer, W.M. (eds.) MKM 2006. LNCS (LNAI), vol. 4108, pp. 2–16. Springer, Heidelberg (2006)

Semantic Relatedness for All (Languages): A Comparative Analysis of Multilingual Semantic Relatedness Using Machine Translation

André Freitas[1(✉)], Siamak Barzegar[2], Juliano Efson Sales[1], Siegfried Handschuh[1], and Brian Davis[2]

[1] Department of Computer Science and Mathematics, University of Passau, Innstrasse 43, ITZ-110, 94032 Passau, Germany
{andre.freitas,juliano-sales,siegfried.handschuh}@uni-passau.de

[2] Insight Centre for Data Analytics, National University of Ireland, Galway, IDA Business Park, Lower Dangan, Galway, Ireland
{siamak.barzegar,brian.davis}@insight-centre.org

Abstract. This paper provides a comparative analysis of the performance of four state-of-the-art distributional semantic models (DSMs) over 11 languages, contrasting the native language-specific models with the use of machine translation over English-based DSMs. The experimental results show that there is a significant improvement (average of 16.7 % for the Spearman correlation) by using state-of-the-art machine translation approaches. The results also show that the benefit of using the most informative corpus outweighs the possible errors introduced by the machine translation. For all languages, the combination of machine translation over the Word2Vec English distributional model provided the best results consistently (average Spearman correlation of 0.68).

Keywords: Multilingual distributional semantics · Machine translation

1 Introduction

Distributional Semantic Models (DSM) are consolidating themselves as fundamental components for supporting automatic semantic interpretation in different application scenarios in natural language processing. From *question answering systems*, to *semantic search* and *text entailment*, distributional semantic models support a scalable approach for representing the meaning of words, which can automatically capture comprehensive associative commonsense information by analysing word-context patterns in large-scale corpora in an unsupervised or semi-supervised fashion [8,18,19].

However, distributional semantic models are strongly dependent on the size and the quality of the reference corpora, which embeds the commonsense knowledge necessary to build comprehensive models. While high-quality texts

E. Blomqvist et al. (Eds.): EKAW 2016, LNAI 10024, pp. 212–222, 2016.
DOI: 10.1007/978-3-319-49004-5_14

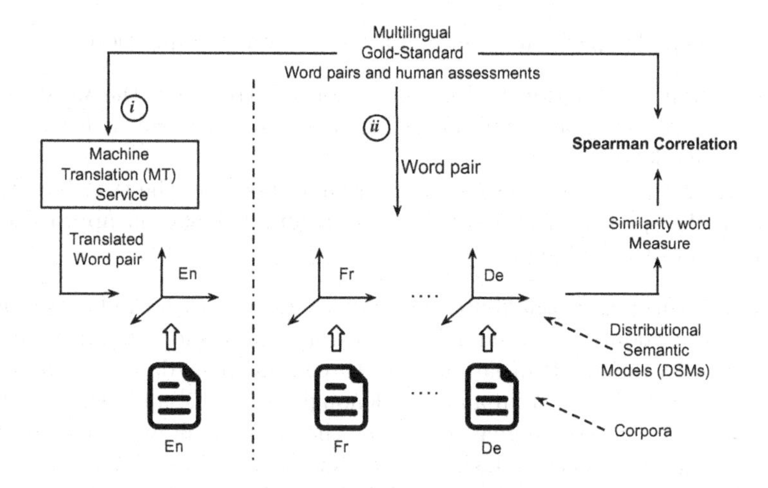

Fig. 1. Depiction of the experimental setup of the experiment.

containing large-scale commonsense information are present in English, such as Wikipedia, other languages may lack sufficient textual support to build distributional models.

To address this problem, this paper investigates how different distributional semantic models built from corpora in different languages and with different sizes perform in computing semantic relatedness similarity and relatedness tasks. Additionally, we analyse the role of machine translation approaches to support the construction of better distributional vectors and for computing semantic similarity and relatedness measures for other languages. In other words, in the case that there is not enough information to create a DSM for a particular language, this work aims at evaluating whether the benefit of corpora volume for English outperforms the error introduced by machine translation.

Given a pair of words and a human judgement score that represents the semantic relatedness of these two words, the evaluation method aims at indicating how close distributional models score to humans. Three widely used word-pairs datasets are employed in this work: Miller and Charles (MC) [14], Rubenstein and Goodenough (RG) [17] and WordSimilarity 353 (WS-353) [7].

In the proposed model the word-pairs datasets are translated into English as a reference language and the distributional vectors are defined over the target end model (Fig. 1). Despite the simplicity of the proposed method based on machine translation, there is a high relevance for the distributional semantics user/practitioner due to its simplicity of use and the significant improvement in the results.

This work presents a systematic study involving 11 languages and four distributional semantic models (DSMs), providing a comparative quantitative analysis of the performance of the distributional models and the impact of machine translation approaches for different models.

In summary, this paper answers the following research questions:

1. Does machine translation to English perform better than the word vectors in the original language (for which languages and for which distributional semantic models)?
2. Which DSMs and languages benefit more and less from the translation?
3. What is the quality of state-of-the-art machine translation approaches for word pairs (for each language)?

Moreover, this paper contributes with two resources which can be used by the community to evaluate multi-lingual semantic similarity and relatedness models: *(i)* a high quality manual translation of the three word-pairs datasets - Miller and Charles (MC) [14], Rubenstein and Goodenough (RG) [17] and WordSimilarity 353 (WS-353) [7] - for 10 languages and *(ii)* the 44 pre-computed distributional models (four distributional models for each one of the 11 languages) which can be accessed as a service[1], together with the multi-lingual approaches mediated by machine translation.

This paper is organised as follows: Sect. 2 describes the related work, Sect. 3 describes the experimental setting; while Sect. 4 analyses the results and provides the comparative analysis from different models and languages, Finally, Sect. 5 provides the conclusion.

2 Related Work

Most of related work has concentrated on leveraging joint multilingual information to improve the performance of the models.

Faruqui and Dyer [6] use the distributional invariance across languages and propose a technique based on canonical correlation analysis (CCA) for merging multilingual evidence into vectors generated monolingually. They evaluate the resulting word representations on semantic similarity/relatedness evaluation tasks, showing the improvement of multi-lingual over the monolingual scenario.

Utt and Pado [20], develop methods that take advantage of the availability of annotated corpora in English using a translation-based approach to transport the word-link-word co-occurrences to support the creation of syntax-based DSMs.

Navigli and Ponzetto [15] propose an approach to compute semantic relatedness exploiting the joint contribution of different languages mediated by lexical and semantic knowledge bases. The proposed model uses a graph-based approach of joint multi-lingual disambiguated senses which outperforms the monolingual scenario and achieves competitive results for both resource-rich and resource-poor languages.

Zou et al. [21] describe an unsupervised semantic embedding (bilingual embedding) for words across two languages that represent semantic information of monolingual words, but also semantic relationships across different languages. The motivation of their works was based on the fact that it is hard to identify

[1] The service is available at http://rebrand.ly/dinfra.

semantic similarities across languages, specially when co-occurrences words are rare in the training parallel text. Al-Rfou et al. [1] produced multilingual word embeddings for about 100 languages using Wikipedia as the reference corpora.

Comparatively, this work aims at providing a comparative analysis of existing state-of-the-art distributional semantic models for different languages as well as analyzing the impact of a machine translation over an English DSM.

3 Experimental Setup

The experimental setup consists of the instantiation of four distributional semantic models (Explicit Semantic Analysis (ESA) [9], Latent Semantic Analysis (LSA) [12], Word2Vec (W2V) [13] and Global Vectors (GloVe) [16]) in 11 different languages - English, German, French, Italian, Spanish, Portuguese, Dutch, Russian, Swedish, Arabic and Farsi.

The DSMs were generated from Wikipedia dumps (January 2015), which were preprocessed by lowercasing, stemming and removing stopwords. For LSA and ESA, the models were generated using the SSpace Package [11], while W2V and GloVe were generated using the code shared by the respective authors. For the experiment the vector dimensions for LSA, W2V and GloVe were set to 300 while ESA was defined with 1500 dimensions. The difference of size occurs because ESA is composed of sparse vectors. All models used in the generation process the default parameters defined in each implementation.

Each distributional model was evaluated for the task of computing semantic similarity and relatedness measures using three human-annotated gold standard datasets: Miller and Charles (MC) [14], Rubenstein and Goodenough (RG) [17] and WordSimilarity 353 (WS-353) [7]. As these word-pairs datasets were originally in English, except for those language available in previous works ([4,5]), the word pairs were translated and reviewed with the help of professional translators, skilled in data localisation tasks. The datasets are available at http://rebrand.ly/multilingual-pairs.

Two automatic machine translation approaches were evaluated: the Google Translate Service and the Microsoft Bing Translation Service. As Google Translate Service performed 16 % better for overall word-pairs translations, this was set as the main machine translation model.

The DInfra platform [2] provided the DSMs used in the work. To support experimental reproducibility, both experimental data and software are available at http://rebrand.ly/dinfra.

4 Evaluation and Results

4.1 Spearman Correlation and Corpus Size

Table 1 shows the correlation between the average Spearman correlation values for each DSM and two indicators of corpus size: # of tokens and # of unique tokens.

Table 1. Correlation between corpus size and different models.

Gold standard	MC		RG		WS353	
	Unique tokens	Tokens	Unique tokens	Tokens	Unique tokens	Tokens
ESA	0.39	*0.48*	0.67	**0.73**	*0.33*	*0.39*
LSA	**0.74**	**0.75**	**0.82**	0.68	**0.64**	0.66
W2V	0.43	0.58	0.71	0.72	0.57	**0.79**
Glove	*0.34*	0.51	*0.51*	*0.61*	0.59	0.63

ESA is consistently more robust (on average) than the other models in relation to the corpus size due the fact that ESA has larger context windows in opposition to the other distributional models. While ESA considers the whole document as its context window, the other models are restricted to five (LSA) and ten (Word2Vec and GloVe) words.

Another observation is that the evaluation of the WS-353 dataset is more dependent on the corpus size, which can be explained by the broader number of semantic relations expressed under the semantic relatedness umbrella.

Table 2 shows the size of each corpus in different languages regarding the number of unique tokens and the number of tokens.

Table 2. The sizes of the corpora in terms of the number of unique tokens and tokens (scale of 10^6).

Lang	Unique tokens	Tokens
en	4.238	902.044
de	4.233	312.380
fr	1.749	247.492
ru	1.766	202.163
it	1.411	178.378
nl	2.021	105.224
pt	0.873	96.712
sv	1.730	82.376
es	0.829	76.587
ar	1.653	46.481
fa	0.925	32.557

4.2 Word-Pair Machine Translation Quality

The second step evaluates the accuracy of state-of-the-art machine translation approa-ches for word-pairs (Table 3). The accuracy of the translation for the WS-353 word pairs significantly outperforms the other datasets. This shows that the higher semantic distance between word pairs (semantic relatedness) has the

Table 3. Translation accuracy.

Dataset/lang	de	fr	ru	it	nl	pt	sv	es	ar	fa
MC	0.48	0.47	0.58	0.42	0.57	0.60	0.55	0.60	0.53	0.38
RG	0.45	0.65	0.53	0.41	0.59	0.51	0.58	0.59	0.43	0.36
WS353	**0.78**	**0.85**	**0.76**	**0.76**	**0.85**	**0.81**	**0.78**	**0.79**	**0.57**	**0.43**

benefit of increasing the contextual information during the machine translation process, subsequently improving the mutual disambiguation process.

For WS-353 the set of best-performing translations has an average accuracy of 80 % (with maximum 85 % and minimum 76 %). This value dropped significantly for Arabic and Farsi (average 50 %).

For MC and RG, the average translation accuracy for the semantic similarity pairs is 51.5 %. This difference may be a result of a deficit of contextual information during the machine translation process. For these word-pairs datasets, the difference between best translation performers and lower performers (across languages) is smaller. Additionally, the final translation accuracy for all languages and all word-pairs datasets is 59 %. French, Dutch and Spanish are the languages with best automatic translations.

4.3 Language-Specific DSMs

In the first part of the experiment, the Spearman correlations (ρ) between the human assessments and the computation of the semantic similarity and relatedness for all DSMs instantiated for all languages were evaluated (Fig. 1 *(ii)*). Table 4 shows the Spearman correlation for each DSM using language-specific corpora (without machine translation), for the three word-pairs datasets.

Table 4. Spearman correlation for the language-specific models.

DS	Models	en	de	fr	ru	it	nl	pt	sv	es	ar	fa	Model AVG.	DS AVG.
MC	ESA	0.69	0.67	0.54	0.66	0.37	0.54	**0.67**	0.37	0.58	0.37	0.56	0.53	**0.56**
	LSA	0.79	**0.70**	0.55	0.63	0.58	0.55	0.41	**0.58**	0.66	**0.46**	0.45	0.56	
	W2V	**0.84**	**0.70**	0.55	0.64	**0.74**	**0.57**	0.37	0.40	**0.74**	0.38	**0.68**	**0.58**	
	Glove	0.69	0.64	**0.64**	**0.76**	0.51	0.55	0.62	0.40	0.65	0.38	0.45	0.56	
RG	ESA	0.80	0.68	0.45	0.63	0.50	0.58	0.51	0.50	0.59	0.36	0.57	0.54	0.53
	LSA	0.72	0.65	0.30	0.51	0.48	0.52	0.30	0.53	0.35	0.35	0.46	0.45	
	W2V	**0.85**	**0.78**	**0.57**	0.64	**0.69**	**0.63**	0.42	**0.57**	**0.64**	**0.36**	0.55	**0.58**	
	Glove	0.74	0.69	0.50	**0.70**	0.59	0.54	**0.52**	0.49	0.61	0.32	**0.59**	0.56	
WS353	ESA	0.50	0.39	0.32	0.44	0.34	0.53	0.44	0.43	0.37	0.26	0.37	0.39	0.41
	LSA	0.54	0.45	0.35	0.40	0.33	0.47	0.39	0.40	0.36	0.28	0.43	0.39	
	W2V	**0.69**	**0.54**	**0.50**	**0.53**	**0.50**	**0.58**	**0.53**	**0.45**	**0.53**	**0.44**	**0.53**	**0.51**	
	Glove	0.49	0.41	0.34	0.42	0.30	0.46	0.38	0.33	0.32	0.26	0.36	0.36	
	Lang AVG	0.70	0.61	0.47	0.58	0.49	0.54	0.46	0.45	0.53	0.35	0.50	0.50	

The comparative language-specific analysis indicates that English is the best-perfor-ming language (0.70), followed by German (0.61). The lowest Spearman correlation was observed in Arabic (0.35). From the tested DSMs, W2V is consistently the best-performing DSM (0.56). The language-specific DSMs achieved higher correlations for MC and RG (0.56 and 0.53, respectively), in comparison to 0.41 for WS-353.

The results for the language-specific DSMs were contrasted to the machine translation (MT) approach, according to the diagram depicted in Fig. 1 *(i)*. The Spearman correlation for the MT-mediated approach are shown in Table 5.

Table 5. Spearman correlation for the machine translation models over the English corpora. *Diff.* represents the difference of machine translation score minus the language specific.

DS	Models	de	fr	ru	it	nl	pt	sv	es	ar	fa	Model AVG.	Diff
MC	ESA-MT	0.55	0.53	0.42	0.38	0.45	0.38	0.48	0.39	0.31	0.58	0.45	−0.08 (−15.1 %)
	LSA-MT	0.61	0.72	0.65	0.67	0.66	0.70	0.74	0.78	0.69	0.75	0.70	0.14 (25.0 %)
	W2V-MT	**0.68**	**0.79**	**0.68**	**0.77**	**0.69**	**0.76**	**0.81**	**0.83**	**0.71**	0.74	**0.75**	**0.17 (29.3 %)**
	GloVe-MT	0.45	0.78	0.67	0.64	0.63	0.56	0.61	0.82	0.69	**0.79**	0.66	0.10 (17.9 %)
RG	ESA-MT	0.62	0.53	0.52	0.61	0.63	0.57	0.56	0.47	0.38	0.71	0.56	0.02 (3.7 %)
	LSA-MT	0.63	0.62	0.59	0.74	0.67	0.64	0.67	0.62	0.55	0.70	0.64	**0.19 (42.2 %)**
	W2V-MT	**0.69**	**0.79**	0.69	**0.78**	0.74	**0.75**	**0.71**	**0.73**	0.57	0.79	**0.72**	0.14 (24.1 %)
	GloVe-MT	0.62	0.77	**0.71**	0.77	**0.78**	0.66	0.66	0.72	**0.65**	**0.80**	0.71	0.15 (26.8 %)
WS353	ESA-MT	0.42	0.45	0.41	0.41	0.44	0.43	0.40	0.35	0.42	0.32	0.40	0.01 (2.6 %)
	LSA-MT	0.51	0.51	0.47	0.48	0.51	0.39	0.51	0.44	0.37	0.43	0.46	**0.07 (17.9 %)**
	W2V-MT	**0.62**	**0.59**	**0.57**	**0.57**	**0.63**	**0.51**	**0.59**	**0.55**	**0.50**	**0.52**	**0.57**	0.06 (11.8 %)
	GloVe-MT	0.45	0.48	0.42	0.43	0.46	0.33	0.42	0.41	0.33	0.37	0.41	0.05 (13.9 %)
	Lang AVG	0.57	0.63	0.57	0.60	0.61	0.56	0.60	0.59	0.52	0.63	0.56	

4.4 Machine Translation Based Semantic Relatedness

Using the MT models, W2V is consistently the best performing DSM (average 0.68), while ESA is consistently the worst performing model (0.47). We can interpret this result by stating that the benefit of using machine translation for ESA does not introduces significant performance improvements in comparison to the language-specific baselines.

The best performing languages are French and Farsi ($\rho = 0.63$). The Spearman correlation variance across languages in the MT models is low, as the impact of the use of the English corpus on the DSM model has a higher positive impact on the results in comparison to the variation of the quality of the machine translation. The results for all languages achieve very similar correlation values.

The impact of the MT model can be better interpreted by examining the difference between the machine translation and the domain-specific models (depicted in Table 6). LSA accounts for the largest average percent improvement (28.4 %) using the MT model, while ESA accounts for the lowest value (−2.9 %). As previously noticed, this can be explained by the sensitivity of these

Table 6. Difference between the language-specific and the machine translation approach. **M. AVG** represents the average of the models and **DS. AVG** represents the average of the datasets.

DS	M	de	fr	ru	it	nl	pt	sv	es	ar	fa	M. AVG	DS. AVG
MC	ESA	−0.18	−0.03	−0.36	0.03	−0.16	−0.44	0.31	−0.32	−0.16	0.03	−0.13	**0.41**
	LSA	−0.13	0.31	0.04	0.16	0.20	0.70	0.27	0.17	0.50	0.68	0.29	
	W2V	−0.02	0.43	0.07	0.05	0.21	**1.04**	1.00	0.13	**0.88**	0.09	0.39	
	GloVe	−0.31	0.22	−0.11	0.25	0.14	−0.10	0.51	**0.26**	0.85	0.75	0.25	
RG	ESA	−0.09	0.19	−0.18	0.21	0.08	0.11	0.12	−0.19	0.06	0.25	0.06	**0.41**
	LSA	−0.03	1.04	0.14	0.52	0.30	**1.15**	0.26	**0.77**	0.57	0.52	0.52	
	W2V	−0.11	0.39	0.08	0.14	0.18	0.76	0.23	0.14	0.59	0.44	0.28	
	GloVe	−0.11	0.55	0.01	0.31	**0.43**	0.28	0.35	0.17	**1.04**	0.36	0.34	
WS353	ESA	0.08	0.40	−0.07	0.18	−0.18	−0.02	−0.07	−0.07	0.60	−0.13	0.07	0.36
	LSA	0.12	0.43	0.19	0.45	0.09	−0.01	0.27	0.21	0.34	0.01	0.21	
	W2V	0.14	0.19	0.09	0.14	0.08	−0.04	0.33	0.04	0.12	0.00	0.11	
	GloVe	0.10	0.41	0.00	0.41	0.00	−0.14	0.28	0.30	0.28	0.04	0.17	
	AVG	0.06	0.52	0.13	0.36	0.23	0.29	0.70	0.22	0.59	0.82		

models to the corpus size due to the dimensional reduction strategy (LSA) or the broader context window (ESA). The remaining models accounted for substantial improvements (W2V = 21.7 %, GloVe = 19.5 %).

Arabic and French achieved the highest percent gains (47 % and 38 %, respectively), while German accounts for worst results (−4 %). These numbers are consistent with the corpus size. For German, the result shows that the corpus volume of the German Wikipedia crossed a threshold size (34 % of the English corpus) above which improvements for computing semantic similarity for the target word-pairs dataset might be marginally relevant, while the translation error accounts negatively in the final result.

The average improvement for the MT over the language specific model for each word-pairs dataset is consistently significant: MC = 20 %, RG = 30 % and WS353 = 14 %.

4.5 Summary

Below, the interpretation of the results are summarised as the core research questions which we aim to answer with this paper:

Question 1: Does machine translation to English perform better than the word vectors in the original language (for which languages and for which distributional semantic models)?

Machine translation to English consistently performs better for all languages, with the exception of German, which presents equivalent results for the language-specific models. The MT approach provides an average improvement of 16.7 % over language-specific distributional semantic models.

Question 2: Which DSMs or MT-DSMs work best for the set of analysed languages?

W2V-MT consistently performs as the best model for all word-pairs datasets and languages, except German, in which the difference between MT-W2V and language-speci-fic W2V is not significant.

Question 3: What is the quality of state-of-the-art machine translation approaches for word-pairs?

The average translation accuracy for all languages and all word-pairs datasets is 59 %. Translation quality varies according to the nature of the word-pair (better translations are provided for word pairs which are semantically related compared to semantically similar word pairs), reaching a maximum of 85 % and a minimum of 36 % across different languages.

For the distributional semantics user/practitioner, as a general practice, we recommend using W2V built over an English corpus, supported by machine translation. Additionally, the accuracy of state-of-the-art machine translation approaches work better for translating semantically related word pairs (in contrast to semantically similar word pairs).

5 Conclusion

This work provides a comparative analysis of the performance of four state-of-the-art distributional semantic models over 11 languages, contrasting the native language-specific models with the use of machine translation over English-based DSMs. The experimental results show that there is a significant improvement (average of 16.7 % for the Spearman correlation) by using off-the-shelf machine translation approaches and that the benefit of using a more informative (English) corpus outweighs the possible errors introduced by the machine translation approach. The average accuracy of the machine translation approach is 59 %. Moreover, for all languages, W2V showed consistently better results, while ESA showed to be more robust concerning lower corpora sizes. For all languages, the combination of machine translation over the W2V English distributional model provided the best results consistently (average Spearman correlation of 0.68).

Future work will focus on the analysis and translation of two other word-pairs datasets: SimLex-999 [10] and MEN-3000 [3].

Acknowledgments. This publication has emanated from research supported by the National Council for Scientific and Technological Development, Brazil (CNPq) and by a research grant from Science Foundation Ireland (SFI) under Grant Number SFI/12/RC/2289.

References

1. Al-Rfou, R., Perozzi, B., Skiena, S.: Polyglot: distributed word representations for multilingual NLP. In: Proceedings of the Seventeenth Conference on Computational Natural Language Learning, pp. 183–192. Association for Computational Linguistics, Sofia, August 2013. http://www.aclweb.org/anthology/W13-3520

2. Barzegar, S., Sales, J.E., Freitas, A., Handschuh, S., Davis, B.: Dinfra: a one stop shop for computing multilingual semantic relatedness. In: Proceedings of the 38th International ACM SIGIR Conference on Research and Development in Information Retrieval, SIGIR 2015, 1027–1028. ACM, New York (2015). http://doi.acm.org/10.1145/2766462.2767870
3. Bruni, E., Tran, N.K., Baroni, M.: Multimodal distributional semantics. J. Artif. Int. Res. **49**(1), 1–47 (2014). http://dl.acm.org/citation.cfm?id=2655713.2655714
4. Camacho-Collados, J., Pilehvar, M.T., Navigli, R.: A framework for the construction of monolingual and cross-lingual word similarity datasets. In: Proceedings of the 53rd Annual Meeting of the Association for Computational Linguistics and the 7th International Joint Conference on Natural Language Processing (ACL-IJCNLP), pp. 1–7. Citeseer (2015)
5. Faruqui, M., Dyer, C.: Community evaluation and exchange of word vectors at wordvectors.org (2014)
6. Faruqui, M., Dyer, C.: Improving vector space word representations using multilingual correlation. In: Proceedings of the 14th Conference of the European Chapter of the Association for Computational Linguistics, pp. 462–471. Association for Computational Linguistics, Gothenburg, April 2014. http://www.aclweb.org/anthology/E14-1049
7. Finkelstein, L., Gabrilovich, E., Matias, Y., Rivlin, E., Solan, Z., Wolfman, G., Ruppin, E.: Placing search in context: the concept revisited. In: Proceedings of the 10th International Conference on World Wide Web, pp. 406–414. ACM (2001)
8. Freitas, A.: Schema-agnositc queries over large-schema databases: a distributional semantics approach. Ph.D. thesis, Digital Enterprise Research Institute (DERI), National University of Ireland, Galway (2015)
9. Gabrilovich, E., Markovitch, S.: Computing semantic relatedness using Wikipedia-based explicit semantic analysis. In: Proceedings of the 20th International Joint Conference on Artifical Intelligence, IJCAI 2007, pp. 1606–1611. Morgan Kaufmann Publishers Inc., San Francisco (2007). http://dl.acm.org/citation.cfm?id=1625275.1625535
10. Hill, F., Reichart, R., Korhonen, A.: Simlex-999: evaluating semantic models with (genuine) similarity estimation. Comput. Linguist. **41**(4), 665–695 (2015)
11. Jurgens, D., Stevens, K.: The s-space package: an open source package for word space models. In: Proceedings of the ACL 2010 System Demonstrations, ACLDemos 2010, pp. 30–35. Association for Computational Linguistics, Stroudsburg (2010). http://dl.acm.org/citation.cfm?id=1858933.1858939
12. Landauer, T.K., Foltz, P.W., Laham, D.: An introduction to latent semantic analysis. Discourse Process. **25**(2–3), 259–284 (1998)
13. Mikolov, T., Chen, K., Corrado, G., Dean, J.: Efficient estimation of word representations in vector space. In: ICLR Workshop Papers (2013)
14. Miller, G.A., Charles, W.G.: Contextual correlates of semantic similarity. Lang. Cogn. Process. **6**(1), 1–28 (1991)
15. Navigli, R., Ponzetto, S.P.: Babelrelate! A joint multilingual approach to computing semantic relatedness. In: AAAI Conference on Artificial Intelligence (2012)
16. Pennington, J., Socher, R., Manning, C.D.: Glove: global vectors for word representation. In: Proceedings of the Empiricial Methods in Natural Language Processing (EMNLP 2014), vol. 12, pp. 1532–1543 (2014)
17. Rubenstein, H., Goodenough, J.B.: Contextual correlates of synonymy. Commun. ACM **8**(10), 627–633 (1965)

18. Sales, J.E., Freitas, A., Davis, B., Handschuh, S.: A compositional-distributional semantic model for searching complex entity categories. In: Proceedings of the Fifth Joint Conference on Lexical and Computational Semantics (*SEM), pp. 199–208 (2016)
19. Turney, P.D., Pantel, P.: From frequency to meaning: vector space models of semantics. J. Artif. Int. Res. **37**(1), 141–188 (2010). http://dl.acm.org/citation.cfm?id=1861751.1861756
20. Utt, J., Pad, S.: Crosslingual and multilingual construction of syntax-based vector space models. Trans. Assoc. Comput. Linguist. **2**, 245–258 (2014)
21. Zou, W.Y., Socher, R., Cer, D.M., Manning, C.D.: Bilingual word embeddings for phrase-based machine translation. In: EMNLP, pp. 1393–1398 (2013)

On Emerging Entity Detection

Michael Färber(✉), Achim Rettinger, and Boulos El Asmar

Karlsruhe Institute of Technology (KIT), Karlsruhe, Germany
{michael.faerber,rettinger}@kit.edu,
boulos.el-asmar@bmw.de

Abstract. While large Knowledge Graphs (KGs) already cover a broad range of domains to an extent sufficient for general use, they typically lack emerging entities that are just starting to attract the public interest. This disqualifies such KGs for tasks like entity-based media monitoring, since a large portion of news inherently covers entities that have not been noted by the public before. Such entities are unlinkable, which ultimately means, they cannot be monitored in media streams. This is the first paper that thoroughly investigates all types of challenges that arise from out-of-KG entities for entity linking tasks. By large-scale analytics of news streams we quantify the importance of each challenge for real-world applications. We then propose a machine learning approach which tackles the most frequent but least investigated challenge, i.e., when entities are missing in the KG and cannot be considered by entity linking systems. We construct a publicly available benchmark data set based on English news articles and editing behavior on Wikipedia. Our experiments show that predicting whether an entity will be added to Wikipedia is challenging. However, we can reliably identify emerging entities that could be added to the KG according to Wikipedia's own notability criteria.

Keywords: Emerging information discovery · Evolving knowledge · Novelty Detection · Entity linking · Text annotation

1 Introduction

Although existing knowledge graphs (KGs) such as DBpedia, Wikidata, and YAGO are already quite powerful in terms of their size, they are inherently incomplete, since they contain concepts and facts of an ever-changing world: constantly, new knowledge needs to be added. Considering Wikipedia, each day over 700 new articles are added to the English Wikipedia which stay permanently.[1] We do regard those articles as *novel entities* w.r.t. Wikipedia, as each article describes one entity which has not been part of Wikipedia before.

A. Rettinger—The research leading to these results has received funding from the European Union Seventh Framework Programme (FP7/2007-2013) under grant agreement no. 611346.

[1] This fact results from our empirical analysis, see Sect. 2.2 for more details.

E. Blomqvist et al. (Eds.): EKAW 2016, LNAI 10024, pp. 223–238, 2016.
DOI: 10.1007/978-3-319-49004-5_15

In this work we attempt to automatically identify such *out-of-KG entities* that are of great importance to numerous time-sensitive tasks which require up-to-date KGs, like semantic media monitoring or automatic speech recognition for TV news. Clearly, not all entities that potentially will get added to Wikipedia should be reported to someone who is interested in breaking news, since entities like *Antonio Ferramolino*, a 16th century Italian architect, were added recently to Wikipedia but not because of a current newsworthy event.

To take this into account we identify a crucial condition that makes an out-of-KG entity a potential candidate for news monitoring tasks: it needs to be *trending* and become *notable* for *the first time*. Those out-of-KG entities which show a notable increase in public interest for the first time are thereby referred to as *emerging*[2]. Hoffart et al. [10] use the same term, however, they do not require out-of-KG entities to be *notable*. *Notability* is an officially specified requirement for novel articles on Wikipedia. In order to determine whether an entity is notable, Wikipedia provides *notability guidelines*[3] to editors. An entity is thereby regarded as notable if it is traceable by reliable secondary literature.

Firstly, this paper presents the first full picture of how missing surface forms (i.e., phrases by which entities are referred to in text) and missing KG-entities impact the task of entity linking. Our empirical analysis of news streams in combination with the editing behavior on Wikipedia reveals that the need for incorporating emerging out-of-KG entities for entity linking occurs frequently, but is least investigated in existing research. Secondly, we create a publicly available benchmark data set and present a machine learning approach to automatically predict emerging out-of-Wikipedia entities. Our assumption is that emergence can be measured by analyzing media streams such as online news. The results show that making predictions about which entities will actually be added to Wikipedia is tricky. However, we are able to identify (actual) emerging entities with high confidence that could be added to the KG according to Wikipedia's own notability criteria. Those entities can be suggested to Wikipedia editors for inserting them into Wikipedia which helps to keep Wikipedia up-to-date with current events.

In summary, we make the following contributions in this paper:

- We describe and formalize the different entity linking challenges arising from out-of-KG entities and surface forms.
- We examine the occurrence and importance of the entity linking challenges regarding emerging entities "in the wild," i.e., based on Wikipedia as KG and annotated English news articles.
- We provide the first public benchmark data set for the *Emerging Entity Detection* challenge.[4]

[2] *Emerging* relates to *trending*: Entities can *emerge* only once. Once they have become *notable*, any (repeated) increase in public interest is just a *trend*.

[3] See https://en.wikipedia.org/wiki/Wikipedia:Notability.

[4] See http://people.aifb.kit.edu/mfa/emerging-entity-detection/.

- We present the first approach for predicting emerging entities based on the history of Wikipedia edits and noun phrases extracted as potential entity mentions from news streams.

The remainder of this paper is organized as follows: In Sect. 2, we analyze the conceptually different challenges for entity linking resulting from out-of-KG entities and out-of-KG surface forms In Sect. 3, the previous work is reported in respect to each challenge. Finally, we introduce our approach for emerging entity prediction in Sect. 4 and conclude in Sect. 5.

2 Entity Linking Challenges Arising from Missing Entities and Missing Surface Forms

We first clarify some terminology:

- An *entity* is a thing which can be uniquely identified via a URI $u \in U$.
- A *Knowledge Graph* is an RDF graph, which consists of a set of RDF triples where each RDF triple (s, p, o) is an ordered set of the following RDF terms: a subject $s \in U \cup B$, a predicate $p \in U$, and an object $U \cup B \cup L$. An RDF term is either a URI $u \in U$, a blank node $b \in B$, or a literal $l \in L$. U, B, and L are pairwise disjoint. In this paper, we do not consider blank nodes.
- A *surface form* is a textual phrase referring to one or several specific entities (e.g., the title of a Wikipedia article). Each entity has none, one or several surface forms attached.
- If a surface form is mentioned in a text, we speak of a *mention* of an entity. The task of linking a mention to a KG entity is referred to as *entity linking*.
- The tuple of a mention and a corresponding entity in a KG is designated by us as *annotation*.
- Entities and surface forms can be present in the KG or not. In the latter case we call them *unknown*, *missing*, or *out-of-KG*. If an out-of-KG entity is trending and notable for the first time, we call it *emerging*. In the moment an entity is inserted into the KG, it is regarded as *novel*.

2.1 Overview of Challenges

Let the following be given:

- KG g at the time t_0 (e.g., Wikipedia at 2015-04-04) with the set of all entities E_{t_0} and for each entity $e \in E_{t_0}$ the set of associated surface forms $S^e_{t_0}$.
- KG g' at time t_1 (e.g., Wikipedia at 2015-05-15) with the set of – in comparison to g – newly added entities $E_{\Delta t}$ since t_0 (i.e., $\Delta t = t_1 - t_0$) and for each entity $e \in (E_{t_0} \cup E_{\Delta t})$ the set of surface forms of e, $S^e_{\Delta t}$, added within Δt. S^e_{t1} is then the set of surface forms for entity e at time t_1.[5]

[5] As we are interested in novel/emerging entities, we do not consider deletions of entities or surface forms within Δt.

- Mention $m \in M$ in a text document given in the time range Δt.
- Function $f : M \rightarrow (E_{t_0} \cup E_{\Delta t})$, indicating the correct entity linking of $m \in M$ to an entity $e \in (E_{t_0} \cup E_{\Delta t})$.

We can then differentiate between the following disjoint challenges for entity linking w.r.t. missing entities and entity surface forms in a KG (in the following called *Challenges*; see also Fig. 1). We thereby first describe each Challenge, before we define it formally for a given mention m.

Challenge 1: Known surface form, known entity. This is the regular task of entity linking, i.e. without any aspect of missing entries. For a given mention in the text, one or several entities exist in the KG whose surface forms match the mention. With the help of a word-sense-disambiguation method, the appropriate entity in the KG is selected for the annotation of the mention. In our example in Fig. 1, "Snowden", the person, is chosen and not the band.

$$\exists e \in E_{t_0} : (m \in S^e_{t_0} \wedge f(m) = e) \wedge \nexists e' \in E_{\Delta t} : m \in S^{e'}_{t_1}$$

Challenge 2: Unknown surface form, known entity. Given the mention in the text, no surface form can be found in the KG that matches the mention and, hence, the mention cannot be linked. However, the entity which should be linked to, already exists in the KG. Missing surface forms regarding this situation can be either small word variations (different (mis)spellings, abbreviations, or substrings; e.g., "Arslonbob" for `:Arslanbob`) or completely new word creations which emerge (e.g., "Kybella" for `:Deoxycholic_acid`).

$$\nexists e \in E_{t_0} : m \in S^e_{t_0} \wedge \exists e' \in E_{t_0} : (m \in S^{e'}_{\Delta t} \wedge f(m) = e')$$

Challenge 3: Known surface form, unknown entity. Here, when using a regular entity linking tool, the given mention in the text might be falsely linked to an existing entity in the KG, since this entity has a surface form which is matching (e.g., "Alphabet"). However, the mention actually refers to an entity which does not exist in the KG yet (e.g., the company `:Alphabet_Inc.`).

$$\exists e \in E_{t_0} : m \in S^e_{t_0} \wedge \exists e' \in E_{\Delta t} : (m \in S^{e'}_{t_1} \wedge f(m) = e')$$

Challenge 4: Unknown surface form, unknown entity. Given the mention in the text, none of the known surface forms of the entities in the KG can be matched and, hence, the mention cannot be linked. Also the entity to be linked to is unknown. Examples are `:Antonio_Ferramolino` and `:41st_G7_summit`.

$$\nexists e \in E_{t_0} : m \in S^e_{t_0} \wedge \exists e' \in E_{\Delta t} : (m \in S^{e'}_{\Delta t} \wedge f(m) = e')$$

2.2 Challenges in the Wild

We can now analyze the above mentioned EL Challenges by monitoring entity mentions in news streams and relating them to the editing behavior in Wikipedia. Due to page limit constraints, we thereby focus primarily on the task of Emerging Entity Detection (i.e., Challenge 4 with emerging entities).

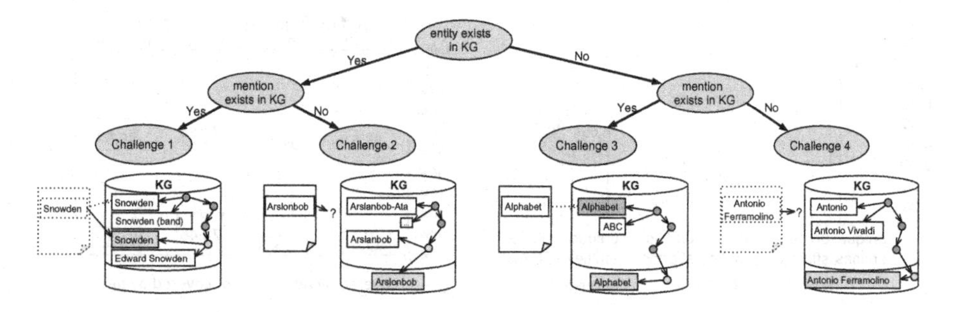

Fig. 1. Four different challenges arising in entity linking tasks when novel entities and novel surface forms start to appear.

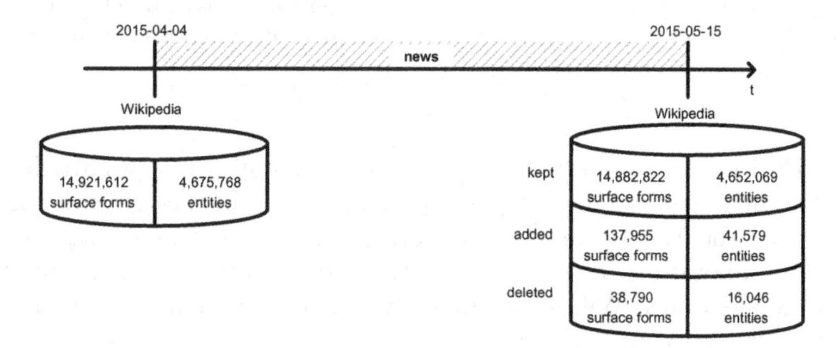

Fig. 2. Timeline with the Wikipedia versions and news used.

Experimental Setup

Wikipedia: Given the Wikipedia states from two different points in time, we first form the set of distinct entities and the set of distinct surface forms for both Wikipedia versions (using the xLiD framework [21] where the surface forms are derived by the title of Wikipedia pages, the redirect pages, the disambiguation pages, and the anchor texts of hyperlinks in Wikipedia). We can now calculate the difference between these two Wikipedia versions which identifies the novel entities and surface forms. The result is depicted in Fig. 2. We can see that 41,579 entities are in the version of May 2015, but not in the earlier version of April 2015. Also, 137,955 surface forms have been added to Wikipedia in this time range. While the major part of these new surface forms belongs to "old" entities, there still are many that correspond to "novel" entities which were added in the time range.

News: To assess the importance of each entity linking challenge (as defined in Sect. 2.1), we need to tap into another data source where traditional entity linking suffers from the mentioned challenges. We choose news articles to investigate how often each entity linking challenge occurs in a real-world news stream. We gather all English news articles from the IJS newsfeed service [17], covering more than 30,000 English news sources within this time range. This results

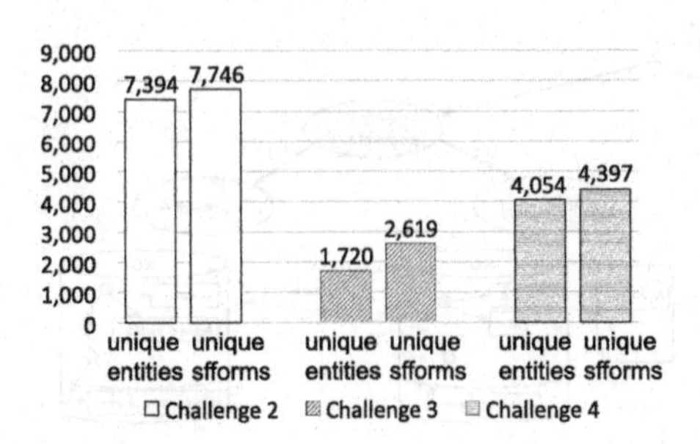

Fig. 3. Number of unique entities and number of unique surface forms used in the annotation. Challenge 1 is not displayed as it neither deals with novel entities nor with new surface forms.

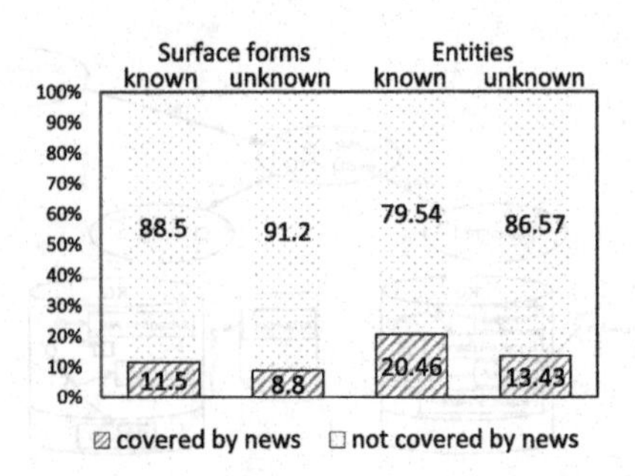

Fig. 4. Proportions to which known and novel entities/surface forms of the KG were also detected in the news, considering the annotations of all Challenges.

in 1,966,540 English news articles in total. We annotate the news articles with Wikipedia entities using the x-LiSA tool [22], a state-of-the-art entity linking system, given the Wikipedia state of May 15, 2015. For this setting, we gain 205,225,526 annotations in total. Given the *diff* of entities and surface forms between the mentioned Wikipedia states, we can now calculate the distribution among the Challenge 1, 2, 3 and 4.

Observations and Discussions

Frequency of Unique Entities and Unique Surface Forms per Challenge: Fig. 3 shows (i) to how many unique novel entities the detected entity mentions link and (ii) how many unique new surface forms were found as mentions. We can observe that Challenge 2 covers more distinct entities than Challenge 4 and that then Challenge 3 follows (always with considerable differences). Apparently, apart from the annotations of Challenge 1, most frequently only new surface forms of already existing KG entities are used in the annotations. This is reasonable since our KG Wikipedia already covers millions of entities and it is likely that a part of these entities get new surface forms which occur as mentions in the news.

Considering Challenge 3 and 4, entities of Challenge 4 are more often linked than entities of Challenge 3. This is comprehensible: Novel entities are more likely to have a surface form which is not existing in the KG so far than having a surface form which is already known. In the latter case, additional entities for existing surface forms are added to the KG, intensifying the ambiguity problem.

Proportion of Named Entities Among Novel Entities: To get a better characterization of the novel entities, we approach the question: "How many novel entities (which had been inserted within the time range) are named entities?" Bunescu and Pasca [2] have developed heuristics for determining whether a Wikipedia entity is a named entity. In order to answer our question, we implemented these

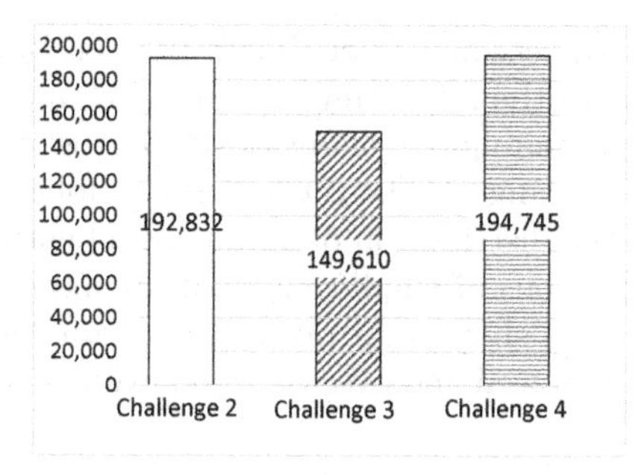

Fig. 5. Number of all annotations w.r.t. the different Entity Linking Challenges for the given time range.

heuristics and gained 33,052 named entities and 8,523 non-named entities.[6] Our evaluation on this classification (given a sample of 300 manually classified, randomly chosen novel entities) revealed an accurracy of 85.67 % for the chosen NER classification method. Note that in the manual evaluation we tended to classify more Wikipedia entities as named entities than the heuristics. For instance, we also considered events which can be given unique names as named entities.[7] In total we can state that focussing on *named* novel entities might be sufficient, especially for emerging entity detection tasks.

Proportion of Emerging Entities and Emerging Surface Forms in the KG: Fig. 4 shows on the right side the ratio of the number of novel entities and surface forms, respectively, detected in the news and being in the KG at time t_1 (and not yet in the KG at time t_0) to the total number of novel entities and surface forms, respectively, being in the KG at time t_1. 13.43 % of the novel entities are found as annotations in the news and 8.80 % of the new surface forms appear as mentions in the news. We can assume that these 13.43 % of the novel entities are the ones which are of highly public interest (i.e., *emerging*), since they occur in the news. On average, each novel entity appears 45.55 times. Mostly, those emerging entities were categorized in Wikipedia at time t_1 under the non-disjoint categories of living people, dead people (especially dying in the given time range Δt), and politicians.

Frequency of Annotations per Entity Linking Challenge: Figure 5 shows the number of all annotations per Challenge for the given time range. Note, that the reported occurrence numbers deliberately include repeated mentions of the same entity or surface form, since here we want to assess the total number of successfully linked mentions in the news. First of all, all Challenges occur considerably

[6] The remaining few entities are not parseable by the Stanford parser.

[7] Given the set of 300 novel entities manually tagged as named entities, 95 of them got classified as of type PERSON, 51 of type LOCATION, 27 of type ORGANIZATION, and 24 of type EVENT (as subtype of MISC).

often. Challenge 1, as expected, happens most frequently (204,675,773 occurrences; not depicted due to the high number), since Wikipedia even at the early point in time already covers millions of entities and grows only by 41,579 novel entities in the given time range.[8] Then, Challenge 4 follows in the occurrence ranking. In contrast to the distribution of unique novel entities and new surface forms (see Fig. 3), annotations of Challenge 4 appear more frequent than annotations of Challenge 2. This shows that our news stream captures novel entities (detected in Challenge 3/4) well and over time (on average 45.55 times per novel entity, see above).

Persistence of Detected Emerging Entities: Considering the annotations where novel entities were found (i.e. Challenges 3 and 4), almost all of those entities still exist in the current Wikipedia version (as of July 2016; above 99 % regarding all Challenges). Those entities seem to be permanently relevant for the KG, which is a strong indicator for the importance of emerging entities and of their detection (see Sect. 4).

2.3 Conclusions

The results of our analysis provide interesting insights into novel Wikipedia entities and surface forms and how they appear in the news. Below is a selection of key findings that we believe are most revealing:

1. Challenge 4 covers most of the novel entities inserted into the KG in Δt. In addition, Challenge 4 occurs – besides Challenge 1, which does not cover any novelty – most frequent regarding the set of all annotations. Thus, when dealing with novel entities, Challenge 4 is the most pressing issue to address.
2. About 13.4 % of the novel Wikipedia-entities are also mentioned in the news. Since those entities start to be mentioned in the news with increasing frequency at a certain point in time (occurring on average 45.55 times), we assume they are *emerging* entities, i.e., of increasing public interest. This clearly shows that *emerging* entity detection is different from *novel* entity detection and should not be treated equally as done by previous work [10].
3. Furthermore, we found out that almost all emerging entities remain in Wikipedia constantly. Together with the item 2 above, i.e. the frequent occurrence of emerging entities in the news, it indicates the great importance of emerging entities for being in the KG and for being detected as early as possible (see Sect. 4).
4. About 75 % of the novel entities are named entities. This indicates that focusing on named entities might be sufficient for many real-world novel entity detection applications. Emerging entities are most frequently living people which are of public interest (e.g., politicians) or people who recently died.

[8] For 11,639 of those 41,579 novel entities, however, only the Wikipedia title or redirects changed (due to typo correction or outsourcing of parts of a page). I.e., on average over 700 entities are inserted into Wikipedia each day which are "really" novel. For the task of Emerging Entity Detection (see Sect. 4), we only consider real novel entities which emerge (i.e., recently gained public interest for the first time).

On our website, we present further results of our analysis, such as the string similarity between the mention and already given surface forms of the target entities.

3 Related Work

In the following, we describe, how the different entity linking challenges w.r.t. novelty have been pursued by the research community. Due to the focus of this paper on Emerging Entity Detection in Sect. 4, we elaborate related work regarding Challenge 4.

3.1 Challenge 1: Linking to in-KG Entities via Known Surface Forms

There is an extensive amount of published work on entity linking (i.e., linking mentions to entries in a KG) and text annotation (entity linking including mention detection for unstructured text). The first approaches on entity linking to Wikipedia have been proposed by Bunescu and Pasca [2] and Cucerzan [5]. In 2008, Milne and Witten [13] built a system including a more sophisticated mention detection step. The annotation of the news texts used in our evaluations is provided by x-LiSA [22].

3.2 Challenge 2: Linking to in-KG Entities via Unknown Surface Forms

Dredze et al. [6] design a system for entity disambiguation taking into account the challenges of *name variations*, *entity ambiguity*, and *absence* of entities in the KG. The authors hence approach the Challenges 1, 2, and 3. They use different features for name variant detection and calculate a similarity score between the entity mention and the KG entity. SVM ranking is used to get the best candidate for each mention. In order to face Challenge 3, they introduce *NIL* as out-of-KG entity to which mentions can always be linked to.

Gottipati and Jiang [9] cover the Entity Linking Challenge 2 besides the traditional entity linking scenario. For that, their system considers not only the entity name for finding the in-KG entity, but also alternative name strings; these strings are gathered (i) from the document containing the mention using a NER tool and (ii) from Wikipedia exploiting page redirects.

3.3 Challenge 3: Linking to Out-of-KG Entities via Known Surface Forms

AIDA, a system for disambiguating named entities, was extended in 2014 [10] so that it can link to out-of-KG entities which share their entity names with in-KG entities. For each mention, besides the in KG-entity candidates, an additional out-of KG entity candidate is introduced which is initially represented by the mention string and later enriched by characteristic keyphrases. Wang et al. [18] also focus on the disambiguation of named entities. They detect so-called *target entities* in the text. These are entities (i) which are not necessarily contained in a KG, but whose names are known and where text documents containing them are available, and (ii) which all come from a so called *target domain* such as "IT companies". They leverage these two aspects for a graph-based model that disambiguates all mentions across all documents collectively. Wu et al. [19] want to classify whether a given mention belongs to an existing KG entity or not, thereby targeting Challenge 3 and 4. The authors use five different spaces to model entities (a contextual, neural embedding, topical, query, and lexical space), but they do not consider the evolution of KGs.

3.4 Challenge 4: Linking to Out-of-KG Entities via Unkown Surface Forms

In this context, it is noteworthy to mention both schema-independent and schema-dependent novel entity detection approaches. All approaches only cover the prediction of whether given mentions (in unstructured text or already extracted) are KG entity candidates and which semantic types these entity candidates can be assigned to. However, they do not focus on *emerging* entities (as entities being of increasing public interest) and also do not correlate their predictions with the actual entity evolutions in a KG (such as the editing behavior in Wikipedia). Thus, to the best of our knowledge, this paper is the first one to define and propose an approach to solve the task of *emerging* entity detection.

Schema-free Novel Entity Extraction: Firstly it is noteworthy to mention that there are the Open Information Extraction approaches such as ReVerb [8] and NELL [3] which provide textual triples and, hence, entity mentions. Those mentions can be used to find out-of-KG entities (targeting Challenge 3 and 4) and to populate the KG [7]. Furthermore, NERC tools and general noun phrase extraction techniques can be used to gain novel entity candidates.

Within the Text Analysis Conference (TAC) tracks and the TREC tracks, the following tracks are related, but are too distant for a comparison with our approach and do not provide a suitable data set: 1. In the *TAC-KBP2015 Tri-lingual Entity Discovery and Linking (EDL) track* [11], besides the ordinary entity linking, non-linkable mentions should be clustered across languages. However, any non-linkable mention is considered as novel. 2. In the *TREC Novelty Detection tracks* [16], the topics (which are events and opinions) are very broad so that they cannot be used as entities. 3. In the *TREC KBA tracks*,[9] the systems had to fill slots on profiles. Like in case of the other mentioned tracks, the task is not to detect *emerging* out-of-KG entities.

The problem of novel entity detection also appears in the area of speech recognition, where it is often refered to as out-of-vocabulary (OOV) problem. Recent systems increase the set of known words by leveraging large external

[9] See http://trec-kba.org/, requested June 26, 2016.

corpora such as the Web [15]. All OOV systems for speech recognition have in common that the OOV words are not assessed w.r.t. relevance, as any utterance needs to be matched, not only emerging entity mentions.

Schema-dependent Novel Entity Detection: Lin and Etzioni [12] introduce the *unlinkable noun phrase problem*: Given a noun phrase that is not linked to Wikipedia as KG, determine whether it is an entity,[10] and if it is, determine its fine-grained entity types. In contrast to us, noun phrases are already given, so that no mention detection step is necessary. Furthermore, Lin et al. do neither consider *emerging* entities nor the evolution of a KG in general. Other works on predicting entity types for out-of-KG entities include HYENA proposed by Yosef et al. [20]. HYENA as first system assigns multiple types to an entity in a hierarchical order by applying a multi-label classifier. Nakashole et al. [14] propose PEARL, a system which assigns entity types to mentions of out-of-Freebase entities with the help of relational patterns.

4 Emerging Entity Detection

In this section, we present the first approach to the task of *emerging entity detection* on the basis of Wikipedia: We propose to train a machine learning model to detect out-of-Wikipedia entities which are emerging, i.e., which are for the first time reaching considerable public interest and are therefore conforming to Wikipedia's own notability requirements. Those entities can then be used for recommending the creation of new Wikipedia articles to Wikipedia editors or for enhancing entity linking in media monitoring systems.

4.1 Used Data Sets

For our analysis shown in Sect. 2.2, we annotated English news articles with the help of x-LiSA [22][11] given a "future" Wikipedia version (2015-05-15), so that we know which mentions refer to novel entities and which mentions just use new surface forms for existing entities. We now use the same annotated corpus for training and testing our machine learning model for *emerging entity detection*.

As emerging entity candidates we use all noun phrases (NPs) which were (i) extracted from the news articles by an implemented noun phrase extraction module (using a slightly extended rule set of [23] on the Part-of-Speech tags gained by the Stanford parser) and which were (ii) not linkable to in-KG entities by the entity linking system x-LiSA [22] given the Wikipedia state of 2015-04-04 (see Fig. 2). By means of the latter, we exclude all noun phrases for which KG entities already exist (i.e., filtering out annotations of Challenge 1 and 3). All noun phrases with annotations of Challenge 4 (i.e., mentions linking to emerging entities) are then labeled as "true", all noun phrases with annotations of

[10] An entity is here understood as "noun phrase that could have a Wikipedia-style article if there were no notability or newness considerations, and which would have semantic types." [12].

[11] Note that any text annotation method for Wikipedia could have been applied here.

Table 1. Number of *true* and *false* target labels for the data sets with different NP series lengths. Only the target labels of the respective last NP per NP series was considered.

NP series length	# *true*	# *false*
2	1754	100071
3	1246	48954
5	831	22066
10	474	8141
20	271	3076
30	169	1751
40	119	1142
50	86	806

Table 2. Examples of true positives and false positives for the data set with NP series length 20.

True positive	False positive
Michael Slager	Elton Simpson
Eric Courtney Harris	Garissa University
Dave Goldberg	Ananta Bijoy Das
Adult Beginners	Joan Kagezi
Dan Fredinburg	Gaioz Nigalidze
LG G4	Russell Begaye
Operation Maitri	Mitchell Santner
Struggle Street	Jose Urena
Oleg Kalashnikov	Severino Gonzalez
Masaan	Operation Fiela

Challenge 2 (i.e., new mentions linking to "old" in-KB entities) or without any Wikipedia annotation as "false". In total, after some initial filtering (e.g., considering only noun phrases with at least three alphanumeric characters) we came up with 15.6M extracted NPs (2.6M unique NPs), extracted from 1.8M English news articles. Note that this data set is highly unbalanced regarding the target labels (ratio true:false is 1:164). In order to reduce the unequal distribution between the classes and the overall data set size, we applied further filtering techniques. For instance, we considered only named entities (using the Stanford NER tagger; see our findings in Sect. 2.2). The reduced data set contained 840,101 NP occurrences and 404,263 unique NPs.

As the overall task is to predict emerging entities as soon as they reach public attention for the first time, we focused on the very first appearances of the NPs in the news stream. We therefore built series of the first n occurrences of each unique NP (with $n = 2, 3, 5, 10, 20, 30, 40, 50$; see Table 1). Based on those NP series, we calculated the features: For each NP occurrence, we extract a number of local noun phrase features (19; e.g., POS tag and suffix of the noun phrase), article features (17; e.g., source of the article), corpus features (7; e.g., slope reg. occurrences of the noun phrase over the last 24h), and global features (12; e.g., Wikipedia PageView slope over the last 24h). A detailed list of all extracted features is available at our website.[12] As most of the features (such as many slope values) are capturing the history of a whole NP group, we only used the the last NP occurrence in each NP series for training and testing.[13]

4.2 Feature Selection and Model Training

To alleviate the imbalance problem, we applied feature selection on the binary and numerical features. Our dimensionality reduction approach for binary

[12] See http://people.aifb.kit.edu/mfa/emerging-entity-detection.

[13] We also experimented with aggregating *all* features for each NP series, but did not yield better evaluation results.

features relies on a variance threshold per target class. For numerical features, the features whose values keep the same range and distribution in the positive class and in the negative class subsets were removed. The most important remaining features after the feature selection process include:

- *PosTag*: Part-of-Speech tag of the NP;
- *hostName*: host name of the article source (e.g., "www.n24.de");
- *npDiffArtsOccurrenceSlope24hSlope*: slope (using linear regression), based on the occurrence number of the NP over the last 24 h;
- *pageViewSlope24h*: slope of the Wikipedia page view values[14] (requests for existing and non-existing Wikipedia pages) reg. the NP over the last 24h;
- *npAsTitleInDEWP*: true if the NP appears as title in the German Wikipedia.

We randomly split our data set and used 75 % for training and 25 % for testing. For training, the distribution among the two classes was equalized by removing instances. The training set was used to fit a Linear SVC model [1,4].[15]

4.3 Evaluation Results

The achieved F_1 scores and accuracy scores for the different data sets corresponding to the different NP series lengths are presented in Table 3. We can recognize an increase regarding the F_1 scores with increasing NP series length (with a maximum of 25.0 at NP series length 40), while the accuracy values roughly remain the same. Table 2 shows some examples of true positive and false positive predictions. Note that the displayed false positive examples were eventually all inserted into Wikipedia, just after the considered time range. This clearly shows that our approach is able to suggest emerging entities to Wikipedia editors before they noticed them.

Table 3. Evaluation results of emerging entity detection.

NP series length	2	3	5	10	20	30	40	50
F1 score (in %)	4.6	6.7	10.7	12.2	23.1	24.7	25.0	19.0
Accuracy (in %)	81.7	81.5	84.2	72.9	82.0	79.4	79.4	71.4

To investigate this further, the top 100 false positive instances (with smallest distance to the hyperplane) were assessed by two independent assessors in order to find out to which extent the recommended NPs are valid Wikipedia emerging entities and thus could be added to Wikipedia. The assessors followed the *notability criteria* given by Wikipedia.[16] In case of varying judgements, the assessors

[14] See http://dumps.wikimedia.org/other/pagecounts-raw/.

[15] We also evaluated machine learning algorithms specialized on imbalanced and time-series data, such as cost-sensitive AdaBoost, cost-sensitive one class classifier and recurrent neural networks. However, this did not yield better results.

[16] See more information on our website.

agreed on a common judgment in a second round. Out of 100 assessed NPs which were classified as emerging entities (using the data with NP series length 20), (i) 37 had not been added to Wikipedia yet[17] but were judged manually as notable,[18] (ii) 20 had not beed added to Wikipedia before and judged manually as not notable, and (iii) 43 had already been added to Wikipedia before, but were not detected as such.[19] If we disregard the mistakes introduced by the entity linking step, we can state a "false false positive" rate of about 65 %. In other words, 65 % of the out-of-KG instances predicted by our approach as emerging entities, actually were feasible emerging entities but not recognized by Wikipedia editors (yet).

5 Conclusions

In this paper, we presented a systematic overview of the different *entity linking* challenges arising from *out-of-KG entities* and *out-of-KG surface forms*. We provided an empirical analysis based on Wikipedia and on annotated English news articles regarding the importance of each of those challenges.

Based on that, we identified *emerging entity detection*, i.e. *trending* entities becoming *notable* for the first time, as the key task to facilitate semantic media monitoring. To address the task, we presented the first trained model for detecting emerging entities. The measured F_1 score lead to the conclusion that a robust prediction of emerging Wikipedia entities is tricky, due to the extreme imbalance in the data. However, this is to a large extend due to Wikipedia missing articles about valid emerging entities. Our approach is verifiably capable of identifying feasible candidate entities which could be added to Wikipedia according to Wikipedia's own notability guidelines. This would improve the up-to-dateness of Wikipedia and semantic media monitoring systems.

References

1. Ben-Hur, A., Horn, D., Siegelmann, H.T., Vapnik, V.: Support vector clustering. J. Mach. Learn. Res. **2**, 125–137 (2002)
2. Bunescu, R., Pasca, M.: Using encyclopedic knowledge for named entity disambiguation. In: Proceedings of the 11th Conference of of the European Chapter of the Association for Computational Linguistics (EACL-06), pp. 9–16, Trento, Italy (2006)
3. Carlson, A., Betteridge, J., Kisiel, B., Settles, B., Hruschka, E., Mitchell, T.: Toward an architecture for never-ending language learning (2010)

[17] Given Wikipedia status of 2015-04-04 as the reference KG.

[18] Some of those entities were inserted later.

[19] Investigations revealed that the already existing Wikipedia entities were not annotated by x-LiSA because no suitable surface form were available for those entities. In most of those cases, the entity was a person and in the news article only the family name was mentioned and extracted. However, in the set of known surface forms from Wikipedia only the full name of the entity was contained. Resolving those issues are left to future work.

4. Cortes, C., Vapnik, V.: Support-vector networks. Mach. Learn. **20**(3), 273–297 (1995)
5. Cucerzan, S.: Large-scale named entity disambiguation based on Wikipedia data. In: Proceedings of the 2007 Joint Conference on EMNLP-CoNLL, pp. 708–716, Prague, Czech Republic. Association for Computational Linguistics, June 2007
6. Dredze, M., McNamee, P., Rao, D., Gerber, A., Finin, T.: Entity disambiguation for knowledge base population. In: Proceedings of the 23rd International Conference on Computational Linguistics, COLING 2010, Stroudsburg, PA, USA, pp. 277–285. Association for Computational Linguistics (2010)
7. Dutta, A., Meilicke, C., Stuckenschmidt, H.: Enriching structured knowledge with open information. In: Proceedings of the 24th International Conference on World Wide Web, WWW 2015, Republic and Canton of Geneva, Switzerland, pp. 267–277 (2015)
8. Fader, A., Soderland, S., Etzioni, O.: Identifying relations for open information extraction. In: Proceedings of the Conference on Empirical Methods in Natural Language Processing, EMNLP 2011, Stroudsburg, PA, USA, pp. 1535–1545. Association for Computational Linguistics (2011)
9. Gottipati, S., Jiang, J.: Linking entities to a knowledge base with query expansion. In: Proceedings of the Conference on Empirical Methods in Natural Language Processing, EMNLP 2011, Stroudsburg, PA, USA, pp. 804–813. Association for Computational Linguistics (2011)
10. Hoffart, J., Altun, Y., Weikum, G.: Discovering emerging entities with ambiguous Names. In: Proceedings of the 23rd International Conference on World Wide Web, WWW 2014, New York, NY, USA, pp. 385–396. ACM (2014)
11. Ji, H., Nothman, J., Hachey, B., Florian, R.: Overview of TAC-KBP2015 tri-lingual entity discovery and linking (2015)
12. Lin, T., Etzioni, O.: No noun phrase left behind: detecting and typing unlinkable entities. In: Proceedings of the 2012 Joint Conference on EMNLP and CoNLL, EMNLP-CoNLL 2012, Stroudsburg, PA, USA, pp. 893–903. ACL (2012)
13. Milne, D., Witten, I.H.: Learning to link with wikipedia. In: Proceedings of the 17th ACM conference on Information and knowledge management, CIKM 2008, New York, NY, USA, pp. 509–518. ACM (2008)
14. Nakashole, N., Tylenda, T., Weikum, G.: Fine-grained semantic typing of emerging entities. In: Proceedings of the 51st Annual Meeting of the Association for Computational Linguistics, pp. 1488–1497 (2013)
15. Parada, C., Sethy, A., Dredze, M., Jelinek, F.: A spoken term detection framework for recovering out-of-vocabulary words using the web. Paragraph **10**(71.24), 323K (2010)
16. Soboroff, I., Harman, D.: Novelty detection: the TREC experience. In: HLT 2005, Stroudsburg, PA, USA, pp. 105–112. ACL (2005)
17. Trampuš, M., Novak, B.: Internals of an aggregated web news feed. In: Proceedings of the Fifteenth International Information Science Conference IS SiKDD 2012, pp. 431–434 (2012)
18. Wang, C., Chakrabarti, K., Cheng, T., Chaudhuri, S.: Targeted disambiguation of ad-hoc, homogeneous sets of named entities. In: Proceedings of the 21st International Conference on World Wide Web, WWW 2012, New York, NY, USA, pp. 719–728. ACM (2012)
19. Wu, Z., Song, Y., Giles, C.L.: Exploring multiple feature spaces for novel entity discovery. In: AAAI 2016, AAAI - Association for the Advancement of Artificial Intelligence, February 2016

20. Yosef, M.A., Bauer, S., Hoffart, J., Spaniol, M., Weikum, G.: HYENA: hierarchical type classification for entity names. In: COLING 2012, pp. 1361–1370 (2012)
21. Zhang, L., Färber, M., Rettinger, A.: xLiD-Lexica: cross-lingual Linked data lexica. In: Proceedings of the Ninth International Conference on Language Resources and Evaluation (LREC 2014), pp. 2101–2105. ELRA (2014)
22. Zhang, L., Rettinger, A.: X-LiSA: cross-lingual semantic annotation. PVLDB **7**(13), 1693–1696 (2014)
23. Zhao, S., Li, C., Ma, S., Ma, T., Ma, D.: Combining POS tagging, lucene search and similarity metrics for entity linking. In: Lin, X., Manolopoulos, Y., Srivastava, D., Huang, G. (eds.) WISE 2013. LNCS, vol. 8180, pp. 503–509. Springer, Heidelberg (2013). doi:10.1007/978-3-642-41230-1_44

Framester: A Wide Coverage Linguistic Linked Data Hub

Aldo Gangemi[1], Mehwish Alam[1(✉)], Luigi Asprino[2], Valentina Presutti[2], and Diego Reforgiato Recupero[3]

[1] Universite Paris 13, Paris, France
{gangemi,alam}@lipn.univ-paris13.fr
[2] National Research Council (CNR), Rome, Italy
luigi.asprino@istc.cnr.it, valentina.presutti@cnr.it
[3] University of Cagliari, Cagliari, Italy
diego.reforgiato@unica.it

Abstract. Semantic web applications leveraging NLP can benefit from easy access to expressive lexical resources such as FrameNet. However, the usefulness of FrameNet is affected by its limited coverage and non-standard semantics. The access to existing linguistic resources is also limited because of poor connectivity among them. We present some strategies based on Linguistic Linked Data to broaden FrameNet coverage and formal linkage of lexical and factual resources. We created a novel resource, Framester, which acts as a hub between FrameNet, WordNet, VerbNet, BabelNet, DBpedia, Yago, DOLCE-Zero, as well as other resources. Framester is not only a strongly connected knowledge graph, but also applies a rigorous formal treatment for Fillmore's frame semantics, enabling full-fledged OWL querying and reasoning on a large frame-based knowledge graph. We also describe Word Frame Disambiguation, an application that reuses Framester data as a base in order to perform frame detection from text, with results comparable in precision to the state of the art, but with a much higher coverage.

Keywords: Frame detection · Framester · FrameNet · Framenet coverage · Knowledge graphs · Frame semantics · Linguistic linked data

1 Introduction

Many resources from different domains are now published using Linked Open Data (LOD) principles to provide easy access to structured data on the web. There are several linguistic resources which are already part of LOD, two of the most important are WordNet [7] and FrameNet [2]. They have already been formalized several times, e.g. in OntoWordNet [12], WordNet RDF [30], FrameNet DAML [22], FrameNet RDF [24], etc. FrameNet allows to represent textual resources in terms of Frame Semantics. The usefulness of FrameNet is however affected by its limited coverage, and non-standard semantics. An evident solution stands on creating valid links between FrameNet and other lexical resources such

E. Blomqvist et al. (Eds.): EKAW 2016, LNAI 10024, pp. 239–254, 2016.
DOI: 10.1007/978-3-319-49004-5_16

as WordNet, VerbNet [19] and BabelNet [23] to create wide-coverage and multilingual extensions of FrameNet. By overcoming these limitations, NLP-based applications such as question answering, machine reading and understanding, etc. would eventually be improved.

This study focuses on a wide coverage resource called "Framester". It is a frame-based ontological resource acting as a hub between linguistic resources such as FrameNet, WordNet, VerbNet, BabelNet, DBpedia, Yago, DOLCE-Zero, and leveraging this wealth of links to create an interoperable *predicate space* formalized according to frame semantics [8], and semiotics [10].

Framester uses WordNet and FrameNet at its core, expands it to other resources transitively, and represents them in a formal version of frame semantics. A frame-detection based application of Framester called as *Word Frame Disambiguation (WFD)* is developed and made available through the *WFD API*. Two evaluations of WFD show that *frame detection by detour* [3] employing large linguistic linked open data is comparable to the state-of-the-art frame detection in precision, and is better in recall.

WFD API uses a simple subset of Framester, which includes a novel set of mappings between frames, WordNet synsets, and BabelNet synsets, and extends frame coverage using semantic relations from WordNet and FrameNet. *WFD* exploits classical Word Sense Disambiguation as implemented in UKB [1] and Babelfy [21], and then uses Framester to create the closure to frames. *WFD* is therefore a new *detour* approach to frame detection and aiming at complete coverage of the frames evoked in a sentence.

This paper is structured as follows: Sect. 2 gives a brief overview of the major existing resources, Sect. 3 details state of the art. Section 4 gives the formal semantics underlying Framester as well as how the resource has been created, while Sect. 5 details the application *WFD* for frame detection based on Framester, along with its evaluation and comparison to the state-of-the-art frame detection algorithm. Finally, Sect. 6 concludes the paper.

2 Linguistic Resources

Some details about the most important linguistic resources forming the core of Framester Cloud are given.

WordNet [7] is a lexical database that groups synonyms into the form of synsets. Each synset is described by a gloss and represents a concept, which is semantically related to other concepts through relations such as hyponymy/hypernymy, meronymy/holonymy, antonymy, entailment, derivation, etc. The conversion of WordNet to RDF has been performed several times; the guidelines and W3C version are described in [31]. OntoWordNet [12] turns the informal WordNet graph into an ontology, representing synsets and the other entities from WordNet as ontology elements (classes, properties, individuals, axioms), and linking them to the DOLCE-Zero foundational ontology[1].

[1] http://www.ontologydesignpatterns.org/ont/d0.owl.

FrameNet [2] containing descriptions and annotations of English words following Frame Semantics (see Sect. 4.1). FrameNet contains *frames*, which describe a situation, state or action. Each frame has semantic roles called *frame elements*. Each frame can be evoked by *Lexical Units (LUs)* belonging to different parts of speech. In version 1.5, FrameNet covers about 10,000 lexical units and 1024 frames. For example in frame **Reshaping** the argument for the role **Deformer** deforms the argument of the role **Patient** in a way that it changes its original shape into a **Configuration** i.e. a new shape. Deformer can also be replaced by a **Cause** i.e., any force or event that causes an effect of changes the shape of the **Patient**.

Lexical units such as **bend**, **crumple**, **crush** etc. are example words, typically used to denote reshaping situations in text, as in the sentence

$$[Hagrid]_{Deformer}\ [rolled]_{lexical\ unit}\ up\ the\ [note]_{Patient}.$$

BabelNet is a wide coverage multilingual graph derived from WordNet, Wikipedia, and several other sources [23]. It is a directed labeled graph consisting of nodes and edges where nodes are the concepts and the edges connect two concepts with a semantic relation such as `is-a`, `part-of` etc.

Predicate Matrix [5] is a lexical resource created by integrating multiple sources containing predicates: WordNet, FrameNet, VerbNet and PropBank. VerbNet (VN) [29] is a broad coverage verb lexicon organized as a hierarchy of verb classes grouped by their sense and their syntactic behaviour. Each verb class contains verb senses, and is associated with thematic roles, and selectional restrictions on the role arguments. Proposition Bank [18] adds semantics to the Penn English Treebank (PTB) by specifying predicate-argument structure. Predicate Matrix uses SemLink [26], a resource containing partial mappings between the existing resources having predicate information as a base, and then extends its coverage via graph-based algorithms. It provides new alignments between the semantic roles from FrameNet and WordNet.

3 State of the Art

The integration between Natural Language Processing (NLP) and Semantic Web under the hat of "semantic technologies" is progressing fast. Most work is however opportunistic: on one hand exploiting NLP algorithms and applications, (typically named-entity recognizers and sense taggers) to populate SW datasets or ontologies, or for creating NL query interfaces, and on the other hand exploiting large SW datasets and ontologies (e.g. DBpedia, YAGO, Freebase [28], etc.) to improve NLP algorithms. For example, large text analytics and NLP projects such as Open Information Extraction (OIE, [6]), Alchemy API,[2] and Never Ending Language Learning (NELL, [15]) recently started trying to ground extracted named entities in publicly available identities such as Wikipedia, DBpedia and

[2] http://www.alchemyapi.com.

Freebase. Most famous, IBM Watson [16] has succeeded in reusing NLP and SW methods in a creative and efficient way. Opportunistic projects for integrating NLP and SW are perfectly fine, but realistic SW applications require a stable semantics when reusing NLP results. At this very moment, that semantics is largely left to the needs of the specific application, and this makes it difficult any comparison between tools or methods.

Standardization attempts are happening since a while, and the recent proposal of Ontolex-Lemon by the OntoLex W3C Community Group[3] will possibly improve resource reuse as Linguistic Linked Data. In addition, platforms exist since a long time which help operational integration of NLP algorithms (GATES, UIMA), or reuse of NLP components as linked data (Apache Stanbol[4], NIF [17], NERD [27], FOX[5]). However, interoperability efforts mainly concentrated on the direct transformation of NLP data models into RDF, so assuming that linguistic entities populate a universe disjoint from the universe of factual data. In the case of W3C OntoLex, a link is established by using so-called "semantics by reference", which allows e.g. to assert that a WordNet synset "references" a class from an existing ontology. In other words, the formal semantics of plain Linguistic Linked Data is delegated to possible mappings that a developer or user wants to make. This approach is conservative and simply avoids the problem of addressing natural language semantics, but has limitations, since it is based on local decisions, which are necessarily arbitrary, and dedicated to a specific task.

On the contrary, a few attempts have been made to formally transform NLP data and lexical resources into regular ontologies and data. On one hand, examples of lexical resources include OntoWordNet [12], FrameNet-OWL [24], FrameBase [28], etc. On the other hand, FRED [13] is a tool that creates formal knowledge graphs (using five-star linked data patterns) from both NLP results and lexical resources.

4 Framester as a Linked Linguistic Predicate Resource

Despite the active development of linguistic linked open data in recent years, there are still few linguistic resources, and they are not linked as intensely as they could be. Figure 1 shows a simplification of the current state of the linguistic resources present in the LOD cloud that are relevant for frame-oriented knowledge. These datasets have heterogeneous schemas that pose inconvenience in their direct and interoperable use.

Framester provides a dense interlinking between existing resources, adds new ones (recently ported to linked data in the context of the Framester project), and provides a homogeneous formalization of those links under the hat of frame semantics. Framester is intended to work as a knowledge graph/linked data hub to connect lexical resources, NLP results, linked data, and ontologies. It is bootstrapped from existing resources, notably the RDF versions of FrameNet

[3] http://www.w3.org/community/ontolex/wiki/Main_Page.
[4] http://stanbol.apache.org.
[5] http://aksw.org/Projects/FOX.html.

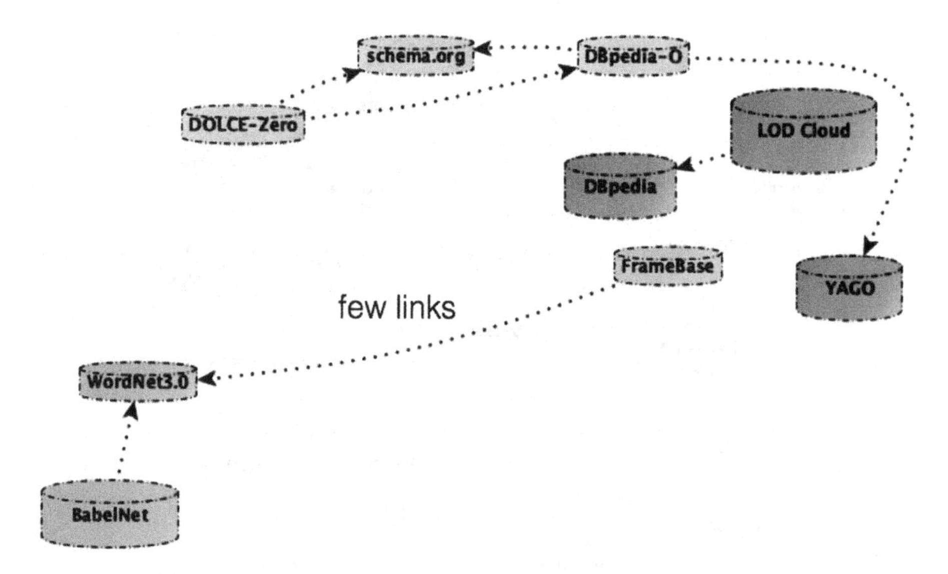

Fig. 1. Current state of Linguistic Linked Data and connections to other resources. Blue, red, green and yellow color represent role-oriented lexical resources, fact-oriented data, wordnet-like lexical resources and ontology schemas respectively. (Color figure online)

[24], OntoWordNet, VerbNet, and BabelNet, by interpreting their semantics as a subset of (a formal version of) Fillmore's frame semantics [8], and semiotics [10], and by reusing or linking to off-the-shelf ontological resources including OntoWordNet, DOLCE-Zero, Yago, DBpedia, etc. A complete depiction of the current state of Framester is shown in Fig. 2. Many resources in the picture, and their linking, are not described in this paper because of limited space. Further details along-with a SPARQL endpoint and a demo of WFD-API are available on-line from http://lipn.univ-paris13.fr/framester/.

The closest resources to Framester are FrameBase and Predicate Matrix. FrameBase is aimed at aligning linked data to FrameNet frames, based on similar assumptions as Framester's: full-fledged formal semantics for frames, detour-based extension for frame coverage, and rule-based lenses over linked data. However, the coverage of FrameBase is limited to an automatically learnt extension (with resulting inaccuracies) of FrameNet-WordNet mappings, and the alignment to linked data schemas is performed manually. Anyway, Framester could be combined with FrameBase (de)reification rules so that the two projects can mutually benefit from their results.

Predicate Matrix is an alignment between predicates existing in FrameNet, VerbNet, WordNet, and PropBank. It does not assume formal semantics, and its coverage is limited to a subset of lexical senses from those resources. The intended meaning of "frames" and "roles" defined in the aligned resources is assumed to be equivalent, though the alignment matrix does not state explicitly the formal conditions, under which such equivalence may hold. An RDF version

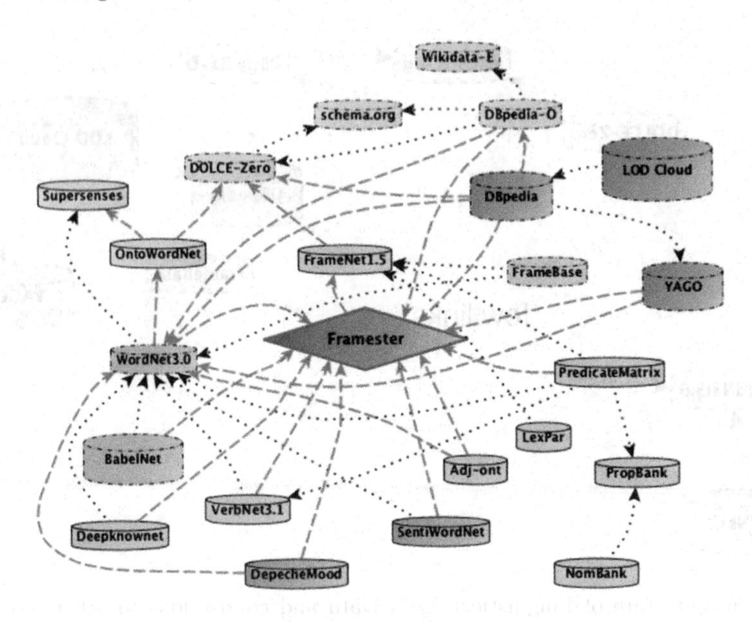

Fig. 2. Framester Cloud. Red color represents the main hub i.e., Framester, Purple represents the links to data sets for Sentiment Analysis. Black and orange arrows represent the existing and Framester specific links respectively. (Color figure online)

of Predicate Matrix has been created in order to add it to the Framester linked data cloud, and to check if those equivalences can be reused in semantic web applications.

4.1 Frame Semantics in OWL

Framester pushes the formalization game further, using the D&S (Descriptions and Situations [11]) knowledge pattern. D&S allows to distinguish the reification of the intension of a predicate (a *description*) from the reification of the extensional denotation of a predicate (a *situation*). A description d can define or reuse *concepts* $c^1, \ldots, c^n$ that can be used to *classify* entities $e^1, \ldots, e^m$ involved in a situation s that is expected to be compatible with d. D&S has been applied in many different ontology design contexts, e.g. proving its flexibility, and eventually being an ideal schema for punning operations in OWL2. As an example, a same set of facts (e.g. a boy pushing another) can be viewed either as an accident, a joke, or voluntary harm: such views are different (intensional) descriptions of different (extensional) situations, consisting of the same entities and relations among them.

D&S perfectly fits the core assumptions of Fillmore's frame semantics, by which a frame is a schema for conceptualizing the interpretation of a natural language text, its denotation (a frame occurrence) is a situation, and the elements (or semantic roles) of a frame are aspects of a frame, which can be either obligatory, optional, inherited, reused, etc. Constructive D&S [9] is an extension of D&S that takes into account a semiotic theory to integrate linguistic and formal semantics. It can therefore support additional frame semantics assumptions such as *evocation* and *semantic typing.*

As described in [24], several recipes can be designed to interpret FrameNet frames and frame elements as OWL classes, object properties, or punned individuals. Both FrameBase and Framester make use of the basic recipe that interprets frames as classes and frame elements as properties. However, Framester goes deeper in providing a two-layered (intensional-extensional) semantics for frames, semantic roles, semantic types, selectional restrictions, and the other creatures that populate the world of lexical resources. The two-layered representation is based on the Descriptions and Situations pattern framework, and exploits OWL2 punning, so enabling both (intensional) navigation in the linked lexical datasets, and the reuse of lexical predicates as extensional classes or properties. The main assumptions for Framester knowledge graphs are as follows:

Frame as a multigrade intensional predicate: A frame is a *multigrade intensional predicate* [25] $f(e, x_1, ..., x_n)$, where f is a first-order relation, e is a (Neo-Davidsonian) variable for any *eventuality* or *state of affairs* described by the frame, and x_i is a variable for any *argument place*, which could admit several *positions* in case multiple entities are expected to be classified in a place. For example in *"Hagrid rolled up a note for Harry"*, multigrade intensional predicate is represented as $Roll(e, Hagrid, note, Harry)$. OWL2 punning allows to represent a frame as either a class $f \sqsubseteq$ `dands`[6]`:Situation` (a subclass of the dands:Situation class, having situations as instances) or as an individual $f \in$ `framester`[7]`:Frame` (an instance of the framester:Frame class) (see Fig. 3).

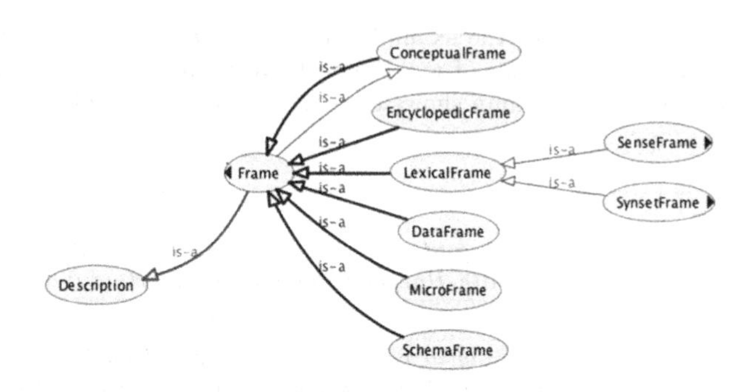

Fig. 3. Framester frame class. (Color figure online)

WordNet synsets are interpreted in a twofold way: as specialized frames, and as semantic types. As equivalence classes of word senses, whose words can evoke

[6] http://www.ontologydesignpatterns.org/cp/owl/descriptionandsituation.owl#.
[7] http://www.ontologydesignpatterns.org/ont/framester/framester.owl#.

one or more frames, they are cloned as instances of `framester:SynsetFrame`, which inherits their semantic roles from the core frames cloned from FrameNet.

Following the OntoWordNet semantics, they are promoted as OWL classes, unary projections of the corresponding synset frames.

Any word or multiword can evoke a frame: this is represented by means of a property chain that connects a word entity to a (punned) frame. A *frame occurrence* (a situation denoted by text or data) $s \in f$ is an instance of f and the entities $\{e, x_1 ... x_n\}$ involved in a situation are individuals. In $Roll(e, Hagrid, note, Harry)$, the frame evoked by the lexical unit "Roll" is the situation i.e., an occurrence of the frame "Reshaping" and the entities $\{e, Hagrid, note, Harry\}$ are the individuals.

Frame Projections include any projections of a frame relation. Assuming frame semantics, each meaning consists of activated frames, whose formal counterparts are multigrade intensional predicates. When only some aspect of that frame is considered, it can be formalized as a (typically unary or binary) projection of a frame relation. Semantic roles as well as co-participation relations are the *binary projections* of a frame. A semantic role is a binary projection $rol(e, x_i)$ of frame f, where e is the reified eventuality i.e., the Neo-Davidsonian variable of a multigrade predicate. A co-participation relation is a binary projection $cop(x_i, x_j)$ of f. Selectional restriction and the semantic type are unary projections of a frame. A selectional restriction is denoted as $res(x_i)$ of f that provides a typing constraint to an argument place. A semantic type $typ(x_i)$ for an external frame f' is reused as one of the domains of f. Figure 4 shows the hierarchy of frame projections. Table 1 shows the examples of each of the frame projections based on the running example.

Table 1. Frame Projections for the example *"Hagrid rolled up the note for Harry."*. The first column keeps the names of the Frame Projections (i.e., Unary and Binary Projections) and the second column shows the corresponding example.

Frame projections	Example
Unary projections	
Semantic type	$Rolls(e, Hagrid, note, Harry) \wedge agent(e, Hagrid) \wedge theme(e, note)$ $\wedge recipient(e, Harry) \wedge Person(Hagrid, Harry) \wedge Text(note)$
Binary projections	
Semantic role	$Rolls(e, Hagrid, note, Harry) \wedge agent(e, Hagrid) \wedge theme(e, note)$ $\wedge recipient(e, Harry)$ semantic roles = {agent,theme,recipient}
Co-participation relation	$rolls(Hagrid, note)$

Due to the expressivity limitations of OWL, some refactoring is needed to represent frame semantics: frames are represented as both classes and individuals, semantic roles and co-participation relations as both (object or datatype) properties and individuals, selectional restrictions and semantic types as both classes and individuals, situations and their entities as individuals. Frames and other predicates are represented as individuals when a schema-level relation is needed (e.g. between a frame and its roles, or between two frames), which cannot be represented by means of an OWL schema axiom (e.g. subclass, subproperty, domain, range, etc.).

Framester Role Hierarchy: Framester preserves the information about the Frame Element inheritance originally present in FrameNet. Additionally, it provides a mapping to generic frame elements which further connects to a more abstract role hierarchy provided by Framester. Figure 4(right) shows the hierarchy of semantic roles as defined in Framester.

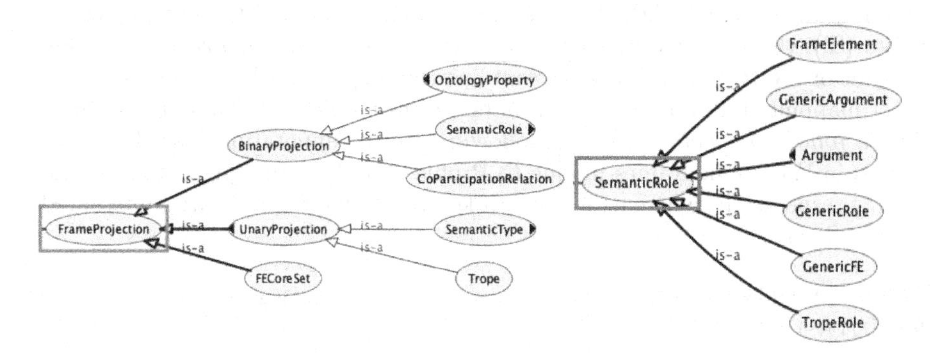

Fig. 4. Hierarchy of (a) Frame projections (left) (b) Semantic roles (right).

4.2 Resource Generation

The extensions to FrameNet were created using the semantic relations already present in WordNet. A set of *base-mappings* was produced by deeply revising existing FrameNet-WordNet mappings (eXtended WordFrameNet [5], FrameBase, and other existing sources found on the Web), and enriching them with new ones. This dataset, called Framester Base, has been manually curated to rectify mapping errors and evocations. Based on these basic mappings further links to other resources were generated. Due to space limitations we only discuss the base mappings. Further extensions were automatically performed based on:

1. WordNet hyponymy relations between noun and verb synsets, where each frame is extended with direct hyponyms of the noun or verb synsets mapped to frames in the Framester Base dataset
2. "Instance-of" relations between WordNet noun synsets
3. Adjective synset similarity
4. Same verb groups including verb synsets
5. Pertainymy relations between adverb synsets and noun or adjective synsets
6. Participle relations between adjective and verb synsets

7. Morphosemantic links between adjective and verb synsets
8. Transitive WordNet hyponymy relations
9. Unmapped siblings of mapped noun or verb synsets
10. Derivational links between different kinds of synsets

The Word Frame Disambiguation Subset. The part of Framester used in the WFD frame detector was bootstrapped by cloning a subset of FrameNet frames (the *core frames*) and its relations, and extending them by means of a manually curated mapping to WordNet synsets. The current experiments used four different Framester profiles to firstly check the impact of automatic extensions on precision and recall of Word Frame Disambiguation API (see next section). The subset of Framester consists of: *(i)* **Base (B)**: just the manually curated mappings, *(ii)* **Direct (D)**: the B profile plus extensions (1) to (7), *(iii)* **Transitive (T)**: the D profile plus extensions (8) to (10) and *(iv)* **FrameNet (F)**: a subset of the B profile that only contains the mappings whose synsets have a direct mapping in FrameNet lexical units. Let us consider the running example, *Hagrid rolled up a note for Harry*, following are the annotations based on each profiles in WFD (the frames unique to Profile D and T are represented in bold and (*) respectively, where as frames evoked by Profile F and B are represented in normal case):

Hagrid [[rolled]$_{\{CauseMotion,\textbf{CauseChange},...\}}$ up]$_{\{\textbf{Reshaping},UndergoChange^*\}}$ a [note]$_{\{Text^*\}}$ for Harry.

5 Word Frame Disambiguation: Evaluation Setting and Results

Word Frame Disambiguation, a framework based on frame detection, has been implemented for evaluation purposes. It is implemented as a pipeline including tokenisation, POS tagging, lemmatization, word sense disambiguation, and finally frame detection by detour using the four WFD profiles. Framester frames have been expanded (when applicable) by using the semantic relations present in FrameNet: `isPerspectivizedIn`, `seeAlso`, `inheritsFrom`, `perspectiveOn` and `uses`.

The four WFD-profiles have been evaluated in a frame detection task, and compared to other sets of mappings (XWFN [5] and FrameBase [28]), as well as to Semafor [4], the state of the art in machine-learning-based frame detection, whose model has been learnt on the annotations of the FrameNet annotated lexicon (see below).

Two textual corpora are used for evaluation: the FrameNet annotated lexicon version 1.5 released in 2010 (78 documents with 170,000 manually annotated sentences), and a corpus (called here the "independent corpus") of 100 heterogeneous texts taken from New York Times news, tweets, Wikipedia definitions, and scientific articles. The texts in the corpora were disambiguated by using two WSD algorithms: (i) Babelfy [21] and (ii) UKB [1]. The word senses provided by the WSD algorithms were then matched against Framester, and the evoked Framester frames were retrieved by following the links provided by the different profiles introduced in Sect. 4.2.

The annotated FrameNet lexicon can be considered a gold standard, since FrameNet developers have a rigorous manual procedure to annotate it. All words that are listed as FrameNet lexemes, and are found in the text, are annotated with exactly one frame. This contrasts with the fact that multiple frames might be evoked by a same word, and that many words that are not FrameNet lexemes can actually evoke a frame.

The independent corpus has been collected for machine reading evaluation purposes [14], and is not a gold standard for frame detection. This means that frame annotations (its ground truth) should be provided from scratch. In this experiment we used the tools intended to be compared, merged their results, asked two experts to judge the correctness of the detected frames, as well as any missing detection, and a third expert to take decisions when the two raters had different opinions.

On one hand, we expected that Semafor would be highly performant on the annotated FrameNet lexicon (since it has been trained on it), and we wanted (Experiment 1) to verify how close we can perform with a detour approach. On the other hand, the second corpus was used to verify (Experiment 2) if any difference in performance between Semafor and detour-based approaches is sensible to the specific Semafor training, or not.

5.1 Experiment 1: FrameNet Annotated Corpus

For Experiment 1, the frames already present in the FrameNet annotated lexicon were used as ground truth.

The performance of Word Frame Disambiguation with all its profiles, as well as Semafor's, were computed, and the results are shown in Table 2: recall obtained for each of the profiles (the values in bold represent the best results). The results were consistent for both the WSD algorithms.

Table 2. Results for different WFD-profiles FN-WN mappings when applied to frame detection against the FrameNet 1.5 full text annotations. Values in **bold** represent the best results.

	UKB				Babelfy			
Framester profiles	Recall	Precision	F_1	New annotations	Recall	Precision	F_1	New Annotations
eXtended WFN	0.511	**0.810**	0.627	832	0.580	**0.820**	0.680	8129
FrameBase	0.719	0.714	0.716	1132	0.621	0.71	0.661	11035
Profile-F	0.688	0.777	0.702	1148	0.673	0.749	0.704	10962
Profile-B	0.671	0.799	**0.729**	1251	0.662	0.780	**0.715**	11661
Profile-D	0.750	0.641	0.690	1929	0.790	0.569	0.660	20382
Profile-T	**0.860**	0.520	0.648	2728	**0.870**	0.444	0.588	26108

There was a significant increase in the newly annotated words in Profile-D and Profile-T as these two profiles extend the coverage of FrameNet. This leads to higher recall for these two profiles. The best recall was obtained for the profile created using transitive hyponymy relation (Profile-T).

The system used as a baseline in our experiments is Semafor [4]. It is a frame-semantic parser, which given a sentence aims at predicting frame-semantic representation using statistical models. As a first step, it extracts targets from the sentences and disambiguates it to a semantic frame. For doing so, it uses semi-supervised learning for frame disambiguation of unseen targets. Then an evoked frame is selected for each predicate.

In the current evaluation, we provide the sentences from the FrameNet 1.5 corpus to Semafor, which generates frame-tagged output and the precision, recall and the $F_1 - measure$ of the system are computed. The results are reported in Table 3. The recall for Framester (Profile-B with Babelfy as disambiguator on BabelNet as target) is .87, higher than Semafor's (.76), as expected, since the coverage of Framester is much wider. On the other hand, the precision of Semafor is very high (.96), but it cannot be compared to Framester on this corpus, since Framester can give multiple frames for a same word, and also annotates the words that are not annotated in the FrameNet corpus: all these annotations would be calculated as false positives, just because the gold standard did not address them.

In order to investigate if the precision of Framester is comparable to Semafor, and if Semafor performs well also on an independent corpus, we have performed the experiment in Sect. 5.2.

Table 3. Results for the baseline (Semafor) on FrameNet 1.5 full text annotations.

	Recall	Precision	$F_1 - Measure$	New Annotations
Semafor	0.76	0.96	0.85	16520

5.2 Experiment 2: Independent Unannotated Corpus

In the second experiment, we wanted to assess the portability of Semafor results out of the training corpus, as well as the accuracy of Framester profiles. We used an independent corpus collected for machine reading evaluation purposes [14]. Frame annotations have been collected by merging the results of all the compared frame detection methods, then asking two experts to judge the correctness of the detected frames, as well as any missing detection, and asking a third expert to take decisions when the two raters had different opinions. The raters were asked to judge the frames detected on a scale including Valid, Metaphorical, or Invalid[8]

[8] Many frames are not really wrong, but they are evoked as metaphorical or metonymical interpretations, e.g. the frame Travelling in a sentence like *Our love traveled distances.*

The inter-rater agreement before the third judgement has been measured by using weighted Cohen's K (WKAPPA) in order to adjust for the different weight of disagreement between absolute differences (valid vs. invalid evocation), and nuanced differences (valid/invalid vs. metaphorical evocation), and its value is 0.532, which is acceptable considering that frame annotation rating is difficult, and semantic annotations in general are accompanied by typically low interrater agreement.

The results are in Table 4, and show the performance of Framester profiles as well as Semafor. As expected, and noticed in Experiment 1, the recall grows significantly with extended profiles, but it's in general lower than with the FrameNet annotated corpus, except for the Profile-T. There is anyway a confirmation that Framester and the detour by WSD approach seems more appropriate for optimizing recall in frame detection. The doubt on the ability of Semafor to be very precise also on an independent corpus is confirmed: Semafor is still precise, but only at .79 against .95 on the corpus used for training. In addition, the best precision for Framester (Profile-B) is almost identical to Semafor's, and both Profile-D and Profile-T outperform Semafor on F1 measure.

Table 4. Results for our resource based on different extensions on the data set from Newspaper. Values in **bold** represent the best results. 'TP' and 'FP' stand for True Positives and False Positives respectively.

	TP	FP	Precision	Recall	F1
eXtended WFN	327	98	0.770	0.277	0.523
FrameBase	434	183	0.703	0.359	0.531
Profile-B	435	126	0.776	0.366	0.571
Profile-D	825	346	0.705	0.622	0.663
Profile-T	1204	664	0.644	**0.781**	**0.713**
Profile-F	452	151	0.750	0.377	0.564
Semafor	365	95	**0.794**	0.334	0.564

6 Conclusion

Framester is a novel linguistic linked data resource. It is based on frame semantics, and provides a whole new set of formally represented and linked lexical resources. Because of its adherence to frame semantics, FrameNet is the entry point for Framester, but it needs a well-built mapping to WordNet, which is at the core of existing lexical resources. Unfortunately, the quality of FrameNet-WordNet mappings is not high, and is largely incomplete.

In this work, we have described a new mapping between FrameNet and WordNet, and shown that this mapping is so good that a simple detour-based frame detector performs comparably to the state-of-the-art, machine-learning-based frame detector.

Ongoing work is about extending the experiments, and making use of the many linked datasets composing Framester with inferences provided by the full frame semantics of Framester's. Abstractive text summarisation, machine understanding and text similarity are some of the tasks that are being addressed.

Acknowledgements. The research leading to these results has received funding from the EFL (Empirical Foundations of Linguistics) LabEx and European Union Horizons 2020 the Framework Programme for Research and Innovation (2014–2020) under grant agreement 643808 Project MARIO Managing active and healthy aging with use of caring service robots.

References

1. Agirre, E., Soroa, A.: Personalizing pagerank for word sense disambiguation. In: Lascarides, A., Gardent, C., Nivre, J., (eds.) EACL 2009, Proceedings of the 12th Conference of the European Chapter of the Association for Computational Linguistics, Athens, Greece, March 30–April 3, 2009, pp. 33–41. The Association for Computer Linguistics (2009)
2. Baker, C.F., Fillmore, C.J., Lowe, J.B.: The Berkeley FrameNet project. In: Boitet, C., Whitelock, P., (eds.), Proceedings of the 36th Annual Meeting of the Association for Computational Linguistics and 17th International Conference on Computational Linguistics, COLING-ACL 1998, August 10–14, 1998, Université de Montréal, Montréal, Quebec, Canada, pp. 86–90. Morgan Kaufmann Publishers/ACL (1998)
3. Burchardt, A., Erk, K., Frank, A.: A WordNet detour to FrameNet. In: Proceedings of the GLDV 2005 Workshop GermaNet II, Bonn (2005)
4. Das, D., Chen, D., Martins, A.F.T., Schneider, N., Smith, N.A.: Frame-semantic parsing. Comput. Linguist. **40**(1), 9–56 (2014)
5. Lopez, M., de Lacalle, E., Laparra, G.R., Matrix, P.: Extending SemLink through WordNet mappings. In: Proceedings of the Ninth International Conference on Language Resources and Evaluation (LREC-2014), Reykjavik, Iceland, 26–31 May 2014, pp. 903–909 (2014)
6. Etzioni, O., Fader, A., Christensen, J., Soderland, S., Mausam, M.: Open information extraction: the second generation. In: IJCAI, pp. 3–10. IJCAI/AAAI. (2011)
7. Fellbaum, C.: WordNet: An Electronic Lexical Database. MIT Press, Cambridge (1998)
8. Fillmore, C.J.: Frame semantics and the nature of language. Ann. N. Y. Acad. Sci. **280**(1), 20–32 (1976)
9. Gangemi, A.: Norms and plans as unification criteria for social collectives. J. Auton. Agents Multi-Agent Syst. **16**(3), 70–112 (2008)
10. Gangemi, A.: What's in a Schema?. Cambridge University Press, Cambridge (2010)
11. Gangemi, A., Mika, P.: Understanding the semantic web through descriptions and situations. In: Meersman et al. [20], pp. 689–706
12. Gangemi, A., Navigli, R., Velardi, P.: The ontowordnet project: extension and axiomatization of conceptual relations in wordnet. In: Meersman et al. [20], pp. 820–838
13. Gangemi, A., Presutti, V., Reforgiato Recupero, D., Nuzzolese, A.G., Draicchio, F., Mongiovì, M.: Semantic web machine reading with FRED, Sermant. Web (2016)

14. Gangemi, A., Recupero, D.R., Mongiovì, M., Nuzzolese, A.G., Presutti, V.: Identifying motifs for evaluating open knowledge extraction on the web. Knowl. Based Syst. **108**, 33–41 (2016)
15. Gardner, M., Talukdar, P.P., Kisiel, B., Mitchell, T.: Improving learning and inference in a large knowledge-base using latent syntactic cues. In: Proceedings of the EMNL 2013 (2013)
16. Gliozzo, A., Biran, O., Patwardhan, S., McKeown, K.: Semantic technologies in IBM Watson. ACL **2013**, 85 (2013)
17. Hellmann, S., Lehmann, J., Auer, S.: Linked-data aware URI schemes for referencing text fragments. In: Teije, A., Völker, J., Handschuh, S., Stuckenschmidt, H., d'Acquin, M., Nikolov, A., Aussenac-Gilles, N., Hernandez, N. (eds.) EKAW 2012. LNCS (LNAI), vol. 7603, pp. 175–184. Springer, Heidelberg (2012). doi:10.1007/978-3-642-33876-2_17
18. Kingsbury, P., Palmer, M.: From TreeBank to PropBank. In: Proceedings of the Third International Conference on Language Resources and Evaluation, LREC, pp. 29–31, 2002, Las Palmas, Canary Islands, Spain. European Language Resources Association, May 2002
19. Kipper, K., Dang, H.T., Palmer, M.S.: Class-based construction of a verb Lexicon. In: Kautz, H.A., Porter, B.W. (eds.), Proceedings of the Seventeenth National Conference on Artificial Intelligence and Twelfth Conference on on Innovative Applications of Artificial Intelligence, July 30 - August 3, 2000, Austin, Texas, USA, pp. 691–696. AAAI Press/The MIT Press (2000)
20. Meersman, R., Tari, Z., Schmidt, D.C.: On the Move to Meaningful Internet Systems 2003: CoopIS, DOA, and ODBASE - OTM Confederated International Conferences CoopIS, DOA, and ODBASE, 3–7 November 2003. LNCS, vol. 2888. Springer, Heidelberg (2011)
21. Moro, A., Raganato, A., Navigli, R.: Entity linking meets word sense disambiguation: a unified approach. Trans. Assoc. Comput. Linguist. (TACL) **2**, 231–244 (2014)
22. Narayanan, S., Fillmore, C.J., Baker, C.F., Petruck, M.R.L.: Framenet meets the semantic web: A DAML+ OIL frame representation. Technology, p. 2000 (2003)
23. Navigli, R., Paolo PonAzetto, S.: BabelNet: the automatic construction, evaluation and application of a wide-coverage multilingual semantic network. Artif. Intell. **193**, 217–250 (2012)
24. Nuzzolese, A.G., Gangemi, A., Presutti, V.: Gathering lexical linked data and knowledge patterns from FrameNet. In: Musen, M.A., Corcho, Ó. (eds.), Proceedings of the 6th International Conference on Knowledge Capture (K-CAP 2011), 26–29 June 2011, Banff, Alberta, Canada, pp. 41–48. ACM (2011)
25. Oliver, A., Smiley, T.: Multigrade predicates. Mind **113**(452), 609–681 (2004)
26. Palmer, M., Semlink: Linking PropBank, VerbNet and FrameNet. In: Proceedings of the Generative Lexicon Conference, Pisa, Italy, GenLex-09 (2009)
27. Rizzo, G., Troncy, R., Hellmann, S., Bruemmer, M.: NERD meets NIF: Lifting NLP extraction results to the linked data cloud. In: LDOW, 5th Workshop on Linked Data on the Web, Lyon, France, 04 2012
28. Rouces, J., de Melo, G., Hose, K.: FrameBase: representing N-Ary relations using semantic frames. In: Gandon, F., Sabou, M., Sack, H., d'Amato, C., Cudré-Mauroux, P., Zimmermann, A. (eds.) ESWC 2015. LNCS, vol. 9088, pp. 505–521. Springer, Heidelberg (2015). doi:10.1007/978-3-319-18818-8_31
29. Schuler, K.K.: Verbnet: a broad-coverage, comprehensive verb Lexicon. Ph.D. thesis, Philadelphia, PA, USA (2005). AAI3179808

30. Van Assem, M., Gangemi, A., Schreiber, G.: Conversion of WordNet to a standard RDF/OWL representation. In: Proceedings of the Fifth International Conference on Language Resources and Evaluation, LREC 2006, Genoa, Italy, pp. 237–242 (2006)
31. van Assem, M., Gangemi, A., Schreiber, G.: RDF/OWL representation of WordNet. World Wide Web Consortium, Working Draft WD-wordnet-rdf-20060619, June 2006

An Investigation of Definability in Ontology Alignment

David Geleta(✉), Terry R. Payne, and Valentina Tamma

Department of Computer Science, University of Liverpool, Liverpool, UK
{D.Geleta,T.R.Payne,V.Tamma}@liverpool.ac.uk

Abstract. The ability to rewrite defined ontological entities into syntactically different, but semantically equivalent forms is an important property of *Definability*. While rewriting has been extensively studied, the practical applicability of currently existing methods is limited, as they are bounded to particular Description Logics (DLs), and they often present only theoretical results. Moreover, these efforts focus on computing single definitions, whereas the ability to find the complete set of alternatives, or even just their signature, can support ontology alignment, and semantic interoperability in general. As the number of possible rewritings is potentially exponential in the size of the ontology, we present a novel approach that provides a comprehensive and efficient way to compute in practice all definition signatures of the feasible (given pre-defined complexity bounds) defined entities described using a DL language for which a particular definability property holds (*Beth definability*). This paper assesses the prevalence, extent and merits of definability over large and diverse corpora, and lays the basis for its use in ontology alignment.

1 Introduction

The ability to rewrite defined ontological entities into syntactically different, but semantically equivalent forms is an important feature of the notion of *Definability*. In particular, *Beth definability* [2,11] is a well-known property from classical logic, that relates the notion of *implicit definability* to the one of *explicit definability*, by stating that every implicitly defined concept is also explicitly definable, in any definitorially complete DL language [21]. For example, given an ontology $\mathcal{O} = \{\mathsf{C} \equiv \mathsf{A} \sqcup \mathsf{B}, \mathsf{A} \sqsubseteq \neg\mathsf{B}, \mathsf{D} \sqsubseteq \exists \mathsf{r}.\top\}$, where the concept C is defined explicitly, i.e. $\mathsf{C} \equiv \mathsf{A} \sqcup \mathsf{B}$, the concept A is defined implicitly under $\mathcal{O}$ by the set of general concept inclusions $\{\mathsf{C} \equiv \mathsf{A} \sqcup \mathsf{B}, \mathsf{A} \sqsubseteq \neg\mathsf{B}\}$. Thus, A can be explicitly defined by the axiom $\mathsf{A} \equiv \mathsf{C} \sqcap \neg\mathsf{B}$.

Definability in general (and Beth definability in particular) have been utilised within DLs to generate syntactically different, albeit semantically equivalent definitions. Known as *rewriting*, this process is primarily used for: (1) extracting equivalent terminology from a general TBox [1]; and (2) finding equivalent query rewritings in ontology-based data access scenarios [18]. These approaches exploit the fact that any defined concept has one or more possible alternative definitions; however they usually focus on finding a single alternate definition; whereas

E. Blomqvist et al. (Eds.): EKAW 2016, LNAI 10024, pp. 255–271, 2016.
DOI: 10.1007/978-3-319-49004-5_17

several ontology engineering tasks would benefit from the ability to identify a complete set. In particular this paper focusses on *ontology alignment* [7], where several approaches have been proposed that successfully align ontologies [3]. However, Stuckenschmidt et al. have argued that existing approaches often fail to compute complex correspondences: typically, systems are only able to identify simple equivalence statements between class or relation names, but often fail to identify richer semantic relation between elements of different ontologies [20]. Thus, the ability to rewrite concept definitions can widen the search space for possible correspondences. This is illustrated by the fact that some alignment mechanisms may not find a simple correspondence for some concept C, but may be able to find a complex correspondence, given the its definition $\mathsf{C} \equiv \mathsf{A} \sqcup \mathsf{B}$.

Determining the complete set of possible definitions of defined concepts is a challenging task, as the number of different definitions is potentially exponential in the size of the ontology. This is problematic for large scale ontologies, such as SNOMED CT [6] or FMA [15]. Existing rewriting algorithms are *language dependent*, and thus different approaches to construct rewritings are used for different types of DL expressivity. Furthermore, even if there was an existing approach for a given language, many rewriting systems provide only a theoretical characterization of the rewriting mechanism, therefore making them less usable in practice. Finally, rewriting *requires a seed signature* to be specified in input, i.e. a restricted vocabulary to be used in defining a given concept. The process of identifying all valid seed signatures is inherently complex, as it requires examining each member of the powerset of the ontology signature and verifying whether it actually implicitly defines a particular entity. Therefore, reducing the search space for these problems is highly desirable.

In this paper we present a pragmatic approach to computing the complete set of rewriting signatures for a given ontology. Our approach exploits Beth definability to identify all possible alternative definitions of defined entities. We present the notion of Beth definability in Sect. 2, and then introduce our novel approach (Sect. 3) that, *in practice*, can efficiently compute the complete set of definition signatures (DS) of defined entities, for any DL language where the Beth definability property holds. Section 4 presents *definition patterns* (DP), that aid comprehension of definability, and also serve as input for a *heuristic-based rewriting approach*, which produces definition axioms without using reasoning services. Section 5 presents an empirical analysis over a wide range of OWL ontologies, and assesses the prevalence of definability and the behaviour of the proposed algorithms for computing definability. Finally, the paper discusses definability for ontology alignment and presents concluding remarks.

2 Beth Definability in Description Logics

The *vocabulary* of a DL ontology[1] consists of the (disjoint) union of the countably infinite sets of concept names, role names and individual names, where an *entity*

[1] In this paper, we assume familiarity with basic notions of Description Logics [1] and the Web Ontology Language [10] (OWL).

e is either a named concept, role, or individual. A *signature* is an arbitrary set of entities; by $\mathsf{Sig}(\mathsf{C})$ we denote the signature of the a complex concept C, while $\mathsf{Sig}(\mathcal{T})$ denotes the signature of a TBox. In this paper, Σ refers to a *definition signature* (DS), i.e. the set of entities that implicitly define a given concept. A DS is used to characterise *implicitly definable* concepts in terms of their explicit definability, by exploiting the *Beth definability theorem.* The theorem, initially studied for first-order logic [2], states that a logical term is *implicitly definable* with respect to a theory if and only if it is also *explicitly definable.* Given that explicit definability implies implicit definability, the Beth definability property holds for some logic language $\mathcal{L}$ if the converse also holds, i.e. if implicit definability implies explicit definability. Consequently, if a term is implicitly defined then it is always possible to define it explicitly. As there are several variants of Beth definability [21], we focus on *Projective Beth definability* which is a stronger formulation [11] with the ability to specify a set of terms, thus permitting us to restrict the vocabulary that can be used in definitions. Beth definability has also been studied in the context of DLs [21], where it has been used to compute explicit definitions based on implicit definitions. We thus assume a general DL language $\mathcal{L}$ for which the Beth definability property holds. We define an explicit definability concept as:

Definition 1 (Explicitly defined concept). *Let* C *be a concept name, and* $\mathcal{T}$ *a TBox, where* $\mathsf{C} \in \mathsf{Sig}(\mathcal{T})$. C *is explicitly defined under* $\mathcal{T}$*, if and only if there is an axiom* $\alpha : \mathsf{C} \equiv \mathsf{D}$*, such that* $\alpha \in \mathcal{T}$*, where* D *is either a concept name in* $\mathcal{T}$*, or a complex concept such that* $\mathsf{Sig}(\mathsf{D}) \subseteq \mathsf{Sig}(\mathcal{T}) \setminus \{\mathsf{C}\}$.

For example, let us consider $\mathcal{T}^{Family}$, a small $\mathcal{ALC}$-TBox describing the family domain, shown in Fig. 1 (upper). The concept Parent, defined by the axioms α_1 and α_2, is the only explicitly defined concept in the ontology. Similarly, we can define implicitly definable concepts:

Definition 2 (Implicitly definable concept). *Let* C *be a concept name,* $\mathcal{T}$ *a TBox, and* Σ *a signature, where* $\mathsf{C} \in \mathsf{Sig}(\mathcal{T})$*, and* $\Sigma \subseteq \mathsf{Sig}(\mathcal{T}) \setminus \{\mathsf{C}\}$. C *is implicitly definable from* Σ *under* $\mathcal{T}$*, if and only if for any two models* $\mathcal{I}$ *and* $\mathcal{K}$ *of* $\mathcal{T}$, $\Delta^{\mathcal{I}} = \Delta^{\mathcal{K}}$*, and for all entity* $P \in \Sigma$, $P^{\mathcal{I}} = P^{\mathcal{K}}$*, then it holds that* $\mathsf{C}^{\mathcal{I}} = \mathsf{C}^{\mathcal{K}}$.

Given the example, it can be seen that both Mother and Father are implicitly defined concepts in $\mathcal{T}^{Family}$, and each has six syntactically different, but semantically equivalent definitions (Fig. 1, lower).

The *number of possible rewritings* of a defined concept, regardless of whether it is explicitly or implicitly defined, is potentially exponential in the size of the ontology. Descriptions of defined concepts (i.e. the right-hand side of a non-primitive concept definition axiom) are built inductively using other, potentially defined concepts. Thus, the number of possible concept rewritings is dependent on the definability of its constituent concepts. As the definability of any defined description member concept is dependent on the definability of its own description, definability is therefore a *recursive notion* [8].

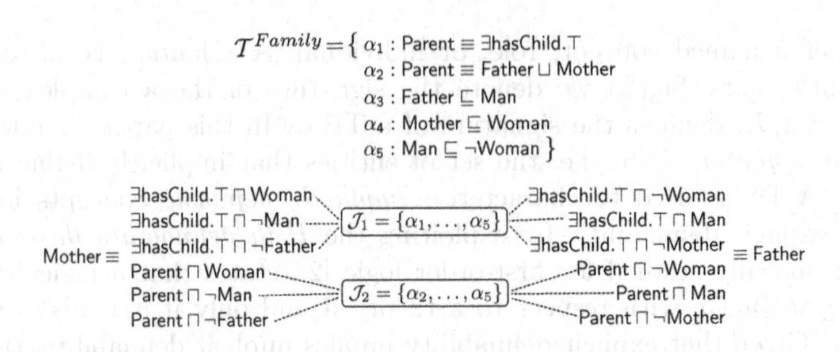

Fig. 1. This small ontology describes a family domain. Concepts Mother and Father are implicitly defined in $\mathcal{T}^{Family}$, hence these are also explicitly definable, as shown by their definition axioms. Each axiom is explained by a justification ($\mathcal{J}_1, \mathcal{J}_2$), denoted with dashed line.

Deciding Definability. A particular concept name C can either be defined *explicitly* or *implicitly* under an ontology, or be *undefined.* Explicit definability is a syntactic notion; deciding whether C is explicitly defined under an ontology is the trivial process of searching the TBox for a concept equivalence axiom whose left-hand side is C, and the potentially complex concept on the right-hand side does not include C (e.g. $\mathsf{C} \equiv \mathsf{D}$ where $\mathsf{C} \notin \mathsf{Sig}(\mathsf{D})$). In contrast, implicit definability is a semantic notion whose detection requires reasoning. Ten Cate et al. have shown that in DLs, testing implicit definability can be reduced to entailment checking [21]; in this article, $\textsc{ImpDef}(\mathsf{C}, \Sigma, \mathcal{T})$ denotes the function that determines whether a concept C is implicitly definable under a TBox $\mathcal{T}$ given a signature Σ). The computational complexity of determining whether a concept is implicitly defined depends on the complexity of the entailment check, which is predicted on the expressivity of the given DL language. Thus it is potentially exponential in the size of the ontology, for expressive DL dialects.

Justifying Definability. It is often difficult for humans to identify the axiom set in a TBox that implies definability. *Justifications* [12] can be used to validate definability and to provide a set of axioms supporting an entailment. A justification $\mathcal{J}$ for an entailment η in an ontology is the ontological fragment in which η holds (i.e. a set of TBox axioms such that $\mathcal{J} \subseteq \mathcal{O}$). A justification is *minimal*, if the entailment in question does not follow from any proper subset of the justification. For example, if we assume $\mathcal{O} = \{\mathsf{A} \sqsubseteq \mathsf{B}, \mathsf{B} \sqsubseteq \mathsf{C}, \mathsf{D} \sqsubseteq \exists \mathsf{r}.\mathsf{C}\}$ and the axiom $\alpha : \mathsf{A} \sqsubseteq \mathsf{C}$, then $\mathcal{O} \models \alpha$ holds as $\{\mathsf{A} \sqsubseteq \mathsf{B}, \mathsf{B} \sqsubseteq \mathsf{C}\} \subseteq \mathcal{O}$; i.e. the entailment is justified[2]. The algorithm checking for implicit definability can be modified to compute not only whether a concept is definable with respect to a given signature, but to also provide its justifications.

Definability of Roles. Determining role definability[3] can also be achieved by using the same method outlined for determining concept definability. However,

[2] Horridge et al. [12] introduced an efficient approach that computes either a single, or all justifications of an entailment.

[3] In this paper, we only focus on concept definability, and omit the description of the approach for determining role definability. However, a full description of the algorithm for deciding role definability is available in [8].

whilst concepts are defined using other concept and role names, roles are defined only in terms of other role names; therefore the entailment check is restricted to the RBox (role axioms), where the definition signature contains role names only.

3 Minimal Definition Signatures (MDSs)

This section describes the approach for finding the *complete set of definition signatures* of a particular defined concept, in any DL language where the Beth definability property holds. A definition signature can be defined as:

Definition 3 (Definition Signature (DS)). *A set of entity names Σ is a definition signature of the concept* C *under a TBox $\mathcal{T}$, if and only if there is a complex concept* D*, such that* $\mathsf{Sig}(\mathsf{D}) \subseteq \Sigma$*, and* $\mathcal{T} \models \mathsf{C} \equiv \mathsf{D}$*, where* $\Sigma \subseteq \mathsf{Sig}(\mathcal{T}) \setminus \{\mathsf{C}\}$.

If a concept C is defined in an ontology, then we can entail that there exists some subset Σ of the ontology signature that implicitly defines C, i.e. members of Σ can be used to construct the right-hand side of a definition axiom for C. We only focus on *acyclic definitions*, as definitions with direct cycles (where the defined concept appears in its corresponding description) have no use in rewriting (as such signatures do not permit the substitution of defined entities), thus are excluded by this definition. As definition signatures may contain *redundant members*, their size could be very large, hence we introduced the notion of *signature minimality*:

Definition 4 (Minimal Definition Signature (MDS)). *A signature Σ is a minimal definition signature of a defined concept* C *under a TBox $\mathcal{T}$, if there exists no other definition signature Σ' such that $\Sigma' \subset \Sigma$.*

The *minimality* property of an MDS refers to minimising the size of the signatures, by eliminating superfluous entities. However, a defined concept may have multiple *unique* MDSs (where the difference of any two MDSs is not an empty set) under an ontology, with the same cardinality. From the definition, it follows that every MDS is also a DS, and any DS may contain at least one, but potentially many MDSs. For example, in the $\mathcal{T}^{Family}$ example (Fig. 1), the signature $\Sigma = \{\mathsf{hasChild}, \mathsf{Man}, \mathsf{Woman}\}$ is a DS of all three defined concepts in the TBox. However, this signature is not a minimal DS of Parent, because it can be defined by the following two MDSs: $\{\mathsf{hasChild}\}$, $\{\mathsf{Mother}, \mathsf{Father}\}$; as formalised by axiom α_1 and α_2, respectively.

Finding MDSs can be computationally expensive, as the number of definitions themselves can be exponential in the size of the ontology. Furthermore, the set of candidate signatures is equivalent to the power set of the TBox signature (excluding the defined concept itself, i.e. $2^{\mathsf{Sig}(\mathcal{T})\setminus\{\mathsf{C}\}}$). In order to reduce this complexity, *modularisation* [16] is used as space reduction mechanism. As modules preserve all entailments with respect to a signature consisting of a concept name, any MDS of a defined concept is contained in the module signature [8].

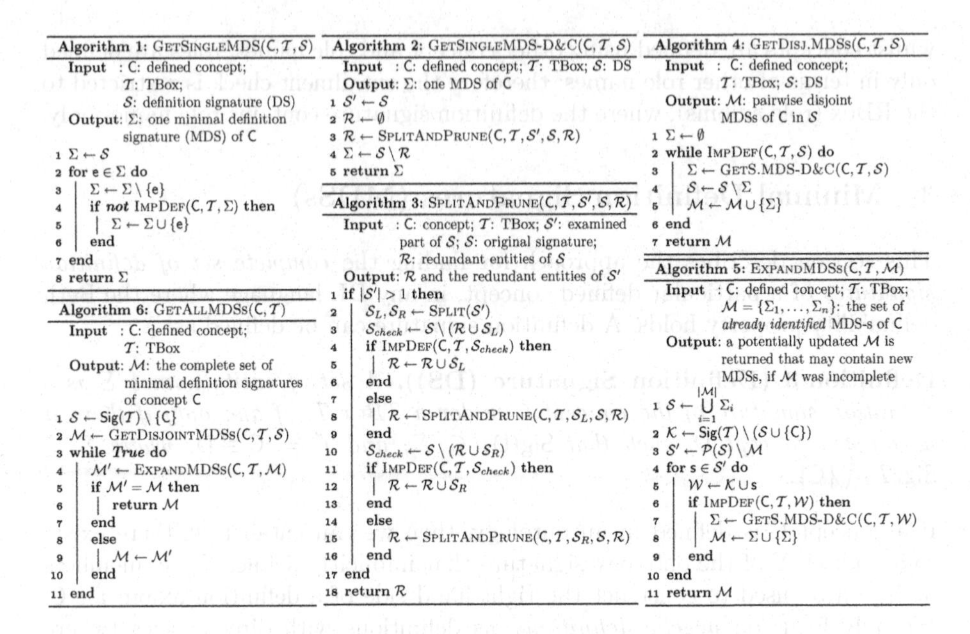

Fig. 2. The algorithms computing Minimal Definition Signatures.

Syntactic locality based modules (LBM) have been shown to be sound approximations of *semantic locality based modules* (that preserve entailments over all the terms that occur in the module) [5], and there are efficient and widely used polynomial time algorithms for extracting syntactic LBMs[4]. As modules can be considerably smaller compared to the original ontology, modularisation is an effective mechanism for reducing the complexity of computing MDSs.

The basic idea behind *computing a single MDS* (Algorithm 1, Fig. 2) is that the input DS (i.e. the signature of the module which describes a given defined concept) is iteratively pruned until it is reduced to a subset that is an MDS. Pruning is achieved by removing a member of the DS and testing the remaining signature to check if it still implicitly defines the given concept. If so, then the entity is redundant as opposed to being required. Algorithm 1 has linear time complexity, as each member of the input is examined exactly once. However, it is worth noting that this *excludes the complexity of the implicit definability check*, which is delegated to an external oracle i.e. a reasoner; thus it always depends on the DL expressivity of the ontology, and that can be exponential in the size of the ontology. Therefore, in order to reduce the overall complexity, we decrease the number of implicit definability checks. By applying a *divide and conquer strategy* (Algorithms 2 and 3) pruning is carried by testing (and removing) entity groups, instead of individual entities. Algorithms 2 and 3 has logarithmic time complexity w.r.t the size of the module signature in the best

[4] The OWL API provides methods for extracting several types of LBMs.

case scenario, polynomial in the worst case[5]. Algorithms 1, 2 and 3 are complete and correct, as if the input is a valid DS, by only removing redundant entities, they reduce it to a minimal DS. Algorithm 4 computes a *set of pairwise disjoint MDSs*, in polynomial time. The input DS is iteratively reduced to an MDS, and the resulting MDS is subtracted from the input signature. This is repeated until the input signature is no longer a DS of the defined concept, thus the algorithm is complete and correct.

In order to find the *complete set of MDSs* (Algorithm 6), the first step requires computing of a set of pairwise disjoint MDSs. After this initial step, any unidentified MDS must overlap with an identified MDS. The rest of the process is delegated to Algorithm 5, which either expands on an existing but incomplete set of MDSs, or confirms its completeness. The expansion process involves iterating through the power set of the union of disjoint MDSs ($\mathcal{S}$, the smallest known set of related signature members of the defined entity), where each subset is combined (into $\mathcal{W}$) with the set of non-MDS entities ($\mathcal{K}$) in the ontology (or module) signature, and tested for validity. Despite its exponential computational complexity, the approach is feasible to use for most real-world ontologies (because $\mathcal{S}$ is typically small), as we show with the empirical evaluation presented in Sect. 5. A more exhaustive description of the algorithms presented above can be found in [8].

4 Definition Patterns

As part of our empirical investigation, we have computed the DSs for numerous ontologies (Sect. 5), and validated them by obtain the corresponding justifications. An explicit concept definition is always formalised as a single axiom, whereas the definition of an implicitly defined concept is derived from an axiom set (i.e. a justification); thus, implicit definitions are often not straightforward to recognise and interpret. By studying the composition of DSs (their cardinality, and the type and number of their member entities) together with their justifications (their size, and the type of their constituent axioms) we have identified a number of *definition patterns*. The patterns aim to generalise the frequent forms of creating definitions; these were inspired by ontology alignment design patterns [17], and by the predefined axiom types in OWL (where the atomic axioms form complex constructs). In addition to the validation and the interpretation of definability, the identifiable definition patterns permit *rule-based definition axiom generation*, i.e. the generation of an explicit definition of a defined entity, according to an inference rule, by processing a given DS and a justification. In contrast to general rewriting methods, this approach is language independent and does not require ontological reasoning (apart from obtaining a single justification, which is achievable in polynomial time [12]). Furthermore, the computational complexity is polynomial in the size of the input. The set of patterns

[5] Both Algorithms 1, 2 and 3 are used in practice, where the former is better suited for computing an MDS from a DS that is the RHS signatures of an explicit definition, as such signature typically contain either none, or only a few redundant members (i.e. reaching the worst-case scenario of Algorithms 2 and 3).

Table 1. Basic definition patterns

Entity	ID	Pattern	Σ (MDS)	Justification		Defined	
				$\|\mathcal{J}\|$	$\mathcal{J}$	Expl.	Impl.
Concept	*1*	Explicit definition	$\|\Sigma\| > 1$	=1	$\{\mathsf{C} \equiv \exists \mathsf{r}.\mathsf{T} \sqcap \mathsf{D}\}$	C	
	2	Explicit synonym	Σ: 1 concept	=1	$\{\mathsf{C} \equiv \mathsf{D}\}$	C, D	
	3	Implicit synonym	Σ: 1 concept	>1	$\{\mathsf{C} \sqsubseteq \mathsf{D}, \mathsf{D} \sqsubseteq \mathsf{C}\}$		C, D
	4	Disjoint union	$\|\Sigma\| > 1$	>1	$\{\mathsf{C} \equiv \mathsf{D}_1 \sqcup \mathsf{D}_2, \mathsf{D}_1 \sqsubseteq \neg \mathsf{D}_2\}$	C	$\mathsf{D}_1, \mathsf{D}_2$
	5	Role domain concept	Σ: 1 role	>1	$\{\mathsf{C} \sqsubseteq \exists \mathsf{r}.\mathsf{T}, \exists \mathsf{r}.\mathsf{T} \sqsubseteq \mathsf{C}\}$		C
	6	Role range concept	Σ: 1 role	>1	$\{\mathsf{T} \sqsubseteq \forall \mathsf{r}.\mathsf{C}, \mathsf{C} \sqsubseteq \exists \mathsf{r}^{-}.\mathsf{T}\}$		C
	7	Synonym role (domain or range)	Σ: 1 role	>1	$\{\mathsf{C} \equiv \exists \mathsf{r}.\mathsf{T}, \mathsf{r} \equiv \mathsf{s}\}$		C
	8	Inverse role (domain or range)	Σ: 1 role	>1	$\{\mathsf{C} \equiv \exists \mathsf{r}.\mathsf{T}, \mathsf{r} \equiv \mathsf{s}^{-}\}$		C
Role	*9*	Explicit definition	$\|\Sigma\| > 1$	=1	$\{\mathsf{r} \equiv \mathsf{s} \circ \mathsf{q}\}$	r	
	11	Explicit synonym	Σ: 1 role	=1	$\{\mathsf{r} \equiv \mathsf{s}\}$	r, s	
	12	Explicit inverse	Σ: 1 role	=1	$\{\mathsf{r} \equiv \mathsf{s}^{-}\}$	r	s
	13	Implicit synonym	Σ: 1 role	>1	$\{\mathsf{r} \sqsubseteq \mathsf{s}, \mathsf{s} \sqsubseteq \mathsf{r}\}$		r, s
	14	Implicit inverse	Σ: 1 role	>1	$\{\mathsf{r} \sqsubseteq \mathsf{s}^{-}, \mathsf{s} \sqsubseteq \mathsf{r}^{-}\}$		r, s

is not exhaustive, i.e. it is not guaranteed to represent all definitions, however as shown by the empirical analysis, it covers a significant portion of cases[6].

Table 1 presents the list of patterns, showing the composition of the minimal definition signature (Σ), the corresponding justification in terms of its size ($|\mathcal{J}|$) and the set of axioms forming the justification. The right column presents the concepts or roles that are explicitly or implicitly defined in a pattern. In some patterns, more than one entity can be defined, for example, in an explicit, or an implicit synonym concept pattern, both concepts C and D are defined, however the actual members of the particular MDS depend on which defined entity the patterns refers to ($\Sigma^{\mathsf{C}} = \{\mathsf{D}\}, \Sigma^{\mathsf{D}} = \{\mathsf{C}\}$). For example, in $\mathcal{T}^{Family}$ (Fig. 1) the set of axioms $\{\alpha_2, \ldots, \alpha_5\}$ form the justification in a *disjoint union* pattern , where the concept Parent is defined explicitly, and the concepts Mother, Father are both defined implicitly by the same MDS $\Sigma = \{\mathsf{Parent}, \mathsf{Mother}, \mathsf{Father}\}$. The definability cases that do not fall into these basic patterns, are arbitrary *combinations* of the basic ones, meaning that the justification of such combination has one or more subset that is the justification (i.e. a minimal set) of some other defined entity which contribute to the definition. Figure 3 presents an example of a pattern combination, where the justification entails the definition of concept Invited_speaker; furthermore, it contains two explicit definitions of the concept Conference_contribution, one explicit definition of Regular_contribution, and a disjoint union pattern of Invited_talk. Although our rule-based approach does not produce a definition axiom for Invited_speaker, it generates an axiom for all the other defined entities, that are characterised by processable patterns.

[6] Further details on patterns, and the axiom generation algorithm are presented in [8].

5 Empirical Analysis

In this section we empirically investigate the occurrence of definability in existing ontologies and the impact it has in supporting semantic interoperability. We also analyse the behaviour of the proposed algorithms to compute the definability of ontological terms. The underlying assumption we make is that the definability status (undefined, or defined: explicitly and/or implicitly) of ontology signature entities, and the number of MDSs of defined entities provide a measure of the usability of an ontology in semantic interoperability[7]. Thus, in order to gain insight on whether, in practice, the use of definability signatures would contribute to ontology alignment in particular, and ontology engineering in general, we have analyzed the *prevalence and the extent of definability* over a wide range of OWL ontologies. Furthermore, we also study the behaviour of the proposed approximations to compute MDSs in terms of run time taken for each of the stages necessary to compute the MDSs. In the first experiment we distinguish between *defined* and *undefined* ontologies (depending on whether they contain at least one defined concept, or no defined entities), and we examine the definability status and type for each entity in all the ontologies of a large and diverse corpus, and considered several characteristics of the defined and the undefined ontologies. A second aim of this empirical analysis is to *characterise the behaviour and assess the practical applicability* of the proposed definability computation algorithms. This is achieved by measuring the processing time of computing MDSs in a corpus of made of 'semantically rich' ontologies, i.e. that contains a large portion of defined entities and MDSs. We remind the reader (Sect. 3), that definability computation is a three step process: first the definability status of each entity is established, next the disjoint MDSs are obtained, finally any potentially unidentified MDS is computed (i.e. the complete set of MDSs). While the first two steps are polynomial (excluding the complexity of the implicit definability check, i.e. an entailment check which depends on the ontology language expressivity), the third step has exponential time complexity.

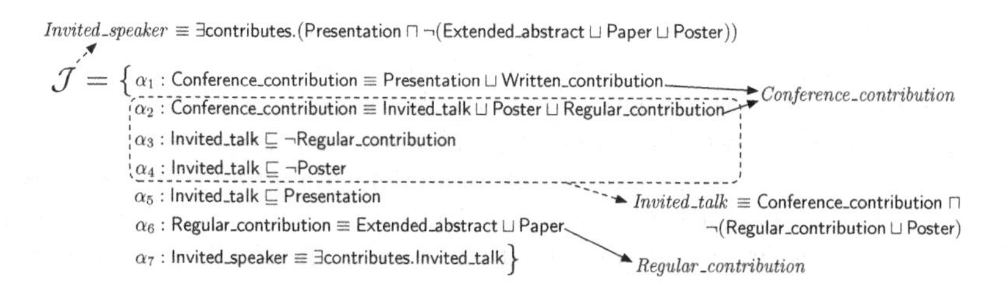

Fig. 3. Example combinations of basic patterns in *Conference.owl*, where respectively, explicit, or implicit definability is denoted with a normal, or a dashed line.

[7] For example, given two versions of an ontology, the one with more defined entities or higher MDS to entity ratio is more valuable, as it may permit the expression of more entities with alignments, that are typically incomplete [7].

Table 2. Comparing defined and undefined ontology properties

(a) OWL Profiles

	EL only	*EL*	*QL only*	*QL*	*RL only*	*RL*	*DL only*	*DL*	*Full*
Defined	1.99%	5.89%	0.81%	3.97%	7.69%	11.60%	48.76%	58.62%	36.17%
Undefined	1.76%	9.36%	1.32%	7.29%	7.22%	15.26%	38.88%	56.60%	42.34%

(b) OWL Constructors

	AL	*C*	*D*	*E*	*F*	*H*	*I*
Defined	82%	28.72%	51.36%	12.33%	15.29%	37.58%	45.40%
Undefined	84%	26.07%	45.02%	9.85%	14.08%	30.13%	37.83%
	N	*O*	*Q*	*R*	*S*	*TRAN*	*U*
Defined	31.04%	29.24%	5.44%	2.84%	18.01%	7.70%	10.83%
Undefined	24.68%	32.62%	4.92%	3.71%	14.02%	6.37%	8.63%

(c) Logical Axioms

	Defined	Undefined
<10	8.94%	25.42%
10-	33.33%	40.74%
101-	50.21%	26.42%
1001-	6.76%	6.14%
10001-	0.76%	1.29%

(d) Ratio of defined entities in ontologies in the corpus

Defined Ratio	*0%*	*<0-10%*	*11-20%*	*21-30%*	*31-40%*	*41-50%*	*51-60%*	*61-70%*	*71-80%*	*81-90%*	*91-100%*
Nb of Ont	50.10%	9.01%	7.87%	7.46%	6.37%	6.63%	3.14%	2.88%	1.41%	1.20%	3.93%

The *experimental framework* used to run this analysis is implemented in Java; the OWL API is used for ontology manipulation and for interacting with the reasoners [13], whilst the OWL Explanation API [12] is used to compute justifications. The framework utilizes both the HermiT [9] and Pellet [19] reasoners[8]. All of the data and software, including computed DSs, definition patterns, and other results are available online[9]. The *evaluation corpus* has been assembled from a variety of OWL ontology datasets, including ontologies in the Manchester Ontology Repository [10], datasets used in the OAEI evaluation challenge[11], and a sizable set of ontologies obtained by crawling the Web [14]. In particular, this evaluation corpus consists of 3576 ontologies of different size, DL expressivity, conceptualized domain, and source of origin (professional, academic etc.). The second experiment uses a sample of 9 ontologies, and is carried out on a 16 GB RAM, 10 core processor machine (with 8 GB memory used by the JVM).

In order to determine which ontology characteristics (if any) are affected by definability, the first experiment examines each concept in all 3576 ontologies with the aim to determine whether they are undefined, or defined either explicitly or implicitly. Therefore, we aim to assess the prevalence of (implicitly) definable concepts in state of the art ontologies. The hypothesis tested in this experiment is that definability occurs in ontology irrespective on the ontology characteristics e.g. size, expressivity etc. The analysis of the entire corpus classifies 1703 ontologies (49.89 %) as defined, i.e. that contain *at least one* defined concept. Out of all concepts, 75.82 % are undefined, 20.74 % are explicitly and 3.44 % are implicitly (only) defined. Table (2.d) shows the proportion of ontologies in the corpus, binned by the ratio of defined to undefined concepts within an ontology. Table (2.c) presents the relative distribution of defined and undefined ontologies, binned by the number of logical axioms; Table (2.a) shows the distribution of OWL profiles in defined and undefined ontologies; and Table (2.b) shows the

[8] HermiT performs faster with most datasets, however Pellet was able to load and process some ontologies that HermiT could not (due to ontologies using datatypes that are not part of the OWL 2 datatype map and no custom datatype definition was given).

[9] http://www.csc.liv.ac.uk/~dgeleta/ontodef.html.

[10] http://owl.cs.manchester.ac.uk/tools/repositories/.

[11] http://oaei.ontologymatching.org.

Table 3. *Cost measured in terms of different characteristics* (time, and number of definability checks $\#Imp$), and *results* ($Def\%$: definable entities in an ontology, $\mathcal{M}$: number of MDSs per defined entity) of the three stages of definability computation

Ontology		$\mathcal{DL}$ expressivity	Logical axioms	$(\mathcal{C} \cup \mathcal{R})$	(1) Definability status			(2) Disjoint MDSs			(3) All MDSs		
ID	Name				Time	#Imp	Def%	Time	#Imp	$\mathcal{M}$	Time	#Imp	$\mathcal{M}$
Conference corpus													
1	cmt	$\mathcal{ALCIN(D)}$	226	86	2.88 s	86	51.16 %	1.85 s	99	1.09	0.54 s	36	1.09
2	conference	$\mathcal{ALCHIF(D)}$	285	109	3.35 s	109	65.14 %	6.63 s	352	1.13	307.54 s	19833	1.54
3	confOf	$\mathcal{SIN(D)}$	196	57	2.92 s	57	15.79 %	2.67 s	195	2.33	8.03 s	573	3.33
4	edas	$\mathcal{ALCOIN(D)}$	739	138	20.84 s	175	28.99 %	28.63 s	397	1.18	23.22 h	1509274	2.80
5	iasted	$\mathcal{ALCIN(D)}$	358	182	16.14 s	182	17.58 %	150.89 s	212	1.25	1058.74 s	1756	1.75
6	sigkdd	$\mathcal{ALEI(D)}$	116	77	3.22 s	77	28.57 %	2.55 s	122	1.50	0.62 s	61	1.55
			384.00	**129.80**	**9.87 s**	**137.20**	**41.45 %**	**38.64 s**	**275.40**	**1.70**	**4.72 h**	**306306.60**	**2.41**
LargeBio corpus													
7	NCI_fma	$\mathcal{ALC}$	9083	6552	5.97 h	6552	29.99 %	222.38 h	229206	1.31	115.76 h	118456	1.31
8	SNOMED_fma	$\mathcal{ALER}$	20243	13431	21.49 h	13431	21.47 %	234.75 h	109980	1.09	756.53 h	384930	1.36
9	SNOMED_nci	$\mathcal{ALER}$	71042	51180	392.95 h	51180	57.87 %	1885.39 h	1145057	1.07	3991.85 h	2619125	1.08
			33456.00	**23721.00**	**140.14 h**	**23721.00**	**36.44 %**	**780.84 h**	**494747.53**	**1.16**	**1621.38 h**	**1040837.00**	**1.25**

OWL constructor usage in defined and undefined ontologies. Apart from some outliers, the results show an even distribution of defined and undefined ontologies, w.r.t. size, and OWL profiles and constructors, thus definability may occur in any type of ontology, regardless of the employed DL language, the size of an ontology, the conceptualised domain of interest, or its origin (source of creation). The only property, which affects the level of definability in an ontology, is, unsurprisingly, the granularity of conceptualisation.

The *second experiment* investigates the feasibility of the algorithms for computing definability by analysing their behaviour and performance. The corpus used in this experiment is made of small ontologies that conceptualise the conference domain (Conference track, OAEI corpus), and 3 vast biomedical ontologies (LargeBio track, OAEI corpus). The aim of this experiment is to assess the time taken by the proposed approximation when computing MDS over a variety of ontologies. Table 3 presents the characteristics of the *sample corpus* (DL expressivity, number of logical axioms, and the number of concepts and role names, ($\mathcal{C} \cup \mathcal{R}$)), and the experiment results. The three numbered partitions show the results of the definability computation steps, where each step is measured in terms of the computation time (this is given either in seconds, or in hours in some cases), and the number of implicit definability checks ($\#Imp$). The first stage establishes the definability status of each concept and role, hence $Def\%$ denotes the ratio of defined entities in the ontology signature. In the other two stages, where first the disjoint MDSs, then all MDSs are computed, $\mathcal{M}$ denotes the MDS to defined entity ratio. In general, as it can be anticipated, the larger, more expressive ontologies take much longer to compute than the smaller, less expressive ones. The definability status and the disjoint MDS computation stages are feasible for both small and large ontologies; whereas obtaining the complete set of MDSs (stage 3) is a considerably more costly operation, in most of the cases[12]. However, despite the cost of the last stage, the difference between the number of MDSs found during stage 2 (on average 1.70 MDSs per defined entity in the small, and 1.16 in the large ontologies) and 3 (2.41, and 1.25 MDSs), in many cases is negligible (0.71, and 0.09 MDSs more per entity). A notable case is the small *edas* ontology, where the first two stages take only 49.47 s to complete, however, the last stage takes 23.22 h, although the MDSs to defined entities ratio has more than doubled (from 1.18 to 2.80). In the *confOf* ontology, the last stage also shows a significant increase, from 2.33 MDSs per entity to 3.33, but in this case the computation time is close (8.03 s) to the sum of the two prior steps (5.59 s). Table 4 provides further information about the computed MDSs (for each ontology in the sample corpus), where the left partition shows the total number of different MDSs in the ontology (|MDS|), the number of MDSs per defined entity, and the cardinality of MDSs; the right partition presents the

[12] The MDS expansion (Algorithm 5) is restricted to computing an MDS union size $\mathcal{S} \leq 20$. This excluded no entities in the Conference, and 441 in the LargeBio corpus.

Table 4. Minimal Definition Signatures and their corresponding Definition Patterns

$\mathcal{O}$	Minimal Definition Signatures (MDSs)									Definition patterns									
	|MDS|	per Def. entity				Cardinality				Concept									Role
		Min	Avg	Med	Max	Min	Avg	Med	Max	1	2	3	4	5	6	7	8	Cmb.	11
Conference corpus																			
1	48	1	1.09	1	3	1	1.06	1	3	9.68 %	0.00 %	0.00 %	0.00 %	29.03 %	19.35 %	3.23 %	0.00 %	0.00 %	38.71 %
2	109	1	1.54	1	14	1	2.07	2	5	2.08 %	0.00 %	0.00 %	0.00 %	8.33 %	4.17 %	0.00 %	0.00 %	2.08 %	83.33 %
3	30	1	3.33	2	8	1	1.43	1	2	0.00 %	0.00 %	0.00 %	0.00 %	46.67 %	3.33 %	0.00 %	0.00 %	43.33 %	6.67 %
4	112	1	2.80	1	34	1	3.05	3	6	7.34 %	0.00 %	0.00 %	21.10 %	5.50 %	5.50 %	0.92 %	0.00 %	21.10 %	38.53 %
5	56	1	1.75	1	5	1	1.68	2	5	26.79 %	1.79 %	7.14 %	0.00 %	0.00 %	0.00 %	21.43 %	0.00 %	14.29 %	28.57 %
6	31	1	1.55	1	6	1	1.13	1	2	2.68 %	0.00 %	0.00 %	3.57 %	3.57 %	0.89 %	0.89 %	0.89 %	62.50 %	25.00 %
	64.33	**1.00**	**2.01**	**1.17**	**11.67**	**1.00**	**1.74**	**1.67**	**3.83**	**8.09 %**	**0.30 %**	**1.19 %**	**4.11 %**	**15.52 %**	**5.54 %**	**4.41 %**	**0.15 %**	**23.88 %**	**36.80 %**
LargeBio corpus																			
7	2583	1	1.32	1	5	2	5.07	4	33	67.52 %	0.00 %	0.00 %	22.45 %	0.00 %	0.00 %	0.00 %	0.00 %	10.03 %	0.00 %
8	3929	1	1.36	1	30	1	4.80	4	720	16.03 %	0.00 %	0.00 %	17.54 %	0.00 %	0.00 %	0.00 %	0.00 %	66.43 %	0.00 %
9	31831	1	1.08	1	5	1	5.59	6	15	16.90 %	0.00 %	0.00 %	15.53 %	0.00 %	0.00 %	0.00 %	0.00 %	67.55 %	0.00 %
	12781.00	**1.00**	**1.25**	**1.00**	**13.33**	**1.33**	**5.15**	**4.67**	**256.00**	**33.48 %**	**0.00 %**	**0.00 %**	**18.51 %**	**0.00 %**	**0.00 %**	**0.00 %**	**0.00 %**	**48.00 %**	**0.00 %**

distribution of MDSs w.r.t. to the corresponding definition patterns[13]. MDS per entity scores show that in most ontologies, about half of the defined entities have only one definition, and there are only a few entities in each ontologies with large number of MDSs. The average cardinality of an MDS is considerably low, with 1.74 entities per MDS in the small, and 5.15 in the large ontologies; however, there are some extreme cases, such as the *SNO_nci* ontology, where one MDS contains 720 entities. In the Conference corpus, 52.51 % of all MDSs correspond to a single entity (i.e. $|\text{MDS}| = 1$) definition pattern. Out of all MDSs in the Conference corpus only 23.88 % of all cases in this corpus are combined, thus for the majority of cases, our rule-based rewriting approach is sufficient; however, in the LargeBio corpus it only covers 52.00 % of all MDSs.

6 Definability and Ontology Alignment

Definability can be exploited in a number of novel contexts; in particular in this paper we argue that MDSs coupled with justification-based explanations [12] can support ontology alignment by identifying seemingly unrelated entities that are used to describe defined entities, and can thus be used to produce (in the case of identifiable patterns) new definition axioms (Sect. 4). Ontology alignment addresses the problem of mapping terms in heterogeneous ontologies and aims to create *alignments*, i.e. sets of correspondences between semantically related entities in different ontologies [7]. Over the past decade more than several alignment approaches have emerged [7]. However, neither state of the art matching systems, nor evaluation measures that assess the quality of alignments, have considered the notion of rewriting. As rewriting permits defined entities to be expressed in syntactically different but semantically equivalent forms, we argue, that an entity is rewritable under an alignment if the entities of its MDS are mapped by the alignment, thus rewriting entails a new type of correspondence, based on the definitions of entities. For instance, given an ontology $\mathcal{O} = \{\mathsf{C} \equiv \mathsf{A} \sqcup \mathsf{B}, \mathsf{B} \sqsubseteq \neg\mathsf{A}\}$, and alignment $\mathcal{A} = \{\langle \mathsf{C}, \mathsf{C}', \equiv\rangle, \langle \mathsf{B}, \mathsf{B}', \equiv\rangle\}$, which maps $\mathcal{O}$ to $\mathcal{O}'$, the implicitly defined concept A is rewritable w.r.t. the alignment, yielding $\langle \mathsf{A}, \mathsf{C}' \sqcap \neg\mathsf{B}', \equiv\rangle$, a definability-based correspondence. This *complex correspondence* describes a relation between a defined entity (or its description) in one ontology, and a complex concept (or role) in an aligned ontology. We suggest that definability-based mappings can potentially (1) increase *alignment coverage* [4] (the ratio of elements of the ontology which are mapped[14]) as they can cover otherwise uncovered entities; (2) increase *coverage retention*, i.e. removing some mappings from an alignment does not necessarily effect its expressive capacity (i.e. coverage) as some entities may be mapped by both an asserted, and a definability-based correspondence; (3) increase *compactness*, i.e. for a given knowledge-based task signature, only a subset of an alignment may be necessary to provide coverage.

[13] Pattern numbers reference Table 1, *Cmb.* denotes those MDS cases that do not correspond to any individual pattern, but to a combination of patterns. Patterns that have no MDS in the sample corpus are omitted for brevity.

[14] An entity e is covered by a mapping c in an alignment $\mathcal{A}$ iff $\{\exists c \in \mathcal{A} | c : \langle e, e', r\rangle\}$.

7 Definability and Ontology Modelling

The empirical analysis presented in previous section has also highlighted how the computation of MDSs can help in identifying modelling errors in ontologies.

We have formalised three types of errors, each of which can be automatically detected[15], but their repairs requires the involvement of an ontology engineer and a domain expert. *(1) Implicit definability by an empty signature:* the only concept in any ontology, which requires no signature for its definition is $\top$. If a named concept is definable by an empty signature, then the ontology is most likely to contain an error, or purposely define the concept as the synonym of $\top$. For example in the *cocus.owl* ontology of the Conference corpus $\mathsf{Person} \equiv \top$. By examining the document, it becomes obvious that this is unintentional, as the ontology contains many other concepts (such as Conference) that are definitely not semantically related to Person. *(2) Unwanted synonym(s):* These occur when two or more concepts, meant to convey different meaning, are wrongly represented as interchangeable synonyms of one another. Figure 4 shows three different ways of defining the concept Anthropometrics_Height. Obviously, A._Height, A._Weight and A._BMI are semantically related, but different concepts. However, in TBox $\mathcal{T}$ where $(\mathcal{J}_1 \cup \mathcal{J}_2 \cup \mathcal{J}_3) \subseteq \mathcal{T}$, these concepts are

$$\mathcal{J}_1 = \{\alpha_1 : \mathsf{Anthropometrics} \sqsubseteq \mathsf{Anthropometrics\_BMI} \sqcap \mathsf{Anthropometrics\_Height} \sqcap \mathsf{Anthropometrics\_Weight}, \; \alpha_2 : \mathsf{Anthropometrics\_Height} \sqsubseteq \mathsf{Anthropometrics}, \; \alpha_3 : \mathsf{Anthropometrics\_Weight} \sqsubseteq \mathsf{Anthropometrics}\} \models \mathit{Anthropometrics\_Height} \equiv \mathsf{Anthropometrics\_Weight}$$

$$\mathcal{J}_2 = \{\alpha_1 : \mathsf{Anthropometrics} \sqsubseteq \mathsf{Anthropometrics\_BMI} \sqcap \mathsf{Anthropometrics\_Height} \sqcap \mathsf{Anthropometrics\_Weight}, \; \alpha_2 : \mathsf{Anthropometrics\_Height} \sqsubseteq \mathsf{Anthropometrics}\} \models \mathit{Anthropometrics\_Height} \equiv \mathsf{Anthropometrics\_BMI}$$

$$\mathcal{J}_3 = \{\alpha_1 : \mathsf{Anthropometrics} \sqsubseteq \mathsf{Anthropometrics\_BMI} \sqcap \mathsf{Anthropometrics\_Height} \sqcap \mathsf{Anthropometrics\_Weight}, \; \alpha_2 : \mathsf{Anthropometrics\_Height} \sqsubseteq \mathsf{Anthropometrics}, \; \alpha_4 : \mathsf{Anthropometrics\_BMI} \sqsubseteq \mathsf{Anthropometrics}\} \models \mathit{Anthropometrics\_Height} \equiv \mathsf{Anthropometrics}$$

Fig. 4. Modelling error: three concepts that should be different, are semantically equivalent to each other (unwanted synonyms in the Bioportal corpus, bp26.owl).

$$\mathcal{J} = \{\alpha_1 : \mathit{Regular\_author} \equiv (\mathsf{Contribution\_1th - author} \sqcup \mathsf{Contribution\_co - author}) \sqcap \underbrace{(\exists \mathsf{contributes.Conference\_contribution})}_{\text{redundant}}, \; \alpha_2 : \mathsf{Contribution\_1th - author} \sqsubseteq \mathsf{Regular\_author}, \; \alpha_3 : \mathsf{Contribution\_co - author} \sqsubseteq \mathsf{Regular\_author}\}$$

$$\alpha_1 \models \mathsf{Regular\_author} \sqsubseteq (\mathsf{Contribution\_1th - author} \sqcup \mathsf{Contribution\_co - author})$$

$$\{\alpha_2, \alpha_3\} \models (\mathsf{Contribution\_1th - author} \sqcup \mathsf{Contribution\_co - author}) \sqsubseteq \mathsf{Regular\_author}$$

$$\mathcal{J} \models \mathit{Regular\_author} \equiv \mathsf{Contribution\_1th - author} \sqcup \mathsf{Contribution\_co - author}$$

Fig. 5. Redundant concepts in explicit definition (Conference corpus, conference.owl).

[15] Error #1 and #2 can also be detected without MDSs, via classifying the ontology and inspecting the resulting tree for unsatisfiable or equivalent concepts, respectively.

defined as equivalent. The correction requires expert knowledge[16]. *(3) Redundant concept(s):* this is not necessarily an error, but a discrepancy between the *intended meaning* (formalised by explicit definition axioms), and the *actual meaning* (the alternative explicit definition, which corresponds to an MDS of the defined concept). Figure 5 presents an example of this case, here Regular_author is defined by axiom α_1, however, its signature is not a minimal, because its subset {Contribution_1th − author, Contribution_co − author} can also be used to define the concept, as it is implied by the justification. An argument can be made, that an explicit concept definition ought to be a succinct representation, meaning that it should only consist of those entities that are necessary to unambiguously describe the concept. However a knowledge engineer may add semantically redundant entities to certain definitions in order to aid human comprehension. This occurs frequently, e.g. in *SNO_nci*, 76.28 % of all MDSs had redundant concepts in explicit definitions, and therefore is not considered an error.

8 Conclusions

In this paper we have presented a novel way to compute the complete set of definition signatures, and introduced a set of new application areas of concept rewriting that motivated the development of this method. In order to justify the viability of these new areas, a large and diverse set of ontologies were subjected to definability computation. This has confirmed the hypothesis that definability is prevalent in any type of ontology, although it is more likely to occur in more expressive, and semantically richer ontologies. Hence the exploitation of MDSs could indeed benefit the previously described application areas. In addition it was shown, that definability computation is feasible for most real world ontologies, and in some cases, it can be useful in dynamic environments as well, due to the fact that a subset of MDSs can be found in polynomial time. However, as the approach does not scale for larger, very expressive ontologies, a more efficient approach should be developed.

References

1. Baader, F.: The Description Logic Handbook: Theory, Implementation, and Applications. Cambridge University Press, Cambridge (2003)
2. Beth, E.W.: On Padoa's method in the theory of definition. Indag. Math. **15**, 330–339 (1953)

[16] Anthropometrics means measurement of the size and proportions of the human body. Axioms $\alpha_2, \alpha_3, \alpha_4$ are correct, as *height, weight* and *BMI* are all type of measurements that make up the general class Anthropometrics, but axiom α_1 is incorrect as *height* and *weight* measurements would share nothing in common, i.e. their intersection would be empty. The correct representation would be to describe Anthropometrics as a disjoint union of these concepts.

3. Cheatham, M., Dragisic, Z., Euzenat, J., Faria, D., Ferrara, A., Flouris, G., Fundulaki, I., Granada, R., Ivanova, V., Jiménez-Ruiz, E., Lambrix, P., Montanelli, S., Pesquita, C., Saveta, T., Shvaiko, P., Solimando, A., dos Santos, C.T., Zamazal, O.: Results of the ontology alignment evaluation initiative 2015. In: Proceedings of the 10th International Workshop on Ontology Matching, pp. 60–115 (2015)
4. David, J., Euzenat, J., Šváb-Zamazal, O.: Ontology similarity in the alignment space. In: Patel-Schneider, P.F., Pan, Y., Hitzler, P., Mika, P., Zhang, L., Pan, J.Z., Horrocks, I., Glimm, B. (eds.) ISWC 2010, Part I. LNCS, vol. 6496, pp. 129–144. Springer, Heidelberg (2010)
5. Del Vescovo, C., Klinov, P., Parsia, B., Sattler, U., Schneider, T., Tsarkov, D.: Empirical study of logic-based modules: cheap is cheerful. In: Alani, H., et al. (eds.) ISWC 2013, Part I. LNCS, vol. 8218, pp. 84–100. Springer, Heidelberg (2013)
6. Donnelly, K.: SNOMED-CT: the advanced terminology and coding system for eHealth. Stud. Health Technol. Inform. **121**, 279 (2006)
7. Euzenat, J., Shvaiko, P.: Ontology Matching, 2nd edn. Springer, Heidelberg (2013)
8. Geleta, D., Payne, T.R., Tamma, V.: Computing minimal definition signatures in description logic ontologies. Technical report, University of Liverpool (2016). https://intranet.csc.liv.ac.uk/research/techreports/tr2016/ulcs-16-003.pdf
9. Glimm, B., Horrocks, I., Motik, B., Stoilos, G., Wang, Z.: HermiT: an OWL 2 reasoner. J. Autom. Reason. **53**(3), 245–269 (2014)
10. Grau, B.C., Horrocks, I., Motik, B., Parsia, B., Patel-Schneider, P., Sattler, U.: OWL 2: the next step for OWL. Web Semant. **6**(4), 309–322 (2008)
11. Hoogland, E., et al.: Definability and interpolation: Model-Theoretic Investigations. Institute for Logic, Language and Computation, Amsterdam (2001)
12. Horridge, M.: Justification based explanation in ontologies. Ph.D. thesis, University of Manchester (2011)
13. Horridge, M., Bechhofer, S.: The OWL API: a Java API for OWL ontologies. Semant. Web **2**(1), 11–21 (2011)
14. Matentzoglu, N., Bail, S., Parsia, B.: A snapshot of the OWL web. In: Alani, H., et al. (eds.) ISWC 2013, Part I. LNCS, vol. 8218, pp. 331–346. Springer, Heidelberg (2013)
15. Rosse, C., Mejino Jr., J.L.: The foundational model of anatomy ontology. In: Burger, A., Davidson, D., Baldock, R. (eds.) Anatomy Ontologies for Bioinformatics. Computational Biology, vol. 6, pp. 59–117. Springer, Heidelberg (2008)
16. Sattler, U., Schneider, T., Zakharyaschev, M.: Which kind of module should I extract? Descr. Log. **477**, 78 (2009)
17. Scharffe, F., Zamazal, O., Fensel, D.: Ontology alignment design patterns. Knowl. Inf. Syst. **40**(1), 1–28 (2014)
18. Seylan, I., Franconi, E., De Bruijn, J.: Effective query rewriting with ontologies over DBoxes. In: IJCAI, vol. 9, pp. 923–929. Citeseer (2009)
19. Sirin, E., Parsia, B., Grau, B.C., Kalyanpur, A., Katz, Y.: Pellet: a practical OWL-DL reasoner. Web Semant.: Sci. Serv. Agents World Wide Web **5**(2), 51–53 (2007)
20. Stuckenschmidt, H., Predoiu, L., Meilicke, C.: Learning complex ontology alignments a challenge for ILP research. In: Proceedings of the 18th International Conference on Inductive Logic Programming (2008)
21. Ten Cate, B., Franconi, E., Seylan, I.: Beth definability in expressive description logics. J. Artif. Intell. Res. (JAIR) **48**, 347–414 (2013)

Alligator: A Deductive Approach for the Integration of Industry 4.0 Standards

Irlán Grangel-González[1,2(✉)], Diego Collarana[1,2], Lavdim Halilaj[1,2], Steffen Lohmann[2], Christoph Lange[1,2], María-Esther Vidal[1,2,3], and Sören Auer[1,2]

[1] Enterprise Information Systems (EIS), Computer Science, University of Bonn, Bonn, Germany
{grangel,collaran,halilaj,vidal,auer}@cs.uni-bonn.de
[2] Fraunhofer Institute for Intelligent Analysis and Information Systems (IAIS), Sankt Augustin, Germany
steffen.lohmann@iais.fraunhofer.de
[3] Universidad Simón Bolívar, Caracas, Venezuela

Abstract. Industry 4.0 standards, such as AutomationML, are used to specify properties of mechatronic elements in terms of views, such as electrical and mechanical views of a motor engine. These views have to be integrated in order to obtain a complete model of the artifact. Currently, the integration requires user knowledge to manually identify elements in the views that refer to the same element in the integrated model. Existing approaches are not able to scale up to large models where a potentially large number of conflicts may exist across the different views of an element. To overcome this limitation, we developed ALLIGATOR, a deductive rule-based system able to identify conflicts between AutomationML documents. We define a Datalog-based representation of the AutomationML input documents, and a set of rules for identifying conflicts. A deductive engine is used to resolve the conflicts, to merge the input documents and produce an integrated AutomationML document. Our empirical evaluation of the quality of ALLIGATOR against a benchmark of AutomationML documents suggest that ALLIGATOR accurately identifies various types of conflicts between AutomationML documents, and thus helps increasing the scalability, efficiency, and coherence of models for Industry 4.0 manufacturing environments.

Keywords: AutomationML · Semantic data integration · Industry 4.0

1 Introduction

In the engineering and manufacturing domain, there is an atmosphere of departure to a new era of digitized production, where traditional industrial engineering methods are synergistically combined with IT and internet technologies, such as cyber-physical systems, sensor networks, big data analytics, and semantic data integration. In different regions, initiatives in these directions are known

E. Blomqvist et al. (Eds.): EKAW 2016, LNAI 10024, pp. 272–287, 2016.
DOI: 10.1007/978-3-319-49004-5_18

under different names, such as *industrie du futur* in France, *industrial internet* in the US or *Industrie 4.0* in Germany. A core vision of these initiatives is to make manufacturing and production more flexible, efficient, and less error-prone by shifting more 'intelligence' to the edge. This shall be achieved by enabling sensors, devices, machines, and storage and transport equipments to directly communicate with each other. To realize this *Industry 4.0* vision, a vast variety of areas related to manufacturing, security, and machine communication need to interoperate by aligning their information models using domain-specific standards.

The *Automation Markup Language* (AutomationML or AML) for exchanging plant engineering information as specified by IEC 62714 [4,9,17,21] is one of the core standards of Industry 4.0. AutomationML can describe plant components and their sub-components from different perspectives, e.g., mechanical or electrical. A key challenge in such settings is *intra-standard interoperability*, i.e., the consistent integration of multiple pieces of information described in AutomationML. To overcome this challenge, we present ALLIGATOR, a deductive approach to integrate AutomationML specifications, and potentially similar document types.

We define an RDF-based representation of AutomationML input documents, aiming to resolve structural semantic inconsistencies, such as granularity of representations, schematic differences, and groupings and aggregations. Based on this semantic representation, we define a set of Datalog rules for identifying conflicts that generate structural semantic inconsistencies. A deductive engine is used to compute the conflicts from the Datalog representations. Conflict resolution is utilized to merge the input documents and produce an integrated AutomationML document.

By automatizing a crucial part of the engineering and modeling processes, ALLIGATOR addresses a key pillar of the Industry 4.0 vision. To the best of our knowledge, ALLIGATOR is the first comprehensive approach for automatically resolving the semantic ambiguity of AutomationML. As a result, the ALLIGATOR approach enhances scalability, efficiency, and coherence of models for Industry 4.0 manufacturing environments. Although our initial implementation and evaluation of the approach focuses on AutomationML, the approach is easily transferable to other Industry 4.0 standardization initiatives. We empirically evaluated the quality of ALLIGATOR against a benchmark of AutomationML documents. The evaluation results suggest that ALLIGATOR accurately identifies various types of conflicts between AutomationML documents.

In summary, this work makes the following contributions:

1. ALLIGATOR, a deductive approach, that combines Deductive Database and Semantic Web technologies for the integration of Industry 4.0 Standards.
2. A set of Datalog rules to characterize semantic heterogeneity types among AutomationML documents.
3. An empirical evaluation that reveals the effectiveness of ALLIGATOR during the integration of AutomationML documents.

The remainder of this paper is structured as follows. Section 2 motivates the problem with a concrete example. Section 3 gives an overview on the background

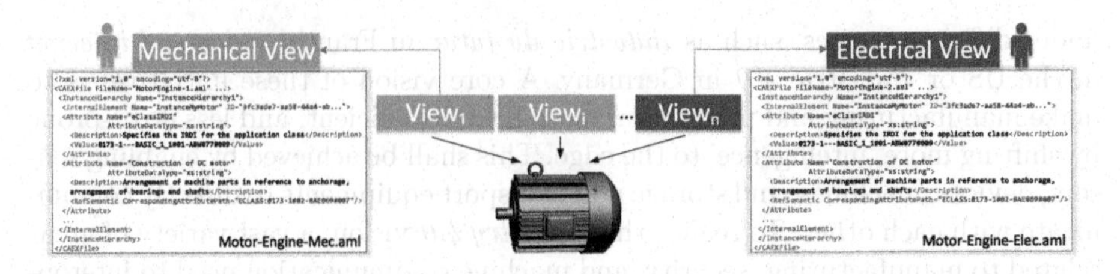

Fig. 1. Motivating example. Results of an engineering process where a motor engine is modeled from different views: a mechanical and an electrical view. Identical elements of the motor engine are defined as different elements in the views, resulting in conflicts between the views.

and introduces the terminology relevant to our approach. Section 4 presents the ALLIGATOR approach, which is evaluated in Sect. 6. Section 7 reviews related work. Section 8 concludes and gives an outlook to future work.

2 Motivating Example

A typical scenario in the mechatronic domain is data exchange between engineering tools during the modeling process. Engineering tools are utilized in different disciplines, such as mechanical and electrical engineering, or systems control. Figure 1 illustrates the results of an engineering modeling process where a *motor engine* is modeled from mechanical and electrical viewpoints. Mechanical engineers design the *motor engine* from the mechanical point of view, whereas electrical engineers model the electrical wiring topology inside the *motor engine*. *AutomationML* is utilized in both views to semantically describe the engine. However, because physical structures in these views are modeled with different properties, *conflicts* might arise when integrating these designs, thus inducing structural semantic inconsistencies.

Figure 2 details the mechanical and electrical views of the motor engine given in Fig. 1. The motor engine is identified as `0173-1#01-AKE162#012` **DC Engine** according to the eCl@ss product classification standard[1]. This reference enables the semantic description of the mechatronic component by pointing to the standard definition of a motor engine in eCl@ss. The AML document `Motor-Engine-Mec.aml` (cf. Fig. 2a) specifies the motor in terms of its construction form as a **DC Engine** (mechanical view). The AML element `RoleClassLib` (lines 2–23) comprises two AML elements `RoleClass`. The first `RoleClass` (lines 4–14) contains AML attributes with references to eCl@ss that semantically describe the engine according to the standard definition of version, classification in the eCl@ss catalog, and the International Registration Data Identifier (IRDI). The second `RoleClass` (lines 15–20) is composed of an AML attribute that

[1] http://www.eclasscontent.com/index.php?id=27022501&version=9_1&language=en&action=det.

defines the construction form of the **DC Engine**; `RefSemantic` (line 18) refers to the eCl@ss standard definition of this AML attribute (`0173-1#02-BAE069#007`).

Figure 2b depicts an AML document that aims at defining the same engine from the electrical viewpoint. As in the mechanical view, the first `RoleClass` (lines 4–14) semantically describes the engine using eCl@ss, while the second `RoleClass` (lines 15–20) defines not the engine as a whole, but a data cable *in* the engine. The `Attribute` in line 16 specifies the data cable and includes the semantic reference to eCl@ss (line 18).

Albeit the structural definition in these views of the **DC Engine** differs in the AML documents, the specification of AutomationML and its eCl@ss integration [19] imply that both descriptions are semantically equivalent. On one hand, the references to eCl@ss indicate that the AML elements between lines 4 and 14 in the two views correspond to the same element in the real world. For example, the specification of AutomationML states that two `RoleClass` elements are semantically equivalent whenever they share the same eCl@ss references for the AML attributes `eClassVersion`, `eClassClassification`, and `eClassIRDI` [19]. However, these views describe different real-world objects, and they should not be defined using `RoleClass` elements in the mechanical and electrical views which are considered semantically identical according to AML. Therefore, these elements are in *conflict*. Accordingly, there are five pairs of conflicting AML elements in this simplified example; each pair of these needs to be merged into one AML element in case the two views are integrated.

Currently, this integration is performed manually by experts, negatively affecting engineering processes. We present Alligator, a deductive framework that exploits the features of logic programming and the RDF data model for representing AML documents, as well as for detecting conflicts whenever AML documents are integrated.

3 Background

AutomationML. AutomationML (Automation Markup Language, IEC 62714) is a standard to exchange information about engineering tools, such as mechanical plant engineering, electrical design, or robot control. AutomationML provides an XML Schema, incorporating three different standards for describing real plant components [20]. At the top level there is the CAEX (IEC 62424) format for plant topology, storing hierarchical object information, properties, and libraries [8]. Secondly, the geometry (mechanical drawings) and kinematics (physical properties, such as force, speed, or torsion) are implemented with COLLADA [3]. Finally, the logic (sequencing, behavior, and control information) is implemented with PLCopen XML (IEC 61131).

AutomationML is built upon four main CAEX concepts: *RoleClassLibrary*, *SystemUnitClassLibrary*, *InterfaceClassLibrary*, and *InstanceHierarchy*. *RoleClassLibrary* specifies vendor independent requirements for the specification of system equipment objects; a *RoleClassLibrary* may comprise several *RoleClass*es, which provide role descriptions of a given class. Such descriptions aim

```
1  <CAEXFile FileName="Motor-Engine-Mec.aml" ...>
2   <RoleClassLib Name="RoleClassLibDCEngine">
3    <Version>1.0.0</Version>
4    <RoleClass Name="eClassClassSpecification">
5     <Attribute Name="eClassVersion">
6      <Value>9.0</Value>
7     </Attribute>
8     <Attribute Name="eClassClassificationClass">
9      <Value>27022501</Value>
10    </Attribute>
11    <Attribute Name="eClassIRDI">
12     <Value>0173-1#01-AKE162#012</Value>
13    </Attribute>
14  </RoleClass>
15  <RoleClass Name="BASIC_27-02-25-01 DC engine
      (IEC)">
16   <Attribute Name="Construction form of DC motor">
17    <Description>Arrangement of machine parts in
        reference to anchorage, arrangement of
        bearings and shafts.</Description>
18    <RefSemantic CorrespondingAttributePath=
       "ECLASS:0173-1#02-BAE069#007"/>
19   </Attribute>
20  </RoleClass>
21 </RoleClassLib>
22 </CAEXFile>
```

(a) Mechanical View

```
1  <CAEXFile FileName="Motor-Engine-Elec.aml" ...>
2   <RoleClassLib Name="RoleClassLibDCEngine">
3    <Version>1.0.0</Version>
4    <RoleClass Name="eClassClassSpecification">
5     <Attribute Name="eClassVersion">
6      <Value>9.0</Value>
7     </Attribute>
8     <Attribute Name="eClassClassificationClass">
9      <Value>27022501</Value>
10    </Attribute>
11    <Attribute Name="eClassIRDI">
12     <Value>0173-1#01-AKE162#012</Value>
13    </Attribute>
14  </RoleClass>
15  <RoleClass Name="27-06 Cable, wire">
16   <Attribute Name="Data cable">
17    <Description>Type of cable for transmission
       of electrical or optical signals on the
       basis for the use in data transmission.
      </Description>
18    <RefSemantic CorrespondingAttributePath=
       "ECLASS:0173-1#02-AKE197#013"/>
19   </Attribute>
20  </RoleClass>
21 </RoleClassLib>
22 </CAEXFile>
```

(b) Electrical View

Fig. 2. Example of AML Documents. A motor engine is semantically described in terms of the eCl@ss standard. Role classes (highlighted in red) model the engine in terms of (a) a construction form and (b) a data cable. Elements of the same type (highlighted in yellow) correspond to conflicts between the views. (Color figure online)

at representing a physical or logical object, e.g., a motor or a robot. The *InterfaceClassLibrary* defines a set of interfaces to describe a plant model. First, it can define relations between the objects of a plant topology. Secondly, it can reference external information, e.g., a 3D description of a motor. The *InstanceHierarchy* describes the plant topology, and defines specific equipment for actual projects. Further, *Attributes* are used to define properties, e.g., length or size, of AML objects, e.g., RoleClasses or Internal Elements. In this paper, we focus on modeling topology information by means of the CAEX format.

AutomationML. Biffl et al. [4] and Kovalenko and Euzenat [11] have characterized mappings to deal with semantic heterogeneity in the engineering domain, and specifically in AutomationML. The authors have identified the following types of semantic heterogeneity: **(M1)** Value processing same properties are not modeled equally, e.g., using different datatypes; **(M2)** Granularity same objects are modeled at different levels of detail; **(M3)** Schematic differences differences in the way how semantics is represented for the same object; **(M4)** Conditional mappings relations between entities exist only if certain conditions occur; **(M5)** Bidirectional mappings relations between entities have to be defined bidirectionally; **(M6)** Grouping and aggregation different semantic modeling criteria are applied to group elements for the same object; and **(M7)** Restrictions on values mandatory values for properties in the object that have to be handled in the mapping process. As a proof of concept, we focus on semantic heterogeneity types, such as granularity (M2), schematic differences (M3), and grouping and aggregation (M6). We selected these types because they present major semantic structural differences to describe similar objects. Additionally, they characterize semantic mappings between two AML elements that can be performed in two ways:

1. *Direct identification* considers two elements to refer to the same entity if the same identifier is used.
2. *Indirect identification* considers two elements to refer to the same entity if both refer to the same identity-providing elements from an external catalog, e.g., *RoleClass* or *Attributes*. For more complex structures as RoleClasses, it is assumed that if the combination of the eCl@ss IRDI, classification level, and version are equal, then the RoleClasses are considered to be the same.

AutomationML Vocabulary. Several approaches exist for adding semantics to the AutomationML language by means of ontologies [1,2,5,6,12,15]. With the exception of the AutomationML ontology[2], designed for the AutomationML Analyzer [16], none of the aforementioned ontologies covers all concepts given in the AutomationML schema. Additionally, they are not available on the web for consulting or querying. Crucial information for ALLIGATOR, such as the mapping with eCl@ass concepts, are not included in the AutomationML Analyzer vocabulary. Therefore, we have developed an RDFS vocabulary describing the main concepts of the AutomationML language.[3] Also, we have included concepts related to the integration with the eCl@ss standard.

4 Our Approach: ALLIGATOR

In this section, we present a formalization of AML documents, as well as the integration problems and proposed solution addressed by the ALLIGATOR approach. Finally, the architecture of ALLIGATOR is described in detail.

4.1 ALLIGATOR Representation of AML Documents

Definition 1 (Alligator Document). *An* ALLIGATOR *document is a tuple* $\Gamma = \langle \theta, V, F \rangle$ *such that* θ *is a set of URIs that identify AML elements,* V *is a set of properties in the AML vocabulary and* F *is an RDF graph composed of triples in* $\theta \times V \times (\theta \cup L)$ *where* L *is a set of literals.*

An ALLIGATOR document $\Gamma = \langle \theta, V, F \rangle$ can represent information from one or several AML documents D_i, where θ is the set of URIs that identify the AML elements in D_i, and the RDF graph F describes the relationships between the AML elements in D_i. In general, V can refer to different vocabularies, e.g., for other standards than AML such as OPC UA, but in this work, we focus on the AML vocabulary.

[2] http://data.ifs.tuwien.ac.at/aml/ontology#.
[3] https://w3id.org/i40/aml/.

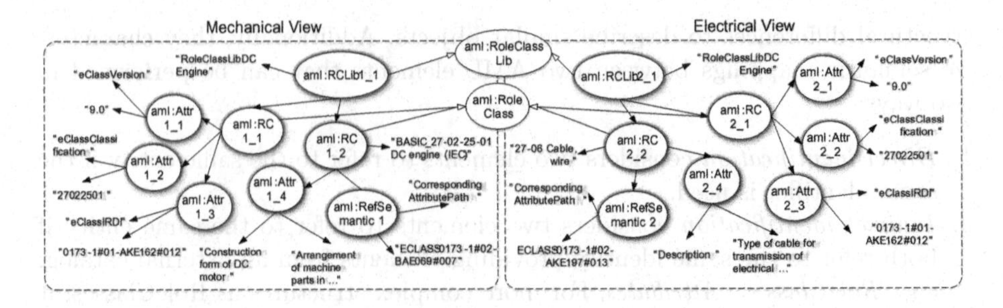

Fig. 3. RDF graph of an ALLIGATOR document. An RDF graph representing AML elements in the union of the mechanical and electrical views in Fig. 2

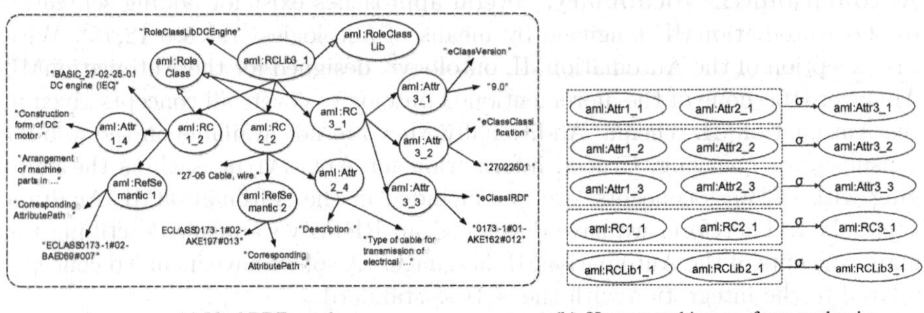

(a) Ideal RDF graph

(b) Homomorphism σ: from nodes in Figure 3 to ideal RDF graph nodes

Fig. 4. Ideal conflict-free ALLIGATOR document. (a) An RDF graph where there is only one RDF resource for the conflicting resources in the mechanical and electrical views of Fig. 6. (b) A homomorphism σ maps conflicting resources in the RDF graph in Fig. 3 to the same resource in the ideal RDF graph.

Example 1. Consider the RDF graph F_1 in Fig. 3. This graph comprises RDF resources representing the AutomationML elements in the mechanical and electrical views shown in Fig. 2; the AutomationML RDF vocabulary is used to describe these resources. An ALLIGATOR document $\Gamma_1 = \langle \theta_1, V, F_1 \rangle$ formally describes this RDF representation of the two views, where θ_1 is the set of the resources in F_1, and V is the AutomationML RDF vocabulary.

Definition 2 (Ideal Alligator Document). *Given an* ALLIGATOR *document $\Gamma = \langle \theta, V, F \rangle$, there is an ideal* ALLIGATOR *document $\Gamma^* = \langle \theta^*, V, F^* \rangle$ such that Γ^* comprises only conflict-free AML elements. Additionally, there is a homomorphism $\sigma : \theta \rightarrow \theta^*$. The RDF ideal graph F^* is defined as follows:*

$$F^* = \{(\sigma(s),\ p,\ \sigma(o)) \mid (s,\ p,\ o) \in F\}$$

Example 2. Consider the RDF graph in Fig. 4a. The ALLIGATOR document $\Gamma^* = \langle \theta^*, V, F^* \rangle$ describes this RDF graph, where θ is the set of RDF resources in the graph, V is the AutomationML RDF vocabulary, and F^* is this RDF graph. Γ^* represents the ideal *conflict-free* ALLIGATOR document of Γ_1. Figure 4b shows a homomorphism σ that maps two conflicting resources in the RDF graph in Fig. 3 to the same resource in Fig. 4a.

Definition 3. *Consider an* ALLIGATOR *document* $\Gamma = \langle \theta, V, F \rangle$*, an ideal conflict-free* ALLIGATOR *document* $\Gamma^* = \langle \theta^*, V, F^* \rangle$*, and a homomorphism* $\sigma : \theta \rightarrow \theta^*$*. A set of conflicts in* Γ *with respect to* Γ^* *and* σ*, conflicts(*$\Gamma \mid \Gamma^*, \sigma$*), corresponds to the set of AML element pairs* (E_i, E_j) *in* $\theta \times \theta$ *such that* E_i *and* E_j *are different but that* σ *maps to the same target AML element in* θ^**:*

$$conflicts(\Gamma \mid \Gamma^*, \sigma) = \{(E_i, E_j) \mid E_i, E_j \in \theta \text{ and } E_i \neq E_j \text{ and } \sigma(E_i) = \sigma(E_j)\}$$

Example 3. Given ALLIGATOR documents Γ_1 and Γ^* from Examples 1 and 2, and the homomorphism σ in Fig. 4b. The set of conflicts($\Gamma_1 \mid \Gamma^*, \sigma$) corresponds to the set of pairs of RDF resources in the RDF graph of Fig. 3 that σ maps to the same resource in the ideal RDF graph (Fig. 4b).

4.2 Problem Definition and Proposed Solution

Given an ALLIGATOR document $\Gamma = \langle \theta, V, F \rangle$, the *AML Conflict Identification* problem determines if a pair (E_k, E_l) of AML elements in θ is conflicting.

Definition 4. *Consider an* ALLIGATOR *document* $\Gamma = \langle \theta, V, F \rangle$*, an ideal conflict-free* ALLIGATOR *document* $\Gamma^* = \langle \theta^*, V, F^* \rangle$*, and a homomorphism* $\sigma : \theta \rightarrow \theta^*$*. The* AML Conflict Identification *problem corresponds to the problem of deciding if* $(E_k, E_l) \in \theta \times \theta$ *belongs to conflicts(*$\Gamma \mid \Gamma^*, \sigma$*).*

Solving the *AML Conflict Identification* problem requires the existence of the ideal conflict-free AML document Γ^* and the homomorphism σ. However, in practice neither Γ^* and σ is known, and ALLIGATOR computes an approximation of the problem. We use *SC(*Γ*)* to refer to the set of pairs (E_k, E_l) that correspond to the solutions of this problem. Once a set *SC(*Γ*)* of conflicting AML elements in F is identified as the solution of the AML *Conflict Identification problem*, the problem of *AML Conflict Resolution* corresponds to the problem of creating an ALLIGATOR document where conflicts in *SC(*Γ*)* are solved.

Definition 5. *Consider an* ALLIGATOR *document* $\Gamma = \langle \theta, V, F \rangle$ *and a set SC(*Γ*) of pairs of conflicting AML elements in* F*. The problem of* AML Conflict Resolution *corresponds to the problem of creating an* ALLIGATOR *document* $\Gamma' = \langle \theta', V, F' \rangle$ *and a homomorphism* $\sigma' : \theta \rightarrow \theta'$*, such that:*

- *For each* (E_i, E_j) *in SC(*Γ*), there is an AML element* E_m *in* θ' *such that* $\sigma'(E_i) = \sigma'(E_j) = E_m$*.*
- $F' = \{(\sigma'(s), p, \sigma'(o)) \mid (s, p, o) \in F\}$*.*

Γ' *represents the* ALLIGATOR *document where pairs of AML elements in SC(*Γ*) are represented as one RDF AML element.*

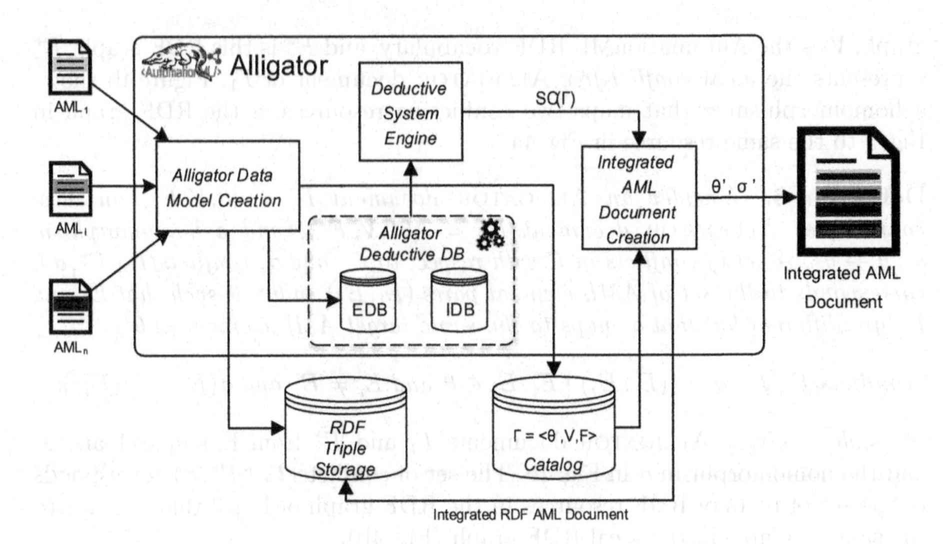

Fig. 5. The ALLIGATOR Architecture. ALLIGATOR receives AML documents and creates an integrated AML document. AML documents are represented as RDF graphs and Datalog predicates (EDB); Datalog intentional rules (IDB) characterize semantic heterogeneity types. A bottom-up evaluation of the Datalog program identifies conflicts between AML documents

We developed ALLIGATOR, an integration tool that relies on deductive database techniques for solving the problems of *AML Conflict Identification* and *AML Conflict Resolution.* Figure 5 depicts the architectural components of ALLIGATOR. Given a set of AML documents, the *Alligator Data Model Creation* component generates an ALLIGATOR document $\Gamma = \langle\theta, V, F\rangle$ that formally describes the union of these input AML documents. Additionally, a set of Datalog extensional facts (EDB) representing the triples in the RDF document F is created. The *Deductive System Engine* relies on the set of Datalog intentional rules (IDB) to compute the set *SC*(Γ) from the Datalog representation of Γ. The set of Datalog intentional rules (IDB) defines different types of semantic heterogeneity that can occur among AML documents that correspond to views of the same mechatronic object definition. *SC*(Γ) is computed as the least minimal fixpoint of the Datalog rules in IDB and the facts in EDB. Further, *SC*(Γ) is utilized by the *Integrated AML Document Creation* component to solve the *AML Conflict Resolution* problem, and to produce an integrated AML document where RDF AML elements in *SC*(Γ) are integrated as one AML element.

4.3 ALLIGATOR Data Model and Deductive System Engine

ALLIGATOR represents AML documents as RDF graphs. AML documents are translated to RDF using Krextor [13], an XSLT-based framework for converting XML to RDF. The RDF AML vocabulary is used to describe AML elements and relations. Further, AML documents are modeled as facts in an extensional database (EDB) of a Datalog program P; for each type of AML element in the AutomationML standard exists an extensional Datalog predicate in P. Rules in the intensional database (IDB) of the Datalog program P characterize types of semantic heterogeneity. Intensional Datalog predicates represent conflicts that can exist between the different AML elements according to the types of semantic heterogeneity. The ALLIGATOR *Deductive System Engine* performs a bottom-up evaluation of P following a semi-naïve algorithm that stops when the least fixed-point is reached [7]. The intensional predicates inferred in the evaluation of P correspond to the pairs of conflicts in the set $SC(\Gamma)$.

5 ALLIGATOR rule-based representation of AutomationML Semantic Heterogeneity

One of the key innovations of ALLIGATOR revolves on the use of a Datalog-rule approach to effectively solve types of semantic heterogeneity. We have developed a set of rules covering the main characteristics of AML. Regarding the attributes, it is possible to determine that, if two attributes refer to the same eCl@ss value, i.e., eCl@ss IRDI, it can be assumed that their semantic meaning is the same. In detail, the AML element `refSemantic` refers to the eCl@ss IRDI using `CorrespondingAttributePath` (cf. Fig. 2 line 18). Thereby, even if two attributes are defined with different names, e.g., *Length* and *StrictLength*, they can still be semantically equivalent whenever they are linked to the same IRDI reference. It is important to remark that these rules have been defined taking into account the AML vocabulary properties. Based on this, the rule in Listing 1 states when two attributes are semantically equivalent.

```
1  sameAttribute(X,Y)    :- hasRefSemantic(X,T) & hasRefSemantic(Y,Z) &
2                           sameRefSemantic(T,Z).
3  sameRefSemantic(X,Y) :- hasCorrespondingAttributePath(X,Z) &
4                           hasCorrespondingAttributePath(Y,Z).
```

Listing 1. Rule 1: Semantic equivalence of two AML attributes

To determine that two `RoleClasses` are semantically equivalent according to their reference to eCl@ss, they have to contain the same version, classification, and IRDI. Based on these three conditions, **Rule 2** (cf. Listing 2) defines two semantically equivalent `RoleClasses`.

```
1  sameRoleClass(X,Y) :- type(X,roleClass) & type(Y,roleClass) & sameEClassIRDI(A,B)&
2                        sameEClassClassification(C,D) & sameEClassVersion(E,F)&
3                        hasAttribute(X,A) & hasAttribute(X,C) & hasAttribute(X,E)&
4                        hasAttribute(Y,B) & hasAttribute(Y,D) & hasAttribute(Y,F).
```

Listing 2. Rule 2: Semantic equivalence of two RoleClasses

Rule 2 relies on simpler rules such as **Rule 3** (cf. Listing 3), which defines the equivalence of two `eClassIRDI` attributes. Similarly, we have defined rules to decide if two values of `eClassVersion` and `eClassClassification` are the same.

```
1  sameEClassIRDI(X,Y) :- hasAttributeName(X,'eClassIRDI') &
2                         hasAttributeName(Y,'eClassIRDI') &
3                         hasAttributeValue(X,Z) & hasAttributeValue(Y,Z).
```

Listing 3. Rule 3: Semantic equivalence of two eClassIRDI AML attributes

These three rules are only examples of the type of rules implemented in ALLIGATOR; the complete set of rules is given on GitHub[4].

6 Empirical Evaluation

We studied the effectiveness of ALLIGATOR in the solution of the problems of *AML Conflict Identification* and *AML Conflict Resolution*. In particular, we assessed the following research questions: (RQ1) Is ALLIGATOR able to identify pairs of conflicting AML elements in AML documents?; (RQ2) Does ALLIGATOR exhibit *equal behavior* whenever different types of semantic heterogeneity occur during the integration of AML documents? The experimental configuration to evaluate these research questions was as follows:

Testbeds. Testbeds were based on the semantic mapping types M2 (granularity), M3 (schematic differences), and M6 (grouping and aggregation), with ten testbeds for each of them, respectively. First, a seed (AML document) was manually created for each testbed according to the type of semantic mapping. Next, we automatically generated two AML documents derived from this seed containing a random number of conflicting AML elements[5]. The generation was performed following a uniform distribution. Testbeds corresponded to pairs of AML documents, and thirty testbeds were evaluated in the study[6].

Gold Standard. To compile a Gold Standard, we relied on the generated testbeds. Formally, the Gold Standard corresponds to an ideal conflict-free ALLIGATOR document $\Gamma^* = \langle \theta^*, V, F^* \rangle$, for each pair of the AML documents in the testbeds. The creation of the conflict-free document as well as the computation of the conflicting elements and different elements was performed manually.

Metrics. We measured the *behavior* of ALLIGATOR in terms of the following metrics:

(a) **Precision** is the fraction of the conflicts identified by ALLIGATOR (i.e., $SC(\Gamma)$) that are conflicts in an AML document (i.e., conflicts$(\Gamma \mid \Gamma^*, \sigma)$).

$$Precision = \frac{|SC(\Gamma) \cap conflicts(\Gamma \mid \Gamma^*, \sigma)|}{|SC(\Gamma)|}$$

[4] https://github.com/i40-Tools/AlligatorRules.
[5] https://github.com/i40-Tools/AMLGoldStandardGenerator.
[6] https://github.com/i40-Tools/HeterogeneityExampleData.

(b) **Recall** is the fraction of the conflicts in an AML document (i.e., conflicts($\Gamma \mid \Gamma^*, \sigma$) that are identified by ALLIGATOR (i.e., $SC(\Gamma)$).

$$Recall = \frac{|SC(\Gamma) \cap conflicts(\Gamma \mid \Gamma^*, \sigma)|}{|conflicts(\Gamma \mid \Gamma^*, \sigma)|}$$

(c) **F-measure** is the harmonic mean of *Precision and Recall.*

Implementation. Experiments were run on a Windows 8 machine with an Intel I7-4710HQ 2.5 GHz CPU and 8 GB 1333 MHz DDR3 RAM. We implemented the *Deductive System Engine* as a meta-interpreter in Prolog that follows the semi-naïve bottom-up evaluation of Datalog programs [7]; we utilized SWI-Prolog version 7.2.3 and the Prolog Development Tool (PDT[7]). An AML extraction module was developed as a part of Krextor to transform AML documents into RDF graphs. This module comprised a set of mapping rules[8] that are executed in Krextor to create RDF graphs using the AML vocabulary. Further, the transformation of the RDF files into Datalog extensional predicates was implemented in Java 1.8. The ALLIGATOR framework, the testbed generator, and the testbeds evaluated in this experiment are publicly available on GitHub[9].

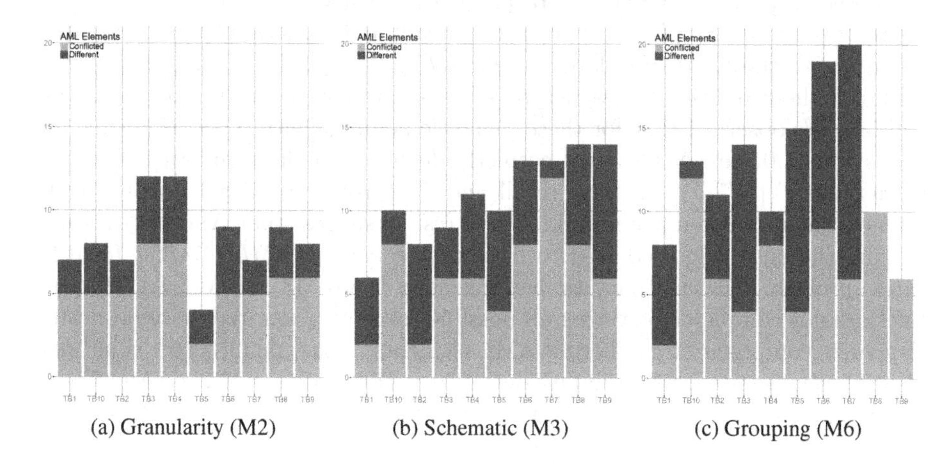

Fig. 6. Size of Integrated AML Documents. Per type of semantic heterogeneity: Granularity (M2), Schematic (M3), and Grouping (M6), the size of the integrated AML documents was reported in terms of the number of conflicts solved (light grey bars), and the different AML elements in the document (dark grey bars). In all the evaluated testbeds, the solved conflicts comprised at least 25 % of the total number of AML elements in the AML document, showing the heterogeneity of the evaluated testbeds

[7] https://sewiki.iai.uni-bonn.de/research/pdt/docs/start.
[8] https://raw.githubusercontent.com/EIS-Bonn/krextor/master/src/xslt/extract/aml.xsl.
[9] https://github.com/i40-Tools/.

Size of the Integrated AML Documents. The goal of this evaluation was to analyze the size of the integrated AML documents with respect to conflicting and different elements. For each type of semantic heterogeneity and testbed of that type, we computed the number of conflicts solved by ALLIGATOR. Further, the number of different AML elements was measured; a different AML element corresponded to an element that appeared in one of the AML documents in the testbed, and was not conflicting with any other AML element. For example, the AML elements in line 15 of the two views in Figs. 2a and 2b are different elements. In consequence, both should be included in the integrated AML document. On the other hand, the AML elements in lines 2, 4, 6, 9, and 12 in both views are pair-wise conflicted AML elements, and each pair should be integrated into only one AML element. Figure 2 reports on the number of *conflicted* and *different* AML elements. We observed that a large number of AML elements in the integrated AML documents result from solving the *Conflict Resolution* problem; being the number of these AML elements at least 25 % of the total elements in the integrated documents. These results illustrated the complexity of the evaluated testbeds, and clearly showed the enhancement assessed by ALLIGATOR during the integration of AML documents.

Effectiveness of ALLIGATOR. The goal of this experiment was to answer our research questions **RQ1** and **RQ2**. ALLIGATOR was run on each of the 30 testbeds to create $SC(\Gamma)$, and precision, recall, and F-measure were computed according to the Gold Standard ($conflicts(\Gamma \mid \Gamma^*, \sigma)$). Table 1 reports on the values of these metrics for each type of semantic heterogeneity, i.e., M2, M3, and M6. We observed that for these semantic heterogeneity types, the value for precision is 1.0, i.e., ALLIGATOR correctly detected all the conflicting elements in $conflicts(\Gamma \mid \Gamma^*, \sigma)$. Further, recall and F-measure are also 1.0 in the testbeds of semantic heterogeneity M2. These results suggest that ALLIGATOR rules capture the knowledge required to *accurately* solve the *AML Conflict Identification* problem. For the semantic heterogeneity types M3 and M6, ALLIGATOR rules are not completely covering all possible conflicts generated between *nested structures* composed of conflicting AML elements. Thus, ALLIGATOR could not identify at most two conflicts in five out of 20 testbeds of type M3 and M6. These results allowed us to positively answer research questions **RQ1** and **RQ2**.

7 Related Work

In the literature, many different approaches are proposed for integrating CAEX documents. In [18], a tool to map two CAEX files is presented. It allows to integrate the AutomationML documents, their respective descriptions, and the modified parts of one file into the other. Further, a mapping algorithm for CAEX files is presented. Nevertheless, the process of mapping is performed manually. Himmler [10] presents a framework to create standardized application interfaces in plant engineering based on AutomationML. The work provides a function-based based standardization framework for the plant engineering domain.

Table 1. Effectiveness of ALLIGATOR. Per semantic heterogeneity type, the effectiveness of ALLIGATOR is reported. In all the testbeds, precision is 1.0. ALLIGATOR exhibits the highest performance in the testbeds of type M2 (F-measure is always 1.0), while in M3 and M6, the F-measure values are at least 0.8

Granularity (M2)										
	TB1	TB2	TB3	TB4	TB5	TB6	TB7	TB8	TB9	TB10
Precision	1.0	1.0	1.0	1.0	1.0	1.0	1.0	1.0	1.0	1.0
Recall	1.0	1.0	1.0	1.0	1.0	1.0	1.0	1.0	1.0	1.0
F-Measure	1.0	1.0	1.0	1.0	1.0	1.0	1.0	1.0	1.0	1.0
Schematic (M3)										
	TB1	TB2	TB3	TB4	TB5	TB6	TB7	TB8	TB9	TB10
Precision	1.0	1.0	1.0	1.0	1.0	1.0	1.0	1.0	1.0	1.0
Recall	1.0	1.0	1.0	1.0	1.0	1.0	0.83	1.0	0.88	0.75
F-Measure	1.0	1.0	1.0	1.0	1.0	1.0	0.90	1.0	0.94	0.85
Grouping (M6)										
	TB1	TB2	TB3	TB4	TB5	TB6	TB7	TB8	TB9	TB10
Precision	1.0	1.0	1.0	1.0	1.0	1.0	1.0	1.0	1.0	1.0
Recall	1.0	1.0	1.0	0.66	1.0	1.0	1.0	1.0	1.0	0.83
F-Measure	1.0	1.0	1.0	0.80	1.0	1.0	1.0	1.0	1.0	0.90

Persson et al. [14] utilize an RDF-based approach to integrate robotized production information modeled with AutomationML. Kovalenko et al. [12] explore how AutomationML can be represented by means of Model-Driven Engineering and the Semantic Web. A small part of an AutomationML ontology is developed, based on the main concepts of the language. Also, the use of rules for consistency checking is proposed, using the Semantic Web Rule Language (SWRL), but no explicit definition of the role of Semantic Web technologies on the integration problem is presented. The *AutomationML Analyzer* [16] is an online tool to browse, query and analyse different AML data by means of Semantic Web technologies; a conceptual design to overcome integration problems in AML is described. All these approaches have the potential to solve specific integration problems for AML. However, they solve rather isolated problems, and a general method capable to automatically integrate AML information from different perspectives is not provided. Contrary, ALLIGATOR combines deductive databases and Semantic Web technologies to effectively integrate documents specified using Industry 4.0 Standards like AML.

8 Conclusions and Future Work

This paper presented ALLIGATOR, a deductive framework for the integration of AML documents. ALLIGATOR relies on Datalog and RDF to *accurately* represent the knowledge that characterizes different types of semantic heterogeneity in AML documents. The results of the empirical evaluation indicate that ALLIGATOR is able to effectively solve the problems of *AML Conflict Identification* and *AML Conflict Resolution*, and exhibits similar behavior for the three studied semantic heterogeneity types, i.e., granularity (M2), schematic (M3), and grouping (M6). In the future, we will empower the ALLIGATOR Deductive System Engine with the expressiveness of Datalog with negation and built-in predicates. Thus, ALLIGATOR will be able to represent other types of semantic heterogeneity in AML, e.g., value processing (M1) and conditional mappings (M4). Further, we plan to extend ALLIGATOR to integrate documents of other Industry 4.0 Standards, such as the OPC-UA machine-to-machine communication protocol.

Acknowledgements. This work has been supported by the German Federal Ministry of Education and Research (BMBF) in the context of the projects *LUCID* (grant no. 01IS14019C), *SDI-X* (no. 01IS15035C), and *Industrial Data Space* (no. 01IS15054).

References

1. Abele, L., Kleinsteuber, M., Hansen, T.: Resource monitoring in industrial production with knowledge-based models and rules. In: PIKM@CIKM, pp. 35–42. ACM (2011)
2. Abele, L., Legat, C., Grimm, S., Muller, A.W.: Ontology-based validation of plant models. In: INDIN, pp. 236–241. IEEE (2013)
3. Barnes, M., Finch, E.L.: COLLADA-Digital Asset Schema Release 1.5.0 specification. Khronos Group, Sony Computer Entertainment Inc (2008)
4. Biffl, S., Kovalenko, O., Lüder, A., Schmidt, N., Rosendahl, R.: Semantic mapping support in AutomationML. In: ETFA, pp. 1–4. IEEE (2014)
5. Björkelund, A., Bruyninckx, H., Malec, J., Nilsson, K., Nugues, P.: Knowledge for intelligent industrial robots. In: AAAI Spring Symposium: Designing Intelligent Robots, vol. SS-12-02. AAAI (2012)
6. Björkelund, A., Malec, J., Nilsson, K., Nugues, P.: Knowledge and skill representations for robotized production. In: Proceedings of the 18th IFAC Congress, Milan (2011)
7. Ceri, S., Gottlob, G., Tanca, L.: What you always wanted to know about datalog (and never dared to ask). IEEE Trans. Knowl. Data Eng. **1**(1), 146–166 (1989)
8. Fedai, M., Epple, U., Drath, R., Fay, D.: A metamodel for generic data exchange between various CAE systems. In: 4th Mathmod Conference, vol. 24, pp. 1247–1256 (2003)
9. Henßen, R., Schleipen, M.: Interoperability between OPC UA and AutomationML. In: Procedia CIRP 25 8th International Conference on Digital Enterprise Technology DET (2014)
10. Himmler, F.: Function based engineering with automationml - towards better standardization and seamless process integration in plant engineering. In: 12 Int. Tagung Wirtschaftsinformatik, WI (2015)
11. Kovalenko, O., Euzenat, J.: Semantic matching of engineering data structures. In: Bill, S., Sabou, M. (eds.) Semantic Web for Intelligent Engineering Applications, Springer (2016)

12. Kovalenko, O., Wimmer, M., Sabou, M., Lüder, A., Ekaputra, F.J., Biffl, S.: Modeling automationml: semantic web technologies vs. model-driven engineering. In: 20th IEEE Conference on Emerging Technologies and Factory Automation, ETFA, pp. 1–4 (2015)
13. Lange, C.: Krextor - an extensible XML→RDF extraction framework. In: Scripting and Development for the Semantic Web (SFSW), CEUR Workshop Proceedings, Aachen, vol. 449, May 2009
14. Persson, J., Gallois, A., Björkelund, A., Hafdell, L., Haage, M., Malec, J., Nilsson, K., Nugues, P.: A knowledge integration framework for robotics. In: 41st International Symposium on Robotics and ROBOTIK 2010 (2010)
15. Runde, S., Dibowski, H., Fay, A., Kabitzsch, K.: A semantic requirement ontology for the engineering of building automation systems by means of OWL. In: ETFA. IEEE (2009)
16. Sabou, M., Ekaputra, F., Kovalenko, O., Biffl, S.: Supporting the engineering of cyberphysical production systems with the automationml analyzer. In: 2016 1st International Workshop on Cyber-Physical Production Systems (CPPS), pp. 1–8. IEEE (2016)
17. Schleipen, M., Gutting, D., Sauerwein, F.: Domain dependant matching of MES knowledge and domain independent mapping of automationml models. In: 2012 IEEE 17th Conference on Emerging Technologies and Factory Automation (ETFA), pp. 1–7. IEEE (2012)
18. Schleipen, M., Okon, M.: The CAEX tool suite - user assistance for the use of standardized plant engineering data exchange. In: 15th IEEE International Conference on Emerging Technologies and Factory Automation, ETFA (2010)
19. Schmidt, N., Lüder, A.: AutomationML and eCl@ss integration (2015)
20. Schmidt, N., Lüder, A.: White paper: automation ML in a nutshell. Technical report (2015)
21. Schmidt, N., Lüder, A., Rosendahl, R., Ryashentseva, D., Foehr, M., Vollmar, J.: Surveying integration approaches for relevance in cyber physical production systems. In: ETFA, pp. 1–8. IEEE (2015)

Combining Textual and Graph-Based Features for Named Entity Disambiguation Using Undirected Probabilistic Graphical Models

Sherzod Hakimov(✉), Hendrik ter Horst, Soufian Jebbara, Matthias Hartung, and Philipp Cimiano

Semantic Computing Group Cognitive Interaction Technology – Center of Excellence (CITEC), Bielefeld University, 33615 Bielefeld, Germany
{shakimov,hterhors,sjebbara,mhartung,cimiano}@cit-ec.uni-bielefeld.de

Abstract. Named Entity Disambiguation (NED) is the task of disambiguating named entities in a natural language text by linking them to their corresponding entities in a knowledge base such as DBpedia, which are already recognized. It is an important step in transforming unstructured text into structured knowledge. Previous work on this task has proven a strong impact of graph-based methods such as PageRank on entity disambiguation. Other approaches rely on distributional similarity between an article and the textual description of a candidate entity. However, the combined impact of these different feature groups has not been explored to a sufficient extent. In this paper, we present a novel approach that exploits an undirected probabilistic model to combine different types of features for named entity disambiguation. Capitalizing on Markov Chain Monte Carlo sampling, our model is capable of exploiting complementary strengths between both graph-based and textual features. We analyze the impact of these features and their combination on named entity disambiguation. In an evaluation on the GERBIL benchmark, our model compares favourably to the current state-of-the-art in 8 out of 14 data sets.

Keywords: Entity disambiguation · Collective entity disambiguation · Named entity disambiguation · Probabilistic graphical models · Factor graphs

1 Introduction

The problem of resolving the real-world reference of entity mentions in textual data, which are already recognized, by linking them to unique identifiers in a knowledge base has received substantial attention in recent years. This *entity disambigation* task is an important first step towards capturing the semantics of textual content.

Earlier approaches to entity disambiguation resolved mentions independently of each other (e.g., DBpedia Spotlight [7], etc.). Recently, several approaches

E. Blomqvist et al. (Eds.): EKAW 2016, LNAI 10024, pp. 288–302, 2016.
DOI: 10.1007/978-3-319-49004-5_19

have been presented that perform *collective entity disambiguation*, attempting to resolve several mentions at the same time within one inference step. Such joint inference approaches can capture dependencies in the choice of identifiers for different mentions.

The features used in entity disambiguation models vary widely. Many approaches rely on features that measure textual coherence. This is typically implemented by a measure of similarity between the context in which a mention appears and the context of the linking candidate. These contexts are of a textual nature and Bag-of-Words (BOW) based similarity as measured by cosine similarity, for instance, can be applied here. A prominent representative of systems using textual coherence is DBpedia Spotlight. Other approaches rely on graph connectivity features exploiting the connectedness between different disambiguation candidates in a knowledge base. Examples of these are the Babelfy [21] and AGDISTIS [26] systems. Finally, recent work has shown the power of using prior probabilities as features on the task. For instance, Tristram et al. [25] have shown that using the PageRank of linking candidates alone can yield quite high results.

Building on these previous results, in this paper we present a novel system that performs collective entity disambiguation by combining all the above-mentioned types of features within one model that is trained discriminatively. In particular, we propose an undirected probabilistic graphical model based on factor graphs. Each factor in the model measures the suitability of the resolution of some mention to a given linking candidate, relying on a set of features that are linearly combined by weights. For inference during training and testing, we rely on a Markov Chain Monte Carlo (MCMC) [2] approach. For training, we rely on the SampleRank [29] algorithm.

We evaluate our approach on standard benchmarking data sets for the entity disambiguation task as available in the GERBIL framework [27]. We show the impact of the features we propose in isolation and in combination. We thus enhance our understanding of the features that work well on the task. Overall, we show that our system outperforms state-of-the-art systems on 8 out of 14 publicly available datasets.

All the data and code used to build our approach are publicly available.[1]

2 Related Work

Given the variety of previous approaches to named entity disambiguation, we structure our discussion of related work according to the features and combinations of features that have been proposed.

One of the first named entity disambiguation systems, DBpedia Spotlight [7], mainly relied on features scoring the textual coherence between the context of the mention and the context of a given linking candidate. Recent approaches include different types of similarities. One example is the approach by Liu et al. [19],

[1] https://github.com/ag-sc/NED.

which considers entity-entity similarity, mention-entity similarity, and mention-mention similarity. The prior probability of a mention is shown to be a strong indicator. A related system is the one of Hoffart et al. [13], which also combines a popularity prior, mention-entity similarity as well as a score of the graph-based coherence between linking candidates. All these features are combined in a linear model. Our approach is related, but extends the feature set used by the above-mentioned approaches, studying in particular the impact of each feature in isolation and in combination.

The connectedness between different linking candidates can also be estimated by the Topic-sensitive PageRank [12] of a linking candidate given another competing candidate or via a random walk over the KB graph, as in Guo and Barbosa [10]. Other approaches relying mainly on graph connectedness include Babelfy [21], TagMe [23] as well as the approaches by Hakimov et al. [11], Alhelbawy and Gaizauskas [1], Usbeck et al. [26], and Jin et al. [15].

By combining different sources of information comprising knowledge about entities, names, context, and the Wikipedia graph in a probabilistic framework, Barrena et al. [3] observe complementary effects between these features. However, they impose strong independence assumptions (i) on the level of features, which essentially renders their model an instance of Naïve Bayes classification, and (ii) on the level of entities as well.

In contrast, we aim at *collective entity disambiguation* in this paper, and frame the task as an inference problem in a probabilistic graphical model to disambiguate all mentions in a text through joint prediction. Previous work on joint entity disambiguation comprises the graph-based approach by Alhelbawy and Gaizauskas [1], for instance: All candidates pertaining to the NEs in the text are represented as nodes in a so-called solution graph that serves as input to a ranking model based on PageRank. As features for the ranking, both an initial confidence (corresponding to prior popularity or mention-entity similarity, respectively) and edge weights in the graph (corresponding to entity-entity coherence) are taken into account. Houlsby and Ciaramita [14] apply a generative probabilistic model, viz. Latent Dirichlet Allocation (LDA; [4]), to the task. They construct a "knowledge base-specific" topic model where each topic corresponds to a Wikipedia article. The word-topic proportions inferred by LDA for each entity mention are directly used in order to link the mention to its most likely Wikipedia concept.

More recently, Ganea et al. [9] and Zwicklbauer et al. [30] have proposed collective entity disambiguation methods as well. Ganea et al. [9] have proposed a joint probabilistic model for collective entity disambiguation that is not trained on any particular data set, but relies on sufficient statistics over all hyperlinks in Wikipedia, considering each anchor text as a mention and the Wikipedia page it refers to as the ground truth entity label. These statistics essentially capture co-occurrence probabilities of mention-entity and entity-entity pairs. Zwicklbauer et al. [30] proposed a method using semantic embeddings of entities for entity disambiguation. They embed entities using Word2Vec [20] by constructing sequences of entities using random walks over the RDF Graph.

Undirected probabilistic graphical models have been successfully applied to a variety of related NLP tasks: Passos et al. [22] propose a method for learning neural phrase embeddings to be applied to Named Entity Recognition by leveraging factor graphs. Singh et al. [24] use factor graphs for cross-document coreference resolution. Our approach differs in that we apply factor graphs for NED while using features specific to the task. We give an overview of how we formulate the NED task with factor graphs in Sect. 3.2.

3 Named Entity Disambiguation with Undirected Factor Graphs

In this work, we present an approach based on imperatively defined factor graphs that addresses Named Entity Disambiguation (NED) with textual and graph-based features. By employing factor graphs, our system is able to disambiguate entity mentions in a document separately and collectively, benefitting from both paradigms. Before we give a formal description of our factor graph approach, we present our candidate retrieval component for retrieving URI candidates for a given entity mention.

3.1 Candidate Retrieval

To reduce the number of possible candidate URIs for a given mention, we implement a retrieval component based on an index that retrieves a subset of k related candidates for this mention. The retrieval component is designed as to provide a high recall, while keeping k as small as possible. Our index is constructed using two different data sources of mention-related surface forms, in particular DBpedia and Wikipedia anchors. In the following, we briefly describe both data sets as well as the generation of our index.

DBpedia Data. We create an index of surface forms of named entities using DBpedia data sets in their 2015-04 version.[2] We collected a set of labeleling properties from these data sets to detect surface forms. All ⟨surface form, URI⟩ pairs are extracted from these data sets while keeping track of the frequency of occurrence of each pair. In addition to label properties, we convert all redirect page URIs into surface forms and pair them with the target page URI. The data set names, label properties, surface form data and all other data sets can be found on our page.

Wikipedia Anchors. We extracted all links in Wikipedia pages and extracted the text mentioned in the anchor and the target link. The text of an anchor refers to the surface form, and the actual link refers to some Wikipedia page (URI). By counting the co-occurrences of ⟨surface form, URI⟩ pairs, we built another table of ⟨surface form, URI, frequency⟩ tuples.

[2] http://wiki.dbpedia.org/Downloads2015-04.

Candidate Retrieval Performance. In order to assess the candidate retrieval performance of our index, we compute the Recall@k, measuring in how many cases we retrieve the correct candidate among the top k results from our index. The results of Recall@k are plotted in Fig. 1 for different values of k. The evaluation is based on the AIDA/CoNLL [13] and MicroPost2014 [6] training sets. We consider three settings: (i) using only the DBpedia table, (ii) using the Wikipedia anchors table, and (iii) We combine both data tables where the frequency of the same ⟨surface form, URI⟩ pairs are summed and the frequency of unique pairs are kept as they are in respective tables. Our results show that the combination approach yields the highest recall. The results also show that considering a number of $k = 10$ represents a reasonable trade-off between recall and efficiency; thus, we rely on this setting in all our experiments. The Recall@10 is 0.934 for the AIDA/CoNLL and 0.814 for the MicroPost2014 training sets, respectively. These figures represent an upper bound in terms of F-Measure for the overall task of entity disambiguation.

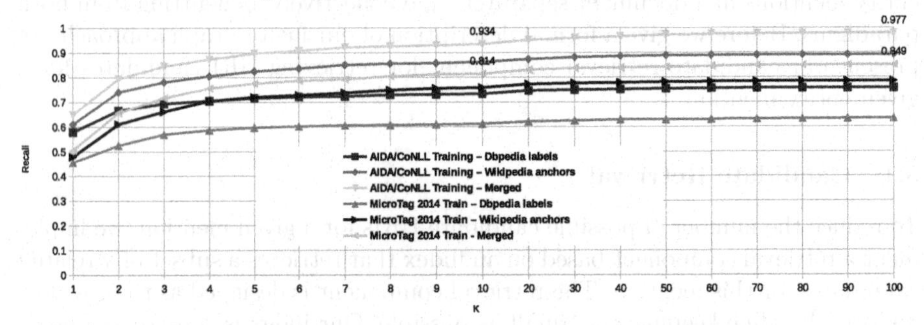

Fig. 1. Recall@k scores for candidate retrieval

3.2 Imperatively Defined Factor Graphs

In this section, we introduce the concept of factor graphs [17], following the notations in [29] and [16]. A factor graph $\mathcal{G}$ is a bipartite graph that defines a probability distribution π. The graph consists of variables V and factors Ψ. Variables can further be divided into sets of *observed* variables X and *hidden* variables Y. A factor Ψ_i connects subsets of observed variables x_i and hidden variables y_i and computes a scalar score based on the exponential of the scalar product of a feature vector $f_i(x_i, y_i)$ and a set of parameters θ_i: $\Psi_i = e^{f_i(x_i,y_i)\cdot\theta_i}$. The probability of the hidden variables given the observed variables is the product of the individual factors:

$$\pi(y|x;\theta) = \frac{1}{Z(x)} \prod_{\Psi_i \in \mathcal{G}} e^{\Psi_i} = \frac{1}{Z(x)} \prod_{\Psi_i \in \mathcal{G}} e^{f_i(x_i,y_i)\cdot\theta_i} \tag{1}$$

where $Z(x)$ is the normalization function. For a given set of observed variables, we generate a factor graph automatically making use of factor templates $\mathcal{T}$. A template $T_j \in \mathcal{T}$ defines the subsets of observed and hidden variables (x, y)

with $x \in X_j$ and $y \in Y_j$ for which it can generate factors and a function $f_j(x, y)$ to generate features for these variables. Additionally, all factors generated by a given template T_j share the same parameters θ_j. With this definition, we can reformulate the conditional probability as follows:

$$\pi(y|x;\theta) = \frac{1}{Z(x)} \prod_{T_j \in \mathcal{T}} \prod_{(x,y) \in T_j} e^{f_j(x,y) \cdot \theta_j} \tag{2}$$

Thus, we define a probability distribution over possible configurations of observed and hidden variables, i.e., assigned URIs. This enables us to explore the joint space of observed and hidden variables in a probabilistic fashion.

Data Representation. In the following, we show how to apply this approach of probabilistic factor graphs to the NED task. We define a document as $d = (\mathbf{w}, \mathbf{a})$ that consists of a sequence of words $\mathbf{w} = (w_1, \ldots, w_{N_w})$ and a set of annotation spans (or entity mentions) $\mathbf{a} = \{a_1, \ldots, a_{N_a}\}$. Documents, words and annotation spans constitute the observed variables X. The assigned URIs of a set of annotation spans $\mathbf{u} = \{u_1, \ldots, u_{N_a}\}$ are considered to be hidden variables Y, where u_i corresponds to annotation span a_i. A disambiguated document, i.e., the collective of words, annotation spans and assigned URIs $(\mathbf{w}, \mathbf{a}, \mathbf{u})$, is referred to as a *configuration* and in the context of sampling as a *state*. Consequently, we can apply Eq. (2) to a disambiguated document to compute its probability given the underlying factor graph. Figure 2 shows a schematic visualization of a disambiguated document along with its factor graph.

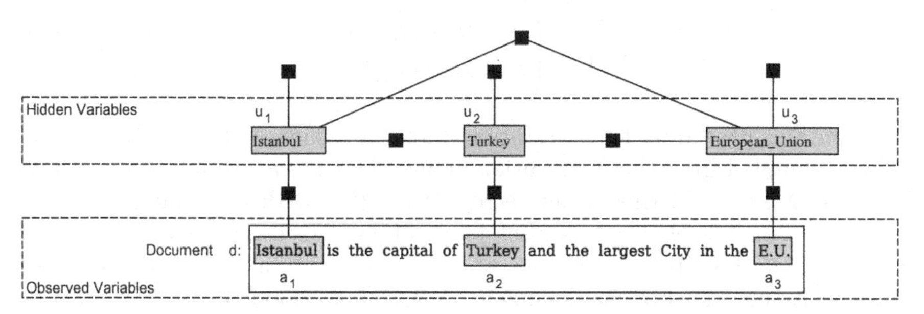

Fig. 2. An exemplary depiction of a factor graph for a disambiguated document with three NEs. The figure shows the division between observed and hidden variables as well as different factors (black boxes) between all variables.

3.3 Inference

This section shows how we infer URIs for a given document using the above formulation of factor graphs. We perform a Markov Chain Monte Carlo (MCMC) sampling procedure [2] that explores the search space of a document in an iterative fashion. The inference procedure performs a local search and can be divided into (i) generating possible successor states for a given state by applying atomic changes, and (ii) selecting a successor state from the set of generated states.

Assuming a document $d = (\mathbf{w}, \mathbf{a})$, our goal is to obtain a configuration (or state) $s^* = (\mathbf{w}, \mathbf{a}, \mathbf{u}^*)$ with the correct URI assignment $\mathbf{u}^*$ for the given annotation spans. For that, we perform an iterative sampling procedure of m steps that performs a local search at each step to find a better disambiguation for a given document.

As a first step, we create an initial state $s_0 = (\mathbf{w}, \mathbf{a}, \mathbf{u_0})$, where URIs $\mathbf{u_0}$ are randomly assigned from the top-k retrieved candidates. This state is used as the starting point of our sampling procedure.

For each annotation span a_i in the state, we retrieve a set of k candidate URIs $Cand(a_i) = \{c_{i1}, \dots, c_{ij}, \dots, c_{ik}\}$ from our candidate retrieval component, using the text of a_i as query. We generate $N_a \cdot k$ modified states that differ from the current state s_t in only a single atomic change. Specifically, the modified state $s'_{ij} = (\mathbf{w}, \mathbf{a}, \mathbf{u}'_{\mathbf{ij}})$ comprises the same observed variables $\mathbf{w}$ and $\mathbf{a}$, but changes exactly one "hidden" assignment of a URI to $\mathbf{u}'_{\mathbf{ij}} = \{u_1, \dots, u_{i-1}, c_{ij}, u_{i+1}, \dots, u_{N_a}\}$, while leaving all other URIs untouched. We consider this pool of generated states to be the collection of all valid states that can be reached from the current state with one atomic change.

Next, we compute the probability of each generated state s'_{ij} using Eq. (2) and obtain a probability distribution over all generated states.[3] We select a single candidate state s'_t by sampling from the distribution of generated states[4] to obtain a potential successor state. We accept the sampled successor state s'_t as our next state s_{t+1} if it has a higher probability than the previous state s_t:

$$s_{t+1} = \begin{cases} s'_t, & \text{if } \pi(s'_t) > \pi(s_t) \\ s_t, & \text{otherwise} \end{cases} \tag{3}$$

Following this procedure for m iterations yields a sequence of states $(s_0, \dots, s_m)$ that are sampled from the distribution defined by the underlying factor graphs. The final state s_m in this sequence represents the predicted configuration s^*. With a reasonable choice of model parameters θ (see Sect. 3.4 below), it is expected that the URI assignments $\mathbf{u}^*$ in s^* constitute a good disambiguation of the entity mentions in the document. A more pseudo-algorithmic description of the inference procedure is given in Algorithm 1 and a schematic visualization of the generation of neighboring states is shown in Fig. 3.

3.4 Learning Model Parameters

In order to optimize parameters θ, we use an implementation of the SampleRank [29] algorithm. The SampleRank algorithm obtains gradients for these parameters from pairs of states (e.g. s_t and s'_t) by observing the individual steps in the

[3] After re-normalizing the probabilities such that $\sum_{s'_{ij}} \pi(s'_{ij}) = 1$.

[4] Our experiments show that a greedy approach that always prefers the state with the highest probability works best.

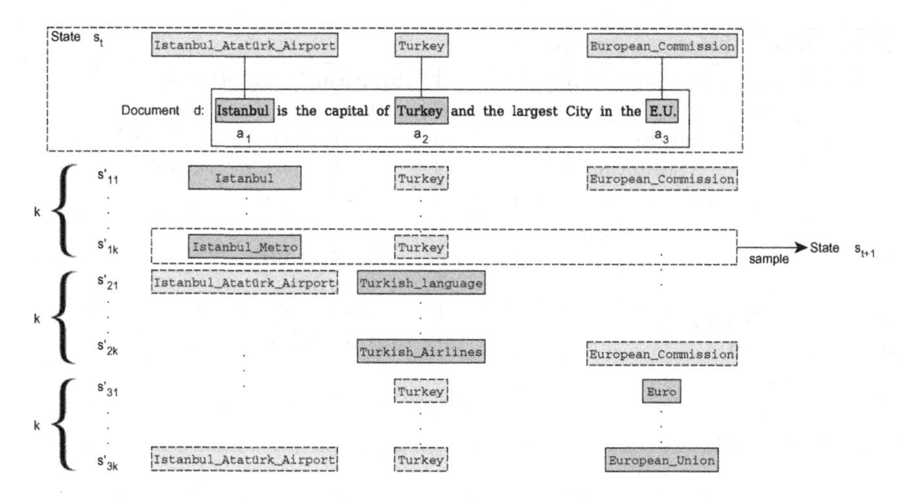

Fig. 3. An exemplary depiction of the sampling procedure. Starting from state s_t we generate states $\{s_{ij}\}$ in its local neighborhood performing only atomic modifications. Specifically, we generate a state for each annotation span and each retrieved candidate URI. Each state is scored according to the current model and the successor state s_{t+1} is selected from these generated states.

inference routine. For that, the algorithm requires a preference function $\mathbb{P}(s', s)$ that indicates which of two states is "objectively" preferred. We implement $\mathbb{P}$ based on an objective function $\mathbb{O}(s)$ that computes a score for a state compared to the ground truth assignments for the respective training document in terms of the ratio of correctly linked entity mentions $N_{a,correct}$ of a state s and the total number of entity mentions N_a in s, i.e., $\mathbb{O}(s) = N_{a,correct}/N_a$. The preference function is thus:

$$\mathbb{P}(s', s) = \begin{cases} 1, & \text{if } \mathbb{O}(s') > \mathbb{O}(s) \\ 0, & \text{otherwise} \end{cases} \tag{4}$$

We modify the original SampleRank algorithm such that we select the best scoring successor state based on the objective function score $\mathbb{O}(.)$ rather than the probability given by the model in Eq. (2). This small modification ensures that the training procedure is guided towards a good solution when updating the model parameters.

The final training algorithm is similar to the inference procedure in Algorithm 1, with two changes, however: Line 5 of the inference algorithm is changed to $s' \leftarrow \arg\max_{s'' \in \{s'_{ij}\}}(\mathbb{O}(s''))$ and an additional call is inserted between line 6 and 7: $\theta \leftarrow \text{UPDATE}(s', s, \theta)$, in order to update the model parameter at each step according to the regular SampleRank algorithm [29].

3.5 Templates

In the following, we describe the templates that are used to automatically instantiate the factors between variables in a configuration and, thus, for the extraction

Algorithm 1. Inference procedure

1: **function** INFERENCE($\mathbf{w}, \mathbf{a}, \mathbf{u_0}$)
2: $s \leftarrow (\mathbf{w}, \mathbf{a}, \mathbf{u_0})$
3: **for** t=1,2 ... m **do**
4: $\{s'_{ij}\} \leftarrow$ NEIGHBORS(s)
5: $s' \leftarrow \arg\max_{s'' \in \{s'_{ij}\}}(\pi(s''))$
6: **if** $\pi(s') > \pi(s)$ **then**
7: $s \leftarrow s'$
8: **else**
9: **break**
10: **end if**
11: **end for**
12: **return** s
13: **end function**

1: **function** NEIGHBORS(s)
2: **for** i=1,2 ... N_a **do**
3: $\{c_j\} \leftarrow$ CANDIDATES($s.a_i$)
4: **for** j=1,2 ... k **do**
5: $s'_{ij} \leftarrow s$
6: $s'_{ij}.u_j \leftarrow c_j$
7: **end for**
8: **end for**
9: **return** $\{s'_{ij}\}$
10: **end function**

of the features that determine the probability of a configuration (see Eq. (2)). Throughout this discussion, we use a_i to denote the ith annotation span, and $Cand(a_i) = \{c_1, \ldots, c_k\}$ to denote the set of entity candidates for a_i. Further, we denote the actually assigned URI for a_i as $u_i \in Cand(a_i)$.

Relative Term Frequency. This template instantiates a factor between each assigned URI u_i and its corresponding annotation span a_i in order to reflect the co-occurrence of u_i and a_i in our index (see Sect. 3.1). The feature value $RTF(a_i, u_i)$ for such a factor is defined by the term-candidate frequency normalized across all candidate URIs that are retrieved for a_i:

$$RTF(a, u) = \frac{freq(a, u)}{\sum_{c \in Cand(a)} freq(a, c)} \quad (5)$$

Edit Similarity. In this template, we add a factor between each annotation span a_i and its assigned URI u_i that reflects the string similarity $l(a_i, u_i)$ between those two based on the Levenshtein distance [18]. We use the maximum length to normalize the string similarity $l(a_i, u_i)$ and it is calculated as follows $v(a_i, u_i) = 1 - \frac{l(a_i, u_i)}{\max(len(a_i), len(u_i))}$ which is added as a feature to the factor. Further, we create n equally distributed thresholds $t_j \in (0, 1]$. For each $t_j \leq v(a_i, u_i)$, an additional boolean feature is added.

Document Similarity. Following [5], we hypothesize a positive impact of the textual context on named entity disambiguation, in particular for NEs that share the same surface form (e.g., *apple*). We represent the content of a document as a bag-of-words vector that is constructed from all of its tokens. Each document is preprocessed by applying tokenization, case normalization, stemming and stop-words removal. Vector components are weighted by their term frequency *tf* and their inverse document frequency *idf*. The latter is computed on the Wikipedia abstract corpus. Given a document d, we denote its document vector as $\boldsymbol{v_d}$. The document vector of an assigned URI u_i is denoted as $\boldsymbol{v_{u_i}}$ which is computed analogously from its corresponding Wikipedia abstract. For each u_i in d, we add a factor to the factor graph whose feature is defined by the cosine similarity of $\boldsymbol{v_d}$ and $\boldsymbol{v_{u_i}}$:

$$cos(\boldsymbol{v}, \boldsymbol{w}) = \frac{\sum_{i=1}^{n} \boldsymbol{v}_i \boldsymbol{w}_i}{\sqrt{\sum_{i=1}^{n} \boldsymbol{v}_i^2}\sqrt{\sum_{i=1}^{n} \boldsymbol{w}_i^2}} \tag{6}$$

Relative Page Rank. The Relative Page Rank template instantiates a factor for each assigned URI u_i to measure its a-priori popularity in Wikipedia. The PageRank scores $PR(u_i)$ are computed on a subgraph of the Wikipedia PageLinks data set[5] excluding all category, disambiguation and file pages. We calculate the PageRank scores based on the approach explained in [21] that uses the random walk algorithm by Das Sarma et al. [8]. We normalize the raw PageRank scores over all $c \in Cand(a_i)$ as described in Eq. (7) and add the relative score $RPR(a_i, u_i)$ as a feature to the factor.

$$RPR(a, u) = \frac{PR(u)}{\sum_{c \in Cand(a)} PR(c)} \tag{7}$$

Topic-Specific PageRank. To measure the degree of coherence between all assigned URIs $(u_1, \ldots, u_{N_a})$ in a state, we introduce the Topic-specific PageRank template. Topic-specific PageRank [12] is computed on the Wikipedia graph using the random walk with restart (RWR) algorithm as described by Moro et al. [21]. Following the notation by Moro et al. [21], we set the RWR parameters as follows: the minimum hit threshold $\eta = 100$, the restart probability $alpha = 0.85$, the number of iterations $n = 1.000.000$ and the transition probability P as uniformly distributed for all neighbor nodes.

For each pair of URIs $p_{ij} = (u_i, u_j)$ where $i \neq j$, we add a new factor to the factor graph connecting u_i and u_j. The feature value for p_{ij} is determined by the sum of the Topic-Specific PageRank values of $TSPR(u_i, u_j)$ and $TSPR(u_j, u_i)$. For pairs where $u_i = u_j$ the feature value is set to 1 in order to encourage repetitions of the same URI.

4 Experiments

In this section, we present our experimental results on different data sets. First, we evaluate the performance of different subsets of features on development data from the AIDA/CoNLL and Micropost2014 data sets in Sect. 4.1. In Sect. 4.2, we compare the performance of our model using the best feature configuration on the GERBIL benchmark [27] (version 1.2.2), addressing the D2KB task in which the named entities are pre-annotated so that only the actual disambiguation, but not the recognition, is evaluated.

[5] http://wiki.dbpedia.org/Downloads2015-04.

Table 1. Micro F_1 scores of models trained on combinations of features RPR (Relative PageRank), RTF (Relative Term Frequency), ES (Edit Similarity), DS (Document Similarity), TSPR (Topic Specific PageRank) as defined in Sect. 3.5

Feature combinations	AIDA/CoNLL Test-A	MicroPost2014 Test
RPR	0.720	0.66
RTF	0.619	0.60
ES	0.500	0.49
DS	0.230	0.29
TSPR	0.725	0.29
RPR + RTF	0.723	0.68
RPR + ES	0.724	0.67
RPR + TSPR	0.747	0.65
RPR + DS	0.720	0.66
RPR + RTF + ES	0.718	**0.70**
RPR + RTF + TSPR	0.747	0.67
RPR + RTF + DS	0.721	0.68
RPR + RTF + ES + DS	0.737	0.69
RPR + RTF + ES + TSPR	**0.781**	0.64
RPR + RTF + ES + TSPR + DS	0.775	0.65

4.1 Model Training and Feature Selection

We trained several models with various combinations of features as defined in Sect. 3.5, using training data from the AIDA/CoNLL [13] and Micropost2014 [6] data sets. We trained and tested models on documents where each annotation has a valid link in a knowledge base, e.g. DBpedia. Each model was trained by iterating 5 times over the training data using the training split of both data sets. Micro F_1 scores for each model are shown in Table 1. Note that the main focus of these experiments is to determine the optimal feature combination. Due to differences in the underlying data splits and evaluation, the results reported in Table 1 are not comparable to official GERBIL results.

The results show that the single best-performing features are the PageRank (RPR) and the Topic-specific PageRank (TSPR) features, which yield an F_1 score of around 0.72. PageRank acts as a strong prior, while TSPR models the connectedness between different linking candidates in a pairwise fashion. Both features are mildly complementary, which can be seen from the fact that they yield the best combination of two features on AIDA/CoNLL, obtaining an F_1 score of 0.747. No combination of three features improves upon this result. The overall best model capitalizes on four features: PageRank, TSPR, relative term frequency and edit similarity. This combination yields an overall performance of $F_1 = 0.78$.

On the MicroPost2014 data set that contains a significantly smaller amount of annotations compared to AIDA/CoNLL (2.1 vs. 20 annotations per document on average), the best model combines PageRank, relative term frequency and edit similarity, yielding an F_1 score of 0.70. Obviosuly, the Topic-specific PageRank is less effective in this text genre, due to its considerably lower degree of connectedness and the lower number of links to explore. Being the only feature that incorporates connectedness of all candidates, the strong individual performance of TSPR on AIDA/CoNLL is indicative of the advantages of a collective disambiguation strategy over an approach that resolves entities independently of one another.

4.2 Comparative Evaluation

In this experiment, we use the best-performing model configurations as determined by feature selection (see Table 1). Our system uses the model with RPR + RTF + ES + TSPR features when the number of annotations in a given document is higher than 3. When the number of annotations is equal or lower than 3, the model with RPR + RTF + ES features is used. We use the top 10 candidates from the Candidate Retrieval component of the system as explained in Sect. 3.1. Returning 100 or more candidates increases the runtime while having no significant improvement on performance.

We compare our system to other state-of-the-art systems on 14 publicly available data sets via GERBIL version 1.2.2 [27]. We implemented a web service called NERFGUN (**N**amed **E**ntity disambiguation by **R**anking with **F**actor **G**raphs over **Un**directed edges), with the two best models after feature selection and submitted to GERBIL. The results of our system[6] and state-of-the-art annotation systems[7] that are integrated into GERBIL are presented in Table 2. Since Ganea et al. [9] and Zwicklbauer et al. [30] evaluated their systems with respect to older versions of GERBIL (version 1.1.4) and these systems do not have submitted a publicly available API to the framework, we cannot fairly compare to them. Thus, we ommited these systems from comparison. This is in particular the case because the evaluation metrics have changed in recent versions of the GERBIL framework. Note that all results presented in Table 2 are based on GERBIL version 1.2.2.

In Table 2 we report Micro F_1 and Macro F_1 measures of compared systems for 14 data sets. Based on Micro F_1 and Macro F_1 measures, NERFGUN obtains the best result on 8 out of 14 data sets. Kea [28] outperforms all systems on the AQUAINT and the DBpedia Spotlight data. On the AQUAINT, the results of our system are comparable to the best annotation system (Micro F_1 0.73 compared to 0.77, Macro F_1 0.72 compared to 0.76, respectively). Babelfy achieves the highest score for the KORE50 data set with 0.74 and 0.71 while NERFGUN obtains 0.44 and 0.40 for Micro F_1 and Macro F_1 measures respectively.

[6] Our results, GERBIL v1.2.2: http://gerbil.aksw.org/gerbil/experiment?id=201604290045.

[7] State-of-the-art annotation systems' results, GERBIL v1.2.2: http://gerbil.aksw.org/gerbil/experiment?id=201604270003.

Table 2. Macro F_1 and Micro F_1 measures for the D2KB task (named entity disambiguation) based on GERBIL v1.2.2; N/A: Not Available. The best scoring system for each data set is highlighted (using **boldface** for the best Micro F_1 result and *italics* for Macro F_1, respectively)

Systems	Measures	ACE2004	AIDA/CoNLL-Complete	AIDA/CoNLL-Test A	AIDA/CoNLL-Test B	AIDA/CoNLL-Training	AQUAINT	DBpediaSpotlight	IITB	KORE50	Microposts2014-Test	Microposts2014-Train	MSNBC	N3-Reuters-128	N3-RSS-500
AGDISTIS	Micro F1	0.63	0.54	0.54	0.52	0.55	0.52	0.27	0.47	0.32	0.67	0.33	0.43	**0.61**	**0.65**
	Macro F1	0.77	0.52	0.49	0.53	0.53	0.51	0.28	0.49	0.3	0.64	0.6	0.61	***0.61***	***0.71***
AIDA	Micro F1	0.14	0.54	0.54	0.52	0.55	0.16	0.19	0.18	0.65	0.3	0.36	0.44	0.44	0.39
	Macro F1	0.44	0.5	0.47	0.5	0.5	0.16	0.17	0.19	0.59	0.28	0.57	0.58	0.38	0.32
Babelfy	Micro F1	0.52	0.66	0.65	0.68	0.66	0.68	0.53	N/A	**0.74**	0.64	0.48	0.51	0.45	N/A
	Macro F1	0.69	0.6	0.59	0.62	0.61	0.68	0.52	N/A	***0.71***	0.59	0.63	0.61	0.39	N/A
DBpedia Spotlight	Micro F1	0.47	0.5	0.48	0.52	0.5	0.53	0.71	0.3	0.45	0.39	0.5	0.49	0.2	0.34
	Macro F1	0.67	0.49	0.47	0.5	0.5	0.52	0.69	0.29	0.41	0.39	0.66	0.61	0.17	0.27
Dexter	Micro F1	0.52	0.51	0.49	0.5	0.52	0.52	0.29	0.21	0.2	0.38	0.41	0.43	0.37	0.36
	Macro F1	0.68	0.48	0.45	0.47	0.48	0.51	0.26	0.21	0.14	0.4	0.59	0.56	0.3	0.31
Entityclassifier.eu NER	Micro F1	0.49	0.5	0.47	0.47	0.51	0.41	0.25	0.14	0.29	0.45	0.41	0.48	0.34	0.37
	Macro F1	0.66	0.48	0.47	0.46	0.48	0.38	0.2	0.16	0.26	0.44	0.6	0.6	0.32	0.34
FOX	Micro F1	0	0.51	0.49	0.47	0.51	0	0.15	0.02	0.29	0.02	0.23	0.32	0.57	0.55
	Macro F1	0.37	0.48	0.44	0.47	0.49	0	0.12	0.02	0.25	0.02	0.5	0.49	0.55	0.59
FREME NER	Micro F1	0.69	0.6	0.59	0.57	0.61	**0.78**	**0.82**	0.43	0.32	0.53	**0.65**	**0.65**	0.42	0.51
	Macro F1	0.81	0.6	0.57	0.59	0.6	***0.78***	***0.83***	0.42	0.3	0.56	***0.78***	***0.76***	0.38	0.48
Kea	Micro F1	0.65	0.62	0.61	0.6	0.63	0.77	0.74	0.48	0.59	**0.7**	0.64	**0.65**	0.44	0.51
	Macro F1	0.76	0.59	0.56	0.59	0.6	0.76	0.73	0.47	0.53	0.67	0.77	0.74	0.39	0.46
NERD-ML	Micro F1	0.56	0.2	0	0.01	0.28	0.59	0.55	0.43	0.32	0.54	0.5	0.51	0.38	0.41
	Macro F1	0.72	0.12	0.01	0.01	0.17	0.57	0.53	0.42	0.26	0.54	0.65	0.62	0.31	0.35
WAT	Micro F1	0.65	0.71	0.7	**0.71**	0.71	0.72	0.66	0.41	0.61	0.65	0.6	0.63	0.44	0.51
	Macro F1	0.77	0.68	0.66	0.68	0.68	0.72	0.68	0.4	0.51	0.62	0.74	0.73	0.37	0.43
xLisa	Micro F1	0.47	0.15	0.41	0.4	0.4	0.42	0.22	**0.57**	0.24	0.45	0.24	0.51	0.43	0.32
	Macro F1	0.63	0.45	0.37	0.33	0.37	0.37	0.22	***0.58***	0.25	0.38	0.24	0.63	0.37	0.29
NERFGUN	Micro F1	**0.73**	**0.72**	**0.71**	**0.71**	**0.72**	0.70	0.49	N/A	0.40	0.65	**0.65**	N/A	0.57	**0.65**
	Macro F1	***0.85***	***0.72***	***0.68***	***0.71***	***0.72***	0.69	0.51	N/A	0.37	***0.79***	0.76	N/A	0.58	0.65

5 Conclusion and Future Work

We have proposed a new approach to collective entity disambiguation that frames the task as a joint inference problem. Our approach relies on an undirected probabilistic graphical model to model dependencies between different factors that score the suitability of assignments of named entities to identifiers in a KB. The model is defined through imperatively defined factor graphs and in particular by templates that 'roll out' the factor graph structure for a given input text by generating corresponding factors. Our model allows to combine and investigate different features in terms of their impact on the task. In particular, our model considers three text-based features, namely (i) term frequency scores, (ii) a similarity measure based on the Levenshtein distance and (iii) the document similarity. Further, our model includes features measuring the degree of connectedness between pairs of linking candidates via the Topic Specific PageRank Score and the PageRank of each linking candidate as a prior. We have shown that a combination of all features with exception of the document similarity feature performs best on the AIDA/CoNLL data sets. Based on Micro F_1 and Macro F_1 measures we outperform well-known annotation systems such as DBpedia Spotlight, Babelfy, WAT and AGDISTIS in 8 out of 14 datasets. In future work, we will extend the approach to also solve the problem of recognition of entities, thus performing named entity recognition and linking within one model in which statistical dependencies between both tasks can be modeled.

Acknowledgements. This work was supported by the Cluster of Excellence Cognitive Interaction Technology 'CITEC' (EXC 277) at Bielefeld University, which is funded by the German Research Foundation (DFG).

References

1. Alhelbawy, A., Gaizauskas, R.J.: Graph ranking for collective named entity disambiguation. In: Proceedings of ACL (Short Papers), Baltimore, MD, pp. 75–80 (2014)
2. Andrieu, C., de Freitas, N., Doucet, A., Jordan, M.I.: An introduction to MCMC for machine learning. Mach. Learn. **50**, 5–43 (2003)
3. Barrena, A., Soroa, A., Agirre, E.: Combining mention context and hyperlinks from Wikipedia for named entity disambiguation. In: Proceedings of ⋆SEM, Denver, CO, pp. 101–105 (2015)
4. Blei, D.M., Ng, A., Jordan, M.: Latent dirichlet allocation. J. Mach. Learn. Res. **3**, 993–1022 (2003)
5. Bunescu, R.C., Pasca, M.: Using encyclopedic knowledge for named entity disambiguation. In: Proceedings of EACL, pp. 9–16 (2006)
6. Cano, A.E., Rizzo, G., Varga, A., Rowe, M., Stankovic, M., Dadzie, A.S.: Making sense of microposts: (# microposts2014) named entity extraction & linking challenge. In: CEUR Workshop Proceedings, vol. 1141, pp. 54–60 (2014)
7. Daiber, J., Jakob, M., Hokamp, C., Mendes, P.N.: Improving efficiency and accuracy in multilingual entity extraction. In: Proceedings of SEMANTICS (2013)
8. Das Sarma, A., Molla, A.R., Pandurangan, G., Upfal, E.: Fast distributed pagerank computation. Theor. Comput. Sci. **561**(Part B), 113–121 (2015). Special Issue on Distributed Computing and Networking
9. Ganea, O.E., Horlescu, M., Lucchi, A., Eickhoff, C., Hofmann, T.: Probabilistic bag-of-hyperlinks model for entity linking. In: Proceedings of WWW (2016)
10. Guo, Z., Barbosa, D.: Robust entity linking via random walks. In: Proceedings of CIKM, Shanghai, China, pp. 499–508 (2014)
11. Hakimov, S., Oto, S.A., Dogdu, E.: Named entity recognition and disambiguation using linked data and graph-based centrality scoring. In: Proceedings of the Workshop on Semantic Web Information Management (SWIM), pp. 1–7 (2012)
12. Haveliwala, T.H.: Topic-sensitive PageRank. In: Proceedings of WWW, pp. 517–526 (2002)
13. Hoffart, J., Yosef, M.A., Bordino, I., Fürstenau, H., Pinkal, M., Spaniol, M., Taneva, B., Thater, S., Weikum, G.: Robust disambiguation of named entities in text. In: Proceedings of EMNLP, Edinburgh, Scotland, UK, pp. 782–792 (2011)

14. Houlsby, N., Ciaramita, M.: A scalable gibbs sampler for probabilistic entity linking. In: de Rijke, M., Kenter, T., de Vries, A.P., Zhai, C.X., de Jong, F., Radinsky, K., Hofmann, K. (eds.) ECIR 2014. LNCS, vol. 8416, pp. 335–346. Springer, Heidelberg (2014)
15. Jin, Y., Kcman, E., Wang, K., Loynd, R.: Entity linking at the tail: sparse signals, unknown entities and phrase models. In: Proceedings of WSDM (2014)
16. Klinger, R., Cimiano, P.: Joint and pipeline probabilistic models for fine-grained sentiment analysis: extracting aspects, subjective phrases and their relations. In: Proceedings of ICDMW, pp. 937–944 (2013)
17. Kschischang, F.R., Frey, B.J., Loeliger, H.A.: Factor graphs and sum product algorithm. IEEE Trans. Inf. Theory **47**(2), 498–519 (2001)
18. Levenshtein, V.I.: Binary codes capable of correcting deletions, insertions, and reversals. Sov. Phys. Dokl. **163**(4), 707–710 (1966)
19. Liu, X., Li, Y., Wu, H., Zhou, M., Wei, F., Lu, Y.: Entity linking for tweets. In: Proceedings of ACL, Sofia, Bulgaria, pp. 1304–1311 (2013)
20. Mikolov, T., Chen, K., Corrado, G., Dean, J.: Efficient estimation of word representations in vector space. arXiv preprint arXiv:1301.3781 (2013)
21. Moro, A., Raganato, A., Navigli, R.: Entity linking meets word sense disambiguation: a unified approach. Trans. Assoc. Comput. Linguist. **2**, 231–244 (2014)
22. Passos, A., Kumar, V., McCallum, A.: Lexicon infused phrase embeddings for named entity resolution. arXiv preprint arXiv:1404.5367 (2014)
23. Piccinno, F., Ferragina, P.: From TagME to WAT. A new entity annotator. In: Proceedings of ACM Workshop on Entity Recognition and Disambiguation, pp. 55–62 (2014)
24. Singh, S., Subramanya, A., Pereira, F., McCallum, A.: Large-scale cross-document coreference using distributed inference and hierarchical models. Proc. ACL **1**, 793–803 (2011)
25. Tristram, F., Walter, S., Cimiano, P., Unger, C.: Weasel. A machine learning based approach to entity linking combining different features. In: Proceedings of ISWC Workshop on NLP and DBpedia (2015)
26. Usbeck, R., Ngonga Ngomo, A.-C., Röder, M., Gerber, D., Coelho, S.A., Auer, S., Both, A.: AGDISTIS - graph-based disambiguation of named entities using linked data. In: Mika, P., et al. (eds.) ISWC 2014, Part I. LNCS, vol. 8796, pp. 457–471. Springer, Heidelberg (2014)
27. Usbeck, R., Röder, M., Ngonga Ngomo, A.C., Baron, C., Both, A., Brümmer, M., Ceccarelli, D., Cornolti, M., Cherix, D., Eickmann, B., et al.: GERBIL. General entity annotator benchmarking framework. In: Proceedings of WWW, pp. 1133–1143 (2015)
28. Waitelonis, J., Sack, H.: Named entity linking in #tweets with kea. In: Proceedings of 6th workshop on Making Sense of Microposts - Named Entity Recognition and Linking (NEEL) Challenge, at WWW2016 (2016)
29. Wick, M., Rohanimanesh, K., Culotta, A., McCallum, A.: SampleRank. Learning preferences from atomic gradients. In: NIPS Workshop on Advances in Ranking, pp. 1–5 (2009)
30. Zwicklbauer, S., Seifert, C., Granitzer, M.: DoSeR - a knowledge-base-agnostic framework for entity disambiguation using semantic embeddings. In: Sack, H., Blomqvist, E., d'Aquin, M., Ghidini, C., Ponzetto, S.P., Lange, C. (eds.) ESWC 2016. LNCS, vol. 9678, pp. 182–198. Springer, Heidelberg (2016). doi:10.1007/978-3-319-34129-3_12

VoCol: An Integrated Environment to Support Version-Controlled Vocabulary Development

Lavdim Halilaj[1,2(✉)], Niklas Petersen[1,2], Irlán Grangel-González[1,2], Christoph Lange[1,2], Sören Auer[1,2], Gökhan Coskun[3], and Steffen Lohmann[2]

[1] Enterprise Information Systems (EIS), University of Bonn, Bonn, Germany
{halilaj,petersen,grangel,langec,auer}@cs.uni-bonn.de
[2] Fraunhofer Institute for Intelligent Analysis and Information Systems (IAIS), Sankt Augustin, Germany
{Lavdim.Halilaj,Niklas.Petersen,Irlan.Grangel-Gonzalez,Christoph.Lange, Soren.Auer,Steffen.Lohmann}@iais.fraunhofer.de
[3] Bayer Business Services, Berlin, Germany
goekhan.coskun@bayer.com

Abstract. Vocabularies are increasingly being developed on platforms for hosting version-controlled repositories, such as GitHub. However, these platforms lack important features that have proven useful in vocabulary development. We present VoCol, an integrated environment that supports the development of vocabularies using Version Control Systems. VoCol is based on a fundamental model of vocabulary development, consisting of the three core activities modeling, population, and testing. We implemented VoCol using a loose coupling of validation, querying, analytics, visualization, and documentation generation components on top of a standard Git repository. All components, including the version-controlled repository, can be configured and replaced with little effort to cater for various use cases. We demonstrate the applicability of VoCol with a real-world example and report on a user study that confirms its usability and usefulness.

Keywords: Vocabulary development · Version control system · Ontology engineering · Integrated development environment · IDE · Git · GitHub · Webhook

1 Introduction

Vocabulary development is currently a major bottleneck for the wide realization of the Semantic Web vision. It requires a significant investment, which is difficult to make by a single person or organization. Identifying the terms and concepts by finding a consensus among the involved stakeholders and defining a shared vocabulary[1] is an effective approach to tackle this problem. However, this

[1] In this work, the term "vocabulary" is used to refer to lightweight ontologies, as they are developed in initiatives like schema.org and defined by the W3C [23].

E. Blomqvist et al. (Eds.): EKAW 2016, LNAI 10024, pp. 303–319, 2016.
DOI: 10.1007/978-3-319-49004-5_20

process, which we refer to as *distributed vocabulary development*, can be quite complex. In fact, the main challenge for vocabulary engineers is to work collaboratively on a shared objective in a harmonic and efficient way, while avoiding misunderstandings, uncertainty, and ambiguity.

On the other hand, *Version Control Systems* (VCS), such as *Subversion* (SVN) or *Git*, are becoming increasingly popular for vocabulary development. In our previous work, we proposed *Git4Voc* [6], a set of best practices which transfer concepts of VCSs to vocabulary development, on the example of Git. We discovered that several aspects of vocabulary development—in particular with regard to revision management, access control, and some governance issues—are already well covered by Git-based version control, especially if developers follow the proposed best practices.

Many of the current vocabulary development activities take place on repository hosting platforms like *GitHub*, *GitLab*, and *BitBucket*. In addition to mere version-controlled (e.g. Git) repositories, these platforms provide features such as change tracking (e.g. diffs), comments, issue tracking, wikis, and notifications. Examples of popular vocabulary projects that are publicly maintained on GitHub include Schema.org, *FOAF*, *BIBO*, *DOAP*, and the *Music Ontology*.[2] However, despite all benefits of developing vocabularies on repository hosting platforms like GitHub, these platforms lack important features that have proven useful in vocabulary development. In particular, they do not provide an integrated environment typically found in systems dedicated to distributed vocabulary development, such as *WebProtégé* [22] or *VocBench* [21].

We designed *VoCol* as a holistic approach for realizing a full-featured vocabulary development environment centered around version control systems. VoCol supports a fundamental round-trip model of vocabulary development, consisting of the three core activities *modeling*, *population*, and *testing*. In the spirit of test-driven software engineering, VoCol allows to formulate queries which represent competency questions for testing the expressivity and applicability of a vocabulary *a priori*. For *a posteriori* testing, it supports the automatic detection of "bad smells" in the vocabulary design by employing SPARQL patterns. For modeling, VoCol integrates a number of techniques facilitating the conceptual work, such as automatically generated documentations and visualizations providing different views on the vocabulary as well as an evolution timeline supporting traceability. For population, VoCol supports the integration of mappings between data sources (e.g., R2RML mappings to relational databases) and the vocabulary. The governance of distributed vocabulary development is supported by the access control as well as the branching and merging mechanisms of the underlying VCS.

As a result, VoCol bridges between the *conceptual* development of vocabularies and the *operational* execution in a concrete IT landscape. The implementation of VoCol is based on a loose coupling, leveraging the webhook method provided by many VCSs with tools and techniques focusing on particular aspects of

[2] See https://github.com/ + schemaorg/schemaorg, foaf/foaf, structureddynamics/Bibliographic-Ontology-BIBO, edumbill/doap, motools/musicontology, among others.

vocabulary development. By proving Vagrant and Docker containers bundling all tools and encapsulating dependencies, VoCol is easily deployable or even usable as-a-service in conjunction with arbitrary VCS installations.

The remainder of this paper is structured as follows: Section 2 introduces the fundamental round-trip model that VoCol is based on and lists requirements that are critical for distributed vocabulary development. Based on the model and requirements, we developed the VoCol system architecture that is presented in Sect. 3. Section 4 introduces an implementation of VoCol that we realized on top of Git. Section 5 reports on a qualitative evaluation of the usefulness and usability of the VoCol environment. Finally, VoCol is compared to related environments for distributed vocabulary development in Sect. 6, before the paper is concluded in Sect. 7.

2 Round-Trip Model and Requirements

Deriving requirements for the envisioned development environment demands the clarification of our understanding of the most fundamental vocabulary development activities. A vocabulary comprises a terminology which is known as *TBox*. The creation of this terminology is realized using a logical formalism during the modeling activities [5]. This comprises the analysis and conceptualization of the domain and the specification of the vocabulary terms, such as classes, properties, and the relationships between them. Once the vocabulary modeling has been completed, the next activity is typically population. It includes the addition of actual data in line with the defined classes and properties, also known as *ABox* [3]. To verify whether the created vocabulary correctly represents the domain, a list of queries can be compiled from *competency questions* [17] and used for testing purposes. Vocabulary engineers may iterate in an incremental fashion between the modeling, population, and querying activities. In fact, these three core activities lead to the conception of *round-trip* development as illustrated in Fig. 1a.

In order to develop an integrated environment that supports the described round-trip development of vocabularies, corresponding requirements have to be identified and addressed accordingly. In our previous work on Git4Voc [6], we identified eleven requirements that are crucial for the successful adaptation of Git to vocabulary development. We gathered these requirements by aggregating insights from the state of the art and our own experiences with developing the vocabularies *MobiVoc* and *SCORVoc* on GitHub.[3] For the design of VoCol, we revised these requirements and grouped them into four categories that need to be addressed by an integrated environment that aims to support full-featured vocabulary development (cf. Fig. 1b). In the following, we briefly summarize the categories and requirements. For a more detailed description, please refer to the Git4Voc paper [6] and referenced works (i.e. [4,8,11,13,14,17,20]).

Collaboration Support: The first category contains requirements that ease collaboration in distributed settings. *R1 Governance:* Stakeholders with

[3] See https://github.com/vocol/mobivoc and https://github.com/vocol/scor.

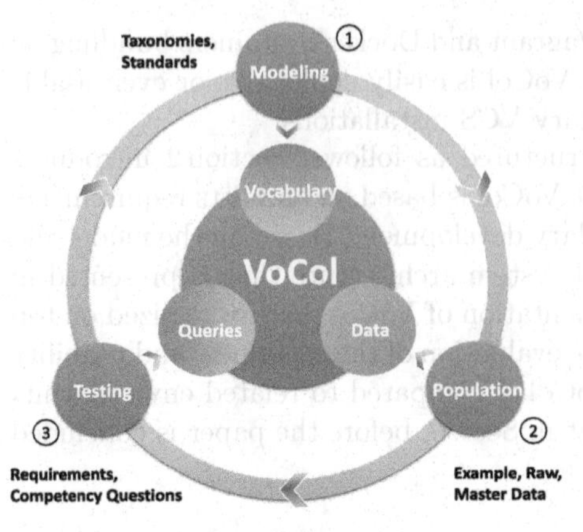

(a) Round-trip model

1. Collaboration Support:
R1 Governance (roles, permissions, etc.)
R2 Communication (issue tracking, notifications, etc.)
R3 Provenance (revision history, semantic diffs, etc.)

2. Quality Assurance:
R4a Syntactic Validation (RDF/OWL compliance, etc.)
R4b Semantic Validation (consistency checking, etc.)
R5 Testing (competency questions, etc.)

3. User Experience:
R6 Documentation (e.g., generated HTML)
R7 Visualization (e.g., node-link diagram)
R8 Editor agnostic (serializations, normalizations, etc.)

4. Vocabulary Deployment:
R9 Machine accessibility (content negotiation, etc.)
R10 Internationalization (multilinguality, etc.)
R11 Querying (e.g., SPARQL endpoint)

(b) Categories and requirements

Fig. 1. (a) Round-trip vocabulary development supported by VoCol; (b) categories and requirements to be addressed by an integrated vocabulary development environment.

different backgrounds and levels of expertise are involved in vocabulary development. Consequently, the definition of roles and permissions is an important requirement [14,20]. *R2 Communication:* The collaborative development of vocabularies is about finding consensus among the different stakeholders. It is essential that they share ideas, make agreements, and discuss issues during the entire development life cycle [13,14]. *R3 Provenance:* It is also crucial to track changes made by the contributors [14]. Each change in the vocabulary reflects the understanding of the domain by the respective stakeholder. In case of disagreements, it is necessary to know which change has been made by whom at which time and for what reason. Furthermore, the development of vocabularies should respond to the evolution of the knowledge domain [20]. Hence, support for detecting and documenting provenance of information and semantic differences between versions is needed during the entire development process.

Quality Assurance: This category comprises requirements for the systematic checking of quality criteria that should be fulfilled by the vocabulary. *R4 Syntax, Semantic, and Constraint Validation:* Syntactic and semantic correctness as well as the application of best practices on designing vocabularies are relevant quality aspects. Providing tool support for these aspects is essential to help contributors in making fewer errors and ultimately increasing the quality of the vocabulary. *R5 Testing:* Competency Questions, i.e., questions the vocabulary must be able to answer, can be translated into queries and used as test cases for the vocabulary [17]. An integrated vocabulary development environment should provide means that allow users to execute such queries efficiently.

User Experience: This category groups requirements for enabling contributors to achieve their objectives effectively and in a user-centered manner. *R6 Documentation:* Domain experts are often team members with little technical expertise in knowledge representation and engineering tools. Thus, presenting the current state of the vocabulary in a human-friendly way is vital. *R7 Visualization:* Visualization is known to have a positive impact on the modeling, exploration, verification, and sense-making of vocabularies [11]. It is particularly helpful for domain experts, but can also provide useful insights for knowledge engineers. *R8 Editor agnostic:* In contrast to software code, vocabularies are conceptual artifacts that can be serialized in different ways. Since contributors can use various editors, which style the syntax differently, support for collaborative vocabulary development should be editor-agnostic and syntax-independent.

Vocabulary Deployment: Finally, there are requirements concerning the deployment of the developed vocabulary that also need to be taken into account by an integrated environment. *R9 Machine accessibility:* An important requirement towards realizing the vision of the web as a global information space is to provide details about the vocabulary terms in a representation that meets the requested type and format [8], thus enabling machines to process the vocabulary correctly. *R10 Internationalization:* The internationalization and localization of vocabularies should also be supported by the environment. The translation of terms into other languages enables a vocabulary to be applicable in different cultures and communities [4]. *R11 Querying:* In order to check whether the developed vocabulary is suitable for a certain use case and to easily retrieve information for a specific task, the environment should support the execution of user-defined queries.

3 System Architecture

In order to implement VoCol as an integrated environment, we developed the system architecture illustrated in Fig. 2. It follows the principles of *Component Based Software Development* (CBSD) [9], which promotes the reuse of components to develop large-scale systems. In other words, it advocates selecting the appropriate *off-the-shelf* components and assembling them into a well-defined software architecture. Following this idea, we composed VoCol from a set of smaller components according to the functionalities they provide. Each of these components is exchangeable and can be replaced by alternatives. In the following, the components are described in detail.

Version Control System: A VCS component is required for the *management of vocabulary changes*. By capturing and storing the changes, various revisions of the vocabulary are created. Contributors should work collaboratively, at best without the need of sharing a common network or the necessity of being always online. In addition, conflicts inevitably arise in environments where multiple contributors are working simultaneously and changing vocabulary terms. The VCS ensures conflict resolution and allows the integration of conflicting changes in an effective and easy way.

Since the VCS is the first component that is aware of changes, we declared it to be the core component of the overall VoCol system. Each additional component that is necessary to support vocabulary development is triggered by the

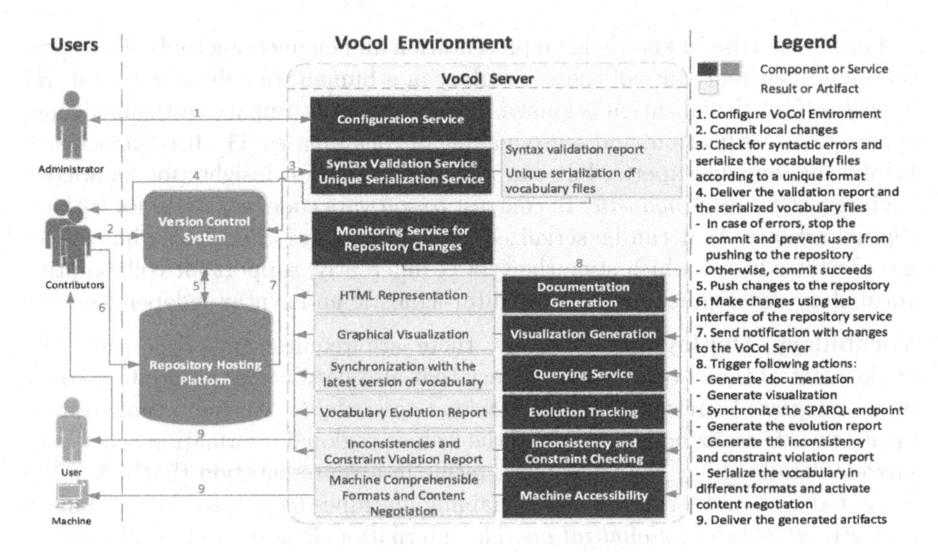

Fig. 2. VoCol architecture and workflow

VCS. We also integrated a repository hosting platform into the VoCol environment (cf. Fig. 2), as it provides low-threshold access to the repository. It acts as the repository storage where the vocabulary files are saved and accessed. Its feature for *Access Control* authenticates users and outputs a permit or a deny message according to the set permissions. Furthermore, using the *Issue Tracker* of the repository hosting platform, contributors are able to discuss the vocabulary by proposing new terms or alternatives for existing ones. In cases where sensitive information should be transmitted, the repository hosting platform can deliver *email notifications* to private user accounts.

Syntax Validation: To ensure that the latest revision of the vocabulary in the VCS is always syntactically correct, VoCol integrates a syntax validation component. In principle, syntax validation could be executed at different stages of the overall workflow. However, with the aim to keep the requirements on the client side at a minimum level, we integrated the syntax validation as a service in the backend. It rejects syntactically incorrect commits and provides a detailed error report in those cases.

Unique Serialization Service: In a distributed environment, contributors use different editors during the development process which may produce different structures of vocabulary files. To avoid this problem, a service integrated into VoCol creates a unique serialization of vocabulary terms before the changes are pushed to the remote repository [7]. Thus, the VCS is prevented from indicating *false-positive* conflicts.

Documentation Generation: A documentation generation service creates an HTML representation of the vocabulary. This permits contributors to easily navigate through the vocabulary by providing a human-friendly overview of it.

Visualization Generation: The integrated visualization component depicts the vocabulary terms and their connections in a graphical way, and allows for the interaction with the visualization. It complements the generated documentation by particularly representing the structure, distribution, and relationships within the vocabulary.

Evolution Tracking: The VCS takes care of maintaining the revision history of the files. To detect *semantic* differences between vocabulary versions, an *evolution tracking* service is integrated into VoCol. It shows which classes and properties have been added, removed, or modified, enabling users to see the vocabulary evolution over time.

Querying Service: VoCol integrates a SPARQL endpoint synchronized with the latest version of the vocabulary. During testing, queries derived from competency questions [17] can be used to verify whether the vocabulary fulfills the domain requirements. These queries are stored in the repository and are preloaded in the query user interface.

Inconsistency and Constraint Checking: After the changes have been pushed to the remote repository, validations of semantic inconsistencies and constraint violation are performed. As a result, two reports with detailed information on respective findings are generated and can be used for corrections.

Machine Accessibility: Using *content negotiation* and dereferenceable URIs, VoCol delivers various machine-comprehensible representations. By specifying the content type in the HTTP header along with the resource URI, the vocabulary can be accessed by different software agents compliant with Linked Data principles.

Monitoring Service: Repository hosting platforms typically expose most of their functionality via web service APIs, so that it can be controlled programmatically. Any change to the repository is delivered as a payload event to a monitoring service listening on VoCol. As a consequence, the services for documentation generation, visualization, evolution tracking, querying, etc. are automatically invoked.

Configuration Service: This service provides a *graphical user interface* to facilitate the configuration of VoCol. The system administrator can choose from various tools for syntax validation and documentation generation. Furthermore, the other services can be activated or deactivated simply by selecting the corresponding checkboxes.

4 Implementation

We use the VCS *Git* at the core of the implementation, together with a set of integrated components providing functionalities for syntax validation, visualization, documentation and evolution report generation, querying, etc.[4]

[4] A live demo of VoCol is available at http://vocol.visualdataweb.org.

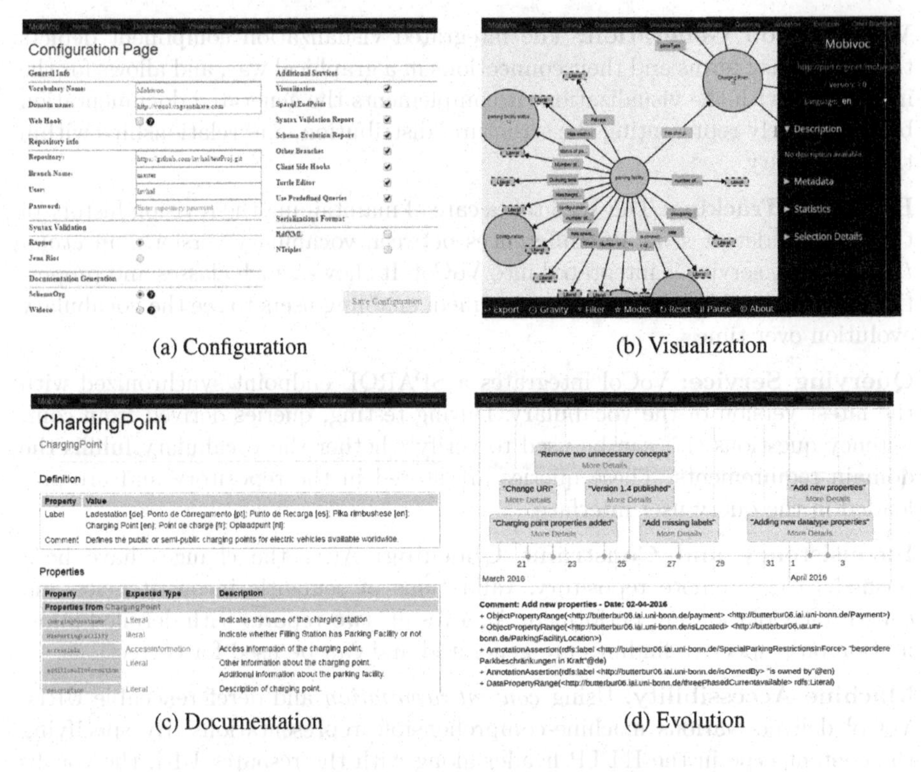

(a) Configuration (b) Visualization

(c) Documentation (d) Evolution

Fig. 3. Screenshots of selected VoCol services

4.1 Configuration

We developed a service that allows the utilization of VoCol for different application scenarios. Using this service, the system administrator configures VoCol by entering the details of the vocabulary repository (i.e., repository URL, user credentials, etc.) in the graphical user interface (cf. Fig. 3a). Next, different tools can be chosen for syntax validation and documentation generation. Via checkboxes, services for visualization, evolution report generation, querying, etc. can be selected for automatic execution from VoCol. The administrator defines the main branch of the repository by entering the value in the *Branch Name* field. For this branch, all selected services will be provided by VoCol. If the option *Monitor Other Branches* is chosen, some of the services are performed on the other branches of the repository too.

Furthermore, the option *Turtle Editor* can be selected to integrate a tool for the online editing of Turtle files into the vocabulary repository [16]. The option *Predefined Queries* indicates that queries defined in files with the extension *.rq* will automatically be loaded into the SPARQL interface. Finally, all serialization formats that VoCol should support via content negotiation can be selected.

VoCol detects the used repository hosting platform (GitHub, BitBucket, etc.) based on the URL entered for the vocabulary repository, and accesses the platform's API to create a *webhook*. This hook contains the address of the VoCol server to which the repository hosting platform will henceforth send information about any push event.

4.2 Client-Side Tasks

Client-side tasks refer to the tasks that are performed before pushing to the repository. To reduce the efforts needed for subsequent corrections, VoCol validates the syntax before pushing the changed files to the repository. An adapted *pre-commit* hook posts vocabulary files that have been changed with tools like *Protégé* or *TopBraid Composer*[5] from the local user repository to the VoCol server. First, the server validates the vocabulary files for syntactic errors. If the validation fails, the user receives a detailed error description, including the file name, the affected lines in the files, and the type of error. If the syntax validation succeeds, a unique serialization of the vocabulary files is created using the SerVCS service [7] we developed on top of the RDF serialization tool Rdf-toolkit[6]. As a result, the vocabulary elements will be serialized in an alphabetic order, which reduces the number of false-positive conflicts indicated by the VCS during the merging process. Additionally, the integrated *TurtleEditor* [16] can be used to edit the vocabulary files directly on the repository hosting platform. Following the idea of a *just-in-time debugger*, this editor implements an instant validator that immediately reports on all found syntax errors. Furthermore, it provides auto-completion of vocabulary terms according to the declared namespaces.

4.3 Server-Side Tasks

Server-side tasks refer to tasks related to the validation and publication of artifacts in human and machine-comprehensible formats that are performed after a Git push event.

Triggering Changes on the Repository: Using the *PubSubHubbub* protocol[7], on each push event, the repository hosting platform delivers a payload with information about the last commit to a server subscribed to it. The *Monitoring Service* implemented in VoCol receives the payload and pulls the vocabulary from the remote repository.

Validation and Error Reporting: Next, the *Syntax Validation* service validates each file for syntax errors using tools like *Rapper* or *Jena Riot*[8]. This task is rerun on the server side to avoid further processing of vocabularies with syntax errors, which can happen if users do not validate the syntax on their commit. If the validation fails, an HTML document is created with detailed information about the errors.

[5] http://protege.stanford.edu, http://www.topquadrant.com/composer/.
[6] https://github.com/edmcouncil/rdf-toolkit.
[7] https://pubsubhubbub.appspot.com.
[8] http://librdf.org/raptor/, https://jena.apache.org/documentation/io/.

Publishing the Artifacts for Humans and Machines: If the syntax validation process is passed successfully, all vocabulary files are merged into a single file. After that, the following tasks are performed automatically; they generate updated artifacts for the evolution report, documentation, and visualization.

Documentation Generation: A human-friendly documentation of the vocabulary is generated using tools such as the documentation generator of Schema.org or *Widoco*[9].

1. Using Schema.org: We developed an HTML generator that creates an RDFa representation for each element of the vocabulary. Next, the content is rendered by the Schema.org tool as one page per resource, which makes the elements dereferenceable. An example of an HTML page generated for a vocabulary term (*ChargingPoint*) is shown in Fig. 3c.
2. Using *Widoco:* A single HTML page listing all elements of the vocabulary is generated by Widoco. This provides the user with a complete overview of the vocabulary that can be easily navigated and searched.

Visualization Generation: The vocabulary is visualized using the web application *WebVOWL* [10]. WebVOWL implements the Visual Notation for OWL Ontologies (VOWL) by graphically representing the vocabulary terms and their relations in a dynamic node-link diagram. An excerpt of a generated visualization is shown in Fig. 3b.

Evolution Tracking: When semantic differences between versions of the vocabulary exist, an evolution report is generated using the tool Owl2vcs [25]. It uses algorithms for structural diffs and three-way merge tools along with OWL 2 direct semantics. The application of direct semantics eliminates problems with blank nodes and allows comparing ontologies axiom by axiom. The report contains each point in time when a new vocabulary revision has been pushed, and lists semantic changes like the *addition*, *removal*, or *modification* of elements, as shown in Fig. 3d.

Machine Accessibility: Machine-comprehensible formats of the vocabulary, such as *Turtle* and *RDF/XML* produced by Rapper, are delivered through a web server configured to perform *content negotiation* according to the best practices for publishing vocabularies[10]. As a result, machines are provided with the latest version of the vocabulary at any time.

Querying Service: An integrated SPARQL endpoint service using *Jena Fuseki* allows performing queries and exporting the results in different formats. This enables users to test whether the vocabulary meets their requirements. Additionally, it checks for the existence of files with the extension *.rq* defining queries. All files found are uploaded to this service by taking the file name as the query name, and the content of the file as the query. Furthermore, we developed a tool that automatically executes queries for constraint violation checking. Some examples

[9] https://github.com/schemaorg/schemaorg/, https://github.com/dgarijo/Widoco.
[10] http://www.w3.org/TR/swbp-vocab-pub/.

Table 1. Examples of predefined queries for constraint checking.

Query name	Expected value	Required
At least one *owl:Ontology* needs to be defined	isNotEmpty	Mandatory
Two resources should not have the same *rdfs:label*	isEmpty	Mandatory
Two resources should not have the same *rdfs:comment*	isEmpty	Mandatory
All resources should have *rdfs:label* and *rdfs:comment* in English	isEmpty	Optional
All resources must not have literals with "foo bar", "lorem" or "ipsum"	isEmpty	Mandatory
All resources should have *rdfs:label* different from *rdfs:comment*	isEmpty	Optional
All resources should have *rdfs:comment* in different languages	isEmpty	Optional
All skos:Concepts should be *skos:inScheme*	isEmpty	Mandatory
All skos:Concepts should have a *skos:broader* statement	isEmpty	Optional

of these predefined queries, that can be easily changed or extended, are listed in Table 1. Whenever the value that is returned after executing the corresponding SPARQL query does not match the value in the "Expected Value" column, this is an indication for constraint violation. The results of this validation process is reported in HTML format. Listing 1 depicts the SPARQL query that checks for missing *rdfs:label* and *rdfs:comment* in English.

```
1 SELECT DISTINCT ?r WHERE { ?r rdf:type ?type .
2         MINUS { ?r rdf:type skos:Concept. }
3         MINUS { ?r rdf:type skos:ConceptScheme. }
4         OPTIONAL { ?r rdfs:label ?label .
5           FILTER( (STRLEN(?label) > 0) && langMatches( lang(?label),'en' ))}
6         OPTIONAL { ?r rdfs:comment ?comment .
7           FILTER((STRLEN(?comment) > 0) && langMatches( lang(?comment),'en'))}
8           FILTER ( !bound(?label) || !bound(?comment) )
9     } ORDER BY ?r
```

Listing 1. Resources should have at least one English *rdfs:label* or *rdfs:comment.*

4.4 Deployment

We deploy the VoCol implementation as VirtualBox and Docker virtual machine images, which can be installed with little effort. The VoCol environment thus works as an isolated solution without affecting the rest of the physical machine. This ensures high portability, allowing the administrator to easily start, stop, move, or share it. With a few additional steps, the VoCol environment can be installed and configured on a clean web server. All implementation details are available on the VoCol website[11].

[11] http://vocol.visualdataweb.org.

5 Evaluation

We are currently applying VoCol in an industrial use case to evaluate its usefulness and effectiveness in a real-world setting. Furthermore, we conducted a qualitative user study to get additional insights into the usefulness and usability of VoCol.

5.1 Industry Application

VoCol is currently applied in an industrial use case to develop vocabularies for describing formally the assets of an enterprise, including how they relate to each other. All of these vocabularies, except the developed *rami* vocabulary[12], are the intellectual property of the industrial partner. We are restricted in the information we can provide here, but would like to share at least some experiences and insights.

A group of seven people contributed in parallel to the development of the vocabularies. While the knowledge engineers conducted most of the formalization, the domain experts participated by creating issues. In total, 46 issues were created ranging from proposals to add, modify, or remove vocabulary terms to VoCol environment issues, such as bug fixes and feature requests. The developed vocabularies currently comprise 151 classes, 93 object properties, 225 datatype properties, and 79 instances.

The loose coupling characteristic of VoCol allowed us to integrate a new component for defining and establishing R2RML mappings between the developed vocabularies and legacy data sources of the industrial partner. By doing so, users were able to execute queries against the legacy systems and receive the results in various representation formats, such as tabular, pie, and bar charts, etc.

VoCol provides very useful and effective support in this use case according to the informal feedback of the involved stakeholders. In particular, the different views on the vocabulary provided by VoCol are considered to be very helpful in getting a better understanding and exploring the state of the art. The easy and comfortable access to all services via one integrated web interface was praised by all stakeholders.

Despite the benefits of VoCol for this use case, one of the drawbacks that we experienced is the lack of a simple form-based editing of vocabulary terms. This prevented domain experts from contributing their ideas directly to the development, but required the continuous involvement of knowledge engineers.

5.2 User Study

We conducted a qualitative user study of VoCol under controlled conditions using the *Concurrent Think Aloud* (CTA) method: Participants were observed and asked to verbalize their thoughts while performing the given tasks [18]. At the beginning of each session, the interviewer gave a general introduction into

[12] http://w3id.org//i40/rami.

VoCol. The interaction with the system as well as comments and suggestions were recorded for later analysis. After completing the tasks, participants had a discussion with the interviewer about their experiences and any difficulties they faced. To measure the usability and ease of use, participants were asked to fill a questionnaire at the end of the interview.

Participants: To ensure that participants represent as closely as possible the targeted user group of the VoCol system, we chose twelve users with different levels of expertise, ranging from basic vocabulary modeling experience to more advanced expertise in knowledge conceptualization and representation.

Tasks and Questionnaire: We designed a set of tasks that comprised all activities of the round-trip development described above (cf. Sect. 2): Starting from the modeling activity, the first task was to define several classes with various numbers of properties. The next task was concerned with the population of the vocabulary, in which users had to create instances based on the defined classes. The tasks were performed on the user machine by committing all changes to the local repository first and later pushing those changes to the remote repository. The SPARQL endpoint was used to execute test queries verifying whether the developed vocabulary met certain criteria. All functionalities provided by VoCol, including the syntax validation before commit and after push events to the remote repository, documentation generation, visualization, etc. were covered in the user study.

In addition, we asked the participants to fill an electronic post-study questionnaire composed of two main sections. The first section contained the USE Questionnaire[13], which uses five-point Likert scales for rating, ranging from 1 (strongly disagree) to 5 (strongly agree). We evaluated four usability dimensions: (1) usefulness; (2) ease of use; (3) ease of learning; and (4) satisfaction. To get more insights into specific areas, we defined three additional questions in the second section of the questionnaire. With these questions, we aimed to get the participants' opinion about: (1) the importance of the individual services integrated into VoCol; (2) negative and positive aspects of the system through an open response question; and (3) possible services to be integrated in the future. The evaluation material is available online.[14]

Results: We obtained the evaluation results by observation, discussions at the end of each session, and the post-study questionnaires. The following are some of the findings that we derived from the analysis of the observation notes and discussions:

- Participants with prior knowledge about VCS, especially with Git, found VoCol very easy to learn and use.
- A few participants expected to see provenance metadata in the browser, i.e., the date and author for each term added to the vocabulary.
- The instant syntax-checking and auto-completion feature of the TurtleEditor was considered very helpful by the majority of the participants.

[13] http://hcibib.org/perlman/question.cgi?form=USE.

[14] https://figshare.com/articles/VoCol_Evaluation_Material/3438371.

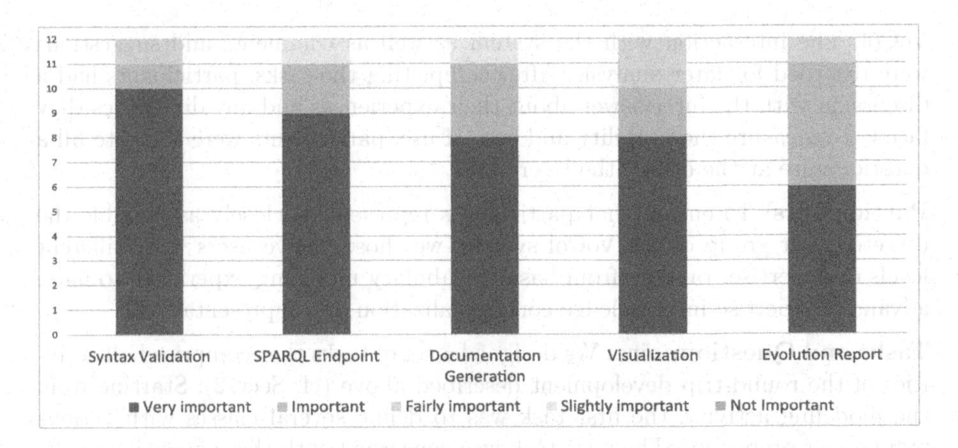

Fig. 4. Importance of the VoCol services according to the study participants.

The results from the USE questionnaire showed that the responders rated their experience with VoCol very high. The average scores received by each dimension are as follows: *usefulness = 4.34*, *ease of use = 3.97*, *ease of learning = 4.35*, and *satisfaction = 4.31*. These scores indicate a high usability of VoCol (nearly all scores are > 4) and correlate with the oral feedback of the participants that VoCol is "easy to learn and use", as well as the informal feedback of the stakeholders form the industrial use case that VoCol provides "very useful and effective support".

Figure 4 shows that each of the services provided by VoCol is of high relevance to the study participants. For instance, 10 of the 12 participants consider *syntax validation* a *very important* service, while the scores for the other services are only slightly lower. Some interesting suggestions made by the participants were: (1) creating a possibility for dynamically adding and removing tools from the user interface; and (2) automatic recommendation of similar vocabularies (e.g., using the LOV[15]API).

6 Related Work

Vocabulary development is an active research topic in the Semantic Web community [15]. One area of research is concerned with the development of web applications that offer low-barrier access to vocabulary development. A well-known approach in this area is *WebProtégé* [22], which is a lightweight version of the Protégé desktop editor. It offers change tracking and collaboration features to support the distributed development of vocabularies, and comes with a customizable user interface that can be adapted to different expertise levels of the users. *VocBench* [21] is a web application targeted at editing SKOS and SKOS-XL thesauri. It supports the workflow management, validation, and publication of vocabularies, and provides a full history of changes as well as a SPARQL query

[15] http://lov.okfn.org/dataset/lov/vocabs.

endpoint. VocBench implements the separation of responsibilities through a role-based access control mechanism, checking user privileges for the different tasks of thesauri editing. *SOBOLEO* [24] also fosters the collaborative editing of SKOS thesauri. It provides a specialized browser to navigate and change the taxonomy and a semantic search engine for annotating web resources. SOBOLEO is used in the domain of social networks and offers tag recommendations for describing people based on existing vocabularies. *TopBraid Enterprise Vocabulary Net* (TopBraid EVN)[16] is a proprietary tool to ease the collaborative creation of SKOS taxonomies and ontologies. It incorporates change audits, role management, concept search capabilities as well as data quality rules to check SKOS and OWL constraints. Moreover, it enables the creation of hierarchy reports through graphical user interfaces. *MoKi* [2] is a collaborative MediaWiki-based tool to support the ontological modeling tailored for business processes. MoKi associates a wiki page, containing both unstructured and structured information, to each entity of the ontology and process model. *PoolParty* [19] provides a web interface for building and managing SKOS thesauri. A user-friendly GUI facilitates the participation of domain experts. It also allows to extract relevant information from external Linked Data sources. *TemaTres*[17] is a web application optimized for SKOS thesauri. It includes an API to access the latest version of the vocabulary, a WYSIWYG editor, and extensive quality assurance support.

The main objective of all these tools is to support the collaborative web-based editing of vocabularies. Although they contain many interesting features for vocabulary development, they are not focused on reusing existing VCSs as a core component of the vocabulary development process. More closely related to VoCol are approaches that aim to extend VCSs with additional features dedicated to vocabulary development.

SVoNt [12] proposes to use Apache Subversion (SVN) as a VCS for the versioning of ontologies. SVoNt uses a separate server to store conceptual changes between different versions of ontologies. These versions are generated as a result of a *diff* operation between the modified and base ontology. SVoNt supports conflict detection and resolution by comparing the structure and semantics of the ontologies. *Ontoology* [1] is a tool for vocabulary development based on Git, similar to the presented VoCol implementation. It generates a documentation using *Widoco*, while an ontology pitfalls report is provided based on the *OOPS* service[18]. Ontoology uses AR2DTool[19] for creating class and taxonomy diagrams. The generated artifacts can become part of the repository after a *pull request* is performed. However, providing a user-friendly client which hides the complexity of the version control system is not in the focus of these works. Thus, these systems are rather suited for ontology development projects that involve purely users with a strong technical background. Furthermore, they do not provide a set of services that is as comprehensive and integrated as that of VoCol.

[16] http://www.topquadrant.com/products/topbraid-enterprise-vocabulary-net/.
[17] http://www.vocabularyserver.com.
[18] http://oops.linkeddata.es.
[19] https://github.com/idafensp/ar2dtool.

7 Conclusions and Future Work

We have presented VoCol, an integrated environment for distributed development of vocabularies based on version control systems. We have defined *distributed vocabulary development* as the process of identifying the main terms and concepts among the involved stakeholders and finding a consensus between them. We argue that the development of an effective and efficient environment for distributed collaboration is the main challenge in this context. The presented VoCol environment supports the identified requirements by extending the functionality of plain version control systems with external tools via the webhook mechanism.

We implemented VoCol on the basis of the widely used VCS Git. Tasks such as content negotiation, documentation and visualization generation, as well as evolution tracking are performed in a fully automated way. In addition, a querying service, synchronized with the latest version of the vocabulary, enables users to execute SPARQL queries. The VoCol environment is easily expandable with other tools to provide additional functionalities. The current implementation of VoCol is tailored to small to medium size vocabularies. However, it can be adjusted for various scenarios by replacing its components with adequate alternatives.

For future work, we plan to implement VoCol also for other VCSs, such as Subversion and Mercurial. Furthermore, we envision an automatic population service that creates data according to the defined terminology of the vocabulary. Finally, we plan to provide VoCol as a service where users can simply subscribe their repositories and benefit from all functionalities.

Acknowledgments. This work has been supported by the German Federal Ministry of Education and Research (BMBF) in the context of the projects *LUCID* (grant no. 01IS14019C), *SDI-X* (no. 01IS15035C) and *Industrial Data Space* (no. 01IS15054).

References

1. Alobaid, A., Garijo, D., Poveda-Villalón, M., Santana-Perez, I., Corcho, Ó.: Ontoology, a tool for collaborative development of ontologies. In: ICBO 2015, CEUR-WS, vol. 1515 (2015)
2. Ghidini, C., Rospocher, M., Serafini, L.: Moki: a Wiki-based conceptual modeling tool. In: ISWC 2010 Posters and Demos, CEUR-WS, vol. 658 (2010)
3. Giuliano, C., Gliozzo, A.M.: Instance-based ontology population exploiting named-entity substitution. In: COLING 2008, ACL, pp. 265–272 (2008)
4. Gracia, J., Montiel-Ponsoda, E., Cimiano, P., Gómez-Pérez, A., Buitelaar, P., McCrae, J.: Challenges for the multilingual web of data. J. Web Semant. **11**, 63–71 (2012)
5. Grüninger, M., Fox, M.S.: Methodology for the design and evaluation of ontologies. In: IJCAI95 Workshop on Basic Ontological Issues in Knowledge Sharing (1995)
6. Halilaj, L., Grangel-González, I., Coskun, G., Lohmann, S., Auer, S.: Git4Voc: collaborative vocabulary development based on git. Int. J. Semant. Comput. **10**(2), 167–192 (2016)
7. Halilaj, L., Grangel-González, I., Vidal, M.E., Lohmann, S., Auer, S.: Proactive prevention of false-positive conflicts in distributed ontology development. In: IC3K 2016, to appear

8. Heath, T., Bizer, C.: Linked data: evolving the web into a global data space. Synth. Lect. Semant. Web: Theor. Technol. **1**(1), 1–136 (2011)
9. Kaur, A., Mann, K.S.: Component based software engineering. Int. J. Comput. Appl. **2**(1), 105–108 (2010)
10. Lohmann, S., Link, V., Marbach, E., Negru, S.: WebVOWL: web-based visualization of ontologies. In: Lambrix, P., Hyvönen, E., Blomqvist, E., Presutti, V., Qi, G., Sattler, U., Ding, Y., Ghidini, C. (eds.) EKAW 2014. LNCS (LNAI), vol. 8982, pp. 154–158. Springer, Heidelberg (2015). doi:10.1007/978-3-319-17966-7_21
11. Lohmann, S., Negru, S., Haag, F., Ertl, T.: Visualizing ontologies with VOWL. Semant. Web **7**(4), 399–419 (2016)
12. Luczak-Rösch, M., Coskun, G., Paschke, A., Rothe, M., Tolksdorf, R.: SVoNt: version control of OWL ontologies on the concept level. In: AST 2010, GI, pp. 79–84 (2010)
13. Noy, N.F., Chugh, A., Liu, W., Musen, M.A.: A framework for ontology evolution in collaborative environments. In: Cruz, I., Decker, S., Allemang, D., Preist, C., Schwabe, D., Mika, P., Uschold, M., Aroyo, L.M. (eds.) ISWC 2006. LNCS, vol. 4273, pp. 544–558. Springer, Heidelberg (2006)
14. Noy, N.F., Tudorache, T.: Collaborative ontology development on the semantic web. In: AAAI Spring Symposium: Semantic Web and Knowledge Engineering, pp. 63–68 (2008)
15. Palma, R., Corcho, O., Gómez-Pérez, A., Haase, P.: A holistic approach to collaborative ontology development based on change management. J. Web Semant. **9**(3), 299–314 (2011)
16. Petersen, N., Coskun, G., Lange, C.: TurtleEditor: an ontology-aware web-editor for collaborative ontology development. In: ICSC 2016, pp. 183–186. IEEE (2016)
17. Ren, Y., Parvizi, A., Mellish, C., Pan, J.Z., van Deemter, K., Stevens, R.: Towards competency question-driven ontology authoring. In: Presutti, V., d'Amato, C., Gandon, F., d'Aquin, M., Staab, S., Tordai, A. (eds.) ESWC 2014. LNCS, vol. 8465, pp. 752–767. Springer, Heidelberg (2014)
18. Russo, J., Johnson, E., Stephens, D.L.: The validity of verbal protocols. Mem. Cogn. **17**, 759–769 (1989)
19. Schandl, T., Blumauer, A.: PoolParty: SKOS thesaurus management utilizing linked data. In: Aroyo, L., Antoniou, G., Hyvönen, E., ten Teije, A., Stuckenschmidt, H., Cabral, L., Tudorache, T. (eds.) ESWC 2010, Part II. LNCS, vol. 6089, pp. 421–425. Springer, Heidelberg (2010)
20. Simperl, E., Luczak-Rösch, M.: Collaborative ontology engineering: a survey. Knowl. Eng. Rev. **29**(01), 101–131 (2014)
21. Stellato, A., Rajbhandari, S., Turbati, A., Fiorelli, M., Caracciolo, C., Lorenzetti, T., Keizer, J., Pazienza, M.T.: VocBench: a web application for collaborative development of multilingual thesauri. In: Gandon, F., Sabou, M., Sack, H., d'Amato, C., Cudré-Mauroux, P., Zimmermann, A. (eds.) ESWC 2015. LNCS, vol. 9088, pp. 38–53. Springer, Heidelberg (2015)
22. Tudorache, T., Nyulas, C., Noy, N.F., Musen, M.A.: WebProtégé: a collaborative ontology editor and knowledge acquisition tool for the web. Semant. Web **4**(1), 89–99 (2013)
23. W3C: Vocabularies (2015). https://www.w3.org/standards/semanticweb/ontology
24. Zacharias, V., Braun, S.: Soboleo - social bookmarking and lighweight engineering of ontologies. In: CKC Workshop at WWW 2007 (2007)
25. Zaikin, I., Tuzovsky, A.: Owl2vcs: Tools for distributed ontology development. In: OWLED 2013, CEUR-WS, vol. 1080 (2013)

Event-Based Recognition of Lived Experiences in User Reviews

Ehab Hassan[1(✉)], Davide Buscaldi[1], and Aldo Gangemi[1,2]

[1] LIPN, Université Paris XIII, CNRS UMR 7030, 93430 Villetaneuse, France
{ehab.hassan,davide.buscaldi,aldo.gangemi}@lipn.univ-paris13.fr
[2] STLab, ISTC-CNR, Rome, Italy

Abstract. User reviews on the web are an important source of opinions on products and services. For a popular product or service, the number of reviews can be large. Therefore, it may be difficult for a potential customer to read all of them and make a decision. We hypothesize and test if lived experiences from reviews may support the confidence of a user in a review. We identify and extract such lived experiences with a novel technique based on machine reading. Our experimental results demonstrate the effectiveness of the technique.

Keywords: Lived experiences extraction · Event extraction · Machine reading · Semantic web · User reviews

1 Introduction

The web has significantly changed how people express themselves and interact with others. Now they can post reviews of products and services in merchant websites, as well as they express their viewpoints and interact with others through blogs and forums. It is now well agreed that user generated content contains valuable information that can be used for real word applications (e-commerce, politics, finance, etc.). As e-commerce is becoming more and more popular, the number of user reviews for a specific product or service may be in hundreds or even thousands. Because of this, taking a decision about commercial offers from a large amount of data on the Web becomes difficult, and takes a lot of time.

Automated solutions for this problem, either from recommender systems [18], commonly exploiting collaborative or content-based filtering, or from user-based ranking systems, have known limitation including provenance assessment, spam detection, and genericity.

Reviews offer (often implicitly) suggestions or opinions on the basis of lived experiences. These reviews are very important in user decisions since they contain non-fictional narrative or stories that people tell about their experience with a product or service. Users can rely on these reviews to project themselves as a potential future consumer, compare their desires and requirements to those of other customers, and make a decision quickly. Lived experiences can give them specific and more interesting information than general opinions, and

E. Blomqvist et al. (Eds.): EKAW 2016, LNAI 10024, pp. 320–336, 2016.
DOI: 10.1007/978-3-319-49004-5_21

provide a larger palette of perspectives than traditional sentiment analysis, since experiences differ among users and hint at their own preferences and reasons for judgment.

In our research, we aim to discover authentic lived experiences in user reviews by hybridizing methods from automated event extraction from text, and the Semantic Web.

In this paper, we describe our efforts to identify, extract, and represent authentic lived experiences extracted from user reviews. Given a set of customer reviews of a product or a service, the task involves four subtasks: (1) identifying reviews which contain lived experiences; (2) for each pertinent review, identifying and extracting the relevant events with their participants, which indicate lived experience contents; (3) representing lived experiences in each review as an event sub-graph containing relevant events and their participants; (4) providing an API to get the extracted subgraphs as linked data.

Section 2 discusses our operational notion of *lived experience.* Section 3 summarizes related work. Section 4 presents the extraction and selection techniques, and introduces a web application system, called LEE (*Lived Experience Extraction*)[1]. Section 5 describes the evaluation performed.

2 What Is a Lived Experience?

We operationally define a *lived user experience* (for the sake of our experiment) as an event mentioned in a review, where the author is among the participants. In addition, we separate lived experiences from generic user opinions, e.g. a situation involving the author giving an opinion about a service or facility is not considered as a lived experience. In other words, we want to detect events in which the author of a review is doing something together with anything associated with a product or service, separating this "quasi-objective" reporting of factuality from any judgment of it. This choice stands on the hypothesis that fake reviews tend to contain opinions that are not associated with actual events. Our hypothesis is supported by spam detection results on so-called *defaming spam* [15]. As an example, let us consider the two following hotel reviews:

1. The view from this hotel's rooms is quite stunning. And that's what make it very special, possibly better than the next door 4 star hotel and than many other hotels in Paris. The bedrooms interior decor is extremely nice. I asked for a room overlooking the pantheon and I got it. My deluxe room was number 32, and was tastefully decorated with a classic and beautiful Pierre Frey wallpaper, and an extra day bed. The bath had bathtub-shower combination and was separated from the toilet. If you book directly through the hotel, you'll get a voucher for a free-drink upon arrival. It was a bit cold at night at some point, maybe because it's March and the heating is not constantly on anymore. Each room has its own heating control, though. Strongly recommended.

[1] https://lipn.univ-paris13.fr/ClientProj/client.jsp.

2. Our stay was absolutely perfect. Its a cool hotel to look at, the design and feel is very trendy and hip. All the staff are terrific, especially the consierge staff-great info and attitude. our room was on the top floor, with great views. Super comfy bed, and neat bathroom. Fantastic, choose this with no hesitation!

According to our definition, the first review contains three lived experiences represented by events where the user is among their participants (1)*I asked for a room overlooking the pantheon and I got it.* (2) *My deluxe room was number 32, and was tastefully decorated with a classic and beautiful Pierre Frey wallpaper, and an extra day bed.* (3) *If you book directly through the hotel, you'll get a voucher for a free-drink upon arrival.* The first lived experience has two event types {*Ask, Get*}, the second and the third have one event type {*Decorate*}, {*Get*}, respectively. All these events have the author as a participant {*I, My, You*}. On the contrary, the other sentences in this review do not represent lived experiences, as they do not contain events *(The bedrooms interior decor is extremely nice)* or they include events, but those events do not have the author as a participant *(The bath had bathtub-shower combination and was separated from the toilet)*.

In the second review, the user writes his opinion in general without telling anything about his lived experiences *(all the staff are terrific, super comfy bed, ...)*. We do not notice any event or action (e.g. what, when, where, how), in which the user was participating. Following our operational definition, we discard this review from the set including lived experiences.

We expect that when choosing a service or product, the decision process assisted by lived experience extraction is quicker and more efficient, since it will be made based on the segments of relevant reviews that mostly stimulate *identification* in the reader. This is supported by results in [5], which show that "*[sites such as] TripAdvisor [...] can involve an "apomediary effect" in which technological features and social identification combine in some circumstances to reduce information to a manageable level*".

Concerning representation, we adopt a neo-Davidsonian modeling of events as modeled in the OWL knowledge graphs extracted by the FRED tool (cf. Sect. 4).

3 Related Work

3.1 Lived Experience Extraction

Lived experiences have been studied mainly in the context of anthropological, historical, and health studies [6,23]. From classical studies in cognitive psychology (e.g. Barsalou) [2], where "autobiographical memories" where immediately associated with computational studies of cognition, the attention moved early to applications in consumer psychology, where a relevant work is [3], which conducted three experiments that showed the importance of personal memories for evaluating either a product or a service, and for understanding the opinion of others. A conclusion was that autobiographical memories involving product or product usage experiences are affectively charged. This finding provides an important clustering factor between event framing, sentiment, and reported facts. In that line of studies, e.g. tourism psychological studies [24] elaborated beyond the affective dimension of lived travel experiences, paying also attention to user expectations, event consequentiality, and memory recollection.

To our knowledge, the only studies devoted specifically to lived experiences from the information extraction perspective concern *personal story extraction* [11,12], and *experience mining* [14].

Personal story extraction defines personal stories as textual discourse that describes a specific series of causally related events in the past, spanning a period of time, where the author or a close associate is among the participants.

Experience mining substantially agrees with personal story extraction in the kind of data and extraction criteria, involving factuality, direct involvement of the author as the experiencer, and the kind of event to be considered.

In this research, we follow these studies, and propose a set of identification criteria in Sect. 4.

3.2 Event Extraction

Event extraction is closely related to our research work. Lived experiences are usually events, since people report about what happened during their experience with a product or service. Previous work in event extraction focused largely on news articles. There are only a few previous works in event extraction from user reviews.

The REES system [1] was evaluated by extracting 100 relation and event types, 61 of which are events, from a news source. An ontology for the relation and events to be extracted in political, financial, business military, and life-related domains was preliminarily defined. REES achieved a 0.70 F-measure. The EVITA event recognition tool [22] has been applied to newswire text. The authors perform identification and tagging of event expression by combining linguistic and statistical methods, and obtain a performance ratio of 0.80 F-measure. The STEP system [4] identifies events (reduced to a classification task) for question answering, with a precision of 0.82 and recall of 0.71. Zhang et al. [26] present a multi-modal framework for semantic event extraction from basketball games. They employ K-means clustering into 9 groups, and list the top 4 rank terms in each group. Van Oorschot et al. [25] extract game events (e.g. goals, fouls) from tweets about football match to automatically generate match summaries. Ritter et al. [21] present an open domain event extraction from Twitter. They propose an approach based on latent variable models to categorize events and classifying extracted events in an open-domain text, with 0.64 F-measure. Due to difference in text size and complexity, this work is hardly comparable to extraction applied to user reviews. Ploeger et al. [19] introduce an automatic event extraction method from various news sources using NLP tools. They extracted 1829 events with 0.71 precision, 0.58 recall, and 0.64 F-Measure.

Unlike most approaches mentioned above, we do not use a predefined list of potentially interesting events, but we use a system of "semantic web machine reading", FRED [10,20], to automatically identify and extract events. A semantic web machine reader is not only open domain and unsupervised, [7] but it also features formal semantics, and attempts to link extracted knowledge to existing data and ontologies. In fact, In a landscape analysis of knowledge extraction tools [8] FRED has got 0.73 precision, 0.93 recall, 0.82 F-measure, and 0.87 accuracy, which is very good for event extraction, specially considering that FRED works in the open domain.

4 Extracting Lived Experiences

Figure 1 is an architectural overview of our lived experiences extraction system. The input to the system is a user review. The output is the event-based lived experiences mentioned in the review.

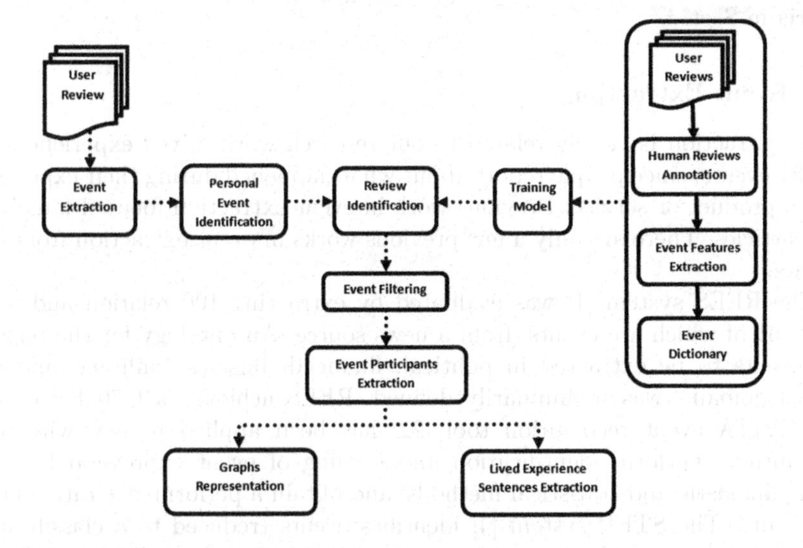

Fig. 1. Event-based Lived Experience Extraction

The LEE system performs lived experiences extraction in three main steps: (1) indicating whether the review contains lived experiences contents or not; (2) if yes, identifying lived experience sentences; (3) representing the results as event sub-graphs. These steps are performed in multiple sub-steps.

Given the input, the system firstly finds event type features which are mentioned in a review. After that, the review is classified as a lived experience review if it contains one or more lived experience content, or not otherwise. For lived experience reviews, the system finds the pertinent events that indicate the main parts of lived experiences. In the last two steps, event participants are identified, and lived experience contents are extracted. Below, we discuss each of the sub-steps.

4.1 Event Extraction

As explained in Sect. 2, we consider events mentioned in user reviews, and have the review author among the participants, as a lived experience. Therefore, a lived experience should involve events and their participants, and the narrator should be among those participants. In this section, we describe our approach to extract events from customer reviews.

Event extraction (EE) can be broadly defined as the creation of specific knowledge concerning facts and situations referred to in some content and/or data: texts, images, video, databases, sensors, etc. In this research, we focus on events expressed by verbs, propositions, common nouns, and named entities (typically proper nouns).

In order to extract events from user reviews, we perform a deep semantic parsing of text which allow to obtain a RDF/OWL knowledge graph representation of the text. We employ a deep variety of machine reading [7], as implemented in the FRED tool[2] [10]. FRED extracts knowledge graphs (formal representation of named entities, senses, taxonomies, relations, events, etc.) from text, resolves it onto Linked Open Data (DBpedia, schema.org, RDF versions of WordNet, FrameNet, VerbNet), and adds data from background knowledge.

Applying a SPARQL query to the semantic graph produced by FRED, we can extract event types with their main participants. From our first example in the introduction, we are able to extract eight event types {*Get, Overlook, Recommend, Ask, Decorate, Have, Make, Separate*}.

4.2 Personal Events Identification

According to our definition in Sect. 2, we assume that events, which have the first or the second person pronoun (i.e. *I, You, We, Me, Us, My, Mine, our, ours, Your, Yours, ...*) as a participant, are the most important elements in lived experiences. Therefore, we keep those events, and use them in the classification task: such events can have the user as a direct participant, i.e. the event has the author as its experiencer. For example, to classify our first review in Sect. 2, we retain 3 event types {*Get, Ask, Decorate*}, which contain the user as a participant, and eliminate 5 event types {*Overlook, Recommend, Have, Make, Separate*}, which do not contain the user among their participants.

4.3 Review Identification

We view lived experience identification as a binary text classification task. We explored the use of machine learning techniques for identifying pertinent reviews, which consist of lived experiences content. This required the development of a set of annotated training examples, where user reviews are assigned "LivedExperience" or "Non-LivedExperience" labels.

[2] http://wit.istc.cnr.it/stlab-tools/fred.

Annotating User Reviews: We follow a traditional text classification approach, in which a corpus of user reviews was hand-annotated (LivedExperience/Non-LivedExperience) to be used as a training and testing set. User reviews would be labeled as "LivedExperience" if an annotator judges that the review contains at least one lived experience.

We did not use any crowdsourcing services such as Amazon's Mechanical Turk (AMT), since in the experiment described in [11], the authors found that the vast majority of annotations produced through the AMT service were either completely random, or were generated by automated response engines, yielding unsuitable results.

We annotated the corpus manually by an expert. Following our definition of what counted as lived experience, we annotated 383 user reviews, assigning the label "LivedExperience" or "Non-LivedExperience" to each. The label "LivedExperience" was assigned to 176 reviews (46 %). To build our classifier, we used 268 reviews as a training set: 134 reviews with the "LivedExperience" class and 134 reviews with the "Non-LivedExperience" class. The remaining 115 reviews (42 lived, 73 non-lived) have been used for the test set.

Development of the Review Classifier: Events that include the author as a participant are used as features to train a Support Vector Machine learning algorithm (SVM). Using our training set, we are able to detect and extract 320 event types: 175 are specific to the reviews containing lived experiences, 50 are specific to non-lived ones, and 95 event types are detected in both classes.

The event types that are homogeneously distributed in the two classes are not useful in our classification task, and have been removed. We considered an event as representative to a class if the probability $P(c|e) \geq \sigma$ where:

$c \in C = \{\text{LivedExperience, Non-LivedExperience}\}$;
e: A generic event;
σ: A threshold that we determined empirically as 0.7 [13].

Applying firstly the classification task on the input reviews allows to perform the extraction process on the reviews, which certainly contain lived experience contents, and to ignore the irrelevant reviews.

4.4 Event Filtering

Many reviews contain either relevant or irrelevant events. We employed a filtering algorithm, which allows to study events by taking into account the impact of their neighbors in the same review to finally select the relevant ones. Event neighbors are supposed to have a big effect on the target event. We considered that two events are neighbors if they are in the same reviews.

The Algorithm 1 takes a review as input, and gives the label (lived for relevant events, non-lived for non-relevant ones) of each event as output.

First, for each target event in each input review, we calculate its distribution in each class of training set reviews used in our classification task. If an event was distributed homogeneously in the two classes, we deleted it from the input review (line 3 to line 9). We considered that the event distribution is homogeneous if it is not higher than 0.7 in one of the two classes of reviews, as empirically established in [13].

Second, we calculate the function $F = \alpha X(\mathbf{e}_j) + \beta Y(\mathbf{e}_j)$ for the remaining events, where α is the value that was weighted to target events, β presents the value that was weighted to event neighbors. We choose α to be equal to the number of event neighbors divided by the total number of lived experience events, and β is 1/the number of lived experience events. The value of α is much larger than the β value since it was weighted to the target event. These values stay the same and do not change for each event in the same review. $X(\mathbf{e}_j)$ represents the bigger distribution value of the target event in the two classes of reviews in our training set. $Y(\mathbf{e}_j)$ is the sum of the bigger distributions values of event neighbors in the two classes of reviews in our training set. We assumed that the distribution value for an event, target or neighbors should be positive if the event is distributed significantly in lived experience reviews from our training set, and negative otherwise.

Finally, the event is classified according to the value of the function F (line 21 to line 24). If F value is higher than 0, the target event represents a lived experience.

Algorithm 1. Event Filtering

Input: $\mathcal{R} = \{R_1, R_2, ..., R_k\}$ such as $R_i = \{\mathbf{e}_1, \mathbf{e}_2, ..., \mathbf{e}_n\}$, α, β
Output: $\mathcal{L} = \{(\mathbf{e}_1, l_1), (\mathbf{e}_2, l_2), ..., (\mathbf{e}_n, l_n)\}$

1 $\alpha = 0$, $\beta = 0$
2 **foreach** *Review* $R_i \in \mathcal{R}$ **do**
3   **foreach** *event* $\mathbf{e}_j \in R_i$ **do**
4     $P(LivedExperience|\mathbf{e}_j) = \frac{frequency(\mathbf{e}_j), \mathbf{e}_j \in TS^+}{frequency(\mathbf{e}_j), \mathbf{e}_j \in TS}$, such as TS^+ is the set of "Lived Experience" reviews of our training set TS
5     $P(Non\text{-}LivedExperience|\mathbf{e}_j) = \frac{frequency(\mathbf{e}_j), \mathbf{e}_j \in TS^-}{frequency(\mathbf{e}_j), \mathbf{e}_j \in TS}$, such as TS^- is the set of "Non Lived Experience" reviews of our training set TS
6   **if** $P(LivedExperience|\mathbf{e}_j) < 0.7$ *and* $P(Non\text{-}LivedExperience|\mathbf{e}_j) < 0.7$ **then**
7   delete $\mathbf{e}_j$ from R_i: $R_i = R_i - \mathbf{e}_j$
8   **end**
9   **end**
10   $\alpha = \frac{|R_i|-1}{|R_i|}$, such as $|R_i|$ is the cardinality of the set of events R_i
11   $\beta = \frac{1}{|R_i|}$
12   **foreach** *event* $\mathbf{e}_j \in R_i$ **do**
13   **if** $P(LivedExperience|\mathbf{e}_j) > 0.7$ **then** $\mu_j = +P(LivedExperience|\mathbf{e}_j)$;
14   **else if** $P(Non\text{-}LivedExperience|\mathbf{e}_j) > 0.7$ **then**
15       $\mu_j = -P(Non\text{-}LivedExperience|\mathbf{e}_j)$
16   Calculate $X(\mathbf{e}_j)$, the larger distribution value of the target event in the two classes of reviews in our training set.
17     $X(\mathbf{e}_j) = \mu_j$
18   Calculate $Y(\mathbf{e}_j)$, the sum of the larger distributions values of event neighbors in the two classes of reviews in our training set
19     $Y(\mathbf{e}_j) = \sum_{\mathbf{e}_v \in R_i/\mathbf{e}_j} X(\mathbf{e}_v)$
20   Calculate $F = \alpha X(\mathbf{e}_j) + \beta Y(\mathbf{e}_j)$
21   **if** $F > 0$ **then return** ($\mathbf{e}_j$, "*LivedExperience*");
22   **else**
23   **return** ($\mathbf{e}_j$, "*NonLivedExperience*")
24   **end**
25   **end**
26 **end**

The review provided as an example in Sect. 1 has three lived experience event types {*Get, Ask, Decorate*}. These events are labeled as lived experience events by the first step in our algorithm, since their distribution in the list of lived experience events in our training set is clearly larger than their distribution in the non-lived experience list. The next step is to calculate the function F to decide whether these events are pertinent or not. Table 1 shows the obtained results. According to their F values, all these events represent lived experiences.

Table 1. Results for the event filtering algorithm.

Event	*P(LivedExp)*	*P(Non-LivedExp)*	*F*
Get	0.852	0.148	$\frac{2}{3} * (0.852) + \frac{1}{3} * (0.778 + 1) = 1.161 > 0$
Ask	0.778	0.222	$\frac{2}{3} * (0.778) + \frac{1}{3} * (0.852 + 1) = 1.136 > 0$
Decorate	1.00	0.00	$\frac{2}{3} * (1) + \frac{1}{3} * (0.852 + 0.778) = 1.21 > 0$

4.5 Event Participant Extraction

A lived experience should involve events and their participants, and the narrator should be among the participants. For each lived experience event, we need to identify its participants, and to build a lived experience graphs. The participants in an event are the arguments of semantic roles (e.g. Agent, Patient, Oblique, Theme) associated with the event.

A complex SPARQL query is used to extract event participants from FRED's knowledge graphs for the input review. For space reasons, we point at http://www-lipn.univ-paris13.fr/~hassan/Lived_Experience_Extraction/ for the query, the dataset, and other material concerning the evaluation described in this paper. Some previous versions of extracting queries have been described in [9].

The SPARQL query generates an event sub-graph containing lived experience events, detected by the filtering algorithm, with their direct or indirect participants. A direct participant $_{d}p_i \ \phi \ e_i$ is an argument of an event e_i. An indirect participant $_{i}p_j \ \psi \ e_i$ is a direct participant of an event e_j that on its turn occurs as a direct participant of e_i (ϕ and ψ are semantic roles).

To improve our extraction method, and keep our lived experience graphs more informative, we extract event participants for the first, second and third degree. In addition, we take advantage of FRED knowledge graphs by extracting *event modifiers* such as *modality, negation, and adverbial qualities*, which enrich relevant events with additional semantic information that enables to distinguish the nuances or polarity of the reported events.

4.6 Extraction of Sentences Containing Lived Experiences

After the previous steps, we are ready to extract lived experience sentences, which consists in the following steps:

– Input Review Segmentation: We divide the input review into its component sentences using punctuation.

- Sentence Ranking: For each relevant event, all review sentences are ranked according to the presence of the relevant event and its participants. We impose two conditions in this step: (1) The sentence must have at least one word evoking an event. (2) The rank should be larger than 2 in order to retrieve an informative sentence. In other words, the sentence should have at least one relevant event, and at least 2 participants, where one of them is the author.
- Finally, the sentence with the highest rank is extracted.

The following shows the lived experience sentences extracted from our first example in Sect. 1:

1. I asked for a room overlooking the pantheon and I got it.
2. My deluxe room was number 32, and was tastefully decorated with a classic and beautiful Pierre Frey wallpaper, and an extra day bed.
3. If you book directly through the hotel, you'll get a voucher for a free-drink upon arrival.

4.7 Lived Experience Graph Representation

We represent lived experiences as event sub-graphs containing relevant events, with their participants, and modifiers. In order to create a compact representation of lived experiences in customer reviews, we encode lived experiences as sequences of events, where each event is represented as an n-tuple of participants. For example, for the first example review, the encoding is as follows:

$$Ask(\text{“}Person'', \text{“}Room''[\text{“}A''])\text{.}$$
$$Get(\text{“}Person'', \text{“}Necessary'', \text{“}Arrival'', \text{“}Book''[\text{“}Direcly''], \text{“}Voucher''[\text{“}A'', \text{“}Free - Drink''])\text{.}$$
$$Decorate(\text{“}Testefully'', \text{“}Bed'', \text{“}Room''[\text{“}Person'', \text{“}Multiple'', \text{“}Deluxe'', \text{“}32'', \text{“}Hotel''[\text{“}This'', \text{“}4''], \text{“}Wallpaper''[\text{“}Beatiful'', \text{“}Classic''])\text{.}$$

Figure 2, shows a diagram of the lived experience graph from that review:

5 Experimental Evaluation

5.1 Dataset

Ott et al. [17] have recently created the first publicly available[3] dataset for deceptive opinion spam research. This dataset contains 800 positive reviews (400 truthful reviews and 400 fake reviews), which have been assigned with 5-stars in the open system ranking and 800 reviews for the negative polarity (400 truthful reviews and 400 fake reviews), which have 1-star in the open system ranking. In this work, we are only interested in the positive truthful reviews which are

[3] Available by request at: http://www.cs.cornell.edu/~myleott/op_spam.

Fig. 2. An event sub-graph representing lived experiences extracted from a user review.

collected from the 20 most popular Chicago hotels on TripAdvisor[4]. We have selected 383 user reviews to be annotated for the training and test set. 17 reviews were excluded because of parsing problems. As we mentioned above, 268 user reviews have been used as a training set for the lived experience identification task, and 115 as test set.

5.2 Review Identification Evaluation

As described, we have used our event dictionary as a collection of features to train a Support Vector Machine (SVM) model with a radial basis function kernel as our classification model. SVM is very well suited for binary classification tasks, and it has the ability to deal with large space features. Since we were looking at entire reviews rather than segments of sentences, we have also considered event

[4] http://www.tripadvisor.com/.

type frequency. Each review is transformed into a feature vector where the i−th component value is the frequency of the i−th event in the review, and then classified as either "LivedExperience" or "Non-LivedExperience".

Our system is evaluated in terms of precision, recall, and F-measure. We performed 10-fold cross-validation on the training set, and yielded good results, see Table 2.

To validate the obtained results, we applied our review classifier to the test set containing 115 user reviews. Table 2 shows the results.

As lived experiences consist of personal events, they should also contain their participants. We assumed that event participants can upgrade the identification task results. By using events and their participants as features in our classifier, we obtained very close results to the previous ones that were obtained using only personal events. However, we noticed a small decrease for both training set and test set, as shown in Table 2.

Table 2. Overall results for review classification using SVM method with four configurations. *All_Events*: use all the personal events as features, $Event_\sigma = 0.7$: delete some common events and use the rest as features, *All_Event_Participant*: use events with their participants as features, and $(Event\text{-}Participant)\sigma = 0.7$: remove the common events and participants, and use the rest as features.

Features	Training set			Test set		
	Precision	Recall	F-measure	Precision	Recall	F-measure
All_Events	84.3 %	84 %	84.1 %	83.4 %	83.5 %	83.4 %
$Event_\sigma = 0.7$	**86.5%**	**85.8%**	**86.1%**	**85.1%**	**85.2%**	**85.2%**
All_Event_Participant	82.7 %	81.7 %	82 %	82.6 %	82.6 %	82.6 %
$(Event\text{-}Participant)\sigma = 0.7$	84.4 %	83.2 %	84 %	79 %	79.1 %	79 %

As shown in this Table, $Event_\sigma = 0.7$ achieves the best results for both the training and test set. The results using all personal events (without deleting the common ones) decrease the performance by about 2 % for the training and test set. Adding participant features to all personal events give good results, but less effectively than the results obtained using personal event features with $\sigma = 0.7$. Deleting common features from the dictionary that contain events and their participants give approximately the same results using all the personal events in the training set, but for the test set we notice some decrease.

Based on this evaluation, we use personal event features in the LEE system (see Sect. 5.4) to identify the input reviews with optimal accuracy. In addition, we can notice that the results in our test set are very close to our original cross-validation evaluation. This allows us to conclude that there is no evidence that the results of our classifier may get worse when applied to different user reviews.

5.3 Comparison to Other Approaches

In order to meaningfully evaluate our model, we have established a reasonable baseline. We chose to compare our method against three baseline models:

Table 3. Overall results for lived experience identification

Model	Training set			Test set		
	Precision	Recall	F-measure	Precision	Recall	F-measure
Personal stories	59.8 %	59.3 %	59.6 %	65.4 %	55 %	60 %
Verbs	66.9 %	66.7 %	66.8 %	75.9 %	75.7 %	75.8 %
Bag-of-words	65 %	63.3 %	64.1 %	69.4 %	69.4 %	69.4 %
$Event_\sigma = 0.7$	**86.5%**	**85.8%**	**86.1%**	**85.1%**	**85.2%**	**85.2%**

(1) The model from [11], called "personal stories", (2) a verb model, and (3) a bag-of-words model.

As we mentioned in Sect. 3.1, The closest study to our approach on personal information identification task is [11]. They employ statistical text classification technology on the content of blog entries to identify personal stories in weblog entries. They investigated several variations of n-gram features (e.g. unigrams and bigrams) to train a Support Vector Machine learning algorithm (SVM). They manually annotated 4252 weblog entries to be used as a training set, and then performed a 10-fold cross validation. Their system achieved precision = 66%, recall = 48 %, and F-Measure = 55 % on this data. To compare our system to their system, we used their pre-trained model [5], and tested it on our training and test set. The achieved results for this set can be shown in Table 3.

The verb model consists in using just verbs, which are extracted from our training set as features for a SVM classifier to discriminate between the two classes of reviews. We chose to compare our method to this model, since most of our event features are represented by verbs. The overall performance on the training set using 10-cross validation and on test set is shown in Table 3.

We also compared our model to another model which only uses bags of words as classifier features. We extracted all the words from our training reviews, and used them to train a Support Vector Machine classifier. We used this model in order to compare our results using only lived experience events, to results obtained using all review contents. Table 3 presents all the results (including our own) on the training set using 10-fold cross validation.

As shown in this table, we outperform other models by nearly 26 % against the personal stories model, about 10 % against the verb model, and 17 % against the bag-of-words model on the test set.

5.4 Lived Experience Extraction Evaluation

We evaluate the Lived Experience Extraction task using 80 reviews from our dataset. We chose reviews that are classified as lived experience reviews by our system, and are equally annotated by the annotator. In total, 790 sentences are used in the evaluation.

[5] https://github.com/asgordon/StoryNonstory.

Table 4. The results of lived experience extraction task.

	Precision	Recall	F-measure
A1	0.55	0.70	0.62
A2	0.48	0.70	0.57
$A1 \cap A2$	0.49	0.74	0.58

Based on those reviews, 2 human evaluators have been asked to manually extract sentences that denote lived experiences. Human evaluators took one week to perform this task. For this reason, we only chose 80 reviews for the evaluation. Inter-annotator agreement among the two judges, computed using Cohen's kappa, is 0.89. According to Landis and Koch [16], this value is in the range (0.81–1.00), which corresponds to "perfect agreement".

For each review, we applied our system to extract lived experience sentences. All the results extracted by our system are compared to the manually extracted results. In addition, we also compared our results to a virtual meta-annotator. The $A1 \cap A2$ meta-annotator extracts lived experience sentences when both annotators believe the sentence to denote a lived experience.

Table 4 gives the precision, recall, and F-measure results of our system. We calculated these values at sentence level. For each sentence in the review, we consider it a "*TP*" if it was labeled as lived experience by both the system and the human annotator, "*FP*" if the system labeled it as a lived experience but not the annotator, "*FN*" if the annotator labeled the sentence as a lived experience, but the system failed to recognize it, and "*TN*" if it was labeled as Non-lived experience by both the system and the human annotator.

$$Precision = \frac{TP}{TP+FP}, Recall = \frac{TP}{TP+FN}$$

The $A1 \cap A2$ meta-annotator found 167 lived experience sentences and 623 Non-lived experience sentences. Using our system, we extract 255 lived experience sentences and 535 Non-lived ones. The "*TP*" is 124 sentences, the "*FP*" is 43 sentences, the "*FN*" is 131, and the "*TN*" is 492.

In the table, column 1 identifies the annotator that we compare our system with. The results indicate that our system is closer to the first annotator than the second one.

The LEE system in available as a web application and RESTful API[6], and provides reusable RDF knowledge graphs e.g. for recommendation services.

6 Conclusion

We have proposed a set of techniques for identifying and extracting lived experiences from user reviews based on data mining, semantic web, and natural language processing methods. The objective is to identify reviews that contain lived

[6] https://lipn.univ-paris13.fr/ClientProj/client.jsp.

experiences by using event features, and then to extract those experiences from relevant reviews. Our experimental results indicate that the proposed techniques are very good.

We believe that this problem will become more and more important, considering the growing amount of user-opinion-based decisions made on the Web. Extracting lived experiences from user reviews is useful for potential users, due to the ability to take into account the cognitive or emotional identification of the reader with the author of a review, either for establishing non-fictional experiences, or for filtering reviews based on the closeness in taste or life habits.

In future work, we intend to create a large-scale comprehensive corpus of lived experience reviews as a *gold-standard*. We aim to link the generated event sub-graphs to contextual knowledge (user ranking, user profiles, etc.), which can enhance these sub-graphs. The sentiment of each event sub-graph can be detected to create a bank of knowledge graphs representing event-oriented lived experiences. In addition, we aim to use lived experience contents to detect fake reviews. Finally, we aim to summarize e.g. hotel reviews based on lived experience contents.

Acknowledgements. I would like to thank Dr. Mehwish Alam for her suggestions. This work is partially supported by a public grant overseen by the French National Research Agency (ANR) as part of the program "Investissements d'Avenir" (reference: ANR-10-LABX-0083).

References

1. Aone, C., Ramos-Santacruz, M.: REES: a large-scale relation and event extraction system. In: Proceedings of the Sixth Conference on Applied Natural Language Processing, pp. 76–83 (2000)
2. Barsalou, L.W.: The content and organization of autobiographical memories. In: Remembering Reconsidered: Ecological and Traditional Approaches to the Study of Memory, pp. 193–243 (1988)
3. Baumgartner, H., Sujan, M., Bettman, J.R.: Autobiographical memories, affect, and consumer information processing. J. Consum. Psychol. **1**, 53–82 (1992)
4. Bethard, S., Martin, J.H.: Identification of event mentions and their semantic class. In: Proceedings of the 2006 Conference on Empirical Methods in Natural Language Processing, pp. 146–154 (2006)
5. Duffy, A.M.: Fellow travellers: what do users trust on recommender websites? A case study of TripAdvisor.com. Ph.D. thesis (2012)
6. Eastmond, M.: Stories as lived experience: narratives in forced migration research. J. Refugee Stud. **20**, 248–264 (2007)
7. Etzioni, O., Banko, M., Cafarella, M.: Machine reading. In: Proceedings of the 21st National Conference on Artificial Intelligence (AAAI) (2006)
8. Gangemi, A.: A comparison of knowledge extraction tools for the semantic web. In: Cimiano, P., Corcho, O., Presutti, V., Hollink, L., Rudolph, S. (eds.) ESWC 2013. LNCS, vol. 7882, pp. 351–366. Springer, Heidelberg (2013)
9. Gangemi, A., Hassan, E., Presutti, V., Reforgiato, D.: Fred as an event extraction tool. In: Proceedings of the Workshop on Detection, Representation, and Exploitation of Events in the Semantic Web, p. 14 (2013)

10. Gangemi, A., Presutti, V., Recupero, D.R., Nuzzolese, A.G., Draicchio, F., Mongiovi, M.: Semantic web machine reading with FRED. In: Sermantic Web (2016). http://www.semantic-web-journal.net/system/files/swj1297.pdf
11. Gordon, A., Swanson, R.: Identifying personal stories in millions of weblog entries. In: Third International Conference on Weblogs and Social Media, Data Challenge Workshop, San Jose, CA (2009)
12. Gordon, A.S.: Story management technologies for organizational learning. In: International Conference on Knowledge Management. In Special Track on Intelligent Assistance for Self-directed and Organizational Learning, Graz, Austria (2008)
13. Hassan, E., Buscaldi, D., Gangemi, A.: Correlating open rating systems and event extraction from text. In: Arik, S., Huang, T., Lai, W.K., Liu, Q. (eds.) ICONIP 2015. LNCS, vol. 9492, pp. 367–375. Springer, Heidelberg (2015). doi:10.1007/978-3-319-26561-2_44
14. Inui, K., Abe, S., Hara, K., Morita, H., Sao, C., Eguchi, M., Sumida, A., Murakami, K., Matsuyoshi, S.: Experience mining: building a large-scale database of personal experiences and opinions from web documents. In: Proceedings of the 2008 International Conference on Web Intelligence and Intelligent Agent Technology. IEEE Computer Society, Washington, DC (2008)
15. Jindal, N., Liu, B.: Opinion spam and analysis. In: Proceedings of the 2008 International Conference on Web Search and Data Mining, pp. 219–230. ACM (2008)
16. Landis, J.R., Koch, G.G.: The measurement of observer agreement for categorical data. Biometrics **33**, 159–174 (1977)
17. Ott, M., Choi, Y., Cardie, C., Hancock, J.T.: Finding deceptive opinion spam by any stretch of the imagination. In: Proceedings of the 49th Annual Meeting of the Association for Computational Linguistics: Human Language Technologies-Vol. 1, pp. 309–319 (2011)
18. Pazzani, M.J., Billsus, D.: Content-based recommendation systems. In: Brusilovsky, P., Kobsa, A., Nejdl, W. (eds.) The Adaptive Web. LNCS, vol. 4321, pp. 325–341. Springer, Heidelberg (2007)
19. Ploeger, T., Kruijt, M., Aroyo, L., de Bakker, F., Hellsten, I., Fokkens, A., Hoeksema, J., ter Braake, S.: Extracting activist events from news articles using existing NLP tools and services. In: Proceedings of the Workshop on Detection, Representation, and Exploitation of Events in the Semantic Web, p. 30 (2013)
20. Presutti, V., Draicchio, F., Gangemi, A.: Knowledge Extraction based on discourse representation theory and linguistic frames. In: ten Teije, A., Völker, J., Handschuh, S., Stuckenschmidt, H., d'Acquin, M., Nikolov, A., Aussenac-Gilles, N., Hernandez, N. (eds.) EKAW 2012. LNCS, vol. 7603, pp. 114–129. Springer, Heidelberg (2012)
21. Ritter, A., Etzioni, O., Clark, S., et al.: Open domain event extraction from twitter. In: Proceedings of the 18th ACM SIGKDD International Conference on Knowledge Discovery and Data Mining, pp. 1104–1112 (2012)
22. Saurí, R., Knippen, R., Verhagen, M., Pustejovsky, J.: Evita: a robust event recognizer for QA systems. In: Proceedings of the Conference on Human Language Technology and Empirical Methods in Natural Language Processing, pp. 700–707 (2005)
23. Treloar, C., Rhodes, T.: The lived experience of hepatitis c and its treatment among injecting drug users: qualitative synthesis. Qual. Health Res. **19**, 1321–1334 (2009)

24. Tung, V.W.S., Ritchie, J.B.: Exploring the essence of memorable tourism experiences. Ann. Tourism Res. **38**, 1367–1386 (2011)
25. Van Oorschot, G., Van Erp, M., Dijkshoorn, C.: Automatic extraction of soccer game events from twitter. In: Proceedings of the Workshop on Detection, Representation, and Exploitation of Events in the Semantic Web, p. 15 (2012)
26. Zhang, Y., Xu, C., Rui, Y., Wang, J., Lu, H.: Semantic event extraction from basketball games using multi-modal analysis. In: 2007 IEEE International Conference on Multimedia and Expo, pp. 2190–2193 (2007)

An Evolutionary Algorithm to Learn SPARQL Queries for Source-Target-Pairs

Finding Patterns for Human Associations in DBpedia

Jörn Hees[1,2(✉)], Rouven Bauer[1,2], Joachim Folz[1,2], Damian Borth[1,2], and Andreas Dengel[1,2]

[1] Computer Science Department, University of Kaiserslautern, Kaiserslautern, Germany
[2] Knowledge Management Department, DFKI GmbH, Kaiserslautern, Germany
joern.hees@dfki.de

Abstract. Efficient usage of the knowledge provided by the Linked Data community is often hindered by the need for domain experts to formulate the right SPARQL queries to answer questions. For new questions they have to decide which datasets are suitable and in which terminology and modelling style to phrase the SPARQL query.

In this work we present an evolutionary algorithm to help with this challenging task. Given a training list of source-target node-pair examples our algorithm can learn patterns (SPARQL queries) from a SPARQL endpoint. The learned patterns can be visualised to form the basis for further investigation, or they can be used to predict target nodes for new source nodes.

Amongst others, we apply our algorithm to a dataset of several hundred human associations (such as "circle - square") to find patterns for them in DBpedia. We show the scalability of the algorithm by running it against a SPARQL endpoint loaded with > 7.9 billion triples. Further, we use the resulting SPARQL queries to mimic human associations with a Mean Average Precision (MAP) of 39.9 % and a Recall@10 of 63.9 %.

1 Introduction

The Semantic Web [1] and its Linked Data [2] movement have brought us many great, interlinked and freely available machine readable RDF [13] datasets, often summarized in the Linking Open Data Cloud[1]. Being extracted from Wikipedia and spanning many different domains, DBpedia [3] forms one of the most central and best interlinked of these datasets.

Nevertheless, even with all this easily available data, using it is still very challenging: For a new question, one needs to know about the available datasets, which ones are best suited to answer the question, know about the way knowledge is modelled inside them and which vocabularies are used, before even attempting

[1] http://lod-cloud.net/.

E. Blomqvist et al. (Eds.): EKAW 2016, LNAI 10024, pp. 337–352, 2016.
DOI: 10.1007/978-3-319-49004-5_22

to formulate a suitable SPARQL[2] query to return the desired information. The noise of real world datasets adds even more complexity to this.

In this paper we present a graph pattern learning algorithm that can help to identify SPARQL queries for a relation $\mathcal{R}$ between node pairs $(s, t) \in \mathcal{R}$ in a given knowledge graph G[3], where s is a source node and t a target node. $\mathcal{R}$ can for example be a simple relation such as "given a capital s return its country t" $\mathcal{R}_{cc}$ or a complex one such as "given a stimulus s return a response t that a human would associate" $\mathcal{R}_{ha}$.

To learn queries for $\mathcal{R}$ from G, without any prior knowledge about the modelling of $\mathcal{R}$ in G, we allow users to compile a ground truth set of example source-target-pairs $\mathcal{GT} \subseteq \mathcal{R}$ as input for our algorithm. For example, for relation $\mathcal{R}_{cc}$ between capital cities and their countries, the user could generate a ground truth list $\mathcal{GT} = \{$(dbr:Berlin, dbr:Germany), (dbr:Paris, dbr:France), (dbr:Oslo, dbr:Norway)$\}$. Given $\mathcal{GT}$ and the DBpedia SPARQL endpoint[4], our graph pattern learner then learns a set of graph patterns $gpl(\mathcal{GT}, G)$ such as:

$$gp_1\text{: } \{\texttt{?source dbo:country ?target}\}$$

$$gp_2\text{: } \{\texttt{?target dbo:capital ?source. ?target a dbo:country}\}$$

In this paper, a graph pattern $gp \in gpl(\mathcal{GT}, G) \subset GP$ is an instance of the infinite set of SPARQL basic graph patterns[5] GP. Each gp has a corresponding SPARQL ASK and SELECT query. We denote their execution against G as ASK(gp) and SELECT(gp). The graph patterns can contain SPARQL variables, out of which we reserve ?source and ?target as special ones. A mapping Φ can be used to bind variables in gp before execution.

The resulting learned patterns can either be inspected or be used to predict targets by selecting all bindings for ?target given a source node s_i:

$$\text{prediction}_{gp}(s_i) = \underset{?target}{\text{SELECT}}(\phi_{?source:=s_i}(gp))$$

For example, given the source node dbr:London the pattern gp_1 can be used to predict dbr:United_Kingdom $\in$ prediction_{gp_1}(dbr:London).

The remainder of this paper is structured as follows: We present related work in Sect. 2, before describing our graph pattern learner in detail in Sect. 3. In Sects. 4 and 5 we will then briefly describe visualisation and prediction techniques before evaluating our approach in Sect. 6.

2 Related Work

To the best of our knowledge, our algorithm is the first of its kind. It is unique in that it can learn a set of SPARQL graph patterns for a given input list of

---

[2] https://www.w3.org/TR/rdf-sparql-query/.

[3] For our purpose G is a set of RDF triples, typically accessible via a given SPARQL endpoint.

[4] http://dbpedia.org/sparql.

[5] https://www.w3.org/TR/sparql11-query/#BasicGraphPatterns.

source-target-pairs directly from a given SPARQL endpoint. Additionally, it can cope with scenarios in which there is not a single pattern that covers all source-target-pairs.

Many other algorithms exist, which learn vector space representations from knowledge graphs. An excellent overview of such algorithms can be found in [17]. We are however not aware that any of these algorithms have the ability of returning a list of SPARQL graph patterns that cover an input list of source-target-pairs.

There are other approaches that help formulating SPARQL queries, mostly in an interactive fashion such as RelFinder [10,11] or AutoSPARQL [15]. Their focus however lies on finding relationships between a short list of entities (not source-target-pairs) or interactively formulating SPARQL queries for a list of entities of a single kind. They cannot deal with entities of different kinds.

Wrt. SPARQL pattern learning, there is an approach for pattern based feature construction [14] that focuses on learning SPARQL patterns to use them as features for binary classification of entities. It can answer questions such as: does an entity belong to a predefined class? In contrast to that, our approach focuses on learning patterns between a list of source-target-pairs for entity prediction: given a source entity predict target entities. To simulate target entity prediction for a single given source with binary classification, one would need to train n classifiers, one for n potential target entities.

In the context of mining patterns for human associations and Linked Data, we previously focused on collecting datasets of semantic associations directly from humans [6,9], ranking existing facts according to association strengths [7,8] and mapping the Edinburgh Associative Thesaurus [12] to DBpedia [5]. None of these previous works directly focused on identifying existing patterns for human associations in existing datasets.

3 Evolutionary Graph Pattern Learner

The outline of our graph pattern learner is similar to the generic outline of evolutionary algorithms: It consists of individuals (in our case SPARQL graph patterns $gp_i \in GP$), which are evaluated to calculate their fitness. The fitter an individual is, the higher its chance to survive and reach the next generation. The individuals of a generation are also referred to as population. In each generation there is a chance to mate and mutate for each of the individuals. A population can contain the same individual (graph pattern) several times, causing fitter individuals to have a higher chance to mate and mutate over several generations.

As mentioned in the introduction, the training input of our algorithm is a list of ground truth source-target-pairs $gtp_i = (s_i, t_i) \in \mathcal{GT}$.

Due to size limitations, we will focus on the most important aspects of our algorithm in the following. For further detail please see our website[6] where you can find the source-code, visualisation and other complementary material.

[6] https://w3id.org/associations.

3.1 Coverage

Before describing the realisation of the components of our evolutionary learner, we want to introduce our concept of coverage.

We say that a graph pattern gp_i covers, models or fulfils a source-target-pair (s_j, t_j) if the evaluation of its SPARQL ASK query returns true:

$$\text{ASK}(\phi_{\texttt{?source}:=s_i,\texttt{?target}:=t_i}(gp))$$

Our algorithm is not limited to learning a single best pattern for a list of ground truth pairs, but it can learn multiple patterns which together cover the list.

We realise this by invoking our evolutionary algorithm in several *runs*. In each run a full evolutionary algorithm is executed (with all its generations). After each run the resulting patterns are added to a global list of results. In the following runs, all ground truth pairs which are already covered by the patterns from previous runs become less rewarding for a newly learnt pattern to cover. Over its runs our algorithm will thereby re-focus on the left-overs, which allows us to maximise the coverage of all ground truth pairs with good graph patterns.

3.2 Fitness

In order to evaluate the fitness of a pattern, we define the following dimensions to capture what makes a pattern "good".

- High *recall*:
  A good pattern fulfils as many of the given ground truth pairs $\mathcal{GT}$ as possible:

$$\text{gt\_matches}_{gp} = |\{(s_i, t_i) \in \mathcal{GT} | \text{ASK}(\phi_{?source:=s_i,?target:=t_i}(gp))\}|$$

$$\text{recall}_{gp} = \frac{\text{gt\_matches}_{gp}}{|\mathcal{GT}|}$$

- High *precision*:
  A good pattern should also be precise. For each individual ground truth pair $(s_i, t_i) \in \mathcal{GT}$ we can define the precision as:

$$\text{precision}_{gp}((s_i, t_i)) = \frac{|\{t_i | t_i \in \text{prediction}_{gp}(s_i)\}|}{|\text{prediction}_{gp}(s_i)|}$$

  The target t_i should be in the returned result list and if possible nothing else. In other words, we are not searching for patterns that return thousands of potentially wrong target for a given source. Over all ground truth pairs, we can define the average precision for *gp* via the inverse of the average result lengths:

$$\text{avg result length}_{gp} = \underset{(s_i,t_i)}{\text{avg}} |\text{prediction}_{gp}(s_i)|$$

$$\text{precision}_{gp} = (\text{avg result length}_{gp})^{-1}$$

- High *gain*:
  A pattern discovered in run r is better if it covers those ground truth pairs $gtp \in \mathcal{GT}$ that aren't covered with high precisions in previous runs ($gp' \in run_q$) already:

$$\text{gain}_{run_r, gp} = \sum_{gtp} \max\{0, \text{precision}_{gp}(gtp) - \max_{\forall q<r: gp' \in run_q} \text{precision}_{gp'}(gtp))\}$$

  Similarly, the potentially remaining gain can be computed as:

$$\text{remains}_{run_r} = \sum_{gtp \in \mathcal{GT}} (1 - \max_{\forall q<r: gp' \in run_q} \text{precision}_{gp'}(gtp))$$

- No *over-fitting*:
  While precision is to be maximised, a good pattern should not *over-fit* to a single source or target from the training input.
- Short *pattern length* and low *variable count*:
  If all other considerations are similar, then a shorter pattern or one with less variables is preferable.
  Note, that this is a low priority dimension. A good pattern is not restricted to a shortest path between `?source` and `?target`. Good patterns can be longer and can have edges off the connecting path (e.g., see gp_2 in Sect. 1).
- Low execution *time & timeout*:
  Last but not least, to have any practical relevance, good patterns should be executable in a short *time*. Especially during the training phase, in which many queries are performed that take too long, we need to make sure to early terminate such queries on both, the graph pattern learner and the endpoint (cf. Sect. 3.7). In case the query was aborted due to a *timeout* and only a partial result obtained, it should not be trusted.

Based on these considerations, we define the *fitness* of an individual graph pattern as a tuple of real numbers with the following optimization directions. When comparing the fitness of two patterns, the fitness tuples for now are compared lexicographically.

1. **Remains** (max): Remaining precision sum remains_{run_r} in the current run r (see Sect. 3.1). Patterns found in earlier runs are considered better.
2. **Score** (max): A derived attribute combining gain with a configurable multiplicative punishment for over-fitting patterns.
3. **Gain** (max): The summed gained precision over the remains of the current run r: $\text{gain}_{run_r, gp}$. In case of timeouts or incomplete patterns the gain is set to 0.
4. **F_1-measure** (max): F_1-measure for precision and recall of this pattern.
5. **Average Result Lengths** (min): avg result length_{gp}.
6. **Recall (Ground Truth Matches)** (max): gt_matches_{gp}.
7. **Pattern Length** (min): The number of triples this pattern contains.
8. **Pattern Variables** (min): The number of variables this pattern contains.

9. **Timeout** (min): Punishment term for timeouts (0.5 for a soft and 1.0 for a hard timeout) (see Sect. 3.7 and gain).
10. **Query Time** (min): The evaluation time in seconds. This is particularly relevant since it hints at the real complexity of the pattern. I.e., a pattern may objectively have a small number of triples and variables, but its evaluation could involve a large portion of the dataset.

3.3 Initial Population

In order to start any evolutionary algorithm an initial population needs to be generated. The main objective of the first population is to form a starting point from which the whole search space is reachable via mutations and mating over the generations. While the initial population is not meant to immediately solve the whole problem, a poorly chosen initial population results in a lot of wasted computation time.

The starting point of our algorithm are single triple SPARQL BGP queries, consisting only of variables with at least a `?source` and `?target` variable, e.g.:

{`?source ?p1 ?v1.`}

While having a small chance of survival (direct evaluation would typically yield bad fitness), such patterns can re-combine (see mating in Sect. 3.4) with other patterns to form good and complete patterns in later generations.

For prediction capabilities, we are searching graph patterns which connect `?source` and `?target`, our algorithm mostly fills the initial population with path patterns of varying lengths l between `?source` and `?target`. Initially such a path pattern purely consists of variables and is directed from source to target:

$$\{\texttt{?source}\ ?p_1\ ?n_1.\ \ldots\ ?n_i\ ?p_{i+1}\ ?n_{i+1}\ .\ \ldots\ ?n_{l-1}\ ?p_l\ \texttt{?target.}\}$$

For example a pattern of desired length of $l = 3$ looks like this:

{`?source ?p1 ?n1. ?n1 ?p2 ?n2. ?n2 ?p3 ?target.`}

As longer patterns are less desirable, they are generated with a lower probability. Furthermore, we randomly flip each edge of the generated patterns, in order to explore edges in any direction.

In order to reduce the high complexity and noise introduced by patterns only consisting of variables, we built in a high chance to immediately subject them to the fix variable mutation (see Sect. 3.5).

3.4 Mating

In each generation there is a configurable chance for two patterns to mate in order to exchange information. In our algorithm this is implemented in a way that mating always creates two children, having the benefit of keeping the amount of individuals the same. Each child has a dominant and a recessive parent. The child will contain all triples that occur in both parents. Additionally, there is a high chance to select each of the remaining triples from the dominant parent and a low chance to select each of the remaining triples from the recessive parent. By this the children have the same expected length as their parents.

Furthermore, as variables from the recessive parent could accidentally match variables already being in the child, and this can be beneficial or not, we add a 50 % chance to rename such variables before adding the triples.

3.5 Mutation

Besides mating, which exchanges information between two individuals, information can also be gained by mutation. Each individual in a population has a configurable chance to mutate by the following (non exclusive) mutation strategies. Currently, all but one of the mutation operations can be performed on the pattern itself (local) without issuing any SPARQL queries. The mutation operations also have different effects on the pattern itself (grow, shrink) and on its result size (harden, loosen).

- **introduce var** select a component (node or edge) and convert it into a variable (loosen) (local)
- **split var** select a variable and randomly split it into 2 vars (grow, loosen) (local)
- **merge var** select 2 variables and merge them (shrink, harden) (local)
- **del triple** delete a triple statement (shrink, loosen) (local)
- **expand node** select a node, and add a triple from its expansion (grow, harden) (local for now)
- **add edge** select 2 nodes, add an edge in between if available (grow, harden) (local for now)
- **increase dist** increase distance between source and target by moving one a hop away (grow) (local)
- **simplify pattern** simplify the pattern, deleting unnecessary triples (shrink) (local) (cf. Sect. 3.7)
- **fix var** select a variable and instantiate it with an IRI, BNode or Literal that can take its place (harden) (SPARQL) (see below)

In a single generation sequential mutation (by different strategies in the order as above) is possible.

We can generally say that introducing a variable loosens a pattern and fixing a variable hardens it. Patterns which are too loose will generate a lot of candidates and take a long time to evaluate. Patterns which are too hard will generate too few solutions, if any at all. Very big patterns, even though very specific can also exceed reasonable query and evaluation times.

Fix Var Mutation. Unlike the other mutations, the fix var mutation is the only one which makes use of the underlying dataset via the SPARQL endpoint G, in order to instantiate variables with an IRI, BNode or Literal. As it is one of the most important mutations and also because performing SPARQL queries is expensive, it can immediately return several mutated children.

For a given pattern gp we randomly select one of its variables `?v` (excluding `?source` and `?target`). Additionally, we sample up to a defined number of source-target-pairs from the ground truth which are not well covered yet (high potential gain). For each of these sampled pairs (s_s, t_s) we issue a SPARQL Select query of the form:

$$\{ \texttt{SELECT distinct ?v \{ VALUES (?source ?target) \{ } (s_s,\ t_s) \texttt{ \} } gp \texttt{ \}} \}$$

We collect the possible instantiations for `?v`, count them over all queries and randomly select (with probabilities according to their frequencies) up to a defined number of them. Each of the selected instantiations forms a separate child by replacing `?v` in the current pattern.

3.6 Selection and Keeping the Population Healthy

After each generation the next generation is formed by the surviving (fittest) individuals from n tournaments of k randomly sampled individuals from the previous generation.

We also employ two techniques, to counter population degeneration in local maxima and make our algorithm robust (even against non-optimal parameters):

- In each generation we re-introduce a small number of newly generated initial population patterns (see Sect. 3.3).
- Each generation updates a hall of fame, which will preserve the best patterns ever encountered over the generations. In each generation a small number of the best of these all-time best patterns is re-introduced.

3.7 Real World Considerations

In the following, we will briefly discuss practical problems that we encountered and necessary optimizations we used to overcome them. We implemented our graph pattern learner with the help of the DEAP (Distributed Evolutionary Algorithms in Python) framework [4].

Batch Queries. The single most important optimization of our algorithm lies in the reduction of the amount of issued queries by using batch queries. This mostly applies to the queries for fitness evaluation (Sect. 3.2). It is a lot more efficient to run several sub-queries in one big query and to only transport the ground truth pairs to the endpoint once (via `VALUES`), than to ask for each result separately.

Timeouts and Limits. Another mandatory optimization involves the use of timeouts and limits for all queries, even if they usually only return very few results in a short time. We found that a few run-away queries can quickly lead to congestion of the whole endpoint and block much simpler queries.

Timeouts are also especially useful as a reliable proxy to exclude too complicated graph patterns. Even seemingly simple patterns can take a very long time to evaluate based on the underlying dataset and its distribution.

Fit to Live Filter. Apart from timeouts we use a filter which checks if mutants and children are actually desirable (e.g., length and variable count in boundaries, pattern is complete and connected), meaning fit to live, before evaluating them. If not, the respective parent takes their place in the new population, allowing for a much larger part of the population to be viable.

Parallelization, Caching, Query Canonicalization and Noise. Two other crucial optimizations to reduce the overall run-time of the algorithm are parallelization and client side caching. Evolutionary algorithms are easy to parallelize via parallel evaluation of all individuals, but in our case the SPARQL endpoint quickly becomes the bottleneck. Ignoring the limits of the queried endpoint will resemble a denial of service attack. For most of our experiments we hence use an internal LOD cache with exclusive access for our learning algorithm. In case the algorithm is run against public endpoints we suggest to only use a single thread in order not to disturb their service (fair use).

Client side caching further helps to reduce the time spent on evaluating graph patterns, by only evaluating them once, should the same pattern be generated by different sequences of mutation and mating operations. To identify equivalent patterns despite different syntactic surface forms, we had to solve SPARQL BGP canonicalization (finding a canonical graph labelling). We were able to reduce the problem to RDF graph canonicalization and achieve good practical run-times with RGDA1 [16].

In the context of caching, one other important finding is that many SPARQL endpoints (especially the widely used OpenLink Virtuoso) often return incomplete and thereby non-deterministic results by default. Unlike many other search algorithms, an evolutionary algorithm has the benefit that it can cope well with such non-determinism. Hence, when caching is used, it is helpful to reduce, but not completely remove redundant queries.

Pattern Simplification. Last but not least, as our algorithm can create patterns that are unnecessarily complex, it is useful to simplify them. We developed a pattern simplification algorithm, which given a complicated graph pattern gp_c finds a minimal equivalent pattern gp_s with the same result set wrt. the `?source` and `?target` variables. The simplification algorithm removes unnecessary edges, such as redundant parallel variable edges, edges between and behind fixed nodes and unrestricting leaf branches.

4 Visualisation

After presenting the main components of our evolutionary algorithm in the previous section, we will now briefly present an interactive visualisation[7]. As the learning of our evolutionary algorithm can produce many graph patterns, the visualisation allows to quickly get an overview of the resulting patterns in different stages of the algorithm.

[7] Also available at https://w3id.org/associations.

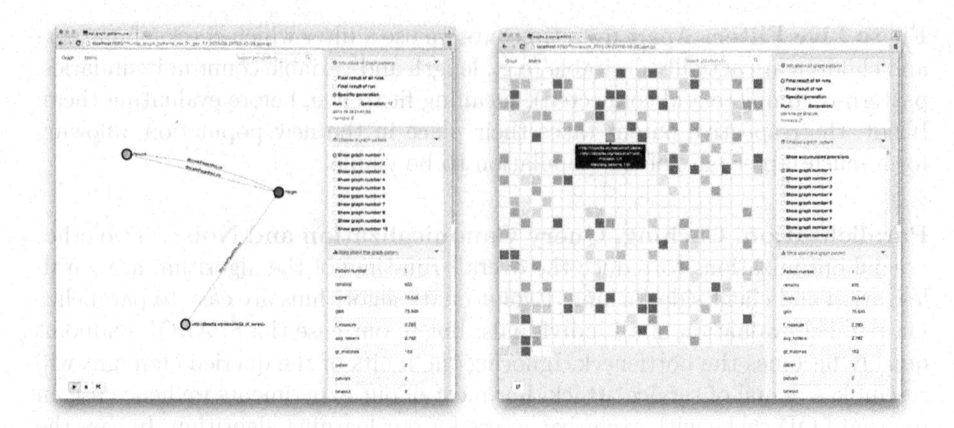

Fig. 1. Visualisation of graph pattern 1 from run 1, generation 12 (left) and the precision vector over all training ground truth pairs of graph pattern 1 (right). Each block represents a (*?source*, *?target*) pair from the ground truth training set. The darker its colour the higher the precision for the ground truth pair.

Figure 1 (left) shows a screen shot of the visualisation of a single learned graph pattern. In the sidebar the user can select between individual generations, the results of a whole run or the overall results (default) to inspect the outcomes at various stages of the algorithm. Afterwards, the individual result graph patterns can be selected. Below these selection options the user can inspect statistics about the selected graph pattern including its fitness, a list of matching training ground truth pairs and the corresponding SPARQL SELECT query for the pattern. Links are provided to perform live queries on the SPARQL endpoint.

At each of the stages, the user can also get an overview of the precision coverage of a single pattern (as can be seen in Fig. 1 (right)) or the accumulated coverage over all patterns.

5 Prediction

As already mentioned in the introduction, the learned patterns can be used to predict targets for a given source. The basic idea is to insert a given source s_i in place of the `?source` variable in each of the learned patterns $gp \in gpl(\mathcal{GT}, G)$ and execute a SPARQL Select query over the `?target` variable (c.f., $\text{prediction}_{gp}(s_i)$ in Sect. 1).

5.1 Query Reduction Technique

While interesting for manual exploration, for practical prediction purposes the amount of learned graph patterns can easily become too large by discovering many very similar patterns that are only differing in minor aspects.

One realisation from visualising the resulting patterns gp, is that we can use their precision vectors wrt. the ground truth pairs to cluster graph patterns. The i-th component of $\boldsymbol{pv}_{gp}$ is defined by the precision value corresponding to the i-th ground truth source-target-pair $stp_i \in \mathcal{GT}$:

$$pv_{gp,i} = \text{precision}_{gp}(stp_i)$$

We employ several standard clustering algorithms on $\boldsymbol{pv}_{gp}$ and select the best patterns cluster($gpl(\mathcal{GT}, G)$) in each cluster as representatives to reduce the amount of queries. By default our algorithm applies all of these clustering techniques and then selects the one which minimises the precision loss at the desired number of queries to be performed during prediction.

In our tests we could observe, that clustering (e.g., with hierarchical scaled euclidean ward clustering) allows us to reduce the number of performed SPARQL queries to 100 for all practical purposes with a precision loss of less than 1 %.

5.2 Fusion Variants

When used for prediction, each graph pattern gp creates an unordered list of possible target nodes $t_j \in \text{prediction}_{gp}(s_i)$ for an inserted source node s_i. We evaluated the following fusion strategies to combine and rank the returned target candidates t_j (higher fusion value means lower rank):

- **target occurrences**: a simple occurrence count of each of the targets over all graph patterns.
- **scores**: sum of all graph pattern scores (from the graph pattern's fitness) for each returned target.
- **f-measures**: sum of all graph pattern F_1-measures (from the graph pattern's fitness) for each returned target.
- **gp precisions**: sum of all graph pattern precisions (from the graph pattern's fitness) for each returned target.
- **precisions**: sum of the actual precisions per graph pattern in this prediction.

By default our algorithm will calculate them all, allowing the user to pick the best performing fusion strategy for their use-case.

6 Evaluation

In order to evaluate our graph pattern learner, we performed several experiments which we will describe in the following.

We ran our experiments against a local Virtuoso 7.2 SPARQL endpoint containing over 7.9 G triples, from many central datasets[8] of the LOD cloud, denoted as G in the following.

[8] Most notably: DBpedia 2015-04 (en, de), Freebase, Yago, Wikidata, GeoNames, DBLP, Wordnet and BabelNet.

6.1 Single Pattern Re-Identification

One of our claims is that our algorithm can learn good SPARQL queries for a relation $\mathcal{R}$ represented by a set of ground truth source-target-pairs $\mathcal{GT}$. In order to evaluate this, we started with simple relations such as "given a capital s return its country t" (see $\mathcal{R}_{cc}$ in Sect. 1). For each $\mathcal{R}$, we used a generating SPARQL query gp_g (such as gp_2 from Sect. 1) to generate $\mathcal{GT} \subset \text{SELECT}(gp_g)$, then executed our graph pattern learner $gpl(\mathcal{GT}, G)$ and checked if gp_g was in the resulting patterns:

$$gp_g \overset{?}{\in} gpl(\mathcal{GT}, G)$$

The result of these experiments is that our algorithm is able to re-identify such simple, readily modelled relations $\mathcal{R}$ in 100 % of our test cases (typically within the first run, so the first 3 min). While this might sound astonishing, it is merely a test that our algorithm can perform the simplest of its tasks: If there is a single SPARQL BGP pattern gp that models the whole training list $\mathcal{GT}$ in G, then our algorithm is quickly able to find it via the fix var mutation in Sect. 3.5. Due to the page limit, we omit further details and instead turn to a more complex relation in the next section.

6.2 Learning Patterns for Human Associations from DBpedia

Two additional claims are that our algorithm can learn a set of patterns, which cover a complex relation $\mathcal{R}$ that is not readily modelled in G, and that we can use the resulting patterns for prediction. Hence, in the following we focus on one such complex relation $\mathcal{R}_{ha}$: human associations. We will present some of the identified patterns and then evaluate the prediction quality.

Dataset. Human associations are an important part of our thinking process. An *association* is the mental connection between two ideas: a *stimulus* (e.g., "pupil") and a *response* (e.g., "eye"). We call such associations *strong associations* if more than 20 % of people agree on the response.

In the following, we focus on a dataset of 727 strong human associations (corresponding to ~ 25.5 K raw associations) from the Edinburgh Associative Thesaurus [12] that we previously already mapped to DBpedia Entities [5]. The dataset contains stimulus-response-pairs such as (dbr:Pupil, dbr:Eye), (dbr:Stanza, dbr:Poetry) and (dbr:Paris, dbr:France).[9]

We randomly split our 727 ground truth pairs into a training set $\mathcal{GT}_{\text{train}}$ of 655 and a test set $\mathcal{GT}_{\text{test}}$ of 72 pairs (10 % random split). All training, visualising and development has been performed on the training set in order to reduce the chance of over-fitting our algorithm to our ground truth.

[9] The full dataset is available at https://w3id.org/associations.

Basic Statistics. We ran the algorithm ($gpl(\mathcal{GT}_{\text{train}}, G)$) on G with a population size of 200, a maximum of 20 generations each in a maximum of 64 runs. The first 5 runs of our algorithm are typically completed within 3, 6, 9, 13 and 15 min. In the first couple of minutes all of the very simple patterns that model a considerable fraction of the training set's pairs are found. With the mentioned settings the algorithm will terminate after around 3 h. It finds roughly 530 graph patterns with a score > 2 (cf. Sect. 3.2).

Notable Learned Graph Patterns. Due to the page limit, we will briefly mention only 3 notable patterns from the resulting learned patterns in this paper. We invite the reader to explore the full results online[10] with the interactive visualisation presented in Sect. 4. The three notable patterns we want to present here are:

{?source gold:hypernym ?target}

{?source dbo:wikiPageWikiLink ?target. ?target dbo:wikiPageWikiLink ?source}

{?source dbo:wikiPageWikiLink ?target. ?v0 skos:exactMatch ?v1. ?v1 dbprop:industry ?target}

The first two are intuitively understandable patterns which typically are amongst the top patterns. The first one shows that human associations often seem to be represented via gold:hypernym in DBpedia (the response is often a hypernym (broader term) for the stimulus). The second one shows that associations often correspond to bidirectionally linked Wikipedia articles. The third pattern represents a whole class of intra-dataset learning by making use of a connection of the ?target to BabelNet's skos:exactMatch.

Prediction and Fusion Strategies Evaluation. As human associations are not readily modelled in DBpedia, it is difficult to assess the quality of the learned patterns *gp* directly. Hence, we evaluate the quality indirectly via their prediction quality on the test-set $\mathcal{GT}_{\text{test}}$.

For each of the $(s_t, t_t) \in \mathcal{GT}_{\text{test}}$ we generate a ranked target list $rtpl_{s_t} = [tp_1, \ldots, tp_n]$ of target predictions tp_i. The list is the result of one of the fusion variants (cf. Sect. 5.2) after clustering (cf. Sect. 5.1). For evaluation, we can then check the rank r_t of t_t in $rtpl_{s_t}$ (lower ranks are better). If $t_t \notin rtpl_{s_t}$, we set $r_t = \infty$.

An example of a ranked target prediction list (for the fusion method *precisions*) for source s_t =dbr:Sled is the ranked list: $rtpl_{\text{dbr:Sled}}$ = [dbr:Snow, dbr:Christmas, dbr:Deer, dbr:Kite, dbr:Transport, dbr:Donkey, dbr:Ice, dbr:Ox, dbr:Obelisk, dbr:Santa_Claus]. In this case the ground truth target t_t =dbr:Snow is at rank $r_t = 1$. As we can see most of the results are relevant as associations to humans. Nevertheless, for the purpose of our evaluation, we will only consider the single t_t corresponding to a s_t as relevant and all other tp_i as irrelevant.

[10] https://w3id.org/associations.

Table 1. Recall@k, MAP and NDCG for our fusion variants and against baselines

	Recall@1	Recall@2	Recall@3	Recall@4	Recall@5	Recall@10	MAP	NDCG
outdeg in	0.000	0.000	0.000	0.000	0.042	0.097	0.029	0.105
outdeg out	**0.069**	**0.125**	**0.153**	0.153	0.167	0.181	0.126	0.209
outdeg bidi	0.014	0.014	0.014	0.014	0.056	0.125	0.045	0.131
indeg in	0.056	0.111	**0.153**	0.167	0.181	**0.306**	0.129	0.207
indeg out	0.056	**0.125**	**0.153**	0.153	0.153	0.194	0.121	0.200
indeg bidi	0.042	0.069	0.111	0.139	0.139	0.194	0.104	0.205
pagerank in	**0.069**	**0.125**	**0.153**	**0.194**	**0.194**	0.292	**0.140**	**0.219**
pagerank out	0.056	0.097	**0.153**	0.153	0.167	0.208	0.117	0.195
pagerank bidi	0.056	0.069	0.111	0.139	0.153	0.236	0.113	**0.219**
hits in	0.014	0.028	0.042	0.069	0.083	0.111	0.046	0.095
hits out	0.056	0.056	0.111	0.125	0.153	0.181	0.102	0.181
hits bidi	0.014	0.042	0.042	0.056	0.069	0.125	0.050	0.110
scores	0.236	0.278	0.375	0.389	0.389	0.556	0.323	0.413
gp precisions	0.250	0.319	0.417	0.500	0.528	**0.639**	0.365	0.457
precisions	0.250	**0.361**	0.444	0.486	0.528	0.625	0.371	0.460
target occs	0.278	0.319	0.458	**0.528**	0.528	0.611	0.381	0.466
f measures	**0.306**	0.347	**0.472**	0.500	**0.542**	0.611	**0.399**	**0.479**

Based on the ranked result lists, we can calculate the Recall@k[11], Mean Average Precision (MAP) and Normalised Discounted Cumulative Gain of the various fusion variants over the whole test set $\mathcal{GT}_{\text{test}}$, as can be seen in Table 1 and Fig. 2.

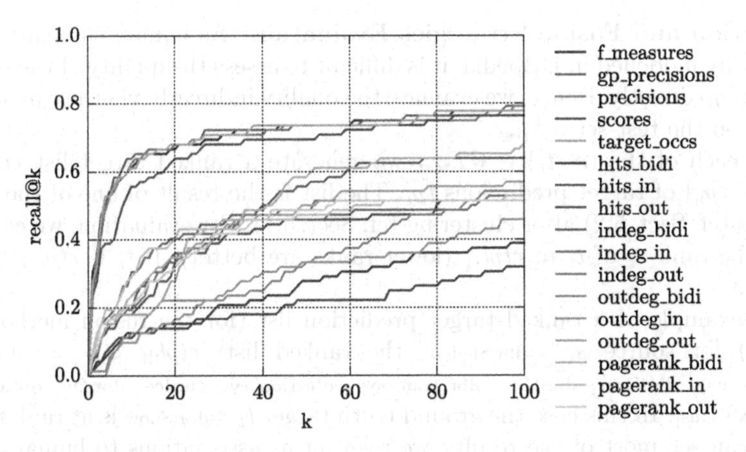

Fig. 2. Recall@k over the different fusion variants and against baselines

[11] We don't provide Precision@k, as it degenerates to Recall@k/k due to the fact that we only have 1 relevant target per result of any (s_t, t_t).

We also calculate these metrics for several baselines, which try to predict the target nodes from the 1-neighbourhood (bidirectionally, incoming or outgoing) by selecting the neighbour with the highest PageRank, HITS score, in- and out-degree [18,19]. As can be seen, all our fusion strategies significantly outperform the baselines.

7 Conclusion and Outlook

In this paper we presented an evolutionary graph pattern learner. The algorithm can successfully learn a set of patterns for a given list of source-target-pairs from a SPARQL endpoint. The learned patterns can be used to predict targets for a given source.

We use our algorithm to identify patterns in DBpedia for a dataset of human associations. The prediction quality of the learned patterns after fusion reaches a Recall@10 of 63.9 % and MAP of 39.9 %, and significantly outperforms PageRank, HITS and degree based baselines.

The algorithm, the used datasets and the interactive visualisation of the results are available online[12].

In the future, we plan to enhance our algorithm to support Literals in the input source-target-pairs, which will allow us to learn patterns directly from lists of textual inputs. Further, we are investigating mutations, for example to introduce FILTER constraints. We also plan to investigate the effects of including negative samples (currently we only use positive samples and treat everything else as negative).

Additionally, we plan to employ more advanced late fusion techniques, in order to learn when to trust the prediction of which pattern. As this idea is conceptually close to interpreting the learned patterns as a feature vector (with understandable and executable patterns to generate target candidates), we plan to investigate combinations of our algorithm with approaches that learn vector space representations from knowledge graphs.

This work was supported by the University of Kaiserslautern CS PhD scholarship program and the BMBF project MOM (Grant 01IW15002).

References

1. Berners-Lee, T., Hendler, J., Lassila, O.: The semantic web. Sci. Am. **284**(5), 34–43 (2001)
2. Bizer, C., Heath, T., Berners-Lee, T.: Linked Data - The Story So Far. Int. J. Semant. Web Inf. Syst. **5**(3), 1–22 (2009)
3. Bizer, C., Lehmann, J., Kobilarov, G., Auer, S., Becker, C., Cyganiak, R., Hellmann, S.: DBpedia - a crystallization point for the Web of Data. Web Semant. Sci. Serv. Agents World Wide Web **7**(3), 154–165 (2009)
4. Fortin, F.A., De Rainville, F.M., Gardner, M.A., Parizeau, M., Gagné, C.: DEAP: evolutionary algorithms made easy. J. Mach. Learn. Res. **13**, 2171–2175 (2012)

[12] https://w3id.org/associations.

5. Hees, J., Bauer, R., Folz, J., Borth, D., Dengel, A.: Edinburgh associative thesaurus as RDF and DBpedia mapping. The Semantic Web. In: ESWC SE. Springer, Heraklion, Crete, Greece (2016)
6. Hees, J., Khamis, M., Biedert, R., Abdennadher, S., Dengel, A.: Collecting links between entities ranked by human association strengths. In: Cimiano, P., Corcho, O., Presutti, V., Hollink, L., Rudolph, S. (eds.) ESWC 2013. LNCS, vol. 7882, pp. 517–531. Springer, Heidelberg (2013)
7. Hees, J., Roth-Berghofer, T., Biedert, R., Adrian, B., Dengel, A.: BetterRelations: using a game to rate linked data triples. In: Bach, J., Edelkamp, S. (eds.) KI 2011. LNCS (LNAI), vol. 7006, pp. 134–138. Springer, Heidelberg (2011). doi:10.1007/978-3-642-24455-1_12
8. Hees, J., Roth-Berghofer, T., Biedert, R., Adrian, B., Dengel, A.: BetterRelations: collecting association strengths for linked data triples with a game. In: Ceri, S., Brambilla, M. (eds.) Search Computing. LNCS, vol. 7538, pp. 223–239. Springer, Heidelberg (2012)
9. Hees, J., Roth-Berghofer, T., Dengel, A.: Linked data games: simulating human association with linked data. In: LWA, pp. 255–260. Kassel, Germany (2010)
10. Heim, P., Hellmann, S., Lehmann, J., Lohmann, S., Stegemann, T.: RelFinder: revealing relationships in RDF knowledge bases. In: Chua, T.-S., Kompatsiaris, Y., Mérialdo, B., Haas, W., Thallinger, G., Bailer, W. (eds.) SAMT 2009. LNCS, vol. 5887, pp. 182–187. Springer, Heidelberg (2009)
11. Heim, P., Lohmann, S., Stegemann, T.: Interactive relationship discovery via the semantic web. In: Aroyo, L., Antoniou, G., Hyvönen, E., ten Teije, A., Stuckenschmidt, H., Cabral, L., Tudorache, T. (eds.) ESWC 2010, Part I. LNCS, vol. 6088, pp. 303–317. Springer, Heidelberg (2010)
12. Kiss, G.R., Armstrong, C., Milroy, R., Piper, J.: An associative thesaurus of English and its computer analysis. In: The Computer and Literary Studies, pp. 153–165. Edinburgh University Press, Edinburgh, UK (1973)
13. Klyne, G., Carroll, J.J.: Resource Description Framework (RDF): Concepts and Abstract Syntax (2004). http://www.w3.org/TR/rdf-concepts/
14. Ławrynowicz, A., Potoniec, J.: Pattern based feature construction in semantic data mining. Int. J. SemWeb Inf. Syst. (IJSWIS) **10**(1), 27–65 (2014)
15. Lehmann, J., Bühmann, L.: AutoSPARQL: let users query your knowledge base. In: Antoniou, G., Grobelnik, M., Simperl, E., Parsia, B., Plexousakis, D., De Leenheer, P., Pan, J. (eds.) ESWC 2011, Part I. LNCS, vol. 6643, pp. 63–79. Springer, Heidelberg (2011)
16. McCusker, J.P.: WebSig: a digital signature framework for the web. Ph.D. thesis, Rensselaer Polytechnic Institute, Troy, NY (2015)
17. Nickel, M., Murphy, K., Tresp, V., Gabrilovich, E.: A review of relational machine learning for knowledge graphs, pp. 1–23 (2015)
18. Reddy, D., Knuth, M., Sack, H.: DBpedia GraphMeasures (2014). http://semanticmultimedia.org/node/6
19. Thalhammer, A., Rettinger, A.: PageRank on Wikipedia: towards general importance scores for entities. In: Know@LOD&CoDeS 2016, CEUR-WS Proceedings (2016)

Things and Strings: Improving Place Name Disambiguation from Short Texts by Combining Entity Co-Occurrence with Topic Modeling

Yiting Ju[1](✉), Benjamin Adams[2], Krzysztof Janowicz[1], Yingjie Hu[3], Bo Yan[1], and Grant McKenzie[4]

[1] STKO Lab, University of California, Santa Barbara, USA
{yju,boyan}@umail.ucsb.edu, janowicz@ucsb.edu

[2] Centre for eResearch, The University of Auckland, Berkeley, New Zealand
b.adams@auckland.ac.nz

[3] Department of Geography, University of Tennessee, Knoxville, USA
yhu21@utk.edu

[4] Department of Geographical Sciences, University of Maryland, College Park, USA
gmck@umd.edu

Abstract. Place name disambiguation is the task of correctly identifying a place from a set of places sharing a common name. It contributes to tasks such as knowledge extraction, query answering, geographic information retrieval, and automatic tagging. Disambiguation quality relies on the ability to correctly identify and interpret contextual clues, complicating the task for short texts. Here we propose a novel approach to the disambiguation of place names from short texts that integrates two models: entity co-occurrence and topic modeling. The first model uses Linked Data to identify related entities to improve disambiguation quality. The second model uses topic modeling to differentiate places based on the terms used to describe them. We evaluate our approach using a corpus of short texts, determine the suitable weight between models, and demonstrate that a combined model outperforms benchmark systems such as DBpedia Spotlight and Open Calais in terms of F1-score and Mean Reciprocal Rank.

Keywords: Place name disambiguation · Natural language processing · LDA · Wikipedia · DBpedia · Linked Data

1 Introduction

Geographic knowledge extraction and management, geographic information retrieval, question answering, and exploratory search hold great promise for various application areas [2,12,19]. From intelligence and media analysis to socio-environmental studies and disaster response, there is demonstrated need to be able to build computational systems that can synthesize and understand

E. Blomqvist et al. (Eds.): EKAW 2016, LNAI 10024, pp. 353–367, 2016.
DOI: 10.1007/978-3-319-49004-5_23

human expressions of information about places and events occurring around the world [8]. Being able to correctly identify geographic references in the abundance of unstructured textual information now available on the Web, in social media, and in other communication media is the first step to building tools for geographic analysis and discovery on these data. Place name, i.e., toponym, disambiguation is key to the comprehension of many texts as place names provide an important context required for the successful interpretation of text [13].

Similar to other named entities, including persons, organizations, and events, place names can be ambiguous. A single place name can be shared among multiple places. To give a concrete example, *Washington* is a place name for more than 43 populated places in the United States alone.[1] Although most of these Washingtons can be accurately located by adding the proper state name or county name, they are all simply referred to as *Washington* in daily conversations, (social) media, photo annotations, and so forth. Figure 1 depicts the distribution of the most common place names for U.S. cities, towns, villages, boroughs, and census-designated places. As shown on the map, these places are distributed across the U.S., indicating that the ambiguity of place names is a widespread phenomenon. It is worth noting that places which share a common name can be of the same or a different type, e.g., the *state* of Washington and the *city* of Washington, Pennsylvania. The situation is even more difficult on a global scale where place names may appear more than 100 times. For example, it takes merely a 45 min car ride to get from Berlin to East London, both located in South Africa. Thus, it is important to devise effective computational approaches to address the disambiguation problem.

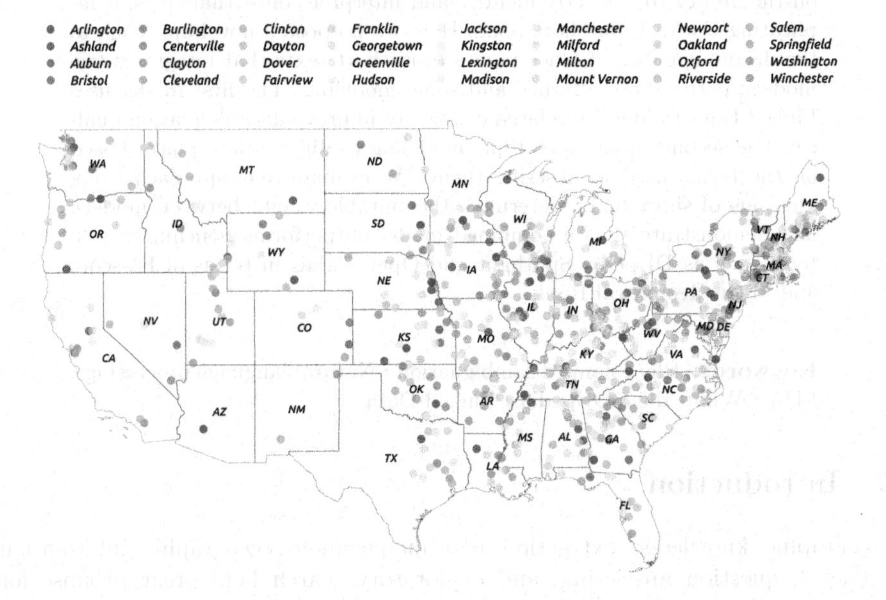

Fig. 1. Distribution of common place names in the US according to Wikipedia.

[1] https://en.wikipedia.org/wiki/Washington.

Given the wide availability of digital gazetteers, i.e., place name dictionaries, such as GeoNames, the Getty Thesaurus of Geographic Names, the Alexandria Digital Library Gazetteer, and Google Places, we assume that the places to be disambiguated are known, i.e. that there is a candidate list of places for any given place name list. After all, unknown places cannot be disambiguated. Thus, we define the task of place name disambiguation as follows: *given a short text which contains a place name and given a list of candidate places that share this name, determine to which specific place the text refers.*

Humans are very good at detecting and interpreting contextual clues in texts to disambiguate place names. Thus, as extension of named entity recognition, place name disambiguation has been tackled using computational approaches that aim at utilizing these contextual clues as well [5,7]. This context typically stems from the terms surrounding the place name under consideration. Typically, short texts from social media, news headlines (and abstracts), captions, and so forth, offer less contextual clues and thus negatively impact disambiguation quality. Consequently, new approaches have to be develop that can extract and interpret other contextual clues.

One such approach is to focus on the detection of surrounding *entities* and use these as contextual clues. Besides the place itself, these entities may include other places, actors, objects, organizations, and events. Examples of such associated entities are landmarks, sports teams, well known figures such as politicians or celebrities, and nearby places that share a common administrative unit [22]. Intuitively, when a text mentions *Washington* along with *Redskins*, an American football team based in Washington, D.C., it is very likely that the *Washington* in the text refers to Washington, D.C., rather than another places called with the same toponym. It has been shown that such a co-occurrence model increases disambiguation quality [11,18].

In addition to entities, implicit *thematic* information buried in the text can also provide contextual evidence to disambiguate place names. Similar to entities, some particular thematic topics are more likely to be mentioned along with a place, which is characterized by those topics. Topic modeling makes it possible to discover topics from the text and match texts with similar topics. Thus, given topics learned from a corpus of texts about candidate places and the topics discovered from the short text under consideration, computing a similarity score between topics representative for the text and for each of the candidate places can provide additional contextual clues [1]. For example, when people are talking about *Washington, DC*, political topics featuring terms such as *conservative*, *policy*, and *liberal* are more likely to be mentioned than when talking about the (small) city of *Washington, Pennsylvania*.

The core distinction between these perspectives is that mentioned entities are explicit information, while thematic information is usually implicit. Both types of information are used as clues by humans to disambiguate a place name. In this paper, we propose a novel approach which integrates *things and strings*, i.e., entity co-occurrence and topic modeling, thereby combining explicit and implicit contextual clues. **The contributions of this work are as follows:**

- We apply topic modeling to place name disambiguation, an approach that has not been taken before.
- We integrate this topic-based model with a reworked version of our previous entity-based co-occurrence model [11] and learn the appropriate weights for this integrated model.
- We compare the integrated model to three well known systems (TextRazor, DBpedia Spotlight, and Open Calais) as baselines and demonstrate that our model outperforms all of them.

2 Related Work

As an extension of named entity disambiguation, place name disambiguation can be conducted using the general approaches from named entity disambiguation. Wikipedia, as a valuable source for ground truth descriptions of named entities, has been used in a number of studies. For example, Bunescu and Pasca [5] trained a vector space model to host the contextual and categorical terms derived from Wikipedia, and employed TF-IDF to determine the importance of these terms. Milne and Witten [17] describes a method for augmenting unstructured text with links to Wikipedia articles. For ambiguous links, the authors proposed a machine learning approach and trained several models based on Wikipedia data. Two named entity disambiguation modules were introduced by Mihalcea and Csomai [16]. One measured the overlaps between context and candidate descriptions, and the other trained a supervised learning model based on manually assigned links in the Wikipedia articles.

For studies specifically focusing on place name disambiguation, Jones and Purves [13] discussed using related places to resolve place ambiguity. Machado et al. [14] proposed an ontological gazetteer which records the semantic relations between places to help disambiguate place names based on related places and alternative place names. In a similar approach, Spitz et al. [22] constructed a network of place relatedness based on English Wikipedia articles. Zhang and Gelernter [24] proposed a supervised machine learning approach to rank candidate places for ambiguous toponyms in Twitter messages that relies on the metadata of tweets and context to a limited extent. In previous work, we leveraged the structured Linked Data in DBpedia for place name disambiguation and demonstrated that a combination of Wikipedia and DBpedia data leads to generally better performance [11].

3 Methodology

The work at hand differs from these previous studies. We apply topic modeling for place name disambiguation and integrate the trained topic model with an entity-based model which captures the co-occurrence relations. Thereby we combine a *things*-based perspective with a *strings*-based perspective.

In the following, we assume that the *surface forms* of place names have been extracted prior to disambiguation, so the primary task of place name disambiguation is to identify the place to which a surface form refers. To accomplish this a list of candidate entities, i.e., places, is selected. In prior work, knowledge bases, such as Wikipedia, DBpedia, and WordNet have been used to obtain candidate entities [6, 10, 15], and here we employ DBpedia as the source of candidate entities. Once a set of candidate places has been identified, the likelihood that the surface form refers to each entity is measured and the disambiguation result is returned if the computed score exceeds a given threshold.

3.1 Entity-Based Co-Occurrence Model

In this section we describe the entity-based co-occurrence method. Wikipedia and DBpedia are used as the sources to train our model. We define the entities from Wikipedia as those words or phrases on a Wikipedia page of the candidate places which have links to another page about these entities. The entities from DBpedia are either subjects or objects of those RDF triples which contain the candidate place entities. Not all RDF triples are selected, but those that fall under the DBpedia namespace, i.e., with prefix *dbp*[2] and *dbo*.[3] While *dbo* provides a cleaner and better structured mapping-based dataset, it does not provide a complete coverage of the original properties and types from the Wikipedia infoboxes. In order to avoid data bias we use both *dbo* and *dbp*. Literals were excluded as well. We treat the subject or object of a triple as a whole, i.e., as an individual entity, instead of further tokenizing it into terms. The harvested entities differ greatly. They include related places (of different types), time zone information, known figures that were born or died at the given place, events that took place there, companies, organizations,[4] sports teams, as well as representative landmarks such as buildings or other physical objects.

Table 1 shows some sample entities for Washington, Louisiana, derived from Wikipedia and DBpedia. It should be noted that there is considerable overlap between place data extracted from Wikipedia and DBpedia. Moreover, some properties such as *population density* in Wikipedia can occur for most or even all candidate places. Such entities which appear frequently but help less to uniquely identify a place will not play a crucial rule in disambiguating the place names.

The entities are assigned weights according to their relative connectivity to the places by means of *term frequency-inverse document frequency (TF-IDF)*. The term frequency of the entity is the number of times the entity appears in Wikipedia and DBpedia, so in this case, it could be 0, 1, and 2. We only count each entity's appearance in a document once, so the term frequency will not be inflated by those entities which are related to many candidate place entities while contribute less to uniquely identify the place. The formula of applying TF-IDF to assign weights to entities is defined in Eqs. 1, 2, and 3.

---

[2] http://dbpedia.org/resource/.

[3] http://dbpedia.org/ontology/.

[4] For example via dbr:FreedomWorks dbp:headquarters dbr:Washington,_D.C.

Table 1. Sample entities for Washington, LA, from Wikipedia and DBpedia

Washington, Louisiana
Wikipedia — St. Landry Parish; Opelousas; Eunice; population density; medianhousehold income; American Civil War; Connecticut; cattle; cow; corn...
DBpedia — United States; Central Time Zone; St. Landry Parish, Louisiana; John M. Parker; KNEX-FM; Louisiana Highway 10...

$$tf(e) = \begin{cases} 0 & e \text{ is not in Wikipedia and DBpedia} \\ 1 & e \text{ is either in Wikipedia or DBpedia} \\ 2 & e \text{ is in both Wikipedia and DBpedia} \end{cases} \quad (1)$$

$$idf(e) = 1 + log(\frac{|E| + 1}{n_e}) \quad (2)$$

$$Weight(e) = TF\text{-}IDF = tf(e) \times idf(e) \quad (3)$$

Here $tf(e)$ defines the term frequency of an entity e, and $idf(e)$ defines the inverse document frequency of e. $|E|$ is the number of all potential candidate places for a surface form, and n_e represents the number of candidate places which contain the entity e. Using TF-IDF entities appearing in multiple candidate places are given lower weights, while entities which are able to uniquely identify a place have more weights. For example, the fact that a place is within the United States becomes irrelevant as it holds for all of them.

We then measure the likelihood that a surface form in a test sentence refers to a candidate place through an entity matching score. To calculate the entity matching score, we first find those entities of the candidate place which also appear in the short text. The weights of matching entities are summed to produce an entity matching score of the candidate place to the surface form in the test sentence. The score is calculated as given in Eq. 4.

$$S_{EC}(s \rightarrow c_i) = \sum_{j=1}^{m} (Weight(e_j) \times I_j) \quad (4)$$

Here m corresponds to the number of entities e for the candidate c_i. I_j is either 1 or 0, referring to whether a matching entity is found in the test for the entity e_j. The candidate place with higher entity matching score is regarded to more likely be the actual place to which the surface form refers. The matching score is the final output of the entity co-occurrence model.

3.2 Topic-Based Model

In this section we introduce the topic-based model. It makes use of the fact that text is *geo-indicative* [1] even without having any direct geographic references.

Hence, even everyday language should be able to provide additional evidence for place name disambiguation. For example, terms such as *humid*, *hot*, *festival*, *poverty*, and even *American Civil War* are more likely to be uttered when referring to Washington, Louisiana than Washington, Maine. The latter rarely experiences hot and humid weather, does not host a popular festival, has substantially less poverty problems compared to its namesake, and did not play a notable role in the civil war. Here we use Latent Dirichlet allocation (LDA) for topic modeling. LDA is a popular unsupervised machine learning algorithm used to discover topics in a large document collection [4]. Each document is modeled as a probability vector over a set of topics, providing a dimensionally-reduced representation of the documents in the corpus.

We use the geo-referenced text from the English Wikipedia as the source material for discovering these thematic patterns. We start with the idea that a collection of texts that describe various features in a local region–such as museums, parks, mountains, architectural landmarks, etc.–give us a foundation for differentiating places referenced in other texts based on thematic, non-geographically specific, terms. For this we need a systematic way to associate the training documents in Wikipedia with well-defined regions. Because administrative regions vary widely in area, they do not provide a good mechanism for aggregation. Instead, our solution is to aggregate the geo-referenced texts in Wikipedia based on an equal area grid over the Earth. This solution means that articles with point-based geo-references are binned together if they spatially intersect with a grid cell, while text related to areal features (such as national parks) can be associated with multiple grid cells.

There are several options for creating a discrete global grid based on an polyhedral simplification of the Earth [21]. In this work we utilize the Fuller icosahedral Dymaxion projection to create a hierarchical triangular mesh [9]. The triangular mesh can be made successively more fine-grained by dividing each triangle into four internal triangles. For place name disambiguation we need grid cells that are fine-grained enough so that two possible places with the same name do not fall within one grid cell. The Fuller projection at hierarchical level 7 (shown in Fig. 2) provides a mesh over the Earth with 327,680 cells with inter-cell distance of 31.81 Km and cell area of 1,556.6 km^2, sufficient to handle most place name disambiguation tasks for meso-scale features like cities.

Once we identified all articles that have geo-references that spatially intersect with a grid cell we can combine all the text to create a *grid document*. For the English Wikipedia the geo-referenced articles intersect with 63,473 grid cells at Fuller level 7. The resulting 63,473 grid documents serve as the training data input for LDA topic modeling. We utilized the MALLET implementation of LDA with hyperparameter optimization, which allows for topics to vary in importance in the generated corpus, and we trained the LDA topic model with 512 topics.

The MALLET toolkit generates an inferencer file for testing new documents against a trained LDA model. For a new document or snippet of text, we use the trained topic model to infer the most likely candidate location based on the inferred mixture of topics. Given a set of candidate locations (i.e., point coordinates) we find the topic mixtures for the grid cells that spatially intersect the

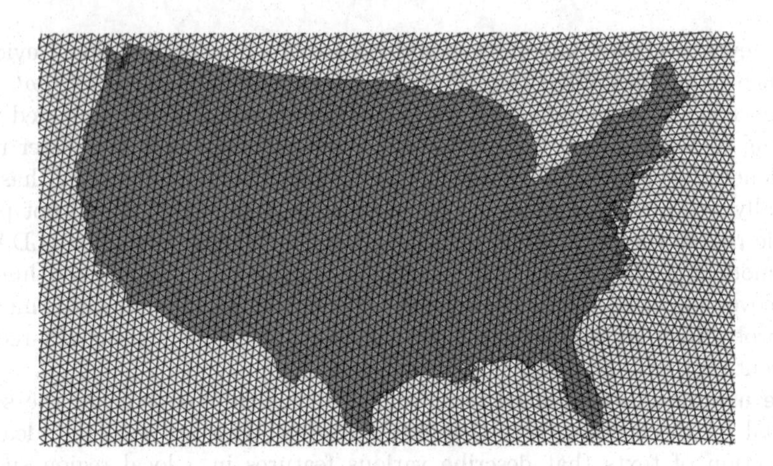

Fig. 2. Level 7 triangular mesh discrete global grid built using Fuller icosahedral Dymaxion projection, shown in U.S. Contiguous Albers projection.

locations and calculate the Jensen-Shannon divergence (Eq. 6) between probability vector representations of the topic mixtures for each candidate and the topic mixture for the new document. The JS divergence is a symmetric measure calculated from the average of the relative entropies (Kullback Leibler divergence, shown in Eq. 5) between two probability vectors (P and Q) and their average, $M = \frac{1}{2}(P + Q)$. The JS divergence is a standard measure of similarity between two probability vectors, and is commonly used for calculating similarity based on topic model results [23]. A lower JS divergence result indicates greater thematic similarity between the new text and the candidate location.

$$KL(P \parallel Q) = \sum_i P(i) \log_2 \frac{P(i)}{Q(i)} \tag{5}$$

$$JS(P \parallel Q) = \frac{1}{2} KL(P \parallel M) + \frac{1}{2} KL(Q \parallel M) \tag{6}$$

3.3 Integrated Model (ETM)

The first model makes use of the co-occurrence of entities as contextual clue to disambiguate place names, while the second model puts emphasis on linguistic aspects, namely co-occurring topics. As argued in the introduction, applying a single model, which extracts partial contextual clues, is often not sufficient to differentiate place names from short texts. Thus, we combine the entity-based model and string-based topic model to an integrated approach called ETM (Entity & Topic Model).

Both the entity co-occurrence model and the topic-based model return a score when comparing each candidate place with each ambiguous place name in a sample text. The scores from these two models are not directly comparable as they involve relative probabilistic measures. To combine the models, we must first standardize the scores of the candidate places for each short text. This results in setting the standardized mean to zero. Scores originally higher than the mean will be positive, and scores originally lower than the mean will be negative. For each candidate place name, the standardized scores from the entity co-occurrence model are then combined with the standardized scores from the topic-based model along with a weighting parameter λ as shown in Eq. 7.

$$S_{ETM}(s \rightarrow c_i) = \lambda S_{ECM}(s \rightarrow c_i) + (1 - \lambda) S_{TM}(s \rightarrow c_i) \tag{7}$$

Here $\lambda \in [0, 1]$, and determines how much each model is weighted in the combined approach. S_{ECM} is the standardized score computed from the entity co-occurrence model for the candidate place name c_i with respect to the surface form s, while S_{TM} is the standardized score from the topic model, namely the JS divergence. S_{ETM} is the score of the combined model, which is the sum of the weighted standardized scores of the two models. Provided that S_{ETM} is the probability of a candidate place which a surface form refers to, the percentile is used as the threshold over which candidate places are returned as the disambiguation result.

4 Evaluation

In this section we evaluate the performance of our proposed ETM and describe the methods through which we gathered the testing corpus and the metrics employed for the evaluation.

4.1 Preparing the Test Corpus

We constructed a text corpus specifically for the evaluation of our place name disambiguation models. The corpus is used to evaluate the performance of the combined ETM and to compare it to existing systems acting as baselines.

Table 2. Three example records of the test corpus extracted from websites.

Oxford, Wisconsin — Located in Marquette County in south-central ~~Wisconsin~~, just minutes west of Interstate 39, Oxford invites you to experience our small town charm along with the area's many year-round outdoor attractions.
Jackson, Montana — The tiny town of Jackson, ~~Montana~~ has made a name for itself as a winter sports destination for the adventurous.
Dayton, Nevada — Since the Native-American tribes in the area were nomadic, this made Dayton the first and oldest permanent non-native settlement in ~~Nevada~~.

To construct the corpus, we first derive ambiguous place names from a list of the most common U.S. place names on Wikipedia.[5] As the list also presents the full place names which could be used to identify the place of interest, we feed the full place names into the Bing Search API,[6] which returns a list of websites related to the place along with URLs. URLs containing "Wikipedia" are filtered out. We then visit the selected websites and extract sentences which contain the full place name. The auxiliary part of the full place name (state or county name) is removed, so the remaining place name is ambiguous. The result of this approach is a set of real-world, i.e., not synthetic, sentences containing ambiguous place names. These sentences comprise our ground truth data.

Sample ground truth sentences are shown in Table 2. The full place name and test sentence are separated by an em-dash, and the auxiliary part of the full place name is removed (shown as *striken* for example purposes). This resulting data contains noise. Some sentences, for instance, contain no meaningful entities or terms that can be categorized into topics, while others seem to be automatically generated from templates. This noise, however, can help evaluate the robustness of our models. In total, the testing corpus consists of 5,500 sentences. The average length of a test sentence is 22.54 words with a median of 19. Note that stop words count towards these statistics, while auxiliary parts of the place name do not.

4.2 Metrics

F-score and Mean Reciprocal Rank (MRR) are used as metrics for the performance evaluation of the place name disambiguation models. The F-score (see Eq. 8) is defined as the harmonic mean of precision and recall [3]. MRR, by comparison, considers the order of the results; see Eq. 9. The reciprocal rank of a test sentence is the inverse of the rank of the correctly identified place name in the list of the candidate places for the surface form.

$$F_1 = 2 \cdot \frac{Precision \cdot Recall}{Precision + Recall} \tag{8}$$

$$MRR = \frac{1}{|Q|} \sum_{r=1}^{|Q|} \frac{1}{rank_i} \tag{9}$$

4.3 Results

In this section, we present the results of our evaluation and compare them to other well recognized named entity disambiguation systems as baselines.

DBpedia Spotlight[7], TextRazor[8], and Open Calais[9] were selected as baseline systems to be compared to ETM. DBpedia Spotlight is based on DBpedia's

[5] https://en.wikipedia.org/wiki/List_of_the_most_common_U.S._place_names.
[6] https://datamarket.azure.com/dataset/bing/search.
[7] https://github.com/dbpedia-spotlight/dbpedia-spotlight.
[8] https://www.textrazor.com/.
[9] http://www.opencalais.com/.

rich knowledge base of structured data [15], which is also employed by our proposed model. Two endpoints of DBpedia Spotlight Web Service (V. 0.7) were used for testing, namely *Annotate* and *Candidates*. The *Candidates* endpoint returns a ranked list of candidates for each recognized entity and concept, while *Annotate* simply returns the best candidate according to the context. TextRazor and Open Calais are two commercial Web services for named entity recognition and named entity disambiguation. Both services offer application programming interfaces (APIs). The TextRazor API returns only one candidate for each entity recognized from the test sentence. Experiments were conducted [20] to compare several named entity disambiguation systems which included DBpedia Spotlight (V. 0.6, confidence = 0, support = 0) and TextRazor. In the experiments, TextRazor demonstrates the best performance in terms of F-score. Open Calais API also returns only one candidate for each recognized entity, while it provides additional *social tags* for each test text instance.

Given that TextRazor and Open Calais do not provide controls on how many candidate places are returned and DBpedia Spotlight relies on *Confidence* and *Support* which are not comparable to percentiles, we choose the highest scores each baseline systems can reach to compare it to our models. For instance, for DBpedia Spotlight, we picked $Confidence = 0.2$ and $Support = 0$, given that this combination of parameter leads to the best overall performance for our setting. From Fig. 3 we can see that Open Calais can obtain relatively higher F-scores and MRR than TextRazor and DBpedia Spotlight on the test corpus. The F-score and MRR of those baseline systems on the testing dataset are shown in Table 3. Compared to these systems, the *individual* performance of the entity-based co-occurrence model and topic-based model do not show a significant improvement, except for the entity co-occurrence model on MRR.

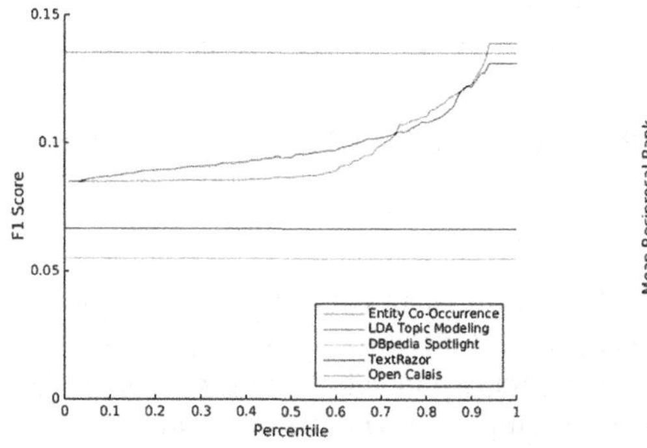

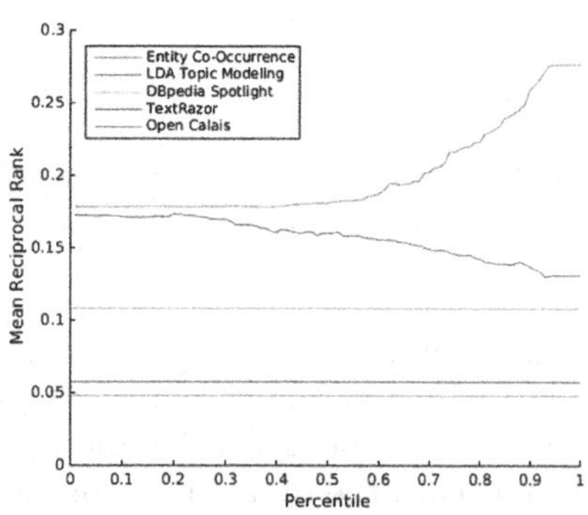

Fig. 3. (left) F-score and (right) Mean Reciprocal Rank for the entity co-occurrence model and the topic model along percentile, and comparison with DBpedia Spotlight, TextRazor, and Open Calais.

Figure 3 also shows how F-score and MRR change along percentiles. Note that the 0.9 at the x-axis refers to the 90th percentile, which means that the candidate places with top 10 % of scores are selected as the disambiguation result. As shown in the plots, when percentile increases, the F-scores of both individual models increase very slightly until the 60th percentile when the scores start increasing dramatically. The MRR for the entity co-occurrence model along percentiles has a similar trend as the F-score, while the MRR for the topic model drops when less candidate places are selected.

ETM, which combines the entity-based co-occurrence model with the topic-based model, demonstrates a significant improvement in terms of F-score and MRR, as shown in Fig. 4. We tested λ values from 0 to 1 with an interval of 0.01 and found that $\lambda = 0.48$ yields the best results on the test dataset. This indicates that both the entity co-occurrence model and the topic model play roughly even roles in ETM for disambiguating place names. At the 94th percentile, the F-score is 0.239, while MRR is 0.239. Note that F-score and MRR are different values though they happen to be rounded to the same value. Out of 5,500 test sentences, 1,315 are correctly disambiguated, given the disambiguation result of 5,509 places. The figures show that both F-score and MRR increase along with percentiles and reach peaks when very low percentage of records are returned as disambiguation results.

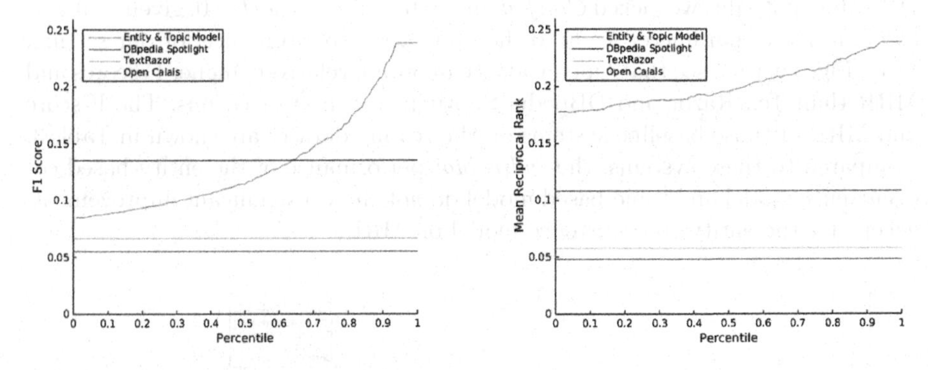

Fig. 4. (left) F-score and (right) Mean Reciprocal Rank for ETM ($\lambda = 0.48$), DBpedia Spotlight, TextRazor, and Open Calais.

In some cases, only one candidate (if available) is taken as the disambiguation results. As stated in the previous paragraph, TextRazor only outputs at most one result, so does DBpedia Spotlight Web Service in the *Annotation* mode. For Open Calais, the disambiguation result is ranked, so the first returned result is taken for this evaluation. When only the candidate places with highest scores are taken, the F-score for ETM reaches 0.238 when λ is set to 0.48. Since always one candidate is picked for each testing sentence, predicted condition positives are the same as condition positives. Thus, the mean reciprocal rank, precision and recall are identical, and they top at 0.238 with λ being 0.48. The change of F-score and Mean Reciprocal Rank for our proposed ETM along λ and its comparison to DBpedia Spotlight, TextRazor, and Open Calais are shown in Fig. 5. As shown in the figure, with the increase of λ, after the peak when λ is around 0.48, F-score and MRR drop mildly until λ approaching 1 when F-score and MRR drop significantly. This implies the entity co-occurrence model plays a more important role for this task, while the topic model still helps to improve the performance. Out of 5500 testing sentences, EMT is able to correctly identify 1311 ambiguous places.

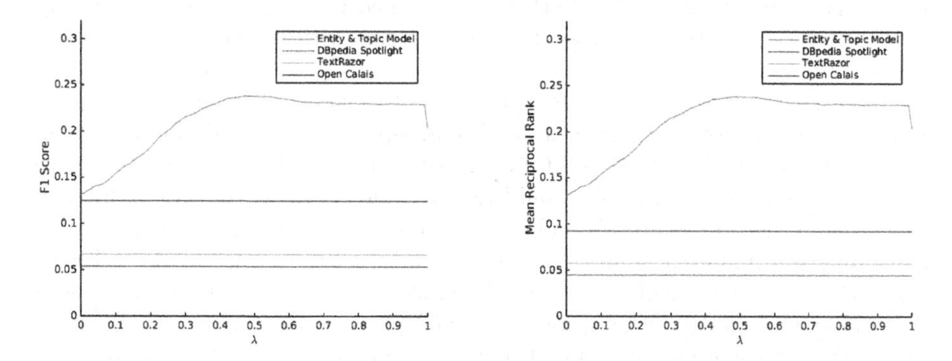

Fig. 5. (left) F-score and (right) Mean Reciprocal Rank for ETM, DBpedia Spotlight, TextRazor, and Open Calais, when only the best candidate entity is taken.

The evaluation of ETM and its comparison to baseline systems are summarized in Table 3. Overall, based on the evaluation, the proposed ETM substantially outperforms existing named entity disambiguation systems in terms of F-score and Mean Reciprocal Rank. The fact that all F-scores are low, is an important reminder for the fact that place name disambiguation from short texts is a difficult task (and that some test sentences did not contain any or only minimal contextual clues).

Table 3. Comparison of systems at best performance in terms of Precision, Recall, F1-Score and Mean Reciprocal Rank (MRR).

Model	Parameters	Precision	Recall	F1-Score	MRR
DBpedia Spotlight	$Confidence = 0.2$; $Support = 0$	0.057	0.053	0.055	0.048
TextRazor	n/a	0.070	0.063	0.067	0.058
Open Calais	n/a	0.148	0.125	0.135	0.108
ETM	$\lambda = 0.48$; 94th percentile	0.239	0.239	0.239	0.239

5 Conclusions and Further Work

In this paper we proposed a novel approach to tackle the challenging task of disambiguating place names from short texts. Place name disambiguation is an important part of knowledge extraction and a core component of geographic information retrieval systems. We have presented two models that are driven by different perspectives, namely an entity-based co-occurrence model and a topic-based model. The first model focuses on the semantic connections between entities and thereby on *things*, while the second model works on the linguistic level by investigating topics associated with places and thereby takes a *string*-based perspective. The integration of both models (called ETM) shows a substantially better performance than the used baseline systems with respect to F-score and MRR.

Nonetheless, there is space for future improvements. For the entity-based model, properties other than those with namespaces of *dbo* and *dbp* have been filtered out. The same is true for literals. Both of these could be added to a future version of ETM, although they would require more work on the used similarity functions in case of the literals and a better alignment to ensure that properties from different namespaces are not mere duplicates. In our work, the ETM is realized as a convex combination of the entity-based co-occurrence model and the topic-based model. Other approaches could be investigated as well. We have used LDA for topic modeling but this is not the only choice that can be used and other approaches will be tested in the future.

As for the experiment, although place entities in our testing corpus have highly ambiguous place names, those places are all some kind of administrative divisions (i.e., cities, towns, villages, etc.) and located within the United States. A potential improvement could be seeking more ambiguous place names from other types of places which are outside of the United States.

Acknowledgement. The authors would like to acknowledge partial support by the National Science Foundation (NSF) under award 1440202 EarthCube Building Blocks: Collaborative Proposal: GeoLink Leveraging Semantics and Linked Data for Data Sharing and Discovery in the Geosciences.

References

1. Adams, B., Janowicz, K.: On the geo-indicativeness of non-georeferenced text. In: International AAAI Conference on Web and Social Media (ICWSM), pp. 375–378 (2012)
2. Adams, B., McKenzie, G., Gahegan, M.: Frankenplace: interactive thematic mapping for ad hoc exploratory search. In: Proceedings of the 24th International Conference on World Wide Web, pp. 12–22. ACM (2015)
3. Bilenko, M., Mooney, R., Cohen, W., Ravikumar, P., Fienberg, S.: Adaptive name matching in information integration. IEEE Intell. Syst. **18**(5), 16–23 (2003)
4. Blei, D.M., Ng, A.Y., Jordan, M.I.: Latent Dirichlet allocation. J. Mach. Learn. Res. **3**(Jan), 993–1022 (2003)
5. Bunescu, R.C., Pasca, M.: Using encyclopedic knowledge for named entity disambiguation. EACL **6**, 9–16 (2006)
6. Cucerzan, S.: Large-scale named entity disambiguation based on Wikipedia data. EMNLP-CoNLL **7**, 708–716 (2007)

7. Fader, A., Soderland, S., Etzioni, O., Center, T.: Scaling Wikipedia-based named entity disambiguation to arbitrary web text. In: Proceedings of the IJCAI Workshop on User-contributed Knowledge, Artificial Intelligence: An Evolving Synergy, Pasadena, CA, USA, pp. 21–26, 2009 (2011)
8. Goodchild, M.F., Glennon, J.A.: Crowdsourcing geographic information for disaster response: a research frontier. Int. J. Digit. Earth **3**(3), 231–241 (2010)
9. Gray, R.W.: Exact transformation equations for Fuller's world map. Cartogr.: Int. J. Geogr. Inf. Geovis. **32**(3), 17–25 (1995)
10. Han, X., Zhao, J., Structural semantic relatedness: a knowledge-based method to named entity disambiguation. In: Proceedings of the 48th Annual Meeting of the Association for Computational Linguistics, pp. 50–59. Association for Computational Linguistics (2010)
11. Hu, Y., Janowicz, K., Prasad, S.: Improving Wikipedia-based place name disambiguation in short texts using structured data from DBpedia. In: Proceedings of the 8th Workshop on Geographic Information Retrieval, p. 8. ACM (2014)
12. Janowicz, K., Hitzler, P.: The digital earth as knowledge engine. Semant. Web **3**(3), 213–221 (2012)
13. Jones, C.B., Purves, R.S.: Geographical information retrieval. Int. J. Geogr. Inf. Sci. **22**(3), 219–228 (2008)
14. Machado, I.M.R., de Alencar, R.O., de Oliveira Campos Jr., R., Davis Jr., C.A.: An ontological gazetteer and its application for place name disambiguation in text. J. Braz. Comput. Soc. **17**(4), 267–279 (2011)
15. Mendes, P.N., Jakob, M., García-Silva, A., Bizer, C., Dbpedia spotlight: shedding light on the web of documents. In: Proceedings of the 7th International Conference on Semantic Systems, pp. 1–8. ACM (2011)
16. Mihalcea, R., Csomai, A.: Wikify!: linking documents to encyclopedic knowledge. In Proceedings of the sixteenth ACM conference on Conference on information and knowledge management, pp. 233–242. ACM (2007)
17. Milne, D., Witten, I.H.: Learning to link with Wikipedia. In: Proceedings of the 17th ACM conference on Information and knowledge management, pp. 509–518. ACM (2008)
18. Overell, S., Rüger, S.: Using co-occurrence models for placename disambiguation. Int. J. Geogr. Inf. Sci. **22**(3), 265–287 (2008)
19. Purves, R., Jones, C.: Geographic information retrieval. SIGSPATIAL Spec. **3**(2), 2–4 (2011)
20. Rizzo, G., van Erp, M., Troncy, R.: Benchmarking the extraction and disambiguation of named entities on the semantic web. In: LREC, pp. 4593–4600 (2014)
21. Sahr, K., White, D., Kimerling, A.J.: Geodesic discrete global grid systems. Cartogr. Geogr. Inf. Sci. **30**(2), 121–134 (2003)
22. Spitz, A., Geiß, J., Gertz, M., So far away, yet so close: augmenting toponym disambiguation and similarity with text-based networks. In: Proceedings of the Third International ACM SIGMOD Workshop on Managing and Mining Enriched Geo-Spatial Data, GeoRich 2016, pp. 2: 1–2: 6. ACM, New York, NY, USA (2016)
23. Steyvers, M., Griffiths, T.: Probabilistic topic models. Handb. Latent Semant. Anal. **427**(7), 424–440 (2007)
24. Zhang, W., Gelernter, J.: Geocoding location expressions in Twitter messages: a preference learning method. J. Spat. Inf. Sci. **2014**(9), 37–70 (2014)

Relating Some Stuff to Other Stuff

C. Maria Keet(✉)

Department of Computer Science, University of Cape Town,
Cape Town, South Africa
mkeet@cs.uct.ac.za

Abstract. Traceability in food and medicine supply chains has to handle stuffs—entities such as milk and starch indicated with mass nouns—and their portions and parts that get separated and put together to make the final product. Implementations have underspecified 'links', if at all, and theoretical accounts from philosophy and in domain ontologies are incomplete as regards the relations involved. To solve this issue, we define seven relations for portions and stuff-parts, which are temporal where needed. The resulting theory distinguishes between the extensional and intensional level, and between amount of stuff and quantity. With application trade-offs, this has been implemented as an extension to the Stuff Ontology core ontology that now also imports a special purpose module of the Ontology of units of Measure for quantities. Although atemporal, some automated reasoning for traceability is still possible thanks to using property chains to approximate the relevant temporal aspects.

1 Introduction

Part-whole relations have been investigated in fields such as ontologies, conceptual modelling, cognitive science, and natural language. The part-whole relation between stuffs like milk, mayonnaise, and alcohol—i.e., uncountable entities other in amounts and indicated in language with mass nouns—or particular amounts of stuff—e.g., the amount of milk in your mug—has been named also part of, but also portion of, piece of, sub quantity of, or ingredient of; e.g., [4,10,12,14,16,18,25]. This already raises the question as to what exactly is going on, and which of those relations are the same or different, so as to be able to choose the right one when developing an ontology or a conceptual model. This becomes crucial in particular when one would want to reason over it for, e.g., traceability in the food chain: the portion in your mug of milk *was* a portion of the amount of milk in the carton, which was again a portion from some batch in the food processing plant, and deriving their relatedness would aid food safety applications in traceability [9,28]. Superficially similar examples are a piece of meat that is contaminated with *E. coli*, yet its fat that would be safe for consumption, and vaporising alcohol from an amount of wine during cooking.

That is, there is a need for making distinctions in how stuffs relate. While it has been recognised that there are differences between parts and portions and stuff and their quantities, this has not yet been fully addressed. Options proposed conflate knowledge at the type and the instance/particular level (the stuff

E. Blomqvist et al. (Eds.): EKAW 2016, LNAI 10024, pp. 368–383, 2016.
DOI: 10.1007/978-3-319-49004-5_24

universal and an amount of it) and the repeatable quantity [14], the temporal dimension has received little attention, or to the extent that it cannot be readily implemented [2,10], and it is not clear whether all those relations are really variations on mereological parthood [18].

We aim to solve these problems by defining seven relations for portions, pieces, and stuff-parts, which are temporalised where needed. In addition, we make a clear distinction between the extensional and intensional levels (amounts vs stuff kinds) that are separate yet represented in the same ontology, and we distinguish portions from quantities. The resulting model with the relations are implemented by extending the Stuff Ontology core ontology of [17] accordingly and importing a special purpose module of the Ontology of units of Measure (OM) for quantities that was developed by domain experts in food [26]. Traceability is then aided by availing of the more precise representation and property chains. The ontologies are available from http://www.meteck.org/stuff.html.

The remainder of the paper first summarises related works (Sect. 2), which is followed by some preliminaries (Sect. 3). Section 4 describes the model and has the formal definitions of the stuff relations, which is subjected to implementation trade-offs in Sect. 5. We discuss in Sect. 6 and conclude in Sect. 7.

2 Related Work

Many ontologies do have at least one relation to relate stuffs specifically. We cover a selection to exemplify the outcome of the assessment, which is that developers of the respective ontologies have struggled with the same questions and either opted for different 'workarounds' or ignored it by overloading parthood. Thereafter we zoom in on the two most recent papers from formal ontology.

2.1 Ontologies as Artefacts

The taxonomy of part-whole relations [18] has a subQuantityOf relation, which the authors admit to be underspecified as lumping together portions of the same kind of stuff and part-stuff of a whole-stuff that are of a different kind of stuff. Elsewhere the latter is also called ingredient [16] and hasSubStuff [17]. The SIO [11] has only has proper part between objects, which may or may not be stuffs, though for liquids, there is also a 'liquid solution component' intended as a specific stuff-part, and mass (a quantity) as 'is attribute of' some 'material entity'. DOLCE-lite (based on [20]) also uses only part, but also has a way to represent the quantity of the amount-of-matter using the has-quality property. BioTop [5] has a temporal part (temporal in the name only, not in the logic, for OWL is atemporal), and therewith one can distinguish descriptively between contiguous and scattered portions, and likewise with portionOf and scatteredPortionOf in the Stuff Ontology [17]. SUMO has piece as "arbitrary parts of Substances" and its super property part for its superclass Object; there is no measure of quantity associated with Substance. It is not much better in domain ontologies that

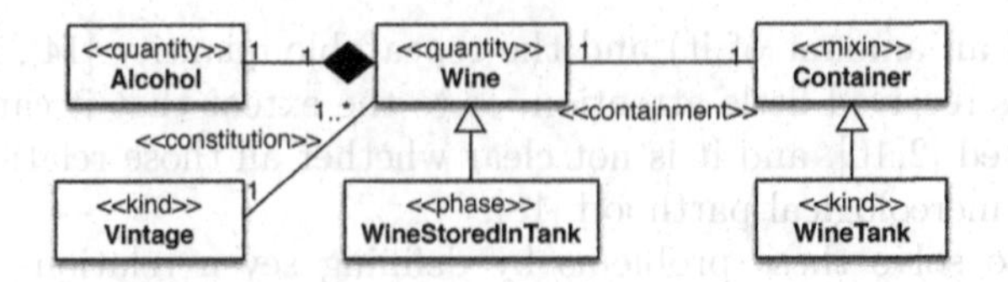

Fig. 1. Guizzardi's example for quantities [14].

typically seek modelling guidance from foundational and core ontologies. For instance, the Environment ontology uses the generic part_of relation from BFO. SNOMED CT's [27] Has active ingredient clearly has to do with a part-stuff of some medicine, but has no domain and range restrictions to enforce it.

A different angle is so-called pedigrees for traceability, notably for medicines, [28] that extends the provenance ontology PROV-O[1]. Currently, it focuses on the amounts and their properties and it states that there is a link between the steps, but not yet what type of relation that is.

2.2 Theoretical Aspects on Representing Stuff Relations

There are multiple papers on relations between stuffs. We discuss in detail only the two most comprehensive proposals on parts and portions for stuffs, as they supersede the others.

Guizzardi [14]'s proposal zooms in on quantities (of stuff) and their parts; an example is shown in Fig. 1. The example is clearly about the extensional level—particular amounts of stuff—though the description of the UML extension less so, it forces all quantities to be in a container, part-quantities are essential, and it is atemporal. The limitation of atemporality is that one cannot fully represent scattered portions, like the glass of wine tapped from the wine in the wine tank, which is further delimited in [14] by the constraint of self-connectedness. For traceability in the food processing chain to ensure food safety, however, this is important; e.g., that the contaminated milk in the bottle on the shelf in the shop is a portion of the batch of milk processed on day x in processing plant y. Conflating the extensional and intensional is tricky with the 1:1 multiplicity between a whole-stuff and a part-stuff. It is the case that for some specific amount of wine, there is one specific amount of alcohol as part of it, and that specific amount of alcohol is part of that specific amount of wine. However, there are more drinks that have alcohol, so if we were to add a class, say, Vodka and a 1:1 association to Alcohol, we have a problem: a same amount of alcohol must be part of both some wine and some vodka, but it cannot be. The underlying issue is that quantification over the relations is different for extensional and intensional parts of stuff, so conflating them will violate either one. Further, while subQuantityOf for particular amounts is indeed essential insofar as it concerns the identity of

[1] https://www.w3.org/TR/prov-o/.

the amount, this may not be the case for universals; e.g., alcohol-free beer is perhaps still beer, decaf coffee still coffee. Finally, it forces a quantity to be in a container, which need not be the case (e.g., a lump of clay).

Donnelly and Bittner [10] do use a temporal mereology for portions of stuff, remain at the extensional level (i.e., no assertions about types of stuff), and with the various summation relations, can differentiate between pure and mixed stuffs. A tricky issue is that they adopted some of Barnett's [4] misconceptions of kinds of stuff. Donnelly & Bittner illustrate"unstructured stuff" ("discrete stuffs" [4], "pure stuffs" [17]) with water. However, by that example, then their and Barnett's "structured stuffs" / "non discrete stuff" is not, which affects the applicability of the summation relations and the feasibility to 'lift' it up to the intensional level. For instance, their examples include "milk, crude oil, graphite, quartz, and wood", but milk and crude oil are homogeneous mixtures, graphite is carbon-only (unstructured pure stuff), quartz without any qualifier (like Amethyst) are just SiO_2 molecules and if water is unstructured, then so must quartz be. Quartz would be structured pure stuff in the Stuff Ontology [17], and therewith obtain the appropriate axioms. Wood is a solid heterogeneous mixture and has its own issues with portions: for a homogeneous mixture, that is easy to establish (freezing or boiling point, sortal weight etc.), but not so for heterogeneous mixtures, due to the compartmentalisation of the different kinds of stuffs that are part of the whole stuff. These issues are not addressed in the formalisation of portions-as-portions (of the same kind of stuff as the whole) and portions-as-parts (of a different kind of stuff as the whole).

Thus, overall, some theoretical advances have been made regarding relating stuff to other stuff, as well as their quantities, but it is incomplete regarding the type/instance issue, the temporal dimension, and relation overloading issues.

3 Preliminaries

Before relating stuffs, three preliminaries have to be outlined: the Stuff Ontology is reused for the intensional level; a brief recap of mereology is included to keep the paper self-contained; and some formalisation considerations are discussed.

The Ontology of Macroscopic Stuff. The Stuff Ontology refines the notion of stuff beyond the mere distinction between pure and mixed stuff [4,7,10], yet in less detail than the philosophy of chemistry [8,24]. Figure 2 includes its four top-level classes: pure and mixed stuff, which is homogeneous or heterogeneous. This is further specialised with classes such as Solution, Suspension, and the Colloids[2], which are all defined classes as well. The same underlying principles have been used as proposed in the philosophy and chemistry literature, including: a granule (also called grain or basis type) of the stuff that is at one finer-grained level than the stuff itself; homogeneous versus heterogeneous matter; and the macroscopic

[2] Colloids are homogeneous mixtures where one phase is evenly dispersed in another; e.g., whipped cream (air in gaseous phase dispersed in cream in liquid phase).

sameness criterion for the least portion, which is the smallest portion that still exhibits the macroscopic properties of that kind of stuff [4,8]. Because its aim was practical usefulness, it is represented in OWL 2 DL, extensively annotated, and available online at http://www.meteck.org/stuff.html.

Parts and Wholes. As the relations between stuff concern parthood relations, we recap here briefly some important aspects of the various mereological theories, following [29]. Part p is a primitive relation, which is reflexive, antisymmetric, and transitive (Eqs. 1-3). Proper parthood, pp, is defined in terms of parthood (Eq. 4), and is irreflexive, asymmetric, and transitive (Eqs. 5-7):

$$\forall x(p(x,x)) \tag{1}$$
$$\forall x,y((p(x,y) \land p(y,x)) \rightarrow x = y) \tag{2}$$
$$\forall x,y,z((p(x,y) \land p(y,z)) \rightarrow p(x,z)) \tag{3}$$
$$\forall x,y(pp(x,y) \equiv p(x,y) \land \neg p(y,x)) \tag{4}$$
$$\forall x,y(pp(x,y) \rightarrow \neg pp(y,x)) \tag{5}$$
$$\forall x \neg (pp(x,x)) \tag{6}$$
$$\forall x,y,z((pp(x,y) \land pp(y,z)) \rightarrow pp(x,z)) \tag{7}$$

Because one needs to consider actual pieces and portions of stuff, i.e., the extensions of stuff universals, an extensional mereology may be of use, which looks at how to exhaustively define an object by its constituent parts, notwithstanding that this has its traps [10]. First, the theory Minimal Mereology (MM) has weak supplementation, saying that every proper part has to be supplemented by some other part (Eq. 8, where o is overlap), or phrased liberally: if a whole has a proper part, then there must be at least two different proper parts.

$$\forall x,y(pp(x,y) \rightarrow \exists z(p(z,y) \land \neg o(z,x))) \tag{8}$$

The alternative is strong supplementation in Extensional Mereology (EM): if an object fails to include another among its parts, then there must be a 'remainder'. EM is highly problematic, especially for colloids, because EM allows non-atomic objects with the same proper parts to be identical, yet sameness of parts tends not to be enough for identity: an amount of air plus an amount of liquid cream one pours into the bowl is surely not the same as whipped cream that one can make of it, yet they have the same parts, or oil and egg yolk versus mayonnaise, and so on. So, EM is a bad idea for stuffs, but MM may be of use.

The total or universal whole is the totality of the quantity of some stuff that exists at some point in time, which is not of interest. For instance, my lemonade I made at time t in Cape Town and your lemonade you made at time t in Bologna are independent and thus certainly not related through parthood. The opposite is either the 'atom'—smallest indivisible part that has no parts—or 'atomless gunk', i.e., infinite divisibility. While infinite divisibility may appear appealing for stuffs, there is, in fact, a relative notion of 'atom': the least portion.

On Formalising It. We have seen that 'pushing' everything there is to say about stuff universals and their amounts into one level—say, first order predicate logic—may be problematic, because things have to be said about stuffs themselves, like that a mixture is composed of at least two different kinds of stuff. This requires quantification over predicates, hence a second order logic. With stuffs being different from objects, one can use a many-sorted logic, so as to quantify over stuffs and over objects [4,10]. However, if one does not want to assert something about its constituent parts, alike a mass-quantity with no declared internal structure [12], then a many-sorted logic is not needed. Here, we delimit the scope to just relations between stuffs, rather than summations, so quantification is over stuffs and their particular amounts only.

Most accounts of portions are either atemporal or are temporal in name only, for it is easier to implement and practically use it. If one wants to be as precise as possible, one cannot avoid the temporal modality at the extensional level, alike in [10]. This lets one distinguish between a portion like the 'upper half of the wine in the tank' and 'the glass of wine just tapped from it' as well as between 'piece' and 'portion': a piece always and only *was* part of the whole, whereas a portion *is or was* part of the whole amount of stuff.

Thus, we end up with a second order logic, where at least the 'first order fragment' of it is temporal. Recalling some basic notation and features of second order logic, we can quantify over predicates, such as $\exists P(P(x) \wedge P(y) \wedge x \neq y)$ meaning 'there exists a property that two distinct entities share in common', and use them as variables; e.g., with, say, *Colour* being a property, then $\neg\exists x \forall P(Colour(P) \rightarrow P(x))$ is the formalisation of 'no object has every colour'. In addition, we will use the usual shorthand notation of $\exists^{\theta x}$ with x an integer > 1 for cardinality constraints beyond simple mandatory/existential quantification, and θ being a comparison operator $\leq$, $=$, or $\geq$. Finally, the temporal modality. A first order LTL with the until and since operators suffices, or just ternaries. We use the latter, using a linear flow of time $\mathcal{T} = \langle \mathcal{T}_p, < \rangle$ where $\mathcal{T}_p$ is a set of time points (indicated with t) and $<$ is a binary precedence relation on $\mathcal{T}_p$ that is assumed to be isomorphic to $\langle \mathbb{Z}, < \rangle$.

4 Relating Stuff

An informal, high-level overview of the various entities and relations is shown in Fig. 2. It is drawn in EER diagram notation so as to avoid the complicating factor of UML's aggregation association, finding meaningful names for the association ends was distracting, and it may make it easier to morph it into a temporal ER, such as ER_{VT} [1], and convert it all to a temporal relational database if one so prefers. Note that inheritance of properties applies, so, among others, also Homogeneous Mixed Stuff has an instantiation relation (because Stuff has) and Portions and Pieces also have a measure of their quantity (because Amount of Stuff has). Those quantities (10ml etc.) have their own representation system, which is summarised into a Quantity extension. Regarding quantities, we concur with other ontologies that quantity kinds are things in their own right, i.e., the

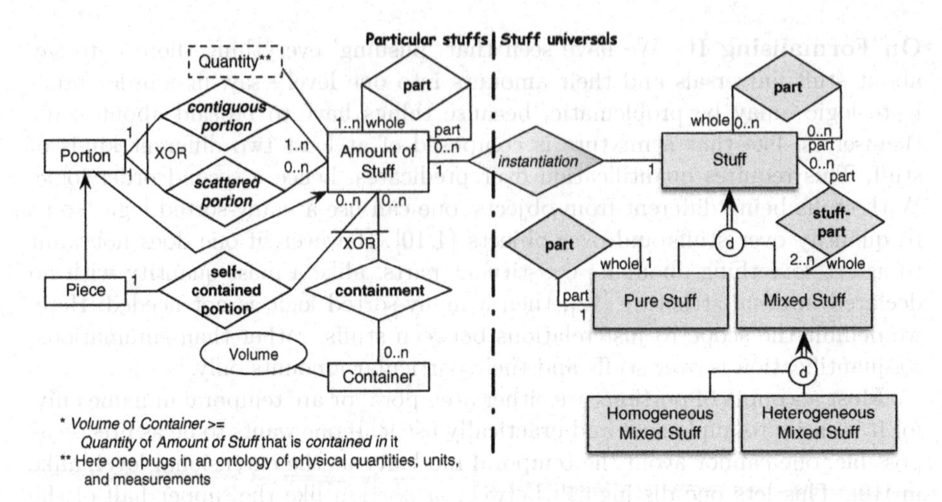

Fig. 2. Stuff relations, depicted informally (and incomplete) using EER Diagram notation, with part-whole relations in bold face and the universals-side in grey.

"quantity as a class" commitment (weight, length etc.) [13,26] rather than as a property/attribute (hasLength etc.) or equating portion and quantity [14], for the identity of a quantity is independent of the entity that 'has' that quantity. For instance, one may rather have a quantity of 1 kg of gold as a present than 1 kg of soil. Put differently, quantities are reusable entities across amounts of matter. Which ontology is then chosen for the quantities, units, and measurements does not matter much.

The model is illustrated in the following example.

Example 1. As example, let us take 'a slice of cake': it is an instance of Piece, as it is a self-contained Amount of Stuff, and it is thus also a Portion. Given that it was cut off from some quantity of cake, it is a scattered portion of cake that is also a Amount of Stuff. The slice (and the cake) as Amount of Stuff is a kind of (instantiation of) a Homogeneous Mixed Stuff. Being a Homogeneous Mixed Stuff, it must have at least two stuff-parts (related through stuff-part), which are Flour, Sugar, Butter, Egg and Vanilla essence, where, e.g., Butter is an Emulsion, which is a homogeneous mixed stuff. A Quantity of 250 g of Butter went in the particular cake the slice came from, which is a stuff-part of the Amount of Stuff that amounts to the whole amount of cake. That amount of butter was in a containment relation to a buttercup that is a Container with a Volume of 500g. Finally, the slice having a left and right contiguous portion, I break it in half and share it with my neighbour. ◇

This example is still incomplete: how much butter did I indulge in when eating my portion of the slice of cake? The whole quantity of cake had as part a quantity of 250 g butter. Let's say the slice is 1/10th of the cake, so, by cake being homogeneous mixed stuff, the slice will also have 1/10th of the butter of

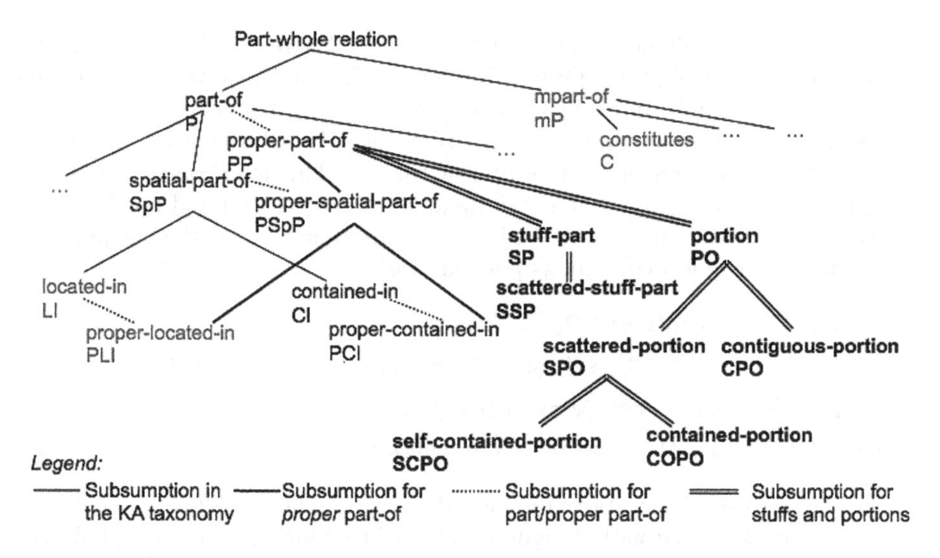

Fig. 3. Section of the basic taxonomy of part-whole relations of Keet and Artale [18] (less and irrelevant sections in grey or suppressed), extended with the stuff relations and their position in the hierarchy.

the whole cake, or 25g. Splitting it into 2 portions, 12.5g of butter was part of the portion of cake I ate. All this can be formally represented with the model in Fig. 2, provided it has the appropriate temporal extensions, which we will address in the remainder of this section.

4.1 Relating Portions

Concerning all those part and portion relations in the example and in Fig. 2, let us start with the axioms of Minimal Mereology for parthood relations part, p, and proper part, pp, which are subsumed by some generic top-relation, pw (part-whole), so we obtain the hierarchy of relations as depicted in Fig. 3. Linking this extended hierarchy to Fig. 2, one can see on the right-hand side of the figure the generic part p with its sub-relationships, and in particular stuff-parts, sp for short, and portions, po. The stuff-part will be discussed in Sect. 4.3. Portions, together with the hierarchy, induces its definition, which is formulated as:

$$\forall x, y \exists^{=1} S(\mathsf{po}(x, y) \leftrightarrow \mathsf{pp}(x, y) \land S(x) \land S(y) \land \mathsf{Stuff}(S)) \tag{9}$$

That is, portions are of the same type of stuff as the whole.

While portion is atemporal, the time dimension has to be introduced to distinguish between scattered (*spo*) and contiguous (*cpo*) portions, for the former *was* a (contiguous) part of the whole portion, whereas the latter *is* a part of the whole portion. The latter being a contiguous part, it then also means that contiguous portion is properly contained in the whole portion, whereas the scattered portion is not. To this end, we take the containment relation from [18], make it proper containment (*pci*), which is included here as (Eq. 10), where R is DOLCE's region, ED is DOLCE's endurant [20], and *has_3D* a shorthand for DOLCE's qualities and qualia to denote something has a physical region. Contiguous portion can then be temporally defined (Eq. 11), so $cpo(x, y, t)$ then reads as "x is a contiguous portion of y at time t". This is in contrast to scattered portion (Eq. 12) that states, informally, that x is a scattered portion of y at time t if it was at some time t'—which is before time t—a contiguous portion of y and at t it is not a contiguous portion of y.

$$\forall x, y(\mathsf{pci}(x, y) \rightarrow \mathsf{pp}(x, y) \wedge \mathsf{R}(x) \wedge \mathsf{R}(y) \wedge \exists z, w(\mathsf{has\_3D}(z, x) \wedge \mathsf{has\_3D}(w, y) \wedge \mathsf{ED}(x) \wedge \mathsf{ED}(y))) \quad (10)$$

$$\forall x, y \exists t(\mathsf{cpo}(x, y, t) \leftrightarrow \mathsf{po}(x, y, t) \wedge \mathsf{pci}(x, y, t)) \quad (11)$$

$$\forall x, y \exists t, t'(\mathsf{spo}(x, y, t) \leftrightarrow \mathsf{cpo}(x, y, t') \wedge \neg \mathsf{cpo}(x, y, t) \wedge t' < t) \quad (12)$$

The last two special relations distinguish between relating self-contained portions that are described with designated pieces like lumps and drops and slices, and relating portions that are housed in a container[3]. The ontological investigation into the entity 'container' is not the scope here, and we appeal to the reader's common sense understanding of it: an object with a cavity such that one can put something in that cavity; e.g., a bottle or a glass that can be filled with an amount of wine, a silo or a bag that store an amount of soy beans. So, if we have a self-contained portion then it is a piece that was part of some amount of stuff (which may be a portion) (Eq.13) and if we have a contained portion, then it is scattered in a container C from the whole amount (e.g., the glass of wine taken from the wine in the wine bottle) (Eq. 14). For the self-contained portion, one cannot say that it is never in a container, for one could have, say, a lump of clay that is put in a sealed container for later use. Therefore, we use only the weak statement that a piece is not necessarily (the "$\neg\Box$") in a container:

$$\forall x, y \exists t, t'(\mathsf{scpo}(x, y, t) \rightarrow \mathsf{spo}(x, y, t) \wedge \neg \Box z(\mathsf{pci}(x, z) \wedge \mathsf{C}(z))) \quad (13)$$

$$\forall x, y \exists t, t'(\mathsf{copo}(x, y, t) \rightarrow \mathsf{spo}(x, y, t) \wedge \Box z(\mathsf{pci}(x, z) \wedge \mathsf{C}(z))) \quad (14)$$

This concludes the specification of the basic set of relations for portions.

4.2 Portions and Pieces

The previous section alluded to one's intuition regarding portions and pieces. While related works do talk about portions, we could not find a formal definition in [7,10,14,17]. Here, we make a first step in that direction, taking the notion of portions from philosophy, in particular the afore-mentioned macroscopic sameness, which implies that a portion of some amount of stuff is of the same type of stuff as the whole amount (Eq. 15), and the least portion would then amount to the equivalent of *Atom*, but then for stuffs. With *Atom* defined as (Eq. 16) (from [29]), the 'least portion type of atom' (LP) then follows from

[3] pieces and portions as objects do differ, which we will discuss in the next subsection.

both (Eq. 17). These can have their temporal counterparts (by adding t, i.e., $Portion(x,t)$, $po(x,y,t)$, and $LP(x,t)$).

$$\forall x \exists^{=1} S(\mathsf{Portion}(x) \leftrightarrow \mathsf{po}(x,y) \wedge S(x) \wedge S(y) \wedge \mathsf{Stuff}(S)) \quad (15)$$

$$\forall x(\mathsf{Atom}(x) \leftrightarrow \neg\exists \mathsf{pp}(y,x)) \quad (16)$$

$$\forall x \exists^{=1}(\mathsf{LP}(x) \leftrightarrow \mathsf{Portion}(x) \wedge \neg\exists \mathsf{po}(y,x) \wedge S(x) \wedge S(y) \wedge \mathsf{Stuff}(S)) \quad (17)$$

Pieces—e.g., a lump of clay, a chip of wood, a drop of blood—are self-contained portions, i.e., they are neither currently contained in the whole amount nor are they necessarily in a separate container:

$$\forall x(\mathsf{Piece}(x,t) \leftrightarrow \mathsf{scpo}(x,y,t)) \quad (18)$$

4.3 Stuff Parts

Stuff part, also called ingredient, was already specified in [17], where at the type level, a pure stuff has as ingredient stuff the same stuff it is, whereas mixed stuffs have at least two other kinds of stuff as part. The issues to examine are whether stuff parts are proper part or just part, temporality, and essentialism.

For pure stuffs as universals, the parthood relation that holds can be considered reflexive, because the domain and range are of the same stuff type, and likewise it is antisymmetric (and obviously transitive); thus, the regular parthood relation p holds. For pure stuffs at the particular level (amounts of pure stuff, like a glass of water), then the parts are of the same type, but it is obviously a smaller amount, so then we obtain proper parthood. One optionally could change the names of part and proper part to other ones to make sure that those two relations only have stuffs as domain and range. However, there is nothing of interest to assert about pure stuffs in that regard as it states the obvious already when it is asserted as being a kind of a pure stuff, like 'gold has as part gold and only gold'. A possible pitfall may be that then on paper there may be confusion, but this ought not to occur in praxis provided one has the taxonomy of part-whole relations imported or DOLCE: if one of the two participants is an object and the other some stuff, then it is a constitution relation by its definition, which would alert the modeller something is amiss, which OntoPartS-2 already does [19].

For mixed stuffs, we end up with proper part both at the universal and particular levels. Because one can say 'interesting' things about mixtures, it does make sense to introduce a separate named relation as a type of proper parthood. For instance, then one can define the part-stuffs a type of mixture is made up of, infer the possible product based on its stuff-ingredients, and play with substitutes in a recipe for case-based reasoning (e.g., soy milk instead of cow's milk, speckled beans instead of kidney beans, etc.); i.e., it serves in automated reasoning. Therefore, the stuff-part, sp and scattered-stuffpart, ssp, relations were added in Fig. 3, which are defined as:

$$\forall \mathsf{x}, \mathsf{y} \exists \mathsf{S}, \mathsf{S}'(\mathsf{sp}(\mathsf{x}, \mathsf{y}) \leftrightarrow \mathsf{pp}(\mathsf{x}, \mathsf{y}) \wedge \mathsf{S}(\mathsf{x}) \wedge \mathsf{S}'(\mathsf{y}) \wedge \mathsf{Stuff}(\mathsf{S}) \wedge \mathsf{Stuff}(\mathsf{S}') \wedge \mathsf{S} \neq \mathsf{S}') \quad (19)$$

$$\forall \mathsf{x}, \mathsf{y} \exists \mathsf{S}, \mathsf{S}'(\mathsf{ssp}(\mathsf{x}, \mathsf{y}, \mathsf{t}) \leftrightarrow \mathsf{pp}(\mathsf{x}, \mathsf{y}, \mathsf{t}') \wedge \mathsf{S}(\mathsf{x}) \wedge \mathsf{S}'(\mathsf{y}) \wedge \mathsf{Stuff}(\mathsf{S}) \wedge \mathsf{Stuff}(\mathsf{S}') \wedge \mathsf{S} \neq \mathsf{S}' \wedge \mathsf{t}' < \mathsf{t}) \quad (20)$$

The *sp* refines the relations hasSubStuff of [17] and sub-quantity-of of [18] into a full definition, as both had only domain and range axioms. It is different from Guizzardi's subQuantityOf [14], in that there is no strong supplementation (recall Sect. 2), it is for designated stuffs and amounts thereof (cf. quantities), and it is not essential. Essentialism at the universal level may apply on a case-by-case basis; e.g., alcohol may be considered to be an essential part of vodka, but not of beer. Whether it is essential at the level of particulars in general, is not entirely clear, unless one defines and identifies a particular amount of stuff as the mereological sum of its quantities (for complications with that, see [10]) and excludes some convoluted corner cases (e.g., distilling the alcohol and putting it back in). Either way, if one assumes that both a portion and a part-stuff are essential to some amount of stuff, then it can be added easily with a temporal logic that has $\top$, $\bot$ and the $\mathcal{U}$ntil and $\mathcal{S}$ince operators [2].

This concludes the sets of relations that relate stuffs to other stuffs.

5 Applying Implementation Trade-Offs

Given the theoretically optimal formal characterisation of parts and portions of stuff presented in the previous section, the next step is to assess how this can be implemented practically with the state of the art technologies. The options are:

1. Use a system that supports a second order language and reasoner and implement it as formalised; e.g., with the *He*terogeneous *t*ool *s*et Hets [23].
2. Squeeze into OWL 2 DL what can be done:
   - Get rid of the second order axioms; either:
   (a) Drop the second order aspects altogether (simply ignore);
   (b) Push that to first order and the first order aspects to instance-level;
   (c) Create two branches in the TBox for the universals and for the particulars, alike in GFO [15].
   - Remove all modal aspects, i.e., the necessity and the temporality, and indicate its intention in the name of the object property only.
3. Use a temporally extended OWL:
   - Second order issue, and options, as above;
   - Choose a temporal trade-off, for the need for temporalising relations already results in an undecidable language [1]; e.g., use concrete domains as workaround, as in tOWL that extends $\mathcal{SHIN}(D)$ [21], or disallow temporal constructors on the right-hand-side of inclusions, as in TQL that extends OWL 2 QL [3].
4. Morph it into a relational database or integrate it with RDF as a Linked Data application such as VacSeen [6].

Option 1 is good for toy examples to verify and validate it with a few examples, but never will be for industry-grade implementations due to the high undecidedability of second order logic. The other three options list several viable usability trade-offs that favour computation over expressiveness, where the ultimate decision lies with the requirements of the use case. Option 2 permits some automated reasoning, but not fully the tracing of some amount of stuff over time. The time dimension is favoured in Option 3, but this is at the cost of, mainly, transitivity and/or qualified cardinality constraints so that one cannot fully represent mixed stuffs, pure stuffs, (solid) heterogeneous mixtures, and colloids, and therewith lose the ability to automatically classify a stuff into its right kind. In addition, TQL and tOWL are preliminary results and are not at the same level of robustness as the technologies of Option 2. Both options, for being in OWL, easily can import the Ontology of units of Measure (OM) that is also represented in OWL and developed by domain experts in food [26]. Finally, one could focus even more on implementation with Option 4, which is good for industry-level applications, but some unenforceable assumptions have to be made regarding its correctness and comprehensiveness and any automated reasoning is limited to what can be done with queries. Therefore, at present, the most straightforward choice seems to be Option 2-c by refining the Stuff Ontology [17] and importing the OM [26], with as future work the data-oriented Option 4 with TQL.

Integrating Quantities. OM is 5 MB and is merged with bibliographic information and FOAF, and including units that are irrelevant to stuff, such as the vase end life of flowers, acceleration, and micro degree Celcius, but also specific ones relevant for food stuffs, such as the lactose_mass_fraction (as stuff-part of, e.g., milk powder). Other models for quantities are also not ideal; e.g. UCUM[4] and EngMath [13] are not available in OWL and QUDT[5] has similar excess baggage as the OM. Therefore, a module was created manually: we reduced the 5MB OM from 25253 axioms (1148 classes, 25 object properties, and 2622 individuals) to 1472 axioms (131 classes, 25 object properties, and 104 individuals) in the 216KB OMmini module. This module was imported into the extended Stuff Ontology, and bridge axioms added. These bridge axioms include alignments, such as om:phenomenon $\equiv$ stuff:PhysicalEndurant, om:'unit of measure' $\sqsubseteq$ Abstract, and om:quantity $\sqsubseteq$ Region, thus also commencing with aligning OM to a foundational ontology—which was still the intention by [26]—as stuff:PhysicalEndurant $\equiv$ dolce:PhysicalEndurant, and likewise for Abstract and Region. Further, the formal counterpart of the dashed 'Quantity**' and 'Container' entity types from Fig. 2 were added; among others:

$$\mathsf{stuff{:}AmountOfStuff} \sqsubseteq \, = 1\, \mathsf{om{:}quantity.om{:}quantity} \tag{21}$$

$$\mathsf{Container} \sqsubseteq \forall \mathsf{containedIn}^{-}.(\mathsf{PhysicalObject} \sqcup \mathsf{AmountOfStuff}) \tag{22}$$

[4] http://www.unitsofmeasure.org/trac.
[5] http://qudt.org/.

$$\mathsf{Portion} \sqsubseteq \exists \mathsf{portionOf.AmountOfStuff} \quad (23)$$

$$\mathsf{Piece} \sqsubseteq \exists \mathsf{isSelfContainedScatteredPortionOf.AmountOfStuff} \quad (24)$$

$$\mathsf{AmountOfMatter} \sqsubseteq \exists \mathsf{instantiation.Stuff} \quad (25)$$

where Eq. 21 then further avails of the quantities from OM, and therewith also AmountOfMatter's subclass Portion and its subclass Piece, and instantiation in Eq. 25 is typed with AmountOfMatter and Stuff, addressing the two-layer issue in the same way as GFO [15]. This resulted in a combined ontology of 1831 axioms (logical axiom count 718), 193 classes, 57 object properties, and 104 individuals) which is in $\mathcal{SROIQ}(\mathcal{D})$, i.e., OWL 2 DL.

Automated Reasoning. For traceability, transitivity of portionOf and property chains yield the most useful results. Take, e.g., the following property chains:

$$\mathsf{scatteredPortionOf} \circ \mathsf{portionOf} \sqsubseteq \mathsf{scatteredPortionOf} \quad (26)$$

$$\mathsf{stuffPart} \circ \mathsf{contiguousPortionOf} \circ \mathsf{SelfContainedScatteredPortionOf} \sqsubseteq \mathsf{scatteredStuffPartOf} \quad (27)$$

$$\mathsf{scatteredPortionOf} \circ \mathsf{scatteredPortionOf} \circ \mathsf{scatteredPortionOf} \sqsubseteq \mathsf{scatteredPortionOf} \quad (28)$$

The chain in Eq. 26 enables one to infer that a scattered portion—say, my glass of wine d.d. 9-7-'16—of a portion (bottle #1234 of organic Pinotage wine) of an amount of matter (cask #3 with wine from wine farm X of Stellar Winery from the 2015 harvest) is a scattered portion of that amount of matter (that cask). Reconsidering the slice of cake from Example 1, the property chain in Eq. 27 can be used to infer that that 12.5 g of butter is a scatteredStuffPartOf the cake: the 12.5 g of butter is a stuffPart of the left-hand side contiguousPortionOf of the slice of cake that, in turn, is a SelfContainedScatteredPortionOf the cake. This same chain in Eq. 27 also can be applied to other use cases; e.g., the amount of alcohol I would consume drinking half a glass of wine is a scatteredStuffPartOf the original amount of wine in the wine bottle. For the pharmaceutical supply chain in [28], we obtain that a portion (on a 'pallet') of the quantity of medicine produced by the manufacturer goes to the warehouse, of which a portion (in a 'case') goes to the distribution centre, of which a portion (as 'items') ends up on the dispensing shelf. Then tracing the customer's portion of medicine can be inferred with Eq. 28. Thus, then one can *infer* the chain of portions in the supply chains, and therewith start tracing it automatically from one amount at home back to the manufacturer (and all the way back to the farm, in case of food).

Note that, because the ontology also has scatteredPortionOf $\sqsubseteq$ portionOf, this combination would result in a cycle and therewith not be a 'regular' RBox, which is not allowed in OWL 2 DL. Making scatteredPortionOf and portionOf siblings does permit the chain. Because DL reasoners do not do much with the hierarchy in the RBox and the semantic differences between these properties—temporality—cannot be represented in OWL anyway, they are made siblings, for the inferences with the property chains are deemed more important. Likewise, due to the declaration of the chains, scatteredStuffPartOf's inverse hasScatterPartStuff is made a sibling of hasPartStuff because the latter was needed more in cardinality constraints for mixtures.

6 Discussion

To the best of our knowledge, this is the first attempt to systematically disentangle the parts and portions, having identified 7 different interactions between stuffs, and named them so for clarity. While the implementation is not a perfect match with the theory presented in Sect. 4, it still has several advantages of the other proposals to date. Notably, (1) there is a clear distinction between the extensional and intensional level; (2) it distinguishes between the (non-repeatable) amounts of stuff and the repeatable quantities; and (3) both are present in the same ontology and immediately usable for ontology development thanks to substantially extending the Stuff Ontology.

A shortcoming is the omission of the temporal dimension, which does not lend itself well for scalable automated reasoning. This is mitigated to some extent by availing of property chains, so that one still can trace a portion to the original amount. This is a limited solution, indeed, but preferable over no such inferences. It might be possible to have it 'both ways' with the Distributed Ontology Language [22]—currently being standardised with OMG—and its technological infrastructure, which breaks up the whole theory into modules based on expressiveness. Then one could have slow automated reasoning where acceptable and fast reasoning where needed. This is an avenue of future work.

The proper treatment of the stuff relations now opens up the opportunities for deployment in the intended use case with food processing, and, for it being a core ontology, also in other domains, such as stuffs in medicine (e.g., pills and vaccines [6,28]) and engineering (e.g., the use cases in [13,16]).

7 Conclusions

Seven relations for portions, pieces, and stuff-parts were defined and formalised in the logic it required, availing both of the temporal dimension and second order. The orchestration with stuffs and amounts of matter make a clear distinction between the extensional and intensional levels (amounts and stuff kinds) and between amount of stuff and its quantity. The implementable components were added to the Stuff Ontology core ontology and a module of the Ontology of units of Measure for the quantities was imported. Some useful automated reasoning was shown to be still possible thanks to property chains.

References

1. Artale, A., Parent, C., Spaccapietra, S.: Evolving objects in temporal information systems. Ann. Math. Artif. Intell. **50**(1–2), 5–38 (2007)
2. Artale, A., Guarino, N., Keet, C.M.: Formalising temporal constraints on part-whole relations. In: Proceedings of KR 2008, pp. 673–683. AAAI Press, Sydney, Australia, 16–19 September 2008 (2008)
3. Artale, A., Kontchakov, R., Wolter, F., Zakharyaschev, M.: Temporal description logic for ontology-based data access. In: Proceedings of IJCAI 2013 (2013)
4. Barnett, D.: Some stuffs are not sums of stuff. Philos. Rev. **113**(1), 89–100 (2004)
5. Beisswanger, E., Schulz, S., Stenzhorn, H., Hahn, U.: BioTop: an upper domain ontology for the life sciences - a description of its current structure, contents, and interfaces to OBO ontologies. Appl. Ontol. **3**(4), 205–212 (2008)
6. Bhattacharjee, P.S., Solanki, M., Bhattacharyya, R., Ehrenberg, I., Sarma, S.: VacSeen: a linked data-based information architecture to track vaccines using barcode scan authentication. In: Proceedings of SWAT4LS 2015, CEUR-WS, vol. 1546, Cambridge, UK, 7–10 December 2015 (2015)
7. Bittner, T., Donnelly, M.: A temporal mereology for distinguishing between integral objects and portions of stuff. In: Proceedings of AAAI 2007, pp. 287–292. Vancouver, Canada (2007)
8. van Brakel, J.: The chemistry of substances and the philosophy of mass terms. Synthese **69**, 291–324 (1986)
9. Donnelly, K.A.-M.: A short communication - meta data and semantics the industry interface: what does the food industry think are necessary elements for exchange? In: Sánchez-Alonso, S., Athanasiadis, I.N. (eds.) MTSR 2010. CCIS, vol. 108, pp. 131–136. Springer, Heidelberg (2010)
10. Donnelly, M., Bittner, T.: Summation relations and portions of stuff. Philos. Stud. **143**, 167–185 (2009)
11. Dumontier, M., et al.: The semanticscience integrated ontology (SIO) for biomedical research and knowledge discovery. J. Biomed. Semant. **5**(1), 14 (2014)
12. Gerstl, P., Pribbenow, S.: Midwinters, end games, and body parts: a classification of part-whole relations. Int. J. Hum.-Comput. Stud. **43**, 865–889 (1995)
13. Gruber, T.R., Olsen, G.R.: An ontology for engineering mathematics. In: Doyle, J., Torasso, P., Sandewall, E. (eds.) Proceedings of KR 1994. Morgan Kaufmann (1994)
14. Guizzardi, G.: On the representation of quantities and their parts in conceptual modeling. In: Proceedings of FOIS 2010, IOS Press, Toronto, Canada (2010)
15. Herre, H., Heller, B.: Semantic foundations of medical information systems based on top-level ontologies. Knowl.-Based Syst. **19**, 107–115 (2006)
16. Höfling, B., Liebig, T., Rösner, D., Webel, L.: Towards an ontology for substances and related actions. In: Fensel, D., Studer, R. (eds.) EKAW 1999. LNCS (LNAI), vol. 1621, pp. 191–206. Springer, Heidelberg (1999)
17. Keet, C.M.: A core ontology of macroscopic stuff. In: Janowicz, K., Schlobach, S., Lambrix, P., Hyvönen, E. (eds.) EKAW 2014. LNCS, vol. 8876, pp. 209–224. Springer, Heidelberg (2014)
18. Keet, C.M., Artale, A.: Representing and reasoning over a taxonomy of part-whole relations. Appl. Ontol. **3**(1–2), 91–110 (2008)
19. Keet, C.M., Khan, M.T., Ghidini, C.: Ontology authoring with FORZA. In: Proceedings of CIKM 2013, pp. 569–578. ACM Proceedings, October 27–November 1, 2013, San Francisco, USA (2013)

20. Masolo, C., Borgo, S., Gangemi, A., Guarino, N., Oltramari, A.: Ontology library. WonderWeb Deliverable D18 (ver. 1.0, 31–12-2003) (2003)
21. Milea, V., Frasincar, F., Kaymak, U.: tOWL: a temporal web ontology language. IEEE Trans. Syst. Man Cybern. **42**(1), 268–281 (2012)
22. Mossakowski, T., Kutz, O., Codescu, M., Lange, C.: The distributed ontology, modeling and specification language. In: Proceedings of WoMo 2013, CEUR-WS, vol. 1081, Corunna, Spain, 15 September 2013 (2013)
23. Mossakowski, T., Maeder, C., Lüttich, K.: The heterogeneous tool set. In: Beckert, B. (ed.) Proceedings of VERIFY 2007. CEUR-WS, vol. 259, pp. 119–135, Bremen, Germany, July 15–16, 2007(2007)
24. Needham, P.: Macroscopic mixtures. J. Philos. **104**, 26–52 (2007)
25. Odell, J.: Advanced Object-Oriented Analysis & Design using UML. Cambridge University Press, Cambridge (1998)
26. Rijgersberg, H., van Assem, M., Top, J.: Ontology of units of measure and related concepts. Semant. Web **4**(1), 3–13 (2013)
27. Snomed, C.T.: Online (version 31-1-2014). http://www.ihtsdo.org/snomed-ct/
28. Solanki, M., Brewster, C.: OntoPedigree: modelling pedigrees for traceability in supply chains. Semant. Web J. **7**(5), 483–491 (2016)
29. Varzi, A.C.: Mereology. In: Zalta, E.N. (ed.) Stanford Encyclopedia of Philosophy. Stanford, fall 2004 edn. (2004). http://plato.stanford.edu/archives/fall2004/entries/mereology/

A Model for Verbalising Relations with Roles in Multiple Languages

C. Maria Keet(✉) and Takunda Chirema

Department of Computer Science, University of Cape Town, Cape Town, South Africa
mkeet@cs.uct.ac.za, CHRTAK003@myuct.ac.za

Abstract. Natural language renderings of ontologies facilitate communication with domain experts. While for ontologies with terms in English this is fairly straightforward, it is problematic for grammatically richer languages due to conjugation of verbs, an article that may be dependent on the preposition, or a preposition that modifies the noun. There is no systematic way to deal with such 'complex' names of OWL object properties, or their verbalisation with existing language models for annotating ontologies. The modifications occur only when the object performs some *role* in a relation, so we propose a conceptual model that can handle this. This requires reconciling the standard view with relational expressions to a positionalist view, which is included in the model and in the formalisation of the mapping between the two. This eases verbalisation and it allows for a more precise representation of the knowledge, yet is still compatible with existing technologies. We have implemented it as a Protégé plugin and validated its adequacy with several languages that need it, such as German and isiZulu.

1 Introduction

Natural language interfaces to ontologies are used both to ameliorate the knowledge acquisition bottleneck and for user interaction with so-called 'intelligent' systems, with the most popular application scenarios in healthcare, weather forecast bulletins, and querying of information systems and question generation in education. This involves mainly knowledge-to-text from OWL files [1,36,37], but also bi-directional in ontology authoring systems [11,14] and the Manchester syntax used since Protégé 4.x. This is done mostly for English, but there are also some works on Latvian [16], Greek [1], and isiZulu [25]. A hurdle for such other languages is the correct 'verbalisation', i.e., a natural language rendering of an axiom, when the name of an OWL object property is not a simple verb in the 3rd person singular. For instance, works for, located in, and is part of all have a dependent preposition, the former two have different verb tenses, and the latter a copulative and noun rather than a regular verb. Regarding the verb tenses, even one tense already raises problems for languages in the Bantu language family, such as isiZulu, which is widely spoken in South Africa. IsiZulu has no single 3rd pers. verb regardless the subject—as in English with, say, 'eats'—but a '3rd

E. Blomqvist et al. (Eds.): EKAW 2016, LNAI 10024, pp. 384–399, 2016.
DOI: 10.1007/978-3-319-49004-5_25

pers.' for each noun class (nc); e.g., if a grandmother (nc1a) 'eats' something is it *udla*, but if an elephant (nc9) 'eats' something it is *idla*. This raises the question of how to model that in an ontology or associated language file, or both.

Further, especially Natural Language Generation (NLG) is expected to take into account prepositions [2]. Prepositions are used in various constructions that may imply a certain relation [35], with the one relevant for ontologies mainly being the dependent prepositions—also called 'deep prepositions' [29] or 'co-verbs' [28]—that in some languages have the preposition associated not with the verb but with the noun. The three principal issues to solve for such prepositions are phonological conditioning, declensions, and noun modifiers. An example of phonological conditioning is preposition contraction in Portuguese, as in *de+a=da* (e.g., *da mesa* 'of the table') [33]. Prepositions may change the article of the noun, as in German and Greek; e.g., the article *der* for *Betrieb* (m.) 'company' together with *arbeitet für* 'works for' results in *arbeitet für<u>den</u> Betrieb*. The lists of prepositions that go with which case are known, yet this has to be encoded somewhere so as to generate the grammatically correct sentence from an ontology. Prepositions may also modify the noun, as happens in Lithuanian and Latvian [16] and in isiZulu and related languages [24]; e.g., the 'of' in 'part of' is handled by the possessive concord for the noun class of 'part' (*ingxenye*, in nc9), *ya-*, that is attached to the object, generating, e.g., *ya + umuntu* = *<u>yo</u>munto* 'of the human' [24]. Although verb conjugation and prepositions could be devolved to the individual language and language-specific implementations, a generic approach that works across languages will facilitate reusability.

To solve these issues, we first take a theoretical approach to achieve a solid foundation conceptually. Both the issue with conjugation and the prepositions can be solved with the so-called positionalist ontological commitment embedded in a representation language, exploiting (1) the *role* an object plays in the relation and (2) the distinction between *relation(ship)* and *relational expression*. As the preposition and its effects on the surface realisation belongs to neither the verb nor the noun *per sé*, the role conveniently can be adorned with such information. The second feature serves as solution to conjugation as well. This is captured at the metamodel layer of the representation language. Therefore, a formal mapping between their respective formalisations in OWL and $\mathcal{DLR}$ is provided, to ensure a rigorous well-founded implementation. The model thus improves both the natural language generation and it provides for a more precise representation of the knowledge. The model has been implemented as a plugin for Protégé. Its adequacy has been validated with isiZulu, chiShona, and German use cases.

In the remainder of the paper we first describe the main language requirements in Sect. 2, which are assessed against related works in Sect. 3. The implementation and validation of the model is described in Sect. 5. We discuss in Sect. 6 and conclude in Sect. 7.

2 Language Requirements and Motivational Use Cases

This section summarises the requirements for verbs, which are straightforwardly problematic, and for prepositions in the context of relating objects, which are challenging on the whole.

2.1 Verbs in IsiZulu and Related Languages

Linguistically, isiZulu (the Zulu language) is a member of the Bantu language family that has a characteristic *noun class system* that categorises each noun into a noun class that determines the agreement with other words in a phrase and exhibits a strong agglutinative character. It determines, among others, the singular/plural form, verb conjugation, and agreement with some prepositions; e.g., *umfundisi* 'student' is in noun class (nc) 1 and its plural, *abafundisi*, is in nc2, and *inja* 'dog' is in nc9 and its plural *izinja* is in nc10. IsiZulu has 17 noun classes. Because of the noun class-driven agreement system, any language annotation model must have some way of processing noun classes.

The nc determines verb conjugation using a subject concord (SC) that is prefixed to the verb stem. Therefore there is no single conjugated verb for 3rd pers. sg./pl., and verbalising an axiom is thus context dependent. That is, for an axiom of the form $C \sqsubseteq \exists R.D$ in an OWL ontology, the noun class of the noun/name of C determines the surface realisation of R. For instance, it is *u-+-dla=udla* 'eats' for *umfundisi* (nc1) and *i-+-dla=idla* 'eats' for *inja* (nc9); the respective plurals are *ba-+-dla=badla* and *zi-+-dla=zidla.* There are only 10 different SCs, as some noun classes have the same one, with 5 variants for the sg. and 5 for the pl. This brings afore the requirements to generate, store, and access those variants somewhere, and to generate or select the right one when verbalising the axiom, and a decision how to name the object property.

Verb negation uses a negative subject concord (NEG SC), which is also determined by the noun class, and the final vowel of the verb stem changes from *-a* to *-i*. So, a 'does not eat' is *aba-+-dli=abadli* for nc1 nouns and *ayi-+-dli=ayidli* for nc9 nouns, and so on for the other noun classes; thus, merging the negation with the verb (as in Japanese [31]). There are 10 different forms of the negated verb for singular and plural nouns.

2.2 Challenges with Prepositions

Prepositions in 'English ontologies' are put together with the verb in the object property (OP) name, yet in multiple other languages they go with, or affect, the noun in the object position in a sentence. The issue is explained easier by referring to a Controlled Natural Language (CNL). For instance, take the axiom of the type as in (A) below in Description Logics (DL) notation, a corresponding *template* (T), and a few examples as *verbalisations* of particular axioms using that template, which generate a *reading* or controlled natural language sentence:

A: $C \sqsubseteq \exists R.D$
T: Each $< C > < R >$ some $< D >$
E1: Each heart is part of some human
E2: Each employee works for some organisation

This works, regardless the nouns and verbs involved. Let us now take the same axiom type (A) for isiZulu, when the verb is 'simple' (teaches, eats, etc.): there is no template but a *pattern* (P) instead (extended from [25]):

P: <QCall for NC_x >onke <pl. of C, is in NC_x > <SC of NC_x > $< R_{root} >$ $< D$ in $\text{NC}_y >$ <RC for NC_y > <QC for NC_y >dwa
E3: Zonke izindlovu zidla ihlamvana elilodwa. 'all elephants eat at least one twig'
E4: Bonke abantu badla ummbila owodwa. 'all humans eat some maize'

Here, the plural of C, *izindlovu*, is in nc10 which has the SC *zi-* and *abantu* 'humans' is in nc2 with SC *ba-* that are added to the verb stem *-dla*. Thus, for patterns, there are variables with any number of terminals that are selected based on some criterion, which is here the noun class of the noun.

Let us extend this now such that R's verb in the ontology would have a (dependent) preposition squeezed in the name, such as 'works *for*' and 'part *of*'. First, a few examples (regardless whether they are ontologically the best way of modelling things), with the preposition component underlined:

E5: zonke izazi zomnyuziki ziyingxenye ye-okhestra elilodwa. 'all musicians are a member of some orchestra'
E6: onke amavazi akhiwe ngobumba 'all vases are constituted of clay'
E7: zonke izincwadi zis emviloph ini eyodwa 'all letters are contained in some envelope'

The *ye-* in E5 is the result of the phonologically conditioned possessive concord for nc9, determined by *ingxenye* 'part': *ya-+i-=ye-*. The 'of' of 'constituted of' in E6 is dealt with by the preposition *nga-* regardless the noun class, but it is also phonologically conditioned (*nga-+u-=ngo-*). The containment in E7 is a locative (spatial), so those rules apply: a locative prefix *e-* and locative suffix, *-ini*, modify the noun *imvilophu* 'envelope' to *emvilophini* '(located/contained) in the envelope'.

This problem is not unique to isiZulu and related languages. Take, for instance, German and again the same axiom type. A template (T), as proposed in [19], reads awkwardly and would be better served by a pattern (P), with "G_C" the gender of C and "IA_{G_D}" the indeterminate article for D's gender:

T: Jeder/s $<$ C $><$ R $>$ mindestens 1 $<$ D $>$
P: <Qall G_C > $<$ C $><$ R $>$ mindestens <IA_{G_D} > $<$ D $>$
E8: T: Jeder/s Arbeiter arbeitet für mindestens 1 Betrieb
P: Jeder Arbeiter arbeitet für mindestens einen Betrieb
'each worker works for at least one company'

noting that the pattern generates a more acceptable sentence. Besides the article, the noun may change as well. For instance, with R a parthood relation, then the 'of' (underlined) in 'part of' can formulated as:

E9: Jedes Herz ist Teil eines Tieres. 'each heart is part of some animal'
E10: Jedes Herz ist ein Teil von mindestens einem Tier. 'each heart is part of at least one animal'

Finally, observe that some verbs with dependent prepositions in English may not be so in other languages, be this a 'co-verb' [28], extended verb [25], or integrated in the noun. For instance, 'part of' in 'part of the body' is *Körperteil* (DE) or *Lichaamsdeel* (NL), the 'for' in 'works for' can modify the verb (*-el-* is added to the verb root *-sebenza*, resulting in the extended stem *-sebenzela* (ZU)), or the preposition is incorporated in the tense ('made by' *-akhiwe* (ZU). Overall, there are gradations from no effect where a preposition can be squeezed in with the verb in naming an OP, to phonological conditioning, to modifying the article of the noun to modifying the noun. So, a preposition does belong neither to the verb nor to the noun uniquely across languages, but, typically, to the *role* that the object plays in the relation described by the verb in the sentence; e.g., it is *yomunto* only if it plays the role of the whole in a part-whole relation like 'heart is part of a human' (*inhliziyo iyingxenye yomuntu* (ZU)).

Thus, we have seen that a '3rd pers. sg.' may be context-dependent, and notions of prepositions may modify the verb or the noun or the article of the noun, or both.

3 Related Works Assessed Against the Requirements

Several approaches have been proposed and used to 'stretch' OWL's object property (OP) usage. We structure them along 5 principal options in two categories from simple to comprehensive and add CNL systems to it, whilst assessing them against the requirements.

'Hacks' in OWL. Although it is well-known that OWL on its own is limited [5], three different workarounds are being used. OPTION 1: IDENTIFIERS. Give the OP a system-generated identifier as 'name' (the IRI), add one or more labels, alike in the OBO ontologies or by overloading the annotation property, and in the application interface layer, such as OBOEdit and Protégé, one has to have an option to select the right label to use in an axiom (e.g., one of badla, idla, adla, zidla, kudla for 'eat' in isiZulu). This separates the ontology component from the natural language. It requires a guarantee that each OP must have at least one label, which is not required by OWL, and it should override the notion of preferred vs. alternative label. OPTION 2: VERB STEM OR INFINITIVE ONLY. Name the OP with the verb stem or its infinitive and conjugate everything as appropriate in the verbalisation interfaces to display it, as in the ACE system [20]. Thus, there is only one IRI for the OP, with as many relational expressions as needed. In isiZulu, an infinitive can also be a noun (e.g., *ukudla* 'to eat' or 'food'). However, once cannot reuse names in the ontology other than for punning [32]. Some noun stems can be classified into multiple noun classes, and the meaning is determined only with the complete word including the prefix (e.g., *umuntu* 'human being' and *ubuntu* 'humanity' have both *-ntu* as stem), so

OWL classes need the complete word for isiZulu and related languages, leaving the verb stem as only option for naming OPs. This, then, assumes an extra rule-based layer for the conjugation and prepositions. OPTION 3: INCLUDE ALL. Name positive and negative verb stems, add all the variants for the positive and negative, i.e., each variant has its own IRI, declare the positives equivalent and the negatives equivalent, and declare the positive and negative stem disjoint. For isiZulu, they are `dla` and `dli` for 'eat' and 'does not eat', equivalences as `badla` $\equiv$ `idla` etc. and `abadli` $\equiv$ `ayidli` etc. (despite that they are essentially synonyms), and disjointness for `dla` $\sqsubseteq \neg$`dli`. This option is only possible in a language where one can express OP equivalence and disjointness. Of the DL-based OWL species, only OWL 2 QL, 2 RL, and 2 DL permit this [32]. Thus, it is not a widely applicable solution. There also will be performance consequences from 'blowing up' the RBox five times in size in the worst case. Further, it conflates the difference between relational expression and relation to the extreme, so it is ontologically a bad choice even if it were to perform well in a particular implementation.

Comprehensive linguistic options outside OWL. Options 1 and 2 require that at least some of the linguistic knowledge be dealt with outside OWL, for which there are two elaborate proposals. OPTION 4: LANGUAGE MODEL. One could use a language model such as *lemon* [30]. Previous work showed that *lemon* was insufficient for the Bantu language family however [9], and the recent W3C community report [https://www.w3.org/2016/05/ontolex/] does not address them: (i) it needs an extension for the noun class information, (ii) it needs to avail of the *lemon* morphology module, and (iii) it was feasible for properties only when the domain and range were fixed and it and its subclasses would have names whose nouns are in the same noun class. Further, LexInfo and ISOcat are used for the linguistic annotation in *lemon*, but they miss both the noun class system information and the system of concordial agreement that requires rules. More generally, descriptive models for annotation are not suited for dealing with rules, for which rule languages exist. This brings us to OPTION 5: GRAMMAR. The grammar rules can be a tailor-made implementation or one can use one of the myriad formal grammars; within CNLs and OWL, there are GF and Codeco [26], possibly together with *lemon* as described in [10]. The OP naming of GF with ACE follows that of ACE (i.e., infinitive). While examples use an 'English ontology' as basis, it could be any with a resource grammar, and subsequently using a translator service either for the terms only as in [3] or to delegate the machine translation to GF [10,15,21]. Translation services are not available for isiZulu, and developing a full resource grammar for GF is unlikely for the foreseeable future, simply because of the limited documentation and investigation into isiZulu grammar. Even then, it still does not resolve the prepositions.

CNL-inspired approaches. Very few works take the simplistic approach of just reusing the name of the relationship or relational expression [19]. Stevens et al. [36] has one rule for processing OP names, being removing "_" (e.g., `derives_from` into derives from), which was feasible because all relations of the Relation Ontology adhere to a restricted naming scheme. In contrast, Hewlett et al. [18] accept incoherent naming and identified seven phrase structure categories of naming OPs in ontologies—(has) NP, V, (is) NP P, (is) VP P, VP NP, is NP, (is) AdjP—and availed of a POS tagger to verbalise them more natural language-like, so that, e.g., a `hasColor` OP verbalises into has a colour. SWAT NL [37] does a similar text-based processing of the OP name. ACE limits the naming scheme of OWL OP names to their infinitive form [20], with the processing happening independent from OWL, as is the case also in [36]. While ACE has a grammar module to do this, the lexical information for NaturalOWL [1] is provided by the domain expert in a Protégé plugin. The separate lexical layer on top of OWL by [1,20,36] have their own data structures rather than a known language model. Another strand of work seeks to link OWL to the Grammatical Framework (GF) [http://www.grammaticalframework.org] with, e.g., AceOwl [15,21]. Overall, there are two extremes in approach: either working with comprehensive top-down annotation frameworks and grammars (e.g., *lemon* [30], GF, Codeco [26]) or a bottom-up approach [1,16,23,36,37]. The few works on languages other than English take, at first at least if not throughout, a bottom-up approach. There are domain-independent solutions for notably Greek [1], Latvian [15,16], and AceWiki was tested with German and Spanish [21], where [15,21] use a 'detour' through GF. Neither of the two recent surveys on NLG and CNLs for OWL address issues of conjugation or prepositions [4,34].

Thus, none of the current approaches caters for the case where there are multiple words for a '3rd pers. sg./pl.' and have flexibility on prepositions.

4 Conceptual Model and Mappings for Relations

In order to obtain the technology-independent model to deal with verb conjugation and prepositions to support also languages other than English, we draw from several sources, which are described first in the preliminaries, after which the model is introduced, and finally the formalisation.

4.1 Preliminaries

From a language viewpoint, it may seem that the pair 'teaches' and 'taught by' or the pair 'works for' and 'employs' are all different relations, for they are different words. This is called the "standard view" on relations in philosophy [13,27]. However, there is only one state of affairs between the professor and the course, or between the worker and the company, respectively, so then there ought to be only one relation for one state of affairs. This is solved by *positionalism*, which relegates 'teaches' etc. to being *relational expressions*, and introducing a different notion of relation(ship). In this case there is one n-ary *relation(ship)* that has n unordered argument places, also called *roles*, in which the objects participate, and to which any number of relational expressions can be attached [13,27]. For instance, a relationship named teaching with the roles [lecturer] and [taught] such that the Professor participates in teaching by playing the [lecturer] role and Course plays the [taught] role.

Positionalism is the underlying commitment of the relational model and a database's physical schema, as well as of the main conceptual modelling languages. It has been employed in Object-Role Modelling (ORM) and its precursor NIAM for the past 40 years [17], UML Class Diagram notation requires *association ends* as roles, and Entity-Relationship (ER) Models have relationship *components* [12]. To illustrate the positionalism, let us take an example in ORM depicted in Fig. 1. It has a binary relationship (ORM fact type) eat with two participating entity types, Lion and Gazelle, where the lion plays the [predator] role and the gazelle plays the [prey] role, and a number of fact type readings, such as *... eats ...*, where the ellipses are filled with the entity types. Together with the fact type reading, it is verbalised as *Each lion eats at least one gazelle.* In the other reading direction, there is no constraint, which is verbalised with 'it is possible' or 'may', so we obtain the sentence *It is possible that a gazelle is eaten by a lion.* The 'by' is only needed when the [prey] role in eat is used to verbalise the axiom. The same mechanism holds for, say, a parthood relationship, with [part] the role that, say, Lecture plays and [whole] the role that Course plays, and a surface reading in both directions may be ... *part of* ... and ... *has part* ...: the 'of' preposition is only used in one reading direction, so is used with one role in that context. Put differently, the preposition is conceptually associated with neither the verb nor the noun, but with the role that an object referred to by the noun plays in the relation.

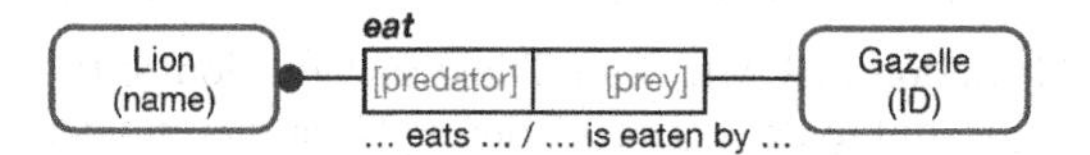

Fig. 1. Example ORM diagram with two entity types, the role names in the role-boxes of the fact type, and the fact type readings below the fact type. The name of the fact type was added to the figure for clarity (typically hidden from view).

An important advantage of positionalism is the separation of relation and reading, though roles are also useful for declaring more precise constraints; e.g., an object may not be allowed to perform two roles at the same time, which cannot easily be asserted with a standard view language. As we shall see, roles are also useful to attach information to for conjugation and prepositions.

There is already a unifying metamodel for the positionalist UML Class Diagrams v2.4.1, EER, and ORM2 [22], which can be extended with an orthogonal component for natural language annotations. This metamodel unifies their language features, the constraints that have to hold when using them, and harmonises their respective terminology. A small extract of this metamodel is depicted in the top-part of Fig. 2: each relationship contains at least two roles, whereas a role is part of exactly one relationship, and each role must have an entity type that plays it (though an entity type does not need to play a role).

4.2 Metamodel for Processing Properties

The task now, then, is to reconcile the positionalism, standard view, and the surface realisations or relational expressions. We first relate the components for positionalism to those of the standard view. This means mapping a relationship with at least two roles that the entities play into a predicate with entity types in a fixed order. These links and types of entities are shown with dashed lines in Fig. 2: it forces an order onto the entities and removes the role elements. If the language has binary relations only, one may simplify this to annotating it with a natural language sentence's nominative and dative/accusative positions. Or, informally: the 'subject' that does the thing and the 'object' that has something done to it, respectively; e.g., the lion (nominative) does the eating and the gazelle (dative) is the one that is eaten, regardless the order of the two elements.

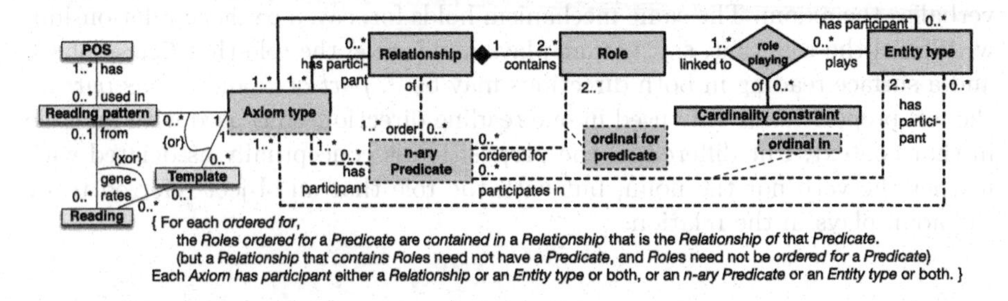

Fig. 2. Simplified depiction in UML Class Diagram notation of the main components (attributes suppressed), linking a section of the unifying metamodel (classes with thick lines; positionalist commitment) to predicates (classes with dashed lines; standard view) and their verbalisation (classes with thin lines).

The second step in model development is to consider whether to show in an ontology development environment or domain experts' interface elements with constraints only, or also the elements themselves as being typed. That is, whether from some actual ontology, it should generate Heart is an Entity type (indicating type of element) and Human has part Heart (without any constraints) as well, or only when they appear in some axiom with constraints. The metamodel in Fig. 2 is permissive of both, through Axiom type. This means that it can take care of those essentially second order statements, like `EntityType(Heart)`, the typing of a relationship (e.g., in DL notation, $\exists \texttt{haspart} \sqsubseteq \texttt{Heart}$ and $\exists \texttt{hasPart}^{-} \sqsubseteq \texttt{Human}$), and those axioms denoting constraints, such as of type $C \sqsubseteq = 1\, R.D$ (e.g., $\texttt{Human} \sqsubseteq = 1\, \texttt{hasPart.Heart}$).

Third, the natural language sentence. This may be split up in a reading *pattern* or *template* and the actual natural language sentences, or *readings*, that are generated from either. The main reason for this is to cater for different natural language grammars. In a 'simple' natural language, such a pattern may well be a straightforward template for the axiom where the nouns for the class and verb for the relation are simply plugged in on the fly, taken from the ontology file.

For grammatically richer languages, the pattern requires additional grammar rules to generate the sentence, as is the case for isiZulu [25], or processing those prepositions (recall Sect. 2). The elements to be plugged into the reading pattern are of a specific POS category, such as noun, verb, possessive concord and so on. This is included on the left-hand side of Fig. 2.

Finally, one can add a myriad of properties or attributes to the classes in Fig. 2, where the main selection of attributes of the classes is included in Fig. 3. These properties are general in the sense of regardless the implementation choices, yet their datatype and value ranges can vary because of that, such as implementing them in a relational database, XML document or linking to the linguistic Linked Open Data cloud. For instance, for tense, case, gender, and grammatical number, it does not matter which language model is chosen as source for interoperability. For the noun class system, it does matter, for no source other than the Noun Class System ontology has sufficient information about noun classes [9], in particular on which noun classes there are and the singular/plural pairs. Note that gender and noun class are optional. To cater for both cases where a preposition is squeezed into the name of the relationship, as is customary for object properties in OWL in English, and to record this separately for languages such as German and isiZulu, both presence/absence of a preposition can be recorded and the actual preposition itself when it does not fit in the relationship's name. Because the latter may not be relevant for some languages, such as English, it is made an optional attribute.

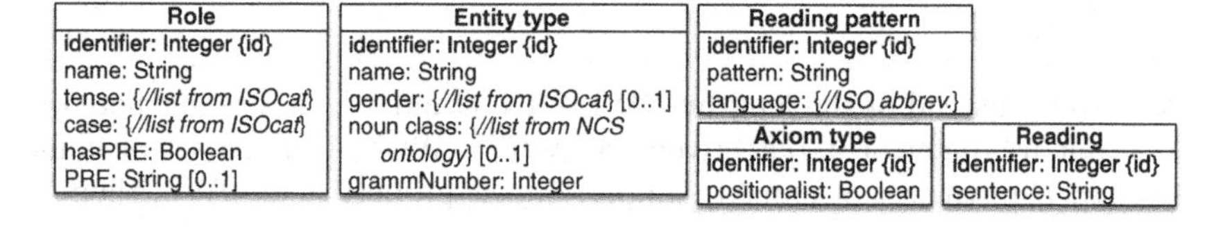

Fig. 3. Several suggested implementation extensions to the metamodel (see text for details).

4.3 Formalisation

Given the conceptual links between the standard view and positionalism, we now specify this formally for the knowledge-to-text case. This means that the bottom-part with the standard view relates with elements from, e.g., OWL, Common Logic, and First Order Predicate Logic, and the top-part with the conceptual modelling languages and its logic-based reconstruction with a positionalist commitment. The latter typically use a language in the $\mathcal{DLR}$ family of Description Logic languages [7], which has been applied first to ER [8] and subsequently in many variants to UML and ORM. What only has to be done is to specify the associations indicated with dashed lines in the UML Class Diagram in Fig. 2. We link the relevant parts of OWL 2 DL to $\mathcal{DLR}$ [8], both of whom have a model-theoretic semantics. The syntax for $\mathcal{DLR}$ is as follows, where P is an atomic relationship and A an atomic entity type (class), based on [8]:

$$R ::= \top_n \mid \mathrm{P} \mid (\$i/n : C) \mid \neg R \mid R_1 \sqcap R_2$$
$$C ::= \top_1 \mid \mathrm{A} \mid \neg C \mid C_1 \sqcap C_2 \mid \exists[\$i]R \mid (\leq k[\$i]R)$$

where i denotes a role (if it is not named, then integer numbers between 1 and n_{max} are used); n is the arity of the relation; the $(\$i/n : C)$ denotes all tuples in $\top_n$ that have an instance of C as their i-th component; k is a nonnegative integer for cardinality constraints). It uses the usual notion of interpretation, where $\mathcal{I} = (\cdot^{\mathcal{I}}, \cdot^{\mathcal{I}})$ and the interpretation function $\cdot^{\mathcal{I}}$ assigns to each concept C a subset $C^{\mathcal{I}}$ of $\Delta^{\mathcal{I}}$ and to each n-ary R a subset $\mathrm{R}^{\mathcal{I}}$ of $(\Delta^{\mathcal{I}})^n$, such that the conditions are satisfied following Table 1.

Table 1. Semantics of $\mathcal{DLR}$ (source: based on [8]).

$\top_n^{\mathcal{I}} \subseteq (\Delta^{\mathcal{I}})^n$	$A^{\mathcal{I}} \subseteq \Delta^{\mathcal{I}}$
$P^{\mathcal{I}} \subseteq \top_n^{\mathcal{I}}$	$(\neg C)^{\mathcal{I}} = \Delta^{\mathcal{I}} \setminus C^{\mathcal{I}}$
$(\neg R)^{\mathcal{I}} = \top_n^{\mathcal{I}} \setminus R^{\mathcal{I}}$	$(C_1 \sqcap C_2)^{\mathcal{I}} = C_1^{\mathcal{I}} \cap C_2^{\mathcal{I}}$
$(R_1 \sqcap R_2)^{\mathcal{I}} = R_1^{\mathcal{I}} \cap R_2^{\mathcal{I}}$	$(\$i/n : C)^{\mathcal{I}} = \{(d_1, \ldots, d_n) \in \top_n^{\mathcal{I}} \mid d_i \in C^{\mathcal{I}}\}$
$\top_1^{\mathcal{I}} = \Delta^{\mathcal{I}}$	$(\exists[\$i]R)^{\mathcal{I}} = \{d \in \Delta^{\mathcal{I}} \mid \exists(d_1, \ldots, d_n) \in R^{\mathcal{I}}.d_i = d\}$
	$(\leq k[\$i]R)^{\mathcal{I}} = \{d \in \Delta^{\mathcal{I}} \mid \|\{(d_1, \ldots, d_n) \in R_1^{\mathcal{I}} \mid d_i = d\| \} \leq k\}$

For OWL, instead of the lengthy OWL 2 DL standard, we present here only the relevant fragment of it (effectively $\mathcal{ALNHI}$). With A in the set of named classes and R in the set of named (simple) object properties in OWL, then:

$$C ::= \top \mid A \mid \forall R.A \mid \exists R.A \mid \leq k\, R \mid \geq k\, R \mid C_1 \sqcap C_2$$
$$R ::= \top_n \mid P \mid P^-$$

The semantics is like for $\mathcal{DLR}$, where "$\exists R.A$" has a semantics $(\exists R.A)^{\mathcal{I}} = \{x \mid \exists y.R^{\mathcal{I}}(x, y) \wedge A^{\mathcal{I}}\}$.

To declare the equivalence mappings, we first use [7,12] for typing of the DL roles/OWL OPs and their DL role components:
Standard view to positionalism:

$$\exists P.C \Longrightarrow \exists[\$1](P \sqcap (\$2/2 : C)) \qquad \exists P^-.C \Longrightarrow \exists[\$2](P \sqcap (\$1/2 : C))$$
$$\forall P.C \Longrightarrow \neg\exists[\$1](P \sqcap (\$2/2 : \neg C)) \qquad \forall P^-.C \Longrightarrow \neg\exists[\$2](P \sqcap (\$1/2 : \neg C))$$

Thus, from standard view to positionalist, we add argument places based on the typing of the relation or the use of the class constructors, by numbering the roles but bearing in mind that they do not have to appear in that order once represented in $\mathcal{DLR}$. In the other direction, we choose the following mapping, which is based on the motivation and algorithm in [12], restricted to binaries only, for OWL has only binary OPs:

Positionalism to standard view:

$$P \sqsubseteq [role]A \sqcap [elor]C \Longrightarrow \begin{array}{l} \exists role.A \sqsubseteq C \\ \exists elor.C \sqsubseteq A \\ role \equiv elor^{-} \end{array}$$

There is one final step to the mappings, which is when there are no domain or range restrictions, as is allowed in ontologies; e.g., there is only some axiom of pattern $C \sqsubseteq \exists R.D$ or $C \sqsubseteq \forall R.D$. This can be linked to a positionalist representation by introducing a property R' as subproperty of R, and make C and D the domain and range of R', and by adding the two roles:

$$C \sqsubseteq \forall R.D \Longrightarrow \begin{array}{l} R' \sqsubseteq [\$1/2]C \sqcap [\$2/2]D \\ R' \sqsubseteq R \end{array}$$

$$C \sqsubseteq \exists R.D \Longrightarrow \begin{array}{l} R' \sqsubseteq [\$1/2]C \sqcap [\$2/2]D \\ R' \sqsubseteq R \\ C \sqsubseteq \exists[\$1/2]R' \end{array}$$

These mappings cover the core possibilities for mappings between a positionalist and standard view logic. DLs were used for the clear link to applications (OWL, Semantic Web technologies), for having a readily available positionalist logic, and for notation convenience, yet it equally well can be cast in other languages, such as plain first order logic and the relational model.

5 Implementation and Testing

We have implemented the model with the mapping as a plugin to Protégé. It was developed in Java and avails of the OWL API for reading the OWL file and it writes into an XML file, which is graphically rendered in the plugin. A screenshot of the plugin is shown in Fig. 4 and it can be downloaded from the project page at http://www.meteck.org/files/geni, together with examples.

Regarding the implemented functionality, it specifically handles the interaction between the standard view OWL and the positionalist elements (Fig. 2, Sect. 4.3) and the annotations/attributes from Fig. 3, plus the additional feature that one can add new linguistic annotation properties. The ISOcat values are used and the noun class numbers were added, which are selectable through drop-down lists. The current version has a relevant subset of the possible axiom types, in particular: the *all-all* (`AllValuesFrom`, $C \sqsubseteq \forall R.D$) and *all-some* (`SomeValuesFrom`, $C \sqsubseteq \exists R.D$) patterns, and in anticipation of the verbaliser, also *subsumption* ($C \sqsubseteq D$), *union* ($C \sqsubseteq E \sqcup D$), *intersection* ($C \sqsubseteq E \sqcap D$) and *complement* ($C \sqsubseteq \neg D$), where C, D and E may be anonymous classes (though the plugin is easier to use with named classes). Each pattern is represented by a single element in the XML for annotations. This enables the user to insert

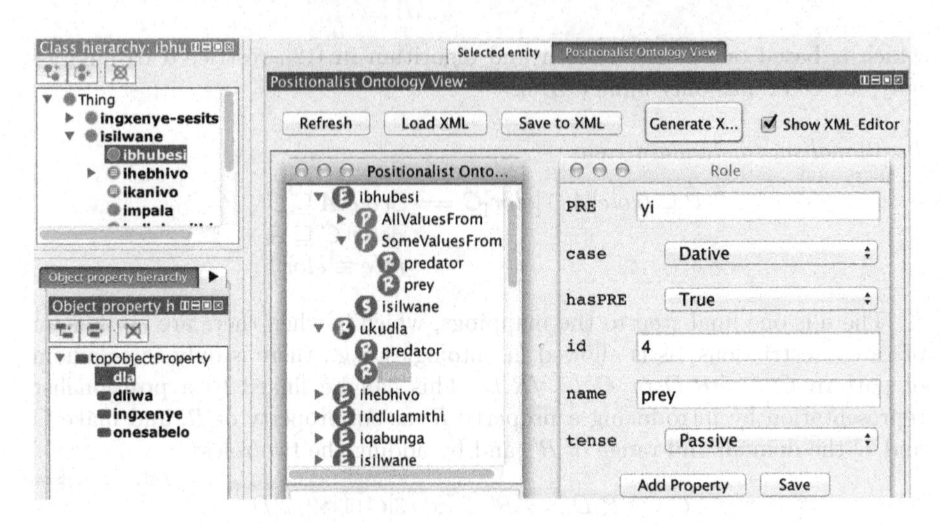

Fig. 4. Screenshot of the plugin with a section of the isiZulu African Wildlife Ontology (left), the positionalist representation (middle), and annotations (right), showing the prey role in the relationship ukudla 'to eat', with passive tense and yi 'by'.

also the desired name in the constructor for verbalisation; e.g., *noma* 'or'. The mapping from the OWL ontology view to the positionalist view is done by the Relationships, which are then used in the all-all and the all-some patterns. The plugin shows this by placing the attribute 'actorName' in the referencing XML element. Verbalisation may then be done by using the noun class of the actor according to the role that the actor is playing in the relationship.

Testing of the model focussed on validation and verification, i.e., on the basis of covering use cases. It was first tested on the positionalism and axiom types functionality. Second, a real modelling scenario was used: a basic isiZulu version of the African Wildlife ontology was created, which includes *ingxenye + ya* 'part of' and *dla + yi* 'eaten by' (see Fig. 4). The German examples from Sect. 2 were modelled in a test ontology. Finally, an ontology about pets was created in chiShona, which has grammar features like isiZulu, that also illustrates naming of intersection (*uyezve*) and complement (*zvisiri*) in anticipation of verbalisation.

6 Discussion

As noted in Sect. 3, currently popular language models, in particular *lemon* [30] and its W3C version, do neither have a way to address noun class information, nor (deep) prepositions other than adding a 'marker' on the *lemon* annotation of an object property. Extending them limits one to a single technology and, moreover, it is still tailored to what in philosophy is called the 'standard view' of relations (roughly: predicates) that do not cater for roles and properties thereof. Also, there was no functional *lemon*-based ontology annotation tool, so it would have to be developed anyway. In contrast, the model proposed here is, by design, technology-independent and the mapping between a logic with standard view commitment and one with a positionalist stance can be implemented for any combination of languages. For instance, one could also link, say, the OWL or Common Logic Interchange Format to the language of UML Class Diagram notation so as to have a better interaction between the logic and conceptual models, thereby enhancing ontology-driven information systems. The proposed model offers a more precise representation of the knowledge, the natural language, and the interaction between the two.

Further, now one can add noun class, case, gender, tense, and prepositions in a simple annotation interface that guarantees syntactic correctness of the XML file, rather than manually writing in text files. These grammar features are present also in other languages, such as Greek [1], Latvian [16], Chinese [28], and languages related to isiZulu, hence, the here presented model can be reused for languages other than the isiZulu focussed on in this paper.

Our next step is to use it with isiZulu and Runyankore so as to generate more correct sentences from the patterns developed by [6,25] and for part-whole relations in particular [24] and evaluate it more comprehensively.

7 Conclusions

A model that reconciles standard view and positionalist commitments was proposed, which is the first precise implementation that maps between representation languages committing to either. Precision was achieved with a formal mapping with OWL and $\mathcal{DLR}$ for logical correctness. The 'roles' (description logic role components) serve as the main vehicle for managing the annotations needed for elaborate conjugation and for prepositions that belong to it. The model with mappings was implemented as a Protégé plugin to validate its adequacy, using examples from isiZulu, chiShona, and German.

Acknowledgments. This work is based on research supported by the National Research Foundation of South Africa (Grant Number 93397).

References

1. Androutsopoulos, I., Lampouras, G., Galanis, D.: Generating natural language descriptions from OWL ontologies: the NaturalOWL system. JAIR **48**, 671–715 (2013)
2. Baldwin, T., Kordoni, V., Villavicencio, A.: Prepositions in applications: a survey and introduction to the special issue. Comput. Linguist. **35**(2), 119–149 (2009)
3. Bosca, A., Dragoni, M., Francescomarino, C.D., Ghidini, C.: Collaborative management of multilingual ontologies. In: Buitelaar, P., Cimiano, P. (eds.) Towards the Multilingual Semantic Web, pp. 175–192. Springer, Berlin (2014)
4. Bouayad-Agha, N., Casamayor, G., Wanner, L.: Natural language generation in the context of the semantic web. Semant. Web J. **5**(6), 493–513 (2014)
5. Buitelaar, P., Cimiano, P., Haase, P., Sintek, M.: Towards linguistically grounded ontologies. In: Aroyo, L., et al. (eds.) ESWC 2009. LNCS, vol. 5554, pp. 111–125. Springer, Heidelberg (2009)

6. Byamugisha, J., Keet, C.M., DeRenzi, B.: Bootstrapping a Runyankore CNL from an isiZulu CNL. In: Davis, B., Pace, G., Wyner, A., Pace, G.J., Pace, G.J., Pace, G.J., Pace, G.J. (eds.) CNL 2016. LNCS, vol. 9767, pp. 25–36. Springer, Heidelberg (2016). doi:10.1007/978-3-319-41498-0_3
7. Calvanese, D., De Giacomo, G.: Expressive description logics. In: The DL Handbook: Theory, Implementation and Applications, pp. 178–218. Cambridge University Press, Cambridge (2003)
8. Calvanese, D., De Giacomo, G., Lenzerini, M., Nardi, D., Rosati, R.: Description logic framework for information integration. In: Proceedings of of KR 1998, pp. 2–13 (1998)
9. Chavula, C., Keet, C.M.: Is lemon sufficient for building multilingual ontologies for Bantu languages? In: Proceedings of OWLED 2014, CEUR-WS, vol. 1265, pp. 61–72, riva del Garda, Italy, 17–18 October 2014
10. Davis, B., Enache, R., van Grondelle, J., Pretorius, L.: Multilingual verbalisation of modular ontologies using GF and *lemon*. In: Kuhn, T., Fuchs, N.E. (eds.) CNL 2012. LNCS, vol. 7427, pp. 167–184. Springer, Heidelberg (2012)
11. Denaux, R., Dimitrova, V., Cohn, A.G., Dolbear, C., Hart, G.: Rabbit to OWL: ontology authoring with a CNL-based tool. In: Fuchs, N.E. (ed.) CNL 2009. LNCS, vol. 5972, pp. 246–264. Springer, Heidelberg (2010)
12. Fillottrani, P.R., Keet, C.M.: Evidence-based languages for conceptual data modelling profiles. In: Morzy, T., Valduriez, P., Ladjel, B. (eds.) ADBIS 2015. LNCS, vol. 9282, pp. 215–229. Springer, Heidelberg (2015)
13. Fine, K.: Neutral relations. Philos. Rev. **109**(1), 1–33 (2000)
14. Fuchs, N.E., Kaljurand, K., Kuhn, T.: Discourse representation structures for ACE 6.6. Technical report, ifi-2010.0010, Department of Informatics, University of Zurich, Switzerland (2010)
15. Gruzitis, N., Barzdins, G.: Towards a more natural multilingual controlled language interface to OWL. In: Proceedings of IWCS 2011, pp. 335–339. ACL, Stroudsburg (2011)
16. Gruzitis, N., Nespore, G., Saulite, B.: Verbalizing ontologies in controlled Baltic languages. In: 4th International Conference on HLT -"The Baltic Perspective", FAIA, vol. 219, pp. 187–194. IOS Press (2010)
17. Halpin, T., Morgan, T.: Information Modeling and Relational Databases, 2nd edn. Morgan Kaufmann, Burlington (2008)
18. Hewlett, D., Kalyanpur, A., Kolovski, V., Halaschek-Wiener, C.: Effective NL paraphrasing of ontologies on the semantic web. In: Proceedings of WS on End-User Semantic Web Interaction, CEUR-WS, vol. 172 (2005)
19. Jarrar, M., Keet, C.M., Dongilli, P.: Multilingual verbalization of ORM conceptual models and axiomatized ontologies. Starlab technical report, Vrije Universiteit Brussel, Belgium, February 2006
20. Kaljurand, K., Fuchs, N.E.: Verbalizing OWL in attempto controlled English. In: Proceedings of OWLED 2007, CEUR-WS, vol. 258, Innsbruck, Austria, 6–7 June 2007
21. Kaljurand, K., Kuhn, T., Canedo, L.: Collaborative multilingual knowledge management based on controlled natural language. Semant. Web **6**(3), 241–258 (2015)
22. Keet, C.M., Fillottrani, P.R.: An ontology-driven unifying metamodel of UML Class Diagrams, EER, and ORM2. DKE **98**, 30–53 (2015)
23. Keet, C.M., Khumalo, L.: Toward verbalizing ontologies in isiZulu. In: Davis, B., Kaljurand, K., Kuhn, T. (eds.) CNL 2014. LNCS, vol. 8625, pp. 78–89. Springer, Heidelberg (2014)

24. Keet, C.M., Khumalo, L.: On the verbalization patterns of part-whole relations in isiZulu. In: Proceedings of INLG 2016, pp. 174–183. ACL, Edinburgh, 5–8 September 2016
25. Keet, C.M., Khumalo, L.: Toward a knowledge-to-text controlled natural language of isiZulu. LRE (2016, in print). doi:10.1007/s10579-016-9340-0
26. Kuhn, T.: A principled approach to grammars for controlled natural languages and predictive editors. J. Logic Lang. Inform. **22**(1), 33–70 (2013)
27. Leo, J.: Modeling relations. J. Phil. Logic **37**, 353–385 (2008)
28. Li, C.N., Thompson, S.A.: Co-verbs in Mandarin Chinese: verbs or prepositions? J. Chin. Linguist. **2**(3), 257–278 (1974)
29. Mathonsi, N.N.: Prepositional and adverb phrases in Zulu: a linguistic adn lexicographic problem. S. Af. J. African Lang. **2**, 163–175 (2001)
30. McCrae, J., et al.: Interchanging lexical resources on the semantic web. LRE **46**(4), 701–719 (2012)
31. McCrae, J., et al.: The Lemon cookbook. Technical report, Monnet Project (2012)
32. Motik, B., Patel-Schneider, P.F., Parsia, B.: OWL 2 web ontology language structural specification and functional-style syntax. W3c recommendation, W3C, 27 October 2009. http://www.w3.org/TR/owl2-syntax/
33. de Oliveira, R., Sripada, S.: Adapting simplenlg for brazilian portuguese realisation. In: Proceedings of INLG 2014, pp. 93–94. ACL, Philadelphia, June 2014
34. Safwat, H., Davis, B.: CNLs for the semantic web: a state of the art. LRE (2016, in print) doi:10.1007/s10579-016-9351-x
35. Schneider, N., Srikumar, V., Hwang, J.D., Palmer, M.: A hierarchy with, of, and for preposition supersenses. In: Proceedings of LAW IX - The 9th Linguistic Annotation Workshop, pp. 112–123, Denver, USA, 5 June 2015
36. Stevens, R., Malone, J., Williams, S., Power, R., Third, A.: Automating generation of textual class definitions from OWL to English. J. Biomed. Sem. **2**(Suppl 2), S5 (2011)
37. Third, A., Williams, S., Power, R.: OWL to English: a tool for generating organised easily-navigated hypertexts from ontologies. In: Poster, Demo Paper at ISWC 2011, Bonn, Germany 23–27 October 2011. Open Unversity, London (2011)

Dependencies Between Modularity Metrics Towards Improved Modules

Zubeida Casmod Khan[1,2](✉) and C. Maria Keet[1]

[1] Department of Computer Science, University of Cape Town, Cape Town, South Africa
zkhan@csir.co.za, mkeet@cs.uct.ac.za

[2] Council for Scientific and Industrial Research, Pretoria, South Africa

Abstract. Recent years have seen many advances in ontology modularisation. This has made it difficult to determine whether a module is actually a good module; it is unclear which metrics should be considered. The few existing works on evaluation metrics focus on only some metrics that suit the modularisation technique, and there is not always a quantitative approach to calculate them. Overall, the metrics are not comprehensive enough to apply to a variety of modules and it is unclear which metrics fare well with particular types of ontology modules. To address this, we create a comprehensive list of module evaluation metrics with quantitative measures. These measures were implemented in the new Tool for Ontology Module Metrics (TOMM) which was then used in a testbed to test these metrics with existing modules. The results obtained, in turn, uncovered which metrics fare well with which module types, i.e., which metrics need to be measured to determine whether a module of some type is a 'good' module.

1 Introduction

A number of techniques for ontology modularisation have been proposed in recent years, such as traversal methods [18], locality-based extraction [9], and partitioning [2,4]. There also have been attempts at analysing which types of modules exist [1], and which types are useful for which purpose, such as that high-level abstraction modules are used for comprehension [15]. There is, however, a disconnect between the two. For instance, if a modeller wants to reuse, say, only the branch of 'social objects' from the DOLCE foundational ontology for an ontology about e-government, then how does the modeller know that the module extracted from DOLCE is a good module, and which one of the modularisation techniques creates the module of the 'best' quality? In fact, it is even unclear how the quality of an ontology module could or should be measured. There are a few studies on evaluation of ontology modules, which focus on a few metrics, such as size, cohesion, coupling, correctness, and completeness [22,24,27], however some of them do not have a defined formula to measure them e.g., intra-module distance [4]. The metrics are not comprehensive enough to apply to the 14 types of modules [15] that exist [11].

E. Blomqvist et al. (Eds.): EKAW 2016, LNAI 10024, pp. 400–415, 2016.
DOI: 10.1007/978-3-319-49004-5_26

Metrics such as size do not fare well with modules created using locality-based techniques [12], while completeness and correctness do not measure well with partition-based modules [22]. This suggests that only specific metrics would be applicable for some type of module to assess the quality of an ontology module. To the best of our knowledge, no one has filled this knowledge gap of how evaluation metrics relate to modules to reveal the module quality.

To solve these problems, we take both existing metrics, where for those that do not have a computable component, we devise a new function, and add three more metrics, totalling to 16. To examine their usefulness, we implemented these metrics into one evaluation tool, the Tool for Ontology Modularity Metrics (TOMM) and examined 189 ontology modules on these metrics. TOMM can be downloaded from http://www.thezfiles.co.za/Modularity/TOMM.zip.

These results from TOMM generated insight into the expected values for evaluation metrics for the different types of modules. We have evidence-based insight about which metrics fare well with which module types. For instance, ontology matching modules fare well with a mix of structural, relational, and information hiding metrics. This insight helps the ontology developer to determine whether a module is of good quality or not.

The remainder of the paper is structured as followed. Related works are summarised in Sect. 2, followed by the evaluation metrics for modules in Sect. 3. The software, TOMM, experimental evaluation, and use-cases is presented in Sect. 4, and a discussion in Sect. 5. Lastly, we conclude in Sect. 6.

2 Related Works

In order to solve the problem of insufficient modularity metrics, we look at the existing work. To start with we use a definition of a module by Khan and Keet stating that a module M is a subset of a source ontology or module M is an ontology existing in a set of modules such that, when combined, make up a larger ontology [15]. In addition, Khan and Keet created a framework for ontology modularity, aimed at guiding the modularisation process [15]. It consists of dimensions for modularity such as use-cases, techniques, types, and properties. The dimensions have been linked to reveal dependencies, and to annotate modules with additional information. It does not include module evaluation. Other works do consider this. For instance, Pathak et al. [22] identified main properties that modules need to satisfy, such as size, correctness, completeness, and evaluated them using existing tools, noting that module correctness is satisfied by most techniques and that completeness and size are difficult to satisfy. They also observed that the logic-based approaches tend to result in modules where completeness is achieved while the graph-based approaches generate modules of smaller size that are not logically complete.

Schlicht and Stuckenschmidt [24] created a set of structural criteria for ontology modules, including connectedness, size, and redundancy of representation, and use quantitative functions to formally measure each criteria value. They argue that structural criteria have an effect on efficiency, robustness and maintainability for the application of semantics-based peer-to-peer systems.

They evaluated the SWOOP and PATO modularity tools on their structural criteria, observing that SWOOP favours modules with a good connectedness over modules with suitable size values, whereas with PATO, a threshold value could be selected such that when the threshold value is increased, so did the size suitability of the module to the detriment of connectedness. Regarding connectedness or cohesion, metrics were introduced by [27], being number of root classes, number of leaf classes, and average depth of inheritance tree of all leaf node. These metrics are not aimed at evaluating the quality of modules, but are rather general for ontologies. In light of this, Oh et al. present new metrics for cohesion to measure the strength of the relations in a module [20], and semantic dependencies were proposed in [5]. Ensan and Du's metrics in [5] use the notion of strong and moderate dependencies between entities. However, there are certain relations in an ontology that are neither strong nor moderate, such as intersections of classes.

3 Evaluation Metrics

The evaluation metrics for modularity was compiled by studying existing literature on modularity. This resulted in 13 metrics from the literature, of which five were short of a metric for quantitative evaluation that have now been devised (indicated with an asterisk), and three new ones have been added (indicated with a double asterisk)[1].

Size. The size metric is a fairly common metric as a modularity evaluation criterion [3,4,20,22,24]. Size refers to the number of entities in a module, $|M|$, which can be subdivided into number of classes, $|C|$, object properties $|OP|$, data properties $|DP|$, and individuals $|I|$. Size is calculated as follows: $Size(M) = |M| = |C| + |OP| + |DP| + |I|$. Note that it excludes the number of axioms, because that is considered a structural criterion.

Appropriateness of module size can be specified by mapping the size of an ontology module to some appropriateness values. Schlicht and Stuckenschmidt [24] propose an appropriate function to measure this, which ranges between 0 and 1, where a module with an optimal size has a value of 1. The function they propose is based on software design principles: since the optimal size of software modules is between 200–300 logical lines of software code, an axiom value of 250 would be the optimal size for an ontology, restricting the module to be between 0 and 500 axioms. The appropriateness equation is defined as follows [24], where x is the number of axioms in the module: $Appropriate(x) = \frac{1}{2} - \frac{1}{2}cos(x.\frac{\pi}{250})$.

Attribute richness is defined as the average number of attributes per class [25]; i.e., each class is defined by a number of axioms with properties describing describing it, which are referred to as attributes: $AR(M) = \frac{|att|}{|C|}$ where att is the number of attributes (OWL properties) of all entities and $|C|$ is the number of classes in the module.

[1] An earlier version of this section was presented at [14] and has now been updated with some corrections, refinements, and better descriptions.

Inheritance richness refers to how the knowledge is distributed across the ontology [25], such as with deep class hierarchies versus one with a flat or horizontal structure with few subclasses; this is calculated as follows: $IR_S(M) = \frac{\sum_{C_i \in C} |H^C(C_1, C_i)|}{|C|}$ where $|H^C(C_1, C_i)|$ is the number of subclasses ($C1$) for a class Ci and $|C|$ is the total number of classes in the ontology.

Cohesion refers to the extent to which entities in a module are related to each other. We use a metric defined in [20]:

$$Cohesion(M) = \begin{cases} \sum_{C_i \in M} \sum_{C_j \in M} \frac{SR(c_i, c_j)}{|M|(|M|-1)} & if |M| > 1 \\ 1 & otherwise \end{cases}$$

where $|M|$ is the number of entities in the module and $|M|(|M|-1)$ represents the number of possible relations between entities in M. The strength of relation for each entity is calculated based on a farness measure.

$$SR(c_i, c_j) = \begin{cases} \frac{1}{farness(i)} & if\ relations\ exist\ between\ c_i\ and\ c_j \\ 0 & otherwise \end{cases}$$

Redundancy has been defined as the duplication of axioms within a set of ontology modules [24]. When a large ontology is partitioned into smaller modules, there are sometimes modules that overlap with regard to shared knowledge. Thus axioms exist in more than one modules. Sometimes this is required for robustness or efficiency. However, these redundant axioms cause difficulty in maintaining the consistency of the modules when modules are to be updated. To measure redundancy in a set of modules, we use: $Redundancy = \frac{(\sum_{i=1}^{k} n_i) - n}{\sum_{i=1}^{k} n_i}$.

Correctness states that every axiom that exists in the module also exists in the original ontology and that nothing new should be added to the module [2,4,16,22], i.e.: $Correctness(M) = M \subseteq O$.

Completeness A module is logically complete if the meaning of every entity is preserved as in the source ontology. The completeness property evaluates that for a given set of entities or signature, every axiom that is relevant to the entity as in the source ontology is captured in the module [2,4,16,22]. $Completeness(M) = \sum_{i}^{n} Axioms(Entity_i(M)) \models Axioms(Entity_i(O))$.

Intra-module distance* d'Aquin et al. [4] introduce the notion of intra-module distance as the distance between entities in a module, which may be calculated by counting the number of relations in the shortest path from one entity to the other, for every entity in the module. The shortest path is calculated based on the entity hierarchy. Based on the description by d'Aquin et al., we refine this to measuring this distance by using Freeman's Farness value [6]. In the field of network centrality, Freeman's Farness value of a node is described as the sum of its distances to all other nodes in the network:

$$Intra\text{-}module\ distance(M) = \sum_{i}^{n} Farness(i) \quad (1)$$

with n the number of nodes in the module and the Farness value defined as $Farness(i) = \sum_{j}^{n} distance_{ij}$ The distance then is measured as the length of the shortest path between entities.

Inter-module distance* in a set of modules has been described as the number of modules that have to be considered to relate two entities [3,4]. Based on this definition, we have created an equation to measure the inter-module distance of a network of modules.

$$IMD = \begin{cases} \sum_{C_i, C_j \in (M_i, M_n)} \frac{NM(C_i, C_j)}{|(M_i, .., M_n)|(|(M_i, .., M_n)| - 1)} & |(M_i, .., M_n)| > 1 \\ 1 & otherwise \end{cases} \tag{2}$$

where $NM(C_i, C_j)$ is the number of modules to consider to relate entities i and j and $|(M_i, .., M_n)|(|(M_i, .., M_n)| - 1)$ represents the number of possible relations between entities in a set of modules $(M_i, ., M_n)$.

Coupling* has been described as a measure of the degree of interdependence of a module [7,19–21]. The coupling value is high if entities in a module have strong relations to entities in other modules. This also means that it will be difficult to update such modules independently because they affect other modules in the system. To measure the coupling of a module, we define our own measure as a ratio of the number of external links (axioms) between a module M_i and M_j, NEL_{M_i,M_j} for n modules in a system to every possible external link between a module M_i and M_j in a system.

$$Coupling(M_i) = \begin{cases} \sum_{i=0}^{n} \sum_{\substack{j=0 \\ i \neq j}}^{n} \frac{NEL_{M_i,M_j}}{|M_i||M_j|} & NEL_{M_i,M_j} > 0 \\ 0 & otherwise \end{cases} \tag{3}$$

where $|M_i|$ is the number of entities in the current module and $|M_j|$ is the number of entities in a related module in the set of n modules. External links in ontology modules depend on what linking language is used.

Encapsulation* d'Aquin et al. mention encapsulation with the notion that "a module can be easily exchanged for another, or internally modified, without side-effects on the application can be a good indication of the quality of the module" [4]. This general idea seems potentially useful for semantic interoperability. There are thus two components to d'Aquin et al.'s encapsulation:

- 'Swappability' of a module, which increases with fewer links to entities in another module in an ontology network; e.g., one can interchange their domain ontologies between foundational ontologies using the SUGOI tool [13].
- Casting it into a measure of knowledge preservation within the given module.

We have designed an equation to calculate the encapsulation of a module in a given a set of modules. For a module, with $n-1$ related, this is measured using the number of axioms in the given module $|Ax_i|$ and the number of axioms that overlap between the given module and related modules, $|Ax_{ij}|$.

$$Encapsulation(M\_i) = 1 - \frac{\sum_{j=1}^{n-1} \frac{|Ax_{ij}|}{|Ax_i|}}{n} \tag{4}$$

Encapsulation values in modules that are equal or close to 1 indicates a good encapsulation value; all or most of the knowledge has been encapsulated and privacy has been completely preserved. Conversely, values that are equal to or close to 0 indicates a poor encapsulation value; none or very little of the knowledge has been encapsulated and privacy has not been preserved.

Independence* Independence evaluates whether a module is self-contained and can be updated and reused separately. In this way, ontology modules can evolve independently. Thus, the semantics of the entire ontology could change without the need for all the modules to be changed. For instance, for the set of Gist foundational ontology modules [17], if information about physical things need to be updated, the relevant module gistPhysicalThing could be updated without needing to alter the remaining modules. In order to determine whether a module is independent, we use two metrics, i.e., the encapsulation and the coupling measure. A module is set to be independent if it has an encapsulation value of 1 and a coupling value of 0. This can be checked using the following code snippet.

$$Ind(M\_i) = \begin{cases} true & Encapsulation(M\_i) = 1 \text{ and } Coupling(M\_i) = 0 \\ false & otherwise \end{cases} \tag{5}$$

where $|M_i|$ is the number of entities in the current module and $|M_j|$ is the number of entities in a related module in the set of n modules.

Relative size** can be defined as the size of the module—i.e., number of classes, properties, and individuals—compared to the original ontology. The relative size of a module strongly influences the result of the module on tasks such as reasoning and maintenance, for if the module extracted is nearly the same as the original one, then not substantial optimisation will be obtained. To compute this, we use the ratio of the size of the module M (i.e., $|M|$) and the original (source) ontology O (i.e., $|O|$):

$$Relative\ size = \frac{|M|}{|O|} \tag{6}$$

Thus, a lower value (between 0 and 1) is better.

Atomic Size** The notion of atoms within ontology modules was first introduced in [26], who define it as a group of axioms within an ontology that have dependencies between each other. Based on the findings from the study, that it is possible to modularise an ontology using atomic decomposition as a method, we propose to measure the size of atoms in ontologies. We define the atomic size as the average size of a group of inter-dependent axioms in a module, and formulate an equation to measure the atomic size of a module by using the sum of all the atoms present in the module, and the size of the ontology.

$$Atomic\ Size(M) = \sum_{i}^{n} \frac{Atom_i}{|M|} \tag{7}$$

Relative Intra-module distance** can be defined as the difference between distances of entities in a module M to a source ontology O. This difference would reveal if the overall distance between the entities in the module has been reduced, and by how many distance units. This is useful in comparing the difference in module size; whether the technique reduces the size considerably. To compare the distances of the original ontology, we compute the farness values for the subset of nodes that exist in a module, which is used to calculate the intra-module distance of the original ontology, and is defined as follows:

$$Relative\ intra\text{-}module\ distance(M) = \frac{Intra\text{-}module\ distance(O)}{Intra\text{-}module\ distance(M)} \tag{8}$$

4 Implementation and Evaluation

We have created a Tool for Ontology Modularity Metrics (TOMM) to evaluate ontology modules. TOMM allows one to upload a module or set of related ontology modules, together with an original ontology (if it exists), and then it computes metrics for the module/s. A screenshot of TOMM's interface is shown in Fig. 1. The metrics are saved as a text file on the user's computer.

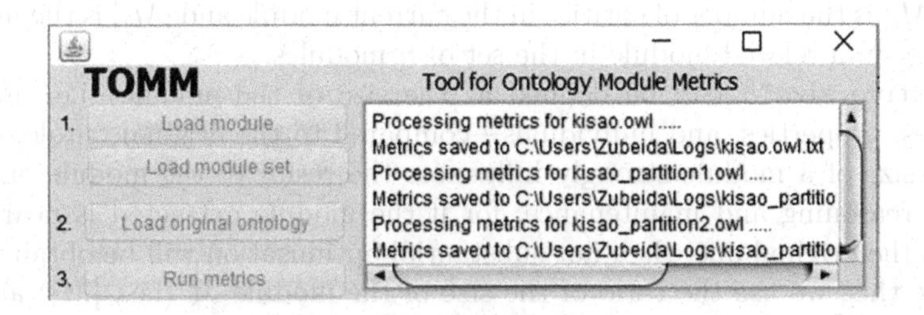

Fig. 1. The interface of TOMM.

4.1 Experimental Evaluation

The purpose of the experiment is to evaluate modules with a set of metrics. We expect that the results will determine how the metrics of a module relate to other factors, such as technique to create them.

Table 1. Averages for a subset of TOMM's metrics; $|T|$ = number of module types, approp. = appropriateness, IMD = intra module distance.

	$\|T\|$	Relative size	Atomic size	Approp.	Relative IMD	Attribute richness	Inheritance richness
T1	13	0.02	5.50	0.34	20.69	0.83	1.48
T2	42	-	5.31	0.64	-	1.45	2.37
T3	7	0.90	6.31	0.11	1.00	0.84	2.30
T4	3	0.02	5.00	0.47	63.66	3.61	1.79
T5	2	0.30	7.20	0.61	1.04	0.87	2.45
T6	10	0.17	2.99	0.30	0.00	0.10	54.32
T7	90	0.01	1.00	0.007	0.00	0.50	1.19
T8	4	0.56	3.64	-	1.02	0.71	3.15
T9	1	1.00	2.89	-	0.00	0.00	2.83
T10	1	0.56	4.21	0.99	1.03	0.00	3.06
T11	3	0.49	3.77	0.89	1.00	0.58	2.44
T12	3	0.42	5.87	0.02	2.17	1.05	2.89
T13	6	1.00	4.33	0.38	1.00	0.73	2.72
T14	1	0.97	5.65	-	1.00	1.78	3.04

Materials and Methods. The method for the experiment is straightforward: (1) take a set of ontology modules; (2) run the TOMM tool for each module; (3) conduct an analysis from the evaluation results for each module. In order to determine which metrics can be used to evaluate which module types, we need to determine how to interpret the values for each metric, which are as follows:

- correctness, completeness and independence are measured as true/false;
- size, no. of axioms, atomic size, intra-module distance. relative intra-module distance, attribute richness, and inheritance richness are measured on a numerical range;
- relative size, appropriateness, cohesion, encapsulation, coupling, and redundancy are measured on a 4-point scale of small (0–0.25), medium (0.25–0.5), moderate (0.51–0.75), and large (0.75–1).

The materials used for the experiment were as follows: Protégé v4.3 [8], TOMM, and a set of ontology modules that serve as the training set. Khan and Keet's set [15] was reused, which contains 189 ontology modules that were collected from ontology repositories and as referenced in the literature. This set contains modules of 14 different types, which are summarised in the appendix. All the test files used for this experimental evaluation can be downloaded from www.thezfiles.co.za/Modules/testfiles.zip.

Results. We ran TOMM for each of the 189 modules and metrics were successfully generated for 188. Due to space limits we include only Table 1 with average values for a subset of the metrics and highlight the notable aspects of the results here; the remaining metric tables are available online together with the test files.

For size, T7 (ontology matching) modules are very small, only 2 % compared to the original ontology. T2 (subject domain) could not be evaluated with the relative size metric as there were no original ontologies. T13 (expressiveness sub-language) is as large as the original ontology. For appropriateness, T10 (entity type abstraction) is the most appropriate at 0.99, meaning that most of the modules have between 200–300 axioms. The relative intra-module distance values determine by how many units (paths between entities) the module has been reduced. T4 (locality) modules were reduced with a high value by 63.65 units followed by T1 (ontology design patterns) by 20.69 units.

The T4 (locality), T8 (optimal reasoning), T9 (axiom abstraction), and T10 (entity type abstraction) modules all hold the correctness metrics; every axiom that exists in the module also exists in the source ontology and nothing new had been added. T1 (ontology design pattern) modules are the only set that all hold the completeness metric; the meaning of every entity in the module is preserved as in the source ontology. For attribute richness, T4 (locality) modules were the richest with a value of 3.61; these modules have on average 3.61 attributes per class. For inheritance richness, T6 (domain coverage) modules had a large value of 54.32 indicating many subclasses per class.

The information hiding and relational criteria only apply to module sets, T2 (subject domain), T6 (domain coverage), T7 (ontology matching), and T8 (optimal reasoning). For encapsulation, T7 (ontology matching) modules had a high value of 1; the knowledge is preserved in the individual modules and they can be changed individually without affecting the other modules in the set. For coupling, most of the modules had 0 values (no links to other modules in the set). The T7 (ontology matching) modules are independent; they are self-contained and also do not contain links to other modules in the set. The experiment uncovered which metrics fare well with which module type as discussed in this section and included in Fig. 2, where for each module type, the metrics and values that fare well with it are stated in bold font.

It is also worthwhile to check whether the techniques used for modularisation have an effect on the quality of the module. For the set of 189 modules in the set, there were four techniques used to generate them: graph partitioning, locality-based modularisation, *a priori* modularisation, and manual methods. The modules that were generated via graph partitioning measured well for the following criteria: relative size (small), encapsulation (large), coupling (small), redundancy (small). The modules generated with locality-based modularity performed well for correctness (all true). *a priori* modules all performed well for encapsulation (large), coupling (small), and redundancy (small). There was no link between the metrics returned by the modules generated by manual methods; all the results differed.

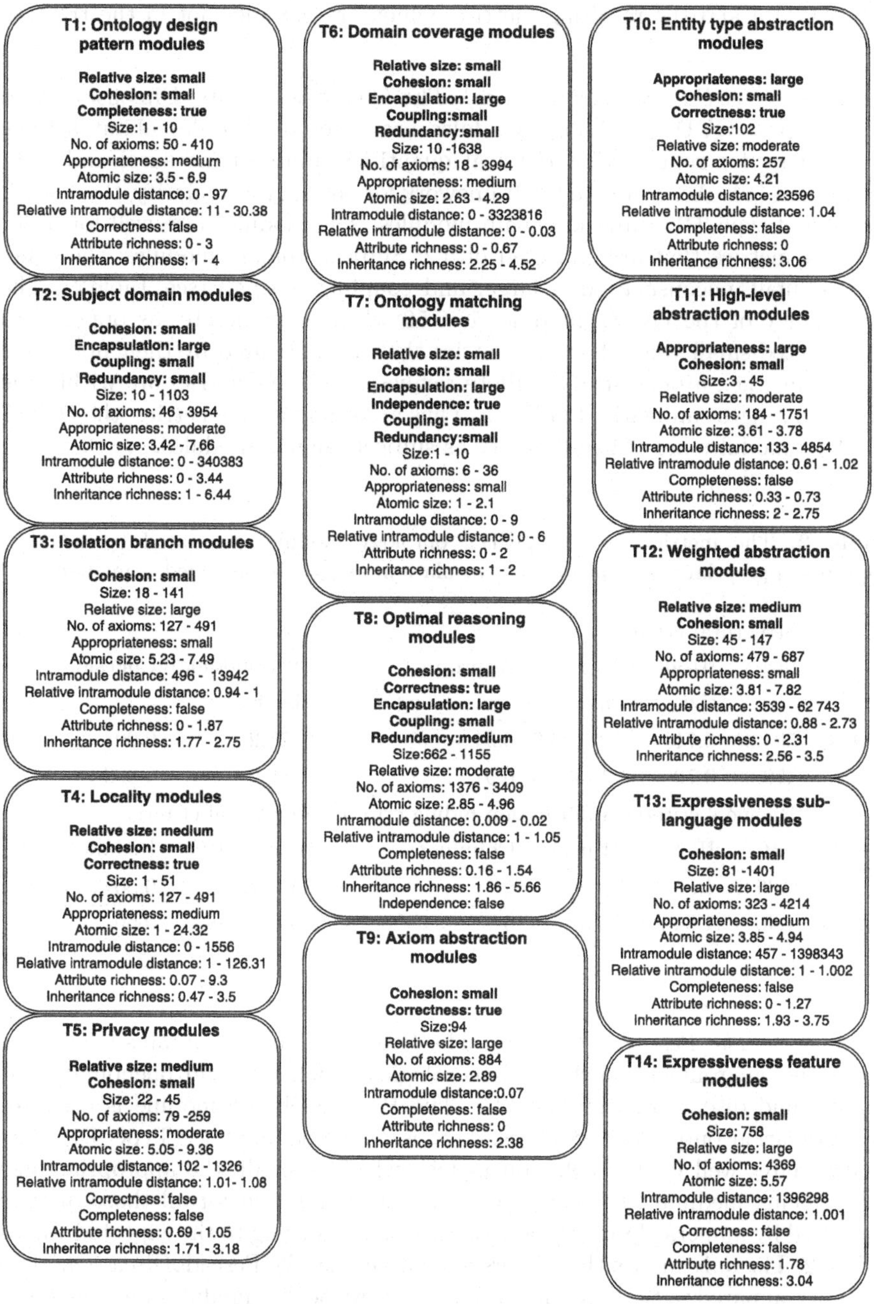

T1: Ontology design pattern modules
Relative size: small
Cohesion: small
Completeness: true
Size: 1 - 10
No. of axioms: 50 - 410
Appropriateness: medium
Atomic size: 3.5 - 6.9
Intramodule distance: 0 - 97
Relative intramodule distance: 11 - 30.38
Correctness: false
Attribute richness: 0 - 3
Inheritance richness: 1 - 4

T2: Subject domain modules
Cohesion: small
Encapsulation: large
Coupling: small
Redundancy: small
Size: 10 - 1103
No. of axioms: 46 - 3954
Appropriateness: moderate
Atomic size: 3.42 - 7.66
Intramodule distance: 0 - 340383
Attribute richness: 0 - 3.44
Inheritance richness: 1 - 6.44

T3: Isolation branch modules
Cohesion: small
Size: 18 - 141
Relative size: large
No. of axioms: 127 - 491
Appropriateness: small
Atomic size: 5.23 - 7.49
Intramodule distance: 496 - 13942
Relative intramodule distance: 0.94 - 1
Completeness: false
Attribute richness: 0 - 1.87
Inheritance richness: 1.77 - 2.75

T4: Locality modules
Relative size: medium
Cohesion: small
Correctness: true
Size: 1 - 51
No. of axioms: 127 - 491
Appropriateness: medium
Atomic size: 1 - 24.32
Intramodule distance: 0 - 1556
Relative intramodule distance: 1 - 126.31
Attribute richness: 0.07 - 9.3
Inheritance richness: 0.47 - 3.5

T5: Privacy modules
Relative size: medium
Cohesion: small
Size: 22 - 45
No. of axioms: 79 -259
Appropriateness: moderate
Atomic size: 5.05 - 9.36
Intramodule distance: 102 - 1326
Relative intramodule distance: 1.01- 1.08
Correctness: false
Completeness: false
Attribute richness: 0.69 - 1.05
Inheritance richness: 1.71 - 3.18

T6: Domain coverage modules
Relative size: small
Cohesion: small
Encapsulation: large
Coupling:small
Redundancy:small
Size: 10 -1638
No. of axioms: 18 - 3994
Appropriateness: medium
Atomic size: 2.63 - 4.29
Intramodule distance: 0 - 3323816
Relative intramodule distance: 0 - 0.03
Attribute richness: 0 - 0.67
Inheritance richness: 2.25 - 4.52

T7: Ontology matching modules
Relative size: small
Cohesion: small
Encapsulation: large
Independence: true
Coupling: small
Redundancy:small
Size:1 - 10
No. of axioms: 6 - 36
Appropriateness: small
Atomic size: 1 - 2.1
Intramodule distance: 0 - 9
Relative intramodule distance: 0 - 6
Attribute richness: 0 - 2
Inheritance richness: 1 - 2

T8: Optimal reasoning modules
Cohesion: small
Correctness: true
Encapsulation: large
Coupling: small
Redundancy:medium
Size:662 - 1155
Relative size: moderate
No. of axioms: 1376 - 3409
Atomic size: 2.85 - 4.96
Intramodule distance: 0.009 - 0.02
Relative intramodule distance: 1 - 1.05
Completeness: false
Attribute richness: 0.16 - 1.54
Inheritance richness: 1.86 - 5.66
Independence: false

T9: Axiom abstraction modules
Cohesion: small
Correctness: true
Size:94
Relative size: large
No. of axioms: 884
Atomic size: 2.89
Intramodule distance:0.07
Completeness: false
Attribute richness: 0
Inheritance richness: 2.38

T10: Entity type abstraction modules
Appropriateness: large
Cohesion: small
Correctness: true
Size:102
Relative size: moderate
No. of axioms: 257
Atomic size: 4.21
Intramodule distance: 23596
Relative intramodule distance: 1.04
Completeness: false
Attribute richness: 0
Inheritance richness: 3.06

T11: High-level abstraction modules
Appropriateness: large
Cohesion: small
Size:3 - 45
Relative size: moderate
No. of axioms: 184 - 1751
Atomic size: 3.61 - 3.78
Intramodule distance: 133 - 4854
Relative intramodule distance: 0.61 - 1.02
Completeness: false
Attribute richness: 0.33 - 0.73
Inheritance richness: 2 - 2.75

T12: Weighted abstraction modules
Relative size: medium
Cohesion: small
Size: 45 - 147
No. of axioms: 479 - 687
Appropriateness: small
Atomic size: 3.81 - 7.82
Intramodule distance: 3539 - 62 743
Relative intramodule distance: 0.88 - 2.73
Attribute richness: 0 - 2.31
Inheritance richness: 2.56 - 3.5

T13: Expressiveness sub-language modules
Cohesion: small
Size: 81 -1401
Relative size: large
No. of axioms: 323 - 4214
Appropriateness: medium
Atomic size: 3.85 - 4.94
Intramodule distance: 457 - 1398343
Relative intramodule distance: 1 - 1.002
Completeness: false
Attribute richness: 0 - 1.27
Inheritance richness: 1.93 - 3.75

T14: Expressiveness feature modules
Cohesion: small
Size: 758
Relative size: large
No. of axioms: 4369
Atomic size: 5.57
Intramodule distance: 1396298
Relative intramodule distance: 1.001
Correctness: false
Completeness: false
Attribute richness: 1.78
Inheritance richness: 3.04

Fig. 2. The set of metrics that can be measured for each module type. Metrics and values in bold font are those which evaluate well for a module type.

Use-Cases. We selected two existing cases of ontology modularisation to evaluate TOMM and the resulting metrics, which are modules not in the training set.

Example 1 (QUDT ontology modules). The Quantities, Units, Dimensions and Data Types (QUDT) ontology modules are a set of modules about science terminology for representing physical quantities, units of measure, and their dimensions [10]. According to the framework for ontology modularity, these modules are of type T2: Subject domain modules. The modules fare well for 3 out of the 4 metrics that are expected of T2: Subject domain modules; the cohesion is small, encapsulation is large, and coupling is small (see Table 2). The redundancy of the QUDT modules is 0.50 which is moderate, as opposed to an expected small value. For the metrics that are measured by their numerical values only, i.e., atomic size, attribute richness, etc., the metrics are within the expected ranges summarised in Fig. 2. Thus according to the metrics, the QUDT ontology modules are of good quality as subject domain modules.

Table 2. The metrics for the QUDT ontology modules generated by TOMM; approp = appropriateness, encap. = encapsulation, avg. = average, med. = median.

	Structural criteria					
	Size	Atomic size	No. of axioms	Approp	Intra module distance	Cohesion
Avg.	595.38	5.71	3112.00	0.91	8577.25	0.008
Med.	479.00	3.70	1443.50	0.91	86.50	0.003
	Richness criteria		Information hiding criteria		Relational criteria	
	AR	IR	Encap	Coupling	Independence	Redundancy
Avg	1.69	1.89	0.99	0	25 % true	0.50
Med	1.40	1.84	0.99	0	75 % false	0.50

Example 2 (The Pescado Ontology). The Pescado ontology contains knowledge about the environment, such as meteorological conditions and phenomena, air quality, and disease information [23]. The PescadoDisease module is a subset of information only about diseases, so it is a locality module (Type T4). The module fares well for the cohesion metric, which is small, the appropriateness value (being medium), the correctness metric (true), and for all those metrics measured by numerical ranges too (see Table 3), according to the expected values of Fig. 2. The only metric that differs is relative size: the PescadoDisease module is small compared to the experimental data where locality modules were medium.

Using TOMM and the use-cases, we were able to evaluate the quality of ontology modules. QUDT and Pescado are different types of modules and therefore different values are expected for their metrics. With both modules, for all

Table 3. The metrics for the Pescado disease ontology generated by TOMM; app = appropriateness.

Structural criteria							
Size	Atomic size	No. of axioms	App.	Intra module distance	Cohesion	Relative size	Relative intra module distance
39.00	3.10	128.00	0.51	158.00	0.16	0.03	10.61
Richness criteria				Logical criteria			
Attribute richness		Inheritance richness		Correctness		Completeness	
0.00		1.67		True		True	

their metrics except one, the values generated by TOMM are as expected for their types; this indicates that the modules are of 'good' quality.

5 Discussion

The list of module metrics that was compiled is a first step in solving the problems regarding the evaluation of ontology modules, and, following from that, knowing how to create a good module. The metrics that are programmed into TOMM allow one to evaluate ontology modules using different metrics such as logical (correctness), structural (relational size), relational (coupling), and so on. Of all the metrics, it was not feasible to include the inter-module distance metric in the program, because these modules were linked using ε-connections, which could not be recognised by the OWL API that was used to develop TOMM. Also, in testing, the 'FMA_subset' module (from T12: weighted abstraction modules) was too large for TOMM to process due to insufficient Java heap space size and increasing the parameters caused the machine to crash. We are looking at running TOMM on a High-Performance Computing Cluster in the future.

We have evaluated modules with TOMM, and analysed their metrics. The results reveal which metrics fare well with which module type, as displayed in Fig. 2; e.g., T1 (ontology design patterns) modules are relatively small compared to the original ontology, and the completeness value is true. For T3, T13, and T14 modules, there is limited associations between them and the metrics. The analysis reveals that they all only fare well for the cohesion metric; all the sets of modules fare well for the cohesion metric.

For the bulk of the modules, T3, T5, T11, T12, T13, and T14 provide good results for structural metrics. Modules of type T1, T4, T9, and T10 have good results for both structural, and logical metrics. Modules of type T2, T6, and T7 have good results for structural, information hiding, and relational metrics, and T8 type of modules have good results for some criteria, structural, logical, relational, and information hiding. Richness criteria only returns a range of numerical values which cannot be mapped to rate values such as small, medium, etc., hence it is unclear what ideal values for such criteria are. Thus, using TOMM to evaluate a module, a user is able to determine whether the module is of 'good' quality. Our approach of evaluating whether a module is of 'good' quality heavily depends on the data from existing modules used in this experiment. The reason for this approach is to offer the developer a practical solution for evaluating modules in Semantic Web applications.

From the assessment on any relation between modularisation technique and ontology module quality metrics, it exhibited a link between the graph partitioning, locality-based, and *a priori* techniques and the metrics; there were certain metrics that were associated with each of the respective techniques. There were four metrics associated with graph partitioning, one with locality, and three with *a priori* techniques. Unsurprisingly, there does not seem to be any clear association between manual modularisation technique and any of the metrics. Perhaps an in-depth qualitative assessment of the manually created modules may reveal what is going on exactly.

The use-case evaluation with the QUDT (of type subject domain modules) and the Pescado-disease modules (of type locality) were promising, showing good modules for their respective types. Others may not fare as well, which time may tell. Most ontology modules we could find are already included in the test set, so that will depend on the modules that are being developed, which, however, can avail of the results presented here to exactly avoid creating a 'bad' module.

6 Conclusion

Five new modularity metrics with measures and three new measures for existing metrics were proposed, making the total to 16 ontology module evaluation metrics. They have been implemented in the TOMM tool to enable scaling up of module evaluation. Our evaluation carried out with 189 modules revealed which metrics work well with which types of modules. This is displayed in bold font in Fig. 2; for each of the 14 module types, the metrics that fare well with them together with the expected values are displayed. It is now possible for an ontology developer to evaluate the quality of a module/set of modules by first classifying its type using the framework for ontology modularity, and then generating its metrics using the TOMM metrics tool. Ontology developers are then able to determine whether their ontology module is of 'good quality' based on comparing the module's metrics to what is expected in Fig. 2.

For future work, we aim to achieve more insight into module evaluation by linking the module evaluation metrics to other characteristics of the modularity framework such as use-cases and properties, to reveal more dependencies. It is also worthwhile to apply the tool to ontology design patterns towards improving pattern quality.

A Appendix: Summarised Types of Ontology Modules

T1 *Ontology design pattern modules* An ontology is modularised by identifying a part of the ontology for general reuse.

T2 *Subject domain modules* A large domain is divided by subdomains present in the ontology.

T3 *Isolation branch modules* A subset of entities from an ontology is extracted but entities with weak dependencies to the signature are not to be included in the module.

T4 *Locality modules* A subset of entities from an ontology is extracted, including all entities that are dependent on the subset.

T5 *Privacy modules* Some information is hidden from an ontology.

T6 *Domain coverage modules* A large ontology is partitioned by its graphical structure and placement of entities in the taxonomy.

T7 *Ontology matching modules* An ontology is modularised for ontology matching into disjoint modules so that there is no repetition of entities.

T8 *Optimal reasoning modules* An ontology is split into smaller modules to aid in overall reasoning over the ontology.

T9 *Axiom abstraction modules* An ontology is modularised to have fewer axioms, to decrease the horizontal structure of the ontology.

T10 *Entity type abstraction modules* An ontology is modularised by removing a certain type of entity e.g., data properties or object properties.

T11 *High-level abstraction modules* An ontology is modularised by removing lower-level classes and only keeping higher-level classes.

T12 *Weighted abstraction modules* An ontology is modularised by a weighting decided by the developer.

T13 *Expressiveness sub-language modules* An ontology is modularised by using a sub-language of a core ontology language.

T14 *Expressiveness feature modules* An ontology is modularised by using limited language features.

References

1. Borgo, S.: Goals of modularity: a voice from the foundational viewpoint. In: Fifth International Workshop on Modular Ontologies (WOMO 2011). Frontiers in Artificial Intelligence and Applications, vol. 230, pp. 1–6. IOS Press, ljubljana, August 2011
2. Cuenca Grau, B., Parsia, B., Sirin, E., Kalyanpur, A.: Modularity and web ontologies. In: 10th International Conference on Principles of Knowledge Representation and Reasoning (KR 2006), pp. 198–209. AAAI Press, Lake District, 2–5 June 2006
3. d'Aquin, M., Schlicht, A., Stuckenschmidt, H., Sabou, M.: Ontology modularization for knowledge selection: experiments and evaluations. In: Wagner, R., Revell, N., Pernul, G. (eds.) DEXA 2007. LNCS, vol. 4653, pp. 874–883. Springer, Heidelberg (2007)
4. d'Aquin, M., Schlicht, A., Stuckenschmidt, H., Sabou, M.: Criteria and evaluation for ontology modularization techniques. In: Stuckenschmidt, H., Parent, C., Spaccapietra, S. (eds.) Modular Ontologies. LNCS, vol. 5445, pp. 67–89. Springer, Heidelberg (2009)
5. Ensan, F., Du, W.: A semantic metrics suite for evaluating modular ontologies. Inf. Syst. **38**(5), 745–770 (2013)

6. Freeman, L.C.: Centrality in social networks conceptual clarification. Soc. Netw. **1**(3), 215–239 (1978)
7. García, J., García-Peñalvo, F.J., Therón, R.: A survey on ontology metrics. In: Lytras, M.D., Ordonez De Pablos, P., Ziderman, A., Roulstone, A., Maurer, H., Imber, J.B. (eds.) WSKS 2010. CCIS, vol. 111, pp. 22–27. Springer, Heidelberg (2010)
8. Gennari, J.H., Musen, M.A., Fergerson, R.W., Grosso, W.E., Crubézy, M., Eriksson, H., Noy, N.F., Tu, S.W.: The evolution of Protégé: an environment for knowledge-based systems development. Int. J. Hum. Comput. Stud. **58**(1), 89–123 (2003)
9. Grau, B.C., Horrocks, I., Kazakov, Y., Sattler, U.: Modular reuse of ontologies: theory and practice. J. Artif. Intell. Res. **31**, 273–318 (2008)
10. Hodgson, R., Keller, P.J.: QUDT-quantities, units, dimensions and data types in OWL and XML (2011). http://www.qudt.org. Accessed September 2011
11. Kalyanpur, A., Parsia, B., Sirin, E., Cuenca Grau, B., Hendler, J.A.: Swoop: a web ontology editing browser. J. Web Semant. **4**(2), 144–153 (2006)
12. Khan, Z.C., Keet, C.M.: The foundational ontology library ROMULUS. In: Cuzzocrea, A., Maabout, S. (eds.) MEDI 2013. LNCS, vol. 8216, pp. 200–211. Springer, Heidelberg (2013)
13. Khan, Z.C., Keet, C.M.: Feasibility of automated foundational ontology interchangeability. In: Janowicz, K., Schlobach, S., Lambrix, P., Hyvönen, E. (eds.) EKAW 2014. LNCS, vol. 8876, pp. 225–237. Springer, Heidelberg (2014)
14. Khan, Z.C.: Evaluation metrics in ontology modules. In: 29th International Workshop on Description Logics (DL 2016), CEUR Workshop Proceedings, vol. 1577, Cape Town, South Africa. CEUR-WS.org, 22–25 April 2016
15. Khan, Z.C., Keet, C.M.: An empirically-based framework for ontology modularisation. Appl. Ontology **10**(3–4), 171–195 (2015)
16. Loebe, F.: Requirements for logical modules. In: First International Workshop on Modular Ontologies (WoMO 2006), CEUR Workshop Proceedings, vol. 232, Athens, Georgia, USA. CEUR-WS.org, 5 November 2006
17. McComb, D.: Gist: the minimalist upper ontology. In: Semantic Technology Conference, San Francisco, CA, 21–25 June 2010
18. Noy, N.F., Musen, M.A.: Specifying ontology views by traversal. In: McIlraith, S.A., Plexousakis, D., van Harmelen, F. (eds.) ISWC 2004. LNCS, vol. 3298, pp. 713–725. Springer, Heidelberg (2004)
19. Oh, S., Ahn, J.: Ontology module metrics. In: International Conference on e-Business Engineering, (ICEBE 2009), pp. 11–18. IEEE Computer Society, Macau, 21–23 October 2009
20. Oh, S., Yeom, H.Y., Ahn, J.: Cohesion and coupling metrics for ontology modules. Inf. Technol. Manag. **12**(2), 81–96 (2011)
21. Orme, A.M., Yao, H., Etzkorn, L.H.: Coupling metrics for ontology-based systems. IEEE Softw. **23**(2), 102–108 (2006)
22. Pathak, J., Johnson, T.M., Chute, C.G.: Survey of modular ontology techniques and their applications in the biomedical domain. Integr. Comput.-Aided Eng. **16**(3), 225–242 (2009)
23. Rospocher, M.: An ontology for personalized environmental decision support. In: Formal Ontology in Information Systems FOIS 2014, pp. 421–42, Rio de Janeiro, Brazil, 22–25 September, 2014

24. Schlicht, A., Stuckenschmidt, H.: Towards structural criteria for ontology modularization. In: First International Workshop on Modular Ontologies, (WoMO 2006), CEUR Workshop Proceedings, vol. 232, Athens, Georgia, USA. CEUR-WS.org, 5 November 2006
25. Tartir, S., Arpinar, I.B., Moore, M., Sheth, A.P., Aleman-Meza, B.: OntoQA: metric-based ontology quality analysis. In: IEEE Workshop on Knowledge Acquisition from Distributed, Autonomous, Semantically Heterogeneous Data and Knowledge Sources, vol. 9 (2005)
26. Vescovo, C.D.: The modular structure of an ontology: atomic decomposition towards applications. In: 24th International Workshop on Description Logics (DL 2011), CEUR Workshop Proceedings, vol. 745, Barcelona, Spain. CEUR-WS.org, 13–16 July 2011
27. Yao, H., Orme, A.M., Etzkorn, L.: Cohesion metrics for ontology design and application. J. Comput. Sci. **1**(1), 107 (2005)

Travel Attractions Recommendation with Knowledge Graphs

Chun Lu[1,2(✉)], Philippe Laublet[2], and Milan Stankovic[1,2]

[1] Sépage, 27 rue du Chemin Vert, 75011 Paris, France
{chun, milstan}@sepage.fr
[2] STIH, Université Paris-Sorbonne, 28 rue Serpente, 75006 Paris, France
philippe.laublet@paris-sorbonne.fr

Abstract. Selecting relevant travel attractions for a given user is a real and important problem from both a traveller's and a travel supplier's perspectives. Knowledge graphs have been used to conduct recommendations of music artists, movies and books. In this paper, we identify how knowledge graphs might be efficiently leveraged to recommend travel attractions. We improve two main drawbacks in existing systems where semantic information is exploited: semantic poorness and city-agnostic user profiling strategy. Accordingly, we constructed a rich world scale travel knowledge graph from existing large knowledge graphs namely Geonames, DBpedia and Wikidata. The underlying ontology contains more than 1200 classes to describe attractions. We applied a city-dependent user profiling strategy that makes use of the fine semantics encoded in the constructed graph. Our evaluation on YFCC100M dataset showed that our approach achieves a 5.3 % improvement in terms of F1-score, a 4.3 % improvement in terms of nDCG compared with the state-of-the-art approach.

Keywords: e-Tourism · Travel attraction · Recommender system · Semantic information · Knowledge graph · Ontology

1 Introduction

The web is today one of the most important sources for travel inspiration and purchase. Selecting relevant travel attractions for a given user is a real and important problem. From a traveler's perspective, "What to do in [destination]" is the most frequent question people type into Google about travel [1], a lot of time and efforts are spent before finding subjectively interesting places to visit [2]. From a travel supplier's perspective, travel attractions represent an important source of revenues. Major travel websites like Expedia[1] and TripAdvisor[2] provide contents about attractions and

[1] https://www.expedia.com/.
[2] https://www.tripadvisor.com/.

E. Blomqvist et al. (Eds.): EKAW 2016, LNAI 10024, pp. 416–431, 2016.
DOI: 10.1007/978-3-319-49004-5_27

commercialize some of them. Some websites are even specialized in this area, such as Peek[3] and Musement[4]. How to sell[5] and what to sell[6] are major concerns of travel suppliers.

In this paper, we present a travel attractions recommender system. The considered scenario is that the system has the knowledge about the attractions the user visited physically in the past in some cities. This knowledge might be collected from users' check-ins on location-based social networks like Foursquare and Yelp, or from geo-tagged photos on photo sharing platforms like Flickr. The system needs to recommend attractions in a new city that the user is going to. For example, in Fig. 1, the user has visited some attractions in Venice and Paris, his/her next destination is Madrid, what attractions might interest him/her?

Fig. 1. Recommendation scenario

Knowledge graphs have been used to recommend music artists [3], movies [4] and books [5]. Our motivation is to identify how knowledge graphs could be efficiently leveraged to recommend travel attractions.

In the related literature, the use of semantic information has been exploited. However, we identified two main drawbacks that we envisage improving: the semantic poorness and the city-agnostic user profiling strategy.

Our hypothesis is that by leveraging a rich travel-domain specific knowledge graph and by applying a city-dependent user profiling strategy, we can improve the user profile and yield better recommendations.

The contributions of this paper are two-fold:

[3] https://www.peek.com/.

[4] https://www.musement.com/.

[5] https://www.tnooz.com/article/real-time-destination-action-tickets-on-a-mobile-as-you-approach-an-attraction/.

[6] https://www.tnooz.com/article/expedia-bets-big-on-tours-and-activities-will-the-industry-win/.

- We present a semi-automatic method with little manual intervention to construct a world scale travel knowledge graph from existing large knowledge graphs namely GeoNames[7], DBpedia[8] and Wikidata[9].
- We present a city-dependent user profiling strategy which makes use of the fine semantics encoded in the constructed graph to better understand travelers' interests.

The remainder of the paper is organized as follows. In Sect. 2, we discuss some related work. In Sect. 3, we present our method of constructing the travel knowledge graph. In Sect. 4, we present our user profiling and recommendation approach. In Sect. 5, we evaluate our system. In Sect. 6, we conclude the paper.

2 Related Work

In the academic literature, the problem being treated in this paper is most closely related to point of interest (POI) recommendations on location-based social networks. There is a large body of work in this area [6]. Generally speaking, existing works make use of several different types of data: social information, contextual data, tags and categories.

In the literature, different vocabularies are used to refer to the same things. We stress that *point of interest* usually refers to the same type of entities as our *travel attractions*, for example Louvre Museum[10]. We may also use *places* and *venues* interchangeably in this paper. The word *category* is used in the *rdf:type* sense, for example Museum[11]. We use *category* when we present an existing work and *type* when we present our approach, with the exception that in Sect. 3 *category* refers to a DBpedia category, for example Category:Museums[12].

Using social information consists of leveraging places that the friends of a user have visited. It has been shown that friends actually have very low overlap of places and that social information contributes little to the recommendation performance [7].

As for contextual data, the geographical information is used because users tend to visit nearby places [8]. The temporal information is used because many users tend to visit different places at different time slots and periodically visit the same places in the same time slot [9]. The use of some other contextual data like the weather and the motion speed has been discussed in [10, 11].

Tags and categories can provide semantic information about POIs. In [12], the authors exploit an aggregated latent Dirichlet allocation model to learn the interest topics of users by mining tags and categories associated to POIs. But it has been pointed out that tags are in many cases missing, wrong or irrelevant for the recommendation purpose, for example, *me and Ann, travel to Europe, Easter 2012* [13].

---

7 http://www.geonames.org/.

8 http://wiki.dbpedia.org/.

9 https://www.wikidata.org/.

10 http://dbpedia.org/resource/Louvre.

11 http://dbpedia.org/ontology/Museum.

12 http://dbpedia.org/page/Category:Museums.

Some authors rely solely on categories to represent user interests. In this paper, we focus on improving user profiling with categories. There are two main drawbacks that we envisage improving: semantic poorness and city-agnostic user profiling strategy.

Semantic poorness. TripAdvisor is a popular travel website providing reviews and travel-related content including travel attractions[13]. There are in total 157 attraction types. TripAdvisor does not allow access to the Content API for purpose of academic research[14]. Foursquare provides a category hierarchy[15] to organize venues in the social network. This hierarchy has been used by many researchers [6, 14, 15]. However, the categorization is not granular enough. For example, the category "Museum" has only 5 subcategories: "Art museum", "Erotic museum", "History museum", "Planetarium", "Science museum". In reality, there are many more different types of museums like military museum, toy museum, food museum, literary museum, archaeological museum and so on. On Foursquare, the Volvo Museum[16] in Gothenburg, Sweden only belongs to the category "Museum". On Wikidata[17], we can know that it is also an "automobile museum" which is a subtype of "museum".

Some authors tried to use semantic information in Wikipedia [13, 16]. However, they only succeeded in retrieving POIs in a limited number of cities and only very general categories. In [16], only 4 cities were retrieved: Toronto, Osaka, Glasgow and Edinburgh, categories like structure, palace, historical, entertainment, museum, zoo were considered. In [13], only 3 cities were retrieved: Pisa, Florence and Rome, only 8 categories were considered: architecture, arts, churches, entertainment, monuments, museums, nature and landmarks.

Extracting POIs and fine semantic information from knowledge graphs is not a trivial task. In Sect. 3, we provide details about how we constructed a world scale travel knowledge graph with data from DBpedia, Geonames and Wikidata. The underlying ontology has more than 1200 attraction types with a high semantic granularity to describe attractions. In the evaluation, we demonstrate that compared to a simple ontology, ours allows to better capture user interests and to yield better recommendations.

City-agnostic user profling strategy. User interests are represented by categories. Each category is assigned a score indicating the degree to which the user likes the category. Frequency-based and time-based are two different strategies which are applied to calculate this score. The intuition behind frequency-based strategy [13, 15] is that the more frequently a user physically visits places of a certain category, the more they like it. The intuition behind time-based strategy [16] is that the longer a user stays physically in places of a certain category, the more he/she likes it. In [16], the authors calculated an approximation of duration based on the taken time of photos.

[13] https://developer-tripadvisor.com/content-api/business-content/categories-subcategories-and-types/.

[14] https://developer-tripadvisor.com/content-api/request-api-access/.

[15] https://developer.foursquare.com/categorytree.

[16] https://foursquare.com/v/volvo-museum/4b9f9e2df964a520432f37e3.

[17] https://www.wikidata.org/wiki/Q3329393.

However, in both cases, the same city-agnostic strategy is applied. The weight of a category depends only on the frequency or the time spent, whatever city the POI is situated in. At the opposite, we propose a city-dependent strategy. The intuition is that a user chooses to travel to a city probably because of its representative attractions. For example, Las Vegas, USA is famous for its casinos, while Geneva, Switzerland is not. Paul visits a casino in Geneva. Mary visits a casino in Las Vegas. Assuming that the frequency/time spent is equal, a city-agnostic approach would conclude that Paul and Mary like the category *casino* with the same degree. With our approach, Paul would have a lower degree than Mary. The data encoded in the constructed knowledge graph allow us to understand the importance of different categories in different cities and thus to develop a city-dependent strategy resulting in a more accurate user profiling. In Sect. 4, we give details about the user profile computation and recommendation.

3 Travel Knowledge Graph

In our past work [17], we constructed a travel destination-centered semantic graph. We gathered travel attraction data via the API of the location-based social network Foursquare. As we stated in Sect. 2, due to the semantic poorness, Foursquare is not satisfying for the problem being treated in this current work and thus is no longer used by us.

We want to construct a travel-domain specific knowledge graph gathering attractions, fine-granular attraction types and cities where attractions are situated.

DBpedia is the most popular dataset on Linked Open Data cloud[18]. On DBpedia, we can find much data about travel attractions. However, extracting a travel knowledge graph is not trivial task. First, in the DBpedia Ontology[19], there is no ontology class which contains all cities, even though the class *dbo:City* exists, however, many cities are not typed with this class for example *dbr:Paris*. Second, there is no ontology class which contains all travel attractions, frequent travel attraction types such as *dbo: Museum, dbo:Monument, dbo:Park* are scattered in the ontology and are not subclasses of a more general class like "*Travel attraction*". Third, travel attractions are often not typed with the aforementioned classes. For example, a famous park in Paris *dbr: Parc_Montsouris* is not typed with *dbo:Park* nor any of its subclasses.

We examined closely other datasets on the Linked Open Data cloud and developed a semi-automatic method which leverages data from GeoNames, DBpedia and Wikidata. We present the 3 main steps of our method (Fig. 2).

3.1 Pool of Travel Attractions

As a first step, we retrieved a pool of travel attractions. We observed that travel attractions are linked by the property *dct:subject* to DBpedia categories which satisfy a certain regular expression.

[18] http://lod-cloud.net/.

[19] http://wiki.dbpedia.org/Downloads2015-10#dbpedia-ontology.

For all DBpedia categories which start with "http://dbpedia.org/resource/Category: Visitor_attractions_in_*", for example, "http://dbpedia.org/resource/Category:Visitor_attractions_in_Barcelona", we ran the following SPARQL query to retrieve the DBpedia entities that are linked to it and its subsumed categories.

select distinct ?POI ?category where {
?POI dct:subject ?category .
?category skos:broader
<http://dbpedia.org/resource/Category:Visitor_attractions_in_Barcelona>}

We then parsed the query results. For each pair result (POI, category), a syntactic verification is conducted, if the *name of the category matches* the regular expression "http://dbpedia.org/resource/Category:(.+?)_(in|of)_(.+)*", we store the POI, the first group of the match as its type string, the third group as its location string. For example, one of the query results is (http://dbpedia.org/resource/Palau_Reial_Major, http://dbpedia.org/resource/Category:Palaces_in_Barcelona), after the verification, we stored the POI "http://dbpedia.org/resource/Palau_Reial_Major", its type string is "Palaces", its location string is "Barcelona". We also stored a hierarchical relationship between each pair of type strings. In the example above, we stored (Palaces, subclassOf, Visitor_attractions).

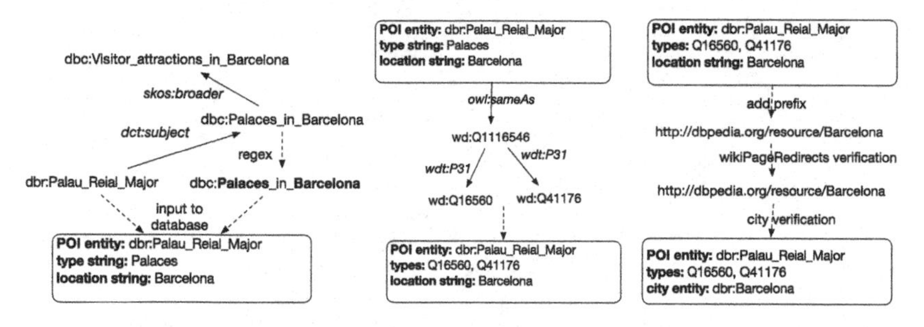

Fig. 2. Illustrations of the method of constructing travel knowledge graph, 3 schemas from left to right corresponding respectively to Sects. 3.1, 3.2 and 3.3

3.2 Types' of Travel Attractions

We examined the type strings stored in Sect. 3.2. Several phenomena make them difficult to use: presence of cycles, for example *Sports_venues* is at the same time parent class and subclass of *Field_hockey_venues*; duplicates with linguistic nuances, for example *Arts_centers* and *Arts_centers*; knowledge organization error, *Music_venues* is subclass of *Sports_venues*.

We observed that Wikidata has a relatively clean schema about POIs. We mapped POI entities to their corresponding entities in Wikidata by using the *owl:sameAs* property in DBpedia. Then for each Wikidata POI entity, we used the property *P31 instance of* to find its type(s).

For each distinct type, we used the property *P279 subclass of* to find all its parent types up to three levels. With the found hierarchical information, we established a subset of Wikidata schema. In this schema, some types are not enough specific such as *Q41176 building*, knowing that there are a certain number of building in a city does not reveal much touristic information. We examined existing travel ontologies and POI categorizations, we manually selected top-level types. This is the only manual intervention in the construction of the travel knowledge graph. In total, we have 1214 classes of which 28 are top-level. We stored our data in the neo4j[20] graph database. Figure 3 is a subset of the ontology with the class *Q33506 museum* and its subclasses. We can observe the richness of the ontology.

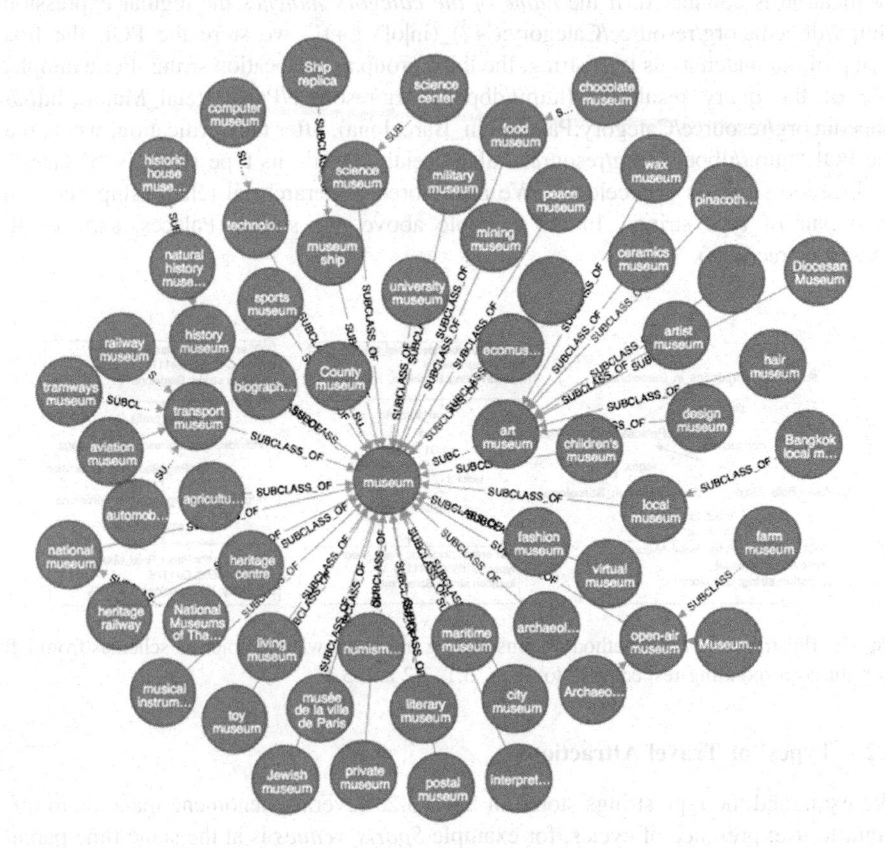

Fig. 3. Museum and its subclasses in the ontology underlying the travel knowledge graph

3.3 Cities of Travel Attractions

We want to link POI entities to city entities. We examined a big amount of location strings obtained in 3.1. We found that except for a small number of exceptions like

[20] http://neo4j.com/.

"City of London, of which only the tower remains", most of their corresponding DBpedia entities use exactly the same string. For each location string, we added the DBpedia prefix "http://dbpedia.org/resource/" to form the entity of the location. For example, the location string "Barcelona" became "http://dbpedia.org/resource/Barcelona". Some corresponding DBpedia entities are redirected to another entity. For example, the location string "Chicago,_Illinois" becomes "http://dbpedia.org/resource/Chicago,_Illinois" after adding the prefix, we then use *dbo:wikipageRedirects* to find the redirection entity "http://dbpedia.org/resource/Chicago".

Locations of different administrative levels can be found, such as cities, regions, countries and continents. Since we want to link travel attractions to cities, we need to identify them. As we stated in the beginning of Sect. 3, it is difficult to verify cities in DBpedia, we mapped all locations that we have in DBpedia to their corresponding entities in GeoNames and Wikidata. We consider that a location is a city if its GeoNames corresponding entity appears in the dump file "cities1000"[21] or if its Wikidata corresponding entity is instance of *Q515 city* or its subsumed classes.

The Fig. 4 is a subgraph illustrating the schema of the knowledge graph.

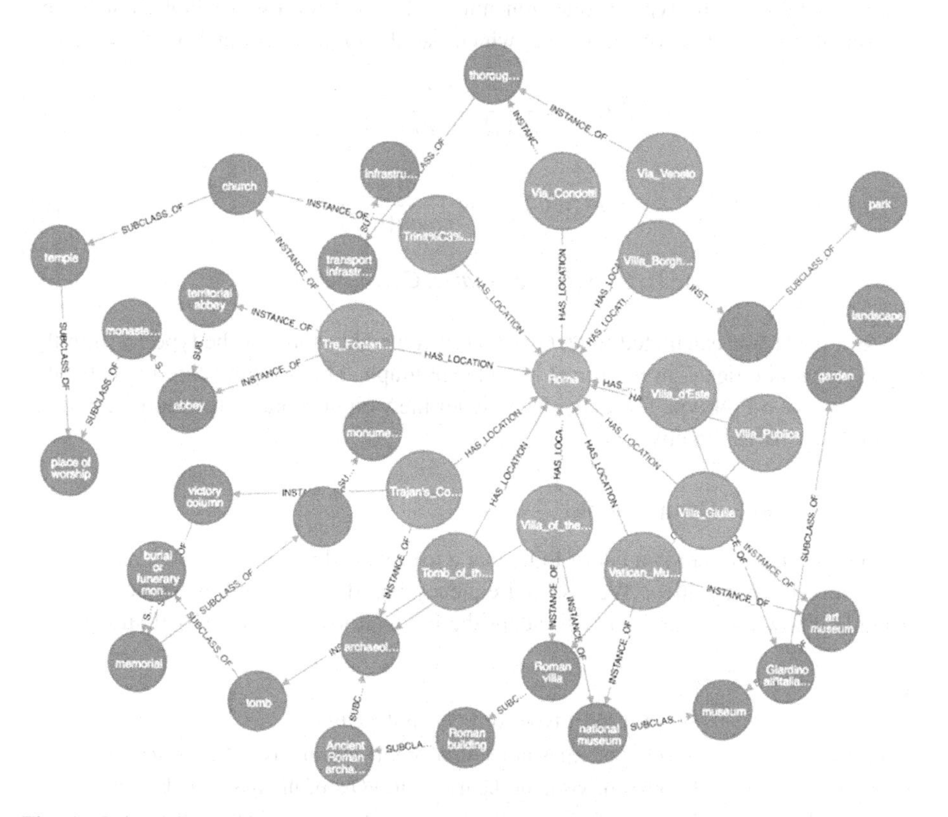

Fig. 4. Subgraph containing travel attractions (green) with their types (red) in the city of Rome (pink) (Color figure online)

[21] http://download.geonames.org/export/dump/.

4 Recommender

4.1 City-Dependent Type Weight

City-dependent implies that the weight of a type should be calculated within the context of its city. We use a TF-IDF-like[22] measure to calculate a weight for each POI type for each city.

$f(t, c)$	Number of distinct POIs of the type t in the city c
T_c	Set of distinct types of POIs in the city c
$\max\{f(t', c) : t' \in T_c\}$	Number of distinct POIs of the type t' which has the highest cardinality in the city c
$\lvert C \rvert$	Number of cities in the travel knowledge graph
$count(t, C) : \lvert f(t, c) > 0 \rvert, t \in T_c, c \in C$	Number of cities which have POIs of the type t

To prevent bias towards cities with bigger number of POIs, we divide the number of distinct POIs of the type in question not by the total number of POIs, but by the number of distinct POIs of the type t' which has the highest cardinality in the city c.

$$tf(t, c) = \frac{f(t, c)}{\max\{f(t', c) : t' \in T_c\}} \tag{1}$$

$$idf(t, C) = \log \frac{|C|}{count(t, C)} \tag{2}$$

$$w_c(t, C) = tf(t, c) * idf(t, C), wherec \in C \tag{3}$$

As a result, we calculated $w_c(t, C)$ which is a weight score for the type t within the city c given all cities C in the travel knowledge graph. This weight score indicates the importance of a POI type for a city. The higher the weight score is, the more important the POI type is to the city.

4.2 User Profile Computation

The input of the system is a list of POIs that a user visited physically in the past in one or several different cities. We use a frequency-based user profiling strategy as an example, knowing that it is easy to adapt these equations to a time-based strategy.

u	User
T_c	Set of distinct types of POIs in the city c
$count(t, C_u)$	Number of cities in the user profile which have POIs of the type t
$visit(u, t, c)$	Number of visits of the user u to POIs of the type t in the city c
$visitAll(u, T_c, c)$	Number of visits of the user u to all types of POIs in the city c

[22] https://en.wikipedia.org/wiki/Tf-idf.

For each city in the user profile, we calculate an interest score for each POI type.

$$I(u,t,c) = \frac{visit(u,t,c) * w_c(t,C)}{visitAll(u,T_c,c)} \tag{4}$$

If there are multiple cities in the user profile and that the same types of POIs are visited in different cities, we aggregate the interest scores of these common types and retain the average score. In the example of Fig. 1, if a user has an interest score of 0.7 for the type *church* in Venice and 0.9 for the same type in Paris, after the aggregation, the user has an interest score of 0.8 for the type *church*.

$$I(u,t,C_u) = \frac{\sum_{c_i} I(u,t,c_i)}{count(t,C_u)}, c_i \in C_u \tag{5}$$

We finally normalized the interest score to the range [0,1].

$$I_{norm} = \frac{I - I_{min}}{I_{max} - I_{min}} \tag{6}$$

4.3 Travel Attraction Scoring

We described our main contributions previously. We do not have the ambition to improve travel attraction scoring method. We are interested in observing, with the same scoring method, if inputting the proposed novelties allows to yield better recommendations compared to existing practices.

In this respect, we adopt the following scoring method which is similar to the one used in [15]. The score of a travel attraction p for a user relies both on the popularity of the attraction and on the user's interest in its types.

The popularity of p is calculated with regards to the set of travel attractions P_c in the city where p is situated. We divide the number of visitors of p by the maximum number of visitors of any other attractions in the same city.

$$Popularity(p,P_c) = \frac{Visitors(p)}{\max\{Visitors(p') : p' \in P_c\}} \tag{7}$$

The final score of a travel attraction for a given user is the sum of its popularity and the user's interest score in its types. As a travel attraction often has multiple types, we use the highest score. For example, Louvre[23] has 3 types: national museum, art museum and museum. If a user has an interest score of 0.5 for national museum, 0.6 for art museum and 0.4 for museum, we would use the score of 0.6.

[23] https://www.wikidata.org/wiki/Q19675.

5 Evaluation

5.1 Experiment Dataset

For the experiments, we use the Yahoo! Flickr Creative Commons 100M (YFCC100M) dataset[24] [18] which contains 100 million public Flickr photos and videos. We constructed a subset of the original dataset for our evaluation use.

5.2 YFCC100M Subset Construction

In this part, we provide detailed description about how we processed the original dataset to obtain our subset. First, we took the file "yfcc100m_dataset". We filtered all the lines where latitude and longitude data were missing and where the accuracy level was below 16 (street level in Flickr). In other words, we retained only geotagged photos and videos with the highest geo-location accuracy. Second, we mapped each photo/video to a POI entity in our travel knowledge graph. We did a pairwise comparison between the geographical coordinates of the photo/video and those of a POI entity. Following [16], if the distance is less than 100 meters, we consider that the photo/video has been taken at the POI entity. If there are several POI entities which are within 100 meters, we choose the nearest one.

One user can post multiple photos in the same POI. We recorded the POI only once because we are only interested in the presence/absence of a POI in his/her history. We also eliminated users who have been to only one city. Because in the considered scenario our system learns user interests from visits in cities which are different from the one for which recommendations are conducted, thus we need more than 1 city in users' histories. For each user, we gathered the cities that he/she has been to in a chronological order. We then eliminated users who have posted photos in less than 5 POIs in the last city in the sequence, because we want to evaluate top-5 performance. In the following table, we show some statistics about the constructed subset which is used for our evaluation.

Mapped photos	2,457,267
Users/Ground truths	3878
Cities	705
Points of interest	106,396
Sequences	3433
Average number of cities in a sequence	5

[24] Yahoo Webscope: http://webscope.sandbox.yahoo.com

We provide more details about the ground truths. A ground truth consists of three things:

- a user NSID (provided by Flickr)
- a travel sequence containing the DBpedia entities of the cities that the user has been to in a chronological order, for example, (http://dbpedia.org/resource/Munich - > http://dbpedia.org/resource/Stockholm - > http://dbpedia.org/resource/New_York_City)
- for each city in the travel sequence, a list of DBpedia entities of the travel attractions that the user has visited in the city

If a travel sequence contains *n* cities, we use the travel attractions visited in the first *n-1* cities to compute a user profile. We then use the profile to generate top-5 recommendations in the *n-th* city. We consider the attractions visited in the *n-th* city as a ground truth. We compare our recommendations with the ground truth.

5.3 Baseline

There are two main novelties in this paper: travel knowledge graph and city-dependent user profiling strategy. These novelties are proposed to improve respectively the semantic poorness and city-agnostic strategy in existing works. By inputting the proposed novelties and the practices in existing works, we can obtain different user profiles. In this evaluation, we want to observe the potential improvements in recommendation by using different user profiles and the same recommendation method described in Sect. 4.

For each of the two drawbacks that we try to improve, we oppose two variants. On the one hand, for the semantic poorness, we oppose *WO* (the whole ontology described in Sect. 3) to *SO* (simple ontology which contains only top-level classes, as existing works use general types), on the other hand, for the user profiling strategy, we oppose *CD* (city-dependent) to *CA* (city-agnostic).

We use these variants to compose four different inputs to calculate four different user profiles, respectively *WOCD, SOCD, WOCA, SOCA*. *SOCA* is our baseline since it uses the two practices of existing works while other inputs use at least one of our proposed novelties.

We then use the same scoring method described in Sect. 4 to generate 4 sets of top-5 recommendations.

5.4 Metrics

We compare recommendations generated by different approaches with the ground truth. Following the ESWC 2014 Linked Open Data-enabled recommender system challenge [19], we used *Precision*@5, *Recall*@5, F_1@5 to assess the relevance of individual recommendations.

$$Precision@5 = \frac{1}{|u|} Precision_u@5 \quad (8)$$

$$Precision_u@5 = \frac{1}{5} \sum_{p=1}^{5} rel_{p,u} \quad (9)$$

$$Recall@5 = \frac{1}{|u|} Recall_u@5 \quad (10)$$

$$Recall_u@5 = \frac{1}{R_u} \sum_{p=1}^{5} rel_{p,u} \quad (11)$$

u is the set of users, R_u is the set of u's ground truth, $rel_{p,u}$ the binary relevance value of the recommended place,

$$rel_{p,u} = \begin{cases} 1, & if\ p\ in\ ground\ truth \\ 0, & otherwise \end{cases} \quad (12)$$

$$F_1@5 = 2 * \frac{Precision@5 * Recall@5}{Precision@5 + Recall@5} \quad (13)$$

We also used the normalized discount cumulative gain (nDCG) to assess the quality of the ranking.

$$DCG_k = \sum_{i=1}^{k} \frac{2^{rel_i} - 1}{log_2(i+1)} \quad (14)$$

$$nDCG_k = \frac{DCG_k}{IDCG_k} \quad (15)$$

k is the number of places that are recommended, $IDCG_k$ is the ideal DCG, in our case, IDCG is achieved when all of the top-5 recommendations are relevant.

5.5 Results and Discussions

In the following table, we report the score of each metric, precision (P), recall (R), F1, nDCG, the standard deviation (σ), the improvement compared with the baseline (Δ), the p-value of Student's t-test.

	WOCD	SOCD	WOCA	SOCA (baseline)
P	0.275	0.261	0.259	0.262
σ	0.263	0.259	0.259	0.261
ΔP	13.5 %	−0.5 %	−1.1 %	
p-value	0.000001	0.597	0.00007	
R	0.167	0.158	0.157	0.158
σ	0.183	0.177	0.177	0.179
ΔR	5.7 %	0 %	−0.6 %	
p-value	0.00000007	0.815	0.007	
F1	0.197	0.187	0.185	0.187
σ	0.201	0.196	0.197	0.198
ΔF1	5.3 %	0 %	−1 %	
p-value	0.00000007	0.753	0.0004	
nDCG	0.291	0.277	0.276	0.279
σ	0.281	0.276	0.276	0.277
ΔnDCG	4.3 %	−0.7 %	−1 %	
p-value	0.00001	0.425	0.00008	

We can clearly observe that WOCD which uses the two novelties proposed in this paper outperforms all other approaches according to all metrics. Compared with the baseline approach SOCA, it achieves a +5.3 % improvement in terms of F1-score, a +4.3 % improvement in terms of nDCG. The improvements over the baseline are statistically significant (p-value < 0.01 for all metrics). This observation proves our hypothesis, that in leveraging a rich travel-domain specific knowledge graph, and applying a city-dependent user profiling strategy, we can improve the user profile and yield better recommendations.

Generally speaking, the standard deviations are rather high. This shows a high variability on different users. In fact, our evaluation with ground truth may be pessimistic. From one hand, there are a big amount of travel attraction candidates, especially in big cities. From the other hand, a user may still prefer a location even if the user has not yet visited it. A user may visit a place and not take any photos. Our method should have better performance than the results reported in this paper. In the future, we envisage conducting a qualitative user study to evaluate the performance of our approach in another setting. WOCD has better performance than SOCD, it shows that with the same city-dependent strategy, using the whole ontology performs better than using the simple ontology. By comparing WOCD and WOCA, we can draw a similar conclusion that the whole ontology yields better recommendations when it is combined with a city-dependent strategy than with a city-agnostic one.

6 Conclusions

In this paper, we identified how knowledge graphs could be efficiently leveraged to solve the travel attractions recommendation problem. Comparing with existing works, we improved two main drawbacks: semantic poorness and city-agnostic user profiling strategy. We described a semi-automatic method to construct a rich world scale travel knowledge graph from existing large knowledge graphs namely Geonames, DBpedia and Wikidata. The underlying ontology contains more than 1200 classes to describe attractions. We presented a city-dependent user profiling strategy which makes use of the fine semantics encoded in the constructed graph. For the evaluation, we processed the YFCC100M dataset and constructed a subset to fit our problem. By inputting four different user profiles generated by four approaches and using the same recommendation method, we showed that our approach achieves statistically significant improvements compared with the baseline approach, a 5.3 % improvement in terms of F1-score and a 4.3 % improvement in terms of nDCG.

The ontology underlying our travel knowledge graph contains rich classes to describe travel attractions' types, however, some attractions are not typed or are typed only with high-level classes. We observed that the type strings that we extract from DBpedia categories, even though difficult to use, provide fine semantic information. As future work, we consider exploring how to use the type strings to complete the semantic annotations of travel attractions in Wikidata. Currently, the ontology is limited to a taxonomy. We envisage taking into account some properties on Wikidata such as P149 architectural style, P1435 heritage status.

References

1. Google: Travel trends: 4 mobile moments changing the consumer journey (2015). https://www.thinkwithgoogle.com/articles/travel-trends-4-mobile-moments-changing-consumer-journey.html
2. Expedia: Custom Research: Exploring the Traveler's Path to Purchase (2014). https://info.advertising.expedia.com/travelerspathtopurchase
3. Passant, A.: dbrec — music recommendations using DBpedia. In: Patel-Schneider, P.F., Pan, Y., Hitzler, P., Mika, P., Zhang, L., Pan, Jeff, Z., Horrocks, I., Glimm, B. (eds.) ISWC 2010. LNCS, vol. 6497, pp. 209–224. Springer, Heidelberg (2010). doi:10.1007/978-3-642-17749-1_14
4. Di Noia, T., Mirizzi, R., Ostuni, V.C., Romito, D., Zanker, M.: Linked open data to support content-based recommender systems. In: Proceedings of the 8th International Conference on Semantic Systems, pp. 1–8. ACM, September 2012
5. Ristoski, P., Loza Mencía, E., Paulheim, H.: A hybrid multi-strategy recommender system using linked open data. In: Presutti, V., et al. (eds.) SemWebEval 2014. CCIS, vol. 475, pp. 150–156. Springer, Heidelberg (2014). doi:10.1007/978-3-319-12024-9_19
6. Bao, J., Zheng, Y., Wilkie, D., Mokbel, M.: Recommendations in location-based social networks a survey. GeoInformatica **19**(3), 525–565 (2015)
7. Cheng, C., Yang, H., King, I., Lyu, M.R.: Fused matrix factorization with geographical and social influence in location-based social networks. In: Twenty-Sixth AAAI Conference on Artificial Intelligence, July 2012
8. Ye, M., Yin, P., Lee, W. C., Lee, D.L.: Exploiting geographical influence for collaborative point-of-interest recommendation. In: Proceedings of the 34th International ACM SIGIR Conference on Research and Development in Information Retrieval, pp. 325–334. ACM, July 2011

9. Yuan, Q., Cong, G., Ma, Z., Sun, A., Thalmann, N.M.: Time-aware point-of-interest recommendation. In: Proceedings of the 36th International ACM SIGIR Conference on Research and Development in Information Retrieval, pp. 363–372. ACM, July 2013
10. Braunhofer, M., Elahi, M., Ricci, F., Schievenin, T.: Context-aware points of interest suggestion with dynamic weather data management. In: Xiang, Z., Tussyadiah, I. (eds.) Information and communication technologies in tourism 2014, pp. 87–100. Springer International Publishing, Cham (2013)
11. Noguera, J.M., Barranco, M.J., Segura, R.J., Martínez, L.: A mobile 3D-GIS hybrid recommender system for tourism. Inf. Sci. **215**, 37–52 (2012)
12. Liu, B., Xiong, H.: Point-of-interest recommendation in location based social networks with topic and location awareness. In: SDM, vol. 13, pp. 396–404, May 2013
13. Brilhante, I.R., Macedo, J.A., Nardini, F.M., Perego, R., Renso, C.: On planning sightseeing tours with TripBuilder. Inf. Process. Manag. **51**(2), 1–15 (2015)
14. Bao, J., Zheng, Y., Mokbel, M.F.: Location-based and preference-aware recommendation using sparse geo-social networking data. In: Proceedings of the 20th International Conference on Advances in Geographic Information Systems, pp. 199–208, November 2012. ACM
15. Chen, C., Zhang, D., Guo, B., Ma, X., Pan, G., Wu, Z.: TripPlanner: personalized trip planning leveraging heterogeneous crowdsourced digital footprints. IEEE Trans. Intell. Transp. Syst. **16**(3), 1259–1273 (2015)
16. Lim, K.H., Chan, J., Leckie, C., Karunasekera, S.: Personalized tour recommendation based on user interests and points of interest visit durations. In: Proceedings of the 24th International Joint Conference on Artificial Intelligence (IJCAI 2015) (2015)
17. Lu, C., Stankovic, M., Laublet, P.: Desperately searching for travel offers? Formulate better queries with some help from linked data. In: Gandon, F., Sabou, M., Sack, H., d'Amato, C., Cudré-Mauroux, P., Zimmermann, A. (eds.) ESWC 2015. LNCS, vol. 9088, pp. 621–636. Springer, Heidelberg (2015). doi:10.1007/978-3-319-18818-8_38
18. Thomee, B., Shamma, D.A., Friedland, G., Elizalde, B., Ni, K., Poland, D., Li, L.J.: YFCC100M: the new data in multimedia research. Commun. ACM **59**(2), 64–73 (2016)
19. Di Noia, T., Cantador, I., Ostuni, V.C.: Linked open data-enabled recommender systems: ESWC 2014 challenge on book recommendation. In: Presutti, V., et al. (eds.) SemWebEval 2014. CCIS, vol. 475, pp. 129–143. Springer, Heidelberg (2014). doi:10.1007/978-3-319-12024-9_17

Making Entailment Set Changes Explicit Improves the Understanding of Consequences of Ontology Authoring Actions

Nicolas Matentzoglu(✉), Markel Vigo, Caroline Jay, and Robert Stevens

The University of Manchester, Oxford Road, Manchester M13 9PL, UK
{nicolas.matentzoglu,markel.vigo,caroline.jay,
robert.stevens}@manchester.ac.uk

Abstract. The consequences of adding or removing axioms are difficult to apprehend for ontology authors using the Web Ontology Language (OWL). Consequences of modelling actions range from unintended inferences to outright defects such as incoherency or even inconsistency. One of the central ontology authoring activities is verifying that a particular modelling step has had the intended consequences, often with the help of reasoners. For users of Protégé, this involves, for example, exploring the inferred class hierarchy.

We explore the hypothesis that making changes to key entailment sets explicit improves verification compared to the standard static hierarchy/frame-based approach. We implement our approach as a Protégé plugin and conduct an exploratory study to isolate the authoring actions for which users benefit from our approach. In a second controlled study we address our hypothesis and find that, for a set of key authoring problems, making entailment set changes explicit improves the understanding of consequences both in terms of correctness and speed, and is rated as the preferred way to track changes compared to a static hierarchy/frame-based view.

Keywords: Ontology engineering · Ontology authoring · Reasoning

1 Introduction

Ontologies are explicit conceptualisations of a domain, and are widely applied in biology, health-care and the public domain. Ontologies are typically represented in a formal representation language such as the Web Ontology Language (OWL), the Open Biomedical Ontologies format (OBO) or the RDF Schema language (RDFS). The central advantage of using such formalisms is their well-defined semantics. Generic reasoning systems can be used to access knowledge in the ontology that is only implied, i.e. not explicitly stated, allowing richer answers to queries, the identification of inconsistent knowledge and improved management of large terminologies through definition-oriented development. There is strong, if mainly anecdotal, evidence that building ontologies using OWL is difficult and error-prone.

E. Blomqvist et al. (Eds.): EKAW 2016, LNAI 10024, pp. 432–446, 2016.
DOI: 10.1007/978-3-319-49004-5_28

Attempts were made to quantify this difficulty [4], but an accurate model of the cognitive complexity of various ontology authoring tasks such as exploration or modelling has yet to be defined. The cognitive complexity of OWL can lead to axioms that do not reflect the intentions of the author. Furthermore, the rich and often complicated semantics of OWL can result in unintended inferences, which are often not made explicit by the authoring tool, and even if they are, are rarely communicated to the author clearly. An interview study recently revealed that many ontology experts *frequently* run the reasoner, sometimes *after every modification*, to detect errors such as unsatisfiable classes and to prevent the spread of errors [12]. Participants also felt that the change evaluation phase, i.e. the phase that determines whether a modelling action had the intended consequences, is not well supported by state of the art development tools. Some ontology authors use DL queries, generated on the fly, to do 'spot checks', others work with competency questions that are crafted upfront to automatically verify the correctness of a change. As the conceptual model of an ontology is, however, not always known upfront, competency question based approaches, perhaps best compared with unit tests in software engineering, unfold their utility later in the engineering process and their coverage of the ontology depends on the user's diligence. Consequently we need the user interface to remove this complexity from the ontologies, support the evaluation of ontologies and to either prevent or detect errors.

In this work, we are concerned with improving the evaluation of modelling actions. In the context of this work, we call the task of evaluating that a particular modelling action has had the desired effect "verification". Verification is a key sub-process of ontology authoring that involves conducting a set of tests, for example to make sure that a definition of a class works as intended and that no unsatisfiable classes were introduced [2]. When developing ontologies with the popular Protégé ontology engineering environment, the verification step is typically realised by invoking the reasoner and exploring the implicit knowledge in the ontology [12], for example by making sure that a particular class has the expected position in the inferred class hierarchy or a freshly introduced property domain restriction results in the expected individual type inferences. We call this approach *static hierarchy/frame-based* (SHFB), where "static" refers to the fact that the inferred hierarchy only reflects a state, without any indication how this state relates to the latest modelling action. We explore the hypothesis that making changes to a number of key entailment sets explicit improves verification compared to the static hierarchy/frame-based approach. Our contributions are as follows:

- We developed the Inference Inspector, a novel Protégé plugin that makes changes to key entailment sets as consequences of modelling actions explicit.
- We conducted an exploratory study to evaluate our Inference Inspector prototype. We find that our approach is better suited for tasks that involve change, such as changing definitions or adding restrictions, and less well suited for tasks that involve the introduction of new entities compared to SHFB and is

well received by users for a number of typical modelling tasks, in particular changing class definitions.
- We conducted a laboratory experiment that confirms our hypothesis. We find that making entailment set changes explicit improves the understanding of consequences both in terms of correctness and speed, and is rated as the preferred way to track changes compared to SHFB.

2 Background and Related Work

Ontology authoring is the creation and maintenance of ontology artefacts represented in a formal knowledge representation language such as OWL, OBO or RDFS. We view an ontology $\mathcal{O}$ as a set of axioms, and α with $\alpha \in \mathcal{O}$ being an axiom in $\mathcal{O}$. The signature of $\mathcal{O}$ is the set of entities across all axioms in $\mathcal{O}$. Typical ontology authoring activities include, but are not limited to, the creation of axioms or annotations. For a detailed discussion of ontology authoring activities see [14]. Research on ontology authoring has experienced a resurgence in recent years [12,14,15]. One reason for this might be the increased utilisation of change-logs for ontology development. WebProtégé for example produces change-logs during ontology authoring, which can form the basis of rich and informative analyses on ontology authoring activities [15].

While ontology authoring is increasingly performed in a programmatic fashion, a large number of ontologies have been built using ontology authoring environments such as Protégé [5] and WebProtégé [11]. Moreover, even if ontologies are created in a programmatic fashion, they are often checked for defects in a visual authoring environment. The work presented here is the continuation of a series of investigations into the processes of ontology authoring [12–14]. The aim of the series is to improve our understanding of ontology authoring processes, in particular to identify typical authoring styles and workflows to guide tool developers to improve their support of those workflows. We identified typical problems during ontology authoring, in particular that Protégé does not cater for all the needs of current authors [12,13]. Ever more sophisticated ontology modelling patterns make the verification of modelling actions difficult. Unintended consequences such as the introduction of unsatisfiable classes, broken definitions (that result in wrong classifications) or wrong inferences on the data level (ABox) are often difficult to spot, which was one of the core incentives for this work. A specially modified version of Protégé that collects interaction events silently during ontology authoring [14], protégéforus, enabled us to study ontology authoring workflows and derive a number of well-founded design suggestions for authoring tools [14]. One of these was *making the changes to the inferred hierarchy explicit*— another major incentive for developing the Inference Inspector.

The existing tool support for ontology authoring activities is still largely inadequate [12]. An example of an early study that established the necessity of presenting explanations for entailments and reporting errors adequately in the context of knowledge representation (KR) systems was McGuiness and Patel-Schneider [8]. Ontology authoring tools continued to receive poor usability ratings [3,6,12] throughout the last 20 years. Examples of unmet demands from users include the ability to compare different versions of the ontology [3] and inadequate debugging support [12]. In particular, making the consequences of modelling actions explicit beyond simply identifying that a defect exists has received little attention. The main effort in this direction came from Denaux et al. [2]. The authors developed a system that provides interactive semantic feedback directly after a change to the ontology. They suggest 6 categories of semantic feedback from the ontology engineering environment, given a single axiom α being added to the ontology: α was already asserted, α was not asserted, but could be inferred, α causes the ontology to be inconsistent, α is novel and the addition results in new implications, α is novel and the addition does not result in new implications, and α causes a concept in the ontology to become unsatisfiable. While Denaux et al. inspired us to produce a better feedback mechanism, their work differs in two fundamental aspects to the research presented here: (1) in Denaux et al. only additions are modelled, i.e., the case that an engineer adds an axiom to the ontology, while we also cover removals, and (2) changes are comprised of a single axiom, while we decided to model sets of additions and removals. Only providing feedback when the reasoner is run returns the responsibility for asking for feedback to the engineer, keeps the interface responsive (reasoning is not required after every step), but also comes with a caveat: given a set of changes, it may not be anymore possible to attribute a particular inference (either lost or gained) to a particular change, thereby putting the burden of identifying the erroneous change back to the engineer. We believe, however, that the gain in responsiveness is worth this caveat, and we can cover some of these shortcomings using justifications, as explained in the next section. The authors evaluate their approach using a task based setting similar to the exploratory study we present later, and find that the feedback was generally considered helpful. However, no formal evaluation was conducted to find out whether the feedback actually led to more accurate modelling. The tool is available online (https://sourceforge.net/projects/entendre/).

3 Inference Inspector: Making the Consequences of Modelling Actions Explicit

We present the Inference Inspector, a Protégé plugin for making the consequences of modelling actions in an ontology explicit. The Inference Inspector is implemented as a plugin for Protégé 5 (5.0.0 at the time of writing). We consider ontologies to be represented in OWL 2 DL, unless otherwise stated. A *modelling action* is defined as a non-empty set of changes $\mathcal{CH}$. A *change* can either be a removal of an axiom α, denoted R_α, or an addition, denoted A_α. For example, given the addition of an axiom α_1: SubClassOf(A, B) and the removal of another axiom α_2: SubClassOf(A, C), the modelling action is defined as $\mathcal{CH} : \{A_{\alpha_1}, R_{\alpha_2}\}$. Axiom *modifications* are always treated as an addition of the revised axiom and a removal. In the previous example, the user might have decided that A should not be subsumed by B, but by C instead, changing the existing SubClassOf(A, B) to SubClassOf(A, C). The Inference Inspector makes changes to a predefined set of key entailment sets explicit. A change to an entailment set is defined as follows. Given an $\mathcal{O}$, a previous version of the ontology $\mathcal{O}'$ and a finite entailment set $\mathcal{E}$, the difference between the respective entailment set of $\mathcal{O}$ and $\mathcal{O}'$, $\mathcal{E}_{\mathcal{O}} \setminus \mathcal{E}_{\mathcal{O}'}$, is called the set of *added inferences w.r.t.* $\mathcal{E}$, and the difference between the entailment set of $\mathcal{O}'$ and $\mathcal{O}$, $\mathcal{E}_{\mathcal{O}'} \setminus \mathcal{E}_{\mathcal{O}}$ is called the set of *removed inferences w.r.t.* $\mathcal{E}$. Given a language $\mathcal{L}$ and an OWL 2 ontology $\mathcal{O}$, the $\mathcal{L}$-entailment set of $\mathcal{O}$, written $\mathcal{E}(\mathcal{O}, \mathcal{L})$, is the set of all axioms in $\mathcal{L}$ that are entailed by $\mathcal{O}$ (entailment set).

There are a number of factors that inform the selection of appropriate entailment sets for presentation [10]. Our approach was fairly practical: the entailments shown should help the user to verify their modelling actions and not be too costly to compute. In order to help the user verifying their modelling actions, the entailment set should be indicative of erroneous or correct modelling and be easily understandable by a typical user. In order to be *indicative of erroneous modelling*, the presented entailments should be verifiable against modelling intentions. Since we cannot directly access the user's modelling intentions, we make a number of simplifying assumptions. Firstly, we consider unsatisfiable classes or ontology inconsistency as bugs, which any user aims to avoid. Secondly, we assume that the majority of users have a mental model of the hierarchical structures of their ontology, including a model of class disjointness, and intend to keep the ontology consistent with that mental model. In other words, we assume that the ontology author knows, for a concept they are modelling, where in the hierarchy it should be situated and which individuals should be members of it, as well as whether it is disjoint from another concept in the domain. Therefore, we primarily serve the modelling intentions of avoiding bugs while producing hierarchies consistent with that mental model. We acknowledge that this assumption is not universally true, as it is, for example, unlikely that any one author of the gene ontology [1], for example, knows all subsumptions between all the concepts it covers. Furthermore, there are other relevant axes that are not covered by our approach, such as partonomy or any class level patterns that are based on object properties, such as existential restrictions. We do, however, believe that the subsumption relation is of central importance in a majority of cases, which is also confirmed by our finding that users look at the class hierarchy 45 % of the time spent editing ontologies with Protégé [14]. Furthermore, the presented entailments must be *easily understandable* by a user, i.e. we do not want to replace a cognitively demanding search, such as a lookup in a large hierarchy, by a cognitively demanding parse, such as an axiom involving deeply nested class expressions. Therefore, our approach only considers entailments that involve named entities, and avoid those that involve complex class expressions such as existential restrictions. Lastly, determining the entailment sets should not be too computationally expensive. Current implementations of subsumption and instantiation perform well in practice [9], despite the high worst case complexity. Computing the full set of disjoint classes, on the other hand, can be more computationally intensive in practice, because reasoners do not implement efficient algorithms for this task. We therefore allow the user to determine whether computing the disjointness relation is worth their while.

We consider the following (groups of) entailment sets: (1) atomic subsumptions between classes, and object and data properties; (2) equivalences between classes, and object and data properties; (3) object property characteristics; (4) disjointness between class names; (5) class assertions and; (6) object property assertions. While (1), (2), (4) and (5) directly correspond to the considerations above, (3) and (6) do not. We include object property assertions in our solution in order to provide a mechanism that allows the user to check whether sub-object property chains (and the object property hierarchy) work as intended. The reason for including object property characteristics was that our extensive experience teaching novice and advanced OWL users has shown that the inheritance or non-inheritance of object property characteristics up or down the object property hierarchy is extremely difficult to understand. For example, making an object property functional makes all its children's properties functional, while the same is not true for transitivity. Although the entailment sets considered here are finite, they are potentially large. To further reduce the amount of information shown to the user, for the three atomic subsumption entailment sets, we consider the transitive reduct, i.e. we query the reasoner only for direct subsumptions.

After the selection of appropriate entailment sets, the second question that is relevant when presenting entailments to users according to Parvizi *et al.* [10] is how they should be ordered. The Inference Inspector implements a configurable system for inference prioritisation. We allow the user to assign a priority to an item from a list of pre-defined inference patterns. Currently, we have implemented five priority levels and 11 inference patterns. The priority levels range from critically important (ontology defects such as unsatisfiability) to unimportant (e.g. asserted axioms).

By default, the Inference Inspector orders the presented consequences by priority. Sometimes, ordering the potentially large number of entailments presented to the user is not enough. In particular, inferences on the ABox (individual) level can be extremely numerous. We therefore employ a grouping strategy for object property assertion axioms and class assertions [10]. By default, we group all axioms of the type ClassAssertion(a,X), where X is a particular class name in the ontology, and ObjectPropertyAssertion(a, b, R), where R is a particular object property name in the ontology. For very large ontologies the list of inferences can be further narrowed down by restricting it to particular entities in the ontology. Justifications for entailments can be computed on demand.

After each reasoner run, a snapshot of the current ontology is created including its inferences. By default users are presented with the consequences of their most recent modelling action (Fig. 1). We define the scope of a single modelling action as the set of all changes that were applied to the ontology between the latest and the previous run of the reasoner. For example, after running the reasoner, the ontology author might add three axioms and remove one. When the user runs the reasoner again, they will be presented with the entailments added and lost since the previous run of the reasoner. From an implementation perspective, this is realised by computing the set difference in accordance with

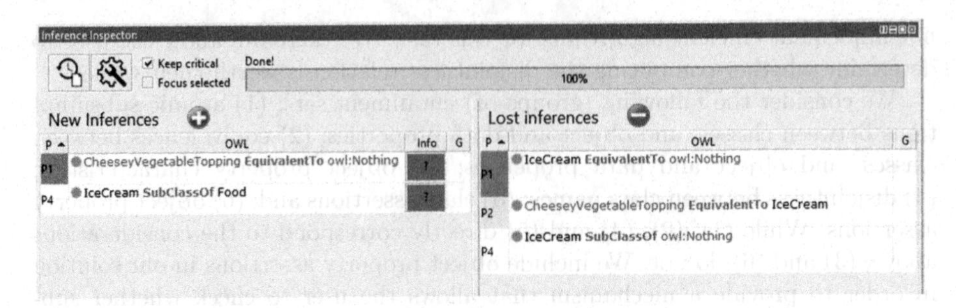

Fig. 1. A snapshot of the inference inspector after the removal of an axiom that made IceCream unsatisfiable. Left: we can see the new position of IceCream in the class hierarchy. Right: we can see respective lost inferences. P1 (Critical), P2 (important) and P4 (not important) are priority levels (P).

the definitions given in the beginning of this section. The Inference Inspector allows the user to compare the current state of the ontology to any snapshot created previously. The first snapshot is always the empty ontology: this means that comparing the current version of the ontology with the empty ontology will always show all inferences (short of those that are explicitly hidden by the user). By default, inferences of critical importance (P1 in Fig. 1) are always shown, no matter which snapshot forms the basis for comparison, but this feature can be switched off, if only the latest changes are of interest. Lastly, the user may restrict the inferences presented, by either: (1) showing only inferences related to the currently selected entity in Protégé or (2) showing only inferences involving entities manually selected in a special entity selector panel. An important caveat of the Inference Inspector implementation is that it relies on the correctness of the reasoner. Reasoners are not always correct [7], and may not support the inference of all the entailment sets considered by the Inference Inspector.

4 Materials and Methods

We conducted two studies to evaluate our approach. The first was an exploratory study performed in the context of an advanced OWL modelling tutorial (E1), intended to evaluate our prototype and isolate modelling actions where authors may benefit from the Inference Inspector. A second, controlled laboratory experiment (E2) validated the hypothesis that making changes to key entailment sets explicit improves modelling performance.

4.1 E1: Prototype Evaluation

Goals. The main goals of this study were to evaluate the Inference Inspector prototype and determine those modelling actions where our approach is likely to add value over existing solutions. The evaluation was designed to be broad and involved rating the Inference Inspector for perceived usefulness and responsiveness, as well as providing feedback on the user interface. From the results we extracted modelling actions where our approach may help, and used this information to design tasks for the second experiment (Sect. 4.2).

Participants. 15 intermediate users of Protégé were recruited in the context of a two day advanced OWL tutorial (see http://ow.ly/pK8P300x9wq). An Amazon Voucher (£10) was given to those that were willing take part. Most participants had successfully completed a beginner level OWL tutorial or had an equivalent experience with OWL. Out of the 15 (9 female) participants, there were 3 students, 5 PhD students, 3 research fellows, 1 data researcher, 1 assistant director for information management, 1 clinician, 1 information architect and 1 bioinformatician. The mean self-reported expertise level (Likert-scale, 1, Novice - 5, Expert) was 2.47 (standard deviation 0.99) for Protégé and and 2.53 (standard deviation 1.06) for OWL.

Experimental Setup. Participants were asked to perform 10 typical ontology authoring tasks in the context of an ontology about pizza (726 axioms,$\mathcal{SHOIN}$) using 5 pre-defined tabs in Protégé. A task typically involved an action, such as adding a definition and running the reasoner, an act of exploration, such as inspecting the changes that occurred, answering one or more control questions about this inspection and finally rating all five tabs for the suitability of performing the task and/or the inspection. The five pre-designed tabs were: a simple list of inferred axioms ("Inferences" view in Protégé), the Protégé "Classes" tab, the Inference Inspector, the Protégé "Individuals" tab and the DL Query tab of the DL Query plugin for Protégé. For navigation purposes, all views showed an asserted class hierarchy on the left hand side, which participants were instructed to ignore when evaluating the suitability of the views for each task. The ten tasks were: (1) understanding the topic of the ontology, (2) identifying unsatisfiable classes, (3) repairing unsatisfiable classes, (4) verifying the repair of an unsatisfiable class, (5–6) verifying the definition of a new class, (7) changing the definition of an existing class, (8) verifying the loosening of a restriction by removing a disjointness axiom, (9) verifying the change of an object property assertion and (10) verifying the addition of a role chain. To avoid participant bias we presented the Inference Inspector simply as a third-party plugin, rather than our own work. Participants were not formally introduced to the Inference Inspector prior to the experiment, but some of the basic functionality was covered as part of the preceding OWL Tutorial. The study took around 50 min.

4.2 E2: Making Changes to Key Entailment Sets Explicit Improves Verification Performance

Goals. The goal of the second study was to verify the following hypothesis: *Making gained and lost entailments explicit improves the user's understanding of consequences of authoring actions compared to a hierarchy/frame-based view.*

Improvement of user understanding was tested through a series of exploration questions and is evaluated using the following metrics:

- Correctness of understanding: (#true positives + #true negatives)/#options
- Speed of understanding: time to completion/correctness
- Ease of understanding: #mouse-click/correctness, scroll time/correctness
- User suitability rating: Likert-scale (0, unusable-4, perfectly usable)
- Breakdown of answer options, i.e. the number of user-selected answers that are correct (true positives), wrong (false positives) and the number of answers not selected by the user that are correct (true negatives) and wrong (false negatives).

The speed score accounts for the correctness of the answer: the more incorrect it is, the higher the penalty on the score.

Participants. 19 (5 female) participants were recruited via word-of-mouth and email-advertisement. The background of the participants ranged from MSc students and intermediate experience to academics and non-academic professionals with a high level of OWL expertise between 22 and 57 years of age (mean 33.28). Out of the 19 participants, there were 5 students, 5 PhD students, 5 academics and 4 non-academics. 4 reported to be involved with ontologies as ontology engineers, 8 as ontology researchers and 6 as ontology tool developers (1 other). The mean self-reported expertise level (Likert-scale, 1, Novice-5, Expert) was 3.53 (standard deviation 0.61) for Protégé and and 3.68 (standard deviation 0.75) for OWL.

Experimental Setup. The controlled study was conducted in a designated usability lab. All participants used the same machine with a 24 in. monitor with Protégé 5.0.0 installed. Protégé was pre-configured with the Inference Inspector and the specially developed Protégé Survey Tool [1] (PST) to administer the survey. Tasks were designed for the following problem areas: tightening conceptualisation (adding restrictions, verifying consequences), loosening conceptualisation (removing restrictions, fixing unsatisfiable classes) and changing conceptualisation (changing class definitions). In order to mitigate the impact of varying user expertise levels, all tasks were designed in pairs, i.e. two very similar tasks were designed with one being tested using the Inference Inspector and the other one being tested using the Classes or the Individuals tab. No task required access to the properties tabs. Tabs were assigned to tasks (1 task of each pair to the Inference Inspector) using a Latin square, and then randomly sampled. The survey contained a total of 14 verification tasks. The TBox focused tasks were presented in a scenario involving an ontology about pizza (604 axioms, $\mathcal{SHOIN}$), and the ABox focused tasks were presented in a scenario involving an ontology about family history (89 axioms, $\mathcal{SHIF}$). The participant was asked to answer 2–3 exploration questions for every task, most of which were of the sort "Did the class hierarchy change?" or "Which are the new subclasses of X?" In order to increase participant focus, all questions were auto-submitted after 60 seconds.

[1] https://github.com/matentzn/protegesurvey.

The questions were designed to be answerable comfortably within that timeout by a reasonably experienced user of Protégé. Answers submitted that way were counted as if they were submitted in the regular way. The study had a maximum duration of 50 min.

5 Results and Discussion

5.1 E1: Prototype Evaluation

Figure 2 shows which views users considered their preferred option to tackle an ontology authoring problem. The participant was allowed to select a single view that was considered the most adequate for addressing a problem. For identifying unsatisfiable classes, at least 5 participants rated the Inference Inspector as the preferred view, compared to 8 who preferred the Classes tab. The Inference Inspector presents unsatisfiable classes clearly to the user, but so does Protégé. The result suggests that at the very least, for this important task, the Inference Inspector is in fact usable. The same argument goes for the repair of unsatisfiable classes, which both views support by allowing the user to delete axioms occurring in a justification. With respect to adding and changing definitions or restrictions, we expected the Inference Inspector to outperform, because it makes the changes explicit and not subject to a potentially complicated search in the class hierarchy. When, however, users were asked to explore the consequences of defining a new named class, they preferred the class hierarchy. This may be because changes with respect to the defined class are made explicit simply by its position in the hierarchy: all sub- and super-classes are new. Only a third of participants preferred the Inference Inspector for this task, possibly because a visual representation of the class hierarchy is easier to understand than a list of axioms. The Inference Inspector did add value, however, when it came to understanding the consequences of a change in the definition of an existing class (i.e. the EquivalentClasses axiom). Seven users preferred the Inference Inspector to six who preferred the class hierarchy, possibly because it is harder to detect a change in the position of a class, than it is to see the introduction of a new one. The first prototype of the Inference Inspector did poorly on problems involving individuals, perhaps because ABox inferences were not ordered or grouped in any way, which was improved for the subsequent study.

Figure 3 shows the distribution of scores, on a 5 point Likert scale, the Inference Inspector received for key usability criteria. Ease of use was the weakest point of the Inference Inspector (mean rating of 2.93). While this can be partially attributed to a lack of familiarity with the tool ("it seems useful but I don't understand it"), participants also found the plugin a "little busy, layout-wise", "overwhelming to use at first", and that there "are maybe too much information/options in the same view" (see next paragraph on free-form feedback). As a consequence of this feedback, we reduced the options shown to the user and adopted features such as the axiom renderer from Protégé to ensure a more familiar look and feel before conducting our second user study. Reliability had a mean rating of 3.47, with at least 4 participants giving it a low rating of 2. The

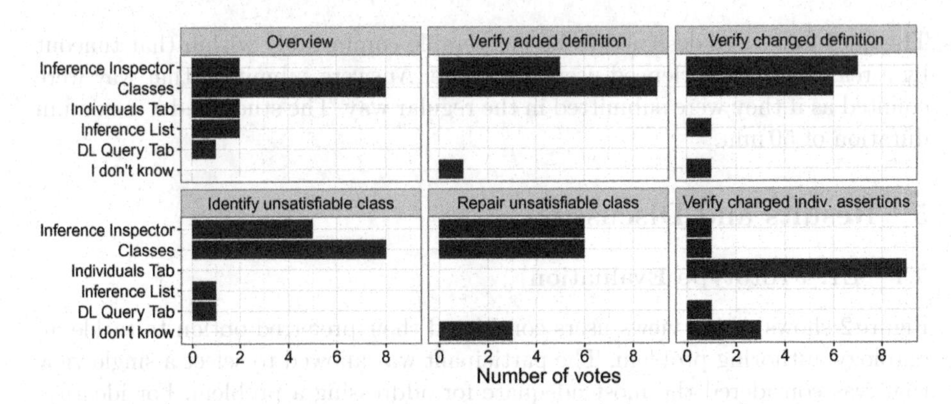

Fig. 2. Preferred view for addressing problem, one vote per participant.

problem with reliability may also have resulted from a lack of familiarity—users may have not been sure what to expect, and therefore been confused by the feedback. Response time was generally rated good (4.27, despite the Inference Inspector being several seconds slower than the reasoner in Protégé. It remains to be seen how well the Inference Inspector scales. Users expressed interest in using the Inference Inspector for their own work (3.87) and would recommend it to others (3.87).

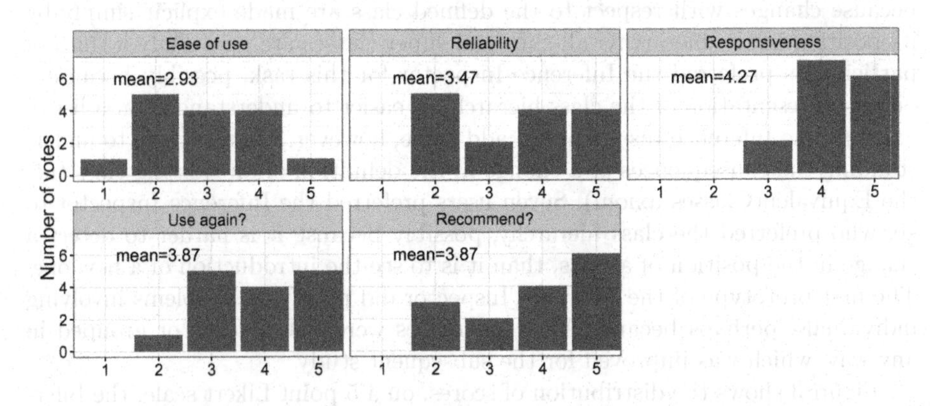

Fig. 3. Ratings of Inference Inspector (1: unusable, 5: perfectly usable)

Study participants were also asked to provide some free-form feedback. They recognised the importance of prioritising and reducing the information shown (independent of whether the Inference Inspector succeeded at this): searching the class hierarchy might be "very tedious in a [..] large ontology [..]." and "some means to filter and sort the output" is required, for example by "prioritizing inferences related to unsatisfiability". Moreover, participants recognised the importance of immediate feedback: "it'd be helpful to have more indication that the repair was successful. As of now, we're looking for the absence of error warnings. If protege could compare the states and flash a message saying 'fixed' [..], it'd be better." This is exactly the sort of feedback we are trying to provide. As the participants were not told about the authors affiliation with the tool, there was little risk of experimenter bias. Unsurprisingly the feedback on the Inference Inspector was mixed. The Protégé Classes tab was rated as favourite for most (TBox related) tasks: "It's the golden standard for all Protégé views", "I felt this view was the most helpful for editing and [..] information." Participants, however, also recognised the potential impact of familiarity bias: "Maybe I like it because I am familiar with it". It is possible the familiarity bias had a significant impact on results, especially since participants were not formally introduced to the Inference Inspector.

5.2 E2: Making Changes to Key Entailment Sets Explicit Improves Verification Performance

We tested the potential for improvement of verification performance provided by the Inference Inspector with respect to (1) a particular range of tasks and (2) the consequences of a change. While we generalise our results for tasks of the kind described in Sect. 4.2, we do not say anything about other kinds of modelling tasks, such as actions that involve datatypes. Secondly, we are interested in understanding the consequences of a change. This means we do not measure the performance of exploration tasks such as "What are the super-classes of A" or "Which one of the following are not sub-classes of A", but instead "Which are the new sub-classes of A" and "Which subclasses were lost to A". The difference is subtle, but important. We use a Wilcoxon signed-rank test to determine whether the difference between the Inference Inspector and the respective Protégé views is statistically significant (at a significance level of $p = 0.05$) for a particular metric.

Exploration tasks were more likely to be performed correctly with the Inference Inspector (80 %), than with the Protégé views (65 %, $p = 0.009$), see Table 1. This provides evidence that our hypothesis, for the specified set of tasks, holds. At a closer look (Fig. 4) we can see that while the problems solved with the Inference Inspector are clustered at the high end of correctness, the ones solved with Protégé are more evenly spread. The same can be observed for the user ratings. We acknowledge, however, that subjective ratings are potentially unreliable due to experimenter bias. Tasks were also performed faster with the Inference Inspector; in the case of exploration tasks, the difference was more than 4.5 sec (mean). However, this difference is not statistically significant ($p = 0.095$). If answer correctness is taken into account (speed), the difference is even greater (and significant, $p = 0.017$). Figure 4 shows how the distribution of task performance time (speed) with the Inference Inspector is shifted to the left. The distribution of the task duration, in particular the high density of short-duration tasks, can be explained by the immediacy with which questions such as "Which classes are unsatisfiable?" or "Did the class hierarchy change?" can be answered with the Inference Inspector. Figure 4 shows how the level to which users needed to scroll is distributed. While there are more tasks that require very little scrolling when using the Inference Inspector, there are also some tasks that require a

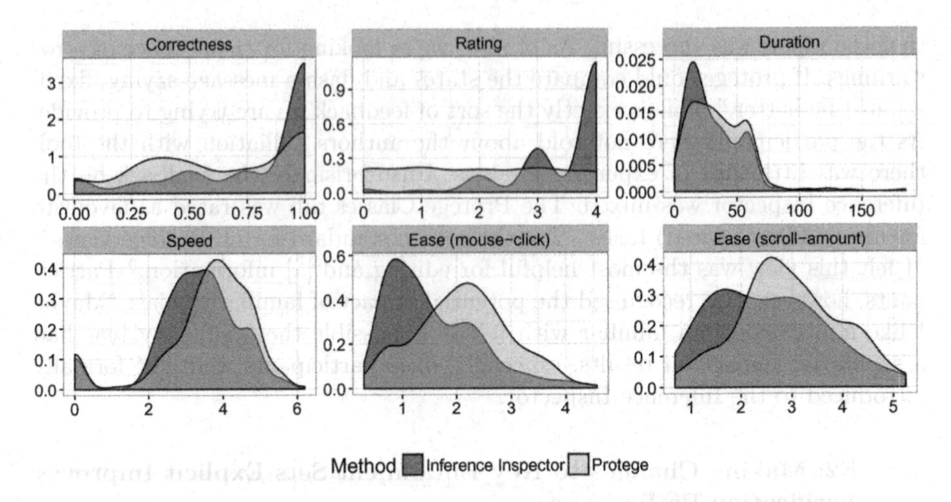

Fig. 4. Kernel density plots for 6 key metrics (x: metric, y: density). Lower three log-rescaled.

lot—for some problems, the Inference Inspector shows the results immediately, for others, searching is required. The difference between the Inference Inspector and Protégé however is not statistically significant ($p = 0.155$). The primary form of navigation in the Protégé hierarchy is expanding the nodes rather than scrolling, which manifests itself typically as mouse-clicks. There were considerably less clicks involved in arriving at a correct answer ($p = 0.015$). Looking at false positives and false negatives gives a more fine grained picture of correctness. False positives are wrong observations, i.e. question options that were false

Table 1. Mean, median and standard deviations for key metrics. II is the Inference Inspector, P is Protégé (depending on the task, either individual or Classes tab). P-values larger than 0.05 indicate that differences are not statistically significant.

Metric	*mean*		*median*		*sd*		*p*
	II	P	*II*	*P*	II	P	
Correctness	0.80	0.65	1.00	0.75	0.32	0.34	*0.009*
Rating	3.51	1.93	4	2	0.82	1.41	*0.003*
Duration	30.25	34.82	23.96	32.13	20.90	23.11	0.095
Speed	43.04	59.31	28.16	43.31	47.04	58.24	*0.017*
Ease (mouse-click)	6.75	11.09	3.00	8.00	10.59	12.03	*0.015*
Ease (scroll-amount)	2.83	5.74	0.00	0.00	13.72	19.38	0.155
False negatives	0.04	0.10	0	0	0.21	0.29	*0.002*
False positives	0.05	0.25	0	0	0.22	0.43	*0.002*
True negatives	0.79	0.56	1	1	0.41	0.50	*0.002*
True positives	0.95	0.73	1	1	0.22	0.44	0.141

but selected by the user. False negative answers are missed answers, that suggest that the view gave the participant a somewhat incomplete picture of the consequences. At first glance, it is surprising that the exploration tasks resulted in more false positive explorations than false negative ones (the converse is true for the Inference Inspector). However, this can be explained by a significant number of binary (Yes/No) questions that were frequently answered wrongly in the case of the Protégé views, for example "Did the class hierarchy change?", which is not always obvious when using the "Classes" tab.

6 Conclusions

Ontologies can be complex systems of axioms, and a modelling action may have consequences throughout the whole system. Being able to apprehend these consequences should be useful in ontology authoring. We presented the Inference Inspector–a tool that shows the consequences of modelling actions–and explored the hypothesis that making changes to key entailment sets explicit improves understanding of the consequences of such actions.

We find that making entailment set changes explicit improves understanding of the consequences of a range of key modelling actions. We provide evidence that the standard static view of an ontology does not adequately support people in understanding the consequences of modelling actions, and should be addressed by current ontology authoring environments. Making the consequences of changes explicit as changes to entailment sets is by no means the only, or necessarily best, way to approach this issue. For example, we can easily imagine solutions that highlight changes to the class hierarchy directly. We hope, however, that our work shows that making changes explicit is a key feature missing from ontology authoring environments based on the static hierarchy/frame-based paradigm and that the Inference Inspector will help ontology authors to verify their modelling choices more easily, thereby improving the ontology authoring experience. The plugin is actively maintained and available at https://github.com/matentzn/inference-inspector. We welcome bug reports and feature requests to be submitted through GitHub's issue tracking system. A demonstration video, along with links to the source code, the tutorial and the results of both studies, can be found at http://ow.ly/pK8P300x9wq.

Acknowledgments. This research has been funded by the EPSRC project WhatIf: Answering "What if..." questions for Ontology Authoring, EPSRC reference EP/J014176/1.

References

1. Consortium, T.G.O.: Gene ontology annotations and resources. Nucleic Acids Res. **41**(D1), D530–D535 (2013)
2. Denaux, R., Thakker, D., Dimitrova, V., Cohn, A.G.: Interactive semantic feedback for intuitive ontology authoring. In: Formal Ontology in Information Systems - Proceedings of the Seventh International Conference, FOIS 2012, Gray, Austra, 24–27 July 2012, pp. 160–173 (2012)

3. Dzbor, M., Motta, E., Buil, C., Gomez, J.M., Görlitz, O., Lewen, H.: Developing ontologies in OWL: an observational study. In: Proceedings of the OWLED 2006 Workshop on OWL: Experiences and Directions, Athens, Georgia, USA, 10–11 November 2006 (2006)
4. Horridge, M., Bail, S., Parsia, B., Sattler, U.: The cognitive complexity of OWL justifications. In: Aroyo, L., Welty, C., Alani, H., Taylor, J., Bernstein, A., Kagal, L., Noy, N., Blomqvist, E. (eds.) ISWC 2011, Part I. LNCS, vol. 7031, pp. 241–256. Springer, Heidelberg (2011)
5. Knublauch, H., Fergerson, R.W., Noy, N.F., Musen, M.A.: The protégé OWL plugin: an open development environment for semantic web applications. In: McIlraith, S.A., Plexousakis, D., Harmelen, F. (eds.) ISWC 2004. LNCS, vol. 3298, pp. 229–243. Springer, Heidelberg (2004)
6. Lambrix, P., Habbouche, M., Pérez, M.: Evaluation of ontology development tools for bioinformatics. Bioinformatics **19**(12), 1564–1571 (2003)
7. Lee, M., Matentzoglu, N., Parsia, B., Sattler, U.: A multi-reasoner, justification-based approach to reasoner correctness. In: Arenas, M., et al. (eds.) ISWC 2015. LNCS, vol. 9367, pp. 393–408. Springer, Heidelberg (2015). doi:10.1007/978-3-319-25010-6_26
8. McGuinness, D.L., Patel-Schneider, P.F.: Usability issues in knowledge representation systems. In: Proceedings of the Fifteenth National Conference on Artificial Intelligence and Tenth Innovative Applications of Artificial Intelligence Conference, AAAI 1998, IAAI 1998, 26–30 July 1998, Madison, Wisconsin, USA, pp. 608–614 (1998)
9. Parsia, B., Matentzoglu, N., Gonçalves, R.S., Glimm, B., Steigmiller, A.: The OWL reasoner evaluation (ORE) 2015 competition report. In: Proceedings of the 11th International Workshop on Scalable Semantic Web Knowledge Base Systems (SSWS-2015), Bethlehem, Pennsylvania, USA, 11 October 2015 (2015)
10. Parvizi, A., Mellish, C., van Deemter, K., Ren, Y., Pan, J.Z.: Selecting ontology entailments for presentation to users. In: KEOD 2014 - Proceedings of the International Conference on Knowledge Engineering and Ontology Development, Rome, Italy, 21–24 October 2014, pp. 382–387 (2014)
11. Tudorache, T., Nyulas, C.I., Noy, N.F., Musen, M.A.: WebProtégé: a collaborative ontology editor and knowledge acquisition tool for the Web. Semant. Web **4**, 89–99 (2013)
12. Vigo, M., Bail, S., Jay, C., Stevens, R.D.: Overcoming the pitfalls of ontology authoring: strategies and implications for tool design. Int. J. Hum.-Comput. Stud. **72**(12), 835–845 (2014)
13. Vigo, M., Jay, C., Stevens, R.: Design insights for the next wave ontology authoring tools. In: CHI Conference on Human Factors in Computing Systems, CHI 2014, Toronto, ON, Canada, April 26–May 01 2014, pp. 1555–1558 (2014)
14. Vigo, M., Jay, C., Stevens, R.: Constructing conceptual knowledge artefacts: activity patterns in the ontology authoring process. In: Proceedings of the 33rd Annual ACM Conference on Human Factors in Computing Systems, CHI 2015, Seoul, Republic of Korea, 18–23 April 2015, pp. 3385–3394 (2015)
15. Wang, H., Tudorache, T., Dou, D., Noy, N.F., Musen, M.A.: Analysis of user editing patterns in ontology development projects. In: Meersman, R., Panetto, H., Dillon, T., Eder, J., Bellahsene, Z., Ritter, N., Leenheer, P., Dou, D. (eds.) ODBASE 2013. LNCS, vol. 8185, pp. 470–487. Springer, Heidelberg (2013)

Data 2 Documents: Modular and Distributive Content Management in RDF

Niels Ockeloen(✉), Victor de Boer, Tobias Kuhn, and Guus Schreiber

The Network Institute, VU University Amsterdam, De Boelelaan 1081, 1081 HV Amsterdam, The Netherlands
{niels.ockeloen,v.de.boer,t.kuhn,guus.schreiber}@vu.nl,
http://www.networkinstitute.org, http://wm.cs.vu.nl

Abstract. Content Management Systems haven't gained much from the Linked Data uptake, and sharing content between different websites and systems is hard. On the other side, using Linked Data in web documents is not as trivial as managing regular web content using a CMS. To address these issues, we present a method for creating human readable web documents out of machine readable web data, focussing on modularity and re-use. A vocabulary is introduced to structure the knowledge involved in these tasks in a modular and distributable fashion. The vocabulary has a strong relation with semantic elements in HTML5 and allows for a declarative form of content management expressed in RDF. We explain and demonstrate the vocabulary using concrete examples with RDF data from various sources and present a user study in two sessions involving (semantic) web experts and computer science students.

Keywords: RDF · Vocabulary · Linked Data · Web documents · Content management · Semantic Web · HTML5

1 Introduction

In this paper we present a novel approach to content management which consists of providing a method for defining and rendering documents in terms of a logical composition of document fragments specified through arbitrary web resources. The approach builds on the semantic notions in HTML5 and on content in the form of Linked Data.

A still growing percentage of websites uses a Content Management System (CMS) to organise and manage their content; more than 44 % in July 2016[1] compared to 40 % two years earlier [16]. There are many CMS implementations, both open source and proprietary, the most popular currently being WordPress[2], Joomla[3] and Drupal[4]. These systems are all imperative software solutions that

[1] http://w3techs.com/technologies/overview/content_management/all, retrieved 5 July 2016.

[2] http://www.wordpress.com.

[3] http://www.joomla.org.

[4] http://www.drupal.org.

E. Blomqvist et al. (Eds.): EKAW 2016, LNAI 10024, pp. 447–462, 2016.
DOI: 10.1007/978-3-319-49004-5_29

have their own specific implementation details and database model. Therefore, sharing fragments of content between these systems is hard, and can only be accomplished by using special convertors or plugins, which either perform an offline migration[5] or depend on popular web feed formats (e.g. RSS) for live content sharing.

Though there are standards considering web documents as a whole, traditionally there have been no clear standards for dealing with fragments of documents in general. However, standards have been developed to share metadata, starting with the Meta Content Framework [9]. MCF was not widely adopted, but can be considered an ancestor of the Resource Description Framework (RDF) [13], now a major cornerstone of the Semantic Web. The number of RDF data sets that constitute the Linked Data Cloud has more than tripled between 2011 and 2014, with over a thousand data sets available containing billions of triples [20]. However, using this data in web documents is not as trivial as doing regular web content management.

In this paper we fold both these problems into one in an effort to solve them with a single solution: a novel way of doing content management expressed in RDF. By expressing all content for a web document in RDF, we provide a way of sharing heterogeneous web content between sites. By doing so, there is essentially no technical difference between regular content and other existing Linked Data. Hence, we also provide a way of including Linked Data in web documents. The main question is how to do this in a way that not only allows for the sharing and re-use of the content itself, but also the effort put into tasks such as data selection and the rendering of selected data into logical document sections.

2 Requirements and General Approach

The uniqueness of our approach is that we do not devise yet another tool to incorporate Linked Data into an existing web platform or system. Instead we turn the technology stack around; using RDF and Linked Data principles as the basis for doing content management on the web. Using this approach, we bring to the web of documents what Linked Data in general brought to the world of databases: eliminating the traditional boundaries for doing content management between web documents coming from different content owners, domains, servers and physical locations.

RDF-based Linked Data [5] has all the necessary properties to form the basis of dealing with fragments of content on the web. Any type of data can be contained, hence also fragments of web content. It has the ability to uniquely identify fragments and the ability to retrieve specific fragments using dereferencing. Furthermore, RDF is well known for its alignment abilities, making a solution expressed in RDF compatible with potential other RDF based approaches to content management. To use RDF as the basis for doing content management

[5] Example of a Joomla to WordPress migration plugin: https://wordpress.org/plugins/fg-joomla-to-wordpress/.

and sharing fragments of content on the web, we formulated the following set of requirements which the solution should meet:

- It should be built upon web standards;
- It should be able to handle heterogeneous Linked Data from multiple sources;
- It should have a separation of concerns with respect to how data is structured, selected and rendered, in the best traditions of the Web;
- It should allow for the sharing of not only content, but also the knowledge and settings involved in selecting and rendering data;
- It should be modular and distributable by design, so that bits and pieces can easily be obtained, combined and re-used on a mix-and-match basis;
- It should have a minimal set of implementation constraints;
- It should be exchangeable with other, similar approaches;
- It should be relatively easy to learn an use for a broad audience of web-contributors such as web developers and content owners;
- It should facilitate easy web page design;
- It should be doable to edit configuration and operation details by hand, in the same sense that HTML, CSS and other fundamentals can be edited by hand *if needed*, even though more elaborate GUIs exist.

To minimise the amount of infrastructure needed and rely on Linked Data standards as much as possible, we employ one of the most basic Linked Data principles to facilitate the actual managing of content: the *dereferencing* of resources. In that approach, the placement of a fragment of content, i.e. an article, is essentially done by creating an RDF triple with the IRI of the article as subject. In order to retrieve the article and render it in the document, the subject IRI is dereferenced. This principle eliminates the difference between using content within one or across sites. Obviously, there are several things we need to know in order to include the article in the document, e.g. which properties of the dereferenced resource to use, how to render the article in HTML and its positioning within the document. In order to express this information, a vocabulary can be used that is interpreted by a *general parsing script*. For this purpose, we defined the Data 2 Documents (d2d) vocabulary. We also implemented a *reference implementation* of the general parsing script to test and evaluate the vocabulary. The script interprets the d2d vocabulary elements, but is not tailored to any specific (type of) website, design or data set. The script is available on GitHub[6]. In order to meet the specified requirements, our approach has the following properties:

- All information regarding data selection, article, section and document composition and rendering are expressed in RDF which facilitates cross-site content sharing and alignment of settings with other RDF-Based systems;
- It is *declarative*, meaning that all knowledge with respect to how document composition should take place is contained in data, rather than functions;
- It provides several *abstraction layers* that are essential to perform proper content management, such as the composition of data into logical chunks of content, i.e. articles and sections, that can be reused across multiple pages and sites. The rendering of these chunks forms another abstraction layer;

[6] https://github.com/data2documents/reference-implementation-php.

- It has a *modular* setup which allows documents to be composed by choosing from various 'article definitions' that define which properties to use from a given RDF resource, which in turn can be rendered by choosing from various 'render definitions' that match the particular 'article definition';
- It is *distributable* as all content and settings are dereferenceable linked data;
- It defines a clear relation between selected data properties and semantic elements in HTML5;
- It provides a declarative template solution to facilitate web page design;
- It poses no restrictions on document structure or layout;

3 The Data 2 Documents Vocabulary

At the heart of our approach lies the d2d vocabulary, that resides in the d2d namespace http://rdfns.org/d2d/. We provide several examples of its use at http://example.d2dsite.net, including site specific content as well as linked data from external sources. A basic 'Hello World' example is given and described in detail, as well as more elaborated examples. All source RDF and template files for all examples can be viewed using an online file browser and editor at http://example.d2dsite.net/browser/. To view the contents of a file, right-click it and choose 'Edit file'. Our project website also uses d2d, and can be found at http://www.data2documents.org.

The d2d vocabulary builds upon and extends notions of HTML5[7], namely the semantic sectioning of content and the separation of a *content layer* and *style layer*. We do so by adding two additional abstract layers: that of *re-usable content*, i.e. beyond a single document, and that of *re-usable rendering* for that content. Furthermore, the d2d vocabulary is aligned with semantic document elements in HTML5 such as 'Article' and 'Section'. According to the HTML5 specification, Sections and Articles can be nested. The difference between the two being that an Article is a 'self-contained' fragment of content that is "independently distributable or reusable, e.g. in syndication".[8] How that syndication is performed in practice, is out of scope for HTML5. Due to its foundation on RDF, this distribution and re-use can be achieved using d2d and dereferencing. When we use the terms 'article' or 'section', we refer to their meaning in HTML5 (Fig. 1).

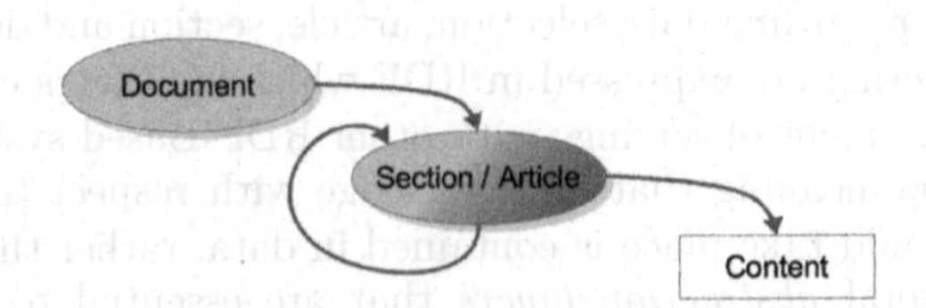

Fig. 1. A web document can be seen as a hierarchy of nested sections and articles

[7] http://www.w3.org/TR/html5-diff/#new-elements.
[8] http://www.w3.org/TR/html5/#the-article-element, retrieved 12 July 2016.

The main design choices for the d2d vocabulary are based upon the requirements discussed above and the aspects that are needed for performing web content management based on RDF. In this setup, each RDF resource can potentially be used as the source for an article or section in a web document. To facilitate this, the main tasks of the d2d vocabulary are to:

- Model the knowledge involved in selecting properties of a given RDF resource that is used as article or section;
- Model the knowledge involved in rendering the selected data properties in HTML, i.e. how to couple selected properties to HTML5 semantics;
- Model the knowledge involved in the creation of 'documents', i.e. document specific properties and the composition of articles and sections;
- Model all the above knowledge in such a way that it can be shared and re-used as small modules for specific RDF resources, articles, sections, etc.

To accomplish these tasks, the d2d vocabulary can be used to define 'Article Definitions', 'Render Definitions' and 'Documents', which are all dereferenceable RDF resources. Actual content that is to be used in the document can also be retrieved using dereferencing, as well as by using SPARQL queries (Fig. 2).

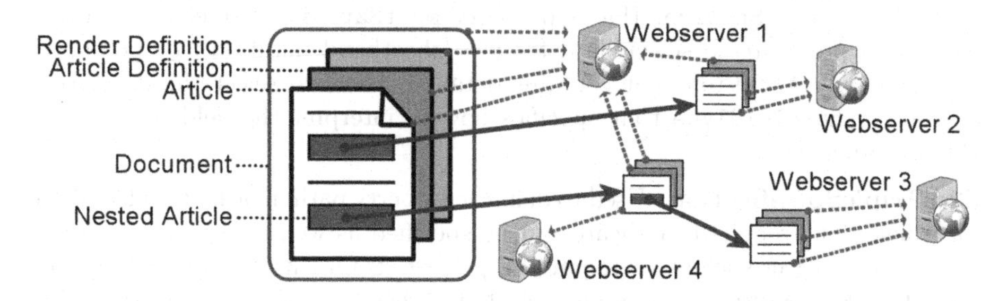

Fig. 2. Symbolic overview of the structuring of a d2d web document. The document contains one root Article, which can potentially be any RDF resource. Which properties form that resource are selected for the article is determined by the Article Definition, and how the article is rendered in HTML5 is determined by the Render Definition. The root article contains nested articles, for which separate definitions are specified. The content as well as the definitions can reside on different web servers, e.g. a content provider can provide them along with the data, or an alternative render definition can be created for a third party article type.

3.1 Main Elements of the D2D Vocabulary

Key concepts of the vocabulary are `Document` and `Section`. `Document` refers to the web document as a whole and consists of a hierarchy of nested `Sections`. One `Section` -or more precisely an `Article` which is defined as a subclass of `Section`- is the root of this hierarchy. Each `Section` contains one or more `Fields` which are small fragments of content that together make up the `Section`. How many `Fields` a `Section` or `Article` has, and of which kind, is specified by a

Section Definition that bundles multiple Field Specifications. How that Section or Article is rendered in HTML is specified by a Render Definition. Below is a description of the most important elements of the d2d vocabulary. A more detailed description can be found at our project website[9].

d2d:Document - Represents the web document as a whole, specifying the main render definitions to use and general document properties such as title and various meta fields. Furthermore it indicates the root article to use.

d2d:Section - Refers to a fragment of content that can be part of a web document. However, in many cases sections are not defined explicitly as external resources may be used to provide content for the section. Instead, within d2d, one can indicate that a selected resource is to be *used* as such within the Field Specification that selects it. d2d:Section has subclasses with additional semantics such as d2d:Article; a self-contained section.

d2d:SectionDefinition - Defines which properties related to a given resource should be used as content for a Section, by bundling a number of Field Specifications. Indicates the RDF classes that it fits to, i.e. the classes that can be used to act as the data source for a Section.

d2d:FieldSpecification - Specifies how data should be selected for a field that is part of a Section. Has a property mustSatisfy that either directly specifies a predicate to select data for use (shorthand notation), or points to a TripleSpecification that contains details on how to select the data. Has a property hasFieldType that specifies how to interpret the field in terms of HTML5 semantics.

d2d:TripleSpecification - Used to define a property path in order to select data for a field. The required predicate can be specified as well as details regarding the selected object such as its required type (e.g. xsd:String or foaf:Person) and role. The role determines how the selected object is used, e.g. as content for the field, as sort key, as SPARQL query, as query endpoint, or as a preferred d2d:SectionDefinition or d2d:RenderDefinition for a nested Section.

d2d:RenderDefinition - Defines how a Section should be rendered. Has a property hasTemplate that can either point to a file containing an HTML5 (sub) template, or a literal holding the actual template. Render definitions are optional; if not specified, fields are rendered in an HTML5 element according to their field type, while additional styling can be applied using CSS.

3.2 Interpretation of the Vocabulary

To provide insight in the vocabulary, we describe the interpretation of a document while referring to the capital letters in Fig. 3. The d2d document resource specifies an RDF resource that is to be used as the root Article (A) for the document through the d2d:hasArticle property. It also specifies one or more 'preferred' Section Definitions that define how specified RDF resources should

[9] http://www.data2documents.org/documentation#vocabulary.

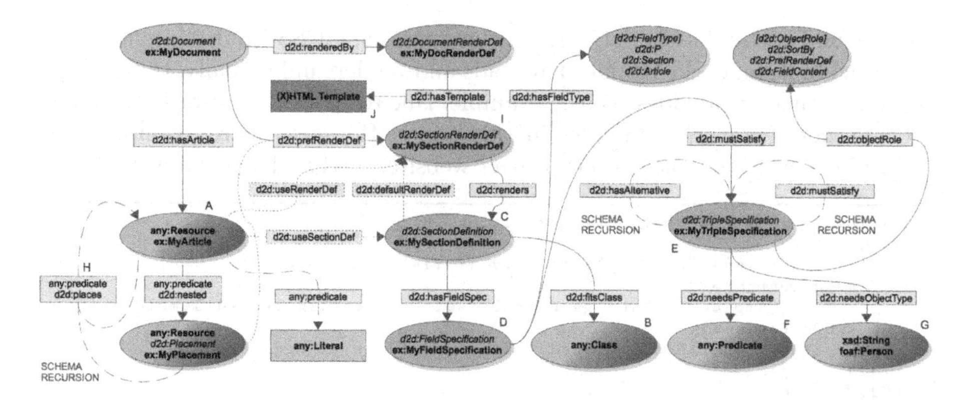

Fig. 3. Schematic overview of how the Data 2 Documents vocabulary is used to select content. Ovals represent RDF resource instances, with their class types indicated in italics. The orange ovals indicate resources that are always specifically created for a d2d web document; hence only the `Document` class itself. The orange/green ovals indicate resources that could be either specifically created for the web document, or be resources in an existing Linked Data set not specifically created for the document. The blue ovals are resources that are instances of d2d classes such as the `Section` and `Render Definitions`. Lastly, the purple ovals are classes from the vocabulary itself that are used to indicate the desired role or field type. A high resolution version of this figure is also available at http://www.data2documents.org/ppre/d2d-schema-0916.png. (Color figure online)

be used as `Article` or `Section`, i.e. which properties should be selected. Exactly which `Section Definition` matches the class of the particular RDF resource is specified through the `d2d:fitsClass` property (B). The matching `Section Definition` (C) contains one ore more `Field Specifications` (D) that specify a particular field for the article. How data for that field should be selected is indicated by the `Triple Specification` (E) that specifies the predicate (F) that the resource *acting as article* should have in order to satisfy the particular field. Optionally, the required type of object for that predicate can be specified (G), and by chaining several `Triple Specifications` a longer property path (H) and/or alternatives can be specified. The data that is to be selected can be a literal, e.g. for a paragraph of text, or yet another resource that is to be processed as a *nested* `Section` or `Article`. Finally, a matching `Render Definition` (I) defines how to render the selected data, optionally using a (sub)template (J) specified by the `Render Definition`. These steps are repeated recursively for each nested `Section` or `Article`, until the complete documents has emerged.

3.3 Declarative Template Solution

In order to gain more control over document rendering, d2d includes a declarative template solution. Specifying templates is optional, as content is already associated with HTML5 semantic elements in the article definition and can be rendered on that basis, while further styling can be done using CSS. (Sub)templates can be nested in a single HTML file that loads and renders individually, to facilitate design. Listing 1.1 provides a basic example. Due to space limitations we cannot provide more extensive examples in this paper. However, more examples and descriptions can be found on our project website, section documentation[10].

```
<d2d:Template d2d:for={article definition IRI}>
  <d2d:Field>
    <!-- Conditional HTML; placed only if field is placed -->
      <d2d:Content>
        <!-- Sample content; gets replaced; Can contain nested Templates -->
      </d2d:Content>
    <!-- Conditional HTML; placed only if field is placed -->
  </d2d:Field>
</d2d:Template>
```

Listing 1.1. Example of d2d HTML template for an article with one field. If no conditional HTML or sample content is needed, just `<d2d:Field />` will suffice.

4 Related Work

Several aspects of our work relate to other scientific work that has been carried out in the past. These aspects include tasks such as the rendering and visualisation of Linked Data, the editing of semantic content, the selection of suitable data using a set of constraints and the use of Linked Data within content management systems.

Vocabularies. Though various vocabularies exist that could be relevant for web content management such as VoID [1] for data set description, PROV [8,15] for describing the provenance of document sections and SHACL[11] for describing and validating RDF Graphs, to the best of our knowledge, no vocabulary exists to the describe the knowledge required to perform actual web content management.

Web Feed Formats. RSS 1[12], 2[13] and Atom[14] are prominent ways of sharing and re-using content across web sites; a process called syndication. Though they can be extended with terms from other name spaces, in practise this only works well for specific domains, e.g. Podcasts[15]. This is due to the fact that many general purpose parsers don't know what to do with such added fields an simply ignore them. In order to process any kind of field without the need of adding explicit support within implementations, a *meta model* is needed to describe how to interpret such fields in a specific context, as is the case in our approach.

Semantic Portals. Semantic MediaWiki [11] allows for semantic annotations to be made within unstructured web content, whereas OntoWiki [2] is form-based and organised as an 'information map' in which each semantic element is

[10] http://www.data2documents.org/documentation#templates.
[11] Shapes Constraint Language: https://www.w3.org/TR/shacl/.
[12] http://web.resource.org/rss/1.0/.
[13] http://www.rssboard.org/rss-specification.
[14] http://tools.ietf.org/html/rfc4287.
[15] RSS extended with 'itunes' elements: http://www.podcast411.com/page2.html.

represented as an editable node. They are tools for authoring semantic content, but are not build specifically to be used as general content management system and do not facilitate the live sharing of semantic content across sites.

Linked Data Rendering and Visualisation. Fresnel [18] is an OWL-based vocabulary that can be used to define *lenses* to select data from a given data graph and *formats* that define how that data should be displayed. Exhibit [12] is an AJAX based publishing framework for structured data that uses an internal data representation format and allows the data to be used in rich web interfaces using templates and pre-defined UI types such as faceted browsing. Callimachus [4] is a template based system that facilitates the management of Linked Data collections and the use of that data in web applications. Uduvudu [14] is a visualisation engine for Linked Data built in JavaScript that lets users describe recurring subgraphs in their data and indicate how these subgraphs should be visualised. Balloon Synopsis [19] is also an approach to include Linked Data in web publications running at the client side, implemented as a jQuery[16] plugin. When it comes to visualising Linked Data in general, Dadzie and Rowe [7] provide a survey of approaches. What makes our approach different is the separation of tasks such as the selection of data elements to form logical documents sections and the rendering of these sections into documents, set up in a way that is completely declarative, modular and distributable in order to facilitate the sharing and re-use of not only the actual content, but also the definitions of what data to select and how to render it in the document. RSLT [17] is a transformation language for RDF data that uses templates associated to resource properties.

Content Management Systems. Drupal has an RDF library to expose information as Linked Data through RDFa [6]. OntoWiki CMS [10] is an extension to OntoWiki that combines several other tools such as the OntoWiki Site Extension and Exhibit to facilitate the use of OntoWiki data in web documents and a dynamic syndication system. The Less template system [3] allows Linked Data that is accessed through dereferencing or SPARQL queries to be used in text-based output formats such as HTML or RSS. It can be used in collaboration with existing content management systems such as WordPress and Typo3[17], however it uses data property names directly in its templates thus provides no separation between data selection and data representation. These plugins and tools focus on the production and consumption of RDF Data within existing systems, but they do not exploit the intrinsic Linked Data properties to facilitate web content management, nor do they store information with respect to the actual content management such as document composition in RDF; it is maintained in implementation specific database systems, which limits interoperability.

5 Evaluation

To evaluate our approach, we performed two experiments: With the first one, we wanted to test the usability and usefulness for expert users, resembling potential

[16] http://jquery.com.
[17] http://typo3.org.

power-users of our system. For this we recruited users who are proficient in Linked Data and Web development techniques. With the second experiment, we wanted to test our approach on users who are technically minded but have limited knowledge and practical experience with Linked Data and Web development. These participants should resemble potential casual users of our system as well as beginners who are trying out these new technologies, but are not yet very familiar with them. For this second experiment, we recruited a larger group of students in Computer Science-related Master programs.

5.1 Design of the Experiments

Both experiments have the same basic structure: The participants first received a brief introduction into the basic concepts of the Data 2 Documents technique. This was done via a general presentation and demo of about 10 min. Then the participants received the detailed instructions, where they were first asked to register and login into a system were each participant was given a separate sub-domain and web server instance with an elementary online source code editor. Figure 5 shows a screenshot of that source code editor. In this editor, the participants found two directories: a read-only directory with a working example[18], and a writable directory that was initially empty and where the participants should create their own website based on the example website[19]. After completing the given tasks, the participants had to fill in a questionnaire asking them about their background, skills (Fig. 4), experiences during the experiment, and their opinions on d2d. Participants were given 9 tasks:[20]

- **Task 1:** Creating a new d2d document by copying from an example
- **Task 2:** Adding a missing introduction article to the new document
- **Task 3:** Changing the title of the introduction article
- **Task 4:** Linking and thereby including an existing FOAF profile as article
- **Task 5:** Creating a new *comment* article and including it in the document
- **Task 6:** Linking and thus including comment articles of other participants

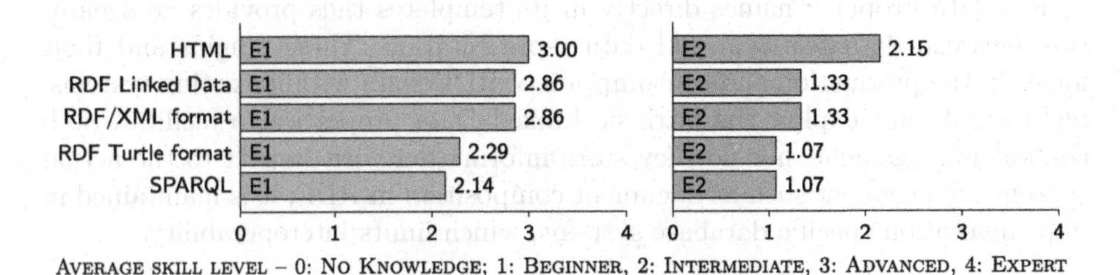

Fig. 4. The average skill level of the participants for experiments E1 and E2

[18] Example site: http://kmvuxx.biographynet.nl.
[19] Participant site after completing the evaluation: http://kmvu03.biographynet.nl.
[20] The detailed instructions can be found here: http://biographynet.nl/assignment/.

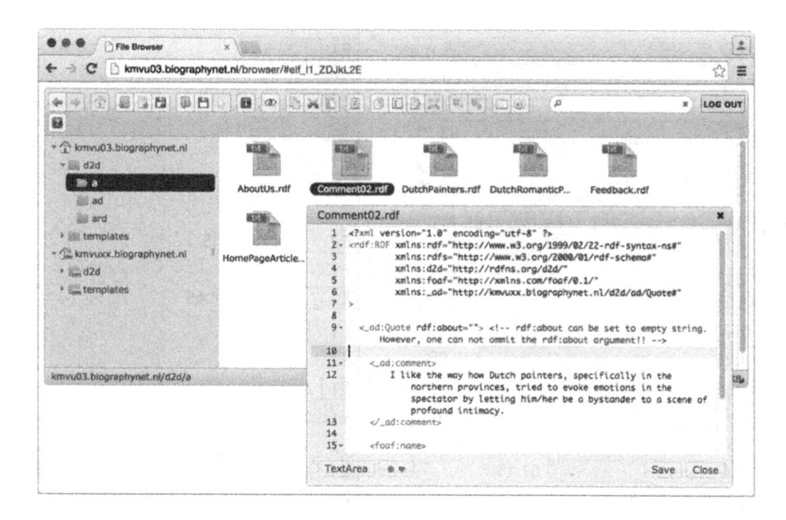

Fig. 5. The online browser and text editor that was used for the experiments. Though all tasks consisted only of elementary text editing, we had to provide a way of browsing and editing the text files, which reside on a web server.

- **Task 7:** Changing an existing listing to show Dutch Painters instead of Prime Ministers of The Netherlands
- **Task 8:** Extending the article definition, render definition and template to show an additional field (creation date) in a listing of paintings
- **Task 9 (optional):** Create a new listing of people using a DBpedia category

For the first experiment, we recruited among employees of VU University Amsterdam as well as their friends who work on general web development and/or Linked Data. The participants of the second experiment were Master students from the *Knowledge and Media* course that was given at the same university during fall 2015.

5.2 Results of the Experiments

For the first experiment, we managed to recruit a relatively small group of 7 participants, but for the second experiment we got a large group of 73 students. Two of the seven participants of the first experiment were female (29 %), which is almost the same rate as for the second experiment, for which 30 % were female. The average age was 35.3 years for the first experiment and 24.5 years for the second. The students of the second experiment were mostly enrolled in the Master programs *Information Science* (76.7 %) and *Artificial Intelligence* (16.4 %).

The left hand side of Fig. 6 shows the rates at which the different tasks were successfully completed by the participants of the two experiments, as revealed by inspecting the resulting files on the server instances after the experiment. These rates are very high at close to or over 85 % for all the tasks up to and including Task 5 for both experiments, and they did not go below 50 % except for the bonus Task 9 (40 %).

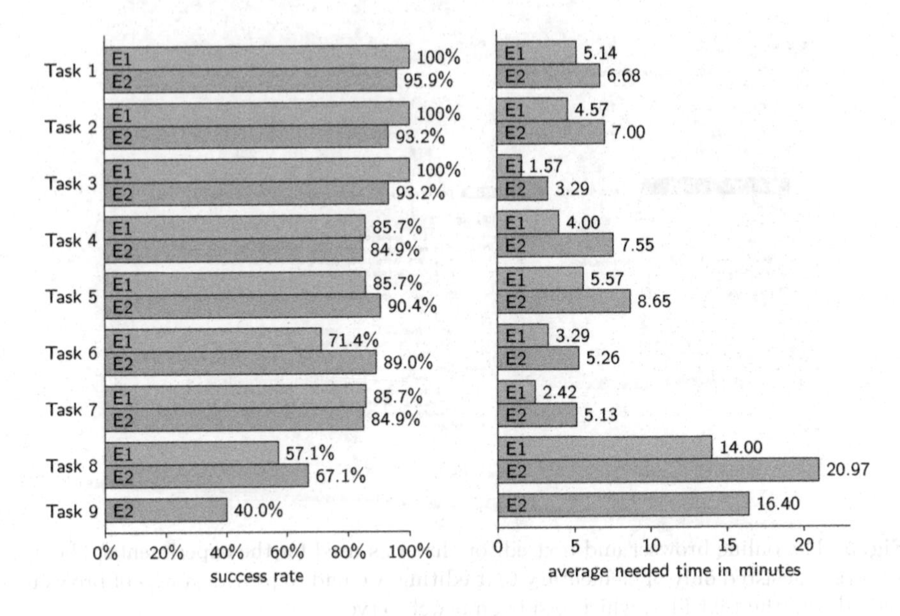

Fig. 6. Success rates (left) and average required time (right) for the different tasks of the two experiments.

For tasks 6 to 8 the success rates are — surprisingly — lower for the experts than for the students, considerably so for tasks 6 and 8. Both groups performed very well, but for some reason the experts were not as good in completing these tasks as the students, even though the latter have a lower skill level.

The solution to this puzzle is presented in the chart on the right hand side of Fig. 6, which plots the average amount of time (in minutes) the participants needed to complete the respective tasks. Over all tasks, the experts spent on average considerably less time than the students. Together with the data for the task success rates presented above, this seems to suggest that the experts did not try as hard as the students. Compared to the students, the experts seem to have been more interested in finishing the experiment in a relatively quick fashion, whereas the students seem to have been more committed to the tasks and were willing to spend more time to complete them.

In absolute terms, the numbers of both settings look very promising. Even the students needed on average only a bit more than one hour to complete the tasks 1 to 8. Taking into account that building or adjusting a website with dynamic and complex content is inherently a challenging task and that users will become more efficient over time when using the same tool, these results seem to indicate that our approach is indeed useful and appropriate.

Table 1 shows an aggregation of the responses of the participants to seven statements (S1–S7) in the questionnaire. They were asked whether or not they agree with these statements on a scale from strongly agree (1) to strongly disagree (5). These statements ask about the participants' opinions with respect

Table 1. Answers from the participants of the two experiments about the degree to which they agree with the given statements about d2d. (* = significant)

Experiment:	E1	E2							
Statement	Avg.	←agree disagree→					Avg.	strong/weak agree (≤ 2)	not disagree (≤ 3)
		1	2	3	4	5		p-value	p-value
S1: "d2d seems to be a suitable approach to perform general Web Content Management such as the creation, sharing and placing of content articles"	2.14	17	40	9	6	1	2.10	* $<10^{-6}$	* $<10^{-12}$
S2: "d2d seems to be a suitable approach to eliminate the traditional boundaries for Content Management between separate web sites, documents, and domains"	1.57	24	29	13	5	2	2.07	* $<10^{-4}$	* $<10^{-12}$
S3: "d2d makes it easy to share content between separate web sites/documents/domains"	1.43	28	22	16	5	2	2.05	* 0.0011	* $<10^{-12}$
S4: "d2d seems to be a suitable approach to use Linked Data in web documents"	1.29	29	29	7	8	0	1.92	* $<10^{-6}$	* $<10^{-11}$
S5: "Manually editing d2d definitions is not significantly harder to do than manually editing HTML"	2.29	25	15	18	13	2	2.34	0.2414	* $<10^{-6}$
S6: "I would consider using d2d, if I have to develop a general website in the future"	3.29	11	22	24	11	5	2.68	0.8254	* $<10^{-6}$
S7: "I would consider using d2d, if I have to develop a website in the future that makes use of Linked Data"	1.71	25	28	9	9	2	2.11	* $<10^{-4}$	* $<10^{-9}$

to the suitability of the approach for different goals (S1–S4), how it compares to plain HTML documents (S5), and whether they would consider using the framework for themselves in the future (S6 and S7). Due to the smaller number of participants, only the average values are shown for experiment 1, but more details are given for the second experiment. All average values are on the *agree* side (which for all statements means in favor of d2d) with the exception of the response from the experts (experiment E1) on statement S6 (but they do strongly agree with the more specific statement S7).

For the second experiment with a larger number of participants, we can make some more detailed analyses. If we lump together *strong agree* and *weak agree* (i.e. ≤ 2), we get a majority of the responses in this area for all statements except S6. To test whether these majorities are not just a product of chance, we can run a statistical test. Our null hypothesis is that when users are asked to select between *strong/weak agree* (≤ 2) on the one hand, and any other option (i.e. *neutral* or *strong/weak disagree*: >2) on the other, they would tend towards the latter or at most have a 50 % chance of selecting *strong/weak agree*. We use a simple one-tailed exact binomial test to evaluate this hypothesis for each of the statements with the data from experiment 2. The results can be seen in Table 1 and they show that we can reject this null hypothesis for all statements except S5 and S6. We have therefore strong statistical reasons to assume that the tendency towards *strong/weak agree* for statements S1, S2, S3, S4, and S7 is not just the product of chance. In a next step, we can soften our previous null hypothesis a bit by adding *neutral* to the lumped-together area, and see whether users have a tendency towards saying *strong/weak agree or neutral* (i.e. ≤ 3). As shown in Table 1, this softer null hypothesis can be rejected for all statements. All data collected and additional charts are available on our project website[21].

6 Conclusions and Future Work

In this paper we described Data 2 Documents (d2d): A vocabulary for doing content management in a declarative fashion, expressed in RDF. The vocabulary is accompanied by a reference implementation that interprets it to create rich web documents. Our results show that participants do not disagree that manually editing d2d definitions is not significantly harder to do than manually editing HTML. We think that this is an impressive result if we consider how much more powerful d2d is, compared to plain HTML. Moreover, as S7 shows, a majority of users would consider using d2d in the future to develop websites making use of Linked Data.

Here, we described Data 2 Documents in its fundamental form, having all content data and definitions as editable XML/RDF files. As future work we plan to develop a sophisticated GUI for working with d2d. But we can do so with the assurance that users are able to fall back on elementary editing skill should it be required. We also plan to port our reference implementation to SWI Prolog[22] for large scale applications to run directly on the ClioPatria[23] semantic web server. A port to Javascript is also planned to browse d2d web documents directly on the client side by requesting the raw RDF data and in order to include d2d defined articles in non-d2d web documents.

We see Data 2 Documents as a first step to bring together the largely separated networks of documents (the Web) and data (Linked Data), for Web users to benefit from the increasing amount of structured data. As such, we think that

[21] http://www.data2documents.org/#evaluation.
[22] http://www.swi-prolog.org.
[23] http://cliopatria.swi-prolog.org.

our approach might form an important step to finally make the vision of the Semantic Web a reality.

Acknowledgments. This work was supported by the BiographyNet(http://www.biographynet.nl) project (Nr. 660.011.308), funded by the Netherlands eScience Center(http://esciencecenter.nl). Partners in this project are the Netherlands eScience Center, the Huygens/ING Institute of the Royal Dutch Academy of Sciences and VU University Amsterdam.

References

1. Alexander, K., Hausenblas, M.: Describing linked datasets-on the design and usage of void, the vocabulary of interlinked datasets. In: Linked Data on the Web Workshop (LDOW 2009). Citeseer (2009)
2. Auer, S., Dietzold, S., Riechert, T.: OntoWiki – a tool for social, semantic collaboration. In: Cruz, I., Decker, S., Allemang, D., Preist, C., Schwabe, D., Mika, P., Uschold, M., Aroyo, L.M. (eds.) ISWC 2006. LNCS, vol. 4273, pp. 736–749. Springer, Heidelberg (2006)
3. Auer, S., Dietzold, S., Riechert, T.: OntoWiki – a tool for social, semantic collaboration. In: Cruz, I., Decker, S., Allemang, D., Preist, C., Schwabe, D., Mika, P., Uschold, M., Aroyo, L.M. (eds.) ISWC 2006. LNCS, vol. 4273, pp. 736–749. Springer, Heidelberg (2006). doi:10.1007/11926078_53
4. Battle, S., Wood, D., Leigh, J., Ruth, L.: The callimachus project: RDFa as a web template language. In: COLD (2012)
5. Bizer, C., Heath, T., Berners-Lee, T.: Linked data - the story so far. Int. J. Semant. Web Inf. Syst. **5**(3), 1–22 (2009)
6. Corlosquet, S., Delbru, R., Clark, T., Polleres, A., Decker, S.: Produce and consume linked data with Drupal!. In: Bernstein, A., Karger, D.R., Heath, T., Feigenbaum, L., Maynard, D., Motta, E., Thirunarayan, K. (eds.) ISWC 2009. LNCS, vol. 5823, pp. 763–778. Springer, Heidelberg (2009)
7. Dadzie, A.S., Rowe, M.: Approaches to visualising linked data: a survey. Semant. Web **2**(2), 89–124 (2011)
8. Groth, P., Gil, Y., Cheney, J., Miles, S.: Requirements for provenance on the web. Int. J. Digit. Curation **7**(1), 39–56 (2012)
9. Guha, R.: Meta content framework. Research report Apple Computer, Englewoods, NJ (1997). http://www.guha.com/mcf/wp.html
10. Heino, N., Tramp, S., Auer, S.: Managing web content using linked data principles-combining semantic structure with dynamic content syndication. In: 2011 IEEE 35th Annual Computer Software and Applications Conference (COMPSAC), pp. 245–250. IEEE (2011)
11. Herzig, D.M., Ell, B.: Semantic MediaWiki in operation: experiences with building a semantic portal. In: Patel-Schneider, P.F., Pan, Y., Hitzler, P., Mika, P., Zhang, L., Pan, J.Z., Horrocks, I., Glimm, B. (eds.) ISWC 2010, Part II. LNCS, vol. 6497, pp. 114–128. Springer, Heidelberg (2010)
12. Huynh, D.F., Karger, D.R., Miller, R.C.: Exhibit: lightweight structured data publishing. In: Proceedings of the 16th International Conference on World Wide Web, pp. 737–746. ACM (2007)
13. Lassila, O., Swick, R.R.: Resource description framework (RDF) model and syntax specification. Technical report (1999). http://www.w3.org/TR/REC-rdf-syntax/

14. Luggen, M., Gschwend, A., Anrig, B., Cudré-Mauroux, P.: Uduvudu: a graph-aware and adaptive UI engine for linked data. In: Proceeding of the LDOW (2015)
15. Moreau, L., Missier, P., Belhajjame, K., B'Far, R., Cheney, J., Coppens, S., Cresswell, S., Gil, Y., Groth, P., Klyne, G., Lebo, T., McCusker, J., Miles, S., Myers, J., Sahoo, S., Tilmes, C.: PROV-DM: the PROV data model. Technical report, W3C (2012). http://www.w3.org/TR/prov-dm/
16. Ockeloen, N.: RDF based management, syndication and aggregation of web content. In: Lambrix, P., Hyvönen, E., Blomqvist, E., Presutti, V., Qi, G., Sattler, U., Ding, Y., Ghidini, C. (eds.) EKAW 2014. LNCS (LNAI), vol. 8982, pp. 218–224. Springer, Heidelberg (2015). doi:10.1007/978-3-319-17966-7_31
17. Peroni, S., Vitali, F.: Templating the semantic web via RSLT. In: Gandon, F., Guéret, C., Villata, S., Breslin, J., Faron-Zucker, C., Zimmermann, A. (eds.) ESWC 2015. LNCS, vol. 9341, pp. 183–189. Springer, Heidelberg (2015). doi:10.1007/978-3-319-25639-9_35
18. Pietriga, E., Bizer, C., Karger, D.R., Lee, R.: Fresnel: a browser-independent presentation vocabulary for RDF. In: Cruz, I., Decker, S., Allemang, D., Preist, C., Schwabe, D., Mika, P., Uschold, M., Aroyo, L.M. (eds.) ISWC 2006. LNCS, vol. 4273, pp. 158–171. Springer, Heidelberg (2006)
19. Schlegel, K., Weißgerber, T., Stegmaier, F., Granitzer, M., Kosch, H.: Balloon synopsis: a jQuery plugin to easily integrate the semantic web in a website. In: CEUR Workshop Proceedings, vol. 1268 (2014)
20. Schmachtenberg, M., Bizer, C., Paulheim, H.: Adoption of the linked data best practices in different topical domains. In: Mika, P., Tudorache, T., Bernstein, A., Welty, C., Knoblock, C., Vrandečić, D., Groth, P., Noy, N., Janowicz, K., Goble, C. (eds.) ISWC 2014, Part I. LNCS, vol. 8796, pp. 245–260. Springer, Heidelberg (2014)

TechMiner: Extracting Technologies from Academic Publications

Francesco Osborne[1(✉)], Hélène de Ribaupierre[1,2], and Enrico Motta[1,2]

[1] Knowledge Media Institute, The Open University, Milton Keynes, UK
{francesco.osborne, enrico.motta}@open.ac.uk
[2] Department of Computer Science, University of Oxford, Oxford, UK
helene.de.ribaupierre@oxford.com

Abstract. In recent years we have seen the emergence of a variety of scholarly datasets. Typically these capture 'standard' scholarly entities and their connections, such as authors, affiliations, venues, publications, citations, and others. However, as the repositories grow and the technology improves, researchers are adding new entities to these repositories to develop a richer model of the scholarly domain. In this paper, we introduce TechMiner, a new approach, which combines NLP, machine learning and semantic technologies, for mining technologies from research publications and generating an OWL ontology describing their relationships with other research entities. The resulting knowledge base can support a number of tasks, such as: richer semantic search, which can exploit the technology dimension to support better retrieval of publications; richer expert search; monitoring the emergence and impact of new technologies, both within and across scientific fields; studying the scholarly dynamics associated with the emergence of new technologies; and others. TechMiner was evaluated on a manually annotated gold standard and the results indicate that it significantly outperforms alternative NLP approaches and that its semantic features improve performance significantly with respect to both recall and precision.

Keywords: Scholarly data · Ontology learning · Bibliographic data · Scholarly ontologies · Data mining

1 Introduction

Exploring, classifying and extracting information from scholarly resources is a complex and interesting challenge. The resulting knowledge base could in fact bring game-changing advantages to a variety of fields: linking more effectively research and industry, supporting researchers' work, fostering cross pollination of ideas and methods across different areas, driving research policies, and acting as a source of information for a variety of applications.

However, this knowledge is not easy to navigate and to process, since most publications are not in machine-readable format and are sometimes poorly classified. It is thus imperative to be able to translate the information contained in them in a free, open

E. Blomqvist et al. (Eds.): EKAW 2016, LNAI 10024, pp. 463–479, 2016.
DOI: 10.1007/978-3-319-49004-5_30

and machine-readable knowledge graph. Semantic Web technologies are the natural choice to represent this information and in recent years we have seen the development of many ontologies to describe scholarly data (e.g., SWRC[1], BIBO[2], PROV-O[3], AKT[4]) as well as bibliographic repositories in RDF [1–3]. However, these datasets capture mainly 'standard' research entities and their connections, such as authors, affiliations, venues, publications, citations, and others. Hence, in recent years there have also been a number of efforts, which have focused on extracting additional entities from scholarly contents. These approaches have focused especially on the biomedical field and address mainly the identification of scientific artefacts (e.g., genes [4], chemical components [5]) and epistemological concepts [6–8] (e.g., hypothesis, motivation, experiments). At the same time, the Linked Open Data cloud has emerged as an important knowledge base for supporting these methods [9–11].

In this paper, we contribute to this endeavour by focusing on the extraction of technologies, and in particular applications, systems, languages and formats in the Computer Science field. In fact, while technologies are an essential part of the Computer Science ecosystem, we still lack a comprehensive knowledge base describing them. Current solutions cover just a little part of the set of technologies presented in the literature. For example, DBpedia [12] includes only well-known technologies which address the Wikipedia notability guidelines, while the Resource Identification Initiative portal [13] contains mainly technologies from PubMed that were manually annotated by curators. Moreover, the technologies that are described by these knowledge bases are scarcely linked to other research entities (e.g., authors, topics, publications). For instance, DBpedia often uses relations such as *dbp:genre* and *dct:subject* to link technologies to related topics, but the quality of these links varies a lot and the topics are usually high-level. Nonetheless, identifying semantic relationships between technologies and other research entities could open a number of interesting possibilities, such as: richer semantic search, which can exploit the technology dimension to support better retrieval of publications; richer expert search; monitoring the emergence and impact of new technologies, both within and across scientific fields; studying the scholarly dynamics associated with the emergence of new technologies; and others. It can also support companies in the field of innovation brokering [14] and initiatives for encouraging software citations across disciplines such as the FORCE11 Software Citation Working Group[5].

To address these issues, we have developed TechMiner (TM), a new approach which combines natural language processing (NLP), machine learning and semantic technologies to identify software technologies from research publications. In the resulting OWL representation, each technology is linked to a number of related research entities, such as the authors who introduced it and the relevant topics.

[1] http://ontoware.org/swrc/.

[2] http://bibliontology.com.

[3] https://www.w3.org/TR/prov-o/.

[4] http://www.aktors.org/publications/ontology.

[5] https://www.force11.org/group/software-citation-working-group.

We evaluated TM on a manually annotated gold standard of 548 publications and 539 technologies and found that it improves significantly both precision and recall over alternative NLP approaches. In particular, the proposed semantic features significantly improve both recall and precision.

The rest of the paper is organized as follows. In Sect. 2, we describe the TechMiner approach. In Sect. 3 we evaluate the approach versus a number of alternative methods and in Sect. 4 we present the most significant related work. In Sect. 5 we summarize the main conclusions and outline future directions of research.

2 TechMiner

The TechMiner (TM) approach was created for automatically identifying technologies from a corpus of metadata about research publications and describing them semantically. It takes as input the IDs, the titles and the abstracts of a number of research papers in the Scopus dataset[6] and a variety of knowledge bases (DBpedia [12], WordNet [15], the Klink-2 Computer Science ontology [16], and others) and returns an OWL ontology describing a number of technologies and their related research entities. These include: (1) the authors who most published on it, (2) related research areas, (3) the publications in which they appear, and, optionally, (4) the team of authors who introduced the technology and (5) the URI of the related DBpedia entity. The input is usually composed by a set of publications about a certain topic (e.g., Semantic Web, Machine Learning), to retrieve all technologies in that field. However, TM can be used on any set of publications.

We use abstracts rather than the full text of publications because we wanted to test the value of the approach on a significant but manageable corpus; in particular, one for which a gold standard could be created with limited resources. In addition, a preliminary analysis revealed that publications which introduce new technologies, a key target of our approach, typically mention them in the abstract.

Figure 1 illustrates the architecture of the system, shows the adopted knowledge bases and lists the features that will be used by the classifier to detect if a candidate is a valid technology. The TM approach follows these steps:

- *Candidate Selection* (Sect. 2.2). TM applies NLP techniques to extract from the abstracts a set of candidate technologies.
- *Candidate Expansion* (Sect. 2.3). It expands the set of candidate technologies by including all the candidates discovered on different input datasets during previous runs which are linked to at least one of the input publications.
- *Publication Expansion* (Sect. 2.4). It expands the set of publications linked to each candidate technology, using the candidate label and the research areas relevant to the associated publications.
- *Candidate Linking* (Sect. 2.5). It applies statistical techniques to link each candidate to its related topics, authors and DBpedia entities.

[6] https://www.elsevier.com/solutions/scopus.

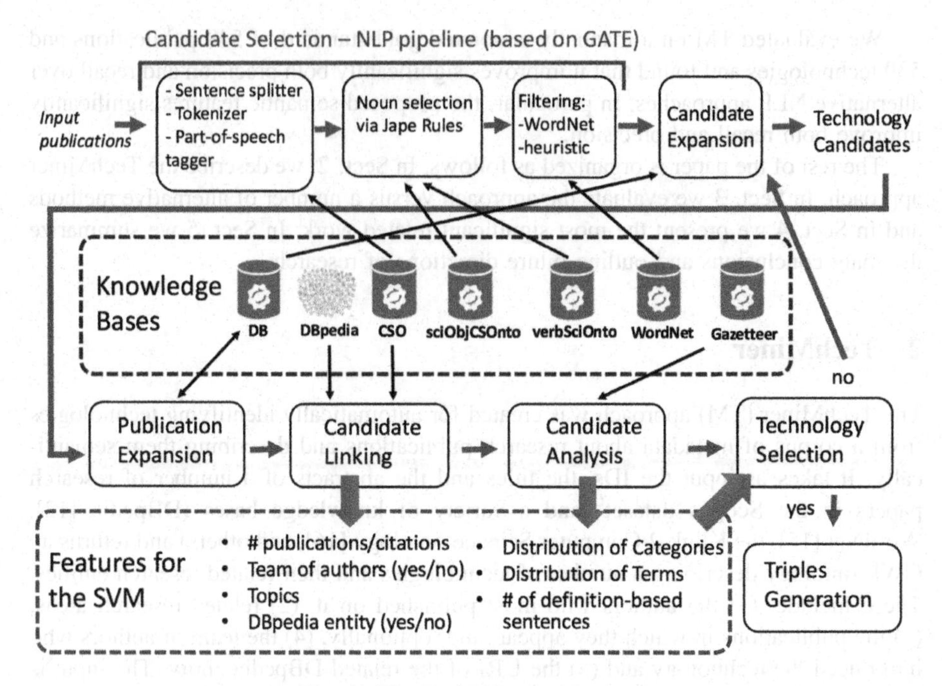

Fig. 1. The TechMiner architecture.

- *Candidate Analysis* (Sect. 2.6). It analyses the sentences in which the candidates appear and derives a weighted distribution of categories and terms.
- *Technology Selection* (Sect. 2.7). It applies a support vector machine classifier for identifying valid technologies. If a candidate is not classified as a technology, TM returns to the Candidate Expansion phase, tries to further expand the set of publications linked to the candidate technology and repeats the analysis.
- *Triple Generations* (Sect. 2.8). It produces the OWL ontology describing the inferred technologies by means of their characteristics and related entities.

In the next sections we shall discuss the background data and each step in details[7].

2.1 Background Data

For supporting the technology extraction task we manually crafted two ontologies: sciObjCSOnto[8] and verbSciOnto[9]. The first was derived from sciObjOnto[10] [17] and defines a number of categories of scientific objects in the Computer Science field and

[7] The ontologies, the JAPE rules and all the materials used for the evaluation is available at http://technologies.kmi.open.ac.uk/rexplore/ekaw2016/techminer/.

[8] http://cui.unige.ch/~deribauh/Ontologies/sciObjCS.owl.

[9] http://cui.unige.ch/~deribauh/Ontologies/verbSciOnto.owl.

[10] http://cui.unige.ch/~deribauh/Ontologies/scientificObject.owl.

their related terms. It contains 47 classes/individuals, and 64 logical axioms and covers concepts such as: algorithm, application, software, implementation, model, approach and prototype. The verbSciOnto ontology was created to represent the verbs usually adopted for describing technologies (e.g., "describe", "develop", "implement"). It contains 26 classes and 67 individuals and 89 logical axioms. Each verb is described with its infinitive, past and present form.

In addition, TM exploits DBpedia, WordNet and the Klink-2 Computer Science Ontology. DBpedia is a well-known knowledge base, which derives from a community effort to extract structured information from Wikipedia and to make this information accessible on the Web. TM uses it to find entities associated to the candidate technologies, with the aim of yielding additional information for the technology extraction process. WordNet[11] is a large lexical database of the English language created by the Princeton University, and is widely used in the NLP field. TM exploits it to filter out generic nouns from the set of candidate technologies.

The Klink-2 Computer Science Ontology (CSO) is a very large ontology of Computer Science that was created by running the Klink-2 algorithm [16] on about 16 million publications in the field of Computer Science extracted from the Scopus repository. The Klink-2 algorithm combines semantic technologies, machine learning and external sources to generate a fully populated ontology of research areas. It was built to support the Rexplore system [18] and to enhance semantically a number of analytics and data mining algorithms. The current version of the CSO ontology includes 17,000 concepts and about 70,000 semantic relationships. The CSO data model[12] is an extension of the BIBO ontology, which in turn builds on the SKOS model[13]. It includes three semantic relationships: *relatedEquivalent*, which indicates that two topics can be treated as equivalent for the purpose of exploring research data (e.g., Ontology Matching, Ontology Mapping), *skos:broaderGeneric*, which indicates that a topic is a sub-area of another one (e.g., Linked Data, Semantic Web), and *contributesTo*, which indicates that the research output of a topic contributes to another (e.g., Ontology Engineering, Semantic Web).

2.2 Candidate Selection

The aim of this first step is to identify a set of candidate technologies from an initial set of publications. To this end, TM processes the text of the abstracts by means of GATE[14], a well-known open source NLP platform, and a number of GATE plugins: OWLIM2, a module for importing ontologies, ANNIE, a component that forms a pipeline composed of a tokenizer, a gazetteer, a sentence splitter and a part-of-speech tagger, and JAPE (Java Annotation Patterns Engine), a grammar language for operating over annotations based on regular expressions.

---

[11] https://wordnet.princeton.edu/wordnet/.

[12] http://technologies.kmi.open.ac.uk/rexplore/ontologies/BiboExtension.owl.

[13] http://www.w3.org/2004/02/skos/.

[14] https://gate.ac.uk/.

The TM approach for identifying the set of candidates performs the following steps: (1) it splits the abstracts into sequences of tokens and assigns them part-of-speech tags (e.g., noun, verb and adverb) using ANNIE; (2) it selects technology candidates from sentences which contain a number of clue terms defined in the sciObjCSOnto ontology (e.g., “algorithm”, “tools”, “API”) and verbs from the verbSciOnto ontology (e.g., “implement”, “create”, “define”) by applying a sequence of JAPE rules; (3) it filters the candidates by exploiting a number of heuristics.

A manual analysis on a variety of sentences about technologies revealed that the technology name can be a proper noun, a common noun or a compound noun, and not necessarily the subject or the object in the sentence. However, sentences about technologies are usually associated with a certain set of verbs and terms. For example, in the sentence: “DAML + OIL is an ontology language specifically designed for its use in the Web” the position of the noun “DAML + OIL” followed by the clue term “language” and subject of “is a”, suggests that DAML + OIL may be the name of a technology.

To identify similar occurrences, TM first uses 6 manually defined JAPE rules to detect a list of candidate nouns or compound nouns which cannot be authors, venues, journals or research topics. It then applies another set of 18 JAPE rules for identifying the sentences that contain both these candidate nouns and the clue terms from the sciObjCSOnto and verbSciOnto ontologies and for extracting the names of candidate technologies.

The rules were created following the methodology introduced in [17, 19] to construct JAPE rules from annotated examples. This approach clusters sentences that have similarities in the sequence of deterministic terms (e.g., terms and verbs described in the ontologies), then replaces these terms with either a JAPE macro or an ontology concept. Non-deterministic terms are instead replaced by a sequence of optional tokens. In this instance, the rules were generated using examples from a dataset of 300 manually annotated publications from Microsoft Academic Search [19]. To improve the recall, we also created some additional JAPE rules to select also nouns that are not associated with any cue terms, but contain a number of syntactic grammatical patterns usually associated with the introduction of technologies.

The resulting candidates are then filtered using the following heuristics. We use WordNet to exclude common names by checking the number of synsets associated to each term contained in a candidate technology. A candidate associated with more than two synsets is considered a general term and gets discarded. However, we took in consideration some relevant exceptions. A preliminary analysis revealed in fact that a large number of technologies in the field of Computer Science are named after common nouns that belong to one or several categories of the Lexicographer Files of WordNet, such as animals (e.g., OWL, Magpie), artefacts (e.g., Crystal, Fedora) and food (e.g., Saffron, Java). Therefore, TM does not exclude the terms in these categories. In addition, we implemented two other heuristics. The first one checks if the term is capitalized or contains uppercase letters (e.g., Magpie, OIL, ebXML) and if so it preserves it even if WordNet suggests that it is a common name. The second one checks the terms that contain hyphens or underscore symbols. If both parts of the term are lower-case (e.g., task-based), they will be analysed separately by the WordNet heuristic, otherwise (e.g., OWL-DL, OWL-s) they will be considered as one word. The current prototype is able to process about 10,000 abstracts in one hour.

2.3 Candidate Expansion

The result of the previous phase is a set of candidate names linked to the publications from which they were extracted. However, the JAPE rules may have failed to recognize some valid technology which is actually mentioned in one of the input papers. Nonetheless, the same technology may have been recognized in previous runs on a different set of initial papers. This happens frequently when examining datasets in different fields. For example, the application "Protégé", may not be recognized when running on a Machine Learning dataset, since the few papers that would mention it may not have triggered the JAPE rules. However, if we already identified "Protégé" by previously analysing a Semantic Web dataset, we can exploit this knowledge to identify the instances of Protégé also in the Machine Learning dataset.

Therefore, in the candidate expansion phase TM enriches the set of candidates by including the technologies discovered during previous runs which were linked to one of the current input papers. This solution takes more time and can introduce some noise in the data, but it is usually able to significantly improve recall without damaging precision too much. We will discuss pros and cons of this solution in the evaluation section.

2.4 Publication Expansion

In this phase, we still may have missed a number of links between candidates and publications. In fact, the full Scopus dataset may have many other publications, not included in the initial dataset, that refer to the candidate technologies. It is thus useful to expand the set of links to collect more data for the subsequent analysis. TM does so by linking to a candidate technology all the papers in the Scopus dataset that mention the candidate label in the title or in the abstract and address the same research area of the set of publications associated to the candidate by the JAPE rules. In fact, taking into account the research area in addition to the label is useful to reduce the risk of confusing different technologies labelled with the same name. TM determines the research areas by extracting the full list of topics associated to the initial papers and finding the lowest common super topic which covers at least 75 % of them according to the CSO ontology. For example, given a candidate technology such as "LODifier" [9], TM will analyse the distribution of topics relevant to the associated papers and may find that most of them are subsumed by the Semantic Web topic, it will then associate the candidate with all the papers that contain the label "LODifier" and are tagged with "Semantic Web" or with one of its sub areas according to CSO, such as "Linked Data" and "RDF".

Finally, the relationships between candidates and publications are saved in a knowledge base and can be used to enrich the set of candidates in the following runs. This process is naturally less accurate than the NLP pipeline and can introduce some incorrect links. However, as discussed in the evaluation, the overall effect is positive since the abundance of links discovered in this phase fosters significantly the statistical methods used in the next steps.

2.5 Candidate Linking

In this phase, TM applies a number of heuristics to link the candidate to related research entities. In particular, it tries to link the candidate with (1) the team of authors who appear to have introduced the technology, (2) related concepts in the CSO ontology, and (3) related entities in DBpedia. The presence and quality of these links will be used as features to decide if the candidate is a valid technology. For example, the fact that a candidate seems to have been introduced by a well-defined team of researchers and is associated to a cohesive group of topics is usually a positive signal.

The authors who first introduce a technology tend to have the highest number of publications about it in the debut year and to be cited for these initial publications. Hence, TM extracts the groups of authors associated to the candidate publications, merges the ones that share at least 50 % of the papers, discards the ones who did not publish in the debut year, and assigns to each of them a score according to the formula:

$$I_{score} = \sum\nolimits_{i=deb}^{cur} \frac{pub_i}{tot\_pub_i}(i+1-deb)^{-\gamma} + \sum\nolimits_{i=deb}^{cur} \frac{cit_i}{tot\_cit_i}(i+1-deb)^{-\gamma} \quad (1)$$

Here pub_i, cit_i, tot_pub_i, tot_cit_i are respectively the number of publications, citations, total publications (for all the papers associated to the candidate) and total citations in the i-th year; deb is the year of debut of the candidate; cur is the current year and γ is a constant > 0 that modules the importance of each year ($\gamma = 2$ in the prototype). Since raw citations follow a power law distribution, we use instead the ratio of publications and citations [20]. Finally, we select the team associated with the highest score, but only if this is at least 25 % higher than the second one. Therefore, only a portion of the technology candidates will be associated with an author's team. Its presence will be used as binary feature in the classification process.

To identify the significant topics, TM extracts the list of keywords associated to the publications and infers from them a set of research areas in the CSO ontology. It does so by retrieving the concepts with the same label as the terms and adding also all their super-areas (the technique is implemented in the Rexplore system and discussed comprehensively in [18]). For example, the term "SPARQL" will trigger the homonym concepts SPARQL and subsequently super-topics such as RDF, Linked Data, Semantic Web and so on.

Finally, TM tries to link the candidate object with entities on DBpedia. It extracts all the sentences in the abstracts and titles which contain the candidate label and annotates them using DBpedia Spotlight [21]. The entity which is associated with at least 25 % more instances than the others is selected as representative of the candidate. If this exists, TM links the candidate to this entity and saves the alternative names, the textual description in English (*dbo:abstract*), and a set of related entities via the *dct:subject* and *rdf:type* relations. The other entities annotated by DBpedia Spotlight will be used for the subsequent linguistic analysis.

2.6 Candidate Analysis

Intuitively a technology should be associated with a semantically consistent distribution of terms related to a specific context (e.g., "tool", "web browser", "plugin", "javascript"). Learning these linguistic signs can help to detect a valid technology. The papers retrieved during the paper expansion phases and the entities retrieved by DBpedia should thus contain a good number of these kinds of terms. Hence, TM scans (1) the abstracts of all related papers, (2) the labels of the entities annotated by DBpedia Spotlight, and, if it exists, (3) the abstract of the linked DBpedia entity and the labels of its related entities for significant terms in an automatically created gazetteer of keywords related to technologies. The gazetteer was built by tokenizing the sentences associated to the annotated technologies in the gold standard from [19] and extracting the terms that were less than 5 tokens away from the technology names. We then removed stop words and selected the most frequent terms from this distribution, ending up with a gazetteer of 500 terms.

TM searches for the significant terms using five different techniques: *(1) co-occurrence*, in which it checks whether the terms occur in the same sentence as the candidate; *(2) proximity-based*, in which it checks whether the terms appear five words before or after a candidate; *(3) definition-based*, in which it checks whether each term *t* appears as part of a definition linguistic pattern, such as 'X is a *t*' or '*t* such as X'; *(4) entity-based*, in which it checks whether the terms appear as part of a linked DBpedia entity; *(5) topic-based*, in which it checks whether the terms appear in the related concepts of the CSO ontology. The result of this process is a distribution of terms, in which each term is associated with the number of times it co-occurred with the candidate label according to the different techniques. We then augment semantically these distributions by including all the concepts from the sciObjCSOnto ontology and assigning to them the total score of the terms which co-occur the most with each concepts label. For example, the category 'application' will co-occur the most with terms such as 'applications', 'prototype', 'system' and so on; hence, it will be assigned the sum of their scores.

The resulting distribution and the information collected in the previous phases are then used as features for selecting the valid technologies from the candidate group.

2.7 Technology Selection

All information collected in the previous phases is then used by TM to decide whether a candidate is a valid technology, by applying a support vector machine (SVM) classifier (adopting a radial basis function kernel) on the set of features extracted in Sects. 2.4 and 2.5, representing both the linguistic signature of the associated papers and the related research entities. We take in consideration the following features (rescaled in the range [− 1, 1]): (i) number of publications and citations; (ii) the presence of an associated team of authors (Sect. 2.4, binary feature); (iii) number of linked research areas in the first, second and third level of the CSO ontology (Sect. 2.4); (iv) presence of a DBpedia entity with the same label (Sect. 2.4, binary feature); (v) distribution of related categories and terms considering each of them as a distinct feature (Sect. 2.5); (vi) number of definition-based sentences addressing the candidate and one of the technology related terms (Sect. 2.5).

When a candidate is classified as a technology, TM saves the related information and proceeds to analyse the next candidate, if it exists. When the candidate fails to be classified as a technology, TM tries to expand the candidate selection by using in the candidate expansion phase the super-topic of the previously high-level topic selected in the CSO. If there are multiple super topics, it selects the one associated with the lowest number of publications. For example, if the first topic was "Semantic Web", the new one will be "Semantics". The process ends when the candidate is classified as a technology, when the root 'Computer Science' is yielded, or after n failed attempts ($n = 2$ in the prototype). The current prototype processes about 2,500 candidate technologies in one hour, taking in account also the queries to external sources (e.g., DBpedia).

2.8 Triple Generation

In this phase, TM generates the triples describing the identified technologies by associating each technology with: (1) the related papers, (2) the number of publications and citations, (3) the team of authors who introduced the technology, (4) the main authors, i.e., the 20 authors with most publications about the technology, (5) the main topics, i.e., the 20 most frequent topics, (6) the categories from sciObjCSOnto (associated with their frequency) and, possibly, (7) the equivalent DBpedia entity.

The output is a fully populated ontology of the technologies identified in the input dataset. To this end, we crafted the TechMiner OWL ontology[15]. Our intention was not to create 'yet another ontology' of the scholarly domain, but to craft a simple scheme for representing our output. For this reason we reused concepts and relationships from a number of well-known scholarly ontologies (including FABIO [22], FOAF[16], CITO, SKOS, SRO[17], FRBR[18]) and introduced new entities and properties only when necessary. The main classes of the TechMiner OWL ontology are *Technology*, *foaf:Person*, to represent the researchers associated to the technology, *Topic* (equivalent to *frbr:concept* and *skos:concept*) and *Category*, representing the category of the technology (e.g., application, format, language).

[15] http://cui.unige.ch/~deribauh/Ontologies/TechMiner.owl.

[16] www.xmlns.com/foaf/0.1/.

[17] http://salt.semanticauthoring.org/ontologies/sro.

[18] http://purl.org/spar/frbr.

3 Evaluation

We tested our approach on a gold standard (GS) of manually annotated publications in the field of the Semantic Web. To produce it, we first selected a number of publications tagged with keywords related to this field (e.g., 'semantic web', 'linked data', 'RDF') according to the CSO ontology. We then created an interface to annotate the abstracts with names and types of technologies. Since recognizing technologies in a field requires a certain degree of expertise, we asked a group of 8 Semantic Web experts (PhD students, postdocs, and research fellows) from The Open University and Oxford University to perform this task. In particular, we asked the annotators to focus on specific technologies which could be identified with a label, and not to consider very common ones, such as "web server". Indeed, we wanted to focus on technologies used or introduced by researches that would usually not be covered by generic knowledge bases. To avoid typos or extremely uncommon labels, we discarded from the output the technologies with labels appearing only once in the full set of 16 million abstracts from the Scopus dataset of Computer Science. The resulting GS includes 548 publications, each of them annotated by at least two experts, and 539 technologies. In this evaluation we focus only on the identification of technologies, and not on the correctness of the links between the technology and other entities (e.g., authors), whose presence is simply used as features for the classification process and will be analysed in future work.

Our aim was to compare the performances of the different techniques discussed in this paper. In particular, we planned to assess the impact of the candidate linking and candidate analysis phases (Sects. 2.5 and 2.6) versus the NLP pipeline, the effect of the semantic features introduced in Sect. 2.6, and the impact of the candidate extension phase (Sect. 2.3). Therefore, we compared the following approaches:

- **NL**: the classic NLP pipeline [19], as discussed in Sect. 2.2, with no additional filters;
- **NLW**: the NLP pipeline which uses WordNet to discard generic terms;
- **TMN**: TM not using semantic features derived by linking the candidates to the knowledge bases (CSO, sciObjCSOnto, DBpedia) nor candidate expansion;
- **TM**: The full TM approach not using candidate expansion;
- **TMN_E:** TMN using candidate expansion;
- **TM_E**: The full TM approach using candidate expansion.

The last four approaches were trained using the gold standard from [19], which counts 300 manually annotated publications from Microsoft Academic Search. TMN_E and TM_E were then applied on a 3,000 publication sample (other than our GS) in the Semantic Web area and learned a total of 8,652 candidates, of which 1,264 were used during the evaluation run, being linked to the initial publications in the GS.

The evaluation was performed by running each approach on the abstracts of the 548 annotated publications in the GS. Since we intended to measure also how the popularity of a technology would affect the outcomes of the approaches, we performed six different tests with each method in which we considered only the technology labels which appeared at least 2, 5, 10, 20, 50 or 100 times in the full set of the Scopus dataset for Computer Science.

We intended to assess both (1) the ability of extracting the technologies from a set of publications, and (2) the ability of yielding a correct set of relationships between those technologies and the publications in which they are addressed. Hence, we computed recall and precision for both tasks. The significance of the results was assessed using non-parametric statistical tests for k correlated data: Wilcoxon's test for $k = 2$ and Friedman's test for $k > 2$.

Table 1 shows the performance of the approaches. We will first discuss the performance of the technology extraction task. The NL method is able to retrieve about half of the technologies with a precision of about 60 %, when considering all labels. The introduction of the WordNet filter (NLW) improves significantly the precision (+12.7 %, p = 0.03), but loses some recall (−4.6 %). TMN is able to further increase precision over NLW (+12.6 %, p = 0.03), lowering the recall to about 44 %. The introduction of the semantic features (TM) improves both precision (+2.1 %, p = 0.03) and recall (+2.4 %, p = 0.03); in particular, TM obtains the best result among all approaches regarding precision (87.6 %) and performs significantly better (p = 0.03) than TMN, NLW and NL regarding F-measure.

Table 1. Precision and recall for the six runs of the six approaches. In bold the best result of each run.

	Technology Recall						Relationship Recall					
Occurrences	**2**	**5**	**10**	**20**	**50**	**100**	**2**	**5**	**10**	**20**	**50**	**100**
NL	54.5%	57.6%	59.4%	59.7%	61.6%	63.1%	50.5%	51.5%	52.2%	52.1%	52.8%	52.8%
NLW	49.9%	52.1%	53.5%	53.9%	56.7%	57.9%	49.2%	50.3%	51.0%	51.3%	53.0%	53.2%
TMN	43.6%	44.8%	45.4%	45.5%	47.4%	48.1%	44.0%	44.7%	45.1%	45.3%	46.7%	46.8%
TM	46.0%	47.6%	48.7%	49.0%	51.9%	52.4%	46.4%	47.3%	48.0%	48.3%	50.2%	50.2%
TMN_E	82.4%	79.1%	77.2%	76.5%	74.4%	72.5%	75.4%	72.3%	70.7%	69.9%	67.9%	66.3%
TM_E	**84.2%**	**81.3%**	**80.4%**	**80.3%**	**80.3%**	**78.1%**	**78.1%**	**75.3%**	**74.3%**	**73.9%**	**73.5%**	**71.5%**
	Technology Precision						**Relationship Precision**					
Occurrences	**2**	**5**	**10**	**20**	**50**	**100**	**2**	**5**	**10**	**20**	**50**	**100**
NL	60.2%	56.5%	55.3%	55.1%	52.8%	49.3%	60.2%	57.7%	56.9%	56.8%	55.9%	54.0%
NLW	72.9%	70.3%	69.8%	71.0%	71.0%	69.6%	74.7%	73.4%	73.3%	74.1%	75.1%	74.4%
TMN	85.5%	84.0%	84.5%	86.3%	86.7%	86.8%	84.5%	83.5%	83.7%	84.6%	85.8%	85.5%
TM	**87.6%**	**87.0%**	**87.0%**	**88.5%**	**88.8%**	**88.4%**	**85.9%**	**85.3%**	**85.2%**	**85.9%**	**86.9%**	**86.3%**
TMN_E	83.6%	80.9%	80.6%	82.0%	80.5%	80.1%	73.7%	70.8%	70.1%	70.2%	69.6%	68.1%
TM_E	86.0%	84.1%	83.5%	84.2%	82.9%	82.0%	76.7%	74.1%	73.4%	73.3%	72.5%	70.8%

The ability of TMN_E and TM_E to consider also pre-learnt candidates yields a massive increase in recall (respectively +38.2 % and +38.8 %), paying a relative small price in precision (−1.6 % and −1.9 %). Once again, the adoption of semantic features increases both precision and recall, yielding no apparent drawbacks. Hence, TM_E performs significantly better than TMN_E regarding F-measure (p = 0.028). In general, TM_E outperforms all the other approaches for recall and F-measure (85.1 %), being able to extract technologies with a recall of 84.2 % and a precision of 86 %.

The approaches that used only the NLP pipeline to identify the candidates (NL, NLW, TMN, TM) improved their recall when considering more popular labels, but also committed more errors. An analysis of the data reveals that this happens mainly because they identify as technologies other kinds of popular named entities (e.g., universities, projects) that, being associated with a great number of publications, have a large chance to be involved in some of the patterns that trigger the JAPE rules. The two solutions that enhance the candidate set (TMN_E, TM_E) suffer from the opposite problem; they tend to perform well when dealing with rare technologies with few occurrences, and not considering them lowers their recall.

Figure 2 shows the F-measure for all the approaches. TM_E yields the best performance (85.1 % when processing all the technologies in the GS), followed by TMN_E, NLW, TMN and NL. The difference between the approaches is significant ($p < 0.0001$).

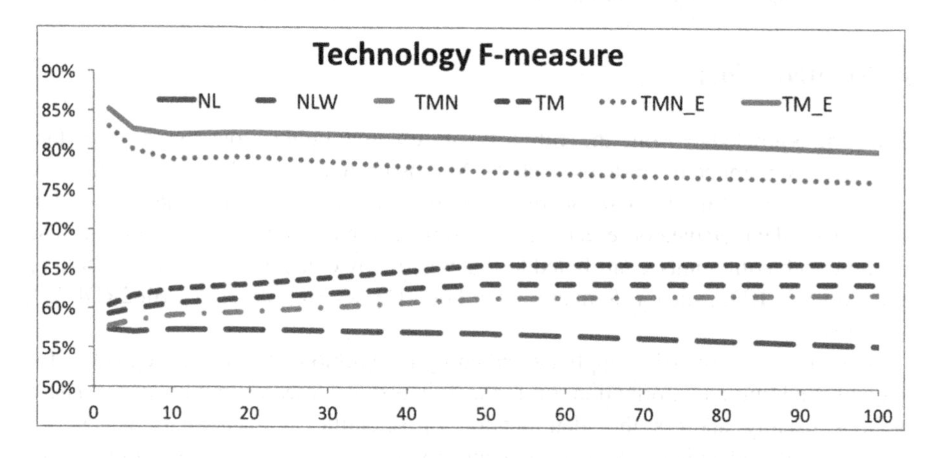

Fig. 2. F-measure of the technology extraction task.

The results regarding the extraction of links between technologies and publications exhibit a very similar dynamic. As before, TM performs best in terms of precision and TM_E in terms of recall. The main difference is that in this test TM_E and TMN_E exhibit a lower precision. This is due to the fact that the method for linking pre-learnt candidates to publications is more prone to error that the NLP pipeline, which links only publications in which it finds a specific linguistic pattern. Figure 3 shows the F-measure regarding relationships. TM_E is again the best solution, followed by the other approaches in the same order as before. The difference among the methods is again highly significant ($p < 0.0001$).

In conclusion, the evaluation shows that the TM approach yields significantly better results than alternative NLP methods and that the introduction of semantic features further improves the performance. The use of pre-learnt candidates introduces a small amount of noise in the set of linked papers, but yields a important increase in recall.

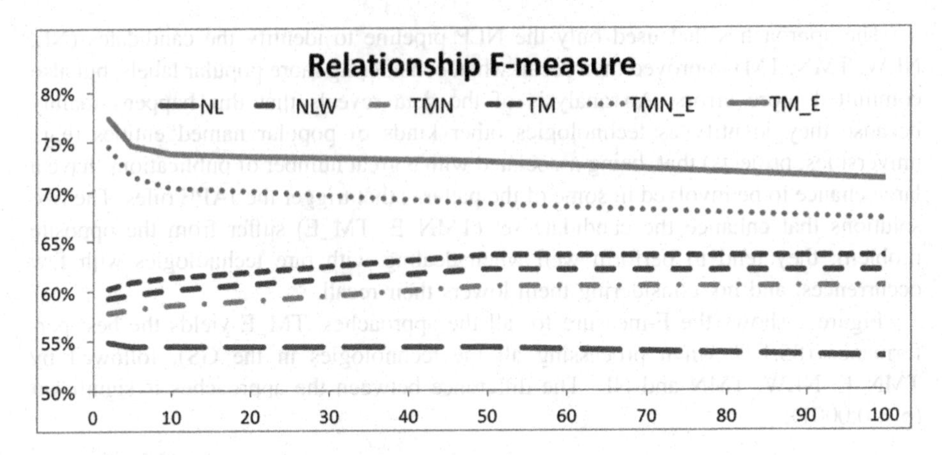

Fig. 3. F-measure of the links between technologies and publications.

4 Related Work

Extracting knowledge from the full text of research publications is an important challenge. A number of systems such as Microsoft Academic Search (academic.research.microsoft.com), Google Scholars (scholar.google.com), and others automatically extract the metadata of research publications and make them available online. The semantic web community contributed to this process by creating a number of scholarly repositories in RDF, such as Semantic Dog Food [1], RKBExplorer [2], Bio2RDF [3], and others.

A number of approaches apply named entity recognition and similar techniques for extracting additional information from the full text of research publications. These methods usually address the identification of scientific artefacts (e.g., genes [4], chemical [5]) and epistemological concepts [6] (e.g., hypothesis, motivation, background, experiment). For example, Groza [7] focused on the identification of conceptualization zones through a novel approach based on the deep dependency structure of the sentences. Ibekwe-Sanjuan and al [23] developed a methodology which combines surface NLP and Machine Learning techniques for identifying categories of information, such as objective, results, conclusion and so on. O'Seaghdha and Teufel [24] addressed instead the identification of the rhetorical zoning (based on argumentative zoning) using a Bayesian latent-variable model. The Dr. Inventor Framework [25] is a publicly available collection of scientific text mining components which can be used to support this kind of tasks.

TM can be classified under the first category, since technologies can be considered scientific objects. As in other methods crafted for this task, it uses a pipeline which includes NLP and machine learning; the main difference is that it focuses on technologies and introduces a number of new statistical and semantic techniques to foster the identification process.

The use of the Linked Open Data cloud for supporting named entity recognition has yielded good results. For example, the LODifier approach [9] combines deep semantic analysis, named entity recognition and word sense disambiguation to extract named entities and to convert them into an RDF representation. Similarly, the AGADISTIS [10] system is a knowledge-base-agnostic approach for named entity disambiguation which combines the Hypertext-Induced Topic Search algorithm with label expansion strategies and string similarity measures. However, this kind of systems can be used only for linking existing technologies to the related entities in knowledge bases, not for discovering new ones. Sateli and Witte [11] presented a method which combines NLP and named entity recognition based on the LOD cloud for identifying rhetorical entities and generating RDF triples describing them. Similarly to TM, they use GATE for NLP and DBpedia Spotlight [21] for linking terms in the publications to DBpedia entities. However, TM uses a classifier to process a number of features derived from the linked research entities.

A number of agencies in the field of innovation brokering and 'horizon scanning' identify new technologies by manually scanning the web [14], leading to high costs and slow throughput. Automatic methods such as TM could bring a dramatic improvement in their workflow, by allowing the selection of a set of candidate technologies with high accuracy. The output produced by TM can also enrich a number of knowledge sources which index technologies, especially considering that, a good number of these, such as Google Patents, cover only patented technologies. As mentioned, DBpedia [12] also includes a number of well-known technologies, even if they are not always described thoroughly. Another interesting resource is the Resource Identification Initiative portal [13], an archive which collects and assigns IDs to a number of scientific objects, including applications, systems and prototypes.

5 Conclusions

We presented TechMiner, a novel approach combining NLP, machine learning and semantic technologies, which mines technologies from research publications and generates an OWL ontology describing their relationships with other research entities. We evaluated TM on a gold standard of 548 publications and 539 technologies in the field of the Semantic Web. The evaluation showed that the use of semantic features significantly improves technology identification, and that the full hybrid method outperforms NLP approaches. These results suggest that using a combination of statistical information derived from the network of relevant of research entities (e.g., authors, topics) and background knowledge offers a competitive advantage in this task.

TM opens up many interesting directions of work. We plan to enrich the approach for identifying other categories of scientific objects, such as datasets, algorithms and so on. This would allow us to conduct a comprehensive study on the resulting technologies, with the aim of better understanding the processes that lead to the creation of successful technologies. We also intend to run our approach on a variety of other research fields and to this end we are testing some methodologies to automatically populate the supporting ontologies with terms automatically extracted from research papers [26]. Finally, since similar experiences in the field of biotechnology [13] highlighted the importance of manually curating this kind of data, we would like to build a pipeline for allowing human experts to correct and manage the information extracted by TechMiner.

Acknowledgements. We thank Elsevier for providing us with access to the Scopus repository of scholarly data. We also acknowledge grant n° 159047 from the Swiss National Foundation.

References

1. Moller, K., Heath, T., Handschuh, S., Domingue, J.: Recipes for semantic web dog food—the ESWC and ISWC metadata projects. In: 6th International Semantic Web Conference, 11–15 November 2007, Busan, South Korea (2007)
2. Glaser, H., Millard, I.: Knowledge-enabled research support: RKBExplorer.com. In: Proceedings of Web Science 2009, Athens, Greece (2009)
3. Dumontier, M., Callahan, A., Cruz-Toledo, J., Ansell, P., Emonet, V., Belleau, F., Droit, A.: Bio2RDF release 3: a larger connected network of linked data for the life sciences. In: 2014 International Semantic Web Conference (Posters & Demos) (2014)
4. Carpenter, B.: LingPipe for 99.99 % recall of gene mentions. In: Proceedings of the Second BioCreative Challenge Evaluation Workshop, vol. 23, pp. 307–309 (2007)
5. Corbett, P., Copestake, A.: Cascaded classifiers for confidence-based chemical named entity recognition. BMC Bioinform. **9**(11), 1 (2008)
6. Liakata, M., Teufel, S., Siddharthan, A., Batchelor, C.R.: Corpora for the conceptualisation and zoning of scientific papers. In: LREC (2010)
7. Groza, T.: Using typed dependencies to study and recognise conceptualisation zones in biomedical literature. PLoS ONE **8**(11), e79570 (2013)
8. de Ribaupierre, H., Falquet, G.: User-centric design and evaluation of a semantic annotation model for scientific documents. In: Proceedings of the 14th International Conference on Knowledge Technologies and Data-driven (2014)
9. Augenstein, I., Padó, S., Rudolph, S.: LODifier: generating linked data from unstructured text. In: The Semantic Web: Research and Applications, pp. 210–224 (2012)
10. Usbeck, R., Ngonga Ngomo, A.-C., Röder, M., Gerber, D., Coelho, S.A., Auer, S., Both, A.: AGDISTIS - graph-based disambiguation of named entities using linked data. In: Mika, P. (ed.) ISWC 2014. LNCS, vol. 8796, pp. 457–471. Springer, Heidelberg (2014). doi:10.1007/978-3-319-11964-9_29
11. Sateli, B., Witte, R.: What's in this paper? Combining rhetorical entities with linked open data for semantic literature querying. In: Proceedings of the 24th International Conference on World Wide Web Companion, pp. 1023–1028 (2015)
12. Bizer, C., Lehmann, J., Kobilarov, G., Auer, S., Becker, C., Cyganiak, R., Hellmann, S.: DBpedia-a crystallization point for the web of data. Web Semant. Sci. Serv. Agents World Wide Web **7**(3), 154–165 (2009)
13. Bandrowski, A., Brush, M., Grethe, J.S., Haendel, M.A., Kennedy, D.N., Hill, S., Hof, P.R., Martone, M.E., Pols, M., Tan, S.C., Washington, N.: The resource identification initiative: a cultural shift in publishing. J. Comparat. Neurol. **524**(1), 8–22 (2016)
14. Scanning Douw, K., Vondeling, H., Eskildsen, D., Simpson, S.: Use of the Internet in scanning the horizon for new and emerging health technologies: a survey of agencies involved in horizon scanning. J. Med. Internet Res. **5**(1), e6 (2003)
15. Fellbaum, C.: WordNet: An Electronic Lexical Database. MIT Press, Cambridge (1998)

16. Osborne, F., Motta, E.: Klink-2: integrating multiple web sources to generate semantic topic networks. In: Arenas, M., et al. (eds.) ISWC 2015. LNCS, vol. 9366, pp. 408–424. Springer, Heidelberg (2015). doi:10.1007/978-3-319-25007-6_24
17. de Ribaupierre, H., Falquet, G.:, An automated annotation process for the SciDocAnnot scientific document model. In: Proceedings of the Fifth International Workshop on Semantic Digital Archives, TPDL 2015 (2015)
18. Osborne, F., Motta, E., Mulholland, P.: Exploring scholarly data with rexplore. In: Alani, H., Kagal, L., Fokoue, A., Groth, P., Biemann, C., Parreira, J.X., Aroyo, L., Noy, N., Welty, C., Janowicz, K. (eds.) ISWC 2013. LNCS, vol. 8218, pp. 460–477. Springer, Heidelberg (2013). doi:10.1007/978-3-642-41335-3_29
19. de Ribaupierre, H., Osborne, F., Motta, E.: Combining NLP and semantics for mining software technologies from research publications. In: Proceedings of the 25th International Conference on World Wide Web (Companion Volume) (2016)
20. Huang, W.: Do ABCs get more citations than XYZs? Econ. Inq. **53**(1), 773–789 (2015)
21. Mendes, P.N., Jakob, M., García-Silva, A., Bizer, C.: DBpedia spotlight: shedding light on the web of documents. In: Proceedings of the 7th International Conference on Semantic Systems, pp. 1–8. ACM (2011)
22. Peroni, S., Shotton, D.: FaBiO and CiTO: ontologies for describing bibliographic resources and citations. Web Semant. Sci. Serv. Agents World Wide Web **17**, 33–43 (2012)
23. Ibekwe-SanJuan, F., Fernandez, S., Sanjuan, E., Charton, E.: Annotation of scientific summaries for information retrieval (2011). arXiv preprint arXiv:1110.5722
24. O'Seaghdha, D., Teufel, S.: Unsupervised learning of rhetorical structure with un-topic models. In: Proceedings of the 25th International Conference on Computational Linguistics (COLING 2014) (2014)
25. Ronzano, F., Saggion, H.: Dr. inventor framework: extracting structured information from scientific publications. In: Japkowicz, N., Matwin, S. (eds.) DS 2015. LNCS (LNAI), vol. 9356, pp. 209–220. Springer, Heidelberg (2015). doi:10.1007/978-3-319-24282-8_18
26. Bordea, G., Buitelaar, P., Polajnar, T.: Domain-independent term extraction through domain modelling. In: The 10th International Conference on Terminology and Artificial Intelligence (TIA 2013), Paris, France (2013)

Ontology Learning in the Deep

Giulio Petrucci[1,2(✉)], Chiara Ghidini[1], and Marco Rospocher[1]

[1] FBK-irst, Via Sommarive, 18, 38123 Trento, Italy
[2] University of Trento, Via Sommarive, 14, 38123 Trento, Italy
{petrucci,ghidini,rospocher}@fbk.eu

Abstract. Recent developments in the area of deep learning have been proved extremely beneficial for several natural language processing tasks, such as sentiment analysis, question answering, and machine translation. In this paper we exploit such advances by tailoring the ontology learning problem as a transductive reasoning task that learns to convert knowledge from natural language to a logic-based specification. More precisely, using a sample of definitory sentences generated starting by a synthetic grammar, we trained Recurrent Neural Network (RNN) based architectures to extract OWL formulae from text. In addition to the low feature engineering costs, our system shows good generalisation capabilities over the lexicon and the syntactic structure. The encouraging results obtained in the paper provide a first evidence of the potential of deep learning techniques towards long term ontology learning challenges such as improving domain independence, reducing engineering costs, and dealing with variable language forms.

1 Introduction

Along the years, the ontology engineering community has been pursuing the ambitious goal to automatically encode increasingly expressive knowledge from text. Even if the fully automatic acquisition of logic-based knowledge is still a long term goal, several approaches have been proposed, achieving remarkable performances but, at the same time, still experiencing some limitations. In a nutshell, approaches that rely on controlled languages such as Attempto [8] pose rigid limitations on the syntax that the text has to adopt, thus constraining the text producers to express knowledge in pre-defined formats and moreover making the approach unsuitable to process already written text. Approaches that instead target expressive knowledge (that is, complex axiom) extraction from written text (e.g., LExO [17]) heavily rely on catalogs of hand-crafted lexico-syntactic patterns that are *rigid* w.r.t. the grammatical structure of the text they can process. The downside of these approaches is that extending a catalog of hand-crafted rules to handle the variability of natural language can be particularly hard, taking also into account that language forms evolve over time, can be domain-specific, and that patterns need to be specifically produced and adapted to the different mother tongues.

In the last years, deep neural networks have been successfully exploited in several Natural Language Processing (NLP) tasks, from the most foundational,

E. Blomqvist et al. (Eds.): EKAW 2016, LNAI 10024, pp. 480–495, 2016.
DOI: 10.1007/978-3-319-49004-5_31

like part-of-speech tagging or semantic role labeling, to the most complex ones, like question-answering, sentiment analysis, and statistical machine translation. Such systems have the advantage to be cheap in terms of feature engineering and can learn how to deal with language variability in a flexible manner.

Stemming from such experiences, we approached ontology learning as a machine transduction task, as described in Sect. 3. We train statistically — i.e. by examples — and in an end-to-end fashion a neural network based system to translate definitory text into Description Logic (DL) [1] formulae. The main contributions of our work are:

- a **general architecture** that enables to formulate the ontology learning problem as a machine transduction task exploiting Recurrent Neural Networks (Sects. 3 and 4). To the best of our knowledge, such approach has never been applied before to ontology learning tasks;
- a **dataset** for a statistical learning based formulation of the OWL complex axioms learning task (Sect. 5.1). The creation of an extensive dataset was necessary as, to the best of our knowledge, no commonly accepted, large-size, dataset exists for this task. Its availability can facilitate the adoption of statistical learning approaches within the ontology learning community;
- a **customisation** of the general architecture for the specific dataset, and its evaluation (Sect. 5.2). The evaluation shows that our model manifest the capability to generalise over sentence structure, as well as, tolerance to unknown words, as discussed in Sect. 6.

Related works are presented in Sect. 2, and concluding remarks are provided in Sect. 7. While our work was performed and evaluated on the specific language model defined by our dataset, the results illustrated in the paper provide a first evidence that deep learning techniques can contribute in a fruitful manner to the ontology engineering community to tackle some of its long term challenges, especially in domain adaptation or extension.

2 State of the Art

In this section, we briefly review the state of the art in ontology learning, with a special focus on those approaches that aim to extract expressive knowledge. For a wider overview, see [5,9,14].

As pointed out in [16], state-of-the-art methods are *able to generate ontologies that are largely informal*, that is, *limited in their expressiveness*. Indeed, tasks like taxonomy construction or factual knowledge extraction (i.e., assertions, entity recognition, and so on) can be largely automatized, thanks to the effectiveness reached by such methods. The success of OWL as the *de facto* standard ontology language for the web paved the way for some advances in automatic extraction of more expressive knowledge, such as terminological axioms or complex class definitions. Some methods and tools have been developed along the years to tackle these problems.

A first collection of tools is based on Controlled Natural Language. The best known one is Attempto Controlled English (ACE) [8], a restricted version of standard English, both in terms of syntax and semantics. By adopting a small set of construction and interpretation rules, sentences written in ACE can be automatically translated in some formal language, including OWL. However, its applicability for ontology learning from generic, available texts is limited, as arbitrary sentences are typically not written in ACE. Furthermore, it works only for English language.

A second collection of tools is based on catalogs of hand-crafted lexico-syntactic patterns. A well known system in this category is LExO [17]. The main idea behind LExO is to transform a set of natural language definitions into a set of OWL axioms as a suggestion to the ontology engineer. Each definition is parsed into a dependency and then transformed into a set of OWL axioms by the application of a set of hand-crafted rules over the syntactic features of the text. The approach has been used in [16] as the foundational idea to build a methodology with the ambitious goal to cover the whole ontology life-cycle management. In particular, a machine learning based approach to determine disjointness of two classes is presented, based on the measurement of their *taxonomic overlap*, i.e. how much is likely an instance of both of them to exist in the ontology: the lower this value, the more two classes are likely to be disjoint. Such metric is estimated starting from their mutual distance in some background taxonomy (like WordNet[1]), the similarity between their lexical contexts, their Pointwise Mutual Information (PMI) over the web and their matching a set of predefined lexico-syntactic patterns. Regarding limitations, pattern-based approaches for ontology learning are *rigid* w.r.t. the grammatical structure of the text they can process: hence, several linguistic phenomena such as conjunctions, negations, disjunctions, quantifiers scope, ellipsis, anaphora, etc., can be particularly hard to parse and interpret. Extending a catalog of hand-crafted rules to handle all such phenomena can be an extremely expensive task, leading to unsustainable costs in engineering, maintaining and evolving the system.

A different approach is taken in [10], where a system for the extraction of $\mathcal{EL}++$ concepts definitions from text is presented. Text fragments involving concepts from the SNOMED CT[2] ontology are matched and their lexical and ontological features are used to train a maximum entropy classifier to predict the axiom describing the involved entities. User feedback can be exploited to adaptively correct the underlying model. However, this approach is tightly bounded to the specific domain considered.

Summing up, state-of-the-art methods for expressive knowledge extraction still experience severe limitations: the approach we propose in this work aims to overcome some of them.

[1] https://wordnet.princeton.edu/.

[2] http://www.ihtsdo.org/snomed-ct/.

3 Ontology Learning as a Transduction Task

The main intuition underpinning our approach is that ontology learning can be seen as a *transduction* process, as defined in [7]: a string from a source language is converted into another string of a target language. In our specific case, we want to convert a sequence of words, that is a *sentence* in natural language, into a sequence of logical symbols, namely a *formula*. As an example, let us consider the following sentence:

$$\text{“}\textit{A bee is an insect that has } 6 \textit{ legs and produces honey.}\text{”} \tag{1}$$

This sentence provides a brief description of a bee and of its main characteristics and can be encoded by means of the following DL formula which, so-to-speak, represents its conversion into the required logical format:

$$\texttt{Bee} \sqsubseteq \texttt{Insect} \sqcap = 6\ \texttt{have.Leg} \sqcap \exists \texttt{produce.Honey} \tag{2}$$

One of the problems we have in converting natural language sentences into logical formulae is the variability of natural language. Consider for instance:

"If something is a bee, then it is an insect
with exactly 6 legs and it also produces honey." (3)

Despite being lexically and syntactically different from (1), (3) provides a description of a bee which is, from the standpoint of an ontological formulation, equivalent to the one provided in (1). Thus, we would like to be able to transform it into the same formula (2). A second, dual problem we have is the fact that sentences often share a similar structure while conveying a completely different meaning. An exemplification of this problem is provided by the sentence:

$$\text{“}\textit{A cow is a mammal that has } 2 \textit{ horns and eats grass}\text{”}. \tag{4}$$

This sentence shares several similarities with sentence (1), and would be translated into a DL formula such as:

$$\texttt{Cow} \sqsubseteq \texttt{Mammal} \sqcap = 2\ \texttt{have.Horns} \sqcap \exists \texttt{eat.Grass} \tag{5}$$

which is, from a structural point of view, very similar to the one in (2). These two phenomena, which can be considered analogous to well known problems of homography (same term different meaning) and synonymity (different terms same meaning) are recurrent when dealing with descriptions (of the world) and the meaning they denote. Being able to deal with them both is one of the challenges that expressive ontology learning has to face. Our approach in dealing with these two problems relies on the following observations:

1. Nouns like `Bee`, `Insect`, `Leg` and `Honey` denote the involved *concepts* while verbs like `have` and `produce` describe relationships among such concepts, or *roles*. From a linguistic standpoint, nouns and verbs – together with adjectives

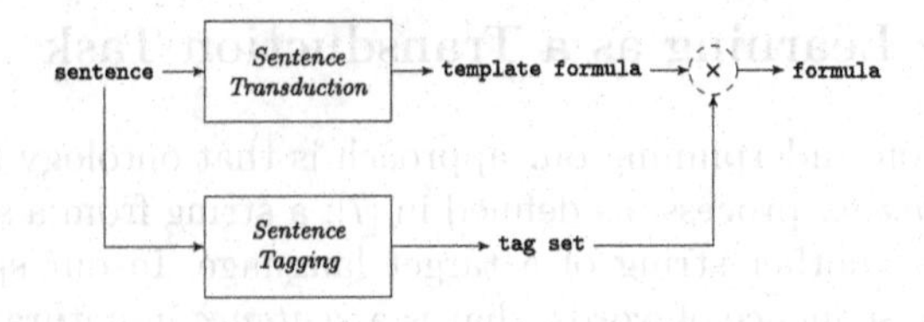

Fig. 1. The whole pipeline

and some adverbs – are *content words*, as they describe the actual content of a sentence. Dually, words such as articles, pronouns, conjunctions, most adverbs, and so on, are considered *function words*, as they express grammatical relationships between words and carry almost no lexical meaning.

2. In the case of different sentences that convey ontological equivalent formulations, such as (1) and (3), we notice that content words are *the same*,[3] while function words change. In the dual case of sentences with similar structures and different meanings (such as (1) and (4)), we note that the sentences share the same function words while content words are radically different.

Stemming from these two observations, our idea is to deal with these two problems by splitting the transduction process of sentences in two parallel phases: a **sentence transduction** phase and a **sentence tagging** phase, as depicted in Fig. 1. The sentence transduction phase focuses on the identification of the logical structure of the formula corresponding to the natural language specification. The output of this phase is a structure that we name *formula template*. The sentence tagging phase focuses on identifying and recognising all the words for what they act in the sentence: a concept, a role, a number, or a generic word.

Going back to our example, the output of the sentence transduction phase for sentences (1), (3), and (4) is the (same) formula template:

$$\mathtt{C_0} \sqsubseteq \mathtt{C_1} \sqcap = \mathtt{N_0}\ \mathtt{R_0}.\mathtt{C_2} \sqcap \exists \mathtt{R_1}.\mathtt{C_3} \tag{6}$$

where C, R and N have been respectively used for concepts, roles and numbers, with a subscript indicating their order. Sentence transduction is therefore the phase which tackles the challenge of identifying a common unifying structure among different natural language sentences that may differ both lexically and syntactically. The output of sentence tagging would instead produce three different structures:

$$A[bee]_{\mathtt{C_0}}\ is\ an\ [insect]_{\mathtt{C_1}}\ that\ [has]_{\mathtt{R_0}}[6]_{\mathtt{N_0}}\ [legs]_{\mathtt{C_2}}\ and\ [produces]_{\mathtt{R_1}}\ [honey]_{\mathtt{C_3}} \tag{7}$$

$$If\ something\ is\ a\ [bee]_{\mathtt{C_0}}\ is\ an\ [insect]_{\mathtt{C_1}}\ that\ [has]_{\mathtt{R_0}}\ exactly\ [6]_{\mathtt{N_0}}\ [legs]_{\mathtt{C_2}}\ and\ it\ also\ [produces]_{\mathtt{R_1}}\ [honey]_{\mathtt{C_3}} \tag{8}$$

$$A\ [cow]_{\mathtt{C_0}}\ is\ a\ [mammal]_{\mathtt{C_1}}\ that[has]_{\mathtt{R_0}}\ [4]_{\mathtt{N_0}}\ [legs]_{\mathtt{C_2}} and\ [eats]_{\mathtt{R_1}}\ [grass]_{\mathtt{C_3}} \tag{9}$$

[3] Possibly after resolving anaphora, coreference, or other linguistic phenomena.

where all the words other than C, R, or N are intended to be tagged with a generic word label, w. This step is close to the slot filling problem, as tackled in [11]: we need to detect the role each word assumes within a certain semantic scope, even a whole sentence. The final step in the pipeline is to combine the outputs of the two phases to obtain the resulting logical formula. Thus, the formula template (6) combined with the tagged sentence (7) will provide formula (2) as a logical conversion of (1); the formula template (6) combined with the tagged sentence (8) will also provide formula (2) as a logical conversion of sentence (3); and finally the formula template (6) combined with the tagged sentence (9) will provide formula (5) as the logical conversion of sentence (4). In the next section we illustrate in detail how RNNs are used to implement the sentence transduction and sentence tagging phases.

4 An RNN-Based Architecture for Ontology Learning

In this section we will provide a brief overview of our computational model for ontology learning, describing the main intuitions behind it. A detailed description of the theoretical framework is provided in [15].

4.1 RNNs and the Gated Recursive Unit Model

Speaking of neural networks, the adjective *recurrent* referred to one of its layers, means that the activation of the layer at time t, say $\mathbf{h}^{\langle t \rangle}$, depends not only on the inputs, say $\mathbf{x}^{\langle t \rangle}$, but also on its previous value, $\mathbf{h}^{\langle t-1 \rangle}$, as in:

$$\mathbf{h}^{\langle t \rangle} = g(\mathbf{x}^{\langle t \rangle}, \mathbf{h}^{\langle t-1 \rangle}; \theta), \tag{10}$$

where g is the so called *cell function* and θ is the set of function parameters to be learnt during the training phase. Dealing with natural language, the t-th timestep is the t-th word in a sentence. The recurrent structure makes such class of models an adequate choice when dealing with sequences, and in particular with natural language, where each word depends on the previous ones.[4] To handle long-term dependencies – syntactic dependencies, speaking of natural language – the cell function can be endowed with some memory effect. Different models have been proposed. Among them, the Gated Recursive Unit (GRU), synthetically depicted in Fig. 2, shows good memory capabilities combined with a higher simplicity w.r.t. other cell functions. The cell (layer unit) state is controlled by two gates: the *reset gate* $\mathbf{r}$ and the *update gate* $\mathbf{z}$, according to the equations in Fig. 3. Intuitively, the update gate balances the amount of information to be kept from the previous state; at the same time such memory can be erased when the reset gate approaches zero. Such structure is repeated for each word. The model parameters to be learnt during the training are $\theta = [\mathbf{W}_r, \mathbf{U}_r, \mathbf{W}_z, \mathbf{U}_z]$. The symbol $\odot$ indicate the Hadamar product.

[4] The goal of our work is to show that the ontology learning task can be tackled using neural network based models trained in a end-to-end fashion. Assessing the best neural network architecture to implement statistical learning for this task is beyond the scope of our work.

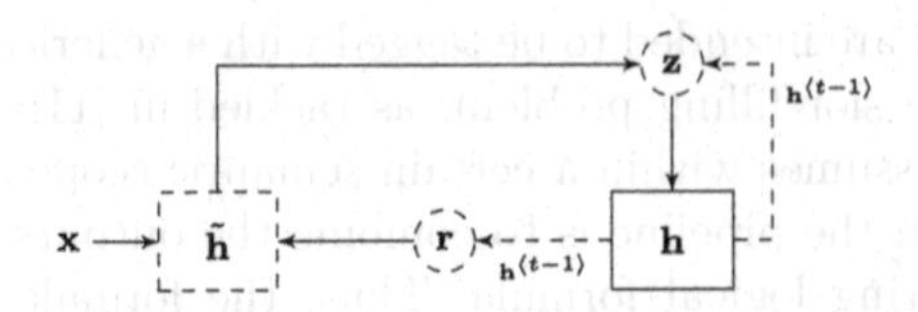

Fig. 2. Gated Recursive Unit

$$\mathbf{r}^{\langle t\rangle} = sigmoid(\mathbf{W}_r\mathbf{x}^{\langle t\rangle} + \mathbf{U}_r\mathbf{h}^{\langle t-1\rangle})$$
$$\mathbf{z}^{\langle t\rangle} = sigmoid(\mathbf{W}_z\mathbf{x}^{\langle t\rangle} + \mathbf{U}_z\mathbf{h}^{\langle t-1\rangle})$$
$$\tilde{\mathbf{h}}^{\langle t\rangle} = tanh(\mathbf{W}_h\mathbf{x}^{\langle t\rangle} + \mathbf{r} \odot \mathbf{U}\mathbf{h}^{\langle t-1\rangle})$$
$$\mathbf{h}^{\langle t\rangle} = \mathbf{z} \odot \mathbf{h}^{\langle t-1\rangle} + (\mathbf{1} - \mathbf{z}) \odot \tilde{\mathbf{h}}^{\langle t\rangle}$$

Fig. 3. GRU equations

4.2 Network Model for Sentence Tagging

The sentence tagging task can be formulated as follows: given a natural language sentence corresponding to some formal representation, we want to apply a tag to each word. The tag identifies the role the word has in the formal representation. We used a regular RNN, a snapshot of which is depicted in Fig. 4. Each word is represented by its index within the vocabulary. The most straightforward representation of such index as a vector would be a one-hot vector: the i-th word in a vocabulary of $|V|$ words will be a vector of $|V|$ components, all with value zero but the i-th, with value 1. Such vector representation is impractical for two main reasons: (i) vectors can become huge, as their dimension is the same of the vocabulary; and (ii) it can be hard to find a meaningful way to compose vectors as they are extremely sparse. Therefore, we map each index within the vocabulary into a low-dimension vector of real numbers, called an *embedding vector.* Embedding vectors act as distributed representations trained to capture all the meaningful features of a word in the context of a given task (see [12]). Moreover, their dimension can be significantly smaller than the number of words in the lexicon, avoiding the *curse of dimensionality*, as in [2]. In our network, each index in the dictionary is associated to an embedding vector. Such vectors are learnt during the training phase. At the t-th step, we feed the network with a *window of words* of width w, i.e. a short sequence of words centered on the t-th word of the sentence. All the embedding vectors corresponding to the words in a window are concatenated in a unique vector and fed into the recurrent layer. We indicate such window as $\mathbf{x}^{\langle t-w;t+w\rangle}$. The activation of such layer is given by Eq. (10). The output of the recurrent layer is then fed into a linear layer that outputs, at each time step, a vector of the same size of the number of possible tags: each component holds a sort of *score* of the corresponding tag. At the timestep t, our network must predict the appropriate tag to apply to the current input word. So we apply a softmax over such scores, modeling a probability distribution across all the possible tags. The tag applied to the t-th word will be the most probable one, i.e. the *argmax* over such output vector $\mathbf{y}^{\langle t\rangle}$. The network is trained minimizing the categorical cross entropy between the expected sequence and the predicted one.

4.3 Network Model for Sentence Transduction

The sentence transduction task can be formulated as follows: given a natural language sentence corresponding to some formal representation, we want to identify

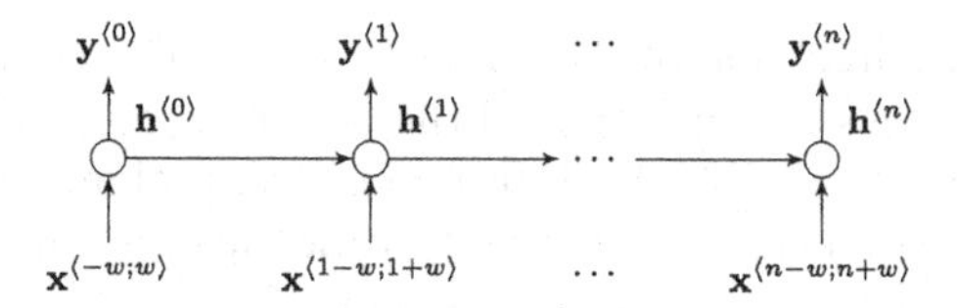

Fig. 4. The RNN used for sentence tagging

the structure of such formal representation, how its concepts and roles are connected, and which connectors are used. The input, the word embeddings, and the output are handled like in the previous model, using single words —represented with $\mathbf{x}^{\langle t \rangle}$— instead of windows in the input layer, and with the output vector that has the same dimension of the catalogue of all the formal representation terms that can be produced. We use also the same training objective, minimizing the categorical cross entropy between the expected sentence and the predicted one.

Both this model and the previous one receive a sequence of words and turn it into another sequence of different symbols. This allows us to use the same training objective and the same training procedure for both architectures. The pivotal difference is that each output symbol of the sentence tagging model corresponds exactly to one word of the input sentence, while this is not the case in the sentence transduction task. We can deal with such situation using two stacked recurrent layers in the so called Recurrent Encoder-Decoder (RED) configuration, a snapshot of which is depicted in Fig. 5.

The main intuition behind such architecture is the following: the first RNN *encodes* the input sequence, so that its hidden state at the last time step —the vector $\mathbf{c}$ in Fig. 5— holds the distributed representation *of the whole sentence.* Such vector is subsequentially fed as a constant input to a second RNN which *decodes* the content of such vector into a new sequence. In this way, the two sequences are at the same time: (i) *tightly coupled* w.r.t. their meaning, as the distributed representation of the whole input sequence is constantly fed into each decoding step; and (ii) *loosely coupled* w.r.t. their structure, since there is no direct correspondence between the input and the output symbols. The cell function of the decoder can be written as:

$$\mathbf{h}^{\langle t \rangle} = g(\mathbf{c}, \mathbf{h}^{\langle t-1 \rangle}; \theta), \tag{11}$$

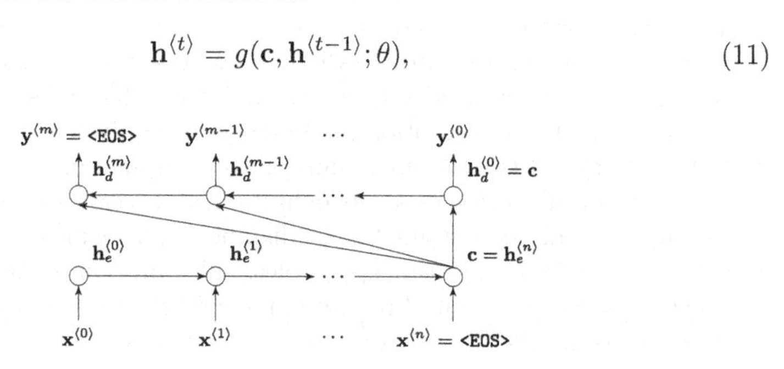

Fig. 5. The RNN Encoder-Decoder network model for sentence transduction

which is a slight simplification of the one used in [4]. The simplification consists in omitting the feedback of the output of the previous step. This helps us to keep the model as simple as possible, both from a conceptual and an implementation point of view (less parameters to deal with, which means a model easier to train and implement) without significant performance loss.

5 Learning Expressive OWL Axioms with RNNs

In this section we show how the general architecture presented in Sect. 4 can be deployed to learn expressive OWL axioms from natural language sentences.

5.1 A Dataset for Learning of OWL Axioms

To the best of our knowledge, there is no commonly accepted dataset for a statistical learning-based formulation of the OWL ontology learning task, especially when focusing on complex class definitions, as we do. Such dataset would consist of pairs of natural language sentences and their corresponding DL axiom, and should be adequately large to enable the training (and evaluation) of Machine Learning based approaches for ontology learning. The creation of such a collection of data would be extremely beneficial for the community and for those, like us, aiming to exploit statistical learning based methods.

As manually building such dataset can be extremely expensive —requiring considerable human-effort to collect, annotate, and validate a large quantity of data— we followed the practice of some notable examples in literature (e.g., [3,6,7,13,18–20]) of verifying the approach over *appropriate* synthetic data.

In details, we started by verbalizing with ACE a set of OWL class definitions in order to have a first seed of definition-like sentences like (1), as typically found in encyclopedias. We extended this seed by manually adding variations of every verbalization and other equivalent structures. So, for the sentence *"all the dogs are mammals"*, we added *"every dog is a mammal"*, *"dogs are mammals"*, *"any dog is also a mammal"* and so on. Or, for *"a bass guitar has at least 4 strings"*, we added *"bass guitars have more than 3 strings"*, *"A bass guitar don't have less than 4 strings"* and so on. Finally, we built a grammar capable to generate all such sentences, with placeholders instead of *concepts*, *roles*, and numbers used in cardinality restriction clauses. All the definitions have a left hand side class description and a right hand side class description. Relations between such class descriptions can be subsumption, disjunction or equivalence. Each side can be itself made of one or two *atomic* definitions, in conjunction or disjunction, which can be a *concept* definition or a *role* definition. Roles can be bound to one or two cardinality constraints, conjuncted or disjuncted. To sum it up, the constructs used in our definitions are concepts, roles, subsumption, disjunction, negation, intersection, union, existential restriction, cardinality restrictions maybe in conjunction or disjunction.[5]

[5] More precisely, the constructs considered correspond to the $\mathcal{ALCQ}$ Description Logic, a well-known, expressive extension of $\mathcal{ALC}$ with qualified number restrictions.

From our grammar, we generated more than 123 millions different *sentence templates*, each of which has been associated to its equivalent *formula template*. We obtained 261189 different formula templates in total.[6] Some examples of sentences and corresponding formulae are in Table 1.

Table 1. Examples of sentence templates and formula templates

Sentence template	Formula template
Anything that R_0 *at least* N_0 C_0 *and is also* C_1 , *is* C_2	$\geqslant N_0R_0.C_0 \sqcap C_1 \sqsubseteq C_2$
Every C_0 *is also something that* R_0 *less than* N_0 *or more than* N_1 C_1	$C_0 \sqsubseteq < N_0R_0.C_1 \sqcup > N_1R_0.C_1$
Anything that R_0 *less than* N_0 *or more than* N_1 C_0 *is* C_1	$< N_0R_0.C_0 \sqcup > N_1R_0.C_0 \sqsubseteq C_1$

5.2 Training and Evaluation

We assessed the capabilities of our approach in learning expressive OWL axioms from text, on various training-test pair configurations generated starting from the dataset described in Sect. 5.1.[7] Next we will describe such training-test pairs and the model configuration in terms of hyper-parameters and training settings. Finally, we will discuss the results of our experiments.

Training and Test Sets. From sentence-formula template couples described in the previous section, we can generate the actual training and test examples for our experiments. Our example is a triple $e = (s, t, f)$ made of:

1. a natural language *sentence* s, namely a sequence of words. A sample of this is sentence (1).
2. a *tag sequence* t corresponding to sentence s. Such tag sequence is obtained mapping each word of sentence s to a tag indicating the role of the word in the sentence. A sample of tag sequence is (7).
3. a *formula template* f corresponding to the translation of s in the target logical language. A sample of formula template is the one in (6).

To turn a sentence template into an actual sentence, we have to fill its placeholder with actual words. *Role* placeholders are filled with verbs, randomly selected from a list of 882. *Concept* placeholder are filled combining words from a list of 1522 adjectives, a first list of 2425 nouns and a second list of 192 nouns. Combinations were allowed only according to 66 patterns. A simplified example of such procedure is presented in Table 2. In our actual dataset, concept definitions can be more complex and involve up to 6 different words, being them nouns, adjectives or the preposition *"of"*.

[6] The list of all the sentence and formula templates is available here: https://drive.google.com/file/d/0B_FaCg6LWgw5Z0UxM2N1dTYwYkU.

[7] The several training, evaluation and test sets used in the experiments are available here: https://drive.google.com/file/d/0B_FaCg6LWgw5ZnBkSEVONWx2YW8.

Table 2. Patterns for concept name generation

Adj.	Noun #1	Noun #2
Cool	Sword	Sharpener
-	Sword	Sharpener
Cool	-	Sharpener
Cool	Sword	-
-	Sword	-
-	-	Sharpener

Table 3. From a sentence template to an example.

Sent. template	*A* $\mathtt{C_0}$ *is a* $\mathtt{C_1}$ *that* $\mathtt{R_0}$ *exactly* $\mathtt{N_0}$ $\mathtt{C_2}$
Sentence	*A bee is a insect that has exactly* $\mathtt{N_0}$ *legs*
Tag sequence	$\mathtt{w\ C_0\ w\ w\ C_1\ w\ R_0\ w\ N_0\ C_2}$
Form. template	$\mathtt{C_0} \sqsubseteq \mathtt{C_1} \sqcap = \mathtt{N_0 R_0 . C_2}$

We applied some basic preprocessing to the text: using only *"a"* as indeterminate article, using only *"do not"* as the negative form, using only singular words (*"Bee"* for *"Bees"*) and the avoiding the third singular person verbal form (*"take"* for *"takes"*). Numbers placeholders, used only in cardinality restrictions, have not been not filled with actual numbers. This missing substitution can be seen as another text preprocessing phase for number identification, which is a task that can be performed easily with a rule based approach. All the generate sentences are *actual natural language sentences* since they are grammatically correct, even if potentially unrealistic from a human point of view – e.g. *"A smoking hurricane is also something that pump at least 18 orange stampede"*.

Transforming such a sentence into a tag sequence is straightforward: each word in the original sentence template is tagged as `w`, while the words filling a placeholder are tagged with the same symbol of the placeholder. Summing up, Table 3 reports all the elements of a training example, starting from a given sentence template. Given a set of examples $\mathcal{T} = \{(s, t, f)\}$, the set of all the pairs of sentences and formulas $\mathcal{T}_F = \{(s, f)\}$ will be a training/test set for the sentence transduction task, while the set of all the pairs of sentences and tag sequences $\mathcal{T}_T = \{(s, t)\}$ will be used for the sentence tagging task.

We generated different training sets, of different dimensions. Fixed the number of training examples, we randomly generated sentence templates from our grammar, turned each of such sample templates into a sentence, and generated the corresponding tag sequence and formula template, emulating the work of a human annotator. In this way, we could test the actual generalisation capabilities of our model from the syntactic point of view. We also randomly marked some words filling the placeholders for concepts and roles as out-of-vocabulary with the `<UNK>` word; for concepts we used the `<UNK>` symbol with 60 % of probability, while for roles we used 20 %. Overall, we re-scanned the whole sentence ensuring that the number of placeholder fillers turned to `<UNK>` was between 20 % and 40 %. In this way we could also test the generalisation capabilities of the model from the lexical point of view. We also generated an evaluation set for each training set, of the same dimension, starting from the same sentence templates, filled with different words. We used such evaluation set to check the network status during the training phase, to be sure that the network was not just

Table 4. Network parameters.

Networks params.	
# words	~5000
# tags	11
# terms	21
Word window	5
dim. embedding	100
dim. hidden (tag.)	200
dim. enc/dec (tra.)	1000

Table 5. Training parameters.

Training parameters	
Training steps	10000
Batch size	50
Learning algo.	AdaDelta
Learning rate	2.0
ρ	0.95
ϵ	10^{-6}
GPU card	Tesla K40
Time (tag.)	~2.5 h
Time (tra.)	~4.5 h

Table 6. Accuracy on the test set

dim.	AT	AF
1000	99.9 %	96.2 %
2000	99.9 %	99.0 %
3000	99.9 %	99.6 %
4000	99.9 %	99.8 %

memorizing the training examples. Finally, for each dataset, we built a larger test set starting from 2 millions of sentence and formula templates generated from our grammar: in this way we ensured that the test sentences are unseen during the training phase. Such templates were turned into the actual test set with the very same procedure followed for the training and the evaluation sets, with the slight difference of increasing the overall minimum probability of out-of-vocabulary words to 30 %. Although the model used for sentence transduction is made of two stacked recurrent layers, they are jointly trained: the first layer produces an embedding of the input sentence which is then decoded into a formula template by the second. Our gold truth is this final formula template. For the sentence tagging model, the gold truth is the output tag sequence.

Experimental Setting and Results. The goal of our experiments was to assess the accuracy of a trained RNN-based architecture in learning expressive $\mathcal{ALCQ}$ axioms from typical definitory sentences. Therefore, we trained and evaluated the proposed architecture on several datasets produced following the procedure described before. For both tasks, tagging and transduction, we defined the network parameters empirically, according to some experiences in literature. We trained both the networks with AdaDelta (see [21]) for 10000 steps, with batches of 50 examples, evaluating the network against the evaluation set every 100 training steps. The network configuration, together with the training parameters and some indication of the training phase duration, are reported in Table 4 and Table 5. The dimensions of the training set (i.e. the amount of annotated examples), together with the results in terms of accuracy in tagging (AT) and in transduction (AF) are reported in Table 6.

We achieve almost 100 % of accuracy for all the datasets in the tagging task and over 96 % of accuracy in the transduction task, thus confirming the feasibility of building accurate RNN-based architectures that learn expressive $\mathcal{ALCQ}$ axioms from typical definitory sentences. We want to remark that none of the test sentences is in the training set, so there is no possibility for the networks to memorize the examples. We remark that in the transduction task, even if the number of possible tag sequences and formula templates is limited, our networks do not classify examples but learn how to *generate* new sequences of symbols, namely a formula template, starting from an initial natural language sentence.

6 Discussion

Reviewing the Ontology Learning state of the art presented in Sect. 2 we can highlight a main intuition: *syntax matters.* Pattern-based approaches, such as LExO, require to manually define rules exploiting the syntactic aspects of the text in order to extract the knowledge it carries. In our work, we pushed such intuition into a totally different direction, with a *learn by examples* approach. We trained a model that *learns* to model encyclopaedic language and its syntactic structures, it *learns* how to parse their occurrences in the text and how to translate them into corresponding logical constructs. Roughly speaking, our tagging model can be seen as a POS tagger, and our transduction model can be seen as a syntactic parser. Both of them extremely *specialized* w.r.t. the type of language of interest. Our model stores in its parameters the embedding of each words in the vocabulary (plus the `<UNK>` word), how to deal with function words, and many other syntactic constraints. Being our model trained in a end-to-end fashion, this *knowledge* —namely the features learnt by the model— remains in the subsymbolic form and is not made explicit.

Our experiments, in which we could achieve extremely high accuracy just annotating 1000 sentences, shows that statistically learning such rules is feasible. Furthermore, our contribution presents several advantages over state-of-the-art pattern-based approaches: (i) it does not require to manually define pattern rules for each possible linguistic natural language variation to be covered, something practically unfeasible; (ii) our model is trained in an end-to-end fashion, from raw text to OWL formulae, without relying on any NLP tool and requiring no feature engineering cost for the input representation; finally, (iii) being our approach purely syntactic, it does not need any domain-specific training: content words of our test set are selected randomly, showing as our model does not rely on their meaning but only on their syntactical features.

Despite being generated starting from sentences in ACE, our system can deal with language variability that goes well beyond controlled English. To confirm this, we generated 4000 random sentence templates and, from them, a set of sentences, filling the various placeholders with a simplified vocabulary compliant with the Attempto Parser Engine (APE)[8]. The APE engine could parse only 13 sentences. A qualitative analysis through the sentences not or incorrectly parsed by the APE service gave us an idea of some of the linguistic phenomena that our system can handle beyond the controlled language. We can roughly split them in two groups:

1. function words that are not parsed by the controlled language but that are actually used in natural language, such as:

[8] http://attempto.ifi.uzh.ch/site/resources/.

   (a) *"anything"* and *"any"* acting as universal quantifier;
   (b) *"at least"*, *"no more than"* or *"exactly"* for cardinality restrictions;
   (c) *"but"* as a conjunction in cardinality restrictions, e.g. *"less than 3 but more than 10"*;
   (d) *"also"*, in the right hand side of an implication, e.g. *"if something is a mammal, it is also an animal"*;
   (e) *"everything that is"* as a quantifier;
   (f) the use of *"some"* to indicate an intersection, e.g. *"some birds are flightless"*;
   (g) *"do not"* as a negation of a role.
2. constructs that are not parsed by the controlled language but are actually used in natural language, such as:
   (a) ellipsis of the demonstrative pronoun *"that"* in conjunction or disjunctions, e.g. *"everything that has a tail and (that) is a dog, also chases cats"*;
   (b) ellipsis of the demonstrative pronoun, role and range concept in conjunction or disjunction of cardinality restrictions, e.g. *"a bass is an instruments that has at least 4 (strings) or (that has) at most 6 strings"*;
   (c) ellipsis of the adverb in cardinality restriction, instead of *"exactly"*, as in *"a bee is an insect that has (exactly) 6 legs"*;
   (d) ellipsis of the quantifier, as in *"dogs are mammals"*.

7 Conclusion and Future Work

In this work we presented an approach for ontology learning where an RNN-based model is trained in a end-to-end fashion to translate definitory sentences into OWL formulae. A GRUs based RE-D is used to transduce a definitory sentence into the corresponding formula template, while a GRUs based RNN maps the proper word to the proper role within such formula template. We trained and tested our approach on a newly created dataset of sentence-formula template pairs, sampling from more than 123M distinct sentence templates and more than 260K distinct formula templates. Our system achieved almost 100 % of accuracy in the sentence tagging task and over 96 % in the sentence transduction task starting from only 1000 training examples. While converting arbitrary natural language text to OWL is still an ambitious, out-of-reach goal, these results give evidence of the capabilities of our approach in translating definition-like sentences to (complex) OWL axioms, while showing good syntactic and lexical generalization capabilities and a reduced tagging effort of 1000 sentences.

The main limitation of our work is that the model has been trained and evaluated on limited amount of data, modeling a *limited* portion of natural language in a sentence-by-sentence (i.e., one sentence is translated to one axiom) fashion. Further validations and evaluation are needed on more realistic data, showing even more language variability. Nonetheless, the encouraging results obtained in the paper pave the way to future extensions and generalizations aiming at tacking some of these limitations. In particular, in our future work we will investigate how to extend our approach (i) to handle portions of knowledge that can be spread across different sentences, overcoming sentence-by-sentence processing, and (ii) to cope with wider language variability, thus covering more realistic encyclopedic text.

References

1. Baader, F., Calvanese, D., McGuinness, D., Nardi, D., Patel-Schneider, P. (eds.): The Description Logic Handbook: Theory, Implementation, and Applications. Cambridge University Press, Cambridge (2003)
2. Bengio, Y., Ducharme, R., Vincent, P., Janvin, C.: A neural probabilistic language model. J. Mach. Learn. Resour. **3**, 1137–1155 (2003)
3. Bowman, S.R., Potts, C., Manning, C.D.: Recursive neural networks for learning logical semantics. CoRR abs/1406.1827 (2014)
4. Cho, K., van Merrienboer, B., Gülçehre, Ç., Bougares, F., Schwenk, H., Bengio, Y.: Learning phrase representations using RNN encoder-decoder for statistical machine translation. CoRR abs/1406.1078 (2014)
5. Cimiano, P., Mädche, A., Staab, S., Völker, J.: Ontology learning. In: Staab, S., Studer, R. (eds.) Handbook on Ontologies, pp. 245–267. Springer, Heidelberg (2009)
6. Graves, A., Wayne, G., Danihelka, I.: Neural turing machines. CoRR abs/1410.5401 (2014)
7. Grefenstette, E., Hermann, K.M., Suleyman, M., Blunsom, P.: Learning to transduce with unbounded memory. CoRR abs/1506.02516 (2015)
8. Kaljurand, K., Fuchs, N.: Verbalizing OWL in Attempto Controlled English. In: OWLED 2007 (2007)
9. Lehmann, J., Voelker, J. (eds.): Perspectives on Ontology Learning. Studies in the Semantic Web. AKA/IOS Press, Berlin (2014)
10. Ma, Y., Syamsiyah, A.: A hybrid approach to learn description logic based biomedical ontology from texts. In: ISWC 2014 Proceedings (2014)
11. Mesnil, G., He, X., Deng, L., Bengio, Y.: Investigation of recurrent-neural-network architectures and learning methods for spoken language understanding. In: INTERSPEECH 2013, pp. 3771–3775 (2013)
12. Mikolov, T., Yih, W., Zweig, G.: Linguistic regularities in continuous space word representations. In: NAACL HLT 2013, pp. 746–751 (2013)
13. Peng, B., Lu, Z., Li, H., Wong, K.: Towards neural network-based reasoning. CoRR abs/1508.05508 (2015)
14. Petrucci, G.: Information extraction for learning expressive ontologies. In: Gandon, F., Sabou, M., Sack, H., d'Amato, C., Cudré-Mauroux, P., Zimmermann, A. (eds.) ESWC 2015. LNCS, vol. 9088, pp. 740–750. Springer, Heidelberg (2015)
15. Petrucci, G., Ghidini, C., Rospocher, M.: Using recurrent neural network for learning expressive ontologies. CoRR abs/1607.04110 (2016)
16. Völker, J., Haase, P., Hitzler, P.: Learning expressive ontologies. In: Ontology Learning and Population: Bridging the Gap between Text and Knowledge, pp. 45–69. IOS Press, Amsterdam (2008)
17. Völker, J., Hitzler, P., Cimiano, P.: Acquisition of OWL DL axioms from lexical resources. In: Franconi, E., Kifer, M., May, W. (eds.) ESWC 2007. LNCS, vol. 4519, pp. 670–685. Springer, Heidelberg (2007)

18. Weston, J., Bordes, A., Chopra, S., Mikolov, T.: Towards AI-complete question answering: a set of prerequisite toy tasks. CoRR abs/1502.05698 (2015)
19. Weston, J., Chopra, S., Bordes, A.: Memory networks. CoRR abs/1410.3916 (2014)
20. Zaremba, W., Sutskever, I.: Learning to execute. CoRR abs/1410.4615 (2014)
21. Zeiler, M.D.: ADADELTA: an adaptive learning rate method. CoRR abs/1212.5701 (2012)

Interest Representation, Enrichment, Dynamics, and Propagation: A Study of the Synergetic Effect of Different User Modeling Dimensions for Personalized Recommendations on Twitter

Guangyuan Piao(✉) and John G. Breslin

Insight Centre for Data Analytics, NUI Galway, IDA Business Park, Lower Dangan, Galway, Ireland
guangyuan.piao@insight-centre.org, john.breslin@nuigalway.ie

Abstract. Microblogging services such as Twitter have been widely adopted due to the highly social nature of interactions they have facilitated. With the rich information generated by users on these services, user modeling aims to acquire knowledge about a user's interests, which is a fundamental step towards personalization as well as recommendations. To this end, researchers have explored different dimensions such as (1) *Interest Representation*, (2) *Content Enrichment*, (3) *Temporal Dynamics* of user interests, and (4) *Interest Propagation* using semantic information from a knowledge base such as DBpedia. However, those dimensions of user modeling have largely been studied separately, and there is a lack of research on the synergetic effect of those dimensions for user modeling. In this paper, we address this research gap by investigating 16 different user modeling strategies produced by various combinations of those dimensions. Different user modeling strategies are evaluated in the context of a personalized link recommender system on Twitter. Results show that *Interest Representation* and *Content Enrichment* play crucial roles in user modeling, followed by *Temporal Dynamics*. The user modeling strategy considering *Interest Representation*, *Content Enrichment* and *Temporal Dynamics* provides the best performance among the 16 strategies. On the other hand, *Interest Propagation* has little effect on user modeling in the case of leveraging a rich *Interest Representation* or considering *Content Enrichment*.

1 Introduction

With the popularity of microblogging services such as Twitter[1], the amount of information available on the Social Web is increasing exponentially. While this information is a valuable resource, its sheer volume limits its value [9]. On the Social Web, as the amount of information available causes information overload for users, the demand for personalized approaches towards information consumption increases. User (interest) modeling aims to analyze user activities on the

[1] https://www.twitter.com.

E. Blomqvist et al. (Eds.): EKAW 2016, LNAI 10024, pp. 496–510, 2016.
DOI: 10.1007/978-3-319-49004-5_32

Table 1. A sample tweet posted by Bob [22]

My Top 3 #lastfm Artists: Eagles of Death Metal(14), The Black Keys(6) & The Wombats(6). http://www.last.fm/user/bob

Social Web in order to provide personalized services for users. To create qualitative and quantitative user models for microblogging services such as Twitter, several design dimensions have been investigated in previous studies.

Interest Representation. The first step of user modeling is to determine how to represent user interests. Several approaches such as *bag-of-words*, *topic models* or *bag-of-concepts* have been used for representing user interests. Take an example from our own recent work (see Table 1 [22]), by using the bag-of-concepts approach, we can assume that the user is interested in DBpedia[2] entities such as `dbpedia`[3] and `dbpedia:The_Wombats` based on a tweet posted by a user named Bob. In addition, we can exploit background knowledge of entities from a Knowledge Base (KB) for extending user interests, e.g., categories of the entities in DBpedia. Throughout the paper, by a *concept* we mean an *entity*, *category* or *class* from a KB (e.g., DBpedia) for representing user interests.

Content Enrichment. As the ideal length of User-Generated Content (UGC) on microblogging services is short[4], there is a need to enrich this short content to better understand the context of it. Embedded links (URLs) in a tweet can be used to enrich the short content, and provide additional information about the tweet. For example, we can follow the link in the sample tweet to retrieve more information about Bob's musical interests. Many sources have shown that a large portion of tweets and retweets contain links[5,6].

Temporal Dynamics. Users might be interested in different topics over time. To capture the dynamics of user interests, some previous studies have used short-term profiles (e.g., considering a user's activities during the last two weeks only), while others have proposed interest decay functions to discount older interests.

Interest Propagation. This dimension exploits cross-domain background knowledge about concepts from a KB such as DBpedia. Based on the concepts directly spotted from UGC, related concepts in the KB can be leveraged for enriching user interest profiles. For instance, Bob (see Table 1) might be interested in `dbpedia:Indie_rock` as he likes indie rock artists such as `dbpedia:The_Black_Keys` and `dbpedia:The_Wombats` based on background knowledge from DBpedia, e.g., `dbpedia:The_Black_Keys` → `dbpedia-owl`[7]`:genre` → `dbpedia:Indie_rock`. Throughout the paper, we

[2] http://wiki.dbpedia.org.
[3] The prefix `dbpedia` denotes http://dbpedia.org/resource/:The_Black_Keys.
[4] http://goo.gl/uewQLu.
[5] http://marketingrelevance.com/news/04/tweet-interesting-information/.
[6] http://goo.gl/RGC16n.
[7] The prefix `dbpedia-owl` denotes http://dbpedia.org/ontology/.

denote the concepts that can be directly extracted from a user's tweets as ***primitive interests*** (e.g., `dbpedia:The_Wombats`), and the concepts that can be propagated from those primitive interests as ***propagated interests*** (e.g., `dbpedia:Indie_rock`).

Although related work reveals many promising insights with respect to those user modeling dimensions, there exists little research on studying the synergetic effect achieved by considering those dimensions together [20]. As those dimensions are not necessarily exclusive of each other, this has in turn motivated us to implement a user modeling framework which can exploit different dimensions at the same time for generating user interest profiles. We then evaluate different user interest profiles generated by different user modeling strategies in the context of a personalized link (URL) recommender system on Twitter.

The contributions of this work are summarized as follows.

- We implemented a user modeling framework, which can incorporate different combinations of four dimensions: (1) *Interest Representation*, (2) *Content Enrichment*, (3) *Temporal Dynamics*, and (4) *Interest Propagation*, to investigate (how) can we combine these different dimensions to retrieve better user interest profiles. To our knowledge, this is the first comprehensive study on these four dimensions.
- We evaluate 16 user modeling strategies generated by different combinations of methods for those four dimensions in the context of link recommendations on Twitter using four different evaluation metrics.

The organization of the rest of the paper is as follows. Section 2 gives some related work, and Sect. 3 describes our user modeling framework. In Sect. 4, we present the experiment setup for our study. Experiment results are presented in Sect. 5. Finally, Sect. 6 concludes the paper with some future work.

2 Related Work

In this section, we provide an overview of some related work from the literature for the aforementioned dimensions in user modeling.

Representation of User Interests. To represent user interest profiles, researchers began by *word-based* approaches such as *bag-of-words* [8,17], *topic modeling* [10]. Degemmis et al. [8] proposed a specific *word-based* approach - using WordNet[8] synsets (which are unordered sets of synonyms) for representing user interests. They showed that their *bag-of-synsets* approach outperformed a *bag-of-words* approach. As *word-based* approaches focus on the words themselves and do not provide semantic information about the words or the relationships among them, a research direction has been proposed over the past few years that uses *concept-based* representations of user interests using a KB from Linked Data form (e.g., Freebase, DBpedia) [4,5,19,23] or using an encyclopedia such as Wikipedia [12,15,16,18]. More recently, we showed that using synsets and

[8] https://wordnet.princeton.edu/.

concepts together for representing user interests can improve the quality of user modeling on Twitter in the context of link recommendations [21].

Enrichment for Short Messages. To better understand the semantics of short messages generated in microblogging services such as Twitter, some researchers have used the content of embedded links (URLs) in short messages to enrich the content [4,13]. In [4], the authors first used URLs in a user's tweets to enrich their content. After that, the user's interest profiles were constructed based on the enriched content. They showed that enriching short content for retrieving user interests enhances the variety and quality of the generated user profiles, and improves the performance of news recommendations.

Dynamics of User Interests. Many methods have been proposed to incorporate the temporal dynamics of user interests based on the hypothesis that the interests of users change over time [2,3,7,19]. For example, Abel et al. [3] studied short-term and long-term user profiles from Twitter for news recommendations. To construct a short-term user profile for a given user, they only used the user's tweets within the last two weeks. On the other hand, a long-term user profile was generated based on the user's entire historical tweets. Another line of work [2,7,19] that incorporates temporal dynamics applies a decay function to the interests of users. The rationale behind the decay function is that higher weights should be given to interests that have occurred recently and lower weights given to older interests.

Interest Propagation using Background Knowledge. There are various related works [19,22,23] that enrich *concept-based* user interest profiles using background knowledge. In [19], the authors built *category-based* user interest profiles by exploring DBpedia categories of entities, e.g., using categories such as `dbc`[9]`:Apple_Inc._executives` to denote user interests if a user is interested in `dbpedia:Steve_Jobs`. Piao et al. [22] proposed a mixed approach that combines the entity- and category-based profiles with the discounting strategy from [19], and proved that the mixed approach performs better than either the entity- or category-based approach. Building on this in a later work [23], the authors showed that by using Concept Frequency - Inverse Document Frequency (CF-IDF) as the weighting scheme and by leveraging different types of information from DBpedia to extend user profiles (i.e., *categories*, and *connected entities via different properties*), the quality of user modeling can be improved.

There are also some studies for user modeling with respect to a specific domain of user interests. For example, Abel et al. [5] proposed using DBpedia to extend user profiles with respect to point of interests (POI), and Nishioka et al. [18] explored different factors of user modeling for modeling user interests with respect to scientific publications in the economic domain. Different from focusing on user interests in a specific domain, our work focuses on user interests extracted from Twitter which are not limited to a specific domain.

While related work reveals several insights regarding each dimension of user modeling, hybrid approaches combining those different dimensions are

[9] The prefix `dbf` denotes http://dbpedia.org/resource/Category:.

considered only to a limited degree. For example, after enriching tweets with the content of embedded links, it would be interesting to explore if interest propagation using background knowledge further improves the quality of user modeling, or if it has little effect or no effect since enough information may already be available from a user's primitive interests.

3 Content-Based User Modeling

In this section, we first introduce user interest profiles as defined in our work, and then present a general process for generating user interest profiles (Sect. 3.1). Subsequently, we provide details of the methods for each of the user modeling dimensions used in the process (Sect. 3.2).

In this work, we use the same definition from [20] to represent the interests of users, which is specified as follows.

Definition 1. *The interest profile of a user $u \in U$ is a set of weighted DBpedia concepts or WordNet synsets, where with respect to a given user u who has an interest $i \in I$, its weight $w(u, i)$ is computed by a certain function w.*

Here, U denotes the set of users, and I denotes the set of concepts in DBpedia and synsets in WordNet, respectively. The weighting scheme $w(u, i)$ measures the importance of a concept with respect to a user. Previous studies showed that using CF-IDF as the weighting scheme provides better performance than using a Concept Frequency (CF) weighting scheme for user modeling in the context of recommender systems [18,23]. Similar to the TF-IDF weighting scheme used in *word-based* user modeling approaches [1], the rationale behind CF-IDF is discounting the weights of concepts appearing frequently in users' interest profiles and increasing the weights of concepts appearing rarely in users' profiles. In the same way, we use the Interest Frequency - Inverse Document Frequency (IF-IDF) as the weighting scheme for our experiments. More formally, it is defined as follows.

- $w_{IF}(u, i) = the\ frequency\ of\ i\ in\ a\ user's\ tweets,$
- $w_{IF-IDF}(u, i) = \underbrace{w_{IF}(u, i)}_{IF} \times \underbrace{\log \frac{M}{m_i}}_{IDF}$

where M is the total number of users, and m_i is the number of users interested in a concept/synset i.

3.1 The Process of Generating User Interest Profiles

Figure 1 presents the process of generating user interest profiles for Twitter considering the aforementioned four different user modeling dimensions. The components with dotted lines are options that can be either can be "enabled" or "disabled" for this user modeling. The process has three major steps:

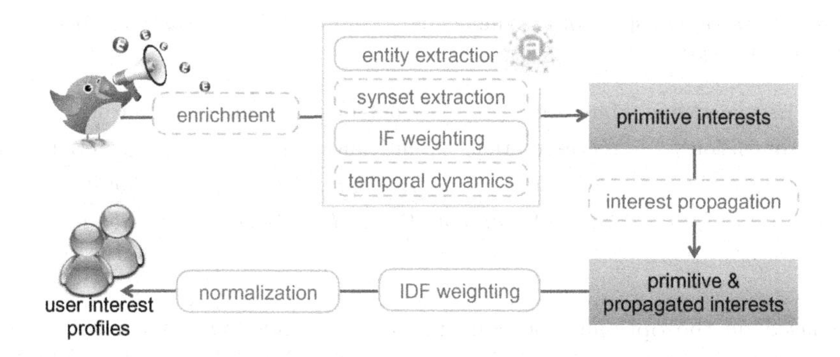

Fig. 1. The process of generating user interest profiles on Twitter

(1) **Primitive interests extraction**. For a given user, we extract all *primitive interests* (DBpedia entities or WordNet synsets) within UGC of a user. If the component *enrichment* is enabled, the content of links embedded in the UGC will also be used for extracting primitive interests.
   - DBpedia entities are extracted using the `Aylien API`[10]. For instance, the API extracts two entities `dbpedia:Microsoft` and `dbpedia:LinkedIn` from the phrase: "*Microsoft to Buy LinkedIn for $26B; LinkedIn to continue as separate brand*". Interest Frequency (IF) is applied to denote the importance of a concept with respect to a user. In addition, it might adhere to strategies for incorporating the *temporal dynamics* of user interests.
   - WordNet synsets can be extracted at the same time as extracting entities. The rationale behinds this is that syntactic information can complement semantic information for generating user interest profiles [21]. For example, given a tweet: "*Just completed a 3.89 km ride. We're gonna need more...*", we can extract synsets such as: s1 = [kilometer, kilometre, km, klick (a metric unit of length equal to 1000 meters (or 0.621371 miles))] and s2 = [drive, ride (a journey in a vehicle (usually an automobile))], which denote the user interests that would be missed if we used a concepts-alone approach.

(2) **Interest propagation.** This component can apply propagation strategies to primitive interests based on background knowledge from DBpedia. The output here is a user interest profile consisting of *primitive interests* as well as *propagated interests.*

(3) **Weighting and normalization**. Finally, the user modeling framework applies Inverse Document Frequency (IDF) to the user interest profile, and further normalizes the profile so that the sum of all weights in the profile is equal to 1: $\sum_{i \in I} w(u,i) = 1$.

[10] http://aylien.com.

Table 2. The design space of user modeling, spanning $2 \times 2 \times 2 \times 2 = 16$ possible user modeling strategies

User modeling dimensions	Interest representation	Content enrichment	Temporal dynamics	Interest propagation
Options	*Concept*	*Enabled*	*Enabled*	*Enabled*
	Synset & concept	*Disabled*	*Disabled*	*Disabled*

Based on the optional components for user modeling (shown with dotted lines in Fig. 1), there are 16 possible strategies which are displayed in Table 2. In the following subsection, we provide details of the methods for each dimension.

3.2 Methods for Each Dimension

Interest Representation: (1) Concept, or (2) Synset & Concept. *Entity recognition* and *synsets extraction* are performed in the first step to extract *primitive interests* from a user's tweets.

Entity recognition in tweets is a challenging task due to the informal nature of and ungrammatical language in tweets. Since our focus in this work is on user modeling and not entity recognition, we have used an existing solution for entity recognition (as does related literature on user modeling).

Different Natural Language Processing (NLP) APIs have been used for DBpedia/Wikipedia entity recognition in the literature. For example, Kapanipathi et al. [12] used the Zemanta API (which is no longer available) after comparing it to other APIs such as DBpedia Spotlight[11], Fattane et al. [24] used tag.me[12], and Piao et al. [23] used the Aylien API, respectively.

Table 3. Evaluation of NLP APIs for DBpedia/Wikipedia entity recognition

API	Precision	Recall	F-measure
Aylien	0.27	0.26	0.26
Alchemy	0.21	0.17	0.19
tag.me	0.12	0.15	0.14

To better investigate the performance of different APIs, we used the Twitter dataset from [14] which contains annotated 1,603 tweets in total where 1,233 of them contain Wikipedia entities. We tested three different NLP APIs: Aylien API, tag.me and Alchemy API[13], which all provide functionality for extracting entities from a given text and representing these with corresponding DBpedia/Wikipedia URIs. A comparative performance is displayed in Table 3. We opted to use the Aylien API for our experiment since it (1) extracts DBpedia entities (*primitive interests*) identified in tweets, and gives their corresponding

[11] http://spotlight.dbpedia.org/rest/annotate, the web service was not accessible at the time of writing this paper.

[12] https://tagme.d4science.org/tagme/.

[13] http://www.alchemyapi.com/.

URIs, (2) it has relatively superior performance to the other APIs as shown in Table 3, and (3) it provides 6,900 calls per day, provided on request for research purposes.

Synset extraction is included in the investigation since concepts from a KB could not express user interests completely [21]. On one hand, there might be new concepts/topics emerging in microblogging services such as Twitter, which cannot be found in a KB. On the other hand, the earlier work [21] showed that using WordNet sunsets and DBpedia concepts together is helpful for improving the quality of user interest profiles. In this regard, in the same way from [21], we adopt a method from [8] which extracts WordNet synsets to build *synset-based* user interest profiles.

Content Enrichment: (1) Enabled, or (2) Disabled. We leverage the content of links embedded in a tweet to enrich the original post content. Based on the selected option for the dimension *Interest Representation*, we apply the same extraction method for the content of embedded links. Therefore, in the case of *concepts* being used for *Interest Representation*, the *concepts* extracted from the content of links embedded in tweets will also be considered as user interests if the *Content Enrichment* dimension option is enabled.

Temporal Dynamics: (1) Enabled, or (2) Disabled. In [23], the authors conducted a comparative study on different interest decay functions [2,6,19] in the context of recommender systems on Twitter. Results showed that those functions have similar performance. We choose a variant of the interest decay function from [6], which performed best overall in the comparative study [23]. This decay function [23] measures the expected weight in terms of an interest i for user k at time t by combining three levels of abstractions using a weighted sum as below:

$$w_{ki}^{t} = \mu_{2week} w_{ki}^{t,2week} + \mu_{2month} w_{ki}^{t,2month} + \mu_{all} w_{ki}^{t,all} \quad (1)$$

where $\mu_{2week} = \mu$, $\mu_{2month} = \mu^2$ and $\mu_{all} = \mu^3$ and $\mu \in [0, 1]$. We set μ as e^{-1} in the same manner as [6,23], for our experiment.

Interest Propagation: (1) Enabled, or (2) Disabled. In [23], the authors also proposed different propagation strategies exploiting different types of background knowledge from DBpedia. Overall, the propagation strategy extending *primitive interests* with categories (Fig. 2(a)) and entities connected via different properties (Fig. 2(b)) in DBpedia, provided the best performance compared to other state-of-art propagation strategies.

As previous studies [19,22] showed that a discounting strategy is required for the extended concepts based on primitive interests, the authors [23] applied a discounting strategy from [22] for the extended categories as follows:

$$CategoryDiscount = \frac{1}{\alpha} \times \frac{1}{\log(SP)} \times \frac{1}{\log(SC)} \quad (2)$$

where: *SP = Set of Pages belonging to the Category, SC = Set of Sub-Categories.* We set the parameter $\alpha = 2$ as in [23]. Thus, an extended category is discounted

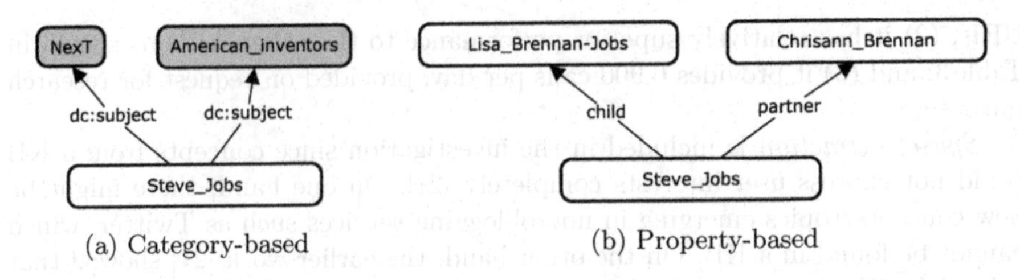

(a) Category-based (b) Property-based

Fig. 2. Three core strategies using DBpedia for extending user interests

heavily if it is a general one, i.e., the category has a great number of pages or subcategories. In addition, the parameter α denotes the discount of the propagated interests from primitive interests. Regarding the property-based extension strategy (Fig. 2(b)), extended entities via different properties are discounted based on the occurrence frequency of a specific property in DBpedia [23]:

$$PropertyDiscount = \frac{1}{\alpha} \times \frac{1}{\log(P)} \tag{3}$$

where: P = *the number of occurrences of a property in the whole DBpedia graph.* The intuition behind *PropertyDiscount* is that entities extended via a property appearing rarely in the DBpedia graph should be given a higher weight than ones extended via a property appearing frequently.

4 Experiment Setup

In the following section, we describe the Twitter dataset used in our experiment (Sect. 4.1), and the evaluation methodology (Sect. 4.2). Subsequently, we present the results using 16 different user modeling strategies in the context of link recommendations on Twitter (Sect. 4.3).

4.1 Twitter Dataset

The dataset used in this experiment is a Twitter dataset from [22], which includes over 340,000 tweets from 480 *active* users on Twitter. An active user denotes that the user published at least 100 Twitter posts [11,15,22]. Table 4 shows the basic statistics about the dataset.

Dataset for link recommendations. In the same way as [23], we further selected users who shared at least one link in their tweets during the previous two weeks, leaving 322 users for our experiment to run upon. We limit our consideration to links having at least four concepts to filter out non-topical links that were automatically generated by third-party applications such as Swarm[14].

[14] https://www.swarmapp.com.

Table 4. Twitter dataset statistics

# of users	480
Total # of tweets	348,554
Average time span of tweets per user (days)	471
Average # of tweets per user	726
Average # of tweets per user per day	7.2

4.2 Evaluation Methodology

We were interested in finding whether combinations of different user modeling dimensions improve the quality of user interest profiles in the context of link recommendations. Therefore, the input to our link recommender system is user interest profiles generated by different user modeling strategies, whereas the output is recommended links (URLs) for users. A lightweight content-based algorithm, like the one used in [5], was applied for recommendations.

Definition 2. *Recommendation Algorithm: given a user profile* P_u *and a set of candidate links* $N = \{P_{i1}, \ldots, P_{in}\}$*, which are represented via profiles using the same vector representation, the recommendation algorithm ranks the candidate items according to their cosine similarity to the user profile.*

We assumed a user was interested in the content of a link if the link was shared by the user in his or her tweets. The ground truth of links was a set of links shared via the user's tweets within the last two weeks, which consists of 3,959 links. Tweets before the last two weeks were used for building user interest profiles. To construct candidate links for recommendations, we further included the links shared by other users but not shared by 322 users in the dataset in addition to the ground truth links from 322 users. The resulting candidate set of links consists of 15,440 distinct links.

The link recommender system measures similarities between a user interest profile and each candidate link, and then provides top-N recommendations based on the similarity scores. We focused on $N = 10$ in our experiment, i.e., the recommendation system would list 10 link recommendations to a user. We measure the quality of recommendations by looking at four different metrics, which were frequently used in the literature [3,5,19,21,23].

- **MRR** The *MRR* (Mean Reciprocal Rank) indicates at which rank the first item *relevant* to the user occurs on average.
- **S@N** The Success at rank N (S@N) stands for the mean probability that a relevant item occurs within the top-N ranked.
- **R@N** The Recall at rank N (R@N) represents the mean probability that *relevant* items are successfully retrieved within the top-N recommendations.
- **P@N** The Precision at rank N (P@N) represents the mean probability that retrieved items within the top-N recommendations are *relevant* to the user.

We set a significance level of alpha = 5 % for all statistical tests. The *bootstrapped paired t-test*[15] was used for testing the significance.

5 Results

In this section, we present the results of experiments using different user modeling strategies in the context of link recommendations. In the following, let *um*(*representation*, *enrichment*, *dynamics*, *semantics*) denote a user modeling strategy where four parameters: *representation*, *enrichment*, *dynamics* and *semantics* represent the four dimensions *Interest Representation*, *Content Enrichment*, *Temporal Dynamics* and *Interest Propagation*, respectively. We use "none" to denote a certain dimension is disabled. For instance, *um*(*concept*, *none*, *none*, *none*) denotes a user modeling strategy using concepts for *Interest Representation* without considering any other dimensions. *um*(*synset* & *concept*, *enrichment*, *none*, *none*) denotes a user modeling strategy using synsets and concepts for *Interest Representation*, and tweets are enriched by the content of embedded links when extracting user interests (i.e., the dimension *Content Enrichment* is enabled).

Table 5 summarizes the recommendation performance using the 16 user modeling strategies in terms of different evaluation metrics. The results are sorted in descending order in terms of MRR. Overall, the best performing strategy is *um*(*synset* & *concept*, *enrichment*, *dynamics*, *none*), which uses DBpedia concepts and WordNet synsets for *Interest Representation*, and considers all other dimensions except *Interest Propagation*. Table 5 shows the importance of (1) *Content Enrichment*, and (2) *Interest Representation* in user modeling. For instance, the strategies enriching tweets with embedded links (1–8 in Table 5) have better performance than the ones without any enrichment (9–16), using the same option for *Interest Representation*. In terms of *Interest Representation* with or without *Content Enrichment*, we observe that using DBpedia concepts with WordNet synsets (1–4 and 9–12) always provides better performance than using concepts alone (5–8 and 13–16). In line with previous work [21], exploiting semantic and lexical knowledge from DBpedia as well as WordNet for *Interest Representation* improves the quality of user modeling.

Table 6 further illustrates statistical differences between the 16 user modeling strategies in terms of MRR. Overall, the results of other evaluation metrics are similar to the MRR and thus omitted for reasons of brevity. The vertical and horizontal dimensions of the table show the comparison between the 16 strategies. As we can see from the table, there are various significant differences between the strategies ($p < .05$, marked in bold font). For example, strategies using concepts and synsets for the dimension *Interest Representation* always significantly outperform strategies using concepts, when other dimensions are kept the same (e.g., 1 and 5). The dimension *Interest Propagation* plays an important role when we use concepts for *Interest Representation* without *Content Enrichment* (13–16). However, when we have a rich interest representation

[15] http://www.sussex.ac.uk/its/pdfs/SPSS_Bootstrapping_22.pdf.

Table 5. Performance of link recommendations using 16 user modeling strategies four different evaluation metrics. The results are sorted in descending order in terms of MRR.

	User Modeling Strategies	MRR	S@10	R@10	P@10
1.	um(synset & concept, enrichment, dynamics, none)	0.3251	0.5062	0.1700	0.1304
2.	um(synset & concept, enrichment, dynamics, propagation)	0.3198	0.4938	0.1654	0.1298
3.	um(synset & concept, enrichment, none, none)	0.3146	0.4876	0.1595	0.1286
4.	um(synset & concept, enrichment, none, propagation)	0.3107	0.4752	0.1534	0.1267
5.	um(concept, enrichment, dynamics, none)	0.2942	0.4193	0.1405	0.1047
6.	um(concept, enrichment, none, none)	0.2886	0.4379	0.1392	0.1062
7.	um(concept, enrichment, dynamics, propagation)	0.2802	0.3975	0.1287	0.0988
8.	um(concept, enrichment, none, propagation)	0.2736	0.4130	0.1332	0.1006
9.	um(synset & concept, none, dynamics, none)	0.2511	0.4255	0.1257	0.0988
10.	um(synset & concept, none, dynamics, propagation)	0.2502	0.4193	0.1259	0.0997
11.	um(synset & concept, none, none, none)	0.2436	0.4068	0.1231	0.0978
12.	um(synset & concept, none, none, propagation)	0.2386	0.3913	0.1179	0.0984
13.	um(concept, none, none, propagation)	0.2083	0.3540	0.0993	0.0820
14.	um(concept, none, dynamics, none)	0.2031	0.3354	0.0927	0.0752
15.	um(concept, none, dynamics, propagation)	0.2024	0.3478	0.0923	0.0795
16.	um(concept, none none, none)	0.1518	0.2609	0.0660	0.0553

(i.e., using concepts and synsets together) or rich content by enrichment, *Interest Propagation* has little effect on the quality of user modeling, i.e., there is no statistical difference between a user modeling strategy with *Interest Propagation* and one without any propagation (1–12). One of the possible reasons might be the rich interest representation, and content is giving sufficient knowledge of user interests. Additionally, the "insufficient quality" of extracted DBpedia entities from tweets using APIs (see the precision in Table 3 in Sect. 3.2), could result in inaccurate interest propagation based on the incorrect entities. This might limit the contribution of propagated interests towards user modeling.

Similar results can be found for temporal dynamics. Although considering *Temporal Dynamics* increases the performance significantly when we use concepts for *Interest Representation* without *Content Enrichment* (13–16), there is no significant difference between strategies with a rich interest representation and rich content (1–12). Nevertheless, we observe that in all of the cases using concepts and synsets for *Interest Representation*, considering the dimension *Temporal Dynamics* provides the best performance (see 1, 9 in Table 5).

To sum up, the two dimensions *Interest Representation* and *Content Enrichment* play significant roles for user modeling, followed by *Temporal Dynamics*. Although the contribution of content enrichment via embedded links might depend on the percentage of embedded links, it is an important and valuable source for enrichment as a large number of tweets are posted with links[16]. The results also show that the *Interest Propagation* dimension had little effect on user modeling when considering different dimensions together, which is different from previous studies considering one or two dimensions [2,19,22,23].

[16] 70 % of one million tweets from U.S. West Coast included links. http://tnw.to/s3R2i.

Table 6. Results of p-values over the 16 user modeling strategies in terms of link recommendations on Twitter (marked in bold font if $p < .05$). Strategies are sorted by MRR results as shown in Table 5.

		2.	3.	4.	5.	6.	7.	8.	9.	10.	11.	12.	13.	14.	15.	16.
1.	um(synset & concept, enrichment, dynamics, none)	.14	.17	.11	**.01**	**.02**	**.00**	**.00**	**.00**	**.00**	**.00**	**.00**	**.00**	**.00**	**.00**	**.00**
2.	um(synset & concept, enrichment, dynamics, propagation)		.35	.21	**.04**	**.04**	**.01**	**.01**	**.00**	**.00**	**.00**	**.00**	**.00**	**.00**	**.00**	**.00**
3.	um(synset & concept, enrichment, none, none)			.24	.10	.05	**.03**	**.01**	**.00**	**.00**	**.00**	**.00**	**.00**	**.00**	**.00**	**.00**
4.	um(synset & concept, enrichment, none, propagation)				.18	.10	**.03**	**.02**	**.00**	**.00**	**.00**	**.00**	**.00**	**.00**	**.00**	**.00**
5.	um(concept, enrichment, dynamics, none)					.31	.05	**.03**	**.02**	**.02**	**.01**	**.01**	**.00**	**.00**	**.00**	**.00**
6.	um(concept, enrichment, none, none)						.26	.05	**.03**	**.02**	**.01**	**.01**	**.00**	**.00**	**.00**	**.00**
7.	um(concept, enrichment, dynamics, propagation)							.26	.10	.08	.05	**.03**	**.00**	**.00**	**.00**	**.00**
8.	um(concept, enrichment, none, propagation)								.13	.13	.07	**.04**	**.00**	**.00**	**.00**	**.00**
9.	um(synset & concept, none, dynamics, none)									.42	.20	.08	**.01**	**.00**	**.00**	**.00**
10.	um(synset & concept, none, dynamics, propagation)										.22	.08	**.01**	**.01**	**.00**	**.00**
11.	um(synset & concept, none, none, none)											.15	**.02**	**.01**	**.01**	**.00**
12.	um(synset & concept, none, none, propagation)												**.04**	**.03**	**.02**	**.00**
13.	um(concept, none, none, propagation)													.32	.27	**.00**
14.	um(concept, none, dynamics, none)														.46	**.00**
15.	um(concept, none, dynamics, propagation)															**.00**
16.	um(concept, none, none, none)															

6 Conclusions

In this paper, we investigated different combinations of four dimensions of user modeling on Twitter: (1) *Interest Representation*, (2) *Content Enrichment*, (3) *Temporal Dynamics of user interests*, and (4) *Interest Propagation*, which have not been studied together. As a result, we end up with 16 different user modeling strategies with all possible combinations (see Table 2). These strategies were evaluated in the context of link recommendations on Twitter. The best-performing strategy is *um(synset & concept, enrichment, dynamics, none)*, which uses DBpedia concepts and WordNet synsets for *Interest Representation* considering *Temporal Dynamics*, with *Content Enrichment*. The results also indicate that *Interest Representation* and *Content Enrichment* are the most important dimensions compared to other dimensions. In future research, we would like to further investigate how different percentages of links in tweets affect the quality of user modeling.

Acknowledgments. This publication has emanated from research conducted with the financial support of Science Foundation Ireland (SFI) under Grant Number SFI/12/RC/2289 (Insight Centre for Data Analytics).

References

1. Abdel-Hafez, A., Xu, Y.: A survey of user modelling in social media websites. Comput. Inf. Sci. **6**(4), 59 (2013)
2. Abel, F., Gao, Q., Houben, G.J., Tao, K.: Analyzing temporal dynamics in Twitter profiles for personalized recommendations in the social web. In: Proceedings of 3rd International Web Science Conference, p. 2. ACM (2011)
3. Abel, F., Gao, Q., Houben, G.-J., Tao, K.: Analyzing user modeling on Twitter for personalized news recommendations. In: Konstan, J.A., Conejo, R., Marzo, J.L., Oliver, N. (eds.) UMAP 2011. LNCS, vol. 6787, pp. 1–12. Springer, Heidelberg (2011)
4. Abel, F., Gao, Q., Houben, G.-J., Tao, K.: Semantic enrichment of Twitter posts for user profile construction on the social web. In: Antoniou, G., Grobelnik, M., Simperl, E., Parsia, B., Plexousakis, D., De Leenheer, P., Pan, J. (eds.) ESWC 2011, Part II. LNCS, vol. 6644, pp. 375–389. Springer, Heidelberg (2011)
5. Abel, F., Hauff, C., Houben, G.-J., Tao, K.: Leveraging user modeling on the social web with linked data. In: Brambilla, M., Tokuda, T., Tolksdorf, R. (eds.) ICWE 2012. LNCS, vol. 7387, pp. 378–385. Springer, Heidelberg (2012)
6. Ahmed, A., Low, Y., Aly, M., Josifovski, V., Smola, A.J.: Scalable distributed inference of dynamic user interests for behavioral targeting. In: Proceedings of 17th International Conference on Knowledge Discovery and Data Mining, pp. 114–122. ACM (2011)
7. Budak, C., Kannan, A., Agrawal, R., Pedersen, J.: Inferring user interests from microblogs. Technical report (2014)
8. Degemmis, M., Lops, P., Semeraro, G.: A content-collaborative recommender that exploits WordNet-based user profiles for neighborhood formation. User Model. User-Adap. Inter. **17**(3), 217–255 (2007)
9. Gauch, S., Speretta, M., Chandramouli, A., Micarelli, A.: User profiles for personalized information access. In: Brusilovsky, P., Kobsa, A., Nejdl, W. (eds.) The Adaptive Web. LNCS, vol. 4321, pp. 54–89. Springer, Heidelberg (2007). doi:10.1007/978-3-540-72079-9_2
10. Harvey, M., Crestani, F., Carman, M.J.: Building user profiles from topic models for personalised search. In: Proceedings of 22nd ACM International Conference on Information and Knowledge Management, pp. 2309–2314 (2013)

11. Jain, P., Kumaraguru, P., Joshi, A.: @i seek 'fb.me': identifying users across multiple online social networks. In: Proceedings of 22nd International Conference on World Wide Web companion, pp. 1259–1268. ACM (2013)
12. Kapanipathi, P., Jain, P., Venkataramani, C., Sheth, A.: User interests identification on Twitter using a hierarchical knowledge base. In: Presutti, V., d'Amato, C., Gandon, F., d'Aquin, M., Staab, S., Tordai, A. (eds.) ESWC 2014. LNCS, vol. 8465, pp. 99–113. Springer, Heidelberg (2014)
13. Kinsella, S., Wang, M., Breslin, J.G., Hayes, C.: Improving categorisation in social media using hyperlinks to structured data sources. In: Antoniou, G., Grobelnik, M., Simperl, E., Parsia, B., Plexousakis, D., De Leenheer, P., Pan, J. (eds.) ESWC 2011, Part II. LNCS, vol. 6644, pp. 390–404. Springer, Heidelberg (2011)
14. Locke, B.W.: Named entity recognition: adapting to microblogging. Ph.D. thesis (2009)
15. Lu, C., Lam, W., Zhang, Y.: Twitter user modeling and tweets recommendation based on Wikipedia concept graph. In: Workshops at the Twenty-Sixth AAAI Conference on Artificial Intelligence (2012)
16. Michelson, M., Macskassy, S.A.: Discovering users' topics of interest on Twitter: a first look. In: Proceedings of Fourth Workshop on Analytics for Noisy Unstructured Text Data, pp. 73–80. ACM (2010)
17. Mislove, A., Viswanath, B., Gummadi, K.P., Druschel, P.: You are who you know: inferring user profiles in online social networks. In: Proceedings of Third ACM International Conference on Web Search and Data Mining, pp. 251–260. ACM (2010)
18. Nishioka, C., Scherp, A.: Profiling vs. time vs. content: what does matter for top-k publication recommendation based on Twitter profiles? In: Proceedings of 16th ACM/IEEE-CS on Joint Conference on Digital Libraries, JCDL 2016, pp. 171–180. ACM, New York (2016)
19. Orlandi, F., Breslin, J., Passant, A.: Aggregated, interoperable and multi-domain user profiles for the social web. In: Proceedings of 8th International Conference on Semantic Systems, pp. 41–48. ACM (2012)
20. Piao, G.: Towards comprehensive user modeling on the social web for personalized link recommendations. In: Proceedings of 2016 Conference on User Modeling Adaptation and Personalization, UMAP 2016, pp. 333–336. ACM (2016)
21. Piao, G.: User modeling on Twitter with WordNet Synsets and DBpedia concepts for personalized recommendations. In: The 25th ACM International Conference on Information and Knowledge Management. ACM (2016)
22. Piao, G., Breslin, J.G.: Analyzing aggregated semantics-enabled user modeling on Google+ and Twitter for personalized link recommendations. In: User Modeling, Adaptation, and Personalization, pp. 105–109. ACM (2016)
23. Piao, G., Breslin, J.G.: Exploring dynamics and semantics of user interests for user modeling on Twitter for link recommendations. In: 12th International Conference on Semantic Systems, pp. 81–88. ACM (2016)
24. Zarrinkalam, F., Kahani, M.: Semantics-enabled user interest detection from Twitter. In: 2015 IEEE/WIC/ACM International Conference on Web Intelligence and Intelligent Agent Technology (2015)

Integrating New Refinement Operators in Terminological Decision Trees Learning

Giuseppe Rizzo[1(✉)], Nicola Fanizzi[1], Jens Lehmann[2], and Lorenz Bühmann[3]

[1] LACAM - Università degli Studi di Bari, Via Orabona 4, 70125 Bari, Italy
{giuseppe.rizzo1,nicola.fanizzi}@uniba.it
[2] Computer Science Institute, University of Bonn,
Römerstr. 164, 53117 Bonn, Germany
jens.lehmann@cs.uni-bonn.de
[3] AKSW- Univerität Leipzig, Augustusplatz 10, 04109 Leipzig, Germany
buehmann@informatik.uni-leipzig.de

Abstract. The problem of predicting the membership w.r.t. a target concept for individuals of Semantic Web knowledge bases can be cast as a concept learning problem, whose goal is to induce intensional definitions describing the available examples. However, the models obtained through the methods borrowed from *Inductive Logic Programming* e.g. Terminological Decision Trees, may be affected by two crucial aspects: the refinement operators for specializing the concept description to be learned and the heuristics employed for selecting the most promising solution (i.e. the concept description that describes better the examples). In this paper, we started to investigate the effectiveness of Terminological Decision Tree and its evidential version when a refinement operator available in DL-Learner and modified heuristics are employed. The evaluation showed an improvement in terms of the predictiveness.

1 Introduction

In the context of the Semantic Web, the effectiveness of the reasoning on the knowledge represented in ontological form through languages derived from Description Logics (DLs) [1] formalism is affected by the inherent incompleteness due to the Open World Assumption.

In the last years, resorting to machine learning methods have shown promising results for tackling this problem, for instance, by inducing predictive models to assess the membership of an individual w.r.t. a given concept for supporting various tasks such as (approximate) query answering and ontology completion. Despite the large availability of inductive methods for solving the problem [2], in this work (and similarly to other previous ones [3–5]) we focused on methods borrowed from *Inductive Logic Programming* (ILP) for solving the concept learning problem. These methods produce intentional definitions that describe the available instances that can be used for classifying them and therefore offering a trade-off between comprehensibility and predictiveness. In these methods, the learning is usually considered as a search process where the best solution as possible (i.e. the most accurate description among the possible ones describing

E. Blomqvist et al. (Eds.): EKAW 2016, LNAI 10024, pp. 511–526, 2016.
DOI: 10.1007/978-3-319-49004-5_33

the instances) is obtained via refinement operators to specialize or generalize the promising concept description, i.e. for obtaining a new concept description which subsumes or is subsumed by the given one. Such methods,e.g. DL-FOIL [6], typically resorts to a *separate-and-conquer* strategy that aims at covering the largest number of positive instances excluding the negative ones. More recently, DL-LEARNER [7] has become a state-of-the-art framework that provides the implementation of various learning algorithms such as CELOE [8] and EL TREE LEARNER (ELTL) [9].

However separate-and-conquer methods suffer of some drawbacks. For instance, such methods learn one concept description at once. In addition, separate-and-conquer approaches tend to consider partial solutions more times yielding inefficient solutions for the learning problem. Finally, these methods may fail to induce the description when the learning problem is hard. On the other hand, *divide-and-conquer* strategies have been exploited to overcome such problems. Among divide-and-conquer solutions, it is possible to mention decision tree models, which have been devised for solving learning problems, also in the context of multi-relational data representations and, in particular, for knowledge bases modeled with Description Logics formalism. Such extensions are called *Terminological Decision Trees* [3]. Also, further extensions, namely *Evidential Terminological Decision Trees*, are able to represent the uncertainty and to handle the presence of tests with uncertain result by resorting to the Dempster-Shafer Theory [4,5,10]. In order to improve the quality of the aforementioned models, there are two crucial aspects that should be investigated: the refinement operator adopted to generate the candidate concept descriptions to be installed as a new node and the heuristics for selecting the best description [11]. Specifically, for both Terminological Decision Trees and their evidential version, the refinement operator used in [3–5] may not generate candidates that discerns the positive instances from the negative ones, likely due to the nature of the employed operator which exploits randomly generated sub-concepts and roles of a knowledge base. As a consequence, the resulting specializations may be not definitely related to the target concept and a large number of both missing values and misclassification cases may be found in the test phase. This problem affects also the values of the heuristic employed for selecting the best concept description: the candidates concepts have similar values of either information gain (in the case of the terminological decision trees) or non-specificity measure (in the case of the evidential terminological decision trees [4]). Moving from this idea, we carried out a preliminary analysis concerning the effectiveness of tree models endowed with another refinement operator and additional measures integrated into the heuristic employed for inducing the models. Specifically, we used a refinement operator adopted by CELOE and implemented in DL-LEARNER and introduced a regularized versions of the heuristics used for the best concept selection which is based on the Jaccard similarity.

The rest of the paper is organized as follows: Section 2 recalls the notion of DL knowledge bases and the refinement operator; Sect. 3 gives some notions about the Terminological Decision Trees and Evidential Terminological Decision Trees and describes the procedure for inducing *Terminological Decision Trees* that includes both the novel refinement operator and a Jaccard-based regularization term in the heuristic exploited for selecting the best concept, Sect. 4 proposes an empirical evaluation in order to understand the effectiveness of the proposed changes in the Decision Tree learning algorithms. Finally, conclusions and further outlooks are reported.

2 Basics

In this section we recall the notions concerning Description Logics and we introduce the class-membership prediction and the concept learning problems. Finally, we briefly provide some notions about the Dempster-Shafer Theory that are used by the extension of terminological decision tree considered in the paper.

2.1 Description Logics and Knowledge Bases

Description Logics (DLs) [1] are a family of knowledge representation languages exploited to model a domain in terms of *concepts* and *roles*. Given a set of atomic concept names $N_C = \{A, B, \cdots\}$ and roles $N_R = \{R, S, \cdots\}$, more complex concept descriptions (usually denoted by the letters $C, D, \cdots$) regarding a set of objects, named *individuals*, can be built by using a set of operators (e.g. complement, conjunction and disjunction between concepts). The set of operators adopted to build the concept descriptions determines the expressiveness of the representation language. In DLs, the knowledge about the domain is intensionally modeled by using a set of inclusion (*subsumption*) axioms between the concepts such as $C \sqsubseteq D$ (C is subsumed by D). Also, the domain can be described by a set of facts concerning the individuals. Such facts are called *concept and role assertions* and they are usually denoted by $C(a)$ and $R(a, b)$. Therefore, a DL *knowledge base* is a couple $\mathcal{K} = (\mathcal{T}, \mathcal{A})$ where $\mathcal{T}$ is the TBox containing the intensional knowledge and $\mathcal{A}$ is the Abox containing the assertions. We will denote the set of individuals occurring in $\mathcal{A}$ by $\mathsf{Ind}(\mathcal{A})$.

Similarly to other first-order logic-based formalisms, the semantics is defined for each concept/role/individual by interpreting them according to the *model-theoretic semantics*. Formally, an interpretation is a couple $\mathcal{I} = (\Delta^{\mathcal{I}}, \cdot^{\mathcal{I}})$ composed by a non-empty set of objects representing the *domain* of the interpretation $\Delta^{\mathcal{I}}$ and an *interpretation function* $\cdot^{\mathcal{I}}$ that maps: (1) each individual a to an object $a^{\mathcal{I}} \in \Delta^{\mathcal{I}}$; (2) each concept C to a subset $C^{\mathcal{I}} \subseteq \Delta^{\mathcal{I}}$; (3) each role R to a subset $R^{\mathcal{I}} \subseteq \Delta^{\mathcal{I}} \times \Delta^{\mathcal{I}}$. The semantics of a complex description, say C is defined by applying recursively the interpretation function to the concepts used to build C. According to the model-theoretic semantics, an interpretation $\mathcal{I}$ *satisfies* an axiom $C \sqsubseteq D$ when $C^{\mathcal{I}} \subseteq D^{\mathcal{I}}$ and an assertion $C(a)$ (resp. $R(a, b)$) when $a^{\mathcal{I}} \in C^{\mathcal{I}}$ (resp. $(a^{\mathcal{I}}, b^{\mathcal{I}}) \in R^{\mathcal{I}}$). $\mathcal{I}$ is a *model* for $\mathcal{K}$ when it satisfies each axiom/assertion α in $\mathcal{K}$ ($\mathcal{I} \models \alpha$). When the axiom α is satisfied w.r.t. these models, we write $\mathcal{K} \models \alpha$. Various reasoning services are available for making new inferences from $\mathcal{K}$, which may involve either the TBox or the ABox. Among them, we recall the *instance-checking* inference service that is crucial from an inductive point of view: given an individual a and a concept description C the goal is determine if $\mathcal{K} \models C(a)$. The *Open World Assumption* (OWA) that is usually made in this context, may affect the ability to prove the truth of either $\mathcal{K} \models C(a)$ or $\mathcal{K} \models \neg C(a)$, as there may be possible to find different interpretations that satisfy either cases.

In the sequel we will denote by $sh \downarrow$ for a concept A (a role R), the set of direct (asserted) sub-classes (resp. sub-roles) of the atomic concept A (resp. role R). Besides, a role R is *applicable* when $\exists A \in N_C$ where $domain(R) \sqsubseteq A$ and there is no A' such that $domain(R) \sqsubseteq A' \sqsubseteq A$. Finally, we denote as $ar(R)$ a concept as $A \in N_C$ where $range(R) \sqsubseteq A$ and there is no A' such that $range(R) \sqsubseteq A' \sqsubseteq A$.

2.2 Class-Membership Prediction and Concept Learning Problem

The task of assessing the membership of an individual w.r.t. a target concept through inductive methods aims at approximating a function from the available instances that allows to determine if an individual is an instance of the concept or not. A possible formalization of the problem, as proposed in [5], is reported below:

Definition 1 (Class-Membership Prediction Problem).

Given

- *a target concept* C;
- *a label set* $\mathcal{L} = \{-1, 0, +1\}$
- *an error threshold* ϵ
- *a training set* $Tr \subseteq \mathsf{Ind}(\mathcal{A})$ *of examples for which – the correct classification value of* $t_{\mathsf{C}}(\cdot) : \mathsf{Ind} \rightarrow \mathcal{L}$ *is known, partitioned into positive, negative and uncertain-membership instances:*
  - $\mathsf{Ps} = \{a \in \mathsf{Ind}(\mathcal{A}) \mid \mathcal{K} \models C(a),\ i.e.\ t_{\mathsf{C}}(a) = +1\}$,
  - $\mathsf{Ns} = \{a \in \mathsf{Ind}(\mathcal{A}) \mid \mathcal{K} \models \neg C(a),\ i.e.\ t_{\mathsf{C}}(a) = -1\}$
  - $\mathsf{Us} = Tr \setminus (\mathsf{Ps} \cup \mathsf{Ns})$, *i.e.* $\{a \in \mathsf{Ind}(\mathcal{A}) : t_{\mathsf{C}}(a) = 0\}$;

Build *a classifier* $h_{\mathsf{C}} : \mathsf{Ind}(\mathcal{A}) \rightarrow \{-1, 0, +1\}$ *for* C *such that*

$$\frac{1}{|Tr|} \sum_{a, \in Tr} \mathbf{1}[h_{\mathsf{C}}(a) = t_{\mathsf{C}}(a)] > 1 - \epsilon$$

where $\mathbf{1}[\cdot]$ *is the* indicator *function returning* 1 *if the argument is true and* 0 *otherwise.*

To this purpose, various methods can be used for approximating this function, e.g. *non-parametric models* [2]. As an alternative, intensional descriptions of the available examples can be produced. Learning such descriptions is usually known as *concept learning problem* [11]. The concept learning problem in the context of a knowledge base can be formalized as follows.

Definition 2 (Concept Learning in DLs).

Given
- *the knowledge base* $\mathcal{K} = \langle \mathcal{T}, \mathcal{A} \rangle$,
- *a target concept* C,
- *the training set* $\mathsf{Tr} = \mathsf{Ps} \cup \mathsf{Ns} \cup \mathsf{Us}$,

Find *a concept description* D *approximating* C*, such that:*
- $\forall a \in \mathsf{Ps}: \ \mathcal{K} \models D(a)$
- $\forall b \in \mathsf{Ns}: \ \mathcal{K} \models \neg D(b)$

Therefore the goal of learning process is to find a concept description that is correct w.r.t. the examples. One could not be interested to a solution that fit perfectly to the training individuals but to induce a description general enough for classifying new individuals. Concept learning can be regarded as a search process in the space of concepts $\mathcal{S}$, which can be explored by imposing a quasi-ordering between DL concepts, i.e. a reflexive and transitive relation and then to use a *refinement operator* which maps a concept onto a set of other concepts. In the following, we consider the subsumption relation $\sqsubseteq$ between concepts as a quasi-ordering relation.

The definition of the refinement operator is reported below:

Definition 3. *Given a quasi-ordered space* $(\mathcal{S}, \sqsubseteq)$*, a downward (resp. upward) refinement operator* ρ *is mapping from* S *to* 2^{S} *such that for any concept description* $C \in \mathcal{S}$ *and* $C' \in \rho(C)$, $C' \sqsubseteq C$ *(resp.* $C \sqsubseteq C'$*).*

2.3 The Dempster-Shafer Theory

One of the models exploited in this paper is a modified version of terminological decision trees endowed with the operators of the Dempster-Shafer Theory (DST) [10]. Therefore, for sake of completeness, we shortly recall the basic notions of this theory used by such predictive models.

The DST is regarded as a generalization of probability theory. In the DST, a domain is usually represented through a *frame of discernement*, denoted by Ω, i.e. a set of mutually and exhaustive hypotheses. For our purposes, the frame of discernment represents the set of admissible membership values w.r.t. the target concept C, i.e. $\Omega = \{-1, +1\}$.

Given the frame of the discernment, a *Basic Belief Assignment* (BBA) can be build, that is a mapping $m : 2^{\Omega} \rightarrow [0, 1]$ such that $\forall A \in 2^{\Omega}\ m(A) > 0$ if $A \neq \emptyset$ and $\sum_{A \in 2^{\Omega}} m(A) = 1$. The value of a BBA function for a set of hypotheses A conveys the amount of belief exactly assigned to A but not to its subsets. In the DST, knowing the BBA allows to determine the *belief* and the *plausibility functions*. The belief function is a mapping $Bel : 2^{\Omega} \rightarrow [0, 1]$ such that $\forall A \in 2^{\Omega}\ Bel(A) = \sum_{B \subseteq A} m(B)$ represents the total amount of belief assigned to A given the available evidences. The plausibility function is a mapping $Pl : 2^{\Omega} \rightarrow [0, 1]$ such that $\forall A \in 2^{\Omega}\ Pl(A) = \sum_{B \cap A \neq \emptyset} m(B)$ and it quantifies the total amount of belief in favor of a set of hypotheses A when further evidences are available.

Other important notions concern the *non-specificity measure* [12] and *the combination rules* [13]. Given a BBA m the non-specificity measure $Ns(m)$

quantifies the degree of imprecision about the knowledge about a set of hypotheses, i.e. $Ns(m) = \sum_{A \subseteq \Omega, A \neq \emptyset} m(A) \log |A|$. A large non specificity measure denotes high uncertainty and imprecision about the available knowledge. As regards the combination rules, they represent operators used to pool BBAs coming from heterogeneous sources of information. The literature proposed various approaches for combining BBAs [13]. Among them, the *Dubois-Prade combination rule* has been adopted in the evidence-based version of a terminological decision tree [14]. The operator pools two BBAs, m_1 and m_2 as follows: $\forall A \in 2^{\Omega}\ m_{12}(A) = \sum_{B \cup C = A} m_1(B) m_2(C)$.

3 Learning Tree Models in DLs

3.1 The Models

The class-membership prediction task can be tackled by inducing either *Terminological Decision Trees* (TDTs) [3] or *Evidential Terminological Decision Trees* (ETDTs) [4].

Definition 4 (Terminological Decision Tree). *Given the knowledge base* $\mathcal{K}$*, a* Terminological Decision Tree *is a binary tree where:*

- *each intermediate node contains a conjunctive concept description* D *that stands for a test;*
- *each leaf contains a label used to denote the (positive/negative) membership w.r.t. the target concept* C
- *the branches correspond, respectively, to the result of the test performed over* D *(resp.* $\neg D$*);*

As illustrated in [3], a TDT can be used to learn concept descriptions and to determine the membership for an unseen individual. However, as argued in [4], when a TDT is used for predicting the class-membership for a new individual, the models cannot assign a definite membership due to intermediate tests with an unknown result. This is similar to the presence of missing values for decision trees targeting attribute-value datasets. In order to take into account this aspect, Evidential Terminological Decision Trees (ETDTs) have been devised [4,5]. They are defined as an extension of the TDTs [3] based on the evidential reasoning [10].

Definition 5 (Evidential Terminological Decision Tree). *Given the knowledge base* $\mathcal{K}$*, an* Evidential Terminological Decision Tree *is a binary tree where:*

- *each intermediate node contains a pair* (D, m) *where* D *is a conjunctive concept description that stands for a test and* m *is used to describe the membership w.r.t.* D*;*
- *each leaf contains both the label and the BBA* m *used to describe the membership w.r.t* C*;*
- *the branches correspond, respectively, to the result of the test performed over* D *(resp.* $\neg D$*);*

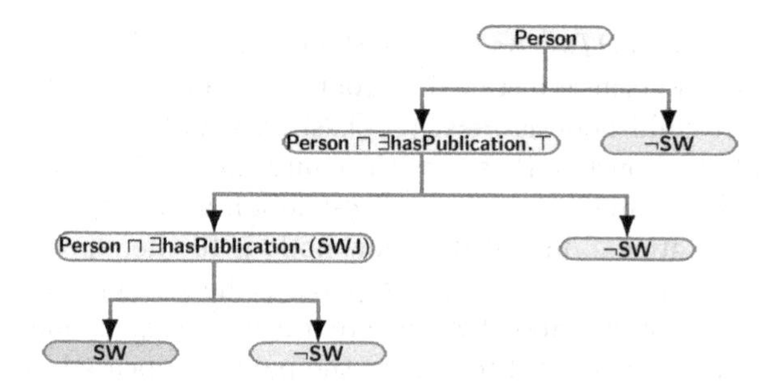

Fig. 1. A TDT for deciding if a person is a researcher that works in the field of the Semantic Web

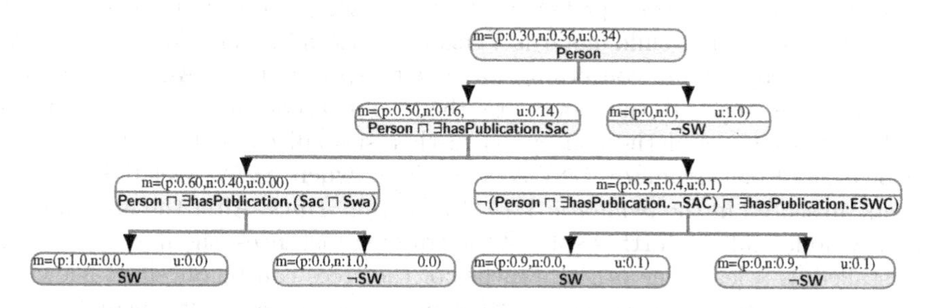

Fig. 2. An ETDT for deciding if a person is a Semantic Web researcher

Figures 1 and 2 report two examples of a TDT and an ETDT that are used for deciding the membership of an individual w.r.t. the target concept *Semantic Web researcher* (SW). The models can be used for deciding if an individual is a researcher whose topic concerns the Semantic Web.

3.2 Training

Given the concept C (used a label to be installed as a leaf) and the training set $\mathsf{Tr} = \langle \mathsf{Ps}, \mathsf{Ns}, \mathsf{Us} \rangle$, the methods for inducing both TDTs and ETDTs apply a *divide-and-conquer* strategy (see [3–5] for further details). The methods perform a recursive partitioning of the training set where, at each level, the individuals are grouped according to the results of some instance-check tests w.r.t. the most promising concepts description. The process is repeated until the instances sorted to a node have the same definite membership w.r.t. C. The concept descriptions that are installed as nodes during the training step are generated by specializing the concept installed as father node and passed as an input for the algorithm. Among the possible candidates, the algorithms select the best description according to a certain heuristic. In the case of ETDTs, the algorithm generates for the current node both a concept and a BBA estimating by using relative frequencies of the positive, negative and uncertain-membership instances routed to the node. Two examples of the learning procedures for TDTs and ETDTs are reported below.

Example 1 (Inducing TDTs). As regards the induction of the TDT reported in Fig. 1, the concept Person is installed as root note. The first refinement that is installed as a left-child node is Person $\sqcap$ $\exists$hasPublication.$\top$, which describes all the instances of the concept Person with a publication. This concept description is obtained by adding an existential restriction as a conjunct. The concept Person $\sqcap$ $\exists$hasPublication.$\top$ installed as new node is further specialized by using the instances with a positive membership w.r.t the concept, resulting in the concept Person $\sqcap$ $\exists$hasPublication.SWJ where a new concept name is introduced, namely SWJ (the concept used to denote the papers appeared in the *Semantic Web Journal*). Again, this concept is installed as left-child node.

Example 2 (Inducing ETDTs). Also the induction of the ETDT reported in Fig. 2 starts from the concept Person. In this case, the first refinement that is installed into the left-child node is Person $\sqcap$ $\exists$hasPublication.SAC, where SAC is a new concept name concerning all papers appeared in SAC proceedings. The instances reached the node are then split according to the instance-check test results, and the concept is further specialized so that the concept Person $\sqcap$ $\exists$hasPublication.SAC $\sqcap$ SWA is obtained, where SWA is related to those papers presented in the *Semantic Web Application Track.* After the installing of the new node and the further split of the training instances, the next node that is installed as a leaf. The other branches of the trees can be obtained likewise. In addition, we can observe that the BBA m assigned to each intermediate node has a decreasing level of non-specificity measure w.r.t. the previous level.

Refinement Operators. As introduced in Sect. 1, refinement operators play a fundamental role for determining the strategy to navigate the concepts space and, in the case of TDTs and ETDTs, for obtaining the candidates concepts to be chosen and installed into the nodes. The examples reported above induce the trees by using the downward refinement operator adopted in [3,4] that generate specializations in one of the following forms: (1) by introducing a new concept name (or its complement as conjunct); (2) by refining a sub-description in the scope of an existential restriction; (3) by refining a sub-description in the scope of an universal restriction. This naïve refinement operator exploits concept names and roles without considering information like the concept hierarchy asserted in a knowledge base. Conversely, the refinement operator implemented in DL-Learner framework (that contains the implementation of various ILP-based learning algorithms) [7] consider this aspect and can be also extended for addressing various DL expressiveness.

Figure 3 describes the refinement operator employed in this work: M_B is the set of the specializations of $\top$ obtained without resorting to disjunction operator that are not disjoint from $B \in \{\top\} \cup N_C$. This means that M_B contains concept in one of the following forms:

- $A \in N_C$ where $A \sqcap B \neq \bot$ and there is no $A' \in N_C$ such that $A' \sqsubseteq A$
- $\neg A \in N_C$ where $\neg A \sqcap B \neq \bot$ and there is no $A' \in N_C$ such that $A \sqsubseteq A'$
- $\forall R.\top$, where R is the most general applicable role for B , i.e. there is no applicable role R' such that $R \sqsubseteq R'$

$$\rho(C) = \begin{cases} \{\top\} \cup \rho_\top(C) & \text{if } C = \top \\ \rho_\top(C) & \end{cases}$$

$$\rho_\top(C) = \begin{cases} \emptyset & \text{if } C = \bot \\ \{C_1 \sqcup C_2 \sqcup \cdots C_n | C_i \in M_B(C)\} & \text{if } C = \top \\ \{A' | A' \in sh\downarrow(A)\} \\ \cup\{A \sqcap D | D \in \rho_B(\top)\} & \text{if } C = A (A \in N_C) \\ \{\neg A' | A' \in sh\uparrow(A)\} \\ \cup\{\neg A \sqcap D | D \in \rho_B(\top)\} & \text{if } C = \neg A (A \in N_C) \\ \{\exists R.E | E = ar_A(R), E \in \rho_A(D)\} \\ \cup\{\exists R.D \sqcap E | E \in \rho_B(\top)\} \\ \cup\{\exists R'.D | R' \in sh\downarrow(R)\} & \text{if } C = \exists R.D \\ \{\forall R.E | E = ar_A(R), E \in \rho_A(D)\} \\ \cup\{\forall R.D \sqcap E | E \in \rho_B(\top)\} \\ \cup\{\forall R.\bot | D = A \in N_C \text{ and } sh\downarrow(A) = \emptyset\} \\ \cup\{\exists R'.D | R' \in sh\downarrow(R)\} & \text{if } C = \forall R.D \\ \{C_1 \sqcap C_2 \sqcap \cdots \sqcap C_{i-1} \sqcap D \sqcap C_{i+1} \cdots C_n | \\ D \in \rho_B(C_i), 1 \le i \le n\} & \text{if } C = C_1 \sqcap \cdots \sqcap C_n \\ \{C_1 \sqcup C_2 \sqcup \cdots \sqcup C_{i-1} \sqcup D \sqcup C_{i+1} \cdots C_n | D \in \rho_B(\top), 1 \le i \le n\} \\ \cup\{(C_1 \sqcup C_2 \sqcup \cdots \sqcup C_{i-1} \sqcup C_i \sqcup C_{i+1} \cdots C_n) \sqcap D | \\ D \in \rho_B(\top), 1 \le i \le n\} & \text{if } C = C_1 \sqcup \cdots \sqcup C_n \end{cases}$$

Fig. 3. The refinement operator available in DL-Learner. Image adapted from [11]

- $\exists R.\top$, where R is the most general applicable role for B

The ρ operator generates the specializations as follows. Firstly, it delegates the refinement process to an operator $\rho_B(\cdot)$, Using the index B allows to exclude the concepts that are disjoint with B. At the beginning $B = \top$. The $\rho_\top(\cdot)$ distinguishes various cases: the simplest cases concern the generation of the refinements for $\bot$ and $\top$. For $\bot$, the specialization process ends by returning an empty set of concepts. In the case of the refinement of $\top$, the operator returns disjunction of concepts C_i where $C_i \in M_B(C)$. Additional cases concern the refinement of an atomic concept A or its negation. For the atomic concept, the refinement operator returns two sets of specializations: the first set contains sub-concepts A' such that $A' \sqsubseteq A$, i.e. $A' \in sh\downarrow(A)$, while the second set contains concepts obtained through the conjunction of the concept A and concepts $D \in \rho_B(\top)$. The case of the complement of an atomic concept is tackled dually to the previous one but the operator generates also refinements in the form $\neg A'$ where $A' \in sh\uparrow(A)$.

The third case concerns the refinement of a concept in the form of an existential restriction $C = \exists R.D$[1]. The operator produces three kinds of refinements: the first one is obtained by replacing the sub-description D with a sub-description E that is a concept subsumed by D and it is not disjoint with the range of the role R; the second kind of refinements is obtained by replacing the sub-description D

[1] The refinement operator was originally devised to consider $\mathcal{ALC}$ expressiveness.

with the one in the form $D \sqcap E$, where E is a refinement contained into the set of specializations of $\top$; the third kind of refinements are obtained by replacing the role R with a sub-role S, i.e. $S \in sh \downarrow (R)$.

The fourth case described in Fig. 3 illustrates the case of a concept in the form of an universal restriction, i.e. $C = \forall R.D$. This case is substantially dual to the case of existential restriction except for the specializations in the form $\forall R.\bot$ generated for the atomic concepts that have no sub-concepts. The last two cases concern concepts in conjunctive and disjunctive forms. In the first case, the refinement operator generates specializations by replacing a sub-description C_i with its refinements obtained by applying recursively the refinement operators. In the second case, the refinement operator produces specializes not only the various concept sub-description C_i (as in the case of conjunctive concept descriptions) but also it adds a new concept D as a conjunct.

Example 3 illustrates a simple example about the generation of the specializations.

Example 3 (ρ refinements). Let the following knowledge base be given:

$$\begin{aligned}\mathcal{K} = \{&\mathsf{Man} \sqsubseteq \mathsf{Person}, \mathsf{Woman} \sqsubseteq \mathsf{Person}, \mathsf{ESWC} \sqsubseteq \mathsf{Publication}\\ &\mathsf{EKAW} \sqsubseteq \mathsf{Publication}, \mathsf{EKAW} \sqcap \mathsf{ESWC} \equiv \bot\\ &\mathsf{domain(hasFirstAuthor)} = \mathsf{Publication},\\ &\mathsf{range(hasFirstAuthor)} = \mathsf{Person} \quad \}\end{aligned}$$

The refinement operator generates the following refinements for $\top$:

$$\begin{aligned}\rho(\top) = \quad &\{\mathsf{Person}, \mathsf{Publication}, \neg\mathsf{Man}, \neg\mathsf{Woman},\\ &\neg\mathsf{EKAW}, \neg\mathsf{ESWC},\\ &\forall\mathsf{hasFirstAuthor}.\top, \exists\mathsf{hasFirstAuthor}.\top, \ldots\}\end{aligned}$$

By using ρ, it is possible to specialize the concept $\mathsf{Publication} \sqcap \exists\mathsf{hasFirstAuthor}.\mathsf{Person}$ generating the following set of concept descriptions:

$$\begin{aligned}\rho(\mathsf{Publication} \sqcap \exists\mathsf{hasFirstAuthor}.\mathsf{Person}) = \quad &\{\mathsf{Publication} \sqcap \exists\mathsf{hasFirstAuthor}.\mathsf{Man}\\ &\mathsf{Publication} \sqcap \exists\mathsf{hasFirstAuthor}.\mathsf{Woman}\\ &\mathsf{EKAW} \sqcap \exists\mathsf{hasFirstAuthor}.\mathsf{Person},\\ &\mathsf{ESWC} \sqcap \exists\mathsf{hasFirstAuthor}.\mathsf{Person}, \ldots \}\end{aligned}$$

Note that the number of the possible specializations that are generated at each step via the refinement operator is *infinite* [11]. To overcome the problem, various strategies can be employed, e.g. by limiting the length of the specializations[2].

---

[2] The length of a concept C, len(C) can be defined inductively as:

- $\mathrm{len}(A) = \mathrm{len}(\top) = \mathrm{len}(\bot) = 1$
- $\mathrm{len}(\neg D) = \mathrm{len}(D) + 1$
- $\mathrm{len}(D \sqcap E) = \mathrm{len}(D \sqcup E) = \mathrm{len}(D) + \mathrm{len}(E) + 1$
- $\mathrm{len}(\exists R.D) = \mathrm{len}(\forall R.D) + 1$.

Heuristics for the Best Candidate Selection. The heuristics used for the concept selection aim at maximizing a purity criterion. This idea, borrowed from the algorithm for the induction of decision trees, is used by TDT. In fact, during the induction of TDTs *information gain* is the criterion used for selecting the best concept description [3]. Instead ETDTs exploits an heuristic based on the minimization of non-specificity measure in order to determine the concept sub-description with the most definite membership [4]. However, both the information gain and the non-specificity measure do not consider aspects such as the complexity of the concept description or the similarity w.r.t. the concept installed into the father node. In the latter case, adding a sub-description that is not similar to the one installed into the father node may increase the risk that most instances are sent along a branch, leading to an error-prone classification model, or that a large number of missing values may be found. To penalize these concept descriptions, we can adopt the idea proposed in [15]: introducing a regularization terms in the information gain/non-specificity measure value. This is basically a *discounting factor* for the purity-measure employed for selecting the concept. As regards the information gain, let C and D two concepts installed into a father and a child node, the regularized version of information gain can be computed as

$$Gain(C,D) = c\left(H(C,\mathsf{Tr}) - \frac{n^l}{n}H\left(D,\mathsf{Ps}^l \cup \mathsf{Ns}^l \cup \mathsf{Us}^l\right) - \frac{n^r}{n}H\left(D,\mathsf{Ps}^r \cup \mathsf{Ns}^r \cup \mathsf{Us}^r\right)\right) \tag{1}$$

where n^l (resp. n^l) is the number of training individuals sent to the left (resp. right) branch, H is the entropy of the concept adopted as a test computed over a set of individuals and $c \in [0,1]$ represents the aforementioned regularization factor. In this paper, the regularization factor takes into account the similarity w.r.t. the concept installed into the father node and it is computed through the Jaccard similarity between the set of the individuals which belong to those concepts. $J(C,D) = \frac{|ret.(C) \cap ret.(D)|}{|ret.(C) \cup ret.(D)|}$ where $ret.(E)$ for a given concept E is the set of individuals which belongs to E. Similarly to the case of information gain, a regularized version of the non specificity measure can be defined.

3.3 Classification

In order to make prediction with the produced models, we consider a ternary classification problem for assessing the membership of an individual [3,4]. The strategy is based on the navigation of tree structure according to the instance-check results. The algorithms start from the root and follows either the left or the positive branch according to the results of the instance check test w.r.t. the concept. The algorithms differ in the strategies exploited for coping with the case of uncertain results w.r.t. the intermediate tests: while the exploration of a TDT is stopped by assigning the uncertain-membership label for the test individual, both branches departing from the node with an uncertain result are navigated in order to reach more leaves when an ETDT is used to classify an individual. In this case, the algorithm collects the BBAs contained into the leaves that are subsequently pooled according to the Dubois-Prade rule [4].

Example 4 (Classification through TDTs). Given the TDT reported in Fig. 1 and a new individual a. Assuming that for this individual the membership w.r.t. the target concept SW is unknown but, according to the available knowledge base, it is an instance of the concept Person and a publication in SWJ exists, the classification algorithm will follow the most-left path of the tree and it will classify the individual as a positive instance. Conversely, Person(a) is entailed from the knowledge base but the it cannot determine if Person $\sqcap$ $\exists$hasPublication.$\top$(a) the classification algorithm stop the traversing of the tree assigning the uncertain-membership value.

Example 5 (Classification through ETDTs). The model proposed in Fig. 2 can be used for classifying an individual a that is an instance of the concept Person. The traversing process checks the membership w.r.t. the intermediate concept description. If neither of the encountered tests is satisfied and the individual is a instance of their complement concept, the algorithm follows the most-right path collecting the BBA of the single leaf and then, computing *Bel* function and assigning the class that corresponds to the hypothesis with the largest belief value. It is straightforward to note that the classification procedure will decide in favor of the negative membership. On the other hand, if an intermediate test with an uncertain result is encountered, e.g. it cannot be determined if a is an instance of either the concept Person $\sqcap$ hasPublication.$\top$ or its complement. In this case, the algorithm explores both the left sub-tree, whose root contains the concept description Person$\sqcap$hasPublication.SAC, and the right branch, whose root contains the concept $\neg$(Person$\sqcap$hasPublication.SAC)$\sqcap$hasPublication.ESWC. Following these branches, the algorithm can collect up to 4 BBAs (if there are further uncertain test results) that are combined according to the Dubois-Prade rule.

4 Empirical Evaluation

In this section we report the settings and the outcomes of an empirical evaluation, where we compared TDTs and ETDTs w.r.t. to other methods implemented in DL-Learner [7].

4.1 Setup

In our experiments, we considered various Web ontologies, whose dimensions and expressiveness are reported in Table 1. Lymph represents an OWL porting of the Lymphography dataset, which is available at the UCI repository (http://archive.ics.uci.edu/ml/). Instead, NTN is an ontology concerning the characters of the New Testament. mutagenesis and carcinogenesis are the porting of the well known datasets typically employed to test ILP methods.

For each ontology, we considered the learning problems available with the DL-Learner release (http://www.dllearner.org). Specifically, for Lymph, we considered the learning problems contained in lymphography_Class2. conf.

Table 1. Ontologies employed in the experiments

Ontology	Expressiv.	# Classes	# Roles	# Individ.
Lymph	$\mathcal{AL}$	53	0	148
NTN	$\mathcal{SHIF}(\mathcal{D})$	47	27	676
MUTAGENESIS	$\mathcal{AL}(\mathcal{D})$	86	5	14145
CARCINOGENESIS	$\mathcal{ALC}(\mathcal{D})$	142	4	22372

Instead, for NTN the learning problem aims at discovering if the ethnicity of an individual is Jewish. Finally, for MUTAGENESIS and CARCINOGENESIS the tasks aim at predicting if a chemical compound is mutagenic and carcinogenic, respectively. In the evaluation, TDTs and ETDTs have been compared against CELOE and ELTL DISJUNCTIVE. For the induction of trees we tested the original models against the new versions endowed with further refinement operators and the Jaccard similarity as a regularization term. As regards the refinement operators, we resort to both the original operator employed in [3,4], the RHO operator available in DL-LEARNER with a maximum length of 2. We used a 10-fold cross validation for assessing the performance of the algorithms.

The performance has been compared in terms of F-measure and other metrics that take into account the Open World Assumption [3,4], which are based on a comparison between inductive classification and the answer of a reasoner (PELLET: http://clarkparsia.com/pellet). The metrics are: (1) *match* (M), i.e. the rate of the test examples for which the inductive model and a reasoner predict the same membership (i.e. +1 vs. +1, −1 vs. −1, 0 vs. 0); (2) *commission*(C), i.e. the rate of the test examples for which predictions are opposite (i.e. +1 vs. −1, −1 vs. +1); (3) *omission* (O), i.e. the rate of test examples for which the inductive method cannot determine a definite membership (−1 or +1) while the reasoner is able to do it; (4) *induction* (I), i.e. rate of test examples where the inductive method can predict a definite membership while it is not logically derivable.

4.2 Outcomes

Table 2 reports the results of the experiments (the improvements due to the new refinement operators are reported by using bold font style). In general, when the refinement operator proposed in [3,4] and TDTs are considered in the experiments, we observed a large omission rate for each ontology. This results can be explained by the difficulty of TDTs to recognize negative instances, likely due to the lack of useful disjointness axioms and the Open World Assumption.

Concerning the experiments with LYMPH ontology, we noticed an improvement w.r.t. the original version of the learning algorithms when we resort to the RHO operator and the regularizer term. The improvement of the match rate and the F-measure in the case of TDTs were really prominent (these improvements were around 28 % and 82 %, respectively). In this case the models were com-

Table 2. Results of the experiments

Ontology	Index	TDT original	TDT regularized+ rho	ETDT original	ETDT regularized+ rho	CELOE	ELTL
Lymph	F_1	18.00 ± 33.27	**100.00 ± 00.00**	63.56 ± 22.38	**70.76 ± 01.55**	87.18 ± 08.29	100.00 ± 00.00
	M%	17.00 ± 19.15	**54.73 ± 01.87**	53.52 ± 03.87	**54.76 ± 01.87**	52.00 ± 03.60	54.77 ± 01.87
	C%	00.00 ± 00.00	00.00 ± 00.00	46.48 ± 03.87	**45.23 ± 01.87**	12.91 ± 07.71	00.00 ± 00.00
	O%	83.00 ± 19.15	**45.23 ± 01.87**	00.00 ± 00.00	00.00 ± 00.00	35.08 ± 05.78	45.23 ± 01.87
	I%	00.00 ± 00.00	00.00 ± 00.00	00.00 ± 00.00	00.00 ± 00.00	00.00 ± 00.00	00.00 ± 00.00
NTN	F_1	30.00 ± 48.31	**100.00 ± 00.00**	40.00 ± 51.64	**100.00 ± 00.00**	100.00 ± 00.00	37.95 ± 05.97
	M%	29.47 ± 47.48	**99.47 ± 01.66**	85.59 ± 12.96	**100.00 ± 00.00**	99.47 ± 01.66	22.85 ± 06.42
	C%	00.00 ± 00.00	00.00 ± 00.00	14.41 ± 12.96	**00.00 ± 00.00**	00.00 ± 00.00	69.27 ± 18.87
	O%	70.53 ± 47.48	**00.53 ± 01.66**	00.00 ± 00.00	00.00 ± 00.00	00.53 ± 01.66	07.90 ± 24.96
	I%	00.00 ± 00.00	00.00 ± 00.00	00.00 ± 00.00	00.00 ± 00.00	00.00 ± 00.00	00.00 ± 00.00
MUTAGENESIS	F_1	00.00 ± 00.00	**70.43 ± 00.02**	70.43 ± 00.17	70.43 ± 00.17	94.00 ± 03.85	70.43 ± 00.17
	M%	00.00 ± 00.00	**54.36 ± 00.20**	54.36 ± 00.20	54.36 ± 00.20	93.03 ± 04.53	54.36 ± 00.20
	C%	00.00 ± 00.00	45.64 ± 00.20	45.64 ± 00.20	45.64 ± 00.20	06.97 ± 04.53	45.64 ± 00.20
	O%	100.00 ± 00.00	**00.00 ± 00.00**	00.00 ± 00.00	00.00 ± 00.00	00.00 ± 00.00	00.00 ± 00.00
	I%	00.00 ± 00.00	00.00 ± 00.00	00.00 ± 00.00	00.00 ± 00.00	00.00 ± 00.00	00.00 ± 00.00
CARCINOGENESIS	F_1	00.00 ± 00.00	**70.51 ± 03.10**	70.46 ± 03.09	**70.51 ± 03.10**	71.48 ± 08.34	66.26 ± 13.26
	M%	00.00 ± 00.00	**54.36 ± 00.20**	54.47 ± 03.70	54.36 ± 00.20	63.42 ± 10.34	49.23 ± 14.82
	C%	00.00 ± 00.00	45.64 ± 00.20	45.53 ± 03.70	45.64 ± 00.20	36.58 ± 10.34	40.58 ± 14.87
	O%	100.00 ± 00.00	**00.00 ± 00.00**	00.00 ± 00.00	00.00 ± 00.00	00.00 ± 00.00	09.09 ± 28.75
	I%	00.00 ± 00.00	00.00 ± 00.00	00.00 ± 00.00	00.00 ± 00.00	00.00 ± 00.00	00.00 ± 00.00

petitive w.r.t. the concepts induced through CELOE and ELTL DISJUNCTIVE. Thanks to the new refinement operator, each tree contained concept descriptions that allowed to discern positive instances and to recognize the negative examples. In addition, we noticed that for this learning problem, a larger number of positive instances were available and this could affect the quality of the trees.

As regards the NTN ontology and the employment of the two refinement operators, the TDTs and ETDTs improved the performance w.r.t. the original versions of the mode only thanks to the RHO operator and the regularizer term. Also, in this case various missing values were found, like the experiments with LYMPH, and the uncertain membership was assigned to test individuals. Consequently, the F-measure was very low: it was only 30 %. On the other hand, resorting to the ETDTs with the original refinement operator improved the F-measure, which was 40 %, and partially mitigated the number of omission cases thanks to the strategy employed for dealing with the missing values. In this case a large number of negative instances have been predicted. With the RHO operator and the regularizer, we observed a significant improvement of the match rate for TDTs, around 70 %. For ETDTs the increase was more limited, but it was still good enough: it was about of 14 %. The improvement in terms of F-measure was very large: it was around 70 % for TDTs and 60 % for ETDTs. This result can be explained by the possibility to set various parameters for RHO operator in order to be fitted w.r.t. the specific learning problem, for instance by setting opportunely the use of data properties. Thanks to the integration of the refinement operator the results are better than the ones obtained by exploiting ELTL DISJUNCTIVE, which induced very poor concept descriptions, and similar to the ones obtained by resorting to CELOE. Finally, in the case of MUTAGENESIS and CARCINOGENESIS ontology, we observed a bad result for the experiments with the original version of TDTs: all test individuals were classified as having an uncertain membership. This was likely due to the expressiveness of the ontology, which is really limited for the refinement operator employed in this experiment. As explained in Sect. 2, the latter considers only concept names and existential or universal restriction obtained from roles that can be found in a knowledge base. But the expressiveness of MUTAGENESIS did not allowed to find this kind of candidate concepts. Besides, no disjointness axiom was found in this ontology. This limit of TDTs is broader when we compared their predictiveness with the one of the original ETDTs: these models were able to reduce the number of omission cases and improve the F-measure. With the RHO operator, the performance of TDTs improved significantly. In fact, the match rate and the F-measure are comparable to the ones obtained via ETDTs, by applying both the original refinement operator and RHO. Similarly to the case of NTN, the performance is better than or as well as the ones obtained through ELTL DISJUNCTIVE although it was worse than the performance of CELOE. This may be due to the fact that CELOE is an accuracy-driven method for inducing concepts which exploits the most promising description for classifying individuals. Conversely, the greedy algorithm employed for growing trees could yield sub-optimal solutions.

5 Conclusions and Extensions

In this work, we integrated the refinements operators available in DL-Learner into the learning algorithms for inducing Terminological Decision Trees and Evidential Terminological Decision Trees. We also proposed to modify the heuristic for selecting the best concept in order to take into account the similarity between a specialization and the concept installed into the father node. An empirical evaluation showed that by modifying the learning algorithms, the resulting models have better performance w.r.t. the original version. Besides the new models can fit with a lower expressivity than the one considered by the original refinement operator. Unfortunately, for some learning problems, the tree models did not outperform other methods proposed in the literature. This work is still preliminary and it can be extended along various directions. Firstly, we can extend the comparison by exploiting further refinement operators and further regularizer terms. In addition, we plan to extend the empirical evaluation by considering also further ontologies and further learning problems in order to investigate the correlation existing between the learning algorithms, the refinement operators and the expressiveness of the ontologies considered in the experiments.

Acknowledgements. This work fulfills the objectives of the PON 02005633489339 project "Puglia@Service - Internet-based Service Engineering enabling Smart Territory structural development" funded by the Italian Ministry of University and Research (MIUR).

References

1. Baader, F., Calvanese, D., McGuinness, D., Nardi, D., Patel-Schneider, P. (eds.): The Description Logic Handbook, 2nd edn. Cambridge University Press, Cambridge (2007)

2. Rettinger, A., Lösch, U., Tresp, V., d'Amato, C., Fanizzi, N.: Mining the semantic web - statistical learning for next generation knowledge bases. Data Min. Knowl. Discov. **24**, 613–662 (2012)
3. Fanizzi, N., d'Amato, C., Esposito, F.: Induction of concepts in web ontologies through terminological decision trees. In: Balcázar, J.L., Bonchi, F., Gionis, A., Sebag, M. (eds.) ECML PKDD 2010, Part I. LNCS, vol. 6321, pp. 442–457. Springer, Heidelberg (2010)
4. Rizzo, G., d'Amato, C., Fanizzi, N., Esposito, F.: Towards evidence-based terminological decision trees. In: Laurent, A., Strauss, O., Bouchon-Meunier, B., Yager, R.R. (eds.) IPMU 2014, Part I. CCIS, vol. 442, pp. 36–45. Springer, Heidelberg (2014)
5. Rizzo, G., d'Amato, C., Fanizzi, N.: On the effectiveness of evidence-based terminological decision trees. In: Esposito, F., Pivert, O., Hacid, M.-S., Rás, Z.W., Ferilli, S. (eds.) ISMIS 2015. LNCS, vol. 9384, pp. 139–149. Springer, Heidelberg (2015). doi:10.1007/978-3-319-25252-0_15
6. Fanizzi, N., d'Amato, C., Esposito, F.: DL-FOIL concept learning in description logics. In: Železný, F., Lavrač, N. (eds.) ILP 2008. LNCS (LNAI), vol. 5194, pp. 107–121. Springer, Heidelberg (2008)
7. Lehmann, J.: Dl-learner: learning concepts in description logics. J. Mach. Learn. Res. (JMLR) **10**, 2639–2642 (2009)
8. Lehmann, J., Auer, S., Bühmann, L., Tramp, S.: Class expression learning for ontology engineering. J. Web Semant. **9**, 71–81 (2011)
9. Lehmann, J., Haase, C.: Ideal downward refinement in the $\mathcal{EL}$ description logic. In: Raedt, L. (ed.) ILP 2009. LNCS (LNAI), vol. 5989, pp. 73–87. Springer, Heidelberg (2010). doi:10.1007/978-3-642-13840-9_8
10. Klir, J.: Uncertainty and Information. Wiley, Hoboken (2006)
11. Lehmann, J., Hitzler, P.: Concept learning in description logics using refinement operators. Mach. Learn. **78**, 203–250 (2010)
12. Smarandache, F., Han, D., Martin, A.: Comparative study of contradiction measures in the theory of belief functions. In: 15th International Conference on Information Fusion, FUSION 2012, Singapore, pp. 271–277, 9–12 July 2012
13. Sentz, K., Ferson, S.: Combination of Evidence in Dempster-Shafer Theory, vol. 4015. Citeseer (2002)
14. Dubois, D., Prade, H.: On the combination of evidence in various mathematical frameworks. In: Flamm, J., Luisi, T. (eds.) Reliability Data Collection and Analysis. Eurocourses, vol. 3, pp. 213–241. Springer, Netherlands (1992)
15. Deng, H., Runger, G.C.: Feature selection via regularized trees. In: The 2012 International Joint Conference on Neural Networks (IJCNN), 2012. IEEE (2012)

SEON: A Software Engineering Ontology Network

Fabiano Borges Ruy[1,3(✉)], Ricardo de Almeida Falbo[1], Monalessa Perini Barcellos[1], Simone Dornelas Costa[1,2], and Giancarlo Guizzardi[1]

[1] Ontology and Conceptual Modeling Research Group (NEMO), Computer Science Department, Federal University of Espírito Santo, Vitória, Brazil
{fabianoruy, falbo, monalessa, gguizzardi}@inf.ufes.br, sidornellas@gmail.com
[2] Computer Department, Federal University of Espírito Santo, Campus Alegre, Alegre, Brazil
[3] Informatics Department, Federal Institute of Espírito Santo, Campus Serra, Serra, Brazil

Abstract. Software Engineering (SE) is a wide domain, where ontologies are useful instruments for dealing with Knowledge Management (KM) related problems. When SE ontologies are built and used in isolation, some problems remain, in particular those related to knowledge integration. The goal of this paper is to provide an integrated solution for better dealing with KM-related problems in SE by means of a Software Engineering Ontology Network (SEON). SEON is designed with mechanisms for easing the development and integration of SE domain ontologies. The current version of SEON includes core ontologies for software and software processes, as well as domain ontologies for the main technical software engineering subdomains, namely requirements, design, coding and testing. We discuss the development of SEON and some of its envisioned applications related to KM.

Keywords: Ontology Network · Ontology Engineering · Software Engineering · Ontology Integration · Knowledge Management

1 Introduction

Software process is a knowledge-intensive process involving many people working in different sub-processes and activities. Moreover, knowledge in Software Engineering (SE) is diverse and organizations have problems to capture, retrieve, and reuse it. An improved use of this knowledge is the basic motivation and driver for Knowledge Management (KM) in SE [1].

Ontologies have been widely recognized as a key enabling technology for KM. They are used for establishing a common conceptualization of the domain of interest to support knowledge representation, integration, storage, search and communication [2]. However, some domains that are the target of KM initiatives are often large and complex. This is the case of SE. If we try to represent the whole domain as a single

E. Blomqvist et al. (Eds.): EKAW 2016, LNAI 10024, pp. 527–542, 2016.
DOI: 10.1007/978-3-319-49004-5_34

ontology, we will achieve a large and monolithic ontology that is hard to manipulate, use, and maintain [3]. On the other hand, representing each subdomain separately would be too costly, fragmented, and again hard to handle.

SE comprises several interrelated subdomains such as Requirements, Design, Coding, Testing, Project Management, and Configuration Management. In one hand, there are few works in the literature that aims at developing ontologies covering wide portions of the SE domain, such as [4–6]. On the other hand, it is easy to find many specific ontologies modeling SE subdomains [7–11]. However, in general, these subdomain ontologies are weakly or even not interrelated, and they are often applied in isolation. In this context, it is important to notice that SE subdomains share concepts, ranging from general (e.g. *Artifact*, *Process*) to ones that are more specific (e.g. *Functional Requirement*, *Test Case*). This striking feature of the SE domain must be considered while representing it. For achieving consistent SE ontologies, concepts and relations should keep the same meaning in any related ontology.

D'Aquin and Gangemi [12] point out a set of characteristics that are presented in "beautiful ontologies", from which we detach the following ones: having a good domain coverage; being modular or embedded in a modular framework; being formally rigorous; capturing also non-taxonomic relations; and reusing foundational ontologies. Most of the existing SE ontologies do not exhibit such characteristics. We believe that an integrated ontological framework, built considering them, can improve ontology-based applications in SE, in particular those related to KM. In such integrated ontological framework, there must be ways for creating, integrating and evolving related ontologies. Thus, we advocate that these ontologies should be built incrementally and in an integrated way, as a network.

An *Ontology Network* (ON) is a collection of ontologies related together through a variety of relationships, such as alignment, modularization, and dependency. A *networked ontology*, in turn, is an ontology included in such a network, sharing concepts and relations with other ontologies [3].

To truly enjoy the benefits of keeping the ontologies in a network, we need to take advantage of the existing resources available in the ON for gradually improving and extending it. Thus, an ON should have a robust base equipped with mechanisms to help its evolution. In our view, an ON should be organized in layers. Briefly, in the background, we need a foundational ontology[1] to provide the general ground knowledge for classifying concepts and relations in the ON. In the center of the ON, core ontologies[2] should be used to represent the general domain knowledge, being the basis for the subdomain networked ontologies. Ideally, these core ontologies should be organized as Ontology Pattern Languages (OPLs) [13] for easing reusing model fragments (ontology patterns) while developing subdomain networked ontologies.

[1] Foundational ontologies span across many fields and model the very basic and general concepts and relations that make up the world, such as object, event, parthood relation etc. [14].

[2] Core ontologies provide a precise definition of structural knowledge in a specific field that spans across different application domains in this field. These ontologies are built based on foundational ontologies and provide a refinement to them by adding detailed concepts and relations in their specific field [15].

Finally, going to the borders, (sub) domain ontologies appear, describing the more specific knowledge.

In this paper, we present SEON, a Software Engineering Ontology Network. SEON provides a well-grounded network of SE reference ontologies[3], and mechanisms to derive and incorporate new integrated subdomain ontologies into the network. The main goals of this work are: to define a layered architecture for SEON that enables supporting network growing; to introduce SEON and its mechanisms to create and integrate SE subdomain ontologies; and to discuss the use of SEON for supporting KM in SE. Due to space limitations, only small portions of SEON are presented here. The current specification is available at nemo.inf.ufes.br/projects/seon, where a machine processable lightweight version implemented in OWL is also available.

This paper is organized as follows. Section 2 discusses SE ontologies. Section 3 presents SEON and how it builds up from foundational to domain ontologies. Section 4 discusses how SEON can be used to support applications of KM in SE. Section 5 discusses related works. Finally, Sect. 6 presents our final considerations.

2 Developing Software Engineering Ontologies

A variety of ontologies have been developed modeling the SE domain. According to Calero et al. [7], these ontologies can be classified as: Generic SE Ontologies, having the ambitious goal of modeling the complete SE body of knowledge; or Specific SE Ontologies, attempting to conceptualize only part (a subdomain) of this discipline. Concerning Generic SE Ontologies, Mendes and Abran [4] propose a SE ontology consisting of an almost literal transcription of the SWEBOK [16] text, with over 4000 concepts. Sicilia and colleagues [5] propose an ontology structure to characterize artifacts and activities, also based on SWEBOK. Wongthongtham and colleagues [6] propose an ontology model for representing the SE knowledge, based on SWEBOK [16] and Sommerville's Software Engineering book [17]. Considering the Specific SE Ontologies, a great number of ontologies is available, representing a variety of SE subdomains, such as Software (e.g. [18, 19]), Software Processes (e.g. [9, 10]), Software Requirements (e.g. [20]), Software Testing (e.g. [8]), and Software Configuration Management (e.g. [11]). For others, see [7].

Some of the specific domain ontologies are developed considering their integration with others [7]. Taking this to the extreme, the combination of ontologies of all SE subdomains would result in an ontology of the complete SE domain. Unfortunately, the reality is that this goal is extremely laborious, not only due to its size, but also due to the numerous problems related to ontology integration and merging [7], such as overlapping concepts, diverse foundational theories, and different representation and description levels, among others. In sum, SE comprises a set of highly interconnected subdomains. This interrelated nature affects any possible representation of the SE domain, and the situations in which it can be applied. Despite of the challenges

[3] A reference ontology is constructed with the goal of making the best possible description of the domain in reality, representing a model of consensus within a community, regardless of its computational properties [18].

involved, an ontological representation covering a large extension of the SE domain remains a desired solution.

In this context, the notion of Ontology Network (ON) [3] applies. The scenario for ONs is radically different from the relatively narrow contexts in which ontologies have been traditionally developed and applied [3]. For instance, the *NeOn Methodology Framework* [3] provides guidance for engineering networked ontologies, making available detailed processes, guidelines and different scenarios for collaboratively building networked ontologies. For building SEON, we have applied some of the NeOn methodological guidelines, in particular those referring to ontology modularization and evaluation, and to the adoption of a pattern-based design approach. Moreover, we complement these valuable guidelines by defining an architecture for SEON that we believe applies for ONs in general. In the next section, we present the SEON architecture and some of the ontologies that comprise it. Besides, we discuss how to develop or integrate new SE subdomain ontologies to SEON.

3 SEON: The Software Engineering Ontology Network

SEON results from several efforts on building ontologies for the SE field. Although SEON itself is a new proposal, the studies and ontologies that our group has developed along the years were important contributions for defining SEON. Hence, SEON rises with three main premises: (i) being based on a well-founded grounding for ontology development; (ii) offering mechanisms to easy building and integrating new SE subdomain ontologies to the network; and (iii) promoting integration by keeping a consistent semantics for concepts and relations along the whole network. SEON architecture is organized considering three ontology generality levels, as Fig. 1 shows.

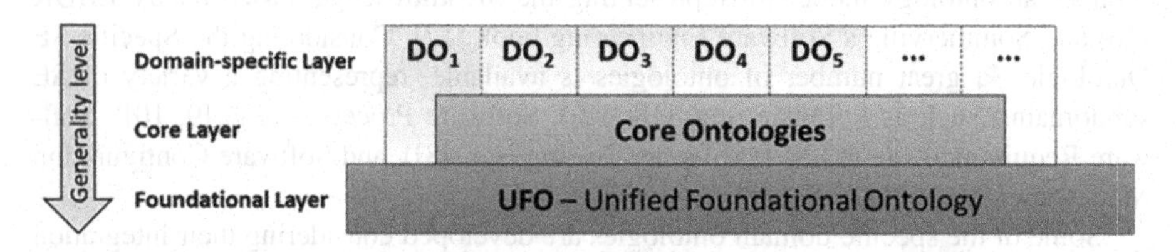

Fig. 1. SEON architecture.

Foundational Layer: at the bottom of SEON, is the Unified Foundational Ontology (UFO), which is developed based on a number of theories from Formal Ontology, Philosophical Logics, Philosophy of Language, Linguistics and Cognitive Psychology. UFO is divided in three parts: an ontology of endurants (objects) [21], an ontology of perdurants (events) [22], and an ontology of social entities [23]. UFO's ontological distinctions are used for classifying SEON concepts, e.g., as objects, actions, commitments, agents, roles, goals and so on. UFO provides the necessary grounding for the concepts and relations of all networked ontologies.

Core Layer: in the center of SEON, there are two core ontologies: the Software Ontology (SwO) and the Software Process Ontology (SPO) [10]. SwO is a core ontology developed based on the work of Wang and colleagues [20], and captures that software products have a complex artifactual nature, being constituted by software artifacts (here called software items) of different nature, namely software systems, programs and code. SPO is a core ontology, also grounded in UFO, aiming at establishing a common conceptualization on software processes. SPO builds upon SwO, and its current version is organized as an Ontology Pattern Language (OPL)[4] [13]. SPO scope is broader, embracing the following aspects of the software process domain: standard, project and performed processes and their activities, artifacts handled, resources used and procedures adopted by activities, team membership, and stakeholders allocation and participation in activities. For dealing with aspects related to organizations (such as team membership), SPO builds upon a core ontology on enterprises [24], which we consider external to SEON. SPO has been developed for more than two decades, and used as basis to develop several ontologies for many SE subdomains (e.g., [8, 11, 25]).

Domain-specific Layer: Over the foundational and core layers, SEON places the domain ontologies. Each networked ontology is grounded in SwO/SPO and also in UFO, and encompasses a SE subdomain (e.g., software requirements, design, configuration management, and measurement). Although not represented in Fig. 1, more specific subdomains ontologies can be developed based on other more general subdomain ontologies. For instance, an ontology on requirements at runtime [26] was developed based on the Reference Software Requirements Ontology.

In a nutshell, the foundational ontology offers the ontological distinctions for the core and domain layers, while the core layer offers the SE core knowledge for building the domain networked ontologies. This way of grounding the ontologies in the network is helpful for engineering the networked ontologies, since it provides ontological consistency and makes a number of modeling decisions easier. The SEON building mechanism also takes advantage of ontology patterns by representing its core layer in a pattern-oriented way. SPO is organized as an OPL, in order to become strongly modular, flexible and reusable. Thus, the ontology engineer can explore alternative models in the design of specific ontologies for the various SE subdomains, select the ontology fragments relevant to the problem in hands and reuse them [13]. Figure 2 shows a fragment of SPO with some of its patterns. This fragment deals with patterns for performed processes and activities; artifacts created, changed or used by those performed activities; and stakeholder participation. Colored blocks are used to delimit each pattern in this figure. During domain ontology development, the ontology engineer selects useful patterns and extends their concepts and relations in the networked ontology (Fig. 4 shows the case for the Requirements Development Process Ontology). In the cases where

[4] An OPL is a network of interconnected Domain-Related Ontology Patterns (DROPs) that provides holistic support for solving ontology development problems for a specific domain. Besides the DROPs, it contains a process guiding how to use and combine them in a specific order, and suggesting patterns for solving the modeling problems in that domain [14].

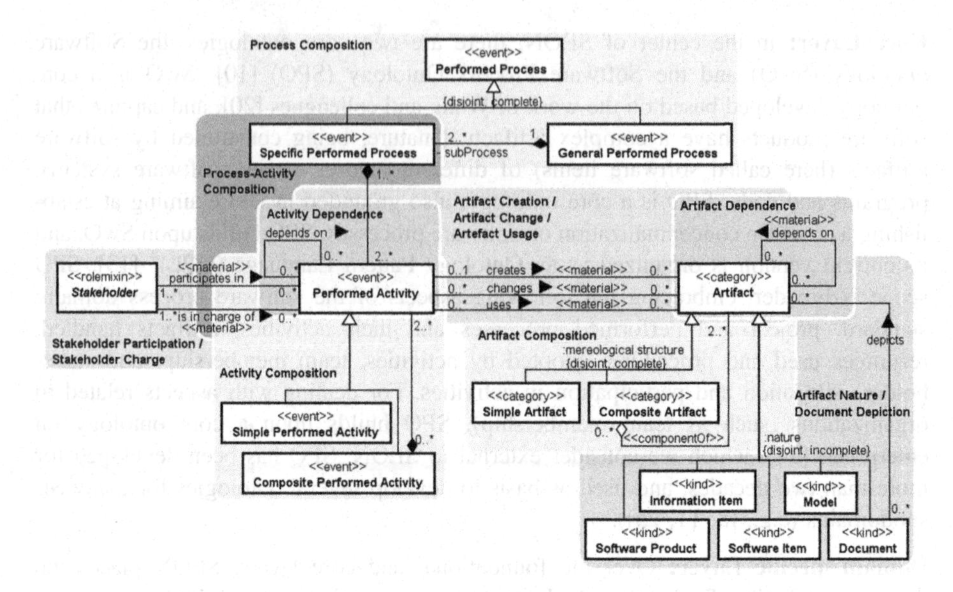

Fig. 2. A portion of SPO, fragmented in patterns (Color figure online).

a domain element is not covered by the core ontologies (SwO/SPO), this domain-specific element should be grounded directly in the foundational ontology (UFO).

The ontology fragments of Fig. 2 and the following are represented in OntoUML. OntoUML is a UML profile that enables making finer-grained modeling distinctions between different types of classes and relations according to the ontological distinctions put forth by the Unified Foundational Ontology (UFO) [21].

By reusing the OPL patterns from the core layer, the development of the domain ontologies is faster and the resulting models are more consistent and uniform [27]. The core ontologies, sustained by the foundational ontology, offer a standardized way for describing all the other elements in the network. Thus, since all the domain networked ontologies inherit the same core and foundational grounds, concepts and relations with the same classification have a common and identifiable background. This is a fundamental aspect for ontology integration.

As networked ontologies are developed and added to SEON, we still need to work on integrating them. Although the domain-specific ontologies share the same conceptual basis, given by the foundational and core ontologies, they still need to be aligned with respect to their specific knowledge, making possible to merge networked ontologies in a meaningful way, by representing information in one ontology in terms of the entities in another [3].

The SEON integration mechanism adopts some alignment guidelines for matching and integrating networked domain ontologies. First, ontologies should be compared looking for equivalent concepts. Since the domain ontologies are produced from the same basis (UFO and SwO/SPO), two concepts can only be considered equivalent if they have the same base type, restricting the search field and speeding up the integration process. Thus, for example, artifacts are compared only with artifacts, performed activities with performed activities, and so on. If concepts have a partial matching, this could mean that one concept is a specialization or a part of another.

Two concepts from distinct ontologies can also have a relationship between them. In this case, it is worth analyzing if there is a relationship to be extended from the base ontologies or a new relationship should be included in the subdomain ontology. From this matching, we can determine the correlation level between the ontologies.

Figure 3 shows the current status of SEON. Each circle represents an ontology. The circles' sizes vary according to the ontologies' sizes in terms of number of concepts, (represented inside the circles in parenthesis). Lines denote links between integrated ontologies, and line thickness represents the coupling level between them in terms of number of relationships between concepts in different ontologies. Blue circles represent the core ontologies; green circles, domain ontologies already integrated to SEON; and gray circles are domain ontologies already developed using UFO and SPO, but not integrated to SEON yet. Due to the wideness and complexity of the SE domain, it requires a continuous and long-term effort. Thus, we expect SEON to continuously evolve, with ontologies being added and integrated incrementally.

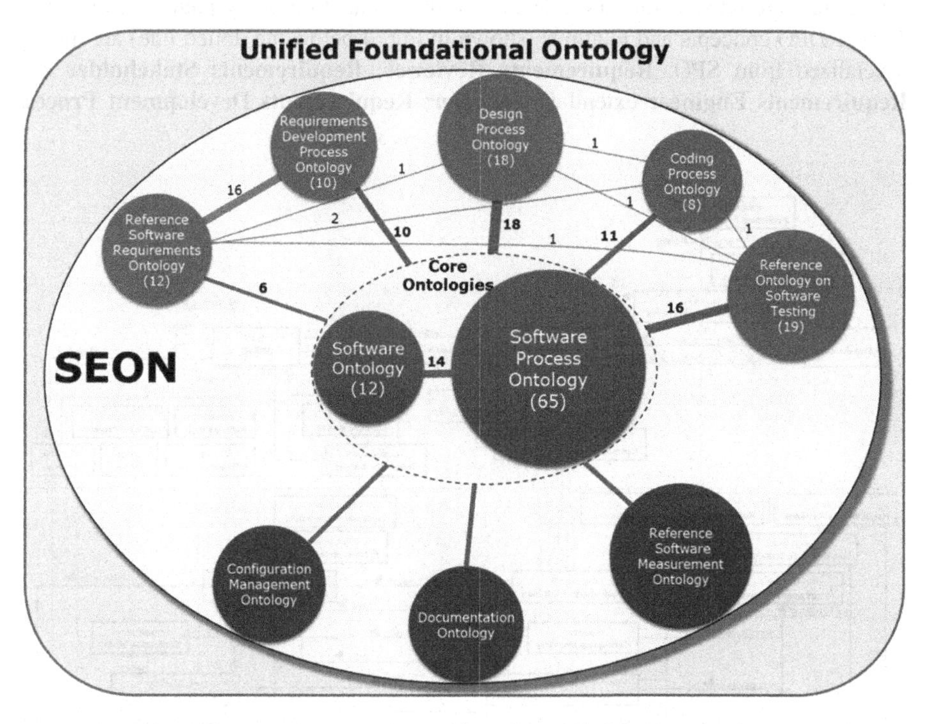

Fig. 3. SEON: The network view (Color figure online).

It is important to notice that, even adopting a layered architecture (see Fig. 1), SEON is a network. Like so, each new added node contributes for the whole network. When a new ontology is added, it should reuse existing elements. Other ontologies, in turn, may be adapted to keep consistency, in order to share the same semantics along the whole network. Even the core ontologies can evolve to adapt or incorporate new patterns discovered when domain ontologies are created or integrated.

Concerning the ontologies that comprise SEON, we should highlight that they have been developed or reengineered using SPO along the time, some of them before SEON conception. Among them, there are ontologies for the following SE subdomains: Software Requirements, Design, Testing [8], Configuration Management [11], Documentation, and Software Measurement [25]. Currently, as Fig. 3 shows, five of them (those addressing the main technical SE subdomains) are integrated to SEON.

The Reference Software Requirements Ontology (RSRO) is centered in the notion of requirement as a goal to be achieved, and addresses the distinction between functional and non-functional requirements, and how requirements are documented in proper artifacts, among others. It is mainly based on SwO.

The other four domain ontologies focus on describing the technical processes, considering its main assets, and extend SPO. The Requirements Development Process Ontology (RDPO), partially shown in Fig. 4, describes the requirements development process. The top of the figure shows the SPO concepts and relations (as presented in Fig. 2) that are reused from the selection of the suitable ontology patterns from SPO. Thus, RDPO concepts and relations (shown in Fig. 4 below the dotted line) are mostly specialized from SPO. **Requirements Reviewer**, **Requirements Stakeholder** and **Requirements Engineer** extend *Stakeholder*; **Requirements Development Process**

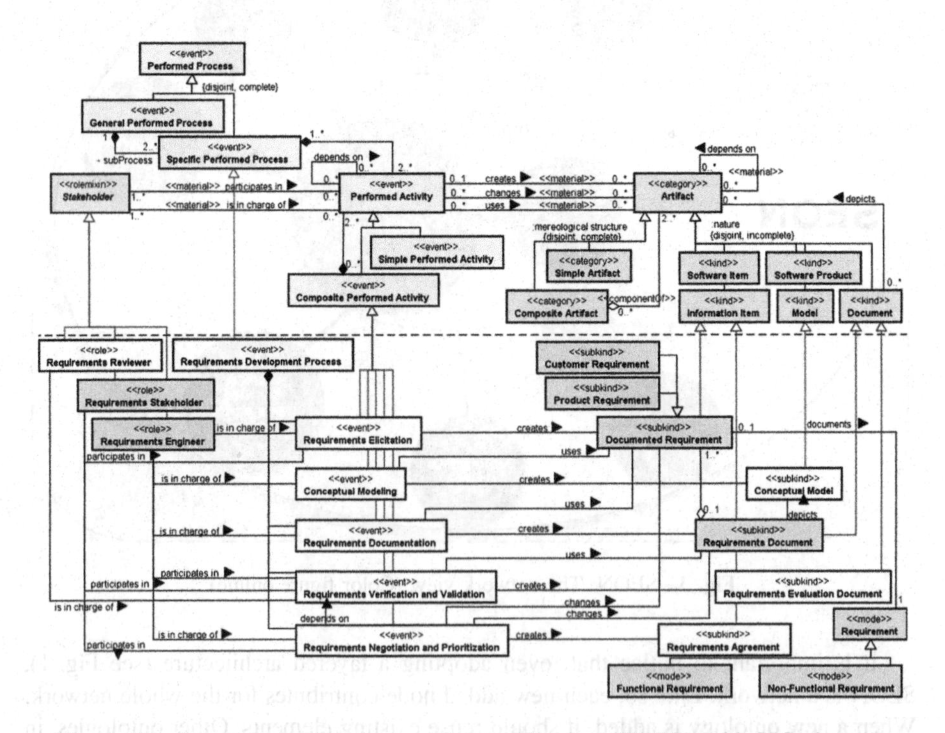

Fig. 4. The Requirements Development Process Ontology (RDPO), with SPO elements.

extends ***Specific Performed Process***, and is decomposed into five ***Composite Performed Activities***. These activities are responsible for producing ***Artifacts*** such as the **Requirements Document**, which depicts **Conceptual Models** and is composed of **Documented Requirements**. The concepts in gray below the line are imported from RSRO and integrated into RDPO.

The Design Process Ontology (DPO) focuses on the architectural and detailed design processes. Concepts and relations in DPO extend the same portion of SPO presented before. The main artifact produced during design is the **Design Document**. This ontology, as the next one, is not shown here due to space limitations.

The Coding Process Ontology (CPO) deals with building the software code based on the requirements and design documents. As the other ontologies, it extends the same portion of SPO. The main CPO product is the **Code *Artifact***.

The Reference Ontology on Software Testing (ROoST) [8] addresses the software testing process. Figure 5 shows a fragment of ROoST, focusing on the testing process and the used and produced artifacts. Although core ontology elements are not presented in Fig. 5, ROoST elements are also specializations from SPO, gotten by reusing patterns [8]. **Test Manager** and **Tester *Stakeholders*** participate in ***Performed Activities*** of the **Testing Process**. **Test Planning** creates the **Test Plan**. The four activities **Test Case Design**, **Test Coding**, **Test Execution** and **Test Result Analysis** are performed at the **Unit**, **Integration** and **System Testing** levels, producing testing process artifacts. For example, **Test Case Design** creates **Test Cases**, using **Test Case Design Inputs**, a role that can be played by ***Artifacts*** used as input for the test case design.

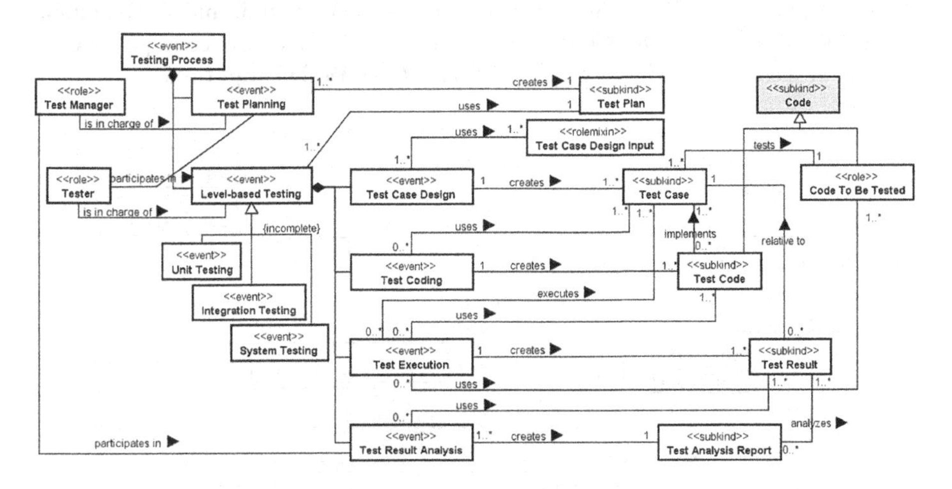

Fig. 5. A fragment of ROoST [8]

Besides the ontologies for the technical SE subdomains already integrated to the network, we expect SEON to continuously grow by adding other SE subdomains ontologies. The SEON integration mechanism has three different ways to incorporate new ontologies into the network, considering the origin of the ontology to be integrated.

In a first situation, consider a new ontology that is created based on UFO and SwO/SPO, and also taking other existing networked ontologies into account. Besides the extensions made from the core ontologies, this ontology tends to use also the related concepts already defined in the other networked ontologies. This situation occurs in RDPO (Fig. 4), which imports concepts from RSRO (the gray ones below the line). This is the best way for increasing SEON, since it reduces modeling and integration efforts, by reusing already defined elements.

The second situation occurs when domain ontologies are developed based on UFO and SwO/SPO, however, independently of the other subdomain networked ontologies. In this situation, although the domain ontology to be integrated to the network shares the same basis of the SEON domain ontologies (UFO/SwO/SPO), some additional integration effort is still required, in order to adapt the common parts focusing on a shared representation. This happened when we integrated ROoST to SEON. ROoST was developed based on SPO and UFO, but disregarding the other domain ontologies already integrated to SEON. This way, while integrating ROoST, we had to align it with the other existing networked ontologies. Figure 6 shows a fragment of the integrated model, encompassing elements from four domain networked ontologies: RSRO, DPO, CPO and ROoST. It shows the activities of coding and test case design, and related artifacts. Most of the concepts and relations shown (as the activities for coding and testing) are just imported from their original ontologies. However, some concepts required further decisions. This is the case of the inputs for the **Test Case Design** activity. The **Test Case Design Input** concept is a general role that can be played by different types of ***Artifacts*** able to be used as inputs for that activity. In this case, the suitable artifacts are the ones used for creating the code (**Requirements Document**, **Design Document**) and the **Code** itself, giving rise to three new concepts, specializations of these three artifacts playing the **Test Case Design Input** role.

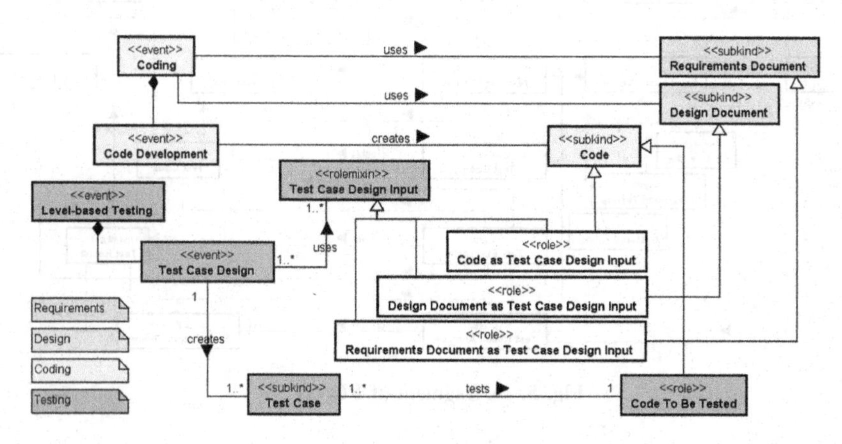

Fig. 6. An integrated SEON fragment.

Finally, the third integration situation happens when external ontologies, developed without taking SwO, SPO or UFO as basis, need to be integrated to SEON. In this case, if we have access to modify the ontology, we need to perform an ontological analysis and reengineering before the integration. By this process, the ontology elements are analyzed and adapted to the UFO distinctions and SwO/SPO domain knowledge. The knowledge represented by the ontology is then preserved, but the representation is adjusted for a better integration into SEON. On the other hand, if the ontology cannot be modified, we have to make the necessary links and adaptations only in the SEON side. In this case, techniques for ontology alignment, as discussed in [3], apply. Currently, we do not have any external ontology integrated to SEON.

4 SEON Envisioned Applications

There are several ontology-based initiatives of applying KM in SE. Some of them adopt centralized KM solutions (e.g., KM systems with large centralized knowledge repositories), such as [28]; others focus on distributed KM solutions, such as the ones using Semantic Wikis [29]. In both cases, ontologies are used to support knowledge representation (e.g., by categorizing or annotating knowledge items), integration, search, and retrieval. However, in KM scenarios spanning different SE subdomains, we need to integrate several ontologies for these subdomains. These are the cases where the benefits of using SEON stand out.

Since the begging of the 2000's, we have been working on ontology-based KM systems to support SE tasks. We started by developing an ontology-based KM infrastructure for a Software Engineering Environment (SEE) [28]. This infrastructure evolved, and more recently, we separate it from the SEE, transforming it in a SE Knowledge Management Portal (SE-KMP). SE-KMP provides general features for managing and assessing knowledge items (including lessons learned, and discussion packages), as well as yellow pages. SE-KMP was extended to manage knowledge related to software testing, in a more specific KM Portal, called TKMP [30]. However, we perceived that, to truly provide benefits for KM in SE, SE-KMP requires integrated ontologies for the several SE subdomains. In fact, this application motivated us to seek for an approach for developing integrated SE domain ontologies, leading to SEON.

In another front of research, we have been working on semantic documentation in SE, by providing an Infrastructure for Managing (SE) Semantic Documents (IMSD) [31]. Semantic documents aim at combining documents and ontologies, and allowing users to access their knowledge in multiple ways. The ultimate goal of semantic documents is not merely to provide metadata for documents, but to integrate documentation and knowledge representation in a way that they use a common structure [32]. We started by annotating requirements documents, and we used a previous versions of the Reference Software Requirements Ontology (RSRO) for this purpose. However, this ontology is not enough to annotate other documents, such as design specifications, source code, and test cases. In particular, for providing information traceability among these artifacts, we need, besides RSRO, other ontologies for design, coding and testing. Moreover, these ontologies need to be integrated. We perceived this clearly when decided to handle in IMSD, besides Requirements Documents, Test Cases. We needed to integrate RSRO to ROoST [8] (the reference ontology on software testing) [31]. With networked ontologies, as in SEON, this effort would not be required. In fact, this scenario of semantic documentation motivated us to start SEON with the set of networked domain ontologies shown in Fig. 3.

Another related scenario for applying SEON is tool integration. Ontologies can be used for semantically integrating heterogeneous tools [11]. Considering different tools working with elements of the same subdomain, an ontology for this subdomain is enough for addressing most of the issues. We have experienced this situation in [11], when we used a Software Configuration Management Ontology for integrating a version control system and a change management tool. However, for more complex contexts, involving several tools, supporting tasks in different SE subdomains, a single ontology is not enough. For instance, to include knowledge items (test cases) in TKMP, we had to import information from different software tools, namely: TestLink, a web-based test management system, and MantisBT, a bug tracking system. To automate this task, we needed to integrate data from both tools.

Addressing ontology integration case by case is an arduous and exhausting task. By integrating these ontologies in SEON, any new initiative that needs to commit to those ontologies could benefit from the efforts already done.

5 Related Works

Regarding the ontologies aiming at covering a large extension of the SE domain [4–6], in general, they present many concepts usually based on acknowledged references such as SE books or reference models (e.g. [16, 17]). Comparing to SEON, the first notable difference is the source for building the ontologies. These Generic SE ontologies use to be based on a few number of sources, in some cases nearing to transcriptions of the referenced source [4]. Contrariwise, each SEON ontology is built based on a set of references, often considering books and standards of the specific (sub)domain. Besides that, the knowledge from the base layers configures as one more source for building the networked ontologies. A second difference regards modularity, since the networked ontologies, even integrated, can be seen, and used, as separated models. Finally, the most important difference regards the mechanisms provided to build SEON incrementally, supported by the foundational and core layers and their patterns. In sum, SEON design considers important characteristics of "beautiful ontologies", as discussed in [12], such as: having a good domain coverage; considering international standards; being modular; being formally rigorous; capturing also non-taxonomic relations; and reusing foundational ontologies.

Concerning SEON ontologies, due to space limitations, it is not possible to contrast all of them with others already published in the literature. Thus, here we decided to compare only SEON's core ontologies (SwO and SPO). Regarding SwO, related work includes the software ontologies presented in [18, 19]. The Core Software Ontology (CSO) [18] detaches. Like SwO, CSO is rigorously formalized, grounded in a foundational ontology (DOLCE, while SwO is grounded in UFO), and was built following a pattern-based approach. Moreover, these two ontologies share concerns related to the polysemy of the concept of software, and in this sense they present similar distinctions. We should highlight that CSO has a broader scope than SwO, addressing concepts of object orientation, such as classes, interfaces and methods, as well as representing workflow information. These aspects are not covered by SwO, because we intend to address them in another networked ontology. The Software Ontology presented in [19] is organized in several modules, addressing aspects related to software and relationships to software process, license, and versions, among others. In this sense, the proposed ontology is related not only to SwO, but also to other SEON's ontologies, such as SPO. It is worthwhile to point out that the Software Ontology presented in [19] is not grounded in a foundational ontology.

Regarding SPO, there are some ontologies on software processes published in the literature. Here, due to space limitations, we compare SPO only with the Ontology for Software Development Methodologies and Endeavors (OSDME) [9]. This choice justifies because OSDME is the basis for an international standard (ISO 24744). OSDME addresses aspects related to process, product and producers. SPO has similar coverage. However, SPO is organized as an Ontology Pattern Language to favor reuse, and is grounded in a foundational ontology (UFO). As discussed in [33], the lack of truly ontological foundations leads to some inconsistencies in OSDME, which are solved in SPO, as discussed in [10]. This reinforces the importance of using a foundational ontology as basis for grounding core and domain ontologies in SEON.

Considering Ontology Networks, in [3], three case studies in the fishery and pharmaceutical domains are presented. Three ONs were developed using NeOn methods and technologies. In general, these ONs are composed of ontologies (expressed in OWL) plus non-ontological resources (such as thesauri). Mappings are an important means to relate the networked ontologies. In two of the studies, the network resources were organized according to the ontologies' types and levels, considering general ontologies (e.g., upper level ontologies or ontologies for time and objects) as independent of the focused domain, and ontologies as references for the domain and basis for providing concepts or relating more specific ontologies. Although the similarities regarding the generality levels, SEON states an architecture with well-defined layers, and it is based on ontological foundations and patterns, facilitating the building and integration of new domain ontologies. We should highlight that SEON's architecture is aligned to the one adopted in the ONIONS Project [34] and further with the ontological architecture proposed by Obrst [35].

6 Final Considerations

Ontologies are a key enabling technology for KM in SE. However, knowledge in SE is diverse and interlinked. For dealing with richer KM scenarios, addressing several SE subdomains, we need integrated ontologies. An ontology network can provide such integrated solution. Thus, in this paper, we presented SEON, a Software Engineering Ontology Network. SEON is designed seeking for: (i) taking advantage of well-founded ontologies (all its ontologies are ultimately grounded in UFO); (ii) providing ontology reusability and productivity, supported by core ontologies organized as Ontology Pattern Languages; and (iii) solving ontology integration problems by providing integration mechanisms. Diverse initiatives can benefit from the use of SEON, especially the ones where the focus is semantic interoperability and that involve a number of related SE subdomains. In this paper, we have explored KM-related scenarios, but SEON can also be used to address other scenarios that demands integrated SE ontologies.

In its current version, SEON includes core ontologies for software and software processes, as well as domain ontologies for the main technical software engineering subdomains, namely requirements, design, coding and testing. Other SE domain ontologies should be developed and integrated to SEON to enlarge its coverage. SEON success criteria relate to how easy SEON grows by incorporating new ontologies, and how successful is to apply SEON to solve integration focused problems (such as standard harmonization and tool integration, besides KM). As ongoing work, we are working on incorporating to SEON already developed SE subdomain ontologies for software documentation, measurement, configuration management and project management. Regarding SEON applications, our research agenda includes: a Standard Harmonization Approach supported by SEON; efforts on using SEON-based annotations in semantic documents, allowing integrating information scattered in multiple documents; and using SEON for semantic integration of SE tools, extending [11].

Acknowledgments. This research is funded by the Brazilian Research Funding Agency CNPq (Processes 485368/2013-7 and 461777/2014-2) and FAPES (Process 69382549/14).

References

1. Rus, I., Lindvall, M.: Knowledge management in software engineering, pp. 26–38. IEEE Software, May/June (2002)
2. O'Leary, D.: Using AI in knowledge management: knowledge bases and ontologies. IEEE Intell. Syst. **13**, 34–39 (1998)
3. Suárez-Figueroa, M.C., Gómez-Pérez, A., Motta, E., Gangemi, A.: Ontology Engineering in a Networked World. Springer Science & Business Media, Heidelberg (2012)
4. Mendes, O., Abran, A.: Issues in the development of an ontology for an emerging engineering discipline. In: First Workshop on Ontology, Conceptualizations and Epistemology for Software and Systems Engineering (ONTOSE). Alcalá Henares, Spain (2005)
5. Sicilia, M.A., Cuadrado, J.J., García, E., Rodríguez, D., Hilera, J.R.: The evaluation of ontological representation of the SWEBOK as a revision tool. In: 29th International Computer Software and Application Conference (COMPSAC), pp. 26–28. Edinburgh, UK (2005)
6. Wongthongtham, P., Chang, E., Dillon, T., Sommerville, I.: Development of a software engineering ontology for multisite software development. IEEE Trans. Knowl. Data Eng. **21**(8), 1205–1217 (2009)
7. Calero, C., Ruiz, F., Piattini, M.: Ontologies for Software Engineering and Software Technology. Springer Science & Business Media, Heidelberg (2006)
8. Souza, E.F., Falbo, R.A., Vijaykumar, N.L.: Using ontology patterns for building a reference software testing ontology. In: 17th IEEE International Enterprise Distributed Object Computing Conference Workshops (EDOCW), pp. 21–30. Vancouver (2013)
9. González-Pérez, C., Henderson-Sellers, B.: An ontology for software development methodologies and endeavours. In: [7] (2006)
10. Bringuente, A.C., Falbo, R.A., Guizzardi, G.: Using a foundational ontology for reengineering a software process ontology. J. Inf. Data Manag. **2**(3), 511 (2011)

11. Calhau, R.F., Falbo, R.A.: An Ontology-based Approach for Semantic Integration. In: 14th IEEE International Enterprise Distributed Object Computing Conference, Vitória, Brazil. Los Alamitos: IEEE Computer Society, pp. 111–120 (2010)
12. d'Aquin, M., Gangemi, A.: Is there beauty in ontologies? Appl. Ontol. **6**(3), 165–175 (2011)
13. Falbo, R.A., Barcellos, M.P., Nardi, J.C., Guizzardi, G.: Organizing ontology design patterns as ontology pattern languages. In: Cimiano, P., Corcho, O., Presutti, V., Hollink, L., Rudolph, S. (eds.) ESWC 2013. LNCS, vol. 7882, pp. 61–75. Springer, Heidelberg (2013). doi:10.1007/978-3-642-38288-8_5
14. Guarino, N.: Formal Ontology and Information Systems. In: Guarino, N. (ed.) Formal Ontology and Information Systems, pp. 3–15. IOS Press, Amsterdam (1998)
15. Scherp, A., Saathoff, C., Franz, T., Staab, S.: Designing core ontologies. Appl. Ontol. **6**(3), 177–221 (2011)
16. Bourque, P., Fairley, R.E.: Guide to the software engineering body of knowledge (SWEBOK (R)): Version 3.0. IEEE Computer Society Press (2014)
17. Sommerville, I.: Software engineering. Addison Wesley, Boston (2004)
18. Oberle, D., Grimm, S., Staab, S.: An ontology for software. In: Staab, S., Studer, R. (eds.) Handbook on Ontologies. International Handbooks on Information Systems, pp. 383–402. Springer, Heidelberg (2009). doi:10.1007/978-3-540-92673-3_17
19. Malone, J., Brown, A., Lister, A.L., Ison, J., Hull, D., Parkinson, H., Stevens, R.: The Software Ontology (SWO): a resource for reproducibility in biomedical data analysis, curation and digital preservation. J. Biomed. Semant. **5**, 25 (2014)
20. Wang, X., Guarino, N., Guizzardi, G., Mylopoulos, J.: Towards an ontology of software: a requirements engineering perspective. In: Proceedings of the 8th International Conference on Formal Ontology in Information Systems, Rio de Janeiro, Brazil, vol. 267, pp. 317–329 (2014)
21. Guizzardi, G.: Ontological Foundations for Structural Conceptual Models: Fundamental research series. Centre for Telematics and Information Technology, Enschede (2005)
22. Guizzardi, G., Wagner, G., Almeida Falbo, R., Guizzardi, R.S.S., Almeida, J.P.A.: Towards ontological foundations for the conceptual modeling of events. In: Ng, W., Storey, V.C., Trujillo, J.C. (eds.) ER 2013. LNCS, vol. 8217, pp. 327–341. Springer, Heidelberg (2013). doi:10.1007/978-3-642-41924-9_27
23. Guizzardi, G., Falbo, R.A., Guizzardi, R.S.S.: Grounding software domain ontologies in the unified foundational ontology (UFO): the case of the ode software process ontology. In: Proceedings of the XI Ibero-American Workshop on Requirements Engineering and Software Environments, pp. 244–251. Recife, Brazil (2008)
24. Falbo, R.A., Ruy, F.B., Guizzardi, G., Barcellos, M.P., Almeida, J.P.A.: Towards an enterprise ontology pattern language. In: Proceedings of the 29th Annual ACM Symposium on Applied Computing - SAC 2014, pp. 323–330 (2014)
25. Barcellos, M.P., Falbo, R.A., Moro, R.D.: A well-founded software measurement ontology. In: 6th International Conference on Formal Ontology in Information Systems 2010, vol. 209, pp. 213–226. IOS Press, Amsterdam (2010)
26. Duarte, B.B., Souza, V.E.S., Leal, A.L.C., Falbo, R.A., Guizzardi, G., Guizzardi, R.S.S.: Towards an ontology of requirements at runtime. In: Proceedings of the 9th International Conference on Formal Ontology in Information Systems, Annecy, France (2016)
27. Ruy, F.B., Reginato, C.C., Santos, V.A., Falbo, R.A., Guizzardi, G.: Ontology engineering by combining ontology patterns. In: Proceedings of the 34th International Conference on Conceptual Modeling (ER 2015), Stockholm, Sweden, pp. 173–186 (2015)
28. Natali, A.C.C., Falbo, R.A.: Knowledge management in software engineering environments. In Proceedings of XVI Brazilian Symposium on Software Engineering, pp. 238–253 (2002)

29. Maalej, W., Panagiotou, D., Happel, H.-J.: Towards effective management of software knowledge exploiting the semantic Wiki paradigm. In: Herrmann, K. Brugge, B. (eds), Software Engineering, GI, LNI, vol. 121, pp. 183–197 (2008)
30. Souza, E.F., Falbo, R.A., Vijaykumar, N.L.: Using lessons learned from mapping study to conduct a research project on knowledge management in software testing. In: 41st Euromicro Conference on Software Engineering and Advanced Applications (2015)
31. Falbo, R.A., Braga, C.E.C., Machado, B.N.: Semantic documentation in requirements engineering. In: 17th Workshop on Requirements Engineering (WER), Pucón, Chile (2014)
32. Eriksson, H.: The semantic-document approach to combining documents and ontologies. Int. J. Hum.-Comput. Stud. **65**(7), 624–639 (2007)
33. Ruy, F.B., Falbo, R.A., Barcellos, M.P., Guizzardi, G.: An ontological analysis of the ISO/IEC 24744 metamodel. In: Proceedings of 8th International Conference on Formal Ontology in Information Systems (FOIS 2o14), Rio de Janeiro, Brazil (2014)
34. Gangemi, A., Pisanelli, D.M., Steve, G.: An overview of the ONIONS project: applying ontologies to the integration of medical terminologies. Data Knowl. Eng. **31**(2), 183–220 (1999)
35. Obrst, L.: Ontological architectures. In: Poli, R., Healy, M., Kameas, A. (eds.) Theory and Applications of Ontology: Computer Applications, pp. 27–66. Springer, Heidelberg (2010)

Discovering Ontological Correspondences Through Dialogue

Gabrielle Santos, Terry R. Payne(✉), Valentina Tamma, and Floriana Grasso

Department of Computer Science, University of Liverpool, Liverpool L69 3BX, UK
{G.Santos,T.R.Payne,V.Tamma,floriana}@liverpool.ac.uk

Abstract. Whilst significant attention has been given to centralised approaches for aligning full ontologies, limited attention has been given to the problem of aligning partially exposed ontologies in a decentralised setting. Traditional ontology alignment techniques rely on the full disclosure of the ontological models that find the "best" set of correspondences that map entities from one ontology to another. However, within open and opportunistic environments, such approaches may not always be pragmatic or even acceptable (due to privacy concerns). We present a novel dialogue based negotiation mechanism that supports the strategic agreement over correspondences between agents with limited or no prior knowledge of their opponent's ontology. This mechanism allows both agents to reach a mutual agreement over an alignment through the selective disclosure of their ontological model, and facilitates rational choices on the grounds of their ontological knowledge and their specific strategies. We formally introduce the dialogue mechanism, and discuss its behaviour, properties and outcomes.

1 Introduction

The emergence of annotated data and sophisticated mechanisms for representing formal data models has promoted the proliferation of novel services and systems. These independent services usually commit to their own knowledge model (*ontologies*) and interoperate in an opportunistic fashion in order to perform some task. However, as data models differ, the extent to which the messages are understood can be restricted; thus approaches are necessary to support semantic reconciliation and thus enable seamless interactions to take place between these services. Usually these approaches rely on reaching some form of agreement on the choice of mappings or *correspondences* to translate between the entities in two ontologies. Whilst the problem of determining the vocabulary to use when integrating heterogeneous knowledge has been investigated by numerous research efforts [2,6,25], they typically require that both ontological models are shared with some party responsible for discovering the correspondences, even though there may be no guarantee that such correspondences exist; thus this is a limiting assumption. Furthermore, privacy has become increasingly pertinent, whereby neither agent is necessarily prepared to disclose its *full* ontology [11,15], *e.g.*, if the knowledge encoded within an ontology is confidential or commercially sensitive.

E. Blomqvist et al. (Eds.): EKAW 2016, LNAI 10024, pp. 543–560, 2016.
DOI: 10.1007/978-3-319-49004-5_35

In this paper, we recast this problem as a form of decentralised negotiation, by exploring how dialogue protocols can be used to determine mappings that satisfy each of the agents requirements and strategies. The use of dialogical models allow the agents to state their position regarding the correctness of some mapping in an asynchronous and distributed fashion, whilst maintaining control over the type of knowledge (class labels vs. ontological model) disclosed. We investigate the issue of reaching an agreement that facilitates the translation of one term from a vocabulary into a corresponding one in a different vocabulary. These translations are not precomputed before any interaction mechanism can be defined, but are rather computed opportunistically (*anytime*) and satisfy the agents requirements and strategies whilst limiting the information exchange only to what is pertinent to support a specific translation. Our main contribution is a dialogue based negotiation mechanism that allows the agents to propose viable lexical mappings and then support these proposals with evidence in the form of ontological fragments, thus collaboratively generating a mutually acceptable partial alignment. These are shared on a per-need basis, and hence the mechanism is purely opportunistic.

This paper is organised as follows: Section 2 introduces the challenge of reconciling heterogeneous knowledge sources, and introduces various constraints that characterise our approach. Section 3 introduces the formalism used throughout the paper, and presents the dialogue protocol and use of arguments to support candidate correspondences. The approach is then illustrated through a walk-through example in Sect. 4, and its theoretical properties are discussed in Sect. 5, before concluding in Sect. 6.

2 Background and Related Work

The ability to reconcile independently developed knowledge sources is crucial in supporting critical decision making in intelligent applications that require the interaction between disparate knowledge sources. *Ontologies* are machine readable specifications of a conceptualisation of some given domain knowledge [13]; they define the entities and the relationships between them that model such knowledge. It is often the case, however, that the agents differ in the vocabularies (ontologies) they assume, thus compromising seamless *semantic interoperability* between dynamic and evolving systems.

Ontology alignment [7] (the creation of sets of mappings between corresponding entities within a pair of ontologies) can support semantic interoperability between knowledge bases, and thus is an essential component for agent communication. However, even in similar domains, the ontologies can be modelled differently using a variety of modelling languages and contrasting assumptions, which can make translating one ontology into another increasingly difficult. For two systems to accurately and successfully communicate, this semantic heterogeneity between ontologies needs to be resolved.

The ontology alignment community has proposed diverse approaches that *align* ontologies in order to find sets of correspondences; however, alignment approaches are typically centralised processes that require full access to both ontologies. Such approaches try to maximise the number of correspondences created (*coverage*) given some objective function, but they are task agnostic, *i.e.* they do not guarantee to provide correspondences that support a given task or set of queries. Even if an alignment can be found, this might not actually support the representation of a joint task [21]. Furthermore, the axioms defined in the ontology may represent proprietary or commercially sensitive knowledge, and an agent may find it strategically important to impose some restriction over access to this knowledge [11,15]. Thus, there is a need for alignment approaches that only generate mappings for the knowledge that is pertinent to some joint task. By structuring the alignment process as a decentralised dialogue, agents can independently determine what axioms they need to expose.

The dialogue based alignment mechanism proposed here is based on the notion of *conversations* as social constructs, where utterances are exchanged in order to achieve some *joint activity* or task [4]; and on the cognitive mechanisms for communication and coordination of activities [14,22]. The dialogue determines whether there is a *common ground* [5] for establishing the alignment, by generating and sharing justifications for each correspondence proposed. An underlying assumption of our dialogue based approach is that it satisfies the principle of least collaborative effort, where participants try to minimise the total effort spent on a conversation, as typically the fewer exchanges required to clarify references, the better this common ground. It also obeys Grice's *Cooperative Principle* [12] by assuming that: (i) the participating agents are truthful; (ii) they make informative contributions as required; and (iii) they keep their interactions terse and do not provide more information than necessary. This principle supports the pertinent sharing of knowledge computed on a per-need basis; and further specification is only applied when the communication becomes ineffective. Previous investigations into *meaning negotiation* have also built upon this principle, whereby ontological reconciliation should be *rational*. This problem was first introduced in [2], where *ontology negotiation* was facilitated through a communication protocol that allowed agents to exchange ontological fragments by successively specifying the meaning of given entities. Other studies have addressed different aspects of ontology negotiation [6,15,18]. Anemone [6] advocated a lazy, minimal protocol whereby agents exchanged logical definitions in an attempt to define a minimal shared ontology with no information loss. However, it assumed that agents had perfect knowledge over the instances of their ontological models (*i.e.* the underlying approach was grounded through an extensional model), which was used to induce a class description covering certain instances.

Other approaches align heterogeneous ontologies through decentralised negotiation mechanisms [15,21] or argumentation [17,18]. In [15], agents selectively exchange details of a priori privately known correspondences, and propose repairs to address any emergent conservativity violations [24], resulting in alignments that are mutually acceptable to both agents without disclosing the full ontological model. Argumentation was used to rationally select correspondences based on the notion of partial-order preferences over their different properties (e.g. structural vs terminological) [18]. This form of correspondence negotiation utilises a course-grained decision mechanism that fails to assess whether or not a correspondence is *acceptable* to each agent (given other mutually accepted correspondences), and assumes that all the correspondences are shared.

3 The Dialogue Mechanism

The dialogue mechanism allows two agents, a *proponent* (a_1) and an *opponent* (a_2), to take turns in exchanging information (through a sequence of *dialogue moves* listed in Table 1) to support a candidate correspondence between the *entities* in their respective ontologies. We assume that an *ontology* $\mathcal{O}$ is modelled as a set of axioms describing theses entities, which consist of *classes* $\mathsf{N_C}$ and their *relations* $\mathsf{N_R}$. As each agent commits to its own ontology $\mathcal{O}$ (*i.e.* agent a_i commits to $\mathcal{O}^{a_i}$), the entities may be *disclosable* or *private*, depending on the strategy or context of the agent. Thus, the aims of the dialogue are to establish an *alignment* (consisting of a set of *correspondences* [7]) for the entities that are *disclosable* (and thus avoid negotiation over any private entity), whereby each agent negotiates over entities in a *disclosable* signature $\Sigma^d = \mathsf{N}^d_\mathsf{C} \cup \mathsf{N}^d_\mathsf{R}$ *i.e.* the set of disclosable class and property names used in $\mathcal{O}$. For clarity, we use Σ in the remainder of this paper to refer to the disclosable signature of an agent.

We assume that the ontologies are represented as an edge-labelled directed graph[1] G, where G is an ordered pair $G = (V, E)$ such that:

- $V \subseteq \mathsf{N_C} \cup L$ is a finite set of vertices (where L is the set of literals);
- $E \subseteq V \times \mathsf{N_R} \times V$ is a ternary relation describing the edges (including labels). As the direction of the edge $e \in E$ represents the 'subsumes' relation ($\sqsubseteq$), two edges are required to represent 'disjoint' ($\perp$) and 'equivalent' ($\equiv$).

We denote with $\pi = \langle s, p, o \rangle$ a subgraph of G (also known as a *triple*) where the disclosable *subject* $s \in \mathsf{N}^d_\mathsf{C}$ and the disclosable *object* $o \in \mathsf{N}^d_\mathsf{C} \cup L$ are vertices, and the disclosable *predicate* $p \in \mathsf{N}^d_\mathsf{R}$ is an edge that relates s to o. We use Π to denote the set of all π.

For two agents to interoperate in an encounter, they need to *align* [7] their respective disclosable vocabulary fragments Σ^{a_1} and Σ^{a_2}, such that the resulting alignment establishes a logical relationship between the disclosable entities belonging to each of the two ontologies. Hence, a correspondence is a mapping between an entity in a source signature (Σ^{a_1}), and a corresponding entity in a target signature (Σ^{a_2}).

Definition 1. *A* **correspondence** *is a triple denoted* $c = \langle e, e', r \rangle$ *such that* $e \in \Sigma^{a_1}$, $e' \in \Sigma^{a_2}$, $r \in \{\equiv, \sqsubseteq, \sqsupseteq, \perp\}$.

We focus on finding concept correspondences, and hence only consider aligning disclosable concept names in $\mathsf{N}^{a_1}_\mathsf{C}$ and $\mathsf{N}^{a_2}_\mathsf{C}$. Furthermore, we only consider *logical equivalence* (as opposed to *subsumption* ($\sqsubseteq$) and *disjointness* ($\perp$)).[2]

---

[1] This is common in ontology alignment approaches [7] and allows us to represent the underlying ontological model irrespectively of the ontology language used (e.g. RDF or OWL).

[2] This assumption does not affect the generality of our approach, and the majority of ontology alignment approaches that align entities only consider equivalence. Extending the dialogue to support the discovery of subsumption relations is the subject of future work.

3.1 Dialogue Protocol

The dialogue protocol comprises a sequence of communicative acts, or *moves* (denoted $\mathcal{M}$), whereby two participating agents take turn to share statements supporting or refuting a candidate correspondence. For every dialogue move, we assume that each agent plays a role; *i.e.* a_1 is either a *sender* x or *recipient* $\hat{x}$ (and conversely, a_2 plays the alternate role, such that they never play the same role concurrently). After each move, the agents swap roles, and thus take turns in acting as sender or recipient. The set of legal *moves*, $\mathcal{T}$, are summarised in Table 1, and their use is illustrated in the walkthrough in Sect. 4. The syntax of each move is of the form $m = \langle x, \tau, e, e', l \rangle$, where τ is the move type such that $\tau \in \mathcal{T}$, and $\mathcal{T} = \{\mathsf{initiate}, \mathsf{propose}, \mathsf{assert}, \mathsf{accept}, \mathsf{reject}, \mathsf{testify}, \mathsf{justify}, \mathsf{fail}, \mathsf{end}\}$; e represents the source entity being discussed (identified within the *initiate* move); e' is the current candidate target entity (*i.e.* the entity that could be mapped to from e); and l represents a list of zero or more additional elements (depending on the type of move). For some moves, it may not be necessary to specify the source entity, the target entity or any additional elements, in which case they will

Table 1. The set $\mathcal{T}$ of legal moves permitted by the dialogue.

Syntax	Description
$\langle x, \mathsf{initiate}, e, \text{nil}, \text{nil} \rangle$	A new source entity e is proposed, with the aim of finding a possible correspondence.
$\langle x, \mathsf{propose}, e, e', \text{nil} \rangle$	A new (*i.e.* not previously disclosed) candidate entity e' is proposed which lexically matches e.
$\langle x, \mathsf{justify}, e, e', \text{nil} \rangle$	A new π is requested to support the candidate correspondence between e and e'.
$\langle x, \mathsf{testify}, e, e', \pi \rangle$	If an undisclosed π is known that supports the candidate correspondence (with the highest ranking predicate), then it is shared; otherwise $\pi = \text{nil}$.
$\langle x, \mathsf{assert}, e, e', A \rangle$	The candidacy of a correspondence between e and e' is asserted, with the supporting argument A containing a subset of disclosed π pairs whose aggregate *neighbourhood similarity* σ_n supports the candidacy. Note that A and σ_n are presented in Sect. 3.3.
$\langle x, \mathsf{accept}, e, e', A \rangle$	The candidacy is accepted if the neighbourhood similarity σ_n of the premise in A is above threshold given the sending agent's similarity metrics.
$\langle x, \mathsf{reject}, e, e', \text{nil} \rangle$	The candidacy is rejected if the neighbourhood similarity σ_n of the premise in A is below threshold given the sending agent's own similarity metrics, and no other supporting evidence is available.
$\langle x, \mathsf{fail}, e, \text{nil}, \text{nil} \rangle$	No further undisclosed candidate entities could be found that lexically match e.
$\langle x, \mathsf{end}, \text{nil}, \text{nil}, \text{nil} \rangle$	The proponent terminates the dialogue.

be empty or unspecified (represented with *nil*). Figure 1 illustrates the different states that can occur during the dialogue, and identifies what moves can be legally taken by which agent. The choice of move is determined by the agent's individual strategy (discussed below).

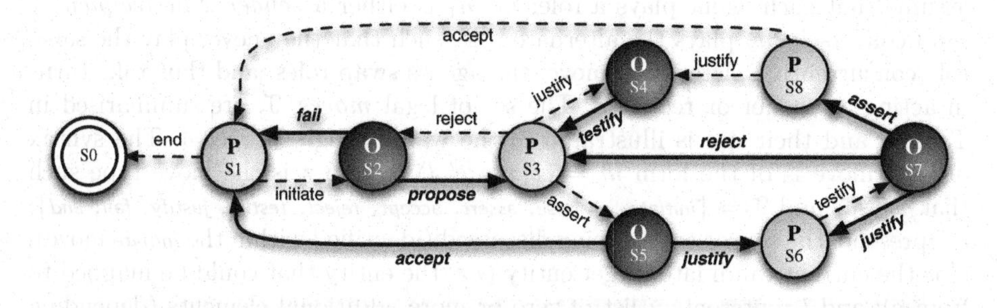

Fig. 1. The dialogue protocol as a state diagram. Nodes indicate the agent whose turn it to utter a move. Moves uttered by the *proponent* are labelled with a light font / dashed edge, whereas those uttered by the *opponent* are labelled with a heavy font / solid edge. The *proponent* always makes the first move (*i.e.* starting from state S1), and the dialogue terminates at state S0.

Both agents manage a public knowledge base, or *Commitment Store*[3] *CS*, which contains a trace of all of the moves uttered by each agent [26]. Each agent manages its own private knowledge base, known as the *Gamma Store*[4] (Γ), that stores private knowledge regarding the ontological structure of the *opponent* that has been garnered through the assertions made in the dialogue. Each of the Gamma Stores contains a partially connected graph, that is:

- either an independent vertex $v_i \in \mathsf{N_C}$ representing a candidate concept from the opponent's ontology for inclusion in a correspondence;
- or the *neighbourhood* of the concept v_i, *i.e.* the subgraph originating from the vertex v_i constructed through the exchange of triples that form a directed path from v_i to support its candidacy.

At each point in the dialogue, an agent selects from one or more moves, depending on its *strategy* which in turn is based on some *objective function* that reflects the agent's current task or objective. Thus, an agent may want to find a maximal alignment (*i.e.* map as many entities as possible) if it is interested in knowledge integration, or find some alignment that maps only those entities that are necessary and sufficient to perform some service [1]. When the proponent has no further entities that it wants to map, it can terminate the dialogue. If the opponent then wishes to explore further correspondences, it can initiate a new

[3] Although the agents maintain individual copies of the *CS*, these will always be identical, and thus we do not distinguish between them.

[4] We distinguish between the sender's Gamma Store, Γ^{x}, and the recipient's store, $\Gamma^{\hat{x}}$.

dialogue and assume the role of *proponent* (*i.e.* the agents can swap these roles). In this paper, we make no assumptions about how the objective function is defined by any specific agent, and whether or not an agent will align all possible entities or terminate early if a sufficient number of entities have been discovered. The only assumptions made are that:

- As the dialogue starts, the agents have no knowledge of their opponent's ontology;
- The agents use their own similarity metrics to assess whether to accept or reject possible correspondences;
- The number of facts about either ontologies that are disclosed to the opponent should be *minimised*.

3.2 Lexical and Structural Similarity

Within the dialogue, the agents try to ascertain a similarity between the shared entities to determine whether or not there is sufficient evidence to justify proposing or accepting a candidate correspondence, given a particular alignment strategy the agent has over a specific task. Many approaches for determining similarity have been proposed, or evaluated in the ontology matching literature [3,7,10,23]. In our approach, the agents can utilise different similarity metrics (e.g. the Jaccard similarity coefficient, or metrics that exploit linguistic resources such as Wordnet [20] to identify synonyms) to determine lexical matchings[5]. However, we make no assumption on the choice of similarity metrics used, nor do we prescribe that the agents have to agree on a common mechanism. Thus, we assume that agents differ in their assessment of the similarity of two labels. A lexical similarity metric is defined formally as:

Definition 2. *The* **lexical similarity metric** *is the function* $\sigma_l : \mathsf{N_C} \times \mathsf{N_C} \rightarrow [0,1]$ *which returns the lexical similarity between the labels of two entity names* $e, e' \in \mathsf{N_C}$*, such that* $\sigma_l(e, e') = 1$ *iff* $e = e'$ *and 0 if the two labels are different.*

This function is used in the initial part of the dialogue to discover those entities in agent a_2's signature that could lexically match an entity in agent a_1's signature (*anchors*). A lexical match is considered *viable* if $\sigma_l(e, e')$ is greater or equal to its threshold ϵ_l.

An important component of the dialogue is how the agents share structural details about the ontology in the neighbourhood of an entity under consideration. For any given entity $e \in \Sigma$, there will be a directed path[6] within the graph that relates e to other entities in its neighbourhood, where the maximum length of the path is bounded by the depth of the ontology. Thus, any triple (π) within this path could be disclosed (*i.e.* shared with the other agent) to provide more details of the entities' local neighbourhood, provided that it forms a path from

[5] See [3] for a good survey of different string similarity metrics.

[6] Given the example in Fig. 3, the neighbourhood of $e = \mathsf{Author}$ would include the triple $\langle \mathsf{Author}, \mathsf{hasInitials}, \mathsf{Initials} \rangle$, but would *not* include the triple $\langle \mathsf{Paper}, \mathsf{hasAuthor}, \mathsf{Author} \rangle$.

the entity itself. Depending on the strategy that an agent may adopt, it may assume a depth-first traversal as opposed to breadth-first when disclosing its triples. Therefore, we assume that each agent utilises a function $\mathsf{rank}(e)$ that generates a strict pre-ordering of triples for a given subject e. This is formally defined as:

Definition 3. *The* **rank function**, $\mathsf{rank} : \mathsf{N_C} \rightarrow \mathfrak{R} \subseteq \Pi$ *returns an ordered list of triples in a path starting at some entity* $e \in \mathsf{N_C}$*, where* $\forall \pi_i, \pi_j \in \mathfrak{R} : \pi_i \succcurlyeq \pi_j$.

An agent can request triples belonging to the local neighbourhood of some entity e' in the other agent's ontology, to support the candidacy of a correspondence. We make no assumptions about how the ranking function is defined by any specific agent, and thus the order in which the triples are ranked. Furthermore, following the *Similarity Flooding* approach [19], we also restrict our attention to paths of length 1, and thus only disclose those triples for which e' is a subject.

As *subject-predicate-object* triples relating to e' are disclosed by one agent, the second agent should try to identify similar localised structures in its own ontology. This may be based purely on the triples themselves, or may also take into account other information that has so far been ascertained or inferred. As with the σ_l function, we make no assumptions about how the similarity function is defined, but simply that there is some function for each agent defined formally as:

Definition 4. *The* **structural similarity metric** *is the function* $\sigma_s : \Pi \times \Pi \rightarrow [0, 1]$ *that returns the structural similarity between two triples* $\pi, \pi' \in \Pi$*, such that* $\sigma_s(\pi, \pi') = 1$ *if the two triples are considered as equivalent, and 0 otherwise.*

3.3 Arguments and Neighbourhood Similarity

The dialogue mechanism utilises arguments that allow the agents to propose candidate correspondences (between the entities in their respective ontologies), and to justify them or refute them on the grounds of some evidential fact. The agents are assumed to be truthful and to cooperate in order to reach an agreement on the best correspondence to use to map two entities from their respective ontologies.

For this reason, agents can only make arguments that assert the validity of a new correspondence that was not previously disclosed, or question its correctness by stating an alternative correspondence for one of the same entities. As each new argument either introduces a new correspondence, or states a new premise for an existing one, there is no possibility of cycles in arguments, and thus the agents will either reach an agreement or they will reject the proposal. The arguments are defined over the language $\mathfrak{L}$, with the same syntactic primitives as defined for the dialogue. Each agent can form arguments about a candidate correspondence c and entities e in the disclosable signature $\Sigma_d^{a_i}$ of their ontology. $\mathfrak{L}$ is the set of formulae ℓ defined by:

$$\ell : := e | c | (\{\pi\}, c)$$

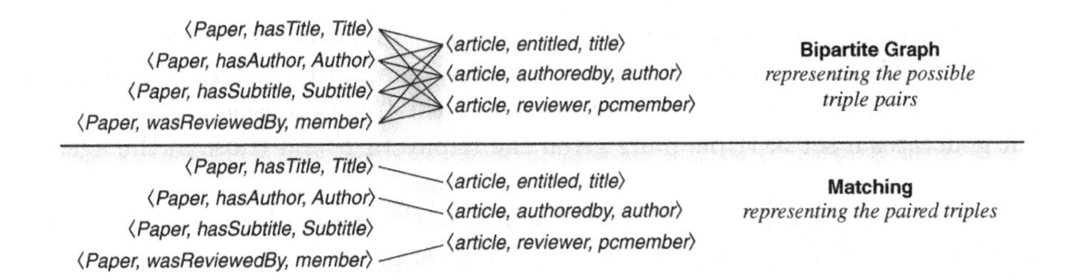

Fig. 2. Possible pairs of triples (top) and a matching (bottom) from the example (Sect. 4).

Hence $\mathfrak{L}$ will contain statements about $\Sigma_d^{a_1}$, $\Sigma_d^{a_2}$ and the correspondences mapping from one signature into the other.

Definition 5. *An* **Argument** *is a pair* $A = (Pr, Cl)$*, where* $Pr \subseteq \mathfrak{L} \cup \{\top\}$ *and* $Cl \in \mathfrak{L}$*. We define* Args*(*$\mathfrak{L}$*) the set of all arguments derivable from the language* $\mathfrak{L}$*.*

In this definition, Pr is the *support* (representing a set of premises of an argument), whilst Cl is the *claim*. Facts (*i.e.* statements with no premises) are represented as $(\top, Cl)$. An argument expresses a relationship between the *claim* and the *support*, such that if the support holds, then the claim must also hold. In our dialogue, the support expresses a *justification* for some neighbourhood similarity (based on a set of related triples) for two entities e and e', and the claim asserts the viability of a correspondence between these two entities, *i.e.* that the correspondence has some evidence of correctness. The support is based upon some *injective matching* between a bipartite graph (Fig. 2 bottom) representing the triples in an agent's own ontology, and those disclosed by the other agent as part of the dialogue, resulting in matched pairs (π, π'). Each π disclosed by one agent will have some similarity to zero or more triples disclosed by its opponent, as illustrated by the example in Fig. 2 (top) between two example sets of triples supporting a correspondence between the entities *Paper* and *article*.

The *neighbourhood similarity* σ_n is computed over the set of all matching (π, π') *pairs* (that form a bipartite graph - Fig. 2, bottom), such that no triple from one ontology is "paired" to more than one triple in the other ontology (*i.e.* finding an *injective*, or *one-to-one* mapping between the sets of triples). Depending on the choice of objective function used [9,16], this can be achieved by finding a *matching* in the graph.

Definition 6. *The* **neighbourhood similarity** *is the function* $\sigma_n : \{(\pi, \pi') \in \Pi \times \Pi \mid \pi \in \Gamma, \pi' \in \mathcal{O}\} \rightarrow [0, 1]$ *that returns an aggregate similarity calculated from a matching generated from the weighted Bipartite graph obtained by calculating all possible structural similarities between the triples in an agent's Gamma Store* Γ *and the triples in the disclosable fragment of the opponent's ontology* $\mathcal{O}^{a_2}$*, such that* $\sigma_n(\pi, \pi') = 1$ *if the neighbourhood is structurally equivalent, and 0 otherwise.*

As we make no assumption w.r.t. the objective function used to generate the matching (other than assuming that a structural similarity metric σ_s is used to generate the similarity of each pair), we define the function $\mathsf{pairing} : \Pi \times \mathcal{O} \rightarrow \Pi$ that generates a set of triple pairs given the triples in Γ and those in the agents ontology $\mathcal{O}$.

For example, assuming the triples in Fig. 3, the agent *Alice* may have disclosed all four triples to *Bob*. Therefore, *Bob* has:

$$\Gamma^{Bob} = \{\langle \mathit{Paper}, \mathit{hasTitle}, \mathit{Title}\rangle, \langle \mathit{Paper}, \mathit{hasAuthor}, \mathit{Author}\rangle, \langle \mathit{Paper}, \mathit{hasSubtitle}, \mathit{Subtitle}\rangle, \langle \mathit{Paper}, \mathit{wasReviewedBy}, \mathit{Member}\rangle\}$$

$$\mathcal{O}^{Bob} = \{\langle \mathit{article}, \mathit{reviewer}, \mathit{pcmember}\rangle, \langle \mathit{article}, \mathit{entitled}, \mathit{title}\rangle, \langle \mathit{article}, \mathit{authoredby}, \mathit{author}\rangle\}$$

By using the structural similarity metric σ_s, the complete set of possible triple pairs in Fig. 2 (left) can be determined. Assuming some objective function, the matching in Fig. 2 (right) can be generated. Thus, we state that:

$$\begin{aligned}\mathsf{pairing}(\Gamma^{Bob}, \mathcal{O}^{Bob}) = \{&(\langle \mathit{Paper}, \mathit{hasTitle}, \mathit{Title}\rangle, \langle \mathit{article}, \mathit{entitled}, \mathit{title}\rangle),\\ &(\langle \mathit{Paper}, \mathit{hasAuthor}, \mathit{Author}\rangle, \langle \mathit{article}, \mathit{authoredby}, \mathit{author}\rangle),\\ &(\langle \mathit{Paper}, \mathit{wasReviewedBy}, \mathit{Member}\rangle, \langle \mathit{article}, \mathit{reviewer}, \mathit{pcmember}\rangle)\}\end{aligned}$$

The premise Pr for the claim by agent a_i for some correspondence c will comprise a subset of pairs from the set $\mathsf{pairing}(\Gamma^{a_i}, \mathcal{O}^{a_i})$, with a corresponding aggregate *neighbourhood similarity* σ_n. Although we make no assumption about how σ_n is defined, it could be based on the structural similarity scores σ_s for each triple pair in Pr. A premise Pr is *acceptable* to an agent if $\sigma_n(Pr)$ is greater or equal to a threshold ϵ_n.

4 Walkthrough Example

We illustrate how two agents utilise the dialogue protocol to find an alignment between the public signatures of their ontologies by means of an example. Two agents, *Alice* and *Bob*, each possess a private ontological fragment (Fig. 3). Both agents implement different structural similarity metrics σ_s, and a subset[7] of the values for different π triple pairs is given in Table 2. For example, the structural similarity[8] σ_s between the triple $\langle \mathit{Paper}, \mathit{hasTitle}, \mathit{Title}\rangle$ and $\langle \mathit{article}, \mathit{entitled}, \mathit{title}\rangle$ for *Alice*, $\sigma_s^{Alice} = 0.70$, whereas for *Bob* the similarity for this pair is $\sigma_s^{Bob} = 0.68$. In the example dialogue (Table 3), we assume that the dialogue has already commenced, resulting in *Alice* accepting the correspondence $\langle \mathit{Author}, \mathit{author}, \equiv\rangle$ in a

[7] Although other similarity pairs have been calculated, these do not appear in the dialogue example (for example, because the distance is lower than those explicitly stated), and thus have not been given for brevity.

[8] These similarity pairs are not generated a priori, but are calculated during the dialogue.

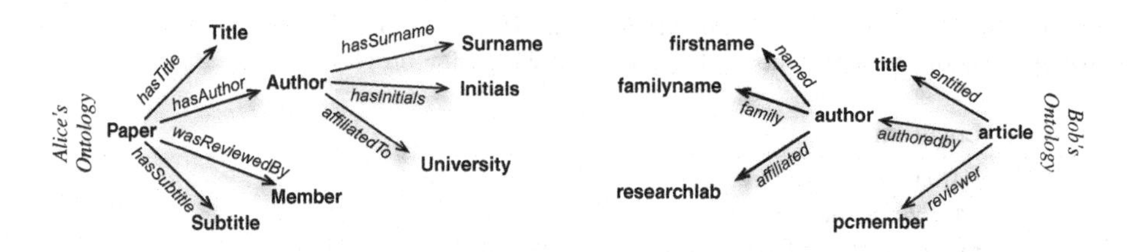

Fig. 3. Two trivial ontology fragments for *Alice* and *Bob* used in the walkthrough example.

Table 2. The structural similarities of possible corresponding triples between *Alice* & *Bob*'s ontologies. Whilst not exhaustive, it lists a subset of triples between the two ontologies.

Alice's π	*Bob*'s π	σ_s^{Alice}	σ_s^{Bob}
⟨*Author*, *hasSurname*, *Surname*⟩	⟨*author*, *family*, *familyname*⟩	0.76	0.72
⟨*Author*, *affiliatedTo*, *University*⟩	⟨*author*, *affiliated*, *researchlab*⟩	0.85	0.86
⟨*Paper*, *hasTitle*, *Title*⟩	⟨*article*, *entitled*, *title*⟩	0.70	0.68
⟨*Paper*, *hasAuthor*, *Author*⟩	⟨*article*, *authoredby*, *author*⟩	0.65	0.61
⟨*Paper*, *hasSubtitle*, *Subtitle*⟩	⟨*article*, *entitled*, *title*⟩	0.68	0.84
⟨*Paper*, *wasReviewedBy*, *Member*⟩	⟨*article*, *reviewer*, *pcmember*⟩	0.66	0.60

previous negotiation round (Moves 1–13; the acceptance of this correspondence is illustrated in Move 13 of Table 3). The order in which the dialogue proponent selects entities for exploration is strategic[9]; for this example, we assume that the first two entities *Alice* explores are (in order): *Author* and *Paper*. We assume a neighbourhood similarity metric $\sigma_n(Pr)$ calculates the average structural similarity $\bar{\sigma}_s$ of the triple pairs in the premise Pr, with a coefficient that increases asymptotically as the cardinality of Pr increases. The metric is defined as $\sigma_n(Pr) = \bar{\sigma}_s \times (1 - \frac{1}{2(|Pr|+1)})$. We also assume a neighbourhood threshold $\epsilon_n = 0.55$ and a lexical threshold $\epsilon_l = 0.75$.

Move 14: Having previously accepted a correspondence for *Author* (Move 13), *Alice* utters a *initiate* move (state S1 in Fig. 1), to explore a possible correspondence for the next entity from her public signature that she wants to align; which in this case is *Paper*.

Move 15: *Bob* identifies *article* as the most similar entity in his ontology to *Paper* with a lexical similarity $\sigma_l^{Bob}(\textit{Paper}, \textit{article}) = 0.82$ (this value is not given in the table). As this is above threshold ϵ_l, he responds with the move ⟨*Bob*, *propose*, *Paper*, *article*, nil⟩.

[9] As mentioned previously, we do not specify here how the strategic choices are made by each agent, but assume some objective function that determines these choices exists.

Table 3. The messages exchanged between *Alice* and *Bob* in the example dialogue fragment (note that the moves 1–12 are not shown for brevity).

Move	Locution
13	⟨*Alice*, *accept*, *Author*, *author*, ((⟨*author*, *family*, *familyname*⟩, ⟨*Author*, *hasSurname*, *Surname*⟩), ((⟨*author*, *affiliated*, *researchlab*⟩, ⟨*Author*, *affiliatedTo*, *University*⟩), ⟨*Author*, *author*, ≡⟩)⟩
14	⟨*Alice*, *initiate*, *Paper*, nil, nil⟩
15	⟨*Bob*, *propose*, *Paper*, *article*, nil⟩
16	⟨*Alice*, *justify*, *Paper*, *article*, nil⟩
17	⟨*Bob*, *testify*, *Paper*, *article*, ⟨*article*, *reviewer*, *pcmember*⟩⟩
18	⟨*Alice*, *justify*, *Paper*, *article*, nil⟩
19	⟨*Bob*, *testify*, *Paper*, *article*, ⟨*article*, *entitled*, *title*⟩⟩
20	⟨*Alice*, *assert*, *Paper*, *article*, ({(⟨*Paper*, *wasReviewedBy*, *Member*⟩, ⟨*article*, *reviewer*, *pcmember*⟩), (⟨*Paper*, *hasTitle*, *Title*⟩, ⟨*article*, *entitled*, *title*⟩)}, ⟨*Paper*, *article*, ≡⟩)⟩
21	⟨*Bob*, *justify*, *Paper*, *article*, nil⟩
22	⟨*Alice*, *testify*, *Paper*, *article*, ⟨*Paper*, *hasAuthor*, *Author*⟩⟩
23	⟨*Bob*, *assert*, *Paper*, *article*, ({(⟨*article*, *reviewer*, *pcmember*⟩, ⟨*Paper*, *wasReviewedBy*, *Member*⟩), (⟨*article*, *entitled*, *title*⟩, ⟨*Paper*, *hasTitle*, *Title*⟩), (⟨*article*, *authoredby*, *author*⟩, ⟨*Paper*, *hasAuthor*, *Author*⟩)}, ⟨*Paper*, *article*, ≡⟩)⟩
24	⟨*Alice*, *accept*, *Paper*, *article*, ({(⟨*article*, *reviewer*, *pcmember*⟩, ⟨*Paper*, *wasReviewedBy*, *Member*⟩), (⟨*article*, *entitled*, *title*⟩, ⟨*Paper*, *hasTitle*, *Title*⟩), (⟨*article*, *authoredby*, *author*⟩, ⟨*Paper*, *hasAuthor*, *Author*⟩)}, ⟨*Paper*, *article*, ≡⟩)⟩

Move 16: *Alice* now knows that ⟨*Paper*, *article*, ≡⟩ is a potential correspondence c (based on *Bob's* lexical similarity claim). She verifies that her lexical similarity for the entity pair is above threshold (in this case $\sigma_l^{Alice}(\mathit{Paper}, \mathit{article}) = 0.79$). As she is aware that the entity *Paper* has a local neighbourhood (i.e. there is at least one π that has *Paper* as its subject), she asks *Bob* to provide some evidence to justify the candidacy of c. At this point, neither agents have support for c; i.e. $Pr = \varnothing$.

Move 17: *Bob* (state S4) generates a strict pre-ordering of the properties for *article*, using the function rank(); i.e. $\mathsf{rank}^{Bob}(\mathit{article}) = \{\mathit{reviewer}, \mathit{entitled}, \mathit{authoredby}\}$. He uses this to determine the next property that has *article* as its *domain* and that has not yet been disclosed (i.e. that has not yet appeared in the commitment store CS). As none of the properties in $\mathsf{rank}^{Bob}(\mathit{article})$ have yet been disclosed, he shares the fact that the highest ranked property *reviewer* relates the two entities *article* and *pcmember*.

Move 18: *Alice* tries to determine if there is sufficient support for c. She realises that ⟨*Paper*, *wasReviewedBy*, *Member*⟩ in her ontology is the most similar triple to the one *Bob* disclosed in move 17, with a similarity $\sigma_s^{Alice} = 0.66$ (Table 2). She calculates that the premise $Pr = \{(\langle \textsf{Paper}, \textsf{wasReviewedBy}, \textsf{Member}\rangle, \langle \textsf{article}, \textsf{reviewer}, \textsf{pcmember}\rangle)\}$ has a neighbourhood similarity $\sigma_n^{Alice} = 0.66 \times (1 - \frac{1}{2(|Pr|+1)}) = 0.66 * 0.75 = 0.495$. She will only *assert* an argument for c if this is above the threshold $\epsilon_n = 0.55$. As this is below threshold, she requests additional evidence to justify c.

Move 19: *Bob's* next highest ranked property that has not been disclosed (*i.e.* does not appear in CS) whose domain is *article*, is the entity *entitled*. Therefore he shares the triple $\langle \textsf{article}, \textsf{entitled}, \textsf{title}\rangle$.

Move 20: *Alice* checks to see if one of her triples is similar to that disclosed by *Bob* in move 19. Although she has two triples that share their highest similarity with *Bob's* disclosed triple, she chooses $\langle \textsf{Paper}, \textsf{hasTitle}, \textsf{Title}\rangle$ as the similarity is higher than $\langle \textsf{Paper}, \textsf{hasSubtitle}, \textsf{Subtitle}\rangle$. She adds this to Pr and calculates the neighbourhood similarity $\sigma_n^{Alice} = (0.66 + 0.7)/2 \times (1 - \frac{1}{2(2+1)}) = 0.68 * 0.8\dot{3} = 0.56$, which (from *Alice's* perspective) is above threshold, Therefore she proposes the argument A for the correspondence $c = \langle \textsf{Paper}, \textsf{article}, \equiv\rangle$, given that:

$$Pr = \{(\langle \textsf{Paper}, \textsf{wasReviewedBy}, \textsf{Member}\rangle, \langle \textsf{article}, \textsf{reviewer}, \textsf{pcmember}\rangle), \\ (\langle \textsf{Paper}, \textsf{hasTitle}, \textsf{Title}\rangle, \langle \textsf{article}, \textsf{entitled}, \textsf{title}\rangle)\}$$

Move 21: Given the argument A for the correspondence c asserted in the previous move, *Bob* (state S5) can make one of two possible moves:

- *accept* the argument A if $\sigma_n^{Bob}(Pr)$ is above threshold, and transition to state S1;
- *justify* the candidacy of c by requesting further support (if other undisclosed properties exist).

In this case, *Bob* calculates that the neighbourhood similarity (from his perspective) is $\sigma_n^{Bob} = (0.60 + 0.68)/2 \times (1 - \frac{1}{2(2+1)}) = 0.64 * 0.8\dot{3} = 0.5\dot{3}$, which is below threshold. However, *Bob* is aware of other triples for the entity *article* that do not appear in Pr, and thus asks *Alice* if she could provide some further evidence to justify c.

Move 22: *Alice* now generates her own strict pre-ordering of the properties for *Paper*, using the function rank(); i.e. $\textsf{rank}^{Alice}(\textsf{Paper}) = \{\textsf{hasTitle}, \textsf{hasAuthor}, \textsf{hasSubtitle}, \textsf{wasReviewedBy}\}$. She shares the triple $\langle \textsf{Paper}, \textsf{hasAuthor}, \textsf{Author}\rangle$ as *hasAuthor* is her highest ranked, non-disclosed property for the domain entity *Paper* (property *hasTitle* was ranked higher but was disclosed in her previous *assert* move).

Move 23: *Bob* recalculates the mean similarity for the new support (inclusive of the triple shared by *Alice* in Move 22): $\sigma_n^{Bob} = (0.60 + 0.68 + 0.61)/3 \times (1 - \frac{1}{2(3+1)}) = 0.63 * 0.875 = 0.551$, which is above threshold. *Bob* is happy to accept the candidacy of c. It is now his turn to *assert* the new argument for c given the new premise Pr.

Move 24: *Alice* confirms that from her perspective, $\sigma_n^{Alice} = (0.66 + 0.7 + 0.65)/3 \times (1 - \frac{1}{2(3+1)}) = 0.67 * 0.875 = 0.59$, which is above threshold, and accepts the argument.

At this point, through co-operation, the agents were able to engage in the joint activity of determining a correspondence between two entities based on the similarity of the local neighbourhood of the entities. Although all of *Bob's* triples were disclosed, *Alice* was able to reach the consensus without revealing knowledge of one of her triples: $\langle$Paper, hasSubtitle, Subtitle$\rangle$, even though from *Bob's* perspective, it was actually more similar to *Bob's* triple $\langle$article, entitled, title$\rangle$ than $\langle$Paper, hasTitle, Title$\rangle$. If in move 20, *Alice* had found that the triple with the highest similarity to $\langle$article, entitled, title$\rangle$ was actually $\langle$Paper, hasSubtitle, Subtitle$\rangle$, then *Bob* would have accepted the support in move 21 (as $\sigma_n^{Bob} = (0.6+0.84)/2 \times (1 - \frac{1}{2(2+1)}) = 0.67 * 0.8\dot{3} = 0.56$, which was above threshold) and fewer properties would have been disclosed.

5 Dialogue Properties

It is customary to analyse dialogue systems in terms of their *soundness*, *completeness* and *termination* properties. Usually these are not considered in isolation, but they are analysed with respect to the compliance shown by the dialogue to the specific agents' strategies. For instance, a *sound* dialogue protocol result can be roughly restated as obtaining a "successful" dialogue results, i.e. verifying that the claim of the dialogue is "acceptable" w.r.t. the adopted strategies [8].

One of the main characteristics of our approach is that the strategy definition is tightly dependent on the specific choices the agents make in terms of similarity measures. Whilst we argue that having a generic framework is a strength of the presented approach as it makes it customisable to suit different interoperability scenarios, this makes it complex to characterise soundness and completeness. Termination is more straightforward to prove as it is independent of the agents' strategies. Regarding soundness it is important to point out that this does not correspond to correctness with respect to a gold standard alignment. Indeed, it is possible to imagine that in a cooperative domain, agents would behave in an intelligent manner in order to be able to influence the outcome of the dialogue and always arrive at the best possible outcome given their internal knowledge and strategies.

The dialogue presented in the previous section allows agents to only put forward new arguments, either by proposing a new correspondence or by providing evidence supporting some candidate correspondence. Once arguments are uttered, they cannot be retracted. The monotonic property of this dialogue helps us to characterise soundness in terms of obtainable outcomes. Indeed, it is possible to clearly identify two possible outcomes of the dialogue, *fail* and *accept*, leading to the dialogue termination at S0, and the state transitions that cause these outcomes to be reached. Either outcomes represent acceptable solutions to the alignment problem, with *fail* explicitly capturing the fact that the agents cannot find a suitable solution within the constraints dictated by their strategies.

The conditions underlying these outcomes are described below, by referring to the states in the diagram in Fig. 1. The pathways to failure are described below:

S2: The proponent initiates the dialogue requesting a match for an entity e (S1), however no entity e' in the opponent signature is a viable match for e, i.e. $\forall e' \in \mathsf{N_C}^{\hat{x}} \sigma_l(e, e') < \epsilon_l$.

S3: Following S1, the opponent responds with an entity e'. The proponent then evaluates the potential correspondence (e, e'): if this is not viable (i.e. $\sigma_l(e, e') < \epsilon_l$) then it rejects it, and the dialogue fails. If the correspondence is viable then the proponent might still request the opponent to provide further evidence supporting this proposal, and hence enter a *justify-testify* loop (S3–S4). If the evidence provided is not deemed sufficient, the proponent can reject the correspondence.

S7: Following S3, the proponent assesses the correspondence proposed, and on finding it suitable she asserts it (S5). This assertion however requires some verification from the opponent, who requests that the proponent provides some supporting evidence for the assertion through a *justify-testify* loop (S6–S7, but this time with the proponent being the opponent, and vice versa). If the opponent deems that the evidence is not sufficient it will reject the assertion made by the proposer and the dialogue will fail.

The pathways for the successful termination of the dialogue are clearly identifiable:

S3: Following S1, the opponent responds with an entity e'. The proponent then evaluates the potential correspondence (e, e') and finds that it satisfies its strategy, and hence asserts the viability of the correspondence requiring further evidence (S3–S4). However the proponent may also require further evidence from the opponent (*justify-testify* loop, S3–S4). If the evidence is deemed sufficient, then the proponent asserts the acceptability of the correspondence from his side. This is then evaluated by the opponent who can confirm the acceptability of the correspondence with respect to its strategy (S6), and the dialogue terminates successfully.

S5: The opponent might also require further evidence (S6, S7) and the if satisfied, it can assert the correspondence as viable from his perspective, and then this is assessed by the proposer (S8) with or without requiring supporting evidence. If the evidence is requested, then it will be assessed and if it is deemed sufficient (according to the agent's internal strategy) then it will be accepted.

Regarding completeness, it is trivial to see that the dialogue is not complete. The dialogue effectively approximates a greedy search over the space of possible correspondences. This approximation is not guaranteed to be complete as solutions will only be accepted by the proponent of an assertion only if it deems the evidence sufficient for the claim. The same assertion will only be accepted by the opponent if it also deems the evidence sufficient and if not it will request further evidence to be put forward. However, this mechanism allows the agents to find a solution without exploring all candidate solutions, hence it satisfies the minimality requirement following Grice's maxims.

Given that the dialogue admits only two possible outcomes, and that it cannot propose correspondences or supporting evidence already proposed it is trivial to show that the dialogue terminates.

Proposition 1. *The negotiation dialogue with the set of moves $\mathcal{M}$ in Sect. 3.1 will always terminate.*

Proof (Sketch). Both agents have finite disclosable signatures, and can only propose one entity to align at a time. Once the entity is proposed, the agents can request that the correspondence is justified in terms of its support; however, this support is also finite, being bounded by the size of the disclosable signature of the ontology. At any point in the dialogue, agents can only add new evidence or assert new correspondences (after having rejected a previous proposal), but are prohibited from revisiting either a correspondence or some evidence previously discussed (i.e. agents can only add to the Commitment Store and not retract from it). If the dialogue does not end before every possible viable correspondence is considered (states S1–S3), then it will end, in the worst case, once the (finite) set of *testify* - *justify* moves providing evidence for the correspondence in the claim have all been made. If no appropriate evidence is provided, then the dialogue will terminate following a *fail* outcome. □

6 Conclusions

We present work on a dialogue based mechanism that allows agents to reach agreement over an alignment between the disclosable entities of their respective ontologies, without the need for prior information of the ontological structures used by either agent, or some centralised machinery. The proponent takes turns to ask questions about a potential correspondence to ascertain if there is sufficient evidence to support it; and the opponent, through introspection accepts, rejects or seeks further or more compelling evidence to support the claim. A dialogue protocol is introduced that allows agents to reach an agreement over mutually acceptable correspondences, and discusses its properties. It is illustrated through an example that shows how the dialogue is used to establish whether two entities in two different ontologies can be mapped, and the formal properties of the dialogue (w.r.t. soundness, completeness and termination) are presented.

References

1. Ankolekar, A., et al.: DAML-S: web service description for the semantic web. In: Horrocks, I., Hendler, J. (eds.) ISWC 2002. LNCS, vol. 2342, p. 348. Springer, Heidelberg (2002)
2. Bailin, S.C., Truszkowski, W.: Ontology negotiation: how agents can really get to know each other. In: Truszkowski, W., Hinchey, M., Rouff, C. (eds.) WRAC 2002. LNCS (LNAI), vol. 2564, pp. 320–334. Springer, Heidelberg (2003). doi:10.1007/978-3-540-45173-0_24

3. Cheatham, M., Hitzler, P.: String similarity metrics for ontology alignment. In: Alani, H., et al. (eds.) ISWC 2013, Part II. LNCS, vol. 8219, pp. 294–309. Springer, Heidelberg (2013)
4. Clark, H.H.: Using Language. Cambridge University Press, Cambridge (1996)
5. Clark, H.H., Schaefer, E.F.: Contributing to discourse. Cogn. Sci. **13**(2), 259–294 (1989)
6. van Diggelen, J., Beun, R.J., Dignum, F., van Eijk, R.M., Meyer, J.J.C.: Anemone: an effective minimal ontology negotiation environment. In: Proceedings of the AAMAS 2006, pp. 899–906 (2006)
7. Euzenat, J., Shvaiko, P.: Ontology Matching. Springer, Heidelberg (2007)
8. Fan, X., Toni, F.: On computing explanations in abstract argumentation. In: ECAI 2014-21st European Conference on Artificial Intelligence, 18–22 August 2014, Prague, Czech Republic - Including Prestigious Applications of Intelligent Systems (PAIS 2014), pp. 1005–1006 (2014)
9. Gale, D., Shapley, L.S.: College admissions and the stability of marriage. Am. Math. Mon. **69**(1), 9–15 (1962)
10. Grau, B.C., Dragisic, Z., Eckert, K., Euzenat, J., et al.: Results of the ontology alignment evaluation initiative 2013. In: Proceedings of the 8th ISWC Workshop on Ontology Matching (OM), pp. 61–100 (2013)
11. Grau, B.C., Motik, B.: Reasoning over ontologies with hidden content: the import-by-query approach. J. Artif. Intell. Res. (JAIR) **45**, 197–255 (2012)
12. Grice, H.P.: Logic and conversation. In: Cole, P., Morgan, J.L. (eds.) Syntax and Semantics, Vol. 3: Speech Acts, pp. 41–58. Academic Press, Cambridge (1975)
13. Gruber, T.R.: Toward principles for the design of ontologies used for knowledge sharing. Int. J. Hum.-Comput. Stud. **43**(5–6), 907–928 (1995)
14. Hulstijn, J.: Dialogue models for inquiry and transaction. University of Twente (2000)
15. Jiménez-Ruiz, E., Payne, T.R., Solimando, A., Tamma, V.: Limiting logical violations in ontology alignment through negotiation. In: Proceedings of the 15th International Conference on Principles of Knowledge Representation and Reasoning, KR 2016 (2016)
16. Kuhn, H.W.: The hungarian method for the assignment problem. Nav. Res. Logist. Q. **2**(1–2), 83–97 (1955)
17. Laera, L., Blacoe, I., Tamma, V., Payne, T., Euzenat, J., Bench-Capon, T.: Argumentation over ontology correspondences in MAS. In: Proceedings of the AAMAS 2007, pp. 1285–1292 (2007)
18. Laera, L., Tamma, V.A.M., Euzenat, J., Bench-Capon, T.J.M., Payne, T.R.: Reaching agreement over ontology alignments. In: Cruz, I., Decker, S., Allemang, D., Preist, C., Schwabe, D., Mika, P., Uschold, M., Aroyo, L.M. (eds.) ISWC 2006. LNCS, vol. 4273, pp. 371–384. Springer, Heidelberg (2006)
19. Melnik, S., Garcia-Molina, H., Rahm, E.: Similarity flooding: a versatile graph matching algorithm and its application to schema matching. In: Proceedings of the 18th International Conference on Data Engineering, San Jose, CA, USA, 26 February - 1 March, 2002, pp. 117–128 (2002). http://dx.doi.org/10.1109/ICDE.2002.994702
20. Miller, G.: WordNet: A Lexical Database for English. ACM Press, New York (1995)
21. Payne, T.R., Tamma, V.: Negotiating over ontological correspondences with asymmetric and incomplete knowledge. In: Proceedings of the AAMAS 2014, pp. 517–524 (2014)
22. Searle, J.R.: Speech Acts: An Essay in the Philosophy of Language, vol. 626. Cambridge University Press, Cambridge (1969)

23. Shvaiko, P., Euzenat, J.: Ontology matching: state of the art and future challenges. IEEE Trans. Knowl. Data Eng. **25**(1), 158–176 (2013)
24. Solimando, A., Jiménez-Ruiz, E., Guerrini, G.: Detecting and correcting conservativity principle violations in ontology-to-ontology mappings. In: Mika, P., et al. (eds.) ISWC 2014, Part II. LNCS, vol. 8797, pp. 1–16. Springer, Heidelberg (2014)
25. Spiliopoulos, V., Vouros, G.A.: Synthesizing ontology alignment methods using the max-sum algorithm. IEEE TKDE **24**(5), 940–951 (2012)
26. Walton, D., Krabbe, E.: Commitment in Dialogue: Basic Concepts of Interpersonal Reasoning. SUNY Series in Logic and Language. State University of New York Press, Albany (1995)

IoT-O, a Core-Domain IoT Ontology to Represent Connected Devices Networks

Nicolas Seydoux[1,2,3(✉)], Khalil Drira[2,3], Nathalie Hernandez[1], and Thierry Monteil[2,3]

[1] IRIT Maison de la Recherche, University of Toulouse Jean Jaurès, 5 allées Antonio Machado, 31000 Toulouse, France
{nseydoux,hernande}@irit.fr
[2] CNRS, LAAS, 7 avenue du Colonel Roche, 31400 Toulouse, France
[3] Univ de Toulouse, INSA, LAAS, 31400 Toulouse, France
{nseydoux,khalil,monteil}@laas.fr

Abstract. Smart objects are now present in our everyday lives, and the Internet of Things is expanding both in number of devices and in volume of produced data. These devices are deployed in dynamic ecosystems, with spatial mobility constraints, intermittent network availability depending on many parameters (e.g. battery level or duty cycle), etc. To capture knowledge describing such evolving systems, open, shared and dynamic knowledge representations are required. These representations should also have the ability to adapt over time to the changing state of the world. That is why we propose IoT-O, a core-domain modular IoT ontology proposing a vocabulary to describe connected devices and their relation with their environment. First, existing IoT ontologies are described and compared to requirements an IoT ontology should be compliant with. Then, after a detailed description of its modules, IoT-O is instantiated in a home automation use case to illustrate how it supports the description of evolving systems.

1 Semantic Interoperability, a Challenge for the IoT

The Internet of Things (IoT) is gaining more and more traction: some projectionists predict up to 50 billion devices connected in the next five to ten years [1]. The Things of the IoT allow to connect the physical world and virtual representations. IoT applications are based on very heterogeneous devices and technologies, and are deployed in domains as diverse as agriculture, domotics[1], smart cities or e-health. Two types of interoperability issues can be identified: syntactic and semantic, brought by the variety of domains and data models [2]. This paper focuses on semantic interoperability, the ability of systems to attribute the same meaning to the data they exchange.

Semantic interoperability is based on shared, unambiguous, machine-understandable vocabularies, which is why semantic web principles and technologies are seen as semantic interoperability providers, as [3] expresses for the specific

[1] Home automation.

E. Blomqvist et al. (Eds.): EKAW 2016, LNAI 10024, pp. 561–576, 2016.
DOI: 10.1007/978-3-319-49004-5_36

domain of IoT. Knowledge expressed in open formats can be shared and reused, and ontologies can evolve to adapt to new contexts or usages. To ensure the reusability of semantic models across projects and domains, good practices in ontology design have been proposed. In IoT projects, many ontologies have been built, but not always according to these guidelines, hence limiting their reusability (see Sect. 3). This is why we propose IoT-O[2], an IoT core-domain modular ontology engineered for reusability and extensibility. IoT-O is also available on the LOV[3], and based on the initial contribution of [4].

IoT systems are strongly bound to their environment, because they are composed of devices in contact with the physical world. Sensors collect data about their environment, and actuators are devices performing actions that have a direct impact on the world: light bulbs, motors, air conditioning, etc. This paper aims at showing how IoT-O can be used as an ontology to semantically describe devices and data in order to make systems aware of their environment, its evolution, and the changes they can bring to it. Such a description allows smart agents to transform their environment thanks to connected actuators, according to the perceptions they have of it through connected sensors.

In the remainder of this paper, Sect. 2 introduces a motivating use case that will serve to instantiate portions of IoT-O. Section 3 presents the design process of IoT-O, and gives an overview of the ontology. Finally, Sect. 4 details how IoT-O is instantiated in the use case.

2 Motivating Use Case

IoT technologies can have a direct impact on the everyday life of citizens, since it connects their physical environment to virtual applications. That is especially relevant in the case of domotics, where the home can be equipped with multiple low-power devices to provide new services. At LAAS-CNRS, the ADREAM project[4] aims at conducting research on smart buildings thanks to an instrumented, energy-positive building. It is equipped with more than 4500 sensing devices, producing up to 500,000 measures a day. Inside the building, there is a mock-up apartment equipped with commercial devices from various vendors. Deployed devices include sensors (temperature, luminosity, humidity, pressure), actuators (fan, space heater, multiple lamps), which communicate using different technologies (phidget, ethernet, zigbee) with gateways connected to a server.

However, small highly distributed devices usually have a limited processing power, which restrict the range of applications they can support. More complex agents can interact with these devices to collect their data and perform advanced processing to provide a higher level service. In our use case, centered on an elderly healthcare scenario, the complex agent is a robot. It is present in the house, and performs tasks such as helping the person in case of fall, moving heavy objects, pushing a wheelchair, fetching objects and bringing medications. Some of these

[2] http://www.irit.fr/recherches/MELODI/ontologies/IoT-O.
[3] http://lov.okfn.org/dataset/lov/vocabs/ioto.
[4] http://www.laas.fr/public/en/adream.

tasks require the robot to know **where the person is** in the apartment. To have this information, the robot can move around the apartment, scan it with its embedded cameras, and through image processing figure out where the person is. However, it requires the robot either to follow the person around all the time, or to scan the apartment completely each time it has to find the person. To make the robot more acceptable to the person, the house can be equipped with an IoT system, collecting information useful to the robot, such as information given by **presence sensors**. Moreover, the connected devices can provide new functionalities to the robot: he can easily interact with connected light switches or sensors. Our use case is composed of two scenarios: the robot must bring pills at fixed hours to the person using the presence sensors to locate her, and the robot must control the temperature in the apartment using temperature sensors and connected fans to improve the comfort of the person (Fig. 1).

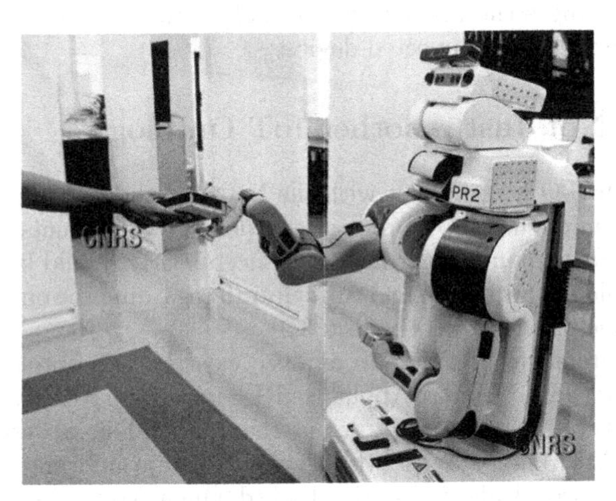

Fig. 1. PR2, the companion robot

In this use case, both syntactic and semantic interoperability are required, among the devices and between the devices and the robot. Syntactic interoperability is ensured using OM2M[5], an open-source horizontal integration platform implementing the oneM2M[6] standard. On top of OM2M, another platform, SemIoTics, enriches the collected data with semantic descriptions, and makes them available to the robot through a REST interface. SemIoTics is driven by a knowledge base capturing knowledge about the devices of the system represented according to our core-domain IoT ontology, and about the environment shared by the robot and the devices (here, the apartment). It is a Java software developed to showcase the role of semantic web technologies in IoT data management, based on Apache Jena. The robot itself is also a semantically enabled agent, it uses a "common sense" ontology and a knowledge base to reason about its 3D environment, as described in [5]. The knowledge specific to the robot relies on

[5] om2m.org.
[6] http://onem2m.org/.

ontologies out of the scope of this paper, but its knowledge base can be extended with any ontology, including IoT-O (as it is done in this paper).

That is why the knowledge described in this paper is implemented in a dedicated knowledge base using IoT-O, ADREAM-Robot[7]: the ontology is shared by the robot and SemioTics, and each system has its own knowledge base. The synchronization between the knowledge bases of the different agents is out of the scope of this paper. The use case focuses on home automation, but IoT-O and our approach are generic enough to be adapted to other domains. For instance, it could be used to support an air quality monitoring system in a smart city, by describing the sensors that collect the data and the services the citizens can subscribe to. The usage of IoT-O and its module in the use case is double: it is used to model the observations about modifications of the apartment, allowing the robot to keep an up-to-date representation of its environment, but also to model the changes the robot wants to make into the apartment through its actions and through the connected devices.

3 IoT-O, Not Just Another IoT Ontology

The design of IoT-O is compliant with the NeOn methodology, presented in [6].

The first step of the NeOn process is to define requirements. We split them in two types: **conceptual**, regarding the concepts that should be present in the ontology (detailed in Sect. 3.1), and **functional**, regarding the ontology structure and design principles (detailed in Sect. 3.2).

These requirements are used to analyze existing IoT ontologies: Semantic Sensor Network (SSN)[8], Smart Appliance REFerence (SAREF)[9], iot-ontology[10], IoT-lite[11], Spitfire[12], IoT-S[13], SA[14] and the oneM2M base ontology[15]. These ontologies are IoT ontologies for which we have found information on the web. Further details are available on the Linked Open Vocabularies for the IoT (LOV4IoT)[16], a recent initiative that lists IoT ontologies, even if they are not referenced on the LOV because they fail to comply with its requirements recalled in [2]. Ontologies related to specific domains impacted by IoT (domotics, agriculture, smart cities...) are out of the scope of this study.

As recommended by NeOn, reusable ontologies that are compliant with parts of the requirements are integrated in our design process. They are analyzed and presented in Sect. 3.3. The core-domain ontology we propose is then described in Sect. 3.4.

[7] https://www.irit.fr/recherches/MELODI/ontologies/Adream-Robot.
[8] http://purl.oclc.org/NET/ssnx/ssn.
[9] http://sites.google.com/site/smartappliancesproject/ontologies.
[10] http://ai-group.ds.unipi.gr/kotis/ontologies/IoT-ontology.
[11] http://iot.ee.surrey.ac.uk/fiware/ontologies/iot-lite.
[12] http://sensormeasurement.appspot.com/ont/sensor/spitfire.owl.
[13] http://personal.ee.surrey.ac.uk/Personal/P.Barnaghi/ontology/OWL-IoT-S.owl.
[14] http://sensormeasurement.appspot.com/ont/sensor/hachem_onto.owl.
[15] http://www.onem2m.org/ontology/Base_Ontology/.
[16] http://www.sensormeasurement.appspot.com/?p=ontologies.

3.1 The Core Concepts of IoT

Conceptual Requirements: These requirements come from an analysis of the IoT domain, driven by the home automation use case introduced in Sect. 2, but not limited to it: the use case is not seen as an end per se, but as an instantiation of the general domain of the IoT. To be reusable in a wide scope of domains, an IoT ontology should contain a set of key concepts. These are representative of IoT systems with no regard to the application domain. This approach facilitates the merging of data collected in different domains for horizontal applications, and allows the ontology to be an extendable core-domain ontology. We distinguish namely:

- **"Device"** and **"software agent"** constitute the two basic components of an IoT system, composed of both physical and virtual elements. The devices can be of two principle types, not mutually exclusive, that are listed below.
- **"Sensor"** are devices acquiring data, and **"observation"** describe the acquisition context and the data collected by the system. These concepts capture the perception the system has of the evolutions of its environment.
- **"Actuator"** are the devices that enable the system to act on the physical world, and **"action"** represents what they can perform. These concepts capture the knowledge the system has on its own abilities to impact its environment, and to make it evolve.
- **"Service":** In many cases, the IoT and the programmable web are very close. Connected devices can be seen as service providers and consumers, and by specifying a notion of service, every aspect of an IoT system can be represented.
- **"Energy":** In the paradigm of pervasive computing, many distributed Things perform computations. Most of these Things being physical devices, a complete modelling of the system will include a description of their energy consumption. Energy management is a crucial topic in IoT systems.
- **"Lifecycle":** Be it data, devices or services, IoT components are all included in different scales of lifecycles. Devices are switched on and off, services are deployed or updated, pieces of data become outdated... The evolution through a set of discrete states representing a lifecycle is an important concept for IoT systems.

Concept Coverage by Existing Ontologies: Table 1 sums up the assessment of existing IoT ontologies regarding the presence of key concepts. One star means that the concept is superficially represented (coarse-grained specialization, few data/object properties), two stars that the requirement is covered, and stars between parentheses indicate that the requirement is met by an included ontology. IoT-O, the ontology we propose, is also included for comparison. Note that we focus on connected device ontologies, and exclude, on purpose, the ontologies SSN is based on, since they are only focused on sensors and observation, which is only a subset of the identified key concepts. We can observe that some of the IoT ontologies cover most of the key concepts but none of them covers them all. Moreover, the different concepts are not represented with the same level of expressivity. In iot-ontology and SAREF, key concepts such as Actuator or Action are present but their representation is limited. For example, an actuator is defined as a device that modifies a property. This is less expressive than what can be expressed for a sensor with SSN which proposes a deep modeling of the sensors and the property they observe, but also of the relations between the sensors and their observations, and of the observations themselves. In eDIANA[17], an ontology referenced by SAREF, some specializations of actuator are given, but the mappings from these specializations to the *saref:Actuator* concept are not available directly. This analysis highlights the fact that an ontology for Actuators and Actions is needed (c.f. Sect. 3.3). This analysis also highlights the failure of existing IoT ontologies in representing correctly all IoT key concepts. As these concepts are not limited to the IoT domain, reusing ontologies dedicated to them (such as SSN for sensor) could help gain in expressivity, as is shown in Sect. 3.2.

Table 1. Key concept coverage in IoT ontologies

	Actuator	Action	Service	Sensor	Observation	Energy	Lifecycle	Device	Software agent
iot-ontology	*	*	**	(**)	(**)		(*)	(**)	**
saref	*	*	**	*		**		**	**
OWL-IoT-S			(**)	(**)	(**)		(*)	(**)	
SA	*		*	(**)	(**)	(**)	(**)	(**)	
iot-lite	*		*	(*)				(*)	
spitfire				(*)	(*)	**		(*)	
ssn				**	**		*	**	
oneM2M			**					*	
IoT-O	**	**	(**)	(**)	(**)	(**)	(**)	(**)	*

3.2 Good Practices for Ontology Design

Functional Requirements: These requirements capture ontology design guidelines and general semantic web good practices in a domain-agnostic fashion.

Reusability: One of the most important aspects of an ontology in such a broad domain as IoT is reusability: if an ontology is ad-hoc to a project, the work done in its definition will not benefit further projects. It is a critical issue that can be solved by different, non-mutually exclusive approaches:

- **Modularization:** as stated in [7], designing ontologies in separated modules makes them easier to reuse and/or extend. IoT applications are related to many various domains, and it is difficult to capture all these application domains in the same ontology. Modular ontologies can be combined together according to specific needs, which is a more scalable approach.

[17] https://sites.google.com/site/smartappliancesproject/ontologies/ediana-ontology.

- **Ontology Design Patterns:** were introduced in [8]. Designing ontologies that respect Ontology Design Pattern (ODP) increases reusability and their potential for alignment, as shown in [9]. ODPs capture modelling efforts: using them is a way to capitalize on previous work, and to take advantage of the maturity of the semantic web compared to the IoT.
- **Reuse of Existing Sources:** avoids redefinition, and prevents from having to align a posteriori the redefined concepts to the existing sources for interoperability. It is a key requirement for interoperability, which is a real issue in heterogeneous systems.
- **Alignment to Upper Ontologies:** Upper-level ontologies define very abstract concepts in a horizontal manner. They articulate very diverse domain-specific ontologies, which is crucial for broad domains like IoT.
- **Compliance with the LOV Requirements:** The LOV[18] is an online vocabulary register that increases visibility of vocabularies, and favours reuse by ensuring the respect of good practices listed in [2].

Level of formalism: To use the full advantages of the semantic description of devices and data, the description should enable reasoning and inference. This choice is motivated by the possibilities it opens:

- Applied to data, it is a way to bring context-awareness, as presented in [10]
- Applied to devices, it enables Thing discovery or self-configuration [11]
- Applied to services it enables automatic composition as in [12]

However, for concrete applications, the model should also by decidable, and in reasonable time, which de facto excludes an OWL-full model: OWL-DL is therefore the best choice. All surveyed ontologies are expressed in OWL-DL.

Table 2. Reusability of IoT ontologies

	Structured by ODP	Modular	Reuses external ontologies	Aligned with upper ontologies	One the LOV	Available online
iot-ontology			*	**	N	Y
saref		**	*		Y	Y
OWL-IoT-S	(*)	*	**	*	N	Y
SA	(*)	*	**	**	N	N
iot-lite					N	Y
spitfire			*	**	Y	N
ssn	**	**	*	**	Y	Y
oneM2M					N	Y
IoT-O	(**)	**	**	**	Y	Y

Assessment of Existing IoT Ontologies: Table 2 shows that the semantic web best practices for reusability are not always followed: some ontologies are not available online, and the majority is not compliant with the requirements

[18] http://lov.okfn.org.

of the LOV. External ontologies are generally not reused, with the exception of SSN. OWL-S, a service ontology is reused in only one case. The other surveyed ontologies propose redefinitions of the service concept. For example, SAREF redefines the concepts present in multiple ontologies, and proposes alignments in an external, textual document. Design patterns have only been used in ontologies importing SSN. Upper ontologies used are DUL[19] (especially used by SSN) and SWEET[20] (for SA). The limited reuse of ontologies shows a lack of federating ontologies, apart from SSN. SSN being a modular ontology compliant with the semantic web good practices, it is possible to say that these guidelines favour reuse. Section 3.3 focuses on such good practices.

3.3 Reused Ontologies for IoT-O

Identification of existing ontologies is included in the NeOn process. Some concepts, which are part of the conceptual requirements are defined by existing ontologies that are imported in IoT-O to avoid redefinition. SSN is a widely used W3C recommended ontology for sensors and observations. To define the notion of service, IoT-O imports Minimal Service Model (MSM), a lightweight service ontology which is generic enough to represent both REST and WSDL services (contrary to OWL-S[21]). The notion of energy consumption dedicated to the IoT is specified in PowerOnt, an ontology referenced by SAREF. The concepts of lifecycle are described using Lifecycle[22], a lightweight vocabulary defining state machines. We extended Lifecycle in the IoT-lifecycle[23] ontology with classes and properties specific to the IoT. Finally, to maximize extensibility and reusability, IoT-O imports DUL[24], a top-level ontology, and aligns all its concepts and imported modules with it.

Focus on SAN: However, no ontology describes the concept of actuator the way SSN describes the concept of sensor. This is why we propose the Semantic Actuator Network (SAN)[25] ontology. Actuators are devices that transform an input signal into a physical output, making them the exact opposite of sensors. SAN is built around Action-Actuator-Effect (AAE)[26], a design pattern we propose, inspired from the Stimulus Sensor Observation (SSO) design pattern described in [13]. Fig. 2 shows a representation of both the AAE and the SSO design patterns. SSN models the state of the world through stimuli converted by sensors into abstract observations, making the system able to be aware of the evolution of its environment. SAN is complementary: it models the transformation of abstract actuations by actuators into real-world effects, leading to the

[19] http://www.ontologydesignpatterns.org/ont/dul/DUL.owl.
[20] http://sweet.jpl.nasa.gov/.
[21] https://www.w3.org/Submission/OWL-S/, more dedicated to WSDL-based services.
[22] http://vocab.org/lifecycle/schema.
[23] https://www.irit.fr/recherches/MELODI/ontologies/IoT-Lifecycle.
[24] http://www.ontologydesignpatterns.org/ont/dul/DUL.owl.
[25] https://www.irit.fr/recherches/MELODI/ontologies/SAN.
[26] http://ontologydesignpatterns.org/wiki/Submissions:Actuation-Actuator-Effect.

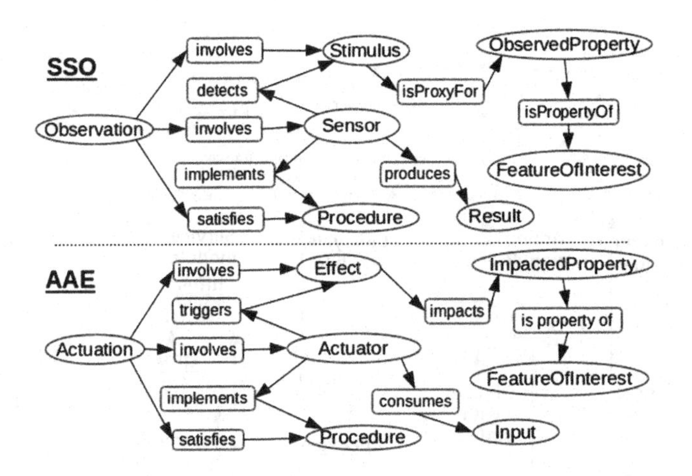

Fig. 2. The SSO and the AAE design patterns, structruring respectively SSN and SAN

representation of the evolution the system brings into its environment. Further details of the main classes of SAN are provided in Sect. 4.4.

3.4 IoT-O, a Modular Core-Domain IoT Ontology

IoT-O, the core-ontology we propose is composed of several modules. IoT-O's architecture is summarized in Fig. 3. The names of the newly created resources are in red and highlighted, the names of the reengineered resources are underlined, and the arrows show dependencies. Solid arrows represent imports, and dashed arrows the reuse of concepts without import.

The Modules of IoT-O:

- The **Sensing module** describes the input data. Its main classes come from SSN: *ssn:Sensor* and *ssn:Observation*. *ssn:Device* and its characteristics (*ssn:-OperatingRange*, *ssn:Deployment...*) provide a generic device description.
- The **Acting module** describes how the system can interact with the physical world. Its main classes come from SAN: *san:Actuator* and *san:Actuation*. It also reuses SSN classes that are not specific to sensing, such as *ssn:Device*.
- The **Lifecycle module** models state machines to specify system life cycles and device usage. Its main classes are *lifecycle:State* and *lifecycle:Transition*.
- The **Service module** represents web service interfaces. Its main classes come from MSM: *msm:Service* and *msm:Operation*. Services produce and consume *msm:Messages*, and RESTful services can be described with hRest.
- **Energy module:** IoT-O's energy module is defined by PowerOnt. It provides the *poweront:PowerConsumption* class, and a set of properties to express power consumption profiles for appliances.

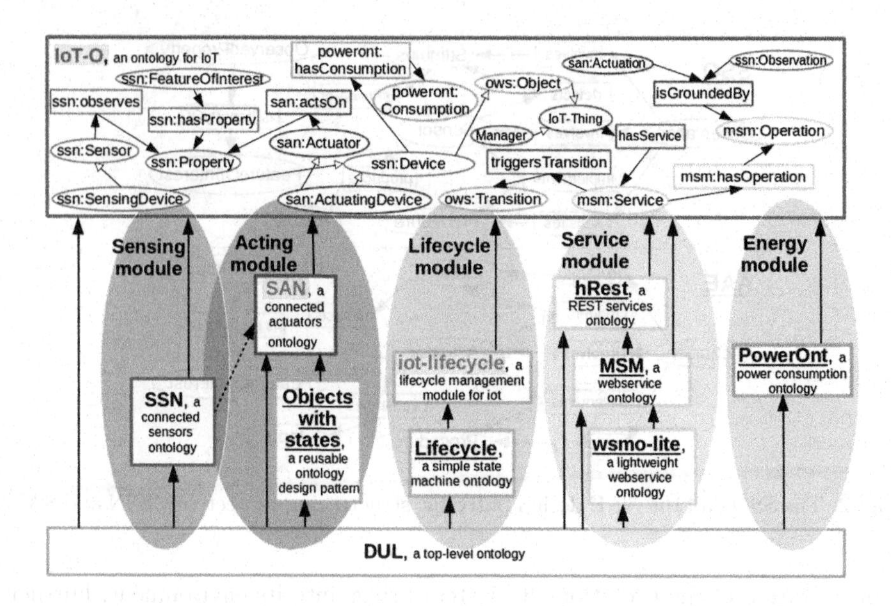

Fig. 3. Overview of IoT-O's architecture (Color figure online)

The Core of IoT-O: IoT-O[27] is both the name of the ontology and of the top module. It gives a conceptualization of the IoT domain, independent of the application, providing classes and relationships to link the underlying modules. Since many concepts are already defined in the modules, IoT-O's core is limited: it defines 14 classes (out of 1126 including all modules), 18 object properties (out of 249) and 4 data properties (out of 78). IoT-O key class is iot-o:IoT_Thing, which can be either an *ssn:Device* or an *iot-o:SoftwareAgent*. The power consumption of *ssn:Devices* is associated to *lifecycle:State* and *poweront:PowerConsumption*. *iot-o:IoT_Thing* is a provider of *msm:Service*, and an *msm:Operation* can have an *iot-o:ImpactOnProperty* on an *ssn:Property*, linking abstract services to the physical world through devices.

As a core domain ontology, IoT-O is meant to be extended regarding specific applicative needs and real-life devices and services. This design, inspired by SSN, makes IoT-O independent of the application.

4 SemIoTics and the Robot: Using IoT-O for Semantic Interoperability

4.1 Implementation of the MAPE-K Loop by the Robot and SemIoTics

[14] describes the concept of autonomic computing, or the control of an entity by an agent thanks to high-level policies and introspective knowledge: the controlling agent and the controlled entity form an autonomic system. The MAPE-K

[27] http://www.irit.fr/recherches/MELODI/ontologies/IoT-O.owl.

loop is a classic control structure in autonomic computing (see Fig. 4), separated in four steps: Monitoring, Analysis, Planning and Execution. The K stands for Knowledge, because the behaviour of the autonomic agent at each step of the loop is guided by a knowledge base, in the general meaning of the term (including but not restricted to the W3C's formalisms of knowledge representation).

In this use case, SemIoTics is performing the Monitoring and the Execution steps when connected devices are involved, and the robot performs the Analysis and the Planning, as well as part of the Monitoring and Execution steps. The robot and SemIoTics have distinct knowledge bases, even if in Fig. 4, a unique knowledge base is represented as the two systems exchange knowledge freely through a rest interface. Consistence issues are not considered in this work, as only one smart agent interacts with the system.

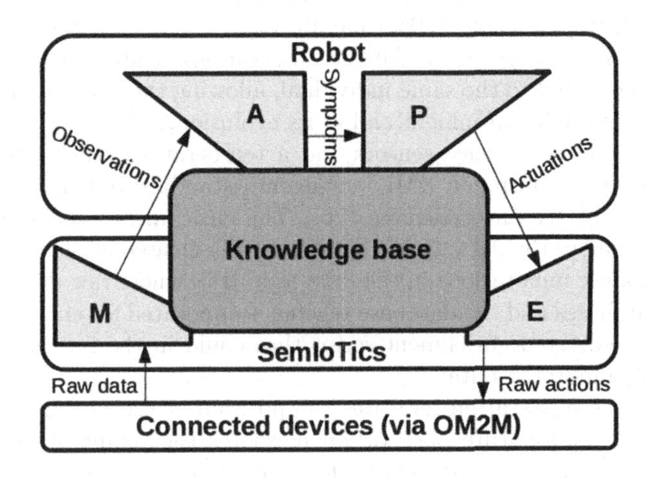

Fig. 4. A representation of the MAPE-K loop, split between the robot and SemIoTics

The process described in Fig. 4 structures the use case: data is first gathered by the sensors, and enriched by SemIoTics. The enriched observations are processed by the robot, which decides to perform actions represented as enriched actuations. These actuations are sent to SemIoTics, which translates them into raw commands for the actuators to perform. In complement to IoT-O, the dogont[28] ontology is used to describe the apartment and the location of devices inside it. Dogont is an ontology identified in the SAREF project, and it is imported by Poweront. We aligned it to IoT-O to integrate it to the use case.

4.2 Monitoring, Where Raw Sensor Data Become Meaningful Observations

The first step of the MAPE-K loop is the monitoring of the controlled system. In the apartment, sensors produce data reflecting their observations. This data

[28] http://elite.polito.it/ontologies/dogont/dogont.html.

is enriched to become a reusable piece of knowledge. Enrichment of sensor data is performed using the SSN ontology, which is in the Sensing module of IoT-O. Each *ssn:Sensor* has an *ssn:Observation* stream composed of *ssn:SensorOutput* whose value is described by *ssn:ObservationValue*. For provenance purposes, a *ssn:SensorOutput* can be linked to its original representation (before enrichment) with the *iot-o:hasRawRepresentation* data property. The sensor's characteristics (*ssn:MeasurementProperty*, the *ssn:Property* of the *ssn:FeatureOfInterest* it observes) are used to enrich the observation as well. IoT-O and SSN are generic ontologies, so they might need to be extended with application-specific modules to be fully functional. Such extension is proposed in the Adream-Robot module[29]. The *ssn:Observation* allow the representation of a characteristic of the environment at a given point in time. The temporality of the sensor measures (and of actuators actions) are represented by a *san:hasDateTime* relations with a http://w3c.org/2006/time#Instant, itself characterized by an *iot-o:hasTimestamp* data property. All the observations related to the same point in time are connected to the same individual, allowing the agent to have a timed representation of its environment and of its evolution.

In our use case, presence sensors and a temperature sensors produce raw observations in the form of XML documents standardized according to the oneM2M Content Instance resource type. The enrichment process requires an approach specific to the data, either by writing a dedicated enrichment script, or by using semantic annotations in the data as in [15], where raw data is stored in relational databases and the database schema is annotated for enrichment. SemIoTics uses a dedicated enrichment script that could in the future be extended by producing annotated data.

The presence observation indicates the position of the person in the apartment, and the temperature observation measures the temperature at a given point in space and time, both in the form of *ssn:ObservationValue* instances. This enriched information is accessed by the robot through SemIoTics' REST interface, and it is used to update the robot's representation of the world. This representation of the world is stored in the robots knowledge base, and used as a context in the Analysis step.

4.3 Analysis: Aggregation of Observations in Abstract Symptoms

In the Analysis step, the robot processes his own representation of the world to determine high-level symptoms that need to be addressed by actions.

In the medication scenario, the robot compares the present time to the time when the medication is due to generate the symptom "Medication must be delivered" if necessary.

In the temperature control scenario, user preferences are represented using the concepts defined in yet another module: Autonomic[30]. *ssn:Property* of the environment controlled by the robot within explicit boundaries expressed in the

[29] https://www.irit.fr/recherches/MELODI/ontologies/Adream-Robot.
[30] http://www.irit.fr/recherches/MELODI/ontologies/Autonomic.

form of *autonomic:PropertyConstraints* are classified as *autonomic:ConstrainedProperty*. In our use case, the *ssn:Property* temperature of the *ssn:FeatureOfInterest* living room air has two constraints, instances of *autonomic:MaximumValue* (25 °C) and *autonomic:MinimumValue* (19 °C). The last *ssn:ObservationValue* of the *autonomic:ConstrainedProperty* is out of the bounds defined by the *autonomic:PropertyConstraint* (26 °C instead of 25), so the temperature is classified by the reasoner as an *autonomic:OutOfBoundsProperty* thanks to custom rules.

4.4 Planning, Where Symptoms Are Used to Create a Plan

In the planing phase, the autonomic agent uses the inferred symptoms and policies defined by the user or by the administrator beforehand to define a series of actions that have to be implemented on the system.

In the medication scenario, the robot uses its representation of its environment to locate the person, as it is kept updated in the monitoring phase thanks to the knowledge produced by the sensors and SemioTics. The robot will plan a trajectory to fetch the medication and to reach the person. In this case, the representation of the trajectory itself is ad-hoc to the robot, and isn't linked to IoT-O. The ontology is used to connect the robots internal representation of the world with the observations collected by the sensors and enriched by SemIoTics, providing semantic interoperability between the robot and SemIoTics. If the robot expresses its needs using the same ontology as SemIoTics, or if their ontologies are aligned, it can seamlessly use elements measured by the sensors to plan its trajectory.

In the temperature control scenario, the description of the actions is performed using SAN, the actuator ontology that also describes the actuators in the system. The agent, with successive queries to the knowledge base, will look for *san:Actuator* instances that *san:actsOn* the *autonomic:OutOfBoundsProperty*, and which *autonomic:ImpactOnProperty* is coherent with the symptom. In the example, since the temperature is too high, the *adream-model:fan* can be used, but also the *adream-model:spaceHeater*, since its *adream-model:turnOff* operation has a *adream-model:NegativeImpact* on the temperature. The orchestration of these actions (if need be) are determined using the Lifecycle module of IoT-O, which represents the devices as state machines by integrating the Objects with States (ows)[31] ontology design pattern. *ssn:Device* (superclass of both *ssn:SensingDevice* and *san:ActuatingDevice*) are objects that *ows:hasState exactly 1 ows:State*, because objects should only be in one state at a time. The *ows:State* is equivalent to the *lifecycle:State* (from the Lifecycle[32] vocabulary, extended by the IoT-Lifecycle[33] ontology), and *lifecycle:State* are connected by *lifecycle:Transition* instances. Thanks to this vision of state machines, stateful transitions (that are only available in certain states of the device) can

[31] http://delicias.dia.fi.upm.es/ontologies/ObjectWithStates.owl.
[32] http://vocab.org/lifecycle/schema.
[33] http://www.irit.fr/recherches/MELODI/ontologies/IoT-Lifecycle.

be represented. Only *msm:Operation* instances that *iot-o:isGroundedBy* a *san:-Actuation* that *iot-lifecycle:triggersTransition* a *lifecycle:Transition* that is a *lifecycle:possibleTransition* of the device current *lifecycle:State* can be called at a given time. For instance, the fan *adream-model:turnOff* operation will only be available if the space heater is on. In our example it is off, so the agent plans to turn on the fan and creates the corresponding *san:ActuationValue*. The selection of devices and their operations is driven by necessity (only the devices impacting the right property are selected), but it can also be driven by policies based on knowledge about the devices intrinsic characteristics expressed with *san:-ActuatingCapability*, composed of *san:ActuatingProperty* that create an actuator profile. It can be used to minimize energy consumption (combined with the Energy module), to optimize reaction time...

4.5 Execution, Where the Plan Is Converted into Actions

In the execution step, the robot implements the planned actions.

For the medication scenario, the robot fetches the medication and brings it directly to the person, it doesn't have to search for her in the apartment. The MAPE-K loop can be repeated while the robot is moving to update the trajectory if the person moves in the house.

For the temperature control scenario, the robot transmits the *san:ActuationValue* that it wants the system to implement to SemIoTics via a REST interface. SemIoTics will handle the transformation of the knowledge into a representation that can be processed by the target device. This translation can be driven by the semantic description of *msm:Operations*, or dedicated annotations as in [16], where XML schemas are annotated for transformation from RDF to XML. SemIoTics uses the semantic description of operations to perform lowering, and perspectives for this technique are presented in Sect. 5. This translation enables the interaction with low-level, constrained devices that are not able to process complex knowledge representations.

5 Conclusion and Future Works

This paper introduces IoT-O, a modular core-domain IoT ontology designed to be compliant with identified requirements. After a detailed presentation of its modules, an instantiation of IoT-O is presented in a home automation use case. IoT-O is used to bring semantic interoperability between SemioTics, a platform enabling semantic access to connected devices, and a robot. SemIoTics and the robot implements the MAPE-K loop, an autonomic computing pattern, and uses IoT-O at each step of the loop to describe knowledge about the connected devices and about the data they produce and consume. The ontology describes the evolving state of the robot's environment through sensor observations, and the capabilities the system offers to impact this environment through the devices.

In this paper, enrichment and lowering techniques (allowing the transformation back and forth from data to knowledge) have been overviewed. Such techniques are essential to include constrained devices into the IoT: enriched data is more reusable than raw data, but it is heavier to exchange and process, so transformation is required between the end devices (sensors, actuators) and the more powerful nodes of the IoT, e.g. gateways, servers and laptops. We are currently working on such an approach. Other perspectives of our work will be to manage data flows over time in order to learn from previous decisions and their consequences to produce explicit knowledge and enrich policies, and synchronization of the distributed knowledge bases of the smart agents: compared to the use case, multiple agents should be able to exchange knowledge about their environment and to maintain coherence between their representations of the world.

References

1. Ganz, F., Puschmann, D., Barnaghi, P., Carrez, F.: A practical evaluation of information processing and abstraction techniques for the internet of things. IEEE Internet Things J. **2**(4), 340–354 (2015)
2. Gyrard, A., Serrano, M., Atemezing, G.A.: Semantic web methodologies, best practices and ontology engineering applied to Internet of Things. In: IEEE 2nd World Forum on Internet of Things (WF-IoT), pp. 412–417. IEEE (2015)
3. Murdock, P.: White paper: semantic interoperability for the web of things (2016)
4. Alaya, M.B., Medjiah, S., Monteil, T., Drira, K.: Toward semantic interoperability in oneM2M architecture. IEEE Commun. Mag. **53**(12), 35–41 (2015)
5. Lemaignan, S.: Grounding the interaction: knowledge management for interactive robots. Ph.D. thesis (2012)
6. del Carmen Suarez de Figueroa Baonza, M.: NeOn methodology for building ontology networks: specification, sheduling and reuse. Ph.D. thesis (2010)
7. Aquin, M.: Modularizing ontologies. In: Suárez-Figueroa, M.C., Gómez-Pérez, A., Motta, E., Gangemi, A. (eds.) Ontology Engineering in a Networked World, pp. 213–233. Springer, Heidelberg (2012)
8. Gangemi, A.: Ontology design patterns for semantic web content. In: Gil, Y., Motta, E., Benjamins, V.R., Musen, M.A. (eds.) ISWC 2005. LNCS, vol. 3729, pp. 262–276. Springer, Heidelberg (2005)
9. Scharffe, F., Euzenat, J., Fensel, D.: Towards design patterns for ontology alignment. In: Proceedings of the 2008 ACM Symposium on Applied Computing - SAC 2008, p. 2321. ACM Press, New York, March 2008
10. Henson, C., Sheth, A., Thirunarayan, K.: Semantic perception: converting sensory observations to abstractions. IEEE Internet Comput. **16**(2), 26–34 (2012)
11. Chatzigiannakis, I., Hasemann, H., Karnstedt, M., Kleine, O., Kröller, A., Leggieri, M., Pfisterer, D., Römer, K., Truong, C.: True self-configuration for the IoT. In: 3rd International Conference on the Internet of Things (IOT) (2012)
12. Han, S.N., Lee, G.M., Crespi, N.: Towards automated service composition using policy ontology in building automation system. In: 2012 IEEE Ninth International Conference on Services Computing, pp. 685–686 (2012)
13. Janowicz, K., Compton, M.: The stimulus-sensor-observation ontology design pattern and its integration into the semantic sensor network ontology. In: Proceedings of the 9th International Semantic Web Conference, 3rd International Workshop on Semantic Sensor Networks, pp. 7–11 (2010)

14. Kephart, J., Chess, D.: The vision of autonomic computing. Computer **36**(1), 41–50 (2003)
15. Le-Phuoc, D., Quoc, H., Parreira, J.X., Hauswirth, M.: The linked sensor middleware-connecting the real world and the semantic web. In: Semantic Web Challenge 2011. Number April 2005, pp. 1–8 (2011)
16. Kopecký, J., Vitvar, T., Bournez, C., Farrell, J.: SAWSDL: semantic annotations for WSDL and XML schema. IEEE Internet Comput. **11**(6), 60–67 (2007)

AutoMap4OBDA: Automated Generation of R2RML Mappings for OBDA

Álvaro Sicilia[1(✉)] and German Nemirovski[2]

[1] ARC Enginyeria i Arquitectura La Salle, Universitat Ramon Llull, Barcelona, Spain
asicilia@salleurl.edu

[2] Business and Computer Science, Albstadt-Sigmaringen-University of Applied Sciences, Albstadt, Germany
nemirovskij@hs-albsig.de

Abstract. Ontology-Based Data Access (OBDA) has become a popular paradigm for the integration of heterogeneous data. The key components of an OBDA system are the mappings between the data source and the target ontology. The great efforts required to create manual mappings are still a significant barrier to adopting the OBDA. Current relational-to-ontology mapping generators are far from providing 100 % of the mappings required in real-world problems. To overcome this issue we present AutoMap4OBDA, a system which automatically generates R2RML mappings based on the intensive use of relational source contents and features of the target ontology. Ontology learning techniques are applied to infer class hierarchies, the string similarity metrics are selected based on the target ontology labels, and graph structures are applied to generate the mappings. We have used the RODI benchmarking suite to evaluate AutoMap4OBDA which outperforms the most advanced state-of-the-art mapping generators.

Keywords: Relational-to-ontology mappings · R2RML · Ontology learning · OBDA

1 Introduction

In recent years, we have witnessed a growing need for businesses to integrate data stored in different formats, using different value scales, distributed in numerous sources, and built upon different schemas. This need arises, on the one hand, in a context of information integration, in particular in interdisciplinary projects, and also in the case of fusions or joint collaborations of companies and institutions. On the other hand, and thanks to the widely available initiatives for open data access such as the Linked Open Data, the quantity of the available data is steadily increasing.

The Ontology-Based Data Access (OBDA) approach has been developed to address this need. In ODBA settings, data queries formulated in terms of an ontology, are rewritten with respect to the native database schema, and forwarded to the database [17]. The main components of an OBDA system are a database – which contains the data –, an ontology – which represents the conceptualization of a domain –, mappings

E. Blomqvist et al. (Eds.): EKAW 2016, LNAI 10024, pp. 577–592, 2016.
DOI: 10.1007/978-3-319-49004-5_37

between the database and the ontology, and the query rewriter which transforms queries into an "understandable" form for the native database management system.

In this context, the development of mappings between the ontology and the database is one of the key issues. The manual mapping in OBDA systems is, nowadays, the most widely adopted solution in academic and industry communities in spite of it being extremely time-consuming and requiring high levels of human expertise [17]. The development of mappings between the ontology and relational data schemas is a process that requires knowledge from a specific domain, for instance medicine, or mechanical engineering, as well as skills in Entity Relationships modeling and ontology design. Finding users with this profile is difficult. Furthermore, creating hundreds of mapping rules manually is error prone. Therefore, creating mappings is still a significant barrier in the adoption of the OBDA approach [13].

Efforts for the automation of the mapping development tasks have been carried out in several studies such as [13]. Sequeda et al. proposed a process for the automated generation of relational-to-ontology mappings as an alternative to the manual process [18]. It starts with a reverse engineering step to derive an ontology from the database schema. Such automatically generated ontology is called putative ontology. It usually has a flat structure as compared to a domain ontology which is created by a community of experts or by a standardization body. In the second step, the putative ontology is aligned with the domain ontology used in an OBDA system by applying ontology matching methods. Thirdly, the alignments between both ontologies are used to derive the final mappings.

As shown in the results of the relational-to-ontology RODI benchmark conducted by Pinkel, current automated mapping generators can address simple mappings. However, all systems failed on advanced tests in which relational databases use design patterns that differ greatly from those used in ontologies [14, 15]. Our analysis – reported in Sect. 5 – shows that the current mapping generators basically rely on the relational schema and do not fully take into account the contents of the database and the features of the target ontology, i.e. the one to be mapped onto the database. We believe that there is room for improvement in those systems if database and ontology features are extensively taken into account.

In this paper we present AutoMap4OBDA[1], a system for the automated generation of R2RML mappings between a relational database and existing target ontology specified in OWL. The main motivation behind the design of this system was to obtain mappings of significantly higher quality compared to the competitors' results. To do so, the database content and features of the target ontology are taken into account during the mapping generation process. Moreover, AutoMap4OBDA has been designed to be used in OBDA scenarios (Sect. 2) and is able to generate fully compliant R2RML mappings without user intervention. The techniques used by AutoMap4OBDA are described in Sect. 3. An evaluation using the RODI benchmark suite has been described in Sect. 4. Finally, AutoMap4OBDA has been compared with the existing systems of the state-of-the-art (Sect. 5).

[1] http://arc.salleurl.edu/automap4obda/.

2 Concepts of an OBDA System

The main purpose of an OBDA system is to provide access to data stored in a relational database by means of queries formulated in terms of a common domain specifics vocabulary encapsulated as an ontology. This approach complies with the requirement of interoperability for systems and applications by facilitating the querying of the relational sources without taking into account their specific schemas.

The core purpose of a relational database is to store data and relations among data in tables. Data is grouped in relations (i.e., tables) of tuples (i.e., table rows), in which each tuple is composed of attributes (i.e., columns). The attributes are defined with a name and a set of permitted values for a particular domain. In this way, a relation is a set of n-tuples where each tuple has the same type of attributes. The data stored in relations should be uniquely identified and linked to related data through attributes.

An ontology is "a formal specification of a shared conceptualization" [3]. The term "shared conceptualization" indicates reaching a consensus among experts whereby the conceptualization represents the related knowledge domain. Ontologies incorporate concepts which are sets, collections, types of objects or kinds of things; object properties that are aspects, properties, features, characteristics that an object can have; and datatype properties that describe types of values. Furthermore, in an ontology items of a knowledge domain are specified by means of axioms.

In terms of R2RML, a declarative language recommended by the W3C for the definition of relational-to-ontology mappings, mappings are declared as a subject map which defines the Internationalized Resource Identifier (IRIs) generated from the logic table which can be defined as a base table, a view (i.e. the result set of a stored query), or a SQL query. Moreover, the data-to-object mappings are declared as predicate and object maps. The subject and objects maps describe how the IRIs should be generated using the columns specified in the logic table and the elements of a target ontology. Generating R2RML mappings consists of two main tasks: first, finding the correspondences between the elements of the database and the target ontology and second, obtaining the SQL views needed to generate the IRIs of the subject and predicate object maps.

3 AutoMap4OBDA: Automated Mappings for OBDA

AutoMap4OBDA is a system that automatically generates R2RML mappings from a relational database and an ontology (further referred as a target ontology). It is invoked from command line passing the connection parameters to the database (e.g., string connection, database name, user, and password), the path to the ontology file, and parameters to enable/disable some features (e.g., Applying ontology learning methods). The goal behind the design of AutoMap4OBDA is to make a relational-to-ontology matcher dependant not only on the database schema, but also on the database content and features of the target ontology. For example, class hierarchies are mined from the instances of the database.

The mapping generation process by AutoMap4OBDA is aligned on the common strategy described in the introduction to this paper: the generating of a putative ontology derived from the database, the alignment of the putative and the target ontology using ontology matching techniques, and the generating of the final R2RML mappings basing on the alignment. The mapping generation process includes two steps more than the common strategy: augmenting the putative ontology and extending the alignment. In particular, the AutoMap4OBDA workflow comprises five steps carried out sequentially. Each step is carried out by a corresponding module (Fig. 1).

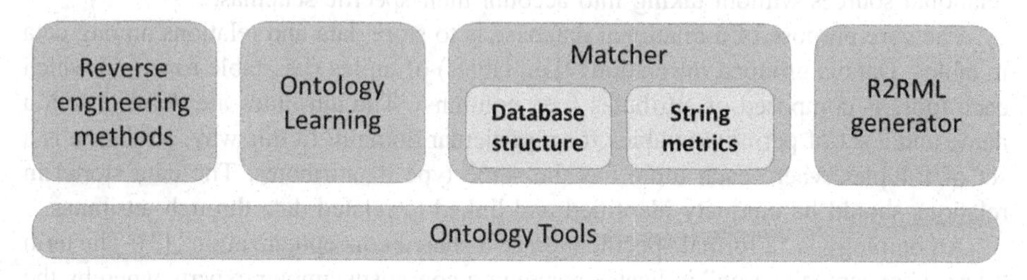

Fig. 1. Architecture of AutoMap4OBDA

- *Reverse engineering methods.* The module contains the functionalities to derive a putative ontology from a relational schema.
- *Ontology learning.* The module contains the method to infer class hierarchies based on the database content.
- *Matcher.* The module contains two sub-modules, one to read the database schema and another one to implement the string similarity metric selection based on labels (i.e., StringAuto and PropString).
- *R2RML generator.* It contains functionalities to generate the R2RML mappings based on the findings of the matcher.
- *Ontology Tools.* It contains functionalities to outline features of the target ontology: maximum entropy, maximum class name length, and maximum number of sub-classes among others.

Before starting the workflow, the target ontology is saturated. That is: specifications of domains and ranges of properties are rewritten in form of axioms.

3.1 Step 1: Generating the Putative Ontology from a Database Schema

The first step comprises generating a putative ontology from a database schema so that each ontology class corresponds to a database table. The name of the table is used to define the class name as well as the class label. Each attribute of a table corresponds to a data property in which the name of the attribute determines the name of the property. If the database table attribute is a foreign key, the domain of the resultant object property will be the class which corresponds to this table and its range will be the class which corresponds to the table whose primary key is referred by the foreign key. Moreover, a class is generated for each attribute and an object property which connects the table class and the attribute class. For each element of the putative ontology (classes, object or data properties) a reference to its origin database element, i.e. a table or a column, is stored as an annotation.

3.2 Step 2: Augmenting the Putative Ontology Basing on the Database Content

In the second step, the putative ontology is augmented based on the database contents. A putative ontology derived from a database schema only, without taking into consideration data values would have a similar structure to the database schema. However, by learning patterns contained in data values the ontology structure can be enriched considerably [4]. In particular, the class hierarchy can be identified by applying data mining methods to the database contents. For example, if a table *Building* is related to a class *Building* of the target ontology, the values of the attribute *Building_Use* can be mined to find subclasses for class *Building* such as *Office* and *Residential.*

One of the main issues in this context is to identify the "categorizing" attributes where subclasses can be extracted from. Cerbah proposed two ways to identify candidates by "categorizing" attributes: Identification of lexical clues in attribute names and filtering through entropy-based estimation of data diversity [4]. In the first option, attributes are selected if their names can reveal a specific role in the table. For example, the attribute *Use* of the table *Building* might determine subclasses of the class *Building*. The problem is that the lexical clues are strongly related to the domain of the database and the language of the contents. Cerbah's suggestion was that the user provided the clues. The second option is based on selecting attributes that have the most balanced distribution of instances. The attributes are characterized using the concept of entropy from information theory which is a measure of the uncertainty of a data source. The attributes with highest entropy – usually primary key attributes – and lowest entropy – attributes with highly repetitive content (up to a single value only) – are usually discarded. The problem with this option is that the discarded attributes may contain candidates to be subclasses. For example, the *Use* attribute of the table *Building* has few instances highly repeated such as *Residential* and *Office*.

In this step we have implemented a set of rules based on the properties of the database and the target ontology in order to select attributes that can be mined to obtain hierarchy classes. In particular, each attribute of a relational table and its values are discarded, that is are not represented within a putative ontology, if they match with one or more of the following rules:

- *The entropy of an attribute is greater than the maximum entropy of the ontology.* By applying this rule we avoid the generation of putative classes whose entropy would be greater than the entropy of any class in the target ontology. We define the entropy of an ontology as the maximum entropy of a class. In order to calculate the entropy of a class, a list of subclasses that can be reached at a certain depth, which is a fraction of the maximum depth of the ontology, is needed. The maximum depth of an ontology is the maximum depth value of all classes. The depth of a class is the maximum number of edges on the longest downward path between the given class and a subclass-leaf (a class which is not a super-class of any class). The entropy of candidate attributes and the entropy of a set of subclasses are calculated according to definitions of Cerbah [4].
- *The number of different values of an attribute is greater than the maximum number of subclasses of any class of the ontology.* By applying this rule we avoid a situation

in which a putative class has more subclasses than the maximum number of subclasses of the corresponding class in the target ontology. The maximum number of subclasses is determined as a number of subclasses of the given class that can be reached at a certain depth. This depth is a fraction of the maximum depth of the ontology obtained in the calculation of the ontology entropy.
- *The length of a value is greater the than maximum class name length in the target ontology plus a factor*. This rule is to avoid generating classes than will not be matched in the next step since the length of the value is much greater than the maximum class name length of the ontology.
- *An attribute value is not a text*. Non text attribute values such as Numbers and Booleans cannot produce class names.
- *An attribute value contains a URL*. A URL cannot be a class name.

The values of attributes that have not been filtered become putative classes that are subclasses of the attribute class. The values of attributes are used as class identifiers and labels.

3.3 Step 3: Matching Using String Similarity Metrics

In this step the classes and data properties of the augmented putative ontology and the target ontology are matched. Full-featured ontology matcher systems use syntactic, semantic, and structural similarity metrics to find correspondences between entities of two ontologies (a source and a target one). Mature tools for ontology alignment can be found in the literature such as LogMap [9], however, when the source ontology is a putative ontology derived from a relational database, those systems may make it difficult to find correspondences between elements of two ontologies. The reason for that is that "hand-made" ontologies, like the target ontology in our settings, usually specify domain knowledge on a high-level of abstraction while putative ontologies derived from relational schemas describe the syntactical structure on a very low level of granularity. Moreover, different design patterns are usually used in ontologies and relational schemas. Therefore, the use of structural metrics may hinder the detection of correspondences.

In this context more simple String-based alignment techniques perform better. These techniques use a predefined string similarity metric. However, as stated by Chetham and Hitzler [5], for some types of ontologies, the performance of different string similarity metrics varies greatly in terms of precision and recall. To address this issue, the authors proposed a set of guidelines to choose the proper metric based on the number of words per entity label after tokenization, on the language of the ontologies and on embedded synonyms. Moreover, the selection takes into account whether the goal is to maximize precision or if it is to recall measures. These guidelines have been implemented by the authors in two matchers named StringAuto [5] and PropString, an entirely string-based approach to aligning properties [6]. AutoMap4OBDA integrates both matchers to match the classes and properties instead of using a full-featured alignment system. The output of this step is a list of correspondences between the classes of the putative and the target ontology and a list of the correspondences between properties of both ontologies.

3.4 Step 4: Extending the Alignment According to the Target Ontology

In the process of generating the putative ontology, mappings are generated between the relational database and the putative ontology. Later, the alignment between the putative ontology and the target ontology is used to update the mappings. Finally, the modified mappings can be used to query the source by means of queries referring to the target ontology. The main drawback of this approach is that the mappings strongly reflect the relational database structure which might not be the same as the structure of the target ontology. This may lead to the generation of incorrect and incomplete mappings.

To overcome this issue, AutoMap4OBDA implements a method to rely on the structure of the target ontology (instead of the putative ontology) to extent the alignment with new correspondences. Thus, it uses the database schema to assure that SQL query can be obtained according to the new correspondences. Once classes of the putative and target ontologies are primarily aligned, additional correspondences between them can be established according to the object properties of the target ontology. This step takes as input correspondences found by the initial matching (step 3). Then, property paths between two classes – of the target ontology – matched in the step 3 are analyzed. These paths are built by the range/domain relations that connect two classes with each other. Such paths can contain arbitrary number of properties and classes connected by these relations. The idea is to find additional correspondences to the classes that are located within such chains. Two classes of the target ontology – previously matched to two putative classes in Step 3 – are taken as input for the step 4 if:

1. There is a property path connecting these two classes within the target ontology,
2. There is a relational path (over a set of foreign keys) between the pair of database tables that correspond to the two putative classes matched to the two classes of the target ontology.

Thanks to the first rule, the connectivity between classes is assured in terms of the target ontology since at least one object property will exist between those concepts. The second rule, assures that a SQL query can be obtained which involves the tables mapped onto the two target ontology classes. This SQL query will be required in Step 5 (described later) to specify the logic table of R2RML mappings. The second rule uses the database schema instead of the putative ontology because the putative ontology had been augmented in Step 2 increasing the difficulty to generate the SQL query.

In order to obtain the property paths, both the database schema and the target ontology are represented as two graph structures. A graph is generated from the database elements where the tables and columns are nodes. The relations of foreign keys are edges between correspondent columns (Algorithm 1). Another graph is derived from the target ontology where the classes and properties are nodes while the edges are the domain and ranges of the properties (Algorithm 2). In both algorithms there is a guarantee that nodes are only inserted once.

```
Algorithm 1: Creating a graph from a relational database schema
INPUT: S = Database schema
OUTPUT: G = Graph
1 For each Table T in S do
2   G.addNode(T.name);
3   For each Column C in T.listOfColumns do
4     If C.isForeignkey then
5       G.addNode(T.name + "." + C.name);
6       G.addEdge(T.name, T.name + "." + C.name);
7       G.addEdge( T.name + "." + C.name, C.foreigntable + "." + C.foreignkey);
8       G.addEdge( C.foreigntable + "." + C.foreignkey, C.foreigntable);
9   return G;
```

```
Algorithm 2: Creating a graph from an ontology
INPUT: O = Ontology; R = Reasoner
OUTPUT: G = Graph
1 For each OWLClass C in O.getClassesInSignature() do
2   G.addNode(C);
3 For each OWLObjectProperty OP in O.getObjectPropertiesInSignature do
4   For each OWLClass CD in R.getObjectPropertyDomains(OP) do
5     For each OWLClass CR in R.getObjectPropertyRanges(OP) do
6       G.addNode(OP);
7       G.addEdge(CD, OP);
8       G.addEdge(OP, CR);
9 return G;
```

To find a path between a pair of tables according to the database schema and a pair of target concepts according to the target ontology an algorithm for finding the shortest path is invoked on the graphs created by Algorithms 1 and 2 accordingly. Although it is not the most performing algorithm, Dijkstra's algorithm was selected because of its ease of implementation [7]. The extending alignment algorithm (Algorithm 3) iterates over matches found in step 3 and invokes Dijkstra's algorithm (*findPathBetween*) to obtain *ontP*, the shortest path between all pairs of classes (i.e., *S* and *T*) of the target ontology using the graph created with Algorithm 2. Moreover, *findPathBetween* is invoked to obtain *dbP*, the shortest path between tables of the database schema using the graph obtained through Algorithm 1. Then, if both paths exist, the components of the *ontP* are processed. The algorithm iterates over the components of the path and checks if the target nodes *nD* have been already matched to the putative classes by looking into the current list of matches *M*. In the case that they have not been matched then they are added to the list of new matches *M'*. The target nodes *nT* are aligned to a putative class from the target nodes. That is, they share the same table and column of the database (i.e., *nD.table* and *nD.column*).

The method applied in this step takes into account two mapping cases (Fig. 2). The first case occurs when the minimum path length between a pair of concepts of the target ontology has more nodes (concepts) than the minimum path length between the aligned concepts of the putative ontology. In this case, extra mappings (i.e., TripleMaps) are generated. The IRIs and SQL queries generated for these extra mappings are the same as those generated for the precedent concept (the leftmost concept in the upper chain on Fig. 2).

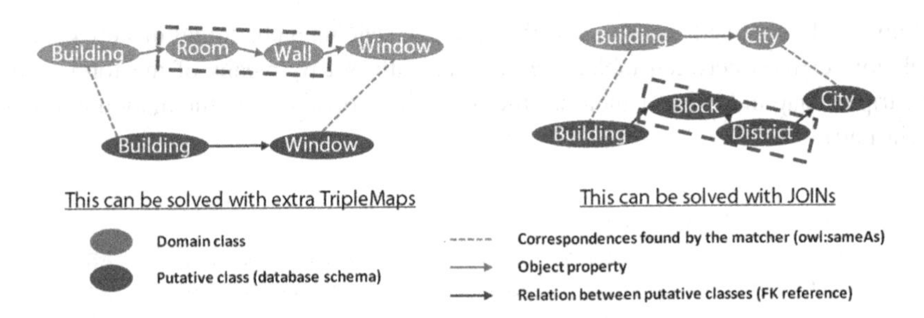

Fig. 2. Mapping generation cases. (1) Less source elements than target elements (Left). (2) More source elements than target elements (Right).

```
Algorithm 3: Extending alignment
INPUT: M = ListOfMatches, dbG = Graph, ontG = Graph
OUTPUT: M' = ListOfNewMatches

1  For each Match S in M do
2    For each Match T in M do
3      ontP = findPathBetween(S.domainname, T.domainname, ontG);
4      dbP = findPathBetween(S.table, T.table, dbG);
5      If ontP.isEmpty() or dbP.isEmpty() then continue;
6        For int i = 0; i < ontP.getNodeList().size; i += 2 do
7          nD = ontG.getNodeSource(ontP[i]);      //domain class
8          op = ontG.getNodeTarget(ontP[i+1]);    //object property
9          nT = ontG.getNodeTarget(ontP[i+2]);    //target class
10         If M.contains(nD) and !M.contains(nT) and !M'.contains(nT) then
11           nT.table = nD.table;
12           nT.column = nD.column;
13           ObjectPropertyList.add(op, nD, nT);
14           M'.add(nT);
15   return M';
```

In the example of Fig. 2, the mappings for the concepts *Room* and *Wall* will have the same IRI and SQL queries as those generated for the mappings of the *Building* concept. The second case occurs when the minimum path length between a pair of concepts of the putative ontology has more nodes (i.e., concepts) than the minimum path length between the aligned concepts of the target ontology. This case is solved by generating an SQL query with multiple join clauses to connect all nodes of the path. Since a path between concepts of the database schema has been found for each pair of connected classes, the SQL query can be obtained by means of joining clauses.

3.5 Step 5: Generating R2RML Mappings

The last step of the AutoMap4OBDA workflow is to generate the R2RML document according to the correspondences found by the matcher. For each correspondence a *TripleMap* is generated which contains a *predicateObjectMap* to define the data properties already aligned. Then, for each object property an extra *TripleMap* is generated. The logical tables of the *TripleMaps* are defined as SQL queries which are automatically obtained according to the database schema paths found in step 4. That is, SQL join clauses between tables are automatically set by means of the foreign keys. The triples map in Listing 1 includes the SQL query obtained for the mapping in Fig. 2 (right part).

```
<mapping1> a rr:TriplesMap;
 rr:logicalTable [rr:sqlQuery "SELECT a.id, d.id FROM Building AS a
   INNER JOIN Block AS b ON a.fkBlock=b.id INNER JOIN District AS c ON
   b.fkDistrict=c.id INNER JOIN City AS d ON c.fkCity=d.id "];
 rr:subjectMap [rr:template ".../building/{a.id}";rr:class ex:Building];
 rr:predicateObjectMap [ rr:predicate ex:hasCity;
  rr:objectMap [rr:template ".../city/{d.id}"]].
```

Listing 1. An example of a R2RML mapping in Figure 2

4 Evaluation

AutoMap4OBDA has been evaluated with relational-to-ontology benchmark suite RODI [14]. RODI offers basic test scenarios from conference, geographical, and oil and gas domains; and mixed scenarios in the conference domain where the database schema has to be matched to an ontology from a different scenario. Each scenario is composed of databases, ontologies, and a set of queries to test if the mappings generated by mapping generating system are performing well. The mappings are evaluated for each scenario by the percentage of successfully answered queries. RODI includes a wide range of relational-to-ontology mapping challenges classified as naming conflicts, structural heterogeneity, and semantic heterogeneity. RODI simulates real-world scenarios by creating different databases with modifications to reproduce design patterns and anti-patterns in databases. For a further explanation of the scenarios addressed by RODI refer to [14, 15].

We have assessed the performance of the state-of-the-art tools described in Sect. 5 related works against AutoMap4OBDA (**AM4O**). The results for BootOX (**B.OX**), IncMap (**IncM.**), ontop, MIRROR (**MIRR.**), COMA ++ (**COMA**), and D2RQ have been obtained by RODI team [15]. While BootOX and IncMap are full relational-to-ontology matching systems, the other mapping generators cannot generate mapping according to an **existing** ontology. To overcome this issue, RODI benchmark includes an ontology matcher (LogMap [9]) to align the putative ontology generated by those systems with the target ontology.

The average execution time of AutoMap4OBDA – in an Intelcore i5 architecture with 10 GB of RAM – has been less than 25 s per scenario for 15 scenarios, 57.96 s for the *Adjusted naming Conference* scenario, and 6.87 min for the *Oil&Gas* whose database has 70 tables with 250 k records and the target ontology has 344 classes, 148 object properties, and 237 data properties.

The results show that AutoMap4OBDA comes out in the top position for eleven out of seventeen scenarios and in second position in three scenarios (Table 1). The scores are based on average of per-test F-measure. The results of AutoMap4OBDA in the mixed scenarios such as *Target ontology: CMT*, *Target ontology: Conference*, and *Target ontology: SIGKDD* is not as good as the other scenarios because the level of semantic heterogeneity is much higher than in basic scenarios.

Table 1. Overall scores of the state-of-the-art tools and AutoMap4OBDA in RODI scenarios (scores based on average of per-test F-measure). Best numbers per scenario in bold.

Scenarios		B.OX	IncM.	ontop	MIRR.	COMA	D2RQ	AM4O
Adjusted naming	CMT	**0.76**	0.45	0.28	0.28	0.48	0.31	0.56
	Conference	0.51	0.53	0.26	0.27	0.36	0.26	**0.56**
	SIGKDD	**0.86**	0.76	0.38	0.30	0.66	0.38	**0.86**
Restructured	CMT	0.41	**0.44**	0.14	0.17	0.38	0.14	0.41
	Conference	0.41	0.41	0.13	0.23	0.31	0.21	**0.54**
	SIGKDD	0.52	0.38	0.21	0.11	0.41	0.28	**0.72**
Combined case	SIGKDD	0.48	0.38	0.21	0.11	0.28	0.28	**0.62**
Missing FK	Conference	0.33	0.41	–	0.17	0.21	0.18	**0.49**
Denormalized	CMT	0.44	0.40	0.20	0.22	–	0.20	**0.52**
GeoData	Classic Rel	0.13	0.08	–	–	–	0.06	**0.44**
Oil&Gas domain	User Queries	0.00	0.00	0.00	0.00	–	0.00	0.00
	Atomic	0.14	0.12	0.10	0.00	0.00	0.08	**0.23**
Target ontology: CMT	Conference	0.20	**0.35**	0.10	0.00	0.00	0.10	0.15
	SIGKDD	0.33	0.33	0.19	0.00	0.14	0.19	**0.38**
Target ontology: Conference	CMT	0.20	0.34	0.05	0.00	0.05	0.05	**0.39**
	SIGKDD	0.13	**0.30**	0.09	0.00	0.04	0.09	0.17
Target ontology: SIGKDD	CMT	0.51	**0.57**	0.19	0.00	0.24	0.26	0.41
	Conference	0.24	**0.44**	0.13	0.00	0.09	0.14	0.19
Average of the tests		0.36	0.37	0.15	0.10	0.20	0.18	**0.43**

In the other scenarios, AutoMap4OBDA outperforms other systems thanks to the methods described in this paper such as extending correspondences according to the target ontology. Indeed, full-featured alignments systems – such as those used in the evaluation of BootOx and ontop among others – have difficulties to match object properties when the structure of the source (putative) ontology is not similar to the structure of the target (domain) ontology. AutoMap4OBDA does not directly match the object properties of the putative and domain ontologies, but the object properties are set by the extending correspondences method described in step 4.

For example, the correspondences illustrated in Fig. 3 can be found by AutoMap4OBDA but not by full-featured alignments systems. In *sigkdd_mixed* scenario, the target ontology has the object property *isCommitteOf* whose domain is *Commitee* and range is *Conference*. Moreover, in the putative ontology the correspondent object

Fig. 3. Example of correspondences found by AutoMap4OBDA in *sigkdd_mixed* scenario

property is *commitee* whose range class is *conferences* and whose domain classes are *best_paper_awards_committs*, *organizing_committees*, and *program_committees*. Those domain classes are subclasses of *committe* class in the domain ontology however an ontology matcher cannot match both object properties. In this case, AutoMap4OBDA fulfills the alignment of the object property in the extending correspondences method once it has aligned the classes correctly.

It is worth mentioning the *GeoData* scenario in which the method of *String similarity metric selection based on target ontology labels for ontology alignment* helped to find much more data properties than the other systems bringing a performance more than three times as high as the following system. No results have been achieved in *Oil&Gas domain User Queries* scenario. This is a real-world scenario where the queries go beyond returning a simple result list to determine whether all objects are of one class. The good result in the *Oil&Gas domain Atomic* scenario – compared with other systems – has been achieved thanks to the ontology learning method applied in step 2 of the AutoMap4OBDA workflow. In this scenario, AutoMap4OBDA was able to find several mappings where the values of the columns are used to set classes of the *subjectMap* as the following example. AutoMap4OBDA outperforms other mapping generators of the state-of-the-art because they cannot generate this kind of mapping in an automated way.

```
<mapping1_201> a rr:TriplesMap;
  rr:logicalTable [ rr:sqlQuery "SELECT pipnpdidpipe FROM pipline WHERE
    pipmedium = 'Oil'" ];
  rr:subjectMap [ rr:template "http://.../oilpipeline/{pipnpdidpipe}";
    rr:class http://sws.ifi.uio.no/vocab/npd-v2#OilPipeline ].
```

Despite the good results AutoMap4OBDA is far from generating a full list of mappings derived from a relational database and a target ontology. Indeed, the average F-measure obtained in the RODI scenarios is 0.43 which is not a remarkable result but it is a step forward in relational-to-ontology mapping generators since it has improved the results by 0.06 with regard the next contender (IncMap) which has an average of 0.37.

5 Related Work

Approaches for ontology creation from existing databases without using any external resource are called direct mapping methods. Sequeda et al. not only surveyed those methods but also standardized the basic transformation rules implemented by those methods [18]. In most systems, the use of direct mapping methods implies that relational tables are mapped to concepts, columns to data properties, and relations implemented by means of foreign/primary keys to an object properties. Apart from the application of transformation rules, direct mapping methods apply reverse engineering techniques to inspect the database schema constraints to unveil its semantics. The implementation of such basic transformation rules is described in [19]. Following this line of work there are D2RQ [2], MIRROR [12] and ontop [16] systems, however they are not fully comparable with AutoMap4OBDA proposed in this paper since they do not generate mappings between a relational database and an existing ontology. These methods generate a putative ontology derived from a relational database and their corresponding mappings.

The prime example of systems that enhance direct mapping methods is RDBToOnto which applies ontology learning methods to identify of semantic patterns in the stored data with the final goal of producing expressive ontologies [4]. Indeed, the ontology learning methods implemented in AutoMap4OBDA are based on the techniques of RDBToOnto. In the same way, the direct mapping methods reported above, RDBToOnto cannot be used directly in OBDA settings.

The closest system to AutoMap4OBDA is BootOX, a relational-to-ontology mapping generator system which applies the similar transformation rules as direct mapping methods and providing support for different OWL profiles [10]. BootOX can produce a set of axioms based on one of the three OWL profiles – QL, RL, and EL – depending on the settings. Moreover, BootOX system includes LogMap as ontology matching system to align a putative ontology – BootOX authors call it a bootstrapped ontology – with a target ontology. The main difference between BootOX and AutoMap4OBDA is that BootOX does not take advantage – in an automatic mode – of the contents of the database to enrich class hierarchy of the putative ontology, instead BootOX proposes users to improve mappings generated in previous step through interaction with the system.

Alternatives to this kind of approach are systems such as COMA ++ which represent the database and the target ontology as directed graphs whose nodes are to be matched [1]. Following this line of work, IncMap automatically maps the target ontology directly to the database using an intermediate graph structures – IncGraphs – and applies a flooding algorithm to merge the graphs aiming at finding the correspondences [13]. In contrast to AutoMap4OBDA, IncMap creates two intermediate graph structures which only rely on the ontology structure and database schema. The graph structures created by AutoMap4OBDA in *extending correspondences according to the target ontology* step are similar to those created by IncMap, however in AutoMap4OBDA graphs data properties and non-foreign key columns are not included as nodes in the graph. Moreover, it does not use ontology learning methods to search further correspondences between the data and classes of the target ontology.

The concept of intermediate graph structured is also present in Karma system, a semi-automatic mapping system which also offers a graphical user interface to verify and correct the mappings [11]. Although Karma inspects database contents to find the correspondences to the target ontology, human intervention is required to generate the final mappings. Therefore, it cannot work in fully automatic mode in the same way as AutoMap4OBDA.

6 Conclusions

We have presented AutoMap4OBDA, a full-featured mapping generator for OBDA scenarios. AutoMap4OBDA produces a putative ontology from a relational database and uses it as an intermediate element in the relational-to-ontology mapping process. AutoMap4OBDA requires a relational database and an ontology specified in OWL as its input. The evaluation presented in this paper confirms that AutoMap4OBDA is a step forward in relational-to-ontology mapping systems. With regards to non-synthetic scenarios such as GeoData and Oil&Gas, the performance of AutoMap4OBDA, requires improvement. Our conclusion of the evaluation is that the generation of relational-to-ontology mappings is a task that cannot be completely automated yet. An expert in the domain where the data originates form still should validate and complement the mappings automatically generated by systems. To do so, we have ensured that mappings generated by AutoMap4OBDA are fully compliant with the R2RML recommendation and that they can be loaded in Map-On, a graphical R2RML mapping editor to support users without ontology engineering and database skills in curation of relational-to-ontology mappings [20].

Providing support for non-relational data sources is an ambitious research line for mapping generators such as AutoMap4OBDA. This requires the addition into AutoMap4OBDA of other ways to extract class hierarchies from non-relational sources (e.g., XML files, graph and NoSQL databases) using ontology learning methods as well as an extension of the Extending correspondences step to handle non-relational sources. Supporting these kinds of sources will lead to using an alternative mapping language such as RDF Mapping Language (RML) [8]. The next step in the AutoMap4OBDA development will be enhancing ontology learning methods to facilitate the inferring of class hierarchies since the results have not been completely satisfactory. A translation feature will be also incorporated to address multi-language scenarios and semantic similarity techniques based on external resources.

Acknowledgements. This work was carried out within the research project ENERSI funded by Ministry of Economy and Competitiveness of the Government of Spain (Reference number RTC-2014-2676-3)

References

1. Aumueller, D., Do, H.H., Massmann, S., Rahm, E.: Schema and ontology matching with COMA++. In: Proceedings of the ACM SIGMOD International Conference on Management of Data, pp. 906–908. ACM (2005)

2. Bizer, C., Seaborne, A.: D2RQ - treating non-RDF databases as virtual RDF graphs. In: 3rd International Semantic Web Conference, vol. 2004. Springer, Heidelberg (2004)
3. Borst, W.N.: Construction of engineering ontologies for knowledge sharing and reuse. Technology, Ph.D. (1997). http://doc.utwente.nl/17864/
4. Cerbah, F.: Mining the content of relational databases to learn ontologies with deeper taxonomies. In: Web Intelligence and Intelligent Agent Technology, pp. 553–557. IEEE (2008)
5. Cheatham, M., Hitzler, Pascal: String similarity metrics for ontology alignment. In: Alani, H., et al. (eds.) ISWC 2013. LNCS, vol. 8219, pp. 294–309. Springer, Heidelberg (2013). doi:10.1007/978-3-642-41338-4_19
6. Cheatham, M., Hitzler, P.: The properties of property alignment. In: 9th International Conference on Ontology Matching, vol. 1317, pp. 13–24. CEUR-WS.Org (2014)
7. Dijkstra, E.W.: A note on two problems in connexion with graphs. Numer. Math. **1**(1), 269–271 (1959)
8. Dimou, A., Sande Vander, M., Colpaert, P., Verborgh, R., Mannens, E., Van De Walle, R.: RML: a generic language for integrated RDF mappings of heterogeneous data. In: 7th Workshop on Linked Data on the Web (2014)
9. Jiménez-Ruiz, E., Cuenca Grau, B.: LogMap: logic-based and scalable ontology matching. In: Aroyo, L., Welty, C., Alani, H., Taylor, J., Bernstein, A., Kagal, L., Noy, N., Blomqvist, E. (eds.) ISWC 2011. LNCS, vol. 7031, pp. 273–288. Springer, Heidelberg (2011). doi:10.1007/978-3-642-25073-6_18
10. Jiménez-Ruiz, E., Kharlamov, E., Zheleznyakov, D., Horrocks, I., Pinkel, C., Skjæveland, M.G., Thorstensen, E., Mora, J.: BootOX: Practical Mapping of RDBs to OWL 2. In: Arenas, M., et al. (eds.) ISWC 2015. LNCS, vol. 9367, pp. 113–132. Springer, Heidelberg (2015). doi:10.1007/978-3-319-25010-6_7
11. Knoblock, C.A., et al.: Semi-automatically mapping structured sources into the semantic web. In: Simperl, E., Cimiano, P., Polleres, A., Corcho, O., Presutti, V. (eds.) ESWC 2012. LNCS, vol. 7295, pp. 375–390. Springer, Heidelberg (2012)
12. de Medeiros, L.F., Priyatna, F., Corcho, O.: MIRROR: automatic R2RML mapping generation from relational databases. In: Cimiano, P., Frasincar, F., Houben, G.-J., Schwabe, D. (eds.) ICWE 2015. LNCS, vol. 9114, pp. 326–343. Springer, Heidelberg (2015)
13. Pinkel, C., Binnig, C., Kharlamov, E., Haase, P.: IncMap: pay-as-you-go matching of relational schemata to OWL ontologies. In: 8th International Conference on Ontology Matching, vol. 1111, pp. 37–48. CEUR-WS.org (2013)
14. Pinkel, C., Binnig, C., Jiménez-Ruiz, E., May, W., Ritze, D., Skjæveland, M.G., Solimando, A., Kharlamov, E.: RODI: a benchmark for automatic mapping generation in relational-to-ontology data integration. In: Gandon, F., Sabou, M., Sack, H., d'Amato, C., Cudré-Mauroux, P., Zimmermann, A. (eds.) ESWC 2015. LNCS, vol. 9088, pp. 21–37. Springer, Heidelberg (2015)
15. Pinkel, C., Binnig, C., Jimenez-Ruiz, E., Kharlamov, E., May, W., Nikolov, A., Skjaeveland, M.G., Solimando, A., Taheriyan, M., Heupel, C., Horrocks, I.: RODI: benchmarking relational-to-ontology mapping generation quality. J. SW (2016). http://www.semantic-web-journal.net/content/rodi-benchmarking-relational-ontology-mapping-generation-quality-0
16. Rodriguez-Muro, M., Rezk, M.: Efficient SPARQL-to-SQL with R2RML mappings. Web Semant.: Sci. Serv. Agents World Wide Web **33**, 141–169 (2015)
17. Savo, D.F., Lembo, D., Lenzerini, M., Poggi, A., Rodríguez-Muro, M., Romagnoli, V., Ruzzi, M., Stella, G.: MASTRO at work: experiences on ontology-based data access. In: DL 2010, pp. 20–31 (2010)

18. Sequeda, J., Garcia-Castro, A., Corcho, O., Tirmizi, S.H., Miranker, D.P.: Overcoming database heterogeneity to facilitate social networks: the Colombian displaced population as a case study. In: 18th International Conference on World Wide Web. ACM (2009)
19. Sequeda, J., Arenas, M., Miranker, D.P.: On directly mapping relational databases to RDF and OWL. In: 21st International Conference on World Wide Web, pp. 649–658. ACM (2012)
20. Sicilia, Á., Nemirovski, G.: Map-on: a web-based editor for visual ontology mapping. J. SW (2016). http://www.semantic-web-journal.net/content/map-web-based-editor-visual-ontology-mapping-0

Word Tagging with Foundational Ontology Classes: Extending the WordNet-DOLCE Mapping to Verbs

Vivian S. Silva(✉), André Freitas, and Siegfried Handschuh

Department of Computer Science and Mathematics,
University of Passau, Innstraße 43, 94032 Passau, Germany
{vivian.santossilva,andre.freitas,siegfried.hasdschuh}@uni-passau.de

Abstract. Semantic annotation is fundamental to deal with large-scale lexical information, mapping the information to an enumerable set of categories over which rules and algorithms can be applied, and foundational ontology classes can be used as a formal set of categories for such tasks. A previous alignment between WordNet noun synsets and DOLCE provided a starting point for ontology-based annotation, but in NLP tasks verbs are also of substantial importance. This work presents an extension to the WordNet-DOLCE noun mapping, aligning verbs according to their links to nouns denoting perdurants, transferring to the verb the DOLCE class assigned to the noun that best represents that verb's occurrence. To evaluate the usefulness of this resource, we implemented a foundational ontology-based semantic annotation framework, that assigns a high-level foundational category to each word or phrase in a text, and compared it to a similar annotation tool, obtaining an increase of 9.05 % in accuracy.

Keywords: Linguistic resources · Semantic annotation · Foundational ontology

1 Introduction

Lexical semantic information is fundamental in many natural language processing and semantic computing applications [2,23,25]. Applications such as question answering and text entailment require complex inferences involving large commonsense knowledge bases. The consumption, interpretation and coordination over large-scale lexical information demands the use of higher level categories capable of generalizing the information without loss in meaning.

The large symbolic word space which is the target of NLP tasks demands strategies to map words to higher level classes which are enumerable and can be used to encode rules and algorithms on the top of these classes. The utility of tagging lies on the potential for encoding generalizations using an enumerable set of categories. These categories can range from simple lexical information, like the

V.S. Silva—CNPq Fellow – Brazil.

E. Blomqvist et al. (Eds.): EKAW 2016, LNAI 10024, pp. 593–605, 2016.
DOI: 10.1007/978-3-319-49004-5_38

grammatical class of a word, to more complex semantic representation, intended to unambiguously state what a concept means in the world. Foundational ontology classes are a good example of semantic representation, composing a set of categories that can determine the most high-level nature of a concept. Additionally, the foundational ontology entities and their connection to logics supports the connection between natural language and reasoning. This connection can help NLP systems to address complex semantic interpretation tasks.

WordNet [6] can be used as a "bridge" between natural language text and higher level semantic representations, including the foundational ontology-based modelling. In an effort to provide the WordNet taxonomy with more rigorous semantics, Gangemi et al. [9] performed the alignment between WordNet upper level synsets and the foundational ontology DOLCE [16]. After a meticulous analysis, the WordNet taxonomy was reorganized to meet the OntoClean [12] methodology requirements, and the resulting upper level nouns were then mapped to DOLCE classes representing their highest level categories. This mapping concentrated on the noun database, since most particulars in DOLCE describe categories whose members are denoted by nouns.

On the other hand, many NLP applications need to deal with events, actions, states, processes and other temporal entities that may not be expressed as nouns, but rather as verbs. Often, verbs are seen as relationships between concepts, and DOLCE in fact provides a well-defined set of properties and axioms that link classes together in a meaningful way, like the properties *performs*, *target*, *instrument*, *makes* or *uses*, among many others. But in natural language a verb can play the role of an entity itself, and a class will be more suitable to represent it than a property.

As an example, consider a rule-based text entailment task where, given the fact "Mary is a mother", known to be true, we want to check whether the fact "Mary gave birth" is also true. Mapping the terms to foundational ontology classes can be done as an intermediary step to reduce the reasoning search space. Here, "give birth" would be better classified as an *action*, while "Mary" can be seen as an *agent*, and "mother" as a *role*. Using a supporting definition (for example, from WordNet) stating that "a mother is a woman who has given birth" and a pre-defined rule asserting that "*if* an agent plays a role *and* the role performs an action, *then* the agent performs the action (while playing the role)", we would have that "*if* Mary plays the role of a mother, *and* a mother performs the action of giving birth, *then* Mary gave birth". As can be seen, the classification of the verb "give birth" as a member of a foundational category is crucial for applying the correct rule and accomplishing the entailment.

This work aims at proposing a semantic annotation model based on foundational ontologies, called FO Tagging, that can be used to enrich text from a knowledge base, bringing valuable information to the execution of natural language processing tasks and semantic computing, and reducing the size of symbolic space they need to deal with. To accomplish this goal, we present an alignment between WordNet verbs and DOLCE, taking as starting point the nouns alignment provided by Gangemi et al. [9]. To identify the correct class for each top level verb, we start by searching for direct links between those verbs and their noun counterparts, that is, a noun that represents an occurrence, or *eventuality* in the Davidsonian logical view [5], of that verb. When there are no such direct links, either a path between the verb and a suitable noun is drawn through indirect links, or a manual evaluation based on the terms present in the synset's gloss is carried out. The DOLCE classes assigned to the top level verbs are then propagated down the taxonomy, resulting in a fully classified verb database.

The paper is organized as follows: Sect. 2 lists some related work regarding the link between linguistic resources and foundational ontologies, as well as semantic annotation approaches. Section 3 presents the basis for the ontological structuring of verbs. Section 4 details the methodology adopted in the alignment followed by the results in Sect. 5 and a quantitative evaluation in Sect. 6. Section 7 draws the conclusions and points to future work.

2 Related Work

The alignment between WordNet noun synsets and the foundational ontology DOLCE performed by Gangemi et al. [9] is probably the most comprehensive attempt to turn WordNet into a conceptually well-grounded ontology. Since DOLCE was developed under a rigorous methodology that ensures the consistency across the taxonomy links, and is oriented towards human cognition and natural language, the resulting mappings and reorganized noun hierarchy can potentially be more useful in practical applications.

To carry out the alignment, the noun synsets taxonomy was analyzed taking into account a set of criteria such as identity, rigidity and unity [12], as well as concepts and individuals differentiation, generality level, among others. The identified inconsistencies were corrected with synsets exclusion or relocation, and the selected top synsets were then mapped to DOLCE classes, adding an upper level descriptive layer to the WordNet ontology.

Another effort to map WordNet concepts to a foundational ontology was presented by Niles and Pease [22], who manually mapped all the WordNet synsets to the Standard Upper Merged Ontology [24]. SUMO can be better defined as a knowledge base rather than a pure foundational ontology, because, differently from DOLCE, it contains many domain specific concepts besides the upper level, domain independent classes. This characteristic leads to a different kind of alignment, where the primary goal was to link the synsets to a similar class in the ontology, that is, a class that has the same meaning, and not a class that represents their upper level category. Only when a similar class was not found in the ontology, a class showing other kind of relationship, like subsumption or instantiation, was chosen and assigned to the synset.

Although all the verb synsets were also mapped to SUMO, the final classification is very heterogeneous. For example[1], one of the senses of the verb *breathe*

[1] Mappings available at https://goo.gl/bflXqx.

("draw air into, and expel out of, the lungs") was mapped to the SUMO concept *Breathing*, considered equivalent in meaning, but the verb *palpebrate* ("wink or blink, especially repeatedly") was mapped to the higher level concept *PhysiologicProcess*, which subsumes it. Even concepts that don't represent temporal entities were assigned to verbs, like the first sense of the verb *sigh* ("heave or utter a sigh; breathe deeply and heavily"), which was mapped to the concept *Organism*, considered equivalent in meaning. The work presented in this paper aims at a more homogeneous and high level classification of the verb synsets, mapping the verbs to domain independent concepts representing only temporal particulars.

Regarding semantic annotation, many tagging approaches have been proposed in the last years, ranging from lexical to semantic annotation features. Lexical annotation, such as the ones based on POS (part-of-speech) tags and syntactic roles [14] are largely employed and serve as a basis for more complex annotation tasks. Besides assigning a tag for each single word in a sentence, lexical annotation can also cover the relationships between words, like syntactic dependency and co-reference [14]. On the other hand, semantic annotation focuses on capturing the meaning of words and the kind of information they carry. Among the most common semantic annotation techniques are the Named Entity Recognition [21], focused on recognizing numeric expressions and entities identified by proper nouns, the Sentiment Annotation [14], which classifies words as positive, negative or neutral for Sentiment Analysis tasks, and Semantic Role Labeling [15], intended to determine the role of entities which refers to a given event.

Foundational ontology-based tagging is a semantic annotation task, as it aims at identifying the most primary meaning of a concept, that is, what its most basic category is. The semantic annotator most closely related to FO tagging is the SuperSense Tagger [1]. SST treats the problem of super sense tagging as a sequential labeling task and implements it as a Hidden Markov Model. The tagset is composed of 41 WordNet high-level noun and verb synsets, called *super senses*. It is also intended to determine the concept's primary category, but the set of super senses can be considered inconsistent, as it mixes higher- and lower-level concepts all together, with overlapping categories that would allow multiple possibilities of classification. For example, a "cake" could be classified as "foods and drinks", but also as "man-made objects", as these super senses overlap, being the first a specialization of the second. FO tagging, in turn, adopts a more rigorous and formal semantic meta-model, pushing the classification to a more stable and conceptually grounded set of categories.

3 Verbs and Ontologies

At first sight, it may seem unsuitable to fit verbs in a classification driven by categories from the DOLCE ontology. An attempt to apply the OntoClean methodology [12] metaproperties and constraints will prove challenging, making it hard to assign rigidity, unity and identity values to verbs, since they are not entities by themselves. On the other hand, the occurrence of a verb in fact represents a temporal entity, in general given by a noun, for which it is possible to assign the aforementioned metaproperties and restrictions. For example, the occurrence of the verb *run* leads to a *running*, *appear* to *apparition*, *leak* to *leakage*, and so on. The proposed approach is to track back the noun denoting a temporal entity that best represents a verb's occurrence, and which is already mapped to DOLCE, and map the verb to the same DOLCE class that was assigned to its noun counterpart.

The adopted model for introducing verbs in the DOLCE-oriented WordNet ontology finds support in the ITP (Intelligent Text Processing) linguistic ontology proposed by Dahlgren [3], which is a content ontology for natural language processing that intends to represent a "world view" based on assumptions about what exists in the world, including verbs, viewed as essential elements in NLP, and how to classify it. Regarding verbs, the ITP ontology follows the Vendlerian approach [26], classifying each verb as an event or state, being events further subdivided into activities, accomplishments and achievements. This is very close to the classification of temporal entities defined in DOLCE, which presents some small variations, detailed in the next Section.

3.1 Perdurants in DOLCE

The most fundamental distinction in DOLCE is that between *endurants* and *perdurants*. Simply put, endurants are entities that are fully present (that is, all their parts are present) at any moment they are present. Contrarily, perdurants are entities that span in time, being only partially present at a given moment, as some of their parts (past or future phases) may not be present at that time. The classification of verbs is restricted to the perdurant branch, which is subdivided into the *stative* and *event* classes, informally described next. A formal description of the DOLCE concepts can be found in [16].

A temporal entity, i.e., a perdurant, is considered *stative* if it is cumulative, and *eventive* otherwise. For example, a "sitting" is stative because it is cumulative, since the sum of two sittings is still a sitting. Stative occurrences are subdivided into *states* and *processes*. A state is an occurrence whose all temporal parts can be described by the same expression used to describe the whole occurrence. A "sitting" is a state because all of its temporal parts are also sittings. Differently, "smoking" is a process, because some temporal parts of a smoking are not smokings themselves. A state can be further specialized into a *cognitive-state*, that is, a state of the (embodied) mind.

Events are subdivided into *accomplishments*, *achievements* and *cognitive-events*. An occurrence is called an achievement if it is atomic, and an accomplishment otherwise. A cognitive-event is an event occurring in the (embodied) mind. Accomplishments are also further subdivided into a series of more specific concepts, but for the purposes of this work only the higher level classes are of interest.

3.2 DOLCE Lite Plus vs. DOLCE Ultra Lite

The alignment between WordNet noun synsets and DOLCE was recently updated to address an entity typing task [11]. Called OntoWordNet (OWN) 2012, this update builds upon OWN [10], an OWL version of WordNet-DOLCE alignment. Besides revising the manual mappings, OWN 2012 also adopts a different version of DOLCE, the DOLCE Ultra Lite Plus [8], which is a simplified version of DOLCE Lite Plus (called simply DOLCE in the rest of this work), intended to make classes and properties names more intuitive and express axiomatizations in a simpler way, among other features. An additional lightweight foundational ontology, called DOLCE Zero (D0), was also developed and integrated into DULplus, generalizing some of its classes.

A substantial difference that can be noted between DLP and the DULplus+D0 resulting ontology (herein called simply DULplus) refers to their hierarchical organization. As mentioned in Sect. 3.1, the main branches in DLP are *endurant* and *perdurant*, and this is the most fundamental distinction that guided the ontology development. DULplus adopts a more relaxed hierarchy, where the distinction between endurants and perdurants is left aside. Instead, there is a highest level class called *Entity*, whose direct subclasses are *Abstract*, *Cognitive Entity*, *Event*, *Information Entity*, *Object*, *Quality*, *System* and *Topic*. The *Event* class is subdivided into *Action* and *Process*, and there is no equivalent to the DLP class *State*. Also, it's not clear whether *Cognitive Entity*, defined as "Attitudes, cognitive abilities, ideologies, psychological phenomena, mind, etc." refers to endurants, perdurants or both.

Clearly identifying which noun synsets in WordNet denote temporal entities is a key point in the methodology adopted in our verb classification, since we consider only perdurants as suitable categories for verbs. Given the above mentioned characteristics of DULplus, even though OWN 2012 is a more recent resource, we opted for using the original WordNet-DOLCE alignment, because we believe that the more rigorous conceptualization expressed in the DLP hierarchy could provide us with a higher quality verb classification.

4 Alignment Methodology

To carry out the WordNet verb synsets alignment to the DOLCE concepts, the noun synsets alignment provided in [9] was used as a reference frame from which the DOLCE classes were transferred to the related verbs, according to the relevant links between word senses in WordNet. The alignment methodology comprises the following steps:

Update and Expansion of Nouns Alignment: as available in [16], 813 noun synsets have been aligned to 50 DOLCE classes. We updated these 813 alignment to bring them to a more recent version of WordNet, since the original ones were done over version 1.6. Using the synset ID mappings provided by Daudé et al. [4], the alignments were migrated from version 1.6 to version 3.0, resulting in 809 aligned synsets (some synsets are excluded or merged from one version to the subsequent one). Then, we expanded the alignments to assign the DOLCE classes to the remaining synsets. The 809 aligned synset are located at the highest levels of the WN hierarchy, then, using the hypernym and instance links recursively, the synsets at the lower levels inherited the DOLCE class from their parent synsets. The final alignment contains 80,897 noun synsets, which corresponds to 98.5 % of the WN 3.0 noun database. The remaining 1.5 % includes the synsets that were not considered in the original DOLCE-WN alignment, and their hyponyms and instances.

Top Level Verbs Selection: similarly to the nouns alignment, the verbs classification was performed over the top level synsets, to be later propagated down the taxonomy through the hyponym links. The WordNet verb taxonomy contains 560 top level synsets, that is, synsets that have no hypernyms, and that were selected as candidates for the direct alignment.

Direct Links: for each of the 560 top level verb synsets, we retrieved all the related word senses given by the *derivationally related form* lexical link. The derivationally related words were then manually filtered in order to identify, among them, the noun that would represent the verb occurrence. In general, these are words whose definition (gloss) starts with expression such as "the act of", "the process of" or "the state of", which are strong evidences that they are classified as perdurants. For example, for the verb *move* ("be in a state of action"), the derivationally related words retrieved were *motion*, *move* and *mover*. In the manual analysis phase, the word *move* ("the act of deciding to do something") was identified as the correct noun counterpart for the verb move, and the DOLCE class *event* associated to it was then also assigned to the verb.

Indirect Links: since not all verbs have a suitable noun counterpart, in many cases no direct link can be found. For the verbs that have no derivationally related form, or none of the derivationally related forms represents the verb occurrence, an indirect path to the appropriate noun was manually searched. This path relies on other kind of links, such as *antonym* and *verb group*. The verb group link acts in a way similar to the derivationally related form, but linking only verbs among them, and the antonym link is also useful because, in general, to be comparable, the verbs need to be of the same kind, leading to the same DOLCE category. This is not always true, as some states, for example *stand still*, have as antonym an event, in this case, *move*, so an additional manual check is required to ensure that the found path is indeed valid. As an example, the verb *ignore* ("be ignorant of or in the dark about"), which has no suitable derivationally related forms, has as antonym the verb *know* ("be cognizant or aware of a fact or a specific piece of information"), which is, in turn, linked to the noun *knowingness* ("having knowledge of"), classified as a *cognitive-event*, so that DOLCE class was directly assigned to the verb *know* and, as a consequence, indirectly linked to the verb *ignore*.

Manual Assignment: finally, for the verbs for which no explicit direct or indirect link to a noun could be found, a careful manual evaluation was carried out. This evaluation has taken into account the implicit relationship to other

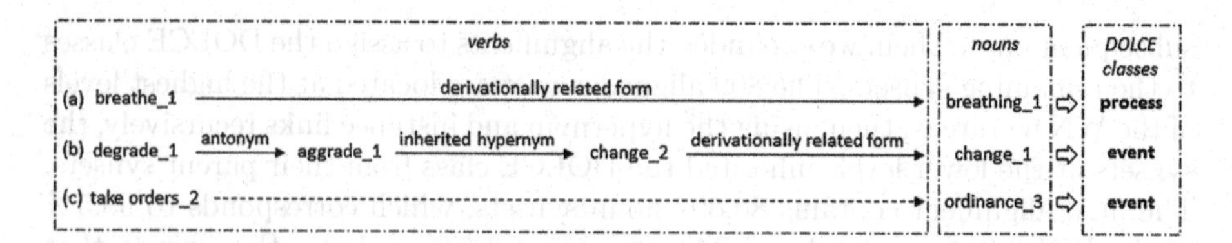

Fig. 1. Examples of WordNet-DOLCE verb mappings obtained by (a) direct link, (b) indirect links and (c) manual assignment. Full lines stand for explicit links in WordNet, while the dashed line represents an implicit relationship. The numbers indicate the word sense in WordNet 3.0

(preferably already classified) verbs, given by the words present in each synset's gloss. Using the gloss to uncover implicit relationships is an important procedure to make the classification as less subjective as possible. For example, verbs having glosses beginning with "be", like *wear* ("be dressed in"), *stay in place* ("be stationary") or *sit* ("be seated") are strong *state* candidates. In other cases, even if the link is not explicit in WordNet, the relationship is very clear, for instance, between the verbs *arch* ("form an arch or curve") and *overarch* ("form an arch over"). Since *arch* was, by inheritance, mapped to the *event* class, and given the high similarity between their glosses (indeed, the second could possibly be a specialization of the first), we could also classify *overarch* as an *event*. In all scenarios, checking if the concept's characteristics described in Sect. 3.1 apply and further analyzing the verb's hyponyms to make sure they also fit in the chosen DOLCE category helped us to reach a consistent classification.

Figure 1 shows some examples of mappings between WordNet verbs and DOLCE concepts, reached through explicit direct and indirect links, and following implicit relationships identified by manual evaluation.

5 Alignment Results

After applying the alignment methodology described in Sect. 4, the 560 top level verb synsets were mapped to five DOLCE classes: *event*, *state*, *process*, *cognitive-event* and *cognitive-state*. These are the same classes used previously to classify all the WordNet noun synsets denoting perdurants. A total of 52.5 % of the synsets were mapped through explicit links, being 36.25 % direct links and 16.25 % indirect links to noun synsets, and 47.5 % through implicit relationships identified by manual analysis. The top level mappings were then propagated down the verb taxonomy using the hypernym-hyponym links, and all the troponyms, as verbs' specializations are called in WordNet, inherited their parent's DOLCE class, resulting in a 100 % mapped verb database.

The adequacy of WordNet hypernym links to effectively represent subsumption relationships is a common concern, but, since we are dealing with very high level categories, the probability of errors in the taxonomy propagation, although not completely eliminated, is considerably reduced. Although not 100 % of the WN hypernym links can be considered correctly assigned, they are intended to represent subordination [18], meaning that synsets are linked in the hierarchy because they somehow show a strong similarity regarding their nature, that is, what they primarily represent in the world, and then tend to converge to the same upper class even if they don't follow a strict subsumption relationship.

Table 1 shows the final distribution of both top level and full taxonomy synsets over the five DOLCE classes. Although it may seem unbalanced towards the *event* class, this is coherent with the noun synsets mapping, where 75 % of the 8,522 perdurant nouns were also classified as *events*. A possible reason for these figures is that most verbs describe actions, and *action*, described as "a perdurant that exemplifies the intentionality of an agent" [16], is indeed a subclass of *accomplishment*, which is a subclass of *event*. The original WordNet-DOLCE nouns alignment opted for a higher level mapping, keeping at the *event* class instead of drilling down more specific concepts, and this choice is also reflected in our final verb mapping.

Table 1. WordNet-DOLCE verb synsets alignment statistics

DOLCE class	Top synsets	Full taxonomy
event	412	12,037
cognitive-event	63	854
state	62	597
process	15	259
cognitive-state	8	20
Total	560	13,767

6 Evaluation

To evaluate the usefulness of the resulting alignments in a semantic annotation task, we run experiments using two datasets: the SemCor dataset [19] and the eXtended WordNet [20]. SemCor is a subset of the Brown corpus [13] which has been manually annotated with WordNet sense numbers. SemCor 3.0 is annotated with WordNet 3.0 synsets, and is divided in three parts: "brown1" and "brown 2", where all nouns, verbs, adjectives and adverbs are annotated, and "brownv", with annotations only for verbs. We used the "brown1" and "brown2" subsets, which make up a total of 20,132 sentences (the original dataset contains 20,138 sentences, but 6 of them are empty sentences). The eXtended WordNet, or XWN, is a resource that provides logical forms for all WordNet synset glosses. Besides the logical forms, XWN also includes word sense disambiguation, being all words present in every gloss annotated with its correspondent WordNet sense number. XWN is divided into four files, one for each grammatical class: "noun", "verb", "adj" and "adv". We used the "noun" and "verb" datasets, since these are the synsets covered in our alignment. To build the sentences, the synset head word, that is, the first word in the synset, was followed by the word "is" and concatenated with the synset gloss, making up a total of 93,197 sentences.

Using the sense number information available in both datasets, we retrieved the synset ID for each word using the JWI API [7], and identified the DOLCE class associated with that ID. All the nouns and verbs in a sentence were labeled with its DOLCE class, and all the adjective, adverbs and stop words received a null label. The full labeled datasets were then used as a gold standard in the semantic annotation task evaluation.

To perform the FO tagging, we opted for the first sense heuristic, or Most Common Sense, as word sense disambiguation technique, using the WordNet's sense ranking to retrieve the most frequent sense for each word. Although being a very straightforward technique, MCS outperforms many WSD systems [17] and is a good alternative for disambiguating commonsense data.

Our semantic annotator then received as input all the SemCor and XWN words/phrases, grouped into sentences, and identified the synset ID using the MCS WSD heuristic, labeling each token with a DOLCE class or with the null label when the synset ID wasn't found in the synset-class mapping. The results were contrasted with the gold standard and compared with two baselines: the random baseline, and the SuperSense Tagger. Although the comparison with SST is only possible for the SemCor dataset, we believe it is worth to show the difference in the results, since, as mentioned in Sect. 2, this is the semantic annotation approach that is most similar to our foundational ontology-based tagging.

The results[2] are summarized in Table 2. The first line shows the accuracy of a baseline that, for each word/phrase, chooses a sense number at random and then assigns the correspondent DOLCE label. The efficacy of the MCS disambiguation method adopted by the FO tagging can be observed by the F1-Score for both datasets, well above the random baseline. When compared to the SST, FO Tagging presents an increase of 9.05 % in the F1-Score for the SemCor dataset.

It is important to emphasize that the goal of the evaluation is not to judge the quality of the alignment itself, but rather to assess how it could be effectively used in an annotation task. Considering the final mapping as a standard to be followed (which is also reflected in the gold standard, built based on it), the bottleneck stands in finding the correct label for a word/phrase when it has more than one label associated to it. The random baseline is relatively high because it does not mean choosing a label among all existing ones, but randomly picking a label only among the ones associated with a given word/phrase. The results,

Table 2. Evaluation results

	XWN			SemCor		
	Precision	*Recall*	*F1-Score*	*Precision*	*Recall*	*F1-Score*
Random	71.82	72.04	71.93	61.52	62.52	62.02
FO Tagging	**89.68**	**89.74**	**89.71**	**86.10**	**86.36**	**86.23**
SuperSense Tagging	-	-	-	76.65	77.71	77.18

[2] Computed by the "conlleval" script, available at http://goo.gl/YL2IBz.

then, show the accuracy of the chosen approach for FO tagging at selecting the most suitable label from the standard mappings set.

7 Conclusion

The previous effort to align WordNet to the foundational ontology DOLCE led to a conceptually more rigorous version of the WordNet noun taxonomy, meant to increase its adequacy as an ontology. We presented an extension to this alignment, using it as a reference frame to map also the verb synsets, using explicit links between word senses and implicit relationships given by the words present in the verbs' glosses to track back the noun that would best represent an occurrence of a given verb, and assign to this verb the same DOLCE class previously associated to its noun counterpart. After aligning the 560 top level verb synsets, the classification was expanded through the taxonomy, resulting in a 100 % aligned WordNet 3.0 verb database.

The resulting alignment was then used in the implementation of a semantic annotation framework, the FO Tagging, which used the Most Common Sense word sense disambiguation technique to identify to which synset each word belongs to and subsequently retrieve the DOLCE label associated with that synset. Compared to the SuperSense Tagger, the most similar semantic annotation tool, FO Tagging shows an increase of 9.05 % in the F1-Score for the SemCor dataset. In addition to the increase in the accuracy, FO Tagging also introduces an ontologically well-grounded set of categories, pushing the classification to a higher level than that provided by the SST tagset.

Besides contributing to expand the benefits brought by the initial noun mapping, the introduction of the verb alignment can also help in the execution of semantic tasks involving natural language processing, like text entailment and question answering, where concepts need to be mapped to a smaller set of categories in order to reduce the reasoning search space. DOLCE classes provide a suitable semantic representation for such tasks, and the evaluation has shown that, even with a straightforward word sense disambiguation technique, with the aid of the WN-DOLCE alignment it is possible to annotate text with a high accuracy. As future work, we intend to try more sophisticated WSD methods to improve the robustness of our semantic annotator, as well as start the analysis of the adjective and adverb databases to expand the alignment also to those synsets. Furthermore, this foundational ontology-based annotation tool will be integrated into a text entailment mechanism currently under development. This mechanism links the text T to the hypothesis H using dictionary definitions as intermediates, and try to determine whether H can be obtained from T by means of a sequence of transformation operations, performed over their foundational representations obtained through FO tagging.

Acknowledgments. This work is in part funded by the SSIX Horizon 2020 project (grant agreement No. 645425).

References

1. Ciaramita, M., Altun, Y.: Broad-coverage sense disambiguation and information extraction with a supersense sequence tagger. In: Proceedings of the 2006 Conference on Empirical Methods in Natural Language Processing, pp. 594–602. Association for Computational Linguistics (2006)
2. Ciaramita, M., Johnson, M.: Supersense tagging of unknown nouns in WordNet. In: Proceedings of the 2003 Conference on Empirical Methods in Natural Language Processing, pp. 168–175. Association for Computational Linguistics (2003)
3. Dahlgren, K.: A linguistic ontology. Int. J. Human-Comput. Stud. **43**(5), 809–818 (1995)
4. Daudé, J., Padró, L., Rigau, G.: Making WordNet mapping robust. Procesamiento del Lenguaje Natural **31** (2003)
5. Davidson, D.: The Logical Form of Action Sentences. University of Pittsburgh Press, Pittsburgh (1967)
6. Fellbaum, C.: WordNet. Wiley Online Library, New York (1998)
7. Finlayson, M.A.: Java libraries for accessing the Princeton WordNet: comparison and evaluation. In: Proceedings of the 7th Global WordNet Conference, Tartu, Estonia (2014)
8. Gangemi, A.: Norms and plans as unification criteria for social collectives. Auton. Agents Multi-Agent Syst. **17**(1), 70–112 (2008)
9. Gangemi, A., Guarino, N., Masolo, C., Oltramari, A.: Sweetening WordNet with DOLCE. AI Magazine **24**(3), 13–24 (2003)
10. Gangemi, A., Navigli, R., Velardi, P.: The OntoWordNet project: extension and axiomatization of conceptual relations in WordNet. In: Meersman, R., Tari, Z., Schmidt, D.C. (eds.) OTM 2003. LNCS, vol. 2888, pp. 820–838. Springer, Heidelberg (2003). doi:10.1007/978-3-540-39964-3_52
11. Gangemi, A., Nuzzolese, A.G., Presutti, V., Draicchio, F., Musetti, A., Ciancarini, P.: Automatic typing of DBpedia entities. In: Cudré-Mauroux, P., Heflin, J., Sirin, E., Tudorache, T., Euzenat, J., Hauswirth, M., Parreira, J.X., Hendler, J., Schreiber, G., Bernstein, A., Blomqvist, E. (eds.) ISWC 2012. LNCS, vol. 7649, pp. 65–81. Springer, Heidelberg (2012). doi:10.1007/978-3-642-35176-1_5
12. Guarino, N., Welty, C.A.: An overview of OntoClean. In: Staab, S., Studer, R. (eds.) Handbook on Ontologies, pp. 201–220. Springer, Heidelberg (2009)
13. Kučera, H., Francis, W.N., et al.: Computational Analysis of Present-Day American English. Brown University Press, Providence (1967)
14. Manning, C.D., Surdeanu, M., Bauer, J., Finkel, J.R., Bethard, S., McClosky, D.: The Stanford coreNLP natural language processing toolkit. In: ACL (System Demonstrations), pp. 55–60 (2014)
15. Martin, J.H., Jurafsky, D.: Speech and Language Processing: An Introduction to Natural Language Processing, Computational Linguistics and Speech Recognition. International Edition. Prentice-Hall (2000)
16. Masolo, C., Borgo, S., Gangemi, A., Guarino, N., Oltramari, A., Schneider, L.: The WonderWeb library of foundational ontologies. WonderWeb deliverable 18. Ontology Library (final) (2003)
17. McCarthy, D., Koeling, R., Weeds, J., Carroll, J.: Finding predominant word senses in untagged text. In: Proceedings of the 42nd Annual Meeting on Association for Computational Linguistics, p. 279. Association for Computational Linguistics (2004)

18. Miller, G.A.: WordNet: a lexical database for English. Commun. ACM **38**(11), 39–41 (1995)
19. Miller, G.A., Leacock, C., Tengi, R., Bunker, R.T.: A semantic concordance. In: Proceedings of the Workshop on Human Language Technology, pp. 303–308. Association for Computational Linguistics (1993)
20. Moldovan, D.I., Rus, V.: Logic form transformation of WordNet and its applicability to question answering. In: Proceedings of the 39th Annual Meeting on Association for Computational Linguistics, pp. 402–409. Association for Computational Linguistics (2001)
21. Nadeau, D., Sekine, S.: A survey of named entity recognition and classification. Lingvisticae Investigationes **30**(1), 3–26 (2007)
22. Niles, I., Pease, A.: Linking lexicons and ontologies: mapping WordNet to the suggested upper merged ontology. In: Proceedings of the IEEE International Conference on Information and Knowledge Engineering, pp. 412–416. IEEE (2003)
23. Pasca, M.A., Harabagiu, S.M.: High performance question/answering. In: Proceedings of the 24th Annual International ACM SIGIR Conference on Research and Development in Information Retrieval, pp. 366–374. ACM (2001)
24. Pease, A., Niles, I., Li, J.: The suggested upper merged ontology: a large ontology for the semantic web and its applications. In: Working Notes of the AAAI-2002 Workshop on Ontologies and the Semantic Web, vol. 28 (2002)
25. Pustejovsky, J., Castano, J., Sauri, R., Rumshinsky, A., Zhang, J., Luo, W.: Medstract: creating large-scale information servers for biomedical libraries. In: Proceedings of the ACL-2002 workshop on Natural language processing in the biomedical domain, vol. 3, pp. 85–92. Association for Computational Linguistics (2002)
26. Vendler, Z.: Linguistics in Philosophy. Cornell University Press, Ithaca (1967)

Locating Things in Space and Time: Verification of the SUMO Upper-Level Ontology

Lydia Silva Muñoz[1](✉) and Michael Grüninger[2]

[1] Computer Science, University of Toronto, 40 St. George Street, Toronto, ON M5S 2E4, Canada
silva@cs.toronto.edu

[2] Mechanical and Industrial Engineering, University of Toronto, 5 King's College Road, Toronto, ON M5S 3G8, Canada
gruninger@mie.utoronto.ca

Abstract. Upper-level ontologies provide an account of the most basic, domain independent entities, such as time, space, objects and processes. They are intended to be broadly reused, among others, during ontology engineering tasks, such as ontology building and integration. Ontology verification is the process by which a theory is checked to rule out its unintended models, and possibly characterize missing intended ones. In this paper, we translate into first-order logic, modularize, and verify the subtheory of location of entities in space and time of the Suggested Upper Merged Ontology (SUMO). As a result, we propose the addition of some axioms that rule out unintended models in SUMO, the correction of others, and make available a modularized version of SUMO characterization of location of entities in time and space represented in standard first-order logic.

Keywords: Spatial location · Time location · SUMO upper level ontology · Ontology verification · Ontology mapping

1 Introduction

Upper-level ontologies, also called *foundational* ontologies, provide an account of the most basic, domain-independent entities such as time, space, objects and processes. Upper-level ontologies are essential for the ontology engineering cycle in activities such as ontology building and integration. They can be used as the foundational substratum on which new ontologies are developed, because they provide some fundamental ontological distinctions, which can help the designer in her task of conceptual analysis, [6]. It is particularly important to understand the models of upper ontologies, because such an understanding makes their ontological commitments explicit. Ontology designers that create new domain-specific ontologies by extension of the upper ontology can then be aware of the ontology that they are using. Among others, upper level ontologies can also be used as oracles for meaning in ontology reconciliation [4].

E. Blomqvist et al. (Eds.): EKAW 2016, LNAI 10024, pp. 606–620, 2016.
DOI: 10.1007/978-3-319-49004-5_39

Ontology verification [5] is the process by which a theory is checked to rule out its unintended models, and possibly characterize missing intended ones. Ontologies which admit unintended models might cause misunderstandings that hinder interoperability because their vocabularies are ambiguously defined. Since foundational ontologies are expected to be broadly reused, their verification is necessary for their correct usage.

In this paper, after translating into standard first-order logic and modularizing the axiomatization of the Suggested Upper Merged Ontology (SUMO) that characterizes the location of entities in space and time, we analyze its ontological commitments, and verify its axiomatization. As a result, we propose the addition of some axioms that rule out unintended models, and the modification of others. We have used the automatic theorem prover Prover9 and model finder Mace4 [7] for the automatic tasks involved in the work that we describe.

2 Ontology Verification

An ontology admits unintended models when it is possible to find features of its conceptualization which are not characterized by its axiomatization. Related to the notion of *ontology verification* is the notion of *ontology mapping*, also called *ontology matching*, and *ontology alignment*, which is concerned with the explicit representation of the existing semantic correspondences among the axiomatizations of different ontologies[1] via *bridge axioms* [4], which are called *translation definitions* in the context of first-order logic. Building a map between two first-order logic ontologies T_1 and T_2 that interprets the first into the second involves translating every symbol of theory T_1 into the language of T_2, translating every sentence of T_1 into the language of T_2, and checking the ability of T_2 to entail every axiom of T_1. The following definition formalizes the notion of relative interpretation between first-order logic theories.

Definition 1. *A map π interprets a theory T_1 into a theory T_2 iff for every sentence α in the language of T_1, $T_1 \models \alpha \Rightarrow T_2 \models \alpha^{\pi}$; being α^{π} the syntactic translation of α into the language of T_2.*

Verifying an ontology T consists of identifying its unintended models and adding to its axiomatization the axioms that rule out those models. In practice, we verify ontologies by comparing the strength of their axiomatizations via *ontology mapping*. The following theorem that follows from [3], introduces a fundamental relation between the models of a theory and the models of the theories that it interprets. Given such a relation, in order to demonstrate that a given theory T_2 can represent every feature that another theory T_1 represents, it suffices to demonstrate that theory T_2 is able to interpret theory T_1.

[1] We assume that an ontology is a set of sentences called *axioms* closed under logical entailment that state the properties that characterize the behaviour of a set of symbols representing constants, relations and functions, called the *signature* of the ontology.

Theorem 1. *If a theory T_1 is interpreted by a theory T_2 by means of a given map π, there is another map δ that sends every model of T_2 into a model of T_1.*

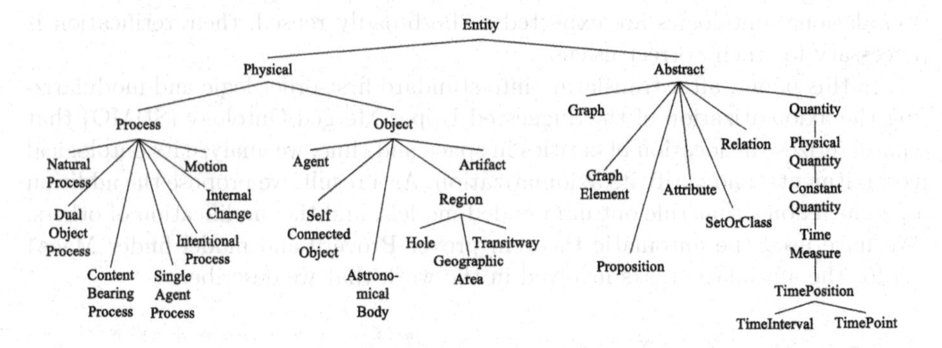

Fig. 1. Top categories of SUMO.

Once an alternative theory which characterizes the intended models of the theory under verification, or which at least characterizes intended features of the theory under verification is found, by building a map that interprets the alternative theory into the theory under verification we demonstrate that the theory under verification does actually characterize those intended features.

3 SUMO

SUMO [10] is a freely available upper level ontology whose partition of top categories is shown in Fig. 1. In addition to the main ontology, which contains about 4000 axioms, SUMO has been extended with a mid-level ontology and a number of domain specific ontologies, all of which account for 20,000 terms and 70,000 axioms. SUMO has been translated into OWL and WordNet [9].

The representation language of SUMO is SUO-KIF[2], a very expressive dialect of KIF[3] with many-sorted features, whose syntax permits higher-order constructions such as predicates that have other predicates, or formulas, as their arguments, and predicates and functions of variable arity [1]. We have built the set of modules used in this paper, after translating SUMO, with loss, into standard first-order logic.

4 SUMO Location of Entities in Time

In this section we review the theories that make possible the representation of the temporal location of entities in SUMO.

[2] http://suo.ieee.org/SUO/KIF/suo-kif.html.
[3] http://logic.stanford.edu/kif/kif.html.

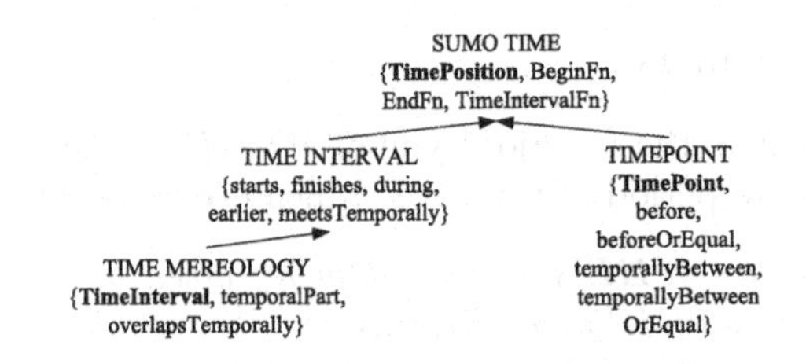

Fig. 2. Theory SUMO TIME. Arrows indicate conservative extensions among subtheories.

4.1 Time Representation

SUMO represents time by means of the subtheory TIME, whose structure of submodules is shown in Fig. 2. This theory was verified in [11]. It includes 3 subtheories:[4] TIME POINT, TIME MEREOLOGY, and TIME INTERVAL. Signature members are shown in the subtheory that introduces them in Fig. 2, and each module is a conservative extension[5] of each connected subtheory below it. These modules respectively characterize a linear ordering between instants of time using the *before* relation, a part-whole relation among intervals of time using the *temporalPart* relation, a mereotopology on intervals of time, and finally TIME accounts for a mereotopology among intervals and instants of time, where function $BeginFn(x)$ and $EndFn(x)$ respectively determine the begin and end time points of time interval x, and $TimeIntervalFn(x, y)$ returns the interval whose begin and end points are respectively x and y.

Definition 2. *Module **TIME** is the subtheory given by the axioms in colore.oor.net/ontologies/sumo/modules/sumo-time.*

4.2 Presence in Time

SUMO assumes that only entities of category *Physical* are present in time. Category *Physical*, shown in Fig. 1, is partitioned into categories *Object* and *Process*. The basic characterization of presence in time, is given in SUMO by subtheory PRESENT, where $time(x, t)$ means that entity x is present at time t.

Definition 3. *Module **PRESENT** is the subtheory given by axioms (1)–(4).*

$$(\forall x, y) time(x, y) \rightarrow Physical(x) \wedge TimePosition(y) \tag{1}$$

$$(\forall x, y) time(x, y) \rightarrow -(time(y, x)) \tag{2}$$

$$(\forall x) Physical(x) \rightarrow \exists y (TimePosition(y) \wedge time(x, y))) \tag{3}$$

$$\begin{aligned}(\forall o) Object(o) \rightarrow ((\exists t_1, t_2)(TimePoint(t_1) \wedge TimePoint(t_2) \wedge before(t_1, t_2) \\ \wedge (\forall t(beforeOrEqual(t_1, t) \wedge beforeOrEqual(t, t_2) \rightarrow time(o, t)))\end{aligned} \tag{4}$$

[4] Available at colore.oor.net/ontologies/sumo/modules.

[5] A theory T' is a *conservative extension* of a theory T if every theorem of T is a theorem of T', and every theorem of T' in the signature of T is also a theorem of T.

4.3 Lifetime of Things

Subtheory WHEN uses the function symbol *WhenFn* to specify the lifespan of objects and the entire period of time over which events occur.

Definition 4. *Module* ***WHEN*** *is the subtheory given by the union of axioms (5)–(7) to subtheories TIME and PRESENT.*[6]

$$(\forall x) Physical(x) \rightarrow TimeInterval(WhenFn(x)) \quad (5)$$

$$(\forall x, t) Physical(x) \rightarrow (temporalPart(t, WhenFn(x)) \leftrightarrow time(x, t)) \quad (6)$$

$$(\forall x, t)(time(x, t) \wedge TimePoint(t)) \leftrightarrow \\ temporallyBetweenOrEqual(BeginFn(WhenFn(x)), t, EndFn(WhenFn(x))) \quad (7)$$

To sum up, SUMO represents time by a rich theory of instants and intervals of time in which instants are organized by a linear ordering, while instants and intervals of time participate in a mereotopology. It is assumed that only entities of category *Physical* exist on time. Function symbol *WhenFn* returns the entire time period during which an entity of category *Physical* exists.

5 SUMO Location of Entities in Space

Based on a mereotopology that relates individuals of category *Object*, and part-whole relation *subProcess*, which relates the temporal parts of individuals of *Process*, SUMO develops the axiomatization that we have grouped in subtheories LOCATION, WHERE, and EVENT LOCATION for characterizing the spatial location of entities of category *Physical*. We review the axiomatization of these subtheories in this section.

5.1 Mereotopology

SUMO represents the notion of mereotopology based on a *ground mereology* [2,12] characterized for individuals of category *Object* by means of relation *part*, and represents the notion of connection among objects, meaning that the objects share at least one point of contact by means of relation *meetsSpatially*.

We have proposed in [8] both extending SUMO with axiom (14) to make possible for SUMO to admit models where every object is not necessarily in relation *part* with another object, and extending SUMO with axiom (22) to characterize the monotonicity of relation *connected* with regard to parthood.

Definition 5. *Module* ***EXTENDED MEREOTOPOLOGY*** *is the theory composed by axioms (8)–(23).*[7]

$$(\forall x, y) part(x, y) \rightarrow Object(x) \wedge Object(y) \quad (8)$$

[6] Axiom (4) is a consequence of axiom (6).

[7] SUMO also defines predicates *properPart*, *overlapsPartially*, and characterizes function symbols *MereologicalProductFn* and *MereologicalDifferenceFn*.

$$(\forall x)Object(x) \rightarrow part(x, x) \tag{9}$$

$$(\forall x, y)part(x, y) \wedge part(y, x) \rightarrow (x = y) \tag{10}$$

$$(\forall x, y, z)part(x, y) \wedge part(y, z) \rightarrow part(x, z) \tag{11}$$

$$(\forall x, y)overlapsSpatially(x, y) \leftrightarrow (\exists z(part(z, x) \wedge part(z, y))) \tag{12}$$

$$(\forall x, y)Object(x) \wedge Object(y) \rightarrow Object(MereologicalSumFn(x, y)) \tag{13}$$

$$\begin{aligned}(\forall x, y, z)Object(x) \wedge Object(y) \rightarrow ((z = MereologicalSumFn(x, y) \rightarrow \\ (\forall p)(part(z, p) \leftrightarrow part(x, p) \wedge part(y, p)))\end{aligned} \tag{14}$$

$$(\forall x)Object(x) \rightarrow connected(x, x) \tag{15}$$

$$(\forall x, y)connected(x, y) \rightarrow Object(x) \wedge Object(y) \tag{16}$$

$$(\forall x, y)connected(x, y) \rightarrow connected(y, x) \tag{17}$$

$$(\forall x, y)meetsSpatially(x, y) \rightarrow meetsSpatially(y, x) \tag{18}$$

$$(\forall x) - (meetsSpatially(x, x)) \tag{19}$$

$$(\forall x, y)connected(x, y) \leftrightarrow (meetsSpatially(x, y) \vee overlapsSpatially(x, y)) \tag{20}$$

$$(\forall x, y)overlapsSpatially(x, y) \rightarrow \neg meetsSpatially(x, y) \tag{21}$$

$$(\forall x, y)part(x, y) \rightarrow \forall z(connected(z, x) \rightarrow connected(z, y)) \tag{22}$$

$$\begin{aligned}(\forall x)SelfConnectedObject(x)) \leftrightarrow (\forall y, z)((MereologicalSumFn(y, z) = x) \\ \rightarrow connected(y, z))\end{aligned} \tag{23}$$

5.2 Spatial Location of Physical Entities

SUMO represents the location of elements of category *Physical* at elements of category *Object*, and also at the particular kind of objects of category *Region* by means of predicates *partlyLocated*, *located*, and *exactlyLocated*. According to SUMO documentation, $partlyLocated(x, y)$ means that entity x is at least partially located at entity y, $located(x, y)$ means that x is *partlyLocated* at y and there is no part of x that is not *located* at y, and, *exactlyLocated* is the actual, minimal location of an Object. Every entity of category *Physical* has in SUMO a spatial location.

Definition 6. *Module* ***PHYSICAL LOCATION*** *is the subtheory of SUMO composed by axioms (24) to (33).*

$$(\forall x, y)partlyLocated(x, y) \rightarrow Physical(x) \wedge Object(y) \tag{24}$$

$$(\forall x, y)Object(x) \wedge partlyLocated(x, y) \rightarrow overlapsSpatially(x, y) \tag{25}$$

$$(\forall x, y)Object(x) \wedge partlyLocated(x, y) \rightarrow (\exists s)(part(s, x) \wedge located(s, y)) \tag{26}$$

$$(\forall x, y)located(x, y) \rightarrow partlyLocated(x, y) \tag{27}$$

$$(\forall x, y)located(x, y) \wedge located(y, x)) \rightarrow (x = y) \tag{28}$$

$$(\forall x, y, z) located(x, y) \wedge located(y, z) \rightarrow located(x, z) \tag{29}$$

$$(\forall x, y) located(x, y) \rightarrow (\forall s)(part(s, x) \rightarrow located(s, y)) \tag{30}$$

$$(\forall y) Physical(y) \rightarrow (\exists l)(located(y, l)) \tag{31}$$

$$(\forall x, r) exactlyLocated(x, r) \rightarrow located(x, r) \tag{32}$$

$$(\forall x, r) exactlyLocated(x, r) \rightarrow \neg(\exists y)(exactlyLocated(y, r) \wedge \neg(y = x)) \tag{33}$$

5.3 WHERE

According to SUMO documentation, entity *Region* represents topographic locations, which encompasses surfaces of objects, imaginary places, and geographic areas. A region can be composed of parts that are not connected with one another, such as, archipelagos. In addition, a region is the only kind of object which can be located at itself, and empty regions are not allowed in SUMO. The subcategories of *Region* in SUMO are *GeographicArea*, *Transitway*, and *Hole*. Function *WhereFn* returns the spatial region where an object is *exactlyLocated* at a given instant of time.

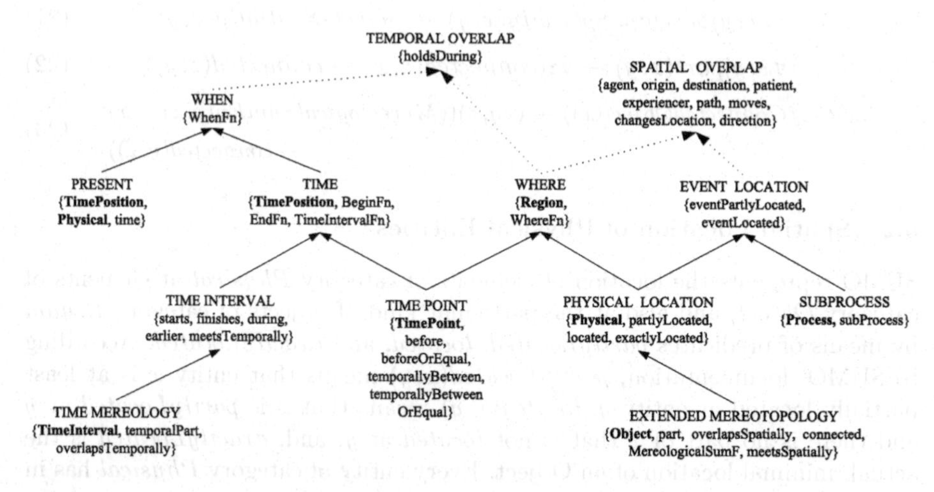

Fig. 3. Modularization of SUMO location of entities in space and time. Solid lines indicate conservative extensions, and dotted lines indicate non conservative extensions.

Definition 7. *Module* ***WHERE*** *is the theory given by the union of axioms (34) to (38) with theory PHYSICAL LOCATION.*[8]

$$(\forall x) Region(x) \rightarrow Object(x) \tag{34}$$

[8] Sentences (37) and (38) result from the translation into standard first-order logic of SUMO higher-order sentence $(\forall x, y, t)(WhereFn(x, t) = y) \leftrightarrow holdsDuring(t, exactlyLocated(x, y))$.

$$(\forall r)Region(r) \rightarrow (\exists y)(located(y, r)) \tag{35}$$

$$(\forall x, y)Physical(x) \wedge TimePoint(y) \rightarrow Region(WhereFn(x, y)) \tag{36}$$

$$\begin{aligned}(\forall x, y, t)Object(x) \wedge TimePoint(y) \\ \rightarrow ((WhereFn(x, t) = y) \rightarrow exactlyLocated(x, y) \wedge Region(y))\end{aligned} \tag{37}$$

$$(\forall x, y, z)exactlyLocated(x, y) \wedge exactlyLocated(x, z) \rightarrow (y = z) \tag{38}$$

We have found that, even though the documentation of SUMO states that "*Region* is the only kind of *Object* which can be located at itself", the formalization of such a condition is not a theorem of SUMO, therefore it admits unintended models where objects that are not regions are actually located at themselves. The following proposition proves our claim. We propose the extension of SUMO with axiom (39) to rule out those unintended models.

Proposition 1. WHERE $\not\models (\forall x)located(x, x) \rightarrow Region(x)$.

Proof. Let S_1 be the union of theory WHERE with sentence $(\exists x)located(x, y) \wedge \neg Region(x)$. Using Mace4 we have built a model of S_1 (See footnote 10). □

$$(\forall x)located(x, x) \rightarrow Region(x) \tag{39}$$

We have also found that SUMO does not enforce that each region *must* be *exactlyLocated* at itself. However, this introduces models where the transitivity of *location* leads to unexpected conclusions. For example, let us suppose that $WhereFn(x, t_1) = y$, and $WhereFn(y, t_2) = z$, due to axioms (37), (32) and (29) it results $located(x, z)$, which should not be concluded as holding at t_1, neither at t_2. This result was an outcome of not enforcing that the exact location of argument y must be itself. The following proposition proves our claim. In order to rule out those unintended models, we propose the addition of axiom (40) to SUMO, and the substitution of axiom (37) by axiom (41).

Proposition 2. WHERE $\not\models (\forall x)Region(x) \rightarrow exactlyLocated(x, x)$.

Proof. Let S_1 be the union of theory WHERE with sentence $(\exists x)Region(x) \wedge \neg exactlyLocated(x, x)$. Using Mace4 we have built a model of S_1 (See footnote 8). □

$$(\forall x)Region(x) \rightarrow exactlyLocated(x, x) \tag{40}$$

$$\begin{aligned}(\forall x, y, t)Object(x) \wedge TimePoint(y) \rightarrow (WhereFn(x, t) = y) \\ \rightarrow exactlyLocated(x, y) \wedge Region(y) \wedge (Region(x) \rightarrow (y = x))\end{aligned} \tag{41}$$

We have found that SUMO admits unintended models where a member of *SelfConnectedObject*[9] can be *exactlyLocated* at a region that is not itself a *SelfConnectedObject*. The following proposition proves our claim. We propose the extension of SUMO with axiom (42) to rule out those unintended models.

[9] Category *SelfConnectedObject* is defined in axiom (23).

Proposition 3. WHERE $\not\models$ $(\forall x,y) exactlyLocated(x,y) \wedge SelfConnectedObject(x) \rightarrow SelfConnectedObject(y)$.

Proof. Let S_1 be the union of theory WHERE with sentence $(\exists x,y) exactlyLocated(x,y) \wedge SelfConnectedObject(x) \wedge \neg SelfConnectedObject(y)$. Using Mace4 we have built a model of S_1.[10] □

$$(\forall x,y) exactlyLocated(x,y) \wedge SelfConnectedObject(x) \rightarrow SelfConnectedObject(y) \tag{42}$$

We have also found that SUMO does not enforce that the parts of an object are *exactlyLocated* at parts of the object in which it is *exactlyLocated*, which leads to the admission of models where this property does not hold. The following proposition proves our claim. We propose the extension of SUMO with axiom (43) to rule out those unintended models.

Proposition 4. WHERE $\not\models$ $(\forall x,y,z,w) part(x,y) \wedge exactlyLocated(x,z) \wedge exactly\ Located(y,w) \rightarrow part(z,w)$.

Proof. Let S_1 be the union of theory WHERE with sentence $(\exists x,y,z,w) part(x,y) \wedge exactlyLocated(y,z) \wedge exactlyLocated(y,w) \wedge \neg part(z,w)$. Using Mace4 we have built a model of S_1 (See footnote 10). □

$$(\forall x,y,z,w) part(x,y) \wedge exactlyLocated(x,z) \wedge exactlyLocated(y,w) \rightarrow part(z,w) \tag{43}$$

We have found that SUMO does not enforce that objects must be *partlyLocated* at the parts of the region where they are *exactlyLocated*. We propose the extension of SUMO with axiom (44) to rule out the involved unintended models.

Proposition 5. WHERE $\not\models$ $(\forall x,y,z) part(x,y) \wedge exactlyLocated(z,y) \rightarrow partlyLocated(z,x)$.

Proof. Let S_1 be the union of theory WHERE with sentence $(\exists x,y,z) part(x,y) \wedge exactlyLocated(z,y) \wedge \neg partlyLocated(z,x)$. Using Mace4 we have built a model of S_1 (See footnote 10).

$$(\forall x,y,z) part(x,y) \wedge exactlyLocated(z,y) \rightarrow partlyLocated(z,x) \tag{44}$$

We have found that SUMO does not define relation *partlyLocated* for objects, but just characterizes it by means of axiom (26). This introduces models where an object that has a part *exactlyLocated* in another object is actually not *partlyLocated* at that object. In addition, there exist models where in spite of an object x being *partlyLocated* at an object y, no part of x is *exactlyLocated* at object y. In order to rule out those unintended models, we propose the substitution of axiom (26) by axiom (45) in SUMO.

[10] Proof available at: colore.oor.net/ontologies/sumo/location/proofs.

Proposition 6. WHERE $\not\models (\forall x, y)Object(x) \rightarrow (partlyLocated(x, y) \leftrightarrow (\exists s)(part(s, x) \wedge exactlyLocated(s, y)))$.

Proof. Let S_1 be the union of theory WHERE with sentence $(\forall x, y)Object(x) \rightarrow (partly\ Located(x, y) \wedge (\forall s)(\neg part(s, x) \vee \neg exactlyLocated(s, y))$. Using Mace4 we have built a model of S_1 (See footnote 10). Let S_2 be the union of theory WHERE with sentence $(\exists x, y, s)Object(x) \wedge part(s, x) \wedge exactlyLocated(s, y) \wedge \neg(partlyLocated(x, y)$. Using Mace4 we have built a model of S_2 (See footnote 10). □

$$(\forall x, y)Object(x) \rightarrow (partlyLocated(x, y) \leftrightarrow (\exists s)(part(s, x) \wedge exactlyLocated(s, y))) \tag{45}$$

Finally, we have found that SUMO does not define *Located* for objects, but just characterizes it by means of axiom (30). This introduces models where an object that is *exactlyLocated* at a part of other object is actually not *located* at that object. In addition, there exist models where in spite of an object x being *located* at an object y, there is no part of y where object x is *exactlyLocated*. In order to rule out those unintended models, we propose the substitution of axiom (30) by axiom (46) in SUMO.

Proposition 7. WHERE $\not\models (\forall x, y)Object(x) \rightarrow (located(x, y) \leftrightarrow (\exists z)(part(z, y) \wedge exactlyLocated(x, z)))$.

Proof. Let S_1 be the union of theory WHERE with sentence $(\exists x, y)Object(x) \wedge located(x, y) \wedge \forall z(part(z, y) \rightarrow \neg exactlyLocated(x, z))$. Using Mace4 we have built a model of S_1 (See footnote 10). Let S_2 be the union of theory WHERE with sentence $(\exists x, y)Object(x) \wedge \forall z(part(z, y) \wedge exactlyLocated(x, z) \rightarrow \neg located(x, y))$. Using Mace4 we have built a model of S_2 (See footnote 10). □

$$(\forall x, y)Object(x) \rightarrow (located(x, y) \leftrightarrow (\exists z)(part(z, y) \wedge exactlyLocated(x, z))) \tag{46}$$

To sum up, function $WhereFn$ returns the spatial region where an entity is *exactlyLocated* at a given instant of time in SUMO, and from that follows that the object is also *located* and *partlyLocated* at such a place. However, since properties *located* and *partlyLocated* are not indexed by a temporal parameter, SUMO can not represent changes on the location of individuals of category object in a systematic way.[11]

[11] Only objects of category $Vehicle$ participating as an *instrument* of a $Translocation$ process have their location related to the objects declared as the *origin* and *destination* of the process respectively at the time where the process begins and ends by means of a higher-order construct, which represents a change of location for objects of category $Vehicle$.

5.4 Spatial Location of Events

SUMO provides a specific characterization for the location of individuals of category *Process* by means of predicates *eventPartlyLocated* and *eventLocated*, which, according to SUMO documentation, respectively represent that a temporal part of the process, or the entire process is located at the indicated object. The following definitions formalize the characterization of the terms *subProcess*, *eventPartlyLocated* and *eventLocated*.

Definition 8. SUMO SUBPROCESS *is the theory given by axioms (47)–(50).*

$$(\forall x, y) subProcess(x, y) \rightarrow Process(x) \wedge Process(y) \tag{47}$$

$$(\forall x) Process(x) \rightarrow subProcess(x, x) \tag{48}$$

$$(\forall x, y) subProcess(x, y) \wedge subProcess(y, x) \rightarrow (x = y) \tag{49}$$

$$(\forall x, y, z) subProcess(x, y) \wedge subProcess(y, z) \rightarrow subProcess(x, z) \tag{50}$$

Definition 9. EVENT LOCATION *is the theory given by axioms (54)–(59).*[12]

$$(\forall x) Physical(x) \leftrightarrow (Process(x) \vee Object(y)) \wedge (Process(x) \rightarrow \neg Object(y)) \tag{51}$$

$$(\forall x)(Physical(x) \leftrightarrow (Object(x) \vee Process(x)) \wedge Process(x) \rightarrow (-Object(x))) \tag{52}$$

$$(\forall x, y) eventPartlyLocated(x, y) \wedge eventPartlyLocated(y, x) \rightarrow (y = x) \tag{53}$$

$$(\forall x, y) eventPartlyLocated(x, y) \rightarrow Process(x) \wedge Object(y) \tag{54}$$

$$(\forall p, l) eventPartlyLocated(p, l) \rightarrow ((\exists s)(subProcess(s, p) \wedge eventLocated(s, l))) \tag{55}$$

$$(\forall x, y) eventPartlyLocated(x, y) \rightarrow \neg eventPartlyLocated(y, x) \tag{56}$$

$$(\forall x, y) eventPartlyLocated(x, y) \rightarrow partlyLocated(x, y) \tag{57}$$

$$(\forall x, y) eventLocated(x, y) \rightarrow eventPartlyLocated(x, y) \tag{58}$$

$$(\forall p, l) eventLocated(p, l) \rightarrow ((\forall s)(subProcess(s, p) \rightarrow eventLocated(s, l))) \tag{59}$$

We have found that there exist models in SUMO where there are processes whose temporary parts are *eventPartlyLocated* at certain objects, but the whole processes themselves are not *eventPartlyLocated* at those objects. The following proposition proves our claim. We propose the extension of SUMO with axiom (60) to rule out those unintended models.

Proposition 8. EVENT LOCATION $\not\models$ $(\forall x, y, z) subProcess(x, y) \wedge eventPartlyLocated(x, z) \rightarrow eventPartlyLocated(y, z)$.

Proof. Let S_1 be the union of theory EVENT LOCATION with sentence $(\exists x, y, z)$ $subProcess(x, y) \wedge eventPartlyLocated(x, z) \wedge \neg eventPartlyLocated$ (y, z). Using Mace4 we have built a model of S_1.[13] □

[12] Axiom (56) follows from the fact that *Object* and *Process* are disjoint categories.
[13] Proof available at: colore.oor.net/ontologies/sumo/location/proofs.

$$(\forall x, y, z) subProcess(x, y) \wedge eventPartlyLocated(x, z) \rightarrow eventPartlyLocated(y, z) \quad (60)$$

In SUMO, every object that has any degree of participation in a process is in relation *involvedInEvent* with the process, and every process that has a *path*, *origin*, or *destination*, which are kinds of *involvedInEvent*, is declared to be partly located at all of them. A *path* relates a motion process x with the object that is the "route along which" the motion occurs, while *origin*, and *destination* account for the locations where a process begins and ends.[14,15]

5.5 Spatio-Temporal Overlap of Related Entities

SUMO characterizes the fact that related individuals of category *Physical* must overlap in space and time. By means of the axiomatization of module TEMPORAL OVERLAP, SUMO guarantees that there exist at least an interval of time in which physical entities coexist with the objects representing their spatial location, and with any other entity with which they are related by part-whole or topological relations.

Definition 10. *Module* ***TEMPORAL OVERLAP*** *is the subtheory given by axioms (61) and (62).*[16,17]

$$(\forall x, y) Physical(x) \wedge Object(y) \wedge (part(x, y) \vee connected(x, y) \vee partlyLocated(x, y)) \rightarrow overlapsTemporally(WhenFn(x), WhenFn(y)) \quad (61)$$

$$(\forall s, p) subProcess(s, p) \rightarrow temporalPart(WhenFn(s), WhenFn(p)) \quad (62)$$

---

[14] Predicates *path*, *origin*, and *destination* are characterized in SUMO as antisymmetric relations whose first and second arguments are respectively individuals of *Process* and *Object*.

[15] We have found that the following SUO-KIF axiom of SUMO, intended to further characterize predicate *origin*, is syntactically incorrect and incomprehensible. We assume that it is a typo.

```
(=> (origin?PROCESS?OBJ)
(eventLocated(WhereFn?PROCESS(BeginFn(WhenFn?PROCESS)))
(WhereFn?OBJ(BeginFn(WhenFn?OBJ)))))
```

[16] The instantiation of the following schema for each binary predicate *rel* of SUMO also contributes to characterizing the temporal overlap of related entities: $(\forall x, y, t) holdsDuring(t, rel(x, y)) \wedge Physical(x) \wedge Physical(y) \rightarrow time(x, t) \wedge time(y, t)$.

[17] Axiom (61) results from the translation into first-order logic of SUO-KIF sentence $(\forall x, y, rel)(BinaryPredicate(rel) \wedge SpatialRelation(rel) \wedge rel(x, y)) \rightarrow overlapsTemporally(WhenFn(x), WhenFn(y))$.

Module SPATIAL OVERLAP guarantees that for at least one instant of time, objects share a location in space with the processes located at them, and with the processes in which they participate. However, SUMO does not represent the actual time during which the participant object and the process were located at the same place, which could be achieved if relations *eventPartlyLocated*, *agent*, *origin*, *destination*, *patient*, *experiencer*, *path*, *moves*, and *changesLocation* were indexed by a temporal parameter.

Definition 11. *Module* ***SPATIAL OVERLAP*** *is the subtheory given by axiom (63).*[18]

$$\begin{aligned}(\forall x,p)(eventPartlyLocated(p,x) \vee agent(p,x) \vee origin(p,x)\\ \vee destination(p,x) \vee patient(p,x) \vee experiencer(p,x) \vee path(p,x)\\ \vee moves(p,x) \vee changesLocation(p,x) \vee direction(p,x)) \wedge Object(x) \rightarrow\\ ((\exists t) verlapsSpatially(WhereFn(p,t),x))\end{aligned} \tag{63}$$

6 Verification Methodology

The methodology used during the verification carried out in this work consisted of deeply analyzing those SUMO theories that represent location of entities in space and time, and also other ontologies that represent the same subject. We compare the ontologies axiomatizations, and determine, based in SUMO axiomatization and documentation, to what extent SUMO is intended, and is able, to represent the commitments of those theories. We have compared SUMO with the theory proposed in [2], and found that SUMO is intended, and by adding some missing axioms also capable, of representing the commitment of such a theory, fact that we demonstrate via ontology mapping.

Let theory $T^{Location}$ be the theory WHERE plus every axiom proposed for extending SUMO in Sect. 5.3. Let $\mathfrak{M}^{intended}$ be the class of intended models of theory $T^{Location}$. We have studied the relation of $T^{Location}$ with the theory given by axioms (64)–(69) from [2], which we define below as $T_{C\&V}$. Although theory $T_{C\&V}$ does not characterize all the features of the class $\mathfrak{M}^{intended}$, given the axiomatization, and documentation of SUMO, every feature that $T_{C\&V}$ characterizes is a feature that the models of class $\mathfrak{M}^{intended}$ should have. Therefore, we have verified the theory WHERE by proving that theory $T^{Location}$ interprets theory $T_{C\&V}$. By proving that, we demonstrate that $T^{Location}$, i.e., SUMO plus its identified missing axioms, is able to characterize all the features of $T_{C\&V}$.

Definition 12. $T_{C\&V}$ *is the theory given by axioms (64)–(69).*

$$(\forall x,y) L(x,y) \rightarrow L(y,y) \tag{64}$$

$$(\forall x,y,z,w) P(x,y) \wedge L(x,z) \wedge L(y,w) \rightarrow P(z,w) \tag{65}$$

[18] Axiom (63) results from the translation into first-order logic of SUO-KIF sentence $(\forall x, rel, p)(CaseRole(rel) \wedge Object(x) \wedge rel(p,x)) \rightarrow ((\exists t) overlapsSpatially(WhereFn(p,t),x))$.

$$(\forall x, y, z, w) C(x, y) \wedge L(x, z) \wedge L(y, w) \rightarrow C(z, w) \tag{66}$$

$$(\forall x, y, z) P(x, y) \wedge L(z, y) \rightarrow PL(z, x) \tag{67}$$

$$(\forall x, y) PL(x, y) \leftrightarrow (\exists z)(P(z, x) \wedge L(z, y)) \tag{68}$$

$$(\forall x, y) WL(x, y) \leftrightarrow (\exists z)(P(z, y) \wedge L(x, z)) \tag{69}$$

Table 1. Translation definitions from $T_{C\&V}$ to $T^{Location}$

$$(\forall x, y) L(x, y) \leftrightarrow Region(y) \wedge exactlyLocated(x, y) \tag{70}$$

$$(\forall x, y) P(x, y) \leftrightarrow part(x, y) \tag{71}$$

$$(\forall x, y) C(x, y) \leftrightarrow connected(x, y) \tag{72}$$

$$(\forall x, y) PL(x, y) \leftrightarrow Object(x) \wedge partlyLocated(x, y) \tag{73}$$

$$(\forall x, y) WL(x, y) \leftrightarrow Object(x) \wedge located(x, y) \tag{74}$$

Theorem 2. *Theory* $T^{Location}$ *interprets theory* $T_{C\&V}$.

Proof. Let us call Δ to the set of translation definitions shown in Table 1. Using Prover9 we have shown that $T^{Location} \cup \Delta \models T_{C\&V}$ (See footnote 13).

7 Conclusions

We have translated into standard first-order logic and modularized the subtheory of the Suggested Upper Merged Ontology (SUMO) that characterizes the location of entities in space and time. We have analyzed its ontological commitments, and verified its axiomatization. As a result, we have proposed the extension of SUMO with a series of axioms that rule out the unintended models that we have identified.

References

1. Benzmüller, C., Pease, A.: Higher-order aspects and context in SUMO. J. Web Sem. **12**, 104–117 (2012)
2. Casati, R., Varzi, A.C.: Parts and Places: The Structures of Spatial Representation. A Bradford book. MIT Press, Cambridge (1999)
3. Enderton, H.B.: A Mathematical Introduction to Logic. Academic Press, Cambridge (1972)

4. Euzenat, J., Shvaiko, P.: Ontology Matching, 2nd edn. Springer, Heidelberg (DE) (2013)
5. Grüninger, M., Hahmann, T., Hashemi, A., Ong, D.: Ontology verification with repositories. In: Formal Ontology in Information Systems, Proceedings of the Sixth International Conference, FOIS 2010, Toronto, Canada, 11–14 May 2010, pp. 317–330 (2010)
6. Guarino, N.: Formal ontology and information systems. In: Formal Ontology in Information Systems - Proceedings of FOIS 1998, Trento, Italy, 6–8 June 1998, pp. 3–15. IOS Press, Amsterdam (1998)
7. McCune, W.: Prover9 and Mace4 (2005–2010). http://www.cs.unm.edu/~mccune/prover9/
8. Silva Muñoz, L., Grüninger, M.: Verifying and mapping the mereotopology of upper-level ontologies. In: Knowledge Engineering and Ontology Development (KEOD), Proceedings of the 8th International Joint Conference on Knowledge Discovery, Knowledge Engineering and Knowledge Management (IC3K), 9–11 November 2016, Porto, Portugal (2016)
9. Niles, I., Pease, A.: Linking lexicons and ontologies: mapping WordNet to the suggested upper merged ontology. In: Proceedings of the 2003 International Conference on Information and Knowledge Engineering (IKE 2003), Las Vegas, Nevada (2003)
10. Niles, I., Pease, A.: Towards a standard upper ontology. In: FOIS 2001: Proceedings of the International Conference on Formal Ontology in Information Systems, pp. 2–9. ACM, New York (2001)
11. Silva Muñoz, L., Grüninger, M.: Mapping and verification of the time ontology in SUMO. In: Formal Ontology in Information Systems - Proceedings of the 9th International Conference, FOIS, 6–9 July 2016, Annecy, France (2016)
12. Achille Varzi, C.: Handbook of spatial logics, chapter spatial reasoning, ontology: parts, wholes, and locations. In: Aiello, M., Pratt-Hartmann, I., Van Benthem, J. (eds.) Handbook of Spatial Logics, pp. 945–1038. Springer, Dordrecht (2007)

Detecting Meaningful Compounds in Complex Class Labels

Heiner Stuckenschmidt, Simone Paolo Ponzetto(✉), and Christian Meilicke

Data and Web Science Group, University of Mannheim, Mannheim, Germany
{heiner,simone,christian}@informatik.uni-mannheim.de

Abstract. Real-world ontologies such as, for instance, those for the medical domain often represent highly specific, fine-grained concepts using complex labels that consist of a sequence of sublabels. In this paper, we investigate the problem of automatically detecting meaningful compounds in such complex class labels to support methods that require an automatic understanding of their meaning such as, for example, ontology matching, ontology learning and semantic search. We formulate compound identification as a supervised learning task and investigate a variety of heterogeneous features, including statistical (i.e., knowledge-lean) as well as knowledge-based, for the task at hand. Our classifiers are trained and evaluated using a manually annotated dataset consisting of about 300 complex labels taken from real-world ontologies, which we designed to provide a benchmarking gold standard for this task. Experimental results show that by using a combination of distributional and knowledge-based features we are able to reach an accuracy of more than 90 % for compounds of length one and almost 80 % for compounds of length two. Finally, we evaluate our method in an extrinsic experimental setting: this consists of a use case highlighting the benefits of using automatically identified compounds for the high-end semantic task of ontology matching.

1 Introduction

Conceptual models of information structures and information flows are a central concept in computer science. They play a crucial role in the design and maintenance of information systems. Besides the classical tasks of creating and evolving conceptual models, the task of identifying mappings between different models as a basis for integrating different systems has become more and more important. The problem of integrating different representations of reality is a long-standing problem in computer science. In particular, it is the core problem of the field of data integration. The database community has developed a variety of methods for identifying matching data elements both on the level of instance and schema data [7]. More recently, the problem of matching elements from different ontologies, namely formal models of an application domain, has been investigated in detail [36]. It has been argued that many matching techniques developed for schema matching can also be applied to ontology matching. However, questions

E. Blomqvist et al. (Eds.): EKAW 2016, LNAI 10024, pp. 621–635, 2016.
DOI: 10.1007/978-3-319-49004-5_40

remain on whether further advances could be achieved by leveraging the formal semantics of ontologies.

Despite much research work in the field, existing approaches to ontology matching still have a number of limitations. For instance, almost all existing methods produce simple one-to-one matches between elements in the representations to be compared [10]. That is, most existing systems rely on the naive assumption that the representations to be compared represent reality at the very same level of granularity. A particular problem that can be observed when trying to match models that describe the domain at different levels of abstraction are situations where the class names describe complex constructs that do not have a direct counterpart in the other model, but their intended meaning can be expressed (or at least approximated) by a logical expression over simpler elements [37]. A complete solution to this problem amounts to developing novel, full-fledged methodologies to ontology matching that cover arbitrary one-to-many mappings. While we envision this as a longer-term goal requiring substantial research efforts, in this paper we provide a first step towards such a solution by addressing the problem of *understanding complex class labels*. More specifically, we focus on the task of *identifying meaningful compounds in complex ontology labels* that might refer to independent classes in a differently structured ontology. This is, to the best of our knowledge, the first attempt to address in detail this problem, which bears nevertheless a strong resemblance with other well-known tasks in Natural Language Processing and Information Retrieval – e.g., syntactic disambiguation of multiword expressions (also known as noun compound bracketing) [2] and query segmentation [3,15, *inter alia*].

1.1 Problem Definition

Real-world ontologies, e.g., those providing semantic models of a highly specialized domain such as the medical one, often provide a description of their fine-grained concepts by means of complex labels that typically require some knowledge of the domain to make sense of. As an example, let us focus on the concept label *natural killer cell receptor 2B4*, which can be found in the Gene Ontology [1]. This label shows properties typical of complex ontology labels. Note that with 'complex' we refer here to the fact that the noun compound exhibits both syntactic and semantic ambiguity. That is, the label could be interpreted in different ways, depending on how its internal syntax is disambiguated. For instance, looking at the first four tokens of our example label, we see that there are at least three ways in which it could be bracketed, and thus interpreted

(1) [natural killer] [cell receptor]
(2) [natural] [killer cell receptor]
(3) [natural killer cell] [receptor]

The first interpretation would be that the label describes the cell receptor of a natural killer. Clearly, this is for humans a quite implausible interpretation of the intended meaning of the label. Nevertheless, the two other possible bracketings

provide us with two equally plausible interpretations, which are both hard to rank as preferred interpretation, even by human subjects. The second possible interpretation, in fact, identifies the natural form of a killer cell receptor, whereas the third one the receptor of a natural killer cell – which is actually the correct interpretation, since 'natural killer cell' is a technical term in immunology. Note that, at a closer look, for semantic applications – such as, for instance, mapping label constituents (i.e., substring) to another resource – we need in practice a task formulation that goes beyond simple bracketing of adjacent noun phrases. First of all, meaningful parts of a label can actually overlap. In our example, these are 'natural killer cell' and 'cell receptor'. The term 'cell' is part of both components and links the two concepts to each other. Beyond that, there are also cases where meaningful compounds consist of terms that are not adjacent in the label. An example is the label 'British Crown colony' where all combinations of terms actually identify a meaningful concept: (i) the 'British Crown', which is in charge of the colony, (ii) 'Crown colony' indicating the property of the colony as belonging to a kingdom, and (iii) 'British Colony', which describes that the colony is or was owned by great Britain.
To provide a workable problem definition, we define criteria for recognizing a meaningful compound within a complex label as follows:

Definition 1. *Given a complex concept label* $l = (l_1, \cdots, l_n)$ *a compound in* l *is a subsequence* $s = (s_1, \cdots, s_m)$ *of* l *where* $m < n$. *A compound* s *in* l *is meaningful if*

- s *is a grammatically correct noun phrase,*
- s *can be the label of a possible concept in some ontology,*
- s *retains a meaningful relation to* l.

Rather than providing a general or exhaustive solution, this definition is inspired by the intended application to ontology matching (Sect. 5). Since the ultimate objective is to find semantic relations to other ontologies, we are interested in parts of the label that can be found as concept labels in other ontologies (requirement 2). Clearly, we are only interested in those concepts that play some part in a complex mapping, and thus have some relation to the complex label (requirement 3). Admittedly, the definition is not unambiguous, so we must rely on human annotations as a reference (Sect. 4.1).

1.2 Contributions

In this paper, we investigate the problem of automatically detecting meaningful compounds in complex class labels as a first step towards complex ontology matching. The contributions of this paper are the following:

- We propose a supervised approach for recognizing meaningful compounds in complex ontology labels[1].

[1] In this work, we focus primarily on labels of length 3: however, our approach can be used in principle with labels of arbitrary lengths.

- We investigate different sets of statistical and knowledge-based features as a basis for the learning approach.
- We create a manually annotated benchmark dataset consisting of about 300 complex three-word labels taken from real-world ontologies.
- We show that, thanks to a combination of statistical and knowledge-based features, we can reach an accuracy of about 90 % for compounds of length one and about 80 % for compounds of length two.
- Based on the results of the experiments, we propose an unsupervised approach for detecting meaningful compounds in labels of arbitrary length.

2 Related Work

Label Analysis. Recently, there has been initial work addressing the analysis and use of complex labels for ontology enrichment and semantic matching. Manaf and others report results of a large scale analysis of the structure of class names on the Semantic Web [21]. They conclude that almost 90 % of all class labels resp. identifiers on the semantic web are actually meaningful in that they provide a natural language description of the intended meaning of the class. More than 96 % of these labels consists of more than one word. Further, they report that complex labels can be parsed syntactically as most labels use camel case syntax or special separators to delimit single words. In our previous work, we have used patterns over linguistic features generated through part-of-speech tagging, syntactic parsing and lexical semantic analysis to detect complex mappings between ontologies [30,31]. In the area of business process modeling, Mendling and others have developed a method for analyzing activity labels based on different modeling styles observed in real world models [20,22]. Other researchers focused instead on domain-specific resources ranging from biomedical ontologies like the Gene Ontology [11] and those found on BioPortal [27], all the way through identifiers found in source code [8].

Noun Phrase Chunking and Compound Bracketing. Two related problems from the field of Natural Language Processing (NLP) are text chunking (also referred to as shallow parsing) and noun compound bracketing. In contrast to full syntactic parsing, text chunking is concerned with the identification of flat, non-overlapping segments of a sentence which identify its basic non-recursive phrases corresponding to major parts-of-speech such as noun, verb and prepositional phrases. Noun phrase chunking is the special problem of identifying basic noun phrases within sentences. Due to the tight relation to full parsing, early approaches relied on established parsing methods [29]. Major advances were made thanks to the organization of a shared task as part of the Conference on Natural Language Learning in 2000 [33]. The participating systems reached an accuracy of over 90 %, with the best performance being reported for a supervised approach based on Support Vector Machines [18]. Further advances were later achieved using better statistical approaches to tagging such as, for instance, Conditional Random Fields [35]. While our task is similar to the chunking problem, identifying 'chunks' in class labels is much harder as labels typically do not have a regular grammatical structure. Similarly, meaningful compound identification is related to the other NLP task of noun compound bracketing, namely the syntactic disambiguation of multiword expressions [2]. For this task the best-performing models are based on a variety of different syntactic and semantic features [24,39]. But while these contributions provide us with useful hints as to which kind of features we need for the task at hand (e.g., N-gram statistics), bracketing is primarily meant as a phrase-internal parsing task: that is, it does not cover cases of meaningful non-adjacent compounds.

Query Segmentation. A problem that is actually closer to our task is that of segmenting web search queries. Keyword queries, in fact, show similarities with class labels as they typically do not have a regular grammatical structure and are often composed of different meaningful compounds (e.g. 'New York budget hotels'). Bergsma and Wang showed that a combination of statistical and linguistic features can be used to learn optimal segmentations from examples with an accuracy ranging between 85 % and 90 % [3]. The results were obtained on a set of 1500 queries sampled from the AOL search query database, a corpus of more than 35 million queries. Zhang and others proposed an unsupervised approach that makes extensive use of background resources like WordNet and Wikipedia to detect potential segments, and applied it to the robust and ad-hoc tracks of TREC reporting good results [41]. However, due to the task-based evaluation approach they opted for, it is not possible to compare their results to the supervised approach of Bergsma and Wang. More recently, Hagen et al. have proposed in [14] a rather light-weight query segmentation method that mostly relies on N-gram statistics from the Google N-gram Corpus [4]. In follow-up work, they show that giving preference to segments that correspond to Wikipedia titles further improves the results [15]. The results reported by Hagen et al. are in the same range as the ones reported by Bergsma and Wang, thus showing that unsupervised approaches can also be competitive.

3 Learning to Detect Meaningful Compounds

We present a method for automatically determining meaningful compounds in complex class labels. Our approach builds upon existing techniques for query segmentation, which are, however, adapted to our specific problem. Following Bergsma and Wang, we propose a supervised approach, and focus in this first initial attempt to explore in detail the feature space for the task at hand.

3.1 Approach

Successful approaches to query segmentation detect segment boundaries based on different features of the neighboring words or, in the case of the unsupervised approach of Hagen et al. [14,15], based on features of all words in the query. This approach does not work for us, as we want to consider all word combinations in a complex label. We solve this problem by regarding each possible word combination as the binary decision problem of determining whether the respective word combination is a meaningful compound or not, and learn a decision function using supervised learning methods. That is, given a concept label, our task is to consider all proper subsequences and decide for each of them whether they are meaningful or not (along the lines of Definition 1). We train the classifier using a wide range of different features. While many features are taken directly from previous work on query segmentation, we go one step further by adding a number of new features more specifically targeted to capture the nature of ontology class labels. We finally arrived at a set of about 80 individual features from different categories, which we now turn to describe in detail.

3.2 Features

Statistical Features: Building on the results of Hagen et al. that show the benefits of N-gram statistics for query segmentation, we use statistical features from large corpora, more specifically the N-gram-based scores for segments (same as proposed by Hagen et al.), as well as features capturing the distributional similarity and relations between words occurring within a label.

Features Based on N-gram Statistics. In [14] the authors propose a measure to estimate the quality of a complete segmentation of a keyword query based on the number of occurrences of a possible segment, normalized by the length of the segment (to account for the power law distribution of N-grams on the web):

$$score(S) = \sum_{s \in S, |s| \geq 2} |s|^{|s|} \cdot count(s)$$

Here S is the complete segmentation consisting of individual segments $s \in S$. Thus the score of a segment is given by $|s|^{|s|} \times count(s)$ where $count(s)$ is the number of occurrences in the N-gram corpus. We use this segment score for all possible word combinations in a class label as feature. Since Hagen et al. treat query segmentation as a global optimization problem, they implicitly consider the relation between the scores of different segments. In order to take this relation into account, we also use the quotient of the scores of all possible compounds as features. We use the Google N-gram corpus [4] to collect statistics for all N-grams up to length 5 and the jWeb1T API [13] to determine their frequency. In [15] the authors show that treating segments that correspond to Wikipedia titles differently improves the results. In the present work, we use empirical evidence from Wikipedia titles as a separate feature (see below), rather than integrating them directly into the N-gram score.

Features Based on Word Similarity. N-gram statistics crucially rely on counting the occurrence of the exact string making up the compound label in very large, i.e., Web-scale corpora. This way, *bank account* is a likely compound, as it frequently occurs in text. However, this is not able to capture that, for instance, *bank* and *account* are strongly associated with each other since they also frequently occur in context, albeit not necessarily in adjacent order – e.g., as in '*open an account in a bank*'. Accordingly, we propose to relax the requirement of exact matching and turn to distributional semantic [38] as a way to estimate the degree of association between each of the compounds' constituents. For each segment s of a complex concept label (of length two), we accordingly compute the pairwise similarity between its tokens. To this end, we use DISCO [17], a freely available toolkit to build semantic spaces from text and compute distributional similarity. In this work, we use both first-order and second-order context vectors [34] to compute the semantic similarity between the compounds' tokens, and use these two similarity scores directly as features for the classifier.

Features Based on Relation Extraction. Open Information Extraction systems such as ReVerb [9] offer another rich source of information to compute the degree of relatedness between the constituents of a compound. Accordingly, we used the ReVerb dataset[2] to compute such a score based on the extraction of relations between sublabels. Given two sublabels, we query for all those triples where one appears in subject position and the other as object, and vice versa. We then count the number of distinct relations that appear in the resulting set of triples in the predicate position, and use this as feature for our classifier. Note that this provides us with an IE-based relatedness score that, in contrast to distributional similarity features, takes explicitly into account the context in which two constituents occur.

Resource-Based Features: Previous work on query segmentation has shown that background knowledge from linguistic resources can significantly improve the identification of meaningful segments. We therefore also include a number of features based on available resources. Following the approach of Zhang and Hagen, we include WordNet and Wikipedia-based features. Since we are concerned here with ontology labels, we also add new, previously unexplored features that are based on the occurrence of words and compounds in the labels of classes, instances and relations of ontologies found on the semantic web.

Wikipedia-Based Features. Successful unsupervised approaches to query segmentation make use of Wikipedia to determine segments that correspond to meaningful concepts. We adopt this approach and test whether combinations of words from a concept label, including the complete label, correspond to a title of a Wikipedia page. The wide coverage of Wikipedia and the fact that Wikipedia pages are created by human editors and are subject to an intellectual revision process make it a very useful source of information about descriptions of meaningful concepts [16]. In order to determine whether a sequence of words corresponds to a Wikipedia title, we use the JWPL Wikipedia API [40].

WordNet-Based Features. WordNet was used as a dictionary in [41] to check whether a word in a query is a proper noun. We adopt and extend this idea. In particular, for each word in a class label, we collect all parts-of-speech

[2] http://openie.cs.washington.edu/.

(PoS) – namely any of noun, verb, adjective or adverb – it can have in WordNet. We consider PoS other than nouns to capture context-specific ambiguity across PoS – e.g., 'light' used as an adjective as in 'light armored vehicle'. We do not attempt at determining the unique exact PoS of the word in context, e.g., using a syntactic parser, as these typically perform badly when applied to small concept labels [26]. PoS of WordNet terms are retrieved using the JWNL API[3].

Ontology-Based Features. We additionally define a set of new, previously unexplored features that are more directly related to the nature of our task. Since, in our case, a meaningful compound consists of a phrase that could appear as a concept name within an ontology, we test for all words in a label whether they occur as the description of an element in existing ontologies available on the Semantic Web. Similar to the case of PoS in WordNet, we do not restrict the search to class names, but also test whether the phrase is used in the descriptions of relations or instances, since this make the candidate less likely to be a meaningful class name. This can be seen as an ontological version of the WordNet-based features described above. Further, for each pair of words in a label, we count the number of ontologies both words occur in. This can be seen as an ontological version of computing word co-occurrence. We use the Watson search engine for ontologies [6] as a tool for accessing available ontologies on the web and computing our features. This approach was inspired by [32], where the authors use Watson as a mechanism to detect background knowledge for ontology matching.

Using Classification Results Within a Bootstrapping Architecture: The different classification tasks that originate from a single label are not independent of each other. Consequently, we first classify shorter compounds and then use the predicted class and the confidence of the classifier for compounds of length n as additional features for classifying compounds of length $n+1$. To this end, we first train base classifiers to decide whether the individual tokens of the label, say 'British Crown Colony' - (referred to as (A) British, (B) Crown and (C) Colony) are meaningful terms on their own. In the next step, the class labels and confidence values of these classifiers are used as features for classifiers that decide whether two-word combinations – i.e., British Crown (AB), Crown Colony (BC) and British Colony (AC), in our case – are meaningful labels themselves.

4 Experiments

4.1 Gold-Standard Dataset

To create a gold-standard for training and evaluating our classifiers, we used the **S**uggested **U**pper **M**erged **O**ntology (SUMO) [25]. SUMO, and its domain ontologies, form a large formal ontology used for research and applications in search, linguistics and reasoning. SUMO contains concepts that describe the

[3] http://sourceforge.net/projects/jwordnet/.

world on a very abstract level, while some of the integrated ontologies cover very specific topics like communication or transportation – the latter, for instance, distinguishing between different types of cargo ships[4].

Analysis of the concept labels found within SUMO revealed that 1579 concepts are described by non-compound labels, whereas 1755 concepts have two-word labels, 635 have three-word labels, and 236 are described using concept labels made up of more than three words. From the whole set of three-word compounds we randomly sampled a subset of 300 labels. These labels cover completely different topics, and range across domains as diverse as from military (e.g., *amphibious assault vehicle*) to medical ones (e.g., *yellow fever virus*)[5].

Given a concept label of the form ABC, three human judges were asked to provide a ground truth by annotating the label's compounds, namely any of A, B, C, AB, BC, or AC, as meaningful or not, based on Definition 1. The final gold standard was created by aggregating the single annotators' judgments based on majority voting. In order to quantify the quality of the annotations and the difficulty of the task we computed the inter-annotator agreement using the kappa coefficient [5] – we use Fleiss' kappa [12]. Our annotators achieved an agreement coefficient κ of .73, .70 and .60 for annotating the two-word compounds AB, BC and AC, respectively. An average agreement of $\kappa = .68$ indicates substantial agreement between annotators, thus corroborating the overall quality of the annotated data, as well as the well-definedness of our task.

4.2 Experimental Setting

We perform experiments using the Rapidminer toolkit [23], version 5. We set up two learning processes: (i) one for classifying single words that uses solely external features of words and word combinations, and (ii) a second one for classifying two-word segments that uses the results of classifying single words, together with external features. For both tasks, we experimented with a number of different learning algorithms. Below, we report results using Support Vector Machines (SVM) and Neural Networks (NN), since these methods showed a significantly better performance than other methods. We use SVM with dot product kernels and NNs with one hidden layer (additional parameters can be found in the process definitions).

Many of our features (e.g., distributional similarity) can be only computed pairwise between different words, and thus require multi-word compounds. Accordingly, we conducted a finer-grained feature analysis using two-word combinations only: in this setting, statistical and knowledge-based features were evaluated separately, in order to quantify the different contribution of background knowledge vs. statistics from large corpora for our task. Given the limited size of our dataset, we employ ten-fold cross validation for all our experiments. For evaluation, we use standard measures of recall, precision and accuracy: below,

[4] SUMO is originally published in the SUMO-KIF format [25]. In our work we use the OWL version available at http://www.ontologyportal.org/.

[5] The gold standard is freely available at https://madata.bib.uni-mannheim.de/57/.

we only report accuracy for each classification task for the sake of brevity. However, all detailed results for our experiments, the Rapidminer processes, and the full feature tables can be found online at https://madata.bib.uni-mannheim.de/57/.

4.3 Results

We present our results in Table 1, where we report accuracy figures for the detection of meaningful, single-word compounds (i.e., A, B or C), as well as two-words (namely, any of AB, BC or AC). Overall, our results for the classification of single-word compounds are generally favorable, with performance figures on average >90 % for both SVMs and neural networks. When looking at the performance on each single token position, we notice the higher results on the rightmost word, namely C: this is because this generally corresponds to the lexical head of the noun phrase[6]. These constituents typically identify, from a semantic point of view, the concept's super-concept, e.g., *amphibious assault vehicles* are *vehicles* (cf. also the head-matching heuristics from [26]) and provide a meaningful concept label in the vast majority of cases. Results on A and B, in contrast, are lower since these tokens are meaningful in a smaller number of cases, which crucially depends on a variety of complex factors, ranging from syntactic – like, for instance, the token having a PoS other than noun (e.g., an adjective, as in "merchant *marine* ship") – through semantic – for example, the single constituent having no meaning related to that of the overall phrase, as in "rift *valley* fever").

Table 1. Results on the identification of meaningful compounds. Performance figures for AB, BC and AC are obtained using the bootstrapping architecture described in Sect. 3, and thus use classification results for A, B and C as additional features.

Feature type learning algorithm	Statistical		Knowledge-based		All	
	SVM	NN	SVM	NN	SVM	NN
A					87.91	**91.21**
B					**90.11**	87.91
C					94.51	**98.90**
Average					90.84	**92.67**
AB	74.34	79.96	79.28	79.95	79.27	**80.27**
BC	70.04	74.35	69.70	**81.84**	80.60	79.27
AC	65.78	63.13	**75.96**	71.37	75.30	74.03
Average	70.05	72.48	74.99	77.72	**78.39**	77.86

[6] The head of a phrase is the word which is grammatically most important in the phrase, since it determines the nature of the overall phrase [28]. For basic non-recursive noun phrases, this typically corresponds to the rightmost noun.

Results on the classification of two-word constituents are lower in that these instances also require in many cases complex decisions integrating heterogeneous features. In general, we note that results on AC are lower that those on AB or BC, which is in line with the higher difficulty of the task highlighted by the lower inter-annotator agreement of our human raters (Sect. 4.1). When looking at the contribution of each single feature group, we note that, in general, knowledge-based features tend to perform better than statistical ones. This is because, while statistical information provides us with better coverage, knowledge-based features are indeed superior for the present task in that they rely on very large amounts of human supervision from large-scale, high-quality semantic resources like Watson, WordNet and Wikipedia. However, the complementarity of both feature types is shown by the overall results – namely those obtained by averaging performance over AB, BC and AC – being obtained when using both statistical and knowledge-based features. We take this to be good news, since it suggests that better performance on this task can be achieved in the future by exploring other heterogeneous knowledge sources, as well as their combination with robust learning algorithms.

5 Use Case

We next analyze whether the detection of meaningful compounds provides us with a valuable knowledge source for the task of matching complex ontology labels. A complete solution for the mapping task itself is beyond the scope of this paper: however, in this work we can already report about some experiments that yield relevant insights. Given a complex compound label, we first apply our method to segment the labels into meaningful parts. We then try to detect a concept with an equivalent or highly similar meaning within a target ontology. Our hunch here is that robust performance on this simplified task indicates that we can use use the results of our segmentation as input to generate partial mappings, which are later used to solve the complex matching task as a whole.

In the following we make use of the same dataset described in Sect. 4.1. Since there exists no evaluation dataset that deals with the problem of complex ontology matching, we formulate a pseudo-matching task as follows. For each compound label from our dataset we remove the corresponding concept from the SUMO ontology. Then we try to anchor this concept back within the target resource. This simulates the task of mapping a concept to an ontology, where an equivalent concept does not exist as named concept. In such a scenario, the concept can be anchored at the right position in the concept hierarchy, or it might be possible to construct an equivalent complex concept description. Let C denote such an concept, let $l(C)$ denote its label, and let $l_m(C) = \{m_1, ..., m_n\}$ denote the set of compounds that have been annotated as meaningful (either from our system or from the human annotators). In our experiments we then aim at creating a mapping for each m_i to one of the concepts in SUMO. In particular, we create a mapping if we find a concept D with $l(D) \cong m_i$, where $\cong$ refers to string equality after normalization. The results of the Ontology Alignment

Table 2. Mapping fragments of a compound label to concepts.

	Baseline	Learning algorithm	Gold standard
Precision	20.1	33.1	31.6
Recall	100	91.6	93.2
F-Measure	33.5	48.6	47.2

Evaluation Initiative have shown that this approach results in highly precise mappings that are often hard to beat in terms of F-measure [10].

In Table 2 we report on the fraction of labels from $l_m(C)$ that can be matched to a concept in SUMO. Performance is computed using standard metrics of precision (fraction of all labels for which a mapping has been generated), recall (fraction of generated mappings compared to the mappings generated by taking all possible sublabels into account) and F_1-measure (the harmonic mean of precision and recall). We compute these scores in three different settings, namely for: (1) a baseline that considers all sublabels as meaningful combinations; (2) the output of our best-performing supervised classifier from Sect. 4.3; (3) the gold standard provided by human annotators (Sect. 4.1), which theoretically provides us with an upper bound for this task. Taking all sublabels into account, we achieve a recall of 100 % (by definition) and a precision of 20.1 %. Using the output of our algorithm yields instead an increased precision of 33.1 %, while maintain recall above 90 %. Overall, we can increase the F-measure from 33.5 % to 48.6 %: we take these as good results with respect the second bullet point in Definition 1 ('it must be a possible concept in some ontology'). Finally, we note that precision and recall change only to a very limited degree when compared against the results of using the gold-standard labels, thus indicating the overall robustness of our approach.

We next analyzed how many mappings generated during our experiments led to a concept that is a superclass of C. This happens for 58.9 % of the instances in the dataset, regardless of whether we use automatically-detected compounds or gold-standard labels. Due to the artificial nature of these experiments – which merely consisting of removing a concept from its place in the reference ontology, as opposed to the full-fledged ontology matching task – we can easily compute these figures in our experimental setting. However, note that in a real matching scenario it is a challenging task to find the right position in the concept hierarchy for a given complex concept label. While in our use case ≈60 % of the generated mappings help us solve the task of attaching the concept to the right place in the target concept hierarchy, the remaining ≈40 % of the mappings establish links to other concepts. Error analysis revealed that these 40 % do not necessarily consist of incorrect mappings. Quite contrary, they might be required to construct complex concept expressions. An example is the concept *fish carrier ship*. The concept *ship* is a superclass of the concept, while the concept *fish* is located in a different branch of the concept hierarchy. A correct mapping would express that a *fish carrier ship* is a *ship* that *carries* the cargo *fish*. That is, this example illustrates the task that needs to be solved for constructing precise equivalence mappings to complex concept descriptions.

6 Conclusions and Future Work

In this paper we presented an approach to detect meaningful compounds within complex ontology class labels. We proposed to view this as a binary classification task, and used a supervised classifier to explore a wide variety of features for solving this problem. Our results indicate that similarly, for instance, to previous results in query segmentation, supervised learning methods offer a viable solution for our task. In particular, they provided us with a complete framework to test many different features and accordingly understand the role and benefits of different knowledge sources. Our best results are obtained by combining statistical and knowledge-rich features, and indicate that future advances could be obtained by additional work on the feature engineering side.

We additionally evaluated the output of our classifier as source for a pseudo ontology matching task with complex class labels. The results indicate that we have to distinguish between two main objectives, in order to solve the challenging problem of matching compound labels. First, we need to identify a concept in the target ontology that is more general than the concept we want to match. So far, we can use our algorithm for detecting meaningful compounds: however, our algorithm cannot determine which of these compounds corresponds to a more general class, i.e., which of the constituents is the head noun. With this additional information we would be able to generate mappings expressing a subsumption relation. Extending our method to detect head nouns would thus be highly beneficial for generating correct subsumption mappings. Second, we have to aim at the construction of complex concept descriptions that are equivalent to the concept denoted by the compound label. This task is obviously much harder than the previously mentioned task. Let us focus again on the example *fish carrier ship* from the previous section. Constructing the equivalent concept description requires more knowledge than identifying the head noun. Moreover, we need to understand which relations hold between those sublabels that have been annotated to be meaningful. For this, relation extraction (which we merely used as a feature in this work) and semantic parsing [19] could prove useful.

With this work we aim at providing a first step towards understanding and solving the problem of matching complex concepts labels. The first results are promising in that our experiments helped us better understand the next steps that need to be taken into account for solving the concrete matching problem. Future work will focus on the open challenge of generating mappings for concepts labeled by compound expression. For generating equivalence mappings, we will turn to analyzing the relation between meaningful sublabels, in order to find an isomorphism between the structures on the linguistic layer and the structures that can be constructed by building complex concept descriptions.

References

1. Ashburner, M., Ball, C.A., Blake, J.A., Botstein, D., Butler, H., Cherry, J.M., Davis, A.P., Dolinski, K., Dwight, S.S., Eppig, J.T., Harris, M.A., Hill, D.P., Issel-Tarver, L., Kasarskis, A., Lewis, S., Matese, J.C., Richardson, J.E., Ringwald, M., Rubin, G.M., Sherlock, G.: Gene ontology: tool for the unification of biology. The gene ontology consortium. Nature Genet. **25**(1), 25–29 (2000)
2. Baldwin, T., Kim, S.N.: Multiword expressions. In: Indurkhya, N., Damerau, F.J. (eds.) Handbook of Natural Language Processing, 2nd edn. CRC Press, Taylor and Francis Group, Boca Raton (2010)
3. Bergsma, S., Wang, Q.: Learning noun phrase query segmentation. In: Proceedings of EMNLP-CoNLL-07, pp. 819–826 (2007)
4. Brants, T., Franz, A.: Web 1T 5-gram version 1. LDC2006T13, Philadelphia, Penn.: Linguistic Data Consortium (2006)
5. Carletta, J.: Assessing agreement on classification tasks: the kappa statistic. Comput. Linguist. **22**(2), 249–254 (1996)
6. d'Aquin, M., Motta, E., Sabou, M., Angeletou, S., Gridinoc, L., Lopez, V., Guidi, D.: Towards a new generation of semantic web applications. IEEE Intell. Syst. **23**(3), 20–28 (2008)
7. Doan, A., Halevy, A.: Semantic-integration research in the database community. AI Mag. **26**(1), 83–94 (2005)
8. Enslen, E., Hill, E., Pollock, L., Vijay-Shanker, K.: Mining source code to automatically split identifiers for software analysis. In: Proceedings of MSR-09, pp. 71–80 (2009)
9. Etzioni, O., Fader, A., Christensen, J., Soderland, S., Mausam, M.: Open information extraction: the second generation. In: Proceedings of IJCAI-11, pp. 3–10. AAAI Press (2011)
10. Euzenat, J., Meilicke, C., Stuckenschmidt, H., Shvaiko, P., Trojahn, C.: Ontology alignment evaluation initiative: six years of experience. In: Spaccapietra, S. (ed.) Journal on Data Semantics XV. LNCS, vol. 6720, pp. 158–192. Springer, Heidelberg (2011)
11. Fernandez-Breis, J.T., Iannone, L., Palmisano, I., Rector, A.L., Stevens, R.: Enriching the gene ontology via the dissection of labels using the ontology pre-processor language. In: Cimiano, P., Pinto, H.S. (eds.) EKAW 2010. LNCS, vol. 6317, pp. 59–73. Springer, Heidelberg (2010)
12. Fleiss, J.L.: Measuring nominal scale agreement among many raters. Psychol. Bull. **76**(5), 378 (1971)
13. Giuliano, C., Gliozzo, A., Strapparava, C.: FBK-irst: lexical substitution task exploiting domain and syntagmatic coherence. In: Proceedings of SemEval-2007 (2007)
14. Hagen, M., Potthast, M., Stein, B., Bräutigam, C.: The power of naive query segmentation. In: Crestani, F., Marchand-Maillet, S., Chen, H.H., Efthimiadis, E., Savoy, J. (eds.) Proceedings of SIGIR-10, pp. 797–798 (2010)
15. Hagen, M., Potthast, M., Stein, B., Bräutigam, C.: Query segmentation revisited. In: Proceedings of WWW-11, pp. 97–106 (2011)
16. Hovy, E., Navigli, R., Ponzetto, S.P.: Collaboratively built semi-structured content and Artificial Intelligence: the story so far. Artif. Intell. **194**, 2–27 (2013)
17. Kolb., P.: DISCO: a multilingual database of distributionally similar words. In: Proceedings of KONVENS-08 (2008)

18. Kudoh, T., Matsumoto, Y.: Chunking with support vector machines. In: Proceedings of NAACL-01, pp. 1–8 (2001)
19. Kwiatkowski, T., Choi, E., Artzi, Y., Zettlemoyer, L.: Scaling semantic parsers with on-the-fly ontology matching. In: Proceedings of EMNLP-13, pp. 1545–1556 (2013)
20. Leopold, H., Smirnov, S., Mendling, J.: Recognising activity labeling styles in business process models. Enterp. Model. Inf. Syst. Architectures **6**(1), 16–29 (2011)
21. Manaf, N.A.A., Bechhofer, S., Stevens, R.: A survey of identifiers and labels in OWL ontologies. In: Proceedings of OWLED 2010 (2010)
22. Mendling, J., Reijers, H., Recker, J.: Activity labeling in process modeling: empirical insights and recommendations. Inf. Syst. **35**(4), 467–482 (2010)
23. Mierswa, I.: Rapid miner. Künstliche Intelligenz **23**(2), 5–11 (2009)
24. Nakov, P., Hearst, M.: Search engine statistics beyond the n-gram: application to noun compound bracketing. In: Proceedings of CoNLL-05, pp. 17–24 (2005)
25. Pease, A.: Ontology: A Practical Guide. Articulate Software Press, Angwin (2011)
26. Ponzetto, S.P., Strube, M.: Taxonomy induction based on a collaboratively built knowledge repository. Artif. Intell. **175**, 1737–1756 (2011)
27. Quesada-Martínez, M., Fernández-Breis, J.T., Stevens, R.: Lexical characterization and analysis of the bioportal ontologies. In: Peek, N., Marín Morales, R., Peleg, M. (eds.) AIME 2013. LNCS, vol. 7885, pp. 206–215. Springer, Heidelberg (2013)
28. Radford, A.: Syntax: A Minimalist Introduction. Cambridge University Press, Cambridge (1997)
29. Ramshaw, L., Marcus, M.: Text chunking using transformation-based learning. In: Proceedings of the Third Workshop on Very Large Corpora, pp. 82–94 (1995)
30. Ritze, D., Meilicke, C., Šváb Zamazal, O., Stuckenschmidt, H.: A pattern-based ontology matching approach for detecting complex correspondences. In: Proceedings of OM-2009 (2009)
31. Ritze, D., Völker, J., Meilicke, C., Šváb Zamazal, O.: Linguistic analysis for complex ontology matching. In: Proceedings of OM-2010 (2010)
32. Sabou, M., d'Aquin, M., Motta, E.: Exploring the semantic web as background knowledge for ontology matching. J. Data Semant. **11**, 156–190 (2000)
33. Sang, E., Buchholz, S.: Introduction to the CoNLL 2000 shared task: chunking. In: Proceedings of CoNLL-00, pp. 127–132 (2000)
34. Schütze, H.: Automatic word sense discrimination. Comput. Linguist. **24**(1), 97–124 (1998)
35. Sha, F., Pereira, F.: Shallow parsing with conditional random fields. In: Proceedings of HLT-NAACL-03, pp. 134–141 (2003)
36. Shvaiko, P., Euzenat, J.: Ontology matching: : state of the art and future challenges. IEEE Trans. Knowl. Data Eng. **25**, 158–176 (2012)
37. Stuckenschmidt, H., Predoiu, L., Meilicke, C.: Learning complex ontology mappings - a challenge for ILP research. In: Proceedings of ILP-08 - Late Breaking Papers (2008)
38. Turney, P.D., Pantel, P.: From frequency to meaning: vector space models of semantics. J. Artif. Intell. Res. **37**, 141–188 (2010)
39. Vadas, D., Curran, J.R.: Parsing noun phrase structure with CCG. In: Proceedings of ACL-08: HLT, pp. 335–343 (2008)
40. Zesch, T., Müller, C., Gurevych, I.: Extracting lexical semantic knowledge from wikipedia and wiktionary. In: Proceedings of LREC-08 (2008)
41. Zhang, W., Liu, S., Yu, C., Sun, C., Liu, F., Meng, W.: Recognition and classification of noun phrases in queries for effective retrieval. In: Proceedings of CIKM-07, pp. 711–720 (2007)

Categorization Power of Ontologies with Respect to Focus Classes

Vojtěch Svátek(✉), Ondřej Zamazal, and Miroslav Vacura

Department of Information and Knowledge Engineering,
University of Economics, W. Churchill Sq. 4, 130 67 Prague 3, Czech Republic
{svatek,ondrej.zamazal,vacuram}@vse.cz

Abstract. When reusing existing ontologies, preference might be given to those providing extensive subcategorization for the classes deemed important in the new ontology (focus classes). The reused set of categories may not only consist of named classes but also of some compound concept expressions that could be viewed as meaningful categories by human ontologist. We define the general notion of focused ontologistic categorization power; for the sake of tractable experiments we then choose a restricted concept expression language and map it to syntactic axiom patterns. The occurrence of the patterns has been verified in two ontology collections, and for a sample of pattern instances their ontologistic status has been assessed by different groups of users.

1 Introduction

Reusing parts of existing semantic web ontologies when designing a new one, or when merely proposing the schema for an RDF dataset to be published, is commonly understood as best practice [6]. With the growing number of ontologies on the semantic web it also becomes more likely to find multiple ontologies covering the given topic. However, mere thematic relevance may not be enough: since the target ontology/schema is to be used in a certain application context, it should exhibit features required in this context. For example, if a reasoner is to be applied on the ontology, its expressiveness should not exceed that expected by the reasoner. In this paper we investigate one another structural feature of an ontology to be potentially reused: its *categorization power*, i.e. its suitability for assigning meaningful categories – not necessarily expressed as *named* classes but possibly in the form of *compound concept expressions* – to individual domain objects (instances). Namely, many tasks related to the management of ontologically described data refer to detailed categorization of objects: companies may provide specific offers to different categories of customers, buyers may only be interested in specific categories of products, and the like. Reusing a categorization structure pre-existing in a widespread vocabulary (or one with potential for future widespread, e.g., cataloged in a respected collection such as LOV[1] [9]) may not only save a part of the design effort but also allow to better interface with

[1] http://lov.okfn.org/.

E. Blomqvist et al. (Eds.): EKAW 2016, LNAI 10024, pp. 636–650, 2016.
DOI: 10.1007/978-3-319-49004-5_41

other applications, e.g., in federated querying or concerted recommendation. We therefore hypothesize that ontologies providing more subcategories for classes important in the given use case – to be called *focus classes* in our approach – would be a more desirable subject of reuse, as whole or in (relevant) part.

To informally introduce the key concepts of our approach (more formally grounded in Sect. 2), let us start with a toy ontology O in Manchester OWL syntax[2] as motivating example:

```
Class: Person
Class: Man        SubClassOf: Person
Class: Woman      SubClassOf: Person
Class: MarriedMan          EquivalentTo: Man and hasSpouse some Thing
Class: ProductivePerson
  EquivalentTo: Person and insuranceCategory some {Enterpreneur,Employed}
Class: Country
ObjectProperty: hasSpouse   Domain: Person    Range: Person
ObjectProperty: bornIn      Domain: Person    Range: Country
ObjectProperty: insuranceCategory     Domain: Person
  Range: {Enterpreneur,Employed,Child,Retired}
DataProperty: zipCode       Domain: Person    Range: string
Individual: UK     Types: Country
Individual: Italy  Types: Country
```

Let us assume we want to build a rich ontology for categorizing persons and need to assess if O is a good reuse candidate. Class `Person` (possibly discovered by lexical search via an ontology search engine) thus becomes our focus class, FC, in O. A simple quantification of the categorization power of O wrt. `Person` could be 4, i.e. the sum of its asserted and inferred subclasses. However, the entities from O can be assembled to many compound expressions containing a subset of `Person` instances, such as: `insuranceCategory value Retired`, `hasSpouse some Thing`, or `Woman and bornIn value Italy`. We can imagine that some of these have only been 'refused entry' to the named class 'elite' (the concept signature of O) due to stringent parsimony or even sloppy modeling. On the other hand, some structurally similar compound expressions are unsuitable for categorizing persons. For example, `bornIn some Thing` does not refine `Person` in any way (all persons are born), while `insuranceCategory value Child and insuranceCategory value Retired` is void. Furthermore, complex conjunctions and especially disjunctions, although possibly containing adequately large subsets of the extent of the focus class, might be mentally too complex to grasp.

For both named subclasses of FC and compound expressions from the former group (for which it would not surprise us to see them transformed to named classes) we propose the term *ontologistic category*. We use this adjective to make distinction from the notion of 'ontological category': while 'ontological' would refer to 'category of beings that exists' (i.e. we cannot deny the existence of categories with complex, unintuitive descriptions or with very small sets of instances), 'ontologistic' refers to a category plausible as reusable domain concept

[2] http://www.w3.org/TR/owl2-manchester-syntax/.

to a human ontologist. Intuitively, we should primarily derive the categorization power of an ontology with respect to FC from the set of ontologistic categories rather than from that of all possible concept expressions.

The approach taken in this paper and reflected in its structure is: to create the overall framing of the focused categorization task (Sect. 2); to choose a restricted (finite and easily manageable) concept expression language and map it to syntactic axiom patterns (Sect. 3); to verify the occurrence of the patterns in ontology collection/s (Sect. 4); to check on a sample of pattern instances if and under what conditions their respective concept expressions are 'ontologistically' plausible (Sect. 5). We also provide an overview of related research (Sect. 6) and summary conclusions with future work prospects.

2 General Model of Ontologistic Categorization Power

Let $PS(FC, O)$ be the set of concept expressions (CEs) that are proper specializations of named class FC with respect to ontology O:

$$PS(FC,O) = \{CE; O \models (CE \sqsubset FC)\}$$

Then *focused ontologistic categorization power* (FOCP) of O with respect to FC could theoretically be defined as

$$FOCP(FC,O) = |CE; CE \in PS(FC,O) \wedge oc(CE)|$$

where the binary function *oc* returns *true* if the CE in its argument is an *ontologistic category* (OC). Obviously, such a definition would be anything but rigorous and operational. First, $PS(FC, O)$ will be infinite in common OWL DL dialects, e.g., considering concept expressions nesting with unlimited depth. Second, the concept of ontologistic category, as outlined in the introduction (and exemplified later in this paper) is fuzzy, context-dependent and subjective. For practical purposes we thus need to (1) restrict the language $\mathcal{L}$ of the CEs, to assure $PS(FC, O)$ finiteness, and (2) approximate the 'typical' result of *oc* (as returned by human oracles in various contexts) by a formula based on measurable features of the CEs. In this paper we only consider boolean features, namely, the presence of *axiom patterns*, mapped on the CEs, in O.

Let an *FC-matching axiom pattern*[3] be a set of OWL axioms with placeholder variables (for concepts, roles and individuals), such that one of them, in the concept position, is the *FC-variable* (to be substituted by FC in pattern matching). For example, a simple axiom pattern could be C `rdfs:subClassOf` FC.

Let a *pattern-CE mapping function m* be a function that takes an n-tuple of entities substituted for variables (other than the FC-variable) in one instantiation of a pattern p and returns the corresponding CE of a certain *type* t, provided (optional) pruning constraints $prun_t$ associated with this type are satisfied. The CE types are understood in the context of the categorization task (having 1-to-1 mapping to patterns) and conform to the CE language $\mathcal{L}$.

[3] From now on simply 'axiom pattern', for simplicity.

The set of patterns used for FOCP computation, together with their mapping functions, should satisfy some requirements: (1) they should assure that the resulting CE is a subconcept of FC (e.g., the 'subclass' pattern above satisfies this trivially), and, (2) the occurrences of two different patterns should not yield the same CE (to assure unique counting). We should further, using a description logic (DL) reasoner and/or syntactic pruning constraints, omit CEs that are either unsatisfiable or equivalent to FC. An open question is whether we should only count logically equivalent (though syntactically different) expressions once. While trivially derivable equivalent concepts such as double negations should be avoided, dissimilar expressions only alignable via complex derivations might have descriptive value of their own. Since the CE language chosen in this paper only covers a small subset of OWL (e.g., without negation and Boolean connectives in general), we tolerate multiple equivalent expressions entering the FOCP computation.

If the above pattern were instantiated as `Man rdfs:subClassOf Person`, an adequate m would take the substitution (`Man`/C) and return the CE `Man`, with type t corresponding to 'named subclass'. No pruning constraints would apply.

Let the *occurrence function* for pattern p wrt. FC, denoted as $Occ(p, FC, O)$, return the number of matches of p in O such that FC is substituted to all occurrences of the FC-variable in p. In the body of Occ the axioms from p are conjunctively interpreted, together with the associated pruning constraints. The matches correspond to the n-tuples submitted to the mapping function m.

The *approximate FOCP* of O with respect to a pattern set $P = \{p_1, ..., p_n\}$ can then be defined as the weighted sum of the Occ function results in O,

$$\widehat{FOCP}(FC, O, \mathcal{L}, P) = Occ(p_1, FC, O) * w_1 + \ ... \ + Occ(p_n, FC, O) * w_n$$

where $w_i \in [0, 1]$ is the weight of pattern p_i indicating the likelihood that its occurrence would produce an OC.

3 Concept Expression Language and Axiom Patterns

For the remainder of this paper we restrict the language of CEs as defined by the following, extremely simple grammar, guaranteeing finiteness for any finite ontology signature:

```
CE := namedClass | simpleExistentialRestriction | valueRestriction
simpleExistentialRestriction := objectProperty 'some' namedClass
valueRestriction := objectProperty 'value' individual
```

When considering the types of CEs corresponding to this ad hoc language (further called $\mathcal{L}_0$), we specifically cater for the top concept (`Thing`) in the role of existential restriction filler. Therefore the CE types are eventually four, as shown in Table 1. The second column displays the CE structure in DL notation; C stands for named concept (class), R for role (object property) and i for individual. We see that t2 and t4 are mere refinements of t3 (generic existential

Table 1. Summary of CE types in $\mathcal{L}_0$

Type	CE in DL	Substituted	Abox path length	Axiom pattern size
t1	C	C	3	1
t2	$\exists R.\top$	R	2	1
t3	$\exists R.C$	R, C	5	3
t4	$\exists R.\{i\}$	R, i	3	4

restriction) in DL terms. The third column indicates which symbols from the CE correspond to variables substituted in the associated axiom pattern. The set of variables in t2 is a subset of those of t3 and t4; if we consider one or more CEs for t3 or t4 with some R then we should also consider the CE for t2 with this R. The fourth column measures the length of Abox path (as sum of resource nodes and predicate edges) connecting the categorized individual with entities ('responsible' for the categorization) substituted for variables from the third column: it is smallest for t2 where only the adjacent edge is applied; t1 and t4 require a whole triple (instantiation or property assertion, respectively), while t3 needs two triples (both assertion and the ensued instantiation to the 'filler' class). The order of the patterns in the table however reflect the increased complexity of their detection in the Tbox using the proposed axiom patterns (fifth column), which we detail in the next subsection.

3.1 Syntactic Axiom Patterns in the Ontology Schema

The CEs in $\mathcal{L}_0$ could in principle be mapped to diverse constellations of axiom patterns with varying expressiveness. However, we implement the patterns primarily in RDFS terms, namely, over `rdfs:subClassOf`, `rdf:type`, `rdfs:domain` and `rdfs:range` axioms, such that all of their arguments are either atomic expressions or variables for which they can be substituted; we so far avoided `rdfs:subPropertyOf` to keep the pattern structure simpler (we will consider adding it in the future). Besides the patterns also address the *pruning* of CEs whose ineligibility follows from the ontology structure. The patterns (p1, .., p4)[4] are pairwise mapped on the previously defined CE types (t1, .., t4), except that for t4 we also supply an additional pattern p5 in which the individual i is not part of the ontology itself but of an associated SKOS codelist. Since the pattern occurrence is to be computed specifically with respect to the entities responsible for categorization (third column in Table 1), we present the patterns in terms of their *occurrence function* $Occ(p\#, FC)$.[5]

[4] We denote these specific patterns using normal font, to differentiate with the superscript notation (p_i) of abstract symbols in Sect. 2.

[5] For simplicity we omit O in the formula; the identity of the ontology follows from the FC it contains. We also avoid the use of DL notation and express the OWL axioms using predicate URIs, to avoid collision with general math notation. The only reference to beyond-RDFS construct is the value restriction for p5.

Pattern p1. CEs of type t1 are simply *subclasses* of FC, i.e. they are matched by the previously mentioned axiom pattern; the occurrence function is the number of these subclasses:

$$Occ(p1, FC) = |\{C;\ C\ \texttt{rdfs:subClassOf}\ FC\}| \tag{1}$$

The subclasses can be both direct or indirect, and possibly even inferred using other kinds of axioms, i.e. they are subclasses of FC in the deductive closure of the ontology computed by a reasoner.

Pattern p2. Next we will consider properties having FC in their *domain*:

$$Occ(p2, FC) = |\{P;\ P\ \texttt{rdfs:domain}\ FC \wedge P \notin prun_2(FC)\}| \tag{2}$$

where $prun_2(FC)$ is the set of properties that have to be pruned as ineligible for this pattern. Again, even cases when FC is *inferred* as domain of P are considered. Since we do not take into account the right-hand side of P, each corresponding CE (of type t2) would contain all instances of FC that appear in the subject of a triple with P as predicate. The corresponding mapping function m thus maps the pattern on the DL expression $\exists P.\top$ (t2 in Table 1, with P substituted for R). $prun_2(FC)$ essentially contains the properties P that appear in an existential restriction[6] $FC \sqsubseteq \exists P.C$; for such properties the CE would contain *all* instances of FC.

Pattern p3. Now we proceed to the *range* of properties with FC in domain and then to the *subclasses of this range*:

$$\begin{aligned} Occ(p3, FC)\ = |\{(P, C);\ \exists D\ (\ P\ \texttt{rdfs:domain}\ FC \wedge P\ \texttt{rdfs:range}_a\ D\ \wedge \\ \wedge\ C\ \texttt{rdfs:subClassOf}\ D\ \wedge\ (P, C) \notin prun_3(FC)\ )\ \}| \end{aligned} \tag{3}$$

where $prun_3(FC)$ is, again, the set of properties that have to be pruned as ineligible for this pattern. The CE (of type t3) would include all instances of FC that appear in the subject of a triple with P as predicate and some i as object such that i is an instance of C. The inferential closure is again used, however, with the exception of the range axiom, which is only considered as asserted (therefore the 'a' index in $\texttt{rdfs:range}_a$) – otherwise not only subclasses of D but also classes having a common superclass with D would be returned as C (since the superclass would become an inferred range of P). The pattern maps on the DL expression $\exists P.C$ (t3). $prun_3(FC)$ contains the pairs (P, C) that appear in an existential restriction $FC \sqsubseteq \exists P.E$ such that $E \sqsubseteq C$; for such properties the CE would contain *all* instances of FC.

Pattern p4. This pattern extends the previous one with an individual that is instance of C:

$$\begin{aligned} Occ(p4, FC) = |\{(P, i);\ \exists C, D\ (\ P\ \texttt{rdfs:domain}\ FC \wedge P\ \texttt{rdfs:range}_a\ D\ \wedge \\ \wedge\ C\ \texttt{rdfs:subClassOf}\ D\ \wedge\ i\ \texttt{rdf:type}\ C\ \wedge\ (P, i) \notin prun_4(FC)\ )\ \}| \end{aligned} \tag{4}$$

[6] The restriction can also be inherited from a superclass or part of a complete definition, or can have the form of a `value` or `self` restriction or of a cardinality restriction that specializes the existential one; analogously for other $prun_n$'s below.

where $prun_4(FC)$ is analogous to the previous variants. The CE (of type t4) would include all instances of FC that appear in the subject of a triple with P as predicate and (the specific individual) i as object. The inferential closure is used as before. The pattern maps on the DL expression $\exists P.\{i\}$ (t4). $prun_4(FC)$ contains the pairs (P, i) that appear in an existential restriction $FC \sqsubseteq \exists P.E$ such that $i \in E$; for such properties the OC would contain *all* instances of FC.

Pattern p5. SKOS is the most widespread alternative to OWL to consider when specifying simpler 'ontological' taxonomies. This variant thus extends the previous one for a specific source of instance i – a SKOS code list (concept scheme):

$$Occ(p5, FC) = |\{(P, i); \exists s \; (P \; \mathtt{rdfs:domain} \; FC \; \wedge \\ \wedge \; P \; \mathtt{rdfs:range}_a \; (\mathtt{skos:Concept} \sqcap \mathbf{value}(\mathtt{skos:inScheme}, s)) \; \wedge \\ \wedge \; i \; \mathtt{skos:inScheme} \; s \; \wedge \; i \; \mathtt{rdf:type} \; \mathtt{skos:Concept} \; \wedge \; (P, i) \notin prun_4(FC)\} \;) \; | \quad (5)$$

where $\mathbf{value}(\mathtt{skos:inScheme}, s)$ shortcuts the DL concept expression $\exists Q.\{s\}$ such that $Q = \mathtt{skos:inScheme}$. The CE (again of type t4) is defined as in Pattern 4. The difference is merely in the selection method for i – rather than instance of a class from the current ontology, it has to be a `skos:Concept` linked to concept scheme s that is in the range of P. The inferential closure is used as before.

With respect to the requirements on axiom patterns from Sect. 2, all patterns assure (p2–p5 via the domain axiom) that the mapped CE is a *specialization* of the FC. It is also easy to see that the patterns, except p4 vs. p5, are *mutually exclusive* since they produce structurally different CEs. (Formally, to assure exclusivity of p2 and p3, D in p3 should not be `owl:Thing`. We however do not anticipate that explicit range axioms would have the default value `Thing`).

4 Survey on Syntactic Pattern Occurrence

The *research questions* to be answered by the analysis were:

1. How many ontologies, and for how many FCs, provide a decent number of 'categorizing' CEs mapped on the patterns from Sect. 3?
2. What are the differences in the occurrence of the individual patterns overall and across different collections?

For our experiments we used two collections of ontologies. First is a small collection from the domain of conference organization, called *OntoFarm*,[7] and the second is the collection from *Linked Open Vocabularies* (LOV) portal.[8] While the former is rather an experimental collection of ontologies (used, among other, in the Ontology Alignment Evaluation Initiative[9]) with heterogeneous styles and relatively rich in axioms, the latter contains real-world (mostly) light-weight ontologies with connection to the Linked Open Data Cloud. In the analysis we

[7] http://owl.vse.cz:8080/ontofarm/.
[8] http://lov.okfn.org/.
[9] http://oaei.ontologymatching.org/.

made use of our *Online Ontology Set Picker framework*[10] to process ontologies from both collections. OntoFarm has 16 ontologies, and for LOV we used January 2016 snapshot where 529 ontologies were available of which we could process successfully 509 at syntactical level.

Due to limited space we only summarize the most important findings of the survey. More detail is in the extended version of this paper and supplementary datasets, both available at http://owl.vse.cz:8080/EKAW2016/.

We counted the occurrences of pattern from Sect. 3 across all classes of all ontologies in the role of FC. We summed up these results at ontology level by identifying 'categorizable FCs' for which the $\widehat{FOCP}$ (with same $w = 1$ for all patterns) reached some threshold τ (1, 3 and 5). Based on the survey, the *research questions* have been answered as follows:

1. Overall, the majority of ontologies have 'interesting' categorization power for *some* of their classes as FCs; however, if we require more than 20–30 % of classes to be 'categorizable', the proportion of ontologies satisfying this requirement is rapidly dropping.
2. There are important differences in the proportion of pattern occurrence, with p2 being most widespread (since it only requires an `rdfs:domain` axiom), followed by p1 and then by p3. For example, in LOV there are 16 % of ontologies in which p2 returns at least 5 categories for 20 % or more of their classes (in the role of FC); for the other patterns there are only between 1–5 % of ontologies satisfying this requirement. A similar ranking holds for the smaller OntoFarm collection, which is structurally richer than (on average) the LOV; however, as appears, the richer axiomatization only allows for a smaller proportion of ontology classes to be categorized, possibly by the effect of CE pruning. p5 is the most rare overall; it only appears in 7 LOV ontologies; however, these ontologies are newer, so we can expect that its occurrence would grow.

5 Ontologistic Categorization Experiment

The CE sets on which the pattern occurrence counts are obtained by automated analysis are mere rough approximations of the true OC sets for the respective FCs. In order to get finer insights, we proceeded to detailed investigation of sample CEs by human 'ontologists', both experts and relative novices (students of relevant subjects). Since we take named (sub)classes as OCs by default, we only examined the CEs of t2, t3 and t4 (we did not further distinguish between the t4 variant returned by pattern p4 and by the 'SKOSsy' p5). The general research questions, this time, were:

1. Is the OC status of CEs correlated with the CE type and/or the background of the human assessor?
2. What is the proportion of clear vs. borderline cases?
3. Which deeper semantic distinctions either lead to negative assessment (even in absence of logical causes for such assessment) or make the decision tricky?

[10] http://owl.vse.cz:8080/OOSP/.

We report on the most interesting results of this effort below.[11]

Initial sampling. As regards the CE sampling for both threads of analysis (expert/novice), we used the same collections as in Sect. 4, i.e. OntoFarm and LOV. From each collection 10 CEs per type have originally been randomly sampled, yielding 80 CEs. After manual removal of duplicates (for OntoFarm as smaller collection) and CEs containing entities with cryptic names without meaning in natural language, 59 CEs remained (28 from OntoFarm and 31 from LOV); there were 17 CEs of t2 (existential restriction with 'filler `Thing`'), 20 of t3 (existential restriction to specific class), and 22 of t4 (value restriction).

Expert ontologist assessment and insights. The analysis has been done by the three authors of the paper, all with 10–20 years of experience in ontological engineering. They first examined the sample of 59 CEs independently and assessed it on the 5-point Likert scale: for each CE X the question "Is X an OC?" was answered as either 'certainly', 'perhaps', 'borderline', 'perhaps not' or 'certainly not'. Then a consensus was sought in a F2F session. The independent assessment had 76 % agreement: in 45 out of 59 cases there was no contradictory assessment (certainly/perhaps yes vs. certainly/perhaps no); we will call these cases *clear positives* (43 cases, incl. the one used twice) and *clear negatives* (3 cases), respectively. The consensus session then yielded a complete consensus on the remaining cases; in 12 out of the 14 'clash' cases the final result was 'yes' (namely, a conceivable situation was formulated in which the CE would be a plausible OC), one case was found dubious due to implausible inference (see the second 'insight' below) and in one case the CE was assumed semantically equivalent to its FC, both resulting into 'no'. Of the seven ultimately negative results, five were of t2, one of t3 and one of t4. Selected general insights into less obvious decisions, with examples, follow (see the supplementary page for complete assessments with commentaries):

- Ontologies tied to software applications, such as some OntoFarm ones (capturing the processes supported by conference software) use object properties to capture relationships that are only relevant within a *short time frame*, e.g., `cmt:finalizePaperAssignment`; a meaningful category of persons would rather refer to their long-term responsibility for paper assignment rather than to the instantaneous action of 'finalizing' it.
- In some cases the use of inferential closure for the filler class in t3 leads to linking relatively *thematically unrelated* entities (especially in the DBpedia ontology), such as in `dbo:beatifiedPlace some dbo:WineRegion` for instances of `dbo:Person`. While this case was found marginally acceptable (there could be some correlation between religiosity and wine production), a similar one, `dbo:headChef some dbo:BaseballPlayer` was rejected not only due to thematic leap but also due odd inference result: the FC was `dbo:Village`, the declared domain of `dbo:headChef` is `dbo:Restaurant`, but the ontology (actually, the 2014 version from the LOV endpoint) enables to infer the axiom `dbo:Restaurant rdfs:subClassOf dbo:Village`.

[11] More detail can be found, again, at http://owl.vse.cz:8080/EKAW2016/.

- Some CEs of t4 are plausible but less useful due to their *inherently limited extent*: for instance, categorizing instances of `geopolitical:area` as `geopolitical:isSuccessorOf value` X, where X is another (geopolitical) area.

Novice ontologist assessment. There were two groups of students involved: Bc-level students in a course on Artificial Intelligence (AI) and MSc-level students in a specialized course on Ontological Engineering (OE). Both courses provided a certain degree of OWL modeling experience (in Protégé and Manchester syntax) prior to this exercise, although OE went into more depth as regards the underlying DL and reasoning. There were 17 AI students and 10 OE students altogether. In both courses the students were first provided with a 30' overview of the notions of CE (in $\mathcal{L}_0$), OC and FC roughly as presented in Sect. 1 of this paper. Then they completed an assignment consisting of 20 atomic tasks, all available in a single sheet of a web questionnaire.[12] In each atomic task the student was required to provide an answer to the question "Is the class CE a meaningful category for categorizing objects of class FC", where FC was a named focus class and CE was a concept expression in Manchester syntax. The answer was again from the 5-point Likert scale, with an additional option '*no judgment, since I don't understand the example*'.

The 59 CEs from the initial sample were randomly divided into three questionnaire versions (one value restriction CE was used twice) to eliminate cribbing; the numbers of returned questionnaires per version were 7, 9 and 11, respectively, with balanced proportion of AI vs. OE students. To avoid protracting and biasing the experiment, the students were instructed to only judge the CEs by the expression itself, i.e. without consulting the respective ontology specification or other external resources. However, specifically for the 'conference' domain of OntoFarm, they were provided with a brief domain glossary (since as students they were not expected to have experience with conference organization matters). In both sessions, 30' sufficed to all students for completing the (20-task) assignment.

We aggregated the results by questionnaire task, and then both by the course and by CE type. The aggregation was carried out by simple summation over the values rescaled to the $[-1; 1]$ interval (i.e. 'certainly' turned to 1, 'perhaps' to 0.5, both 'borderline' and 'no judgment' to 0, etc.), and then normalized by dividing by the number of students on the task. This way, for example, a task assigned to eight students, with the responses 'certainly', 'perhaps', 'borderline' and 'perhaps not', all present twice, yields the normalized sum (NS) of $(2 * 1 + 2 * 0.5 + 2 * 0 + 2 * -0.5)/8 = 0.25$.

A short digest of the results follows:

- The average NS over all 60 tasks was 0.07, i.e. rather low, although positive. Of the 60 NS values, 28 were positive, 5 zero and 27 negative. The values strictly below 0.25 and above -0.25, possibly viewed as 'borderline aggregates', were 34 (57 %).

[12] The questionnaire was in Czech. Its English translation is available from the paper web page http://owl.vse.cz:8080/EKAW2016/.

- The cases[13] with highest positive and lowest negative values are in Table 2; the type is listed in the third column. We see that cases with highest positive polarity tend to achieve higher absolute values than cases with highest negative polarity, and that t4 dominates the upper end of the spectrum. Interestingly, the negative cases correspond each to a different type and also have different semantic roots: the village with baseball player head-chef was already discussed before (distant and dubious inference), the 'conference in city' one deals with a seemingly mandatory property leading to $OC \equiv FC$ (here, however, the experts' consensual opinion diverged: how about future editions not yet having a location, or virtual conferences?), and the 'day followed by Friday' only holds for one individual, in turn.
- The average NS was higher for the OE students (0.12) than for the AI students (0.04), which might be attributed to more developed 'ontologistic thinking' of the latter. The *inter-task* variance, indicating the tendency towards giving uneven values (averaged over the students filling the same task) across the questionnaire, was about the same (0.16) for both courses. However, the *intra-task* variance, indicating the degree of agreement within the students filling the same task, was higher for the AI students (2.51) than for the OE students (2.12), i.e. the rating of the latter was more coherent.
- The average NS was highest for t4 (0.21, with 15 positives, 1 zero and 7 negatives), lower for t3 (0.02, with with 9 positives, 3 zero and 8 negatives) and lowest for t2 (−0.05, with with 4 positives, 1 zero and 12 negatives).

In comparison with the 'expert ontologist' assessment:

- The students gave a significantly lower score: only about a half of CEs are viewed as OCs, compared to 88 % (52/59) by the final consensus of experts. This can be explained by their lower ability to figure out specific situations in which less obvious CEs might become plausible.
- If we apply the same method of average NS computation on the initial assessment of experts the proportion of 'borderline aggregates' between −0,25 and 0,25 is only 14 % (in contrast to 57 % for the students' values).
- There is agreement on less frequent OC status of t2 (i.e. lower reliability of pattern p2). Out of the 17 respective CEs, as mentioned above, only 4 were viewed as OCs by students and 12 by the experts (who in turned judged all CEs of other types, except two, as OCs).
- As regards the case-by-case comparison between students and experts, there is also correlation in the sense that the 43 experts' clear positives obtained a positive average NS from students (0.14), while the 14 initially 'clash' cases obtained a slightly negative average NS (−0.07) and the 3 negative cases obtained a clearly negative average NS (−0.24).

[13] Most namespace prefixes used can be expanded using the prefix.cc service. Prefixes unlisted by this service follow: p-act=http://purl.org/procurement/public-contracts-activities#, p-aut=http://purl.org/procurement/public-contracts-authority-kinds#, p1=http://www.loc.gov/premis/rdf/v1, p1-sm=http://id.loc.gov/vocabulary/preservation/storageMedium#, sigkdd=http://oaei.ontologymatching.org/2016/conference/data/sigkdd.owl.

Table 2. CEs with highest and lowest average NS of student scores

FC	Expression	Type	Avg.NS
ofrd:FridgeFreezer	ofrd:styleOfUnit value ofrd:SingleDoor	4	0.91
gr:BusinessEntity	pco:mainActivity value p-act:GeneralServices	4	0.86
gr:BusinessEntity	pco:authorityKind	4	0.61
	value p-aut:LocalAuthority		
akt:Generalized-Transfer	akt:information-transfer-medium-used	4	0.59
	value akt:Email-Medium		
p1:Storage	p1:hasStorageMedium value p1-sm:mag	4	0.59
fabio:Item	fabio:isStoredOn value fabio:web	4	0.50
...	...	...	...
dbo:Village	dbo:headChef some dbo:BaseballPlayer	3	−0.50
sigkdd:Conference	sigkdd:City_of_conference some Thing	2	−0.56
gr:DayOfWeek	gr:hasNext value gr:Friday	4	−0.56

As regards the *research questions* from the start of this section: ad (1) the OC status assessment strongly depends both on the CE type and the expert/novice distinction; ad (2) most cases are clear for experts but not for novices; (3) semantic distinctions leading to negative or inconclusive assessment might be related to temporality, inference over distant paths or inherently small cardinality of some concepts (referring to the exemplified insights above).

A conclusion to be made with respect to FOCP computation is that t4 (value restriction) and to some degree t3 (existential restriction) might successfully complement t1 (named class) in the role of OC. As regards the relatively poor performance of p2, it might be premature to completely abandon it at this phase, since for some models it might yield the only unnamed CEs, as the analysis from Sect. 4 indicates, and its contribution to the overall categorization power should still be considered. A naïve weighted approximate FOCP formula for $\mathcal{L}_0$ and $P = \{p1, p2, p3, p4\}$, derived from the students' assessment, could be

$$\widehat{FOCP}(FC, O, \mathcal{L}_0, P) = Occ(p1, FC, O) * 1.0 + Occ(p2, FC, O) * 0.3$$
$$+Occ(p3, FC, O) * 0.5 + (Occ(p4, FC, O) + Occ(p5, FC, O)) * 0.7$$

where the numerical weight coefficients roughly correspond to the ratio of CEs with positive average NS, per type.

6 Related Work

Since we are unaware of prior work on precisely the same topic, we reference to related research that only overlaps with ours at the abstract level (mere notion of "classification power"), systematically applies other kinds of metrics on ontologies, or addresses similar application-level goals (ontology reuse or transformation) by other means.

The term classification/categorization power previously appeared in many scientific texts, however, rarely as a rigorously defined notion. For example, on many occasions, automated classifiers (typically, machine-learning-based) are reported to have certain 'classification power' with respect to classes from an ontology, which is merely an informal circumscription of measures such as accuracy or error rate. The 'power' also clearly pertains to the classifier and not to the ontology itself. Partially relevant is the analysis made by Giunchiglia and Zaihrayeu [3], who categorized 'lightweight' ontologies with respect to two dimensions: complexity of labels (simple noun phrases vs. use of connectives and prepositions) and use of 'intersection' operator allowing to combine atomic entities of different nature (e.g., the atomic concepts 'Italy' and 'vacation' implicitly combine into 'vacation in Italy'). Maximal 'classification power' is obtained when both explicitly complex labels and implicit concept combinations are allowed. This however only applies to classifying documents extrinsic to the ontology, since 'intersection' of concepts of different nature is not coherent with the set-theoretic semantics of DL. Overall, their 'classification power' is a global property of the method by which the ontology has been built. In contrast, our notion of FOCP applies to individuals intrinsic to the DL world of the ontology and is calculated with respect to a focus class. Under the restrictions assuring finiteness of the CO set, FOCP can be expressed as a numerical value.

As regards the analysis of ontology repositories in terms of various aggregated features and metrics (logical-structural, graph, lexical etc.), there has recently been renewed interest, following up with the early work of Tempich and Volz [8] (aiming to build a benchmark for testing ontology tools). A large scale study of OWL ontology metrics has been carried out by Matentzoglu et al. [5]. However, the categorization power of ontologies has not been, to our knowledge, studied (never mind with the flavor presented here).

Our own ongoing work on the PURO modeling language [7] deals with various options how the same "background" state of affairs can be expressed in OWL. PURO structurally resembles OWL but relaxes some of its modeling constraints. A library of transformation patterns allows to proceed from one PURO model to alternative OWL ontologies in different encoding styles. An example relevant to our case is the notion of *enterpreneur*, which is likely to be expressed as type in PURO, but could be translated to relationship (`insuranceCategory`) restricted to the `Entrepreneur` individual in OWL (i.e. a CO based on a compound concept expression), assuming we prefer a style using object properties with "codelist individuals". Analogously, *born in* may possibly be a relation in PURO but can be translated not only to OWL property restrictions but also to named classes such as `PersonBornInUK`, assuming we prefer an "encapsulating" encoding style used, e.g., in the DBpedia Ontology.[14] Modeling in PURO and applying the transformation patterns may thus make hidden COs explicit in the domain. A similar account of alternative "typecasting" (but with smaller coverage) has been given by Krisnadhi et al. [4].

[14] http://wiki.dbpedia.org/services-resources/ontology.

The broad context of our research, the task of ontology reuse, has been studied by Schaible et al. [6]: the users expressed their preferences on reuse strategy in a survey. The results indicate that reusing multiple entities from the same vocabulary (even if some of them are by themselves less popular than analogous entities from other vocabularies) may often be preferred; this corroborates the relevance of our approach to measuring the categorization power of ontologies with respect to focus classes. Reuse support [2] is also systematically sought by the maintainers of LOV [9], primarily at keyword relevance level; we are in contact with them and will seek to integrate our complementary approaches.

7 Conclusions and Future Work

Ontologies are an important means of subcategorizing entities already known to belong to a general focus class, and the scope of subcategories need not be confined to named classes, especially in the linked data world, which is relatively 'property-centric'. High categorization power of an ontology for certain classes might serve as argument for *reusing* this ontology when building a new one, or preparing a dataset scheme, if the notions corresponding to these classes are important in the modeled domain. Additionally, plausible compound concept expressions can be *transformed* to named ones if needed. We provided, to our knowledge, the first systematic study on the categorization power of OWL ontologies covering several compound concept expression patterns. The main contributions of the paper are: (1) formulation of the problem and description of the pattern set; (2) empirical analysis of two ontology collections for (syntactical) pattern occurrence; (3) in-depth ontologistic analysis of a sample of pattern occurrences, carried out both by ontologistic experts (paper authors) and two groups of slightly trained students.

While the presented research focused on the general principles and empirical analysis, aspects of it have also been implemented into our OOSP tool, allowing to recommend ontologies with respect to their estimated FOCP.[15] The estimation currently does not distinguish between the CE types; it is however going to be tuned by lowering the weight for instances of patterns exhibiting lower proportion of plausible OCs, as indicated in Sect. 5.

As future work we plan to investigate if inclusion of additional types of concept constructors into the $\mathcal{L}$ language could still yield relevant results (true OCs) while keeping the computational complexity tractable. One candidate is the *inverseOf* predicate: in some 'modeling styles' only one of the pair of mutually inverse properties is included in the ontology and some categorization tasks might then be carried out against the direction of such a relationship.

In addition to the structural aspect of the CEs, we may also leverage on their *lexical* aspects. Presence of suffixes such as `-Category`, `-Type` or `-Kind` in either

[15] This work has been published separately as a demo paper [10]. The demo paper only contains a minimalist informal explanation of the notions of FC, CE and OC; apart from that the content of the demo paper is disjoint with the current submission.

property or filler class names may indicate that the filler individual is actually a type (note the highly scoring case with `pco:authorityKind` in Table 2).

We also envisage to align this schema-oriented analysis to *data-oriented* one. If a representative sample of data is available for an ontology, we can of course derive the OC status of CEs from the numbers of FC instances satisfying them (or not), which can be obtained by a simple SPARQL query. This empirical analysis will also allow us to refine the reliability estimate of the patterns (in combination with ontologistic assessment as in this paper) and clues used in the schema-oriented approach, which is usable even if instance data are not available.

In longer term, the goal is to embed a properly tuned ontology/fragment recommender into our OBOWLMorph OE tool [1] and also make available via an API to third-party tools such as the ProtégéLOV Plugin [2].

Acknowledgment. Ondřej Zamazal has been supported by the CSF grant no. 14-14076P, "COSOL – Categorization of Ontologies in Support of Ontology Life Cycle". Vojtěch Svátek has been supported by the the UEP IGA F4/28/2016 project.

References

1. Dudáš, M., Hanzal, T., Svátek, V., Zamazal, O.: OBOWLMorph: starting ontology development from PURO background models. In: Tamma, V., Dragoni, M., Gonçalves, R., Ławrynowicz, A. (eds.) OWLED 2015. LNCS, vol. 9557, pp. 14–20. Springer, Heidelberg (2016). doi:10.1007/978-3-319-33245-1_2
2. García-Santa, N., Atemezing, G.A., Villazón-Terrazas, B.: The ProtégéLOV plugin: ontology access and reuse for everyone. In: Gandon, F., Guéret, C., Villata, S., Breslin, J., Faron-Zucker, C., Zimmermann, A. (eds.) ESWC 2015. LNCS, vol. 9341, pp. 41–45. Springer, Heidelberg (2015). doi:10.1007/978-3-319-25639-9_8
3. Giunchiglia, F., Zaihrayeu, I.: Lightweight ontologies. In: Liu, L., Tamer Özsu, M. (eds.) Encyclopedia of Database Systems, pp. 1613–1619. Springer, Heidelberg (2009)
4. Krisnadhi, A.A., Hitzler, P., Janowicz, K.: On the capabilities and limitations of OWL regarding typecasting and ontology design pattern views. In: Tamma, V., Dragoni, M., Gonçalves, R., Ławrynowicz, A. (eds.) OWLED 2015. LNCS, vol. 9557, pp. 105–116. Springer, Heidelberg (2016). doi:10.1007/978-3-319-33245-1_11
5. Matentzoglu, N., Bail, S., Parsia, B.: A snapshot of the OWL Web. In: Alani, H., et al. (eds.) ISWC 2013. LNCS, vol. 8218, pp. 331–346. Springer, Heidelberg (2013). doi:10.1007/978-3-642-41335-3_21
6. Schaible, J., Gottron, T., Scherp, A.: Survey on common strategies of vocabulary reuse in linked open data modeling. In: Presutti, V., d'Amato, C., Gandon, F., d'Aquin, M., Staab, S., Tordai, A. (eds.) ESWC 2014. LNCS, vol. 8465, pp. 457–472. Springer, Heidelberg (2014). doi:10.1007/978-3-319-07443-6_31
7. Svátek, V., Homola, M., Kluka, J., Vacura, M.: Metamodeling-based coherence checking of OWL vocabulary background models. In: OWLED 2013, Montpellier (2013)
8. Tempich, C., Volz, R.: Towards a benchmark for semantic Web reasoners - an analysis of the DAML ontology library. In: EON-2003 (2003)
9. Vandenbussche, P.-Y., Vatant, B.: Linked open vocabularies. ERCIM News, 21–22 (2014)
10. Zamazal, O., Svátek, V.: Ontology search by categorization power. In: Workshop SumPre 2016 at ESWC (2016)

Selecting Optimal Background Knowledge Sources for the Ontology Matching Task

Abdel Nasser Tigrine(✉), Zohra Bellahsene, and Konstantin Todorov

LIRMM/University of Montpellier, Montpellier, France
{Tigrine,Bella,Todorov}@lirmm.fr

Abstract. It is a common practice to rely on background knowledge (BK) in order to assist and improve the ontology matching process. The choice of an appropriate source of background knowledge for a given matching task, however, remains a vastly unexplored question. In the current paper, we propose an automatic BK selection approach that does not depend on an initial direct matching, can handle multilingualism and is domain independent. The approach is based on the construction of an index for a set of BK candidates. The couple of ontologies to be aligned is modeled as a query with respect to the indexed BK sources and the best candidate is selected within an information retrieval paradigm. We evaluate our system in a series of experiments in both general-purpose and domain-specific matching scenarios. The results show that our approach is capable of selecting the BK that provides the best alignment quality with respect to a given reference alignment for each of the considered matching tasks.

1 Introduction

Over the past years, the web has been continuously evolving from a web of documents to a web of data, following the principles of data and knowledge representation, publishing and linking. The Linked Open Data project[1], the web of data most successful initiative to date, comprises nowadays hundreds of datasets over several domains of life and science. While information is expressed by the help of RDF (Resource Description Framework) statements, knowledge about the domains of interest is given in the form of ontologies, which provide common vocabularies to name classes of things (concepts) and relations between these classes, defining in an explicit manner their semantics. Ontologies, expressed in RDFS, OWL or SKOS, can be simple sets of terms, thesauri or more complex structured vocabularies with logical expressions that allow for the inference of new facts.

It occurs often that ontologies, describing similar or equivalent domains of knowledge, are expressed differently. These differences, referred to as ontology heterogeneities, can occur in terms of terminology (choosing different names to refer to the same concepts and relations), structure or semantics (relating classes

[1] http://linkeddata.org.

E. Blomqvist et al. (Eds.): EKAW 2016, LNAI 10024, pp. 651–665, 2016.
DOI: 10.1007/978-3-319-49004-5_42

in different ways, giving different intensions to information) or simply in terms of syntax (choosing different formal representations). In order to unlock the potential of the web of data and foster the creation of a veritable information network, heterogeneous ontologies have to be linked together by explicitly declaring the equivalence relations between their entities (classes and properties). The field of ontology matching has taken the challenge of proposing solutions that allow to automatically discovering the correspondence between ontological elements in the presence of one or more of the heterogeneities cited above. As a result of almost 20 years of research and practice, many approaches and systems have been developed, capable of aligning highly heterogeneous ontologies [1].

It has been shown in recent publications [2,3] that the ontology matching process can benefit largely from the use of Background Knowledge (*BK*). *BK* is understood as any external reference knowledge that can facilitate the matching process, given in the form of large general-purpose ontologies or well-established knowledge graphs (such as DBPedia or YAGO), domain specific ontologies, or the web at large. We outline three main advantages of the use of *BK* when aligning ontologies. In the first place, as observed by [4], there is always an inherent semantic gap between two ontologies, coming from the missing semantic context of their acquisition. *BK* can help close that gap, as shown in [5]. In the second place, and even more importantly, the ontology matching process is heavy and costly: most of the existing matching tools are complex engineering artefacts comprising a sophisticatedly orchestrated pipeline of matching modules, mapping filtering and semantic verification components. In a recent study [3], we have shown that an appropriately chosen *BK* source can help significantly lighten the overall matching process. Finally, in specific domains there is a clear need for specific reference knowledge, since the commonly used external knowledge sources, such as WordNet, fail to provide the semantic information that is needed to discover correctly the correspondences between domain specific concepts.

While there is little doubt about the benefits of using *BK* for ontology matching, outlined as one of the challenges for the field by [1], an important question remains largely unanswered: how to select an optimal *BK* source for a given ontology matching task out of a set of known *BK* sources? We understand "optimal" as the source that provides the best quality of the alignment produced for two ontologies. In this paper, we attempt to answer this question by proposing an approach for the automatic selection of a *BK* source for a given ontology matching task. We situate the problem in an information retrieval framework. The set of known *BK*s is indexed by using the well-known vector space model while the two ontologies to be aligned are represented as a query document. The comparison between the ontologies and the *BK*s is based on their content, but also on their structure. We elaborate on the different choices of a similarity measure for this task. Particularly, we show that the commonly used cosine similarity is not the best choice in this scenario and we propose the use of correlation-based similarity measures. The selection system that implements this approach has the properties of being *fully automatic*, *domain independent* and *multilingual*, as well as being entirely *dissociated from the alignment process*.

In order to evaluate the proposed approach, we carry out experiments on benchmark data coming from the Ontology Alignment Evaluation Initiative (OAEI). We used as background knowledge sources a mixed set of domain specific and general purpose knowledge graphs. The results show that our approach guarantees the selection of the optimal *BK* source with respect to each of the matching tasks that has been performed in terms of the quality of the produced alignment by using the selected *BK*.

The paper is structured as follows. The following section introduces the BK selection problem by focusing on requirements and criteria for selection. Section 3 describes our approach that we support experimentally in Sect. 4. Several related results are discussed and compared to our method in Sect. 5 before we conclude in Sect. 6.

2 The Background Knowledge Selection Problem

Ontology matching is the process of automatically discovering semantic correspondences between the entities of two ontologies that are assumed to cover overlapping domains of knowledge. The result of the ontology matching process is a set of pairs of cross-ontology entities (names of concepts or properties) called an *alignment*, where the entities of each pair are bound by a given semantic relation (most commonly equivalence) and each pair of entities is assigned a confidence value indicating the strength of this relation [6].

As pointed out in the introduction, using an appropriate background knowledge source (hereafter, *BK* for short) in the matching process can help improve the results and, potentially, decrease the complexity of the process. A *BK* is understood as any piece of external information that can be used in order to improve the matching quality. According to [1], one can consider as *BK* for ontology matching a large range of external sources, such as linked data, domain specific corpora of schema and alignments, domain specific or general purpose ontologies, dictionaries and thesauri, lexical databases. Since, on the one hand, the choice of *BK* has a direct impact on the results and, on the other hand, – there is a multitude of available *BK* sources, researchers in the field have recognized the need for an approach to select automatically the optimal *BK* for a given matching task.

2.1 Criterion of Optimality of a *BK*

There is a large set of *BK* sources that can be considered, some of them not even known to the user. For that reason, it is important to frame formally the criterium that defines an optimal *BK*. Since the aim of using a *BK* is to provide good quality matching results no matter the provenance and nature of the *BK*, its choice has to be motivated by the maximization of the alignment quality. Suppose that for a given matching task (a pair of ontologies to be aligned) there exists a reference alignment, either provided by an expert or produced automatically. The best *BK* for this matching task, that we call *optimal BK*, is selected as the one that produces the best alignment with respect to the reference in terms of F-measure.

Let $\Gamma = \{BK_1, ..., BK_n\}$ be a set of known BKs. Let s, t be two ontologies (s for *source* and t for *target*) and $\mathcal{A} = \{A_1, ..., A_n\}$ – a set of alignments of s and t each using a different BK from Γ, given as $A_i = (s, t, A_{ref}, F_i)$, where A_i is produced by using BK_i, A_{ref} is a reference alignment for s and t and F_i is the corresponding F-measure score of A_i with respect to A_{ref}, $\forall i = 1, ..., n$.

Definition 1 (Optimal BK). *We define the* **optimal** *background knowledge source for the task of matching t and s as the background knowledge source $BK_o \in \Gamma$, which corresponds to an alignment from $\mathcal{A}$ with maximal F-measure value. In case multiple BKs produce alignments with maximal F-measure values, the one with the lowest number of entities is defined as optimal.*

This optimality criterion can be verified only in the presence of a reference alignment. Since this criterion plays a role in the process of conception, development and configuration of a BK selection systems, which is tested on ontologies with existing reference alignments, we consider that this definition of optimality is fair.

Note that the optimality criterion is inherently semantic, although this remains implicit in the definition. The BK-based ontology matching systems are designed in such manner that the BK that maximizes the F-measure value of an alignment of two ontologies is the one that is closest in content to these ontologies. This is the reason why our selection method is based on content similarity between the BK sources and the input ontologies. In our experiments, we show that a BK selected by the help of this method is optimal in the sense of Definition 1 (see Sect. 4).

As a final remark, note that we do not base the selection criterium on the improvement provided by a BK as compared to a direct matching. Our assumption is that an user is looking for a BK because she is not satisfied with the results achieved by a direct matching, or in other words, we assume that the best BK-assisted alignment is better than the best direct one for the particular matching task (although not in general).

2.2 Requirements to an Automatic BK Selection System

We outline the requirements that, in our view, a BK selection system has to meet.

1. BK **type independence.** The system has to be able to take into account BKs expressed or serialized differently (as long as there exists a parser able to extract the textual information and structure from the BKs) as well as BKs of different semantic nature (thesauri, ontologies, lexical databases, corpora).
2. **Domain independence.** The system should be able to propose a BK for a pair of ontologies of any given domain of knowledge.
3. **Multilingualism.** The system has to be able to select a BK that assists the alignment of cross-lingual ontologies.

4. **Optimality.** It should be guaranteed (experimentally during the system conception and tuning) that the system returns an optimal BK with respect to a matching task in the sense of Definition 1.

3 An Information Retrieval Approach to Automatic Selection of *BK*

We situate the BK selection problem in an information retrieval framework. We consider Γ as a corpus of documents and a given pair of ontologies s and t to be aligned – as a query document in the form of a set of terms. The corpus Γ is indexed in order to represent its content by using standard information retrieval techniques, that we explain further on. One of the particularities of our approach consists in the fact that we construct a common index for all known BKs, independent on their domain and focus. As we shall see, in this way we reply to the first two requirements given in the previous section. Additionally, the effort of indexing large ontologies is performed only once, which contributes to the efficiency of the approach. Note also, that the query is given in the form of a unique document representing the pair of ontologies (and not one document per ontology), because we want to retrieve the background knowledge that is common for the two and therefore allows for their reconciliation. In the following, we first explain how we transform the BK sources and the query ontologies to documents. Then we describe the indexing process and present a set of similarity measures that we use in our approach for retrieving the optimal BK for a given matching task. The overall process is depicted in Fig. 1.

3.1 Modeling Ontologies and *BK*s as Structure-Content Documents

We map a BK source to a text document that we call a BK-document in a manner that allows for taking both the content and the structure of the knowledge sources into account. The term extraction method consists in the following. Each token of the labels of the concepts of a given BK becomes a term in the corresponding BK-document. In order to preserve the information relevant to the BK structure, we add a given term, appearing in the label of a given concept A, to the BK-document every time when it appears also in labels of sub- or super-classes within the BK hierarchy or, when it appears in the label of a concept in the domain or the range of a property of A or, more generally speaking, when it appears in the label of a concept related to A by a relation of any kind (synonymy, subsumption, *etc.*).

To illustrate this idea, take a part of a BK where the classes {*Author, Administrator, PhD Student, Professor*} are all in a `subclassOf` relation with the class *Person*. Without the use of the structure, the resulting BK-document would be d_{BK} = {*Person, Author, Administrator, PhdStudent, Professor*} with all the terms having the same weight with respect to their frequency of occurrence. However, the term *Person* seems to be more important than the other terms in the BK as it is a label of their common superclass. Not considering

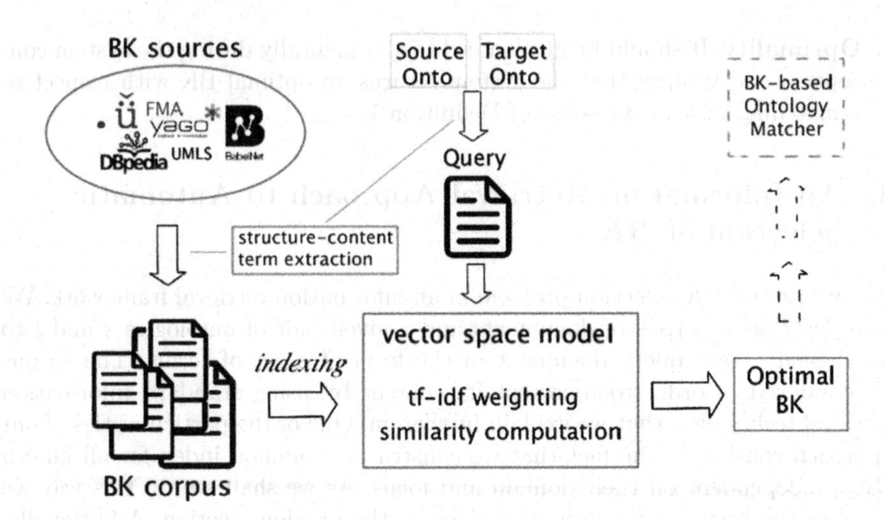

Fig. 1. The *BK* indexing and selection process. The elements in dashed lines are not part of the selection workflow.

this semantic information affects considerably the computation of weights in the indexing phase. By using the structural information, the *BK*-document becomes d'_{BK}={*Person, Person, Person, Person, Person, Author, Administrator, PhD-Student, Professor*}.

We model in this manner all known *BK*s in Γ, resulting in the corpus $D_\Gamma = \{d_{BK_1}, d_{BK_2}, ..., d_{BK_n}\}$. We proceed in the same manner in order to represent a pair of ontologies, t and s, as a query document, denoted by $q_{t,s}$ by creating a document for each ontology as shown above and then merging the two documents into a single one.

3.2 Indexing

Prior to indexing, we apply a standard set of text preprocessing methods, such as normalization of characters and spaces, removing diacritics or accents, deleting numbers, punctuations and stop words, tokenization and lemmatization.

We index the documents in D_Γ by using the well-known vector model. We build an indexation matrix M, which has *BK*-document vectors as rows and index term vectors as columns. Standardly, the index terms are the terms collected from the *BK*-documents without repetition after preprocessing. We denote the set of index terms by C_t. Each element w_{ij} of M corresponds to the weight of the term t_j with respect to d_{BK_i}.

In order to compute the weights w_{ij}, we use the well-known TF-IDF (Term Frequency – Inverse Document Frequency) weighting scheme that captures the importance of a term both within a single document and in a collection of documents. For a term t_j and a BK-document d_{BK_i}, the weight w_{ij} is calculated as follows:

$$w_{ij} = tf(t_j, d_{BK_i}) * log\left(\frac{n}{df(t_j)}\right), \tag{1}$$

where $tf(t_j, d_{BK_i})$ is the frequency of occurrence of the term t_j in the document d_{BK_i} and $df(t_j)$ is the number of documents containing t_j and n is the number of documents in D_Γ.

3.3 Retrieving the Optimal BK

We need a similarity measure of some kind, which allows the system to evaluate the relatedness of a given BK source to the input ontologies query. In information retrieval, the most commonly applied similarity measure is the cosine similarity, denoted by *cosine*, given as the normalized dot product of two document vectors. We noticed that in certain cases parameter free measures of (rank) correlation can be applied more successfully than the cosine similarity measure for our particular problem. A measure of correlation expresses the degree of dependence of two random variables. It takes values between −1 and 1, assuming strong correlations for high positive values, strong anti-correlations for high negative values and no dependence for values close to 0. We introduce two of the most popular choices for correlation measures and show how they can serve as similarity measures for two documents [7]. Note that the correlation measures to be presented, just as the cosine measure, are based on the dot product of vectors. In that sense, the cosine similarity can be seen as a correlation, just as the correlations can be seen as similarity measures.

Pearson Correlation. The Pearson's coefficient r between two variables X and Y is calculated from their covariance $cov_{X,Y}$ and their standard deviations σ_X and σ_Y in the following way:

$$r_{X,Y} = \frac{cov_{X,Y}}{\sigma_X * \sigma_Y} = \frac{\sum_i (x_i - m_X)(y_i - m_Y)}{\sqrt{\sum_i (x_i - m_X)^2}\sqrt{\sum_i (y_i - m_Y)^2}}, \tag{2}$$

where m_X and m_Y are the means of X and Y.

In our case, each of the variables X and Y corresponds to either a BK-document or a query document that all live in the same space and are therefore representable by the same type and number of features (i in (2) takes values from 1 to the number of index features). The sets of values $\{x_i\}$ and $\{y_i\}$ correspond to the values of the $tf * idf$ vectors in our vector model. Pearson is used to test the linear dependency of variables.

Spearman Correlation. Spearman's coefficient provides a measure of the correlation of two variables X and Y represented as lists of statistical rankings.

In contrast to Pearson, it is applied when a general monotonic relationship is expected between the variables. Formally, it is given as

$$\rho_s = 1 - \frac{6\sum_i d_i^2}{n^3 - n}, \tag{3}$$

where d_i is the difference between the rank of the ith observation of the variable X and the rank of ith observation of the variable Y, $d_i = rank(x_i) - rank(y_i)$. We transform the $tf * idf$ values to ranks by using the partial order on real numbers. In case of equal values, we assign equal ranks.

BK Selection. For a given corpus of indexed BK-documents D_Γ, a query $q_{s,t}$ and a similarity measure $\sigma \in \{cosine, r, \rho_s\}$, the selected BK is the one whose BK-document maximizes the similarity between the query and the documents in D_Γ, given as

$$D_{BK_{s,t}} = \underset{D_{BK} \in D_\Gamma}{\arg\max}\, \sigma(D_{BK}, q_{s,t}). \tag{4}$$

In case more than one BKs provide a maximal similarity to the query, the one with lowest number of entities is selected.

In case where an ontology matching system uses more than one BK at a time, we can recommend the top-K BKs, with respect to their similarity to the input ontologies. These sources can be further combined – a problem that is not addressed in this paper. Where no optimal BK is found in the set of BKs, we can easily apply threshold limits to avoid this scenario. In our study, we have considered that an optimal BK always exists in the set of known BKs. We show experimentally in Sect. 4 that the selected background knowledge source $BK_{s,t}$ corresponding to the BK-document $D_{BK_{s,t}}$ is optimal in the sense of Definition 1.

4 Evaluation and Results

In order to evaluate our approach, we have conducted experiments on data coming from the OAEI[2] of year 2015. Our aim is to test if our approach selects the optimal BK among a set of BK sources in two scenarios: selecting a domain specific BK or selecting a general purpose BK. We take a set of BKs Γ = {Yago, DBpedia, BabelNet, DBnary, FMA, Doid, Uberon}, described below. The set Γ includes general purpose sources (Yago, DBpedia, BabelNet and DBnary), as well as several domain specific anatomy and biomedical BKs (FMA, Doid, Uberon).

As described in Sect. 3, the selected BK for each task has to conform to the optimality criterion given in Definition 1. In order to verify this criterion, we need a F-measure score produced as a result of the comparison of an alignment given by an ontology matching system that uses BK in its process and a reference alignment. For the totality of our experiments, we have used the system

[2] http://oaei.ontologymatching.org/.

LYAM++ [3] developed in our research group, which is entirely based on BK and has also shown to perform well on the OAEI MultiFarm track on which it participated last year [8]. LYAM++ does not use complex matching methods with regard to the BK. For instance, LYAM++ uses the *hasSynonyms* relation present in a given BK to match between two concepts.

Note again that the aim of these experiments is not to show the quality of the ontology matching tools, but to evaluate the performance of the BK-selection approach proposed in this paper. The pairs of ontologies to be aligned, as well as the BKs in Γ are modeled as documents and indexed as described in the previous section. In the retrieval phase, we have tested the performance of the three similarity measures given in Sect. 3 - the commonly used cosine similarity and the two parameter-free correlation coefficients (Pearson and Spearman).

4.1 The OAEI Tracks

The **Anatomy track** aims at discovering alignments between a human anatomy ontology, part of the NCI Thesaurus[3] and a mouse anatomy ontology. This track is considered as a large-scale matching task because the input ontologies are of a large size and very rich semantically. The **Large Biomedical track** aims at aligning three large bio-medical ontologies, namely FMA, SNOMED and the NCI Thesaurus. Since FMA appears in a query couple, we did not consider this ontology as background knowledge source for this track. The Anatomy and the Large biomedical tracks have been selected to make sure that at each experiment our approach selects from Γ the right BK for the right domain, respectively anatomy and biomedicine.

The **Conference track** contains a dataset of about 15 ontologies from the scientific publication field, together with reference alignments. We have used with the Conference dataset to test if the approach selects the optimal general purpose BK among the mixed set of BKs Γ. Note that, although the Conference data are specific to the scientific publishing domain, they can be considered as general purpose due to the type of concepts that are used to describe this domain, often dealing with common sense knowledge.

Finally, the **MultiFarm track** is derived from the Conference track data by translating the conference ontologies into several different languages with the aim to challenge the performance of cross-lingual ontology matching tools. In our scenario, this track is appropriate for testing whether the selection procedure is able of choosing an optimal multilingual knowledge source for aligning cross-lingual ontologies.

4.2 BK Sources

We give a quick overview of the BK sources used in this evaluation. BabelNet [9] is a multi-lingual semantic network and an ontology that has been built by merging different encyclopedic and linguistic resources. The integration of these

[3] https://ncit.nci.nih.gov/ncitbrowser/.

resources has been conducted automatically. BabelNet appears to be an appropriate choice of a *BK* for the *MultiFarm* track. DBnary[4] [10] is a multilingual lexical database extracted from Wiktionary, which is also a potentially good candidate for the MultiFarm track. DBpedia [11] is a large multilingual knowledge graph extracted from Wikipedia, covering a multitude of areas such as music, films, people, places, *etc.* YAGO [12] (Yet Another Great Ontology) is a another large multilingual general purpose knowledge graph extracted from Wikipedia, GeoNames[5] and WordNet. Both YAGO and DBpedia can be considered as candidates for a large variety of general purpose ontology matching tasks. Doid[6] is an open source ontology for the integration of biomedical data associated with human diseases. In our evaluation setting, Doid can be potentially used for aligning Biomedical or Anatomy-related ontologies. FMA (Foundational Model of Anatomy) [13] is the reference ontology for the anatomy field. Similarity, UBERON [14] is a multi-species anatomy ontology. Both ontologies can be used for the Anatomy ontology matching task.

4.3 Results Presentation and Analysis

The results from this series of experiments are presented in Tables 1 and 2. For each *BK*, we provide the average similarity scores obtained by each of the three similarity measures described in Sect. 3 and the corresponding average F-measure values obtained in the *BK*-based alignment. The average values are computed over all pairs of ontologies in each track. The best score achieved by each similarity measure, as well as the highest F-measure value are highlighted. The two correlation coefficients take values in the $[-1, 1]$ interval, which explains the negative numbers. We are interested in strong correlations, corresponding to high similarity. Recall that according to our criterion, the *BK* selected by a similarity measure is optimal if it guarantees the highest F-measure.

As it can be seen from the results in Tables 1 and 2, our approach systematically selects the optimal *BK*, independently on the track, by using the Spearman correlation and very conclusively so with similarity scores around and above 0.5. The performance of the selection method is flawed when using the cosine similarity, which is the common choice for a similarity measure in an information retrieval setting, but ranks only second best in our experiments. Namely, it fails to detect the optimal *BK* on the Anatomy and the Conference tracks and on the other tracks its outcome does not always help to come up with a clear cut decision.

We explain this observation by the fact that the Spearman measure is based on ranks instead of real values and on monotonic dependencies between the two vectors. Precisely, a small variation in the corresponding values of two vectors influences the cosine similarity negatively, while this is less so in case of Spearman. Spearman seems to be also better suited to dealing with highly sparse data

[4] http://kaiko.getalp.org/about-dbnary/.
[5] http://www.geonames.org/.
[6] http://do-wiki.nubic.northwestern.edu/do-wiki/index.php.

Table 1. Anatomy and large biomedical tasks

	Anatomy				Large Biomedical			
	Cosin	Pearson	Spearman	F-m	Cosin	Pearson	Spearman	F-m
BabelNet	0.03	0.06	−0.76	0.05	0.0	0.0	−0.66	0.0
DBnary	0.02	0.07	−0.72	0.0	0.02	0.02	−0.51	0.0
DBpedia	0.01	0.01	−0.70	0.0	0.05	0.06	−0.60	0.0
Doid	0.10	0.15	−0.14	0.66	0.15	**0.14**	0.40	0.53
Uberon	0.26	**0.33**	**0.40**	**0.79**	**0.16**	**0.14**	**0.5**	**0.60**
YaGo	0.01	0.01	−0.20	0.0	0.03	0.03	−0.22	0.0
FMA	**0.30**	**0.33**	0.20	0.46	-	-	-	-

Table 2. Conference and multifarm tasks

	Conference				MultiFarm			
	Cosine	Pearson	Spearman	F-m	Cosine	Pearson	Spearman	F-m
BabelNet	0.28	−0.08	**0.73**	**0.61**	**0.30**	**0.39**	**0.44**	**0.49**
DBnary	0.06	−0.01	−0.34	0.46	0.16	−0.20	0.22	0.29
DBpedia	0.1	−0.02	0.12	0.12	0.01	0.03	0.09	0.10
Doid	0.10	0.01	−0.05	0.0	0.0	0.0	0.0	0.0
FMA	0.08	−0.05	−0.07	0.0	0.0	0.0	0.0	0.0
Uberon	0.11	0.0	0.05	0.0	0.0	0.0	0.0	0.0
YaGo	**0.30**	**0.11**	0.31	0.10	0.10	0.09	−0.09	0.11

(it is most of the times the case that the query document corresponding to two input ontologies contains much less terms than any of the BK-documents in the selection pool).

As for the Pearson correlation coefficient, Spearman appears to largely outperform that, as well. This is mainly due to the fact that the Pearson correlation measure is suited for testing linearity of the variables, while Spearman assumes monotonic behavior of the rank pairs. We draw the reader's attention to the fact on the Anatomy track Pearson yields a maximal value for two different BKs – UBERON and FMA. The BK selected by this measure is UBERON, the smaller of the two sources, conforming to our selection condition given at the end of Sect. 3. UBERON is well known as the perfect BK for the anatomy matching task at OAEI. However, if UBERON did not exist in the set of BKs the approach would have selected FMA as the optimal BK because FMA has a higher correlation value than the other BKs.

Finally, we note that the choice of a similarity measure is more important in the presence of multiple similar in terms of domains BKs, as this is the case in the OAEI experiments, where the only similarity measure that remains unflawed is Spearman.

5 Related Work

We provide an overview of related approaches to ontology matching with regard to two aspects of this task: (1) the use of background knowledge and (2) the automatic selection of background knowledge.

5.1 Ontology Matching Using Background Knowledge

The idea of using background knowledge (BK) for enhancing ontology matching task is not new and has been successfully adopted in several matching approaches. Although not directly relevant to our study, which focuses on the *BK* selection process, we summarize here the main groups of relevant approaches.

An intuitive idea is to rely on reusing existing mappings in order to improve the mappings produced by a system. Several approaches [15–17] follow this paradigm. The main drawback of this group of methods is the fact that they depend heavily on the quality of the re-used mappings and, hence, on the performance of the ontology matching techniques that have been used to produce them.

Another approach consists in using a corpus, which can be seen as a rich collection of data elements and their data types, relationships between elements, sample data instances and other information that can be used to discover mappings between entities [18,19]. Furthermore, domain specific ontologies are often seen as quality sources of background knowledge. In [20], the alignment process takes place in two steps: *anchoring* and *driving relations*. Anchoring consists in matching the concepts of the source and target ontologies to the concepts of the reference knowledge using standard ontology matching techniques. Relations between source and target concepts are derived by checking if their corresponding anchored concepts are related.

A group of approaches relies on the web in order to discover (by crawling) automatically relations between the input ontologies entities that may exist in various knowledge sources distributed on the web [4]. This is particularly useful when the needed background knowledge is spread among different sources. More recently, several approaches have been proposed that rely on general purpose knowledge graphs, such as Yago and DPBedia. It has been shown that such sources are particularly useful for aligning cross-lingual ontologies [3,5].

5.2 Automatic Selection of Background Knowledge

To the best of our knowledge, there are only a few approaches that have addressed the question of automatic *BK* selection.

In [21], an automatic background knowledge selection approach has been proposed for the particular task of matching biomedical ontologies based on the notion of mappings gain (MG). MG is used to estimate the individual usefulness of background knowledge sources and is defined as a function of the improvement of the number of correct mappings by using a given *BK* source as compared to a direct mapping of two ontologies. The *BK* providing the highest MG is selected.

The authors have shown experimentally the correlation between the mapping gain and the F-measure in most of the matching tasks that they perform.

The closest to our work is reported in [22]. In this approach, a local repository is built from a set of ontologies, to be used as background knowledge sources. These resources are indexed by extracting concept names, comments and labels. In addition and separately, structural features are taken into account. Querying the repository is performed for a given source and target ontology, modeled as sets of key words. If a suitable BK is not found for the given ontologies in the repository, the method searches the web for appropriate BK ontologies. In the likely case of returning more than one ontology, all found ontologies are used for the matching independently and a unified result set is produced. As reported, the adopted strategy aims at selecting the BK that maximizes the F-measure score for a given matching task, although no experimental results are presented to illustrate and support this selection criterium.

Positioning. We present the major points, which differentiate our technique as compared to the two approaches described above.

Although also relying on information retrieval techniques, in our method, the query is represented as a single document, built in the same way as the BK-documents. This allows to avoid the complex weighted similarity computation in [22], depending on the setting of two parameters. Instead of creating two classes of features (one for terms and one for structure), we embed the structure of the BKs *and* of the input ontologies in their respective textual representations (see Sect. 3), which has the potential to produce a more compact index.

We make clear distinction between query ontologies and BKs in the evaluation phase. Indeed, the evaluation presented in [22] is highly biased by the fact that the authors use the same type of ontologies as queries and as BK (for example, if aligning two ontologies from the Benchmark track of OAEI, the authors would include the rest of the Benchmark ontologies as BK candidates), which will lead to having in the repository always a very useful BK source at hand. This is however, hardly a realistic scenario. In contrast, we use as BKs well-established and widely used knowledge graphs that are much likely to be called upon as BK sources for solving a real-life alignment problem. In that line of thought, our approach is generic and can handle different domains, contrarily to [21].

In both [21,22] the BK selection appears to be strongly coupled with the actual matching procedure. One of the main motivations of work is to dissociate completely the BK selection process from the alignment procedure.

In contrast to our work, the results presented in [22] are not reproducible, because no information is given regarding the selected BKs, neither about the similarity measure that has been used in the process. In turn, no direct experimental comparison to the approach is possible.

Note that a direct comparison to these approaches was not possible due to the difficutly to reproduce the scenarios presneted in these papers on the one hand and on the other hand – the difficulty to apply our scenario to these approaches that do not adresse the multi-domain or multilngual selection problems.

6 Conclusion and Future Work

In this work, we have addressed the problem of automatic selection of background knowledge sources for the task of ontology matching. We propose an approach using information retrieval techniques implemented in an automatic domain independent system that can handle multilingual input ontologies. We build an index for a set of known, well-established in the semantic web field knowledge sources, that are often used as BK for aligning ontologies. For a pair of ontologies to be matched, the selection process is a result of querying the indexed corpus for semantically similar BK sources from the indexed data. We provide an in-depth empirical study and we show that in certain cases the standard choice of a cosine similarity is not the most optimal and parameter free correlation measures can help discriminate better between close in terms of domains BK sources. We define an optimality criterion of selection based on the quality of the matching and we show experimentally that our approach satisfies this criterion. Contrarily to state of the art approaches, our technique has the advantage of not being based on a preliminary direct matching between the input ontologies.

In the future, we plan to work on optimizing the selection process by improving the quality of the index features. In that respect, we will consider the task of BK preselection for a particular domain, to be applied prior to the selection algorithm. We also plan to improve our selection criterion in order to take into account the trade-off between optimality and efficiency. Finally, we will investigate the benefits of the development of a BK selection method to assist the instance matching and link discovery processes.

References

1. Shvaiko, P., Euzenat, J.: Ontology matching: state of the art and future challenges. IEEE Trans. Knowl. Data Eng. **25**(1), 158–176 (2013)
2. Faria, D., Pesquita, C., Santos, E., Palmonari, M., Cruz, I.F., Couto, F.M.: The agreementmakerlight ontology matching system. In: Meersman, R., Panetto, H., Dillon, T., Eder, J., Bellahsene, Z., Ritter, N., De Leenheer, P., Dou, D. (eds.) ODBASE 2013. LNCS, vol. 8185, pp. 527–541. Springer, Heidelberg (2013)
3. Tigrine, A.N., Bellahsene, Z., Todorov, K.: Light-weight cross-lingual ontology matching with LYAM++. In: Debruyne, C., Panetto, H., Meersman, R., Dillon, T., Weichhart, G., An, Y., Ardagna, C.A. (eds.) ODBASE. LNCS, vol. 9415, pp. 527–544. Springer, Cham (2015). doi:10.1007/978-3-319-26148-5_36
4. Sabou, M., d'Aquin, M., Motta, E.: Exploring the semantic web as background knowledge for ontology matching. J. Data Semant. **11**, 156–190 (2008)
5. Todorov, K., Hudelot, C., Geibel, P.: Fuzzy and cross-lingual ontology matching mediated by background knowledge. In: Bobillo, F., Carvalho, R.N., Costa, P.C.G., d'Amato, C., Fanizzi, N., Laskey, K.B., Laskey, K.J., Lukasiewicz, T., Nickles, M., Pool, M. (eds.) URSW 2011-2013. LNCS, vol. 8816, pp. 142–162. Springer, Heidelberg (2014)

6. Ngo, D.H., Bellahsene, Z., Todorov, K.: Opening the black box of ontology matching. In: Cimiano, P., Corcho, O., Presutti, V., Hollink, L., Rudolph, S. (eds.) ESWC 2013. LNCS, vol. 7882, pp. 16–30. Springer, Heidelberg (2013). doi:10.1007/978-3-642-38288-8_2
7. Conover, W.J.: Practical Nonparametric Statistics. Wiley, Hoboken (1998)
8. Tigrine, A.N., Bellahsene, Z., Todorov, K.: Lyam++ results for OAEI 2015. In: OM at ISWC (2015)
9. Navigli, R., Ponzetto, S.P.: Babelnet: the automatic construction, evaluation and application of a wide-coverage multilingual semantic network. Artif. Intell. **193**, 217–250 (2012)
10. Sérasset, G.: DBnary: wiktionary as a lemon-based multilingual lexical resource in RDF. Semant. Web **6**(4), 355–361 (2015)
11. Auer, S., Bizer, C., Kobilarov, G., Lehmann, J., Cyganiak, R., Ives, Z.G.: DBpedia: a nucleus for a web of open data. In: Aberer, K., Choi, K.-S., Noy, N., Allemang, D., Lee, K.-I., Nixon, L.J.B., Golbeck, J., Mika, P., Maynard, D., Mizoguchi, R., Schreiber, G., Cudré-Mauroux, P. (eds.) ASWC 2007 and ISWC 2007. LNCS, vol. 4825, pp. 722–735. Springer, Heidelberg (2007)
12. Suchanek, F.M., Kasneci, G., Weikum, G.: Yago,: a core of semantic knowledge. In: WWW, pp. 697–706 (2007)
13. Rosse, C., Mejino, J.: A reference ontology for biomedical informatics: the foundational model of anatomy. J. Biomed. Inf. **36**(6), 478–500 (2003)
14. Haendel, M., Balhoff, J.P., Bastian, F.B., Blackburn, D.C., Blake, J.A., Bradford, Y., Comte, A., Dahdul, W.M., Dececchi, T., Druzinsky, R.E., Hayamizu, T.F., Ibrahim, N., Lewis, S.E., Mabee, P.M., Niknejad, A., Robinson-Rechavi, M., Sereno, P.C., Mungall, C.J.: Unification of multi-species vertebrate anatomy ontologies for comparative biology in uberon. J. Biomed. Semant. **5**, 21 (2014)
15. Do, H.H., Rahm, E.: COMA - a system for flexible combination of schema matching approaches. In: VLDB, pp. 610–621 (2002)
16. Groß, A., Hartung, M., Kirsten, T., Rahm, E.: GOMMA results for OAEI 2012. In: Workshop on Ontology Matching (2012)
17. Saha, B., Stanoi, I., Clarkson, K.L.: Schema covering: a step towards enabling reuse in information integration. In: ICDE, pp. 285–296 (2010)
18. Madhavan, J., Bernstein, P.A., Chen, K., Halevy, A.Y., Shenoy, P.: Corpus-based schema matching. In: Proceedings of IJCAI-03 Workshop on Information Integration on the Web (IIWeb-03), pp. 59–63 (2003)
19. Madhavan, J., Bernstein, P.A., Doan, A., Halevy, A.Y.: Corpus-based schema matching. In: Proceedings of the 21st International Conference on Data Engineering, ICDE, pp. 57–68 (2005)
20. Aleksovski, Z., Klein, M., ten Kate, W., van Harmelen, F.: Matching unstructured vocabularies using a background ontology. In: Staab, S., Svátek, V. (eds.) EKAW 2006. LNCS(LNAI), vol. 4248, pp. 182–197. Springer, Heidelberg (2006). doi:10.1007/11891451_18
21. Faria, D., Pesquita, C., Santos, E., Cruz, I., Couto, F.: Automatic background knowledge selection for matching biomedical ontologies. PLoS ONE **9**(11), e111226 (2014). doi:10.1371/journal.pone.0111226
22. Quix, C., Roy, P., Kensche, D.: Automatic selection of background knowledge for ontology matching. In: SWIM, p. 5 (2011)

Considering Semantics on the Discovery of Relations in Knowledge Graphs

Ignacio Traverso-Ribón[1(✉)], Guillermo Palma[2], Alejandro Flores[4], and Maria-Esther Vidal[2,3]

[1] FZI Research Center for Information Technology, Karlsruhe, Germany
traverso@fzi.de
[2] Universidad Simón Bolívar, Caracas, Venezuela
[3] University of Bonn and Fraunhofer, Bonn, Germany
[4] University of Maryland, College Park, USA

Abstract. Knowledge graphs encode semantic knowledge that can be exploited to enhance different *data-driven* tasks, e.g., query answering, data mining, ranking or recommendation. However, knowledge graphs may be incomplete, and relevant relations may be not included in the graph, affecting accuracy of these data-driven tasks. We tackle the problem of *relation discovery* in a knowledge graph, and devise $\mathcal{KOI}$, a semantic based approach able to discover relations in portions of knowledge graphs that comprise *similar* entities. $\mathcal{KOI}$ exploits both datatype and object properties to compute the similarity among entities, i.e., two entities are similar if their datatype and object properties have similar values. $\mathcal{KOI}$ implements graph partitioning techniques that exploit similarity values to discover relations from knowledge graph partitions. We conduct an experimental study on a knowledge graph of TED talks with state-of-the-art similarity measures and graph partitioning techniques. Our observed results suggest that $\mathcal{KOI}$ is able to discover missing edges between related TED talks that cannot be discovered by state-of-the-art approaches. These results reveal that combining semantics encoded both in the similarity measures and in the knowledge graph structure, has a positive impact on the *relation discovery* problem.

Keywords: Relation discovery · Semantic similarity · Graph partitioning

1 Introduction

Following Linked Data initiatives and exploiting features of Semantic Web technologies, large volumes of data are publicly available in the form of knowledge graphs usually described using the RDF data model, e.g., DBpedia[1] or YAGO[2]. Simultaneously, data-driven applications that rely on knowledge graphs are progressively increasing [5]. However, as traditional semi-structured data, knowledge

[1] http://wiki.dbpedia.org/.
[2] https://yago-knowledge.org.

E. Blomqvist et al. (Eds.): EKAW 2016, LNAI 10024, pp. 666–680, 2016.
DOI: 10.1007/978-3-319-49004-5_43

graphs may be incomplete, either because relations among graph entities were unknown at the time the graph was created, or because the knowledge graph creation process failed to completely identify all existing relations. This situation encourages the development of techniques for the discovery of missing relations.

Discovering relations in knowledge graphs requires the analysis of both the semantics encoded in the knowledge graph, and the connectivity or structure of the represented relations. However, the majority of the state-of-the-art approaches are based either on the structure of the graph [2,10], or on properties of the knowledge graph entities [7,16]. Although some approaches combine both types of knowledge [21], they do not take into account domain semantics encoded in semantic similarity measures to discover missing relations [15].

In this paper we propose $\mathcal{KOI}$, an approach for relation discovery in knowledge graphs that considers the semantics of both entities represented in the knowledge graph and their neighborhoods. $\mathcal{KOI}$ receives as input a knowledge graph, and encodes the semantics about the properties of graph entities and their neighbors in a bipartite graph. Entity neighbors correspond to ego-networks, e.g., the friends of a person in a social network or the set of TED talks related to a given TED talk. $\mathcal{KOI}$ partitions the bipartite graph into parts of highly similar entities connected to also similar ego-networks. Relations are discovered in these parts following the *homophily* prediction principle, which states that entities with similar characteristics tend to be related to similar entities [13]. Intuitively, the *homophily* prediction principle allows for relating two entities $t1$ and $t2$ whenever they have similar datatype and object property values (neighborhoods).

We evaluate the behavior of $\mathcal{KOI}$ in a knowledge graph of TED talks[3]; we crafted this knowledge graph by crawling data from the official TED website (http://www.ted.com/). We compare relations discovered by $\mathcal{KOI}$ with two baselines of relations identified by the METIS [9] and *k-Nearest Neighbors* (KNN) algorithms. We empowered KNN with statistic and semantic similarity measures (Sect. 6.3). Experimental outcomes suggest the following statements: (i) Semantics encoded in similarity measures and knowledge graph structure enhances the performance of relation discovery methods; and (ii) $\mathcal{KOI}$ outperforms state-of-the-art approaches, obtaining higher values of precision and recall.

To summarize, the contributions of this paper are as follows:

- $\mathcal{KOI}$, a relation discovery method that implements graph partitioning techniques and relies on semantics encoded in similarity measures and graph structure to discover relations in knowledge graphs;
- A knowledge graph describing TED talks crafted from the TED website; and
- A empirical evaluation on a real-world knowledge graph of TED talks to analyze the performance of $\mathcal{KOI}$ with respect to state-of-the-art approaches.

This paper comprises six additional sections. Section 2 motivates our approach with an example, and Sect. 3 introduces preliminary definitions. We explain our approach in Sect. 4 and the related work in Sect. 5. Section 6 reports on experimental results and describes the crafted TED knowledge graph. Section 7 concludes and presents future work ideas.

[3] https://github.com/itraveribon/TED_KnowledgeGraph.

2 Motivating Example

In this section, we provide an example to motivate the problem of knowledge discovery tackled in this paper. We show an example of relation discovery between TED talks publicly available in the TED website. TED talks are described through textual properties, e.g., title, abstract or tags, and their relations with other talks in order to provide recommendations to the users. Relations between talks are defined by TED curators manually, which corresponds to a time expensive task and prone to omissions. Therefore, it would be helpful to have automatic methods able to ease the relation discovery and other curation tasks. We check the TED website in 2015 and 2016, and compare both versions of the website in order to detect relations between talks that are only represented in the newer version of the website. In total, we observe 62 relations that are included in 2016 but are not present in the 2015 version, i.e., TED curators do not discover these relations until 2016. One example is the relation between talks *The politics of fiction*[4] and *The shared wonder of film*[5]. Both talks are present in both versions of the website. However, only in 2016 is possible to find a relation between them. Thus, we can conclude that there are missing relations between TED talks in the 2015 version of the website. An approach able to discover these relations automatically would alleviate the effort of curators and improve the quality (completeness) of the data. Though the relation between *The politics of fiction* and *The shared wonder of film* is not included in the 2015 website, the rest of knowledge regarding to these talks allows for intuiting a high degree of relatedness between them. We observe that both talks have keywords or tags in common as *Culture* or *Storytelling*. We also find some expressions in their abstracts or descriptions, that though do not match exactly, are clearly related such as *identity politics* and *cultural walls*, or *film* and *novel*. Moreover, if their sets of related TED talks are compared, we observe they share two related talks, *The clues to a great story*[6] and *The mistery box*[7]. Thus, related talks have properties in common. $\mathcal{KOI}$ relies on this observation and exploits entity properties to discover missing relations between these entities.

3 Preliminaries

In this section we present definitions required to understand our approach.

Definition 1 *(RDF Triple* [1]*).* Let U be a set of RDF URI references, B a set of Blank nodes, and L a set of RDF literals. A tuple $(s, p, o) \in UB \times U \times UBL$ is an RDF triple, where s is called subject, p predicate and o object.

Definition 2 *(Knowledge graph* [18]*).* Given a set T of RDF triples, a knowledge graph is a pair $G = (V, E)$, where $V = \{s|(s,p,o) \in T\} \cup \{o|(s,p,o) \in T\}$ is a set of entities and $E = \{(s,p,o) \in T\}$ a set of relations.

[4] http://www.ted.com/talks/elif_shafak_the_politics_of_fiction.
[5] http://www.ted.com/talks/beeban_kidron_the_shared_wonder_of_film.
[6] http://www.ted.com/talks/andrew_stanton_the_clues_to_a_great_story.
[7] http://www.ted.com/talks/j_j_abrams_mystery_box.

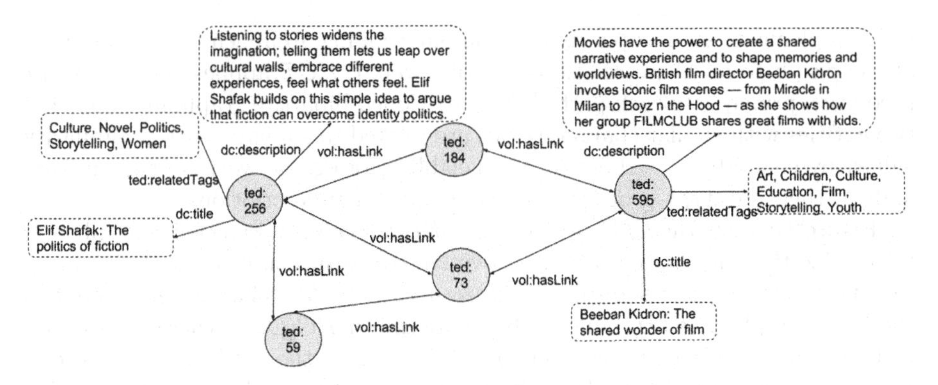

Fig. 1. Portion of a knowledge graph of TED talks. Nodes represent TED talks, while dashed squares represent datatype property values.

Figure 1 shows a portion of a knowledge graph describing TED talks. The predicate *vol:hasLink* connects related talks, while the rest of predicates correspond to datatype properties and connect talks with string literals.

Definition 3 *(Ego-Network).* Let $G = (V, E)$ be a knowledge graph and $L = \{p \mid (s, p, o) \in E\}$ be a set of predicates. Given an entity $v_i \in V$ and a predicate $r \in L$, the ego-network of v_i according to r is defined as the set of entities connected to v_i through an edge with predicate r: ego-net$(v_i, r) = \{v_j \mid (v_i, r, v_j) \in E\}$.

The ego-network of the entity *ted:256* with respect to the predicate *vol:hasLink* (Fig. 1) is formed by entities *ted:59*, *ted:73*, and *ted:184*.

4 Our Approach: $\mathcal{KOI}$

4.1 Problem Definition

Let $G' = (V, E')$ and $G = (V, E)$ be two knowledge graphs. G' is an *ideal* knowledge graph that contains all the existing relations between entities in V. G is the *actual* knowledge graph, which contains only a portion of the relations represented in G', i.e., $E \subseteq E'$. Let $\Delta(E', E) = E' - E$ be the set of relations existing in the ideal graph that are not represented in the actual knowledge graph G, and $G_{\text{comp}} = (V, E_{\text{comp}})$ the *complete* knowledge graph, which contains a relation for each possible combination of entities and predicates $E \subseteq E' \subseteq E_{\text{comp}}$.

Given a relation $e \in \Delta(E_{\text{comp}}, E)$, the *relation discovery problem* consists of determining if $e \in E'$, i.e., if a relation e corresponds to an existing relation in the ideal graph G'.

4.2 Our Solution

We propose $\mathcal{KOI}$, a relation discovery method for knowledge graphs that considers semantics encoded in similarity measures and the knowledge graph structure.

$\mathcal{KOI}$ implements an unsupervised graph partitioning approach to identify parts of the graph from where relations are discovered. $\mathcal{KOI}$ applies the *homophily* prediction principle to each part of the partitioned bipartite graph, in a way that two entities with similar characteristics are related to similar entities. Similarity values are computed based on: *(a)* the neighbors or ego-networks of two entities, and *(b)* their datatype property values (e.g., textual descriptions).

Figure 2 depicts the $\mathcal{KOI}$ architecture. $\mathcal{KOI}$ receives a knowledge graph $G = (V, E)$ like the one showed in Fig. 1, two similarity measures S_v and S_u, and a set of constraints S. As result, $\mathcal{KOI}$ returns a set of relations discovered in the input graph G. $\mathcal{KOI}$ builds a bipartite graph $BG(G, r)$ where each entity in V is connected with its ego-network according to the predicate r. Figure 3a contains the bipartite graph built from the knowledge graph in Fig. 1 according to predicate *vol:hasLink*. By means of a graph partitioning algorithm and the similarity measures S_v and S_u, $\mathcal{KOI}$ identifies graph parts containing highly similar entities with highly similar ego-networks, i.e., similar entities that are highly connected in the original graph. According to the *homophily* prediction principle, $\mathcal{KOI}$ produces candidate missing relations inside the identified graph parts. Figure 3b represents with red dashed lines the set of candidate discovered relations. Only those relations that satisfy a set of constraints S are considered as discovered relations. Listing 1.1 shows an example of constraints and Fig. 4 includes the corresponding *score* values for each candidate relation.

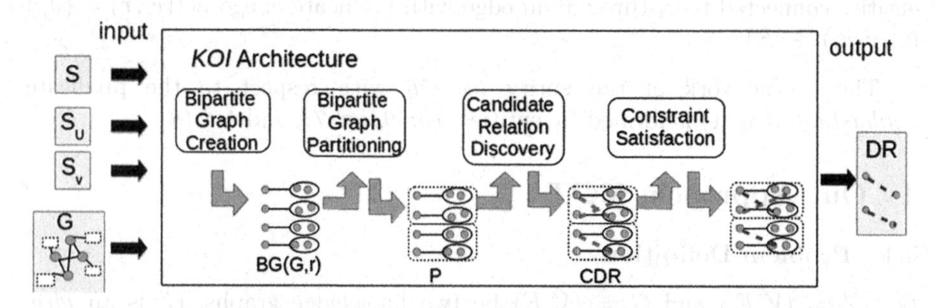

Fig. 2. $\mathcal{KOI}$ Architecture. $\mathcal{KOI}$ receives a knowledge graph G, two similarity measures S_v and S_u, and a set S of constraints. $BG(G, r)$ is a bipartite graph and represents relations between entities in G and their corresponding ego-networks built in terms of r. A graph partitioning algorithm is used to partition $BG(G, r)$ into a set P of parts; each part corresponds to a portion of $BG(G, r)$ where both entities and ego-networks are highly similar. Parts in P are used to identify candidate discovered relations *CDR* (red edges). Then, a constraint satisfaction outputs in *DR* the relations in *CDR* that meet the constraints in S (green edges). (Color figure online)

Bipartite Graph Creation. Determining the membership of each relation $e \in \Delta(E_{\text{comp}}, E)$ in E' is expensive in terms of time due to the large amount of relations included in $\Delta(E_{\text{comp}}, E)$, and may produce a large amount of false positives. $\mathcal{KOI}$ leverages from the homophily intuition to tackle this problem by finding highly similar portions of the graph, i.e., portions including entities with similar ego-networks and similar datatype property values. In order to consider at the same time both similarities, $\mathcal{KOI}$ builds a bipartite graph where each entity is associated with its ego-network. The objective is to find a partitioning of this graph, such that each part contains highly similar entities and highly similar ego-networks. Thus, the $\mathcal{KOI}$ graph partitioning problem is an optimization problem where these two similarities are maximized on entities of each part.

Definition 4 *($\mathcal{KOI}$ Bipartite Graph).* Let $G = (V, E)$ be a knowledge graph and $L = \{p \mid (s, p, o) \in E\}$ be a set of predicates. Given a predicate $r \in L$, the $\mathcal{KOI}$ Bipartite Graph of G and r is defined as $BG(G, r) = (V \cup U(r), E_{BG}(r))$, where $U(r) = \{\text{ego-net}(v_i, r) \mid v_i \in V\}$ is the set of ego-networks of entities in V, and $E_{BG}(r) = \{(v_i, u_i) \mid v_i \in V \land u_i = \text{ego-net}(v_i, r)\}$ is the set of edges that associate each entity with its ego-network.

Figure 3a shows a $\mathcal{KOI}$ bipartite graph for the knowledge graph in Fig. 1.

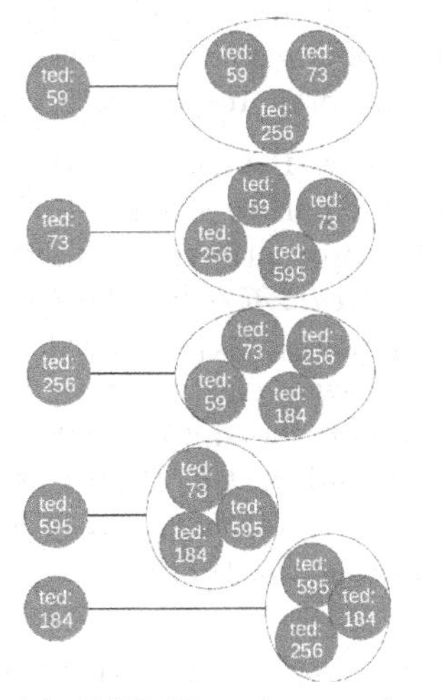

(a) $\mathcal{KOI}$ Bipartite graph generated from Figure 1. Each entity is connected to its ego-network.

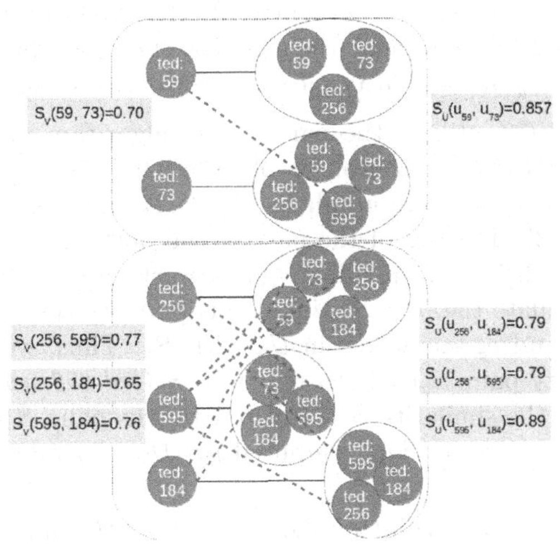

(b) Partition found by the $\mathcal{KOI}$ Bipartite Graph in Figure 3a. Dashed squares represent partitions and red dashed edges candidate relations of our approach. S_v and S_u are entity and ego-network similarity measures, respectively.

Fig. 3. Example of $\mathcal{KOI}$ Graphs. A $\mathcal{KOI}$ bipartite graph and its partitioning (Color figure online)

Bipartite Graph Partitioning. To identify portions of the knowledge graph where the *homophily* prediction principle can be applied, the bipartite graph $BG(G, r)$ is partitioned in a way that entities in each part are highly similar (i.e., similar datatype properties) and connected (i.e., have similar ego-networks).

Definition 5 *(A Partition of a $\mathcal{KOI}$ Bipartite Graph).* Given a $\mathcal{KOI}$ bipartite graph $BG(G,r) = (V \cup U, E_{BG})$, a partition $P(E_{BG}) = \{p_1, p_2, ..., p_n\}$ satisfies the following conditions:

- Each part p_i contains a set of edges $p_i = \{(v_x, u_x) \in E_{BG}\}$,
- Each edge (v_x, u_x) in E_{BG} belongs to one and only one part p of $P(E_{BG})$, i.e., $\forall p_i, p_j \in P(E_{BG}), p_i \cap p_j = \emptyset$ and $E_{BG} = \bigcup_{p \in P(E_{BG})} p$.

Definition 6 *(The Problem of $\mathcal{KOI}$ Bipartite Graph Partitioning).* Given a $\mathcal{KOI}$ bipartite graph $BG(G,r) = (V \cup U, E_{BG})$, and similarity measures S_v and S_u for entities in V and ego-networks in U. The problem of $\mathcal{KOI}$ Bipartite Graph Partitioning corresponds to the problem of finding a partition $P(E_{BG})$ such that Density($P(E_{BG})$) is maximized, where:

- Density($P(E_{BG})$)=$\sum_{p \in P(E_{BG})}$(partDensity(p)), and
- $\text{partDensity}(p) = \overbrace{\frac{\sum_{v_i, v_j \in V_p} [v_i \neq v_j] S_v(v_i, v_j)}{|V_p|(|V_p| - 1)}}^{(A)} + \overbrace{\frac{\sum_{u_i, u_j \in U_p} [u_i \neq u_j] S_u(u_i, u_j)}{|U_p|(|U_p| - 1)}}^{(B)}$

where component (A) represents the similarity between entities in edges of part p and (B) represents the similarity between the corresponding ego-networks. S_v and S_u are similarity measures for entities and ego-networks, respectively.

$\mathcal{KOI}$ utilizes the partitioning algorithm proposed by Palma et al. [15] to solve the optimization problem of partitioning a $\mathcal{KOI}$ bipartite graph.

The bipartite graph in Fig. 3a is partitioned into two parts represented in Fig. 3b. Entities of the part in the bottom are $V_p = \{\textit{ted:256, ted:595, ted:184}\}$ and their corresponding ego-networks are $U_p = \{u_{256}, u_{595}, u_{184}\}$. In order to calculate the *partDensity* of this part, we compare pair-wise entities in V_p with S_v and ego-networks in U_p with S_u. Thus, we compute the similarity S_v for entity pairs $S_v(\textit{ted:256, ted:595})$, $S_v(\textit{ted:256, ted:184})$, and $S_v(\textit{ted:595, ted:184})$, and the similarity S_u for ego-networks pairs $S_u(u_{256}, u_{595})$, $S_u(u_{256}, u_{184})$, and $S_u(u_{595}, u_{184})$. The computed *partDensity* value is in this case 0.775.

Candidate Relation Discovery. $\mathcal{KOI}$ applies the *homophily* prediction principle in the parts of a partition of a $\mathcal{KOI}$ bipartite graph, and discovers relations between entities included in the same part.

Definition 7 *(Candidate relation).* Given two knowledge graphs $G = (V, E)$ and $G_{comp} = (V, E_{comp})$. Let $BG(G,r) = (V \cup U, E_{BG})$ be a $\mathcal{KOI}$ bipartite graph. Let $P(E_{BG})$ be a partition of E_{BG}. Given a part $p = \{(v_x, u_x) \in E_{BG}\} \in P(E_{BG})$, the set of candidate relations $CDR(p)$ in part p corresponds to the set of relations $\{(v_i, r, v_j) \in E_{comp}\}$ such that v_j is included in some ego-network u_x and edges (v_i, u_i) and (v_x, u_x) are contained in the partition p.

In Fig. 3b candidate relations are represented as red dashed lines. One example is the relation (*ted:59, vol:hasLink, ted:595*). This candidate relation is discovered due to the presence of *ted:59* and *ego-net(ted:73, vol:hasLink)* in the same partition and the inclusion of the entity *ted:595* in the ego-network *ego-net(ted:73, vol:hasLink)*.

```
ASK {
  { SELECT count(?uk) as ?repetitions
        count(?vunion) as ?union
        count(?vinter) as ?intersection
        ?intersection/?union
           * ?repetitions as ?score
       WHERE {
     vj dc:isPartOf ?uk.
     ?vk dc:relation ?uk.0
     vi dc:relation ?ui.
     ?ui dc:isPartOf ?p.
     ?uk dc:isPartOf ?p.
     {?vunion dc:isPartOf ?uk} UNION
            {?vunion dc:isPartOf ?ui}.
     ?vinter dc:isPartOf ?uk.
     ?vinter dc:isPartOf ?ui.
     }
 }
   FILTER (?score > THETA)
}
```

Listing 1.1. SPARQL specification of a constraint on the product between the similarity of the ego-networks and the amount of times a relation is discovered.

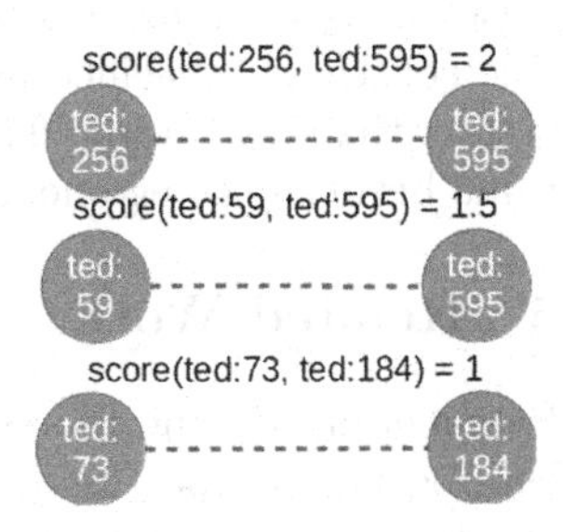

Fig. 4. Application of the relation constraint described in Listing 1.1 for the candidate relations (red dashed edges) found in Fig. 3b.

Constraint Satisfaction. A relation constraint is a set of RDF constraints that states conditions that must be satisfied by a candidate discovered relation in order to become a discovered relation, i.e., relations belonging to the ideal knowledge graph. RDF constraints are expressed using the SPARQL language as suggested by Lausen et al. [11] and Fischer et al. [3]. Only the candidate relations that fulfill relation constraints are considered as discovered relations.

Definition 8 *(Discovered Relations).* Given a set of candidate relations CDR and a set of relation constraints S, the set of discovered relations DR is defined as the subset of candidate relations that satisfy the given contraints $DR(CDR, S) = \{(v_i, r, v_j) \mid (v_i, r, v_j) \in CDR \wedge \forall rc \in S \Rightarrow satisfy(rc(v_i, r, v_j))\}$.

Although an upper bound for the problem of checking if a constraint is satisfied by a candidate discovered relation is PSPACE-complete [17], because number of constraints is smaller than the size of the knowledge graph, the complexity of this decision problem can be expressed in terms of data, and is LOGSPACE for SPARQL [11,17].

Listing 1.1 illustrates a constraint that states a condition for a candidate discovered relation $cdr = (v_i\ r\ v_j)$ to become a discovered relation. Whenever the candidate discovered relation $cdr = (v_i\ r\ v_j)$ is identified in several parts of a partition P, the number of times that cdr appears is taken into account, as well as the similarity between the ego-network of v_i and the ego-networks where v_j is included. To determine if the constraint is satisfied, a *score* is computed and the value of this score has to be greater than a threshold θ_i. The score is defined as the product of the number of times a relation is discovered and the similarity between corresponding ego-networks. For each discovered relation, Fig. 4 contains the value of the corresponding score described in Listing 1.1. Relation (*ted:256*, *vol:hasLink*, *ted:595*) gets the highest value for this score being discovered four times in Fig. 3b. Moreover, the similarity between ego-networks *ego-net*(*ted:595,vol:hasLink*) and *ego-net*(*ted:184*, *vol:hasLink*) is 0.5. The constraint, specified as an ASK query, is held if at least one score value is greater than the threshold θ. Therefore, we consider only the maximum similarity value between the ego-networks.

5 Related Work

Palma et al. [15] and Flores et al. [4] present approaches for relation discovery in heterogeneous bipartite graphs. Palma et al. present semEP, a semantic-based graph partitioning approach that finds the minimal partition of a weighted bipartite graph with highest density. semEP utilizes parts in the same way $\mathcal{KOI}$ does, in order to find missing relations. However, they consider entities as isolated elements and do not consider their ego-networks during the partitioning process. esDSG [4] performs similarly than semEP, i.e., given a weighted bipartite graph, esDSG identifies a subgraph that is highly dense and comprise highly similar entities. Again, ego-networks are not considered.

Researchers of the social network field study the structure of friendship induced graphs, and define the concept of ego-network as the set of entities that are at one-hop distance to a given entity. Epasto et al. [2] reports on high quality results in the friend suggestion task by analyzing the ego-networks of the induced knowledge graphs. In this case, the discovery of the relations is based purely on the ego-network of the entities and no datatype property value is considered.

Redondo et al. [7] propose an approach to discover relations between video fragments based on visual information and background knowledge extracted from the Web of data in form of annotations. Like [4,15] entities or video fragments are considered as isolated elements in the knowledge graph, and the similarity is computed as the number of coincident annotations between two video fragments.

Sachan and Ichise [21] discover relations between authors in a co-author network extracted from dblp. Their approach is based on the dense subgraph approach. They consider the connections in the knowledge graph and some features of the authors and from the papers like the keywords. However, the comparison of such features relies on the syntactic level, and the semantics is omitted.

Kastrin et al. [10] present an approach to discover relations among biomedical terms. They build a knowledge graph with such terms with the help of SemRep [20], a tool for recovering semantic propositions from the literature. In this case, it is not only important the existence of the relation, but also the type of the relation. Unlike $\mathcal{KOI}$, they only consider the graph topology, discarding semantic knowledge encoded in datatype properties.

Nunes et al. [14] link entities based on the number of co-occurrences in a text corpus and distance, measured in number of hops, between the entities in a knowledge graph. Unlike $\mathcal{KOI}$, this approach needs a corpus labeled with entities and only takes into account the object properties, omitting the semantics encoded in datatype properties.

6 Empirical Evaluation

6.1 Knowledge Graph Creation

In this section we describe the characteristics of the crafted TED knowledge graph and its links to external vocabularies. This knowledge graph is built from a real-world dataset of TED talks and playlists[8].

The knowledge graph of TED talks consists of 846 talks and 125 playlists (15/12/2015). Playlists are described with a title and the set of included TED talks. Each TED talk is described with the following set of datatype properties:

- *dc:title* (Dublin Core vocabulary) represents title of the talk;
- *dc:creator* models speaker;
- *dc:description* represents abstract;
- and *ted:relatedTags* corresponds to set of related keywords.

Apart from the datatype properties, TED talks are connected to playlists that include them through the object property *ted:playlist*. A *vol:hasLink* (Vocabulary Of Links[9]) object property connects each pair of talks that are together in at least one playlist. We crawled the playlists available in the TED website[10]. Playlists contain sets of TED talks that usually address similar topics. TED playlists are created and maintained by curators, who decide if a certain video may or may be not included in a certain playlist.

Additionally, we enriched the knowledge graph by adding similarity values between each pair of entities. We computed four similarity measures (TFIDF, ESA [6], Doc2Vec [12], and Doc2Vec Neighbors) using as input the concatenation of datatype properties title, description and related tags. ESA similarity values were computed using the public ESA endpoint[11], and Doc2Vec (D2V) values were obtained training the *gensim* implementation [19] with the pre-trained Google News dataset[12]. Doc2Vec Neighbors (D2VN) is defined as:

$$\text{D2VN}(v_1, v_2) = \frac{\sqrt{S_v(v_1, v_2)^2 + S_u(\text{ego-net}(v_1, r), \text{ego-net}(v_2, r))^2}}{\sqrt{2}},$$

where r corresponds to *vol:hasLink* and S_v and S_u are defined as follows:

$$S_v(v_1, v_j) = Doc2Vec(v_i, v, j) \tag{1}$$

$$S_u(V_1, V_2) = \frac{2 * \sum_{(v_i,v_j) \in WEr} \text{Doc2Vec}(v_i, v_j)}{|V_1| + |V_2|} \tag{2}$$

[8] Data collected on 15/2/2015 and 22/04/2016.
[9] http://purl.org/vol/ns/.
[10] http://www.ted.com/playlists.
[11] http://vmdeb20.deri.ie:8890/esaservice.
[12] Google pre-trained dataset: https://goo.gl/flpokK.

where WEr represents the set of edges included in the 1-1 maximal bipartite graph matching following the definition of Schwartz et al. [22].

Unlike the knowledge graph created by Taibi et al. [23], our knowledge graph of TED talks includes information about the playlists, the relations between TED talks, and four similarity values for each pair of talks (TFIDF, ESA, Doc2Vec, and Doc2Vec Neighbors). The knowledge graph of TED talks is publicly available at https://goo.gl/7TnsqZ.

6.2 Experimental Configuration

We empirically evaluate the effectiveness of $\mathcal{KOI}$ to discover missing relations in the 2015 TED knowledge graph, which is based on a real-world dataset. We compare $\mathcal{KOI}$ with METIS [9] and *k-Nearest Neighbors* (KNN) empowered with four similarity measures: TFIDF, ESA, Doc2Vec, and Doc2Vec Neighbors.

Research Questions: We aim at answering the following research questions: (**RQ**1) Does semantics encoded in similarity measures affect the relation discovery task? In order to answer this question we compare four similarity measures, one statistical-based measure (TFIDF) and three semantic similarity measures (ESA [6], Doc2Vec [12], and Doc2Vec Neighbors). Doc2Vec Neighbors considers both, the semantics encoded in datatype properties and the structure of the graph by taking into account the ego-networks. (**RQ**2) Is $\mathcal{KOI}$ able to outperform common discovery approaches as METIS or KNN?

Implementation: We implemented $\mathcal{KOI}$ in Java 1.8 and executed the experiments on an Ubuntu 14.04 64 bits machine with CPU: Intel(R) Core(TM) i5-4300U 1.9 GHz (4 physical cores) and 8GB RAM. In order to perform a fair evaluation, we used the library *WEKA* [8] version 3.7.12 to split the dataset following the 10-fold cross-validation strategy. The cross-validation was performed over the set of relations among TED talks. In order to discover relations using the METIS solver version 5.1[13], we apply METIS on a KOI Bipartite Graph with the same similarity measures S_u and S_v above specified for $\mathcal{KOI}$. METIS returns a partitioning of the given graph, and we produce candidate discovered relations as explained in Sect. 4. In order to perform a fair comparison, the same constraint (Listing 1.1) is applied for the results of both, $\mathcal{KOI}$ and METIS.

Evaluation Metrics: For each discovery approach, we compute the following metrics: (i) *Precision*: Relation between the number of correctly discovered relations and the whole set of discovered relations. (ii) *Recall*: Relation between the number of correctly discovered relations and the number of existing relations in the dataset. (iii) *F-Measure*: harmonic mean of precision and recall. Values showed in Tables 1 and 2 are the average values over the 10-folds. Moreover, we draw the F-Measure curves for $\mathcal{KOI}$ and METIS and calculate the Precision-Recall Area Under the Curve (AUC) coefficients (Table 3).

[13] http://glaros.dtc.umn.edu/gkhome/metis/metis/download.

6.3 Discovering Relations with K-Nearest Neighbors

In our first experiment, we discover relations in the graph using the K-Nearest Neighbors (KNN) algorithm under the hypothesis that highly similar TED talks should be related. Given a talk, we discover a relation between it and its K most similar talks. This experiment evaluates the impact of considering semantics encoded in domain similarity measures during the relation discovery task (**RQ1**).

Table 1 reports on the results obtained by four similarity measures: TFIDF, ESA [6], Doc2Vec [12], and Doc2Vec Neighbors. The first three similarity measures only consider knowledge encoded in datatype properties. On the other hand, Doc2Vec neighbors compares two entities considering the knowledge located in datatype properties and the structure of the graph by taking into account the ego-networks. Results obtained with the first three similarity measures suggest that Doc2Vec and ESA, which are semantic similarity measures, are able to outperform TFIDF, which does not take into account semantics. Doc2Vec obtains the highest F-measure value (0.196) with $K = 13$, which is significantly better than the maximum values obtained by ESA (0.137) and TFIDF (0.133). Thus, we can conclude that considering semantics encoded in Doc2Vec has a positive impact in the relation discovery task with respect to ESA and TFIDF. Results obtained with the Doc2Vec Neighbors indicate that knowledge encoded in ego-networks is of great value and that combining it with the knowledge encoded in datatype properties allows for obtaining a higher F-measure value (0.285) than the other three similarity measures.

Table 1. Effectivenness of KNN. D2V = Doc2Vec, D2VN = Doc2Vec Neighbors. D2VN presents the best results with an F-measure of 0.285 for $K = 4$. Relevance of the knowledge encoded in ego-networks is reported

	Precision				Recall				F-Measure			
K	TFIDF	ESA	D2V	D2VN	TFIDF	ESA	D2V	D2VN	TFIDF	ESA	D2V	D2VN
2	0.219	0.251	0.300	0.558	0.036	0.042	0.048	0.156	0.06	0.07	0.08	0.244
3	0.203	0.240	0.286	0.424	0.050	0.060	0.069	0.212	0.077	0.092	0.107	0.283
4	0.191	0.220	0.267	0.322	0.061	0.072	0.085	0.257	0.089	0.104	0.123	0.285
5	0.179	0.205	0.255	0.254	0.072	0.083	0.101	0.288	0.098	0.113	0.138	0.27
6	0.172	0.196	0.243	0.208	0.083	0.094	0.115	0.322	0.106	0.121	0.148	0.253
7	0.165	0.187	0.235	0.175	0.092	0.104	0.129	0.35	0.112	0.127	0.158	0.233
8	0.158	0.177	0.227	0.149	0.101	0.111	0.142	0.373	0.117	0.129	0.165	0.233
9	0.153	0.169	0.217	0.128	0.110	0.120	0.152	0.391	0.12	0.132	0.168	0.193
10	0.147	0.160	0.212	0.113	0.118	0.126	0.165	0.41	0.123	0.133	0.175	0.177
11	0.143	0.154	0.207	0.1	0.124	0.133	0.177	0.422	0.133	0.135	0.18	0.171
12	0.139	0.149	0.200	0.089	0.132	0.140	0.186	0.434	0.128	0.137	0.181	0.147
13	0.134	0.144	0.195	0.08	0.138	0.146	0.184	0.442	0.128	0.136	0.196	0.135
14	0.131	0.138	0.190	0.072	0.145	0.151	0.205	0.45	0.129	0.136	0.186	0.125

Table 2. Comparison of $\mathcal{KOI}$ and METIS. Values of θ correspond to the value of variable THETA of the constraint in Listing 1.1

	KOI			METIS		
θ	Precision	Recall	F-Measure	Precision	Recall	F-Measure
0	0.068	0.645	0.123	0.011	0.732	0.022
0.7	0.563	0.47	0.512	0.218	0.553	0.31
0.8	0.633	0.424	0.507	0.234	0.518	0.319
0.9	0.646	0.374	0.473	0.251	0.478	0.326
1	0.678	0.356	0.466	0.271	0.458	0.337
1.1	0.728	0.343	0.465	0.305	0.439	0.356
1.2	0.776	0.334	0.466	0.336	0.428	0.373
1.3	0.808	0.321	0.46	0.367	0.413	0.385
1.4	0.853	0.304	0.448	0.392	0.393	0.389
1.5	0.867	0.287	0.431	0.41	0.378	0.39
1.6	0.887	0.265	0.408	0.432	0.357	0.388

6.4 Effectiveness of $\mathcal{KOI}$ Discovering Relations

We executed $\mathcal{KOI}$ using the definitions of S_v and S_u in Eqs. 1 and 2, respectively. We compare $\mathcal{KOI}$ with respect to METIS [9] using the relation constraint constraint defined in Listing 1.1.

Table 2 contains the obtained results with $\mathcal{KOI}$ and METIS. The highest F-measure value is 0.512 and is obtained by $\mathcal{KOI}$ with $\theta = 0.7$. This F-measure value is higher than the one obtained with KNN and Doc2Vec Neighbors (0.285) and also higher than the maximum value obtained by METIS (0.39). We also observe that the parameter θ, which corresponds to *THETA* in Listing 1.1, can be configured depending on the respective importance of precision and recall. Lower values of θ deliver high values of recall, while high values of θ deliver high values of precision. Figure 5 shows the F-Measure curve for values of $\theta \in [0, 2]$. $\mathcal{KOI}$ is able to get higher F-Measure values for almost all θ values. We also computed the

Table 3. Area Under the Curve coefficients for $\mathcal{KOI}$, KNN Doc2Vec Neighbors and METIS

Approach	AUC	F-Measure
$\mathcal{KOI}$	**0.396**	**0.512**
METIS	0.244	0.39
KNN D2VN	0.223	0.285

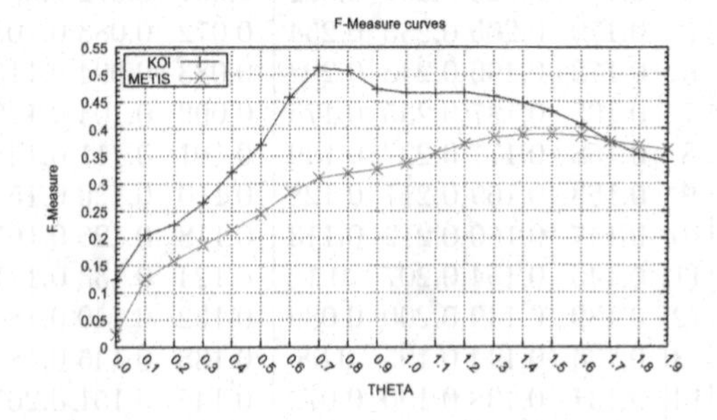

Fig. 5. F-Measure curves of $\mathcal{KOI}$ and METIS. Area under the curve indicates quality of the approaches

Precision-Recall Curve for $\mathcal{KOI}$, METIS and KNN Doc2Vec Neighbors. Table 3 shows that $\mathcal{KOI}$ gets a higher AUC value (0.396) than METIS (0.244) and KNN (0.223).

7 Conclusions and Future Work

In this paper we present $\mathcal{KOI}$, an approach that exploits semantics and graph structure information in order to discover missing relations in a knowledge graph. $\mathcal{KOI}$ considers semantics encoded in entities and their ego-networks to identify relations between entities with similar datatype properties and similar ego-networks. Reported experimental results suggest that $\mathcal{KOI}$ outperforms state-of-the-art approaches that: (i) do not consider semantics (KNN TFIDF), (ii) do not identify graph portions containing highly similar entities (KNN D2VN and METIS). In the future, we plan to extend $\mathcal{KOI}$ to take into account domain specific knowledge in graphs of more specific domains, e.g., social network, financial, or clinical data. Further, we plan to extend $\mathcal{KOI}$ to consider the relevance or importance of the entities in ego-networks, as well as to discover relations between different types of entities, e.g., drugs and proteins.

Acknowledgements. This work is supported by the German Ministry of Education and Research within the SHODAN project (Ref. 01IS15021C) and the German Ministry of Economy and Technology within the ReApp project (Ref. 01MA13001A).

References

1. Arenas, M., Gutierrez, C., Pérez, J.: Foundations of RDF databases. In: Tessaris, S., Franconi, E., Eiter, T., Gutierrez, C., Handschuh, S., Rousset, M.-C., Schmidt, R.A. (eds.) Reasoning Web. LNCS, vol. 5689, pp. 158–204. Springer, Heidelberg (2009)
2. Epasto, A., Lattanzi, S., Mirrokni, V., Sebe, I.O., Taei, A., Verma, S.: Ego-net community mining applied to friend suggestion. VLDB Endow. **9**(4), 324–335 (2015)
3. Fischer, P.M., Lausen, G., Schätzle, A., Schmidt, M.: RDF constraint checking. In: EDBT/ICDT 2015 Joint Conference (2015)
4. Flores, A., Vidal, M., Palma, G.: Exploiting semantics to predict potential novel links from dense subgraphs. In: 9th Alberto Mendelzon International Workshop on Foundations of Data Management (2015)
5. Fundulaki, I., Auer, S.: Linked open data - introduction to the special theme. ERCIM News **2014**(96) (2014)
6. Gabrilovich, E., Markovitch, S.: Computing semantic relatedness using Wikipedia-based explicit semantic analysis. In: IJCAI, vol.7 (2007)
7. García, J.L.R., Sabatino, M., Lisena, P., Troncy, R.: Detecting hot spots in web videos. In: ISWC Poster and Demo Track. CEUR-WS.org (2014)
8. Hall, M., Frank, E., Holmes, G., Pfahringer, B., Reutemann, P., Witten, I.H.: The weka data mining software: an update. ACM SIGKDD Explor. Newsl. **11**(1), 10–18 (2009)
9. Karypis, G., Kumar, V.: A fast and high quality multilevel scheme for partitioning irregular graphs. SIAM J. Sci. Comput. **20**(1) (1998)

10. Kastrin, A., Rindflesch, T.C., Hristovski, D.: Link prediction on the semantic MEDLINE network - an approach to literature-based discovery. In: Džeroski, S., Panov, P., Kocev, D., Todorovski, L. (eds.) DS 2014. LNCS, vol. 8777, pp. 135–143. Springer, Heidelberg (2014)
11. Lausen, G., Meier, M., Schmidt, M.: Sparqling constraints for RDF. In: 11th International Conference on Extending Database Technology, EDBT. ACM (2008)
12. Le, Q.V., Mikolov, T.: Distributed representations of sentences and documents. CoRR, abs/1405.4053 (2014)
13. Liben-Nowell, D., Kleinberg, J.: The link-prediction problem for social networks. J. Am. Soc. Inf. Sci. Technol. **58**(7), 1019–1031 (2007)
14. Pereira Nunes, B., Dietze, S., Casanova, M.A., Kawase, R., Fetahu, B., Nejdl, W.: Combining a co-occurrence-based and a semantic measure for entity linking. In: Cimiano, P., Corcho, O., Presutti, V., Hollink, L., Rudolph, S. (eds.) ESWC 2013. LNCS, vol. 7882, pp. 548–562. Springer, Heidelberg (2013). doi:10.1007/978-3-642-38288-8_37
15. Palma, G., Vidal, M.-E., Raschid, L.: Drug-target interaction prediction using semantic similarity and edge partitioning. In: Mika, P., et al. (eds.) ISWC 2014, Part I. LNCS, vol. 8796, pp. 131–146. Springer, Heidelberg (2014)
16. Pappas, N., Popescu-Belis, A.: Combining content with user preferences for ted lecture recommendation. In: 11th International Workshop on Content Based Multimedia Indexing. IEEE (2013)
17. Pérez, J., Arenas, M., Gutierrez, C.: Semantics and complexity of SPARQL. ACM Trans. Database Syst. **34**(3), 30–43 (2009)
18. Pirró, G.: Explaining and suggesting relatedness in knowledge graphs. In: Arenas, M., et al. (eds.) ISWC 2015. LNCS, vol. 9366, pp. 622–639. Springer, Heidelberg (2015). doi:10.1007/978-3-319-25007-6_36
19. Řehůřek, R., Sojka, P.: Software framework for topic modelling with large corpora. In: LREC 2010 Workshop on New Challenges for NLP Frameworks. ELRA (2010). http://is.muni.cz/publication/884893/en
20. Rindflesch, T.C., Kilicoglu, H., Fiszman, M., Rosemblat, G., Shin, D.: Semantic medline,: an advanced information management application for biomedicine. Inf. Serv. Use **31**(1–2), 15–21 (2011)
21. Sachan, M., Ichise, R.: Using semantic information to improve link prediction results in network datasets. Int. J. Eng. Technol. **2**(4), 71–76 (2010)
22. Schwartz, J., Steger, A., Weißl, A.: Fast algorithms for weighted bipartite matching. In: Nikoletseas, S.E. (ed.) WEA 2005. LNCS, vol. 3503, pp. 476–487. Springer, Heidelberg (2005)
23. Taibi, D., Chawla, S., Dietze, S., Marenzi, I., Fetahu, B.: Exploring TED talks as linked data for education. Br. J. Educ. Technol. **46**(5), 1092–1096 (2015)

ACRyLIQ: Leveraging DBpedia for Adaptive Crowdsourcing in Linked Data Quality Assessment

Umair ul Hassan[1(✉)], Amrapali Zaveri[2], Edgard Marx[3], Edward Curry[1], and Jens Lehmann[4,5]

[1] Insight Centre for Data Analytics, National University of Ireland, Galway, Ireland
{umair.ulhassan,edward.curry}@insight-centre.org
[2] Stanford Center for Biomedical Informatics Research, Stanford University, Stanford, USA
amrapali@stanford.edu
[3] AKSW Group, University of Leipzig, Leipzig, Germany
emarx@informatik.uni-leipzig.de
[4] Computer Science Institute, University of Bonn, Bonn, Germany
jens.lehmann@cs.uni-bonn.de
[5] Knowledge Discovery Department, Fraunhofer IAIS, Sankt Augustin, Germany
jens.lehmann@iais.fraunhofer.de

Abstract. Crowdsourcing has emerged as a powerful paradigm for quality assessment and improvement of Linked Data. A major challenge of employing crowdsourcing, for quality assessment in Linked Data, is the cold-start problem: how to estimate the reliability of crowd workers and assign the most reliable workers to tasks? We address this challenge by proposing a novel approach for generating test questions from DBpedia based on the topics associated with quality assessment tasks. These test questions are used to estimate the reliability of the new workers. Subsequently, the tasks are dynamically assigned to reliable workers to help improve the accuracy of collected responses. Our proposed approach, ACRyLIQ, is evaluated using workers hired from Amazon Mechanical Turk, on two real-world Linked Data datasets. We validate the proposed approach in terms of accuracy and compare it against the baseline approach of reliability estimate using gold-standard task. The results demonstrate that our proposed approach achieves high accuracy without using gold-standard task.

1 Introduction

In recent years, the *Linked Data* paradigm [7] has emerged as a simple mechanism for employing the Web for data and knowledge integration. It allows the publication and exchange of information in an interoperable way. This is confirmed by the growth of Linked Data on the Web, where currently more than 10,000 datasets are provided in the Resource Description Format (RDF)[1]. This vast

[1] http://lodstats.aksw.org.

E. Blomqvist et al. (Eds.): EKAW 2016, LNAI 10024, pp. 681–696, 2016.
DOI: 10.1007/978-3-319-49004-5_44

amount of valuable interlinked information gives rise to several use cases to discover meaningful relationships. However, in all these efforts, one crippling problem is the underlying data quality. Inaccurate, inconsistent or incomplete data strongly affects the consumption of data as it leads to unreliable conclusions. Additionally, assessing the quality of these datasets and making the information explicit to the publisher and/or consumer is a major challenge.

To address the challenge of *Linked Data Quality Assessment* (LDQA), crowdsourcing has emerged as a powerful mechanism that uses the "wisdom of the crowds" [9]. An example of a crowdsourcing experiment is the creation of LDQA tasks, then submitting them to a crowdsourcing platform (e.g. Amazon Mechanical Turk), and paying for each task that the workers perform [6,16,21]. Crowdsourcing has been utilized in solving several problems that require human judgment include LDQA. Existing research has focused on using crowdsourcing for detecting quality issues [21], entity linking [6], or ontology alignments [14,16]. A major challenge of employing crowdsourcing for LDQA is to accurate responses for tasks while considering *reliability* of workers [3,5,8,15]. Therefore, it is desirable to find the most reliable workers for the tasks.

In this paper, we study the problem of *adaptive task assignment* in crowdsourcing specifically for quality assessment in Linked Data. In order to make appropriate assignments of tasks to workers, crowdsourcing systems currently rely on the estimated reliability of workers based on their performance on previous tasks [10,15]. Some approaches rely on expectation-maximization style approaches to jointly estimate task responses and reliability of workers after collecting data from large number of workers [11,15]. However, it is a difficult problem to estimate the reliability of new workers. In fact, existing crowdsourcing systems have been shown to exhibit a *long-tail* phenomena where the majority of workers have performed very few tasks [10]. The uncertainty of worker reliability leads to low *accuracy* of the aggregated tasks responses. This is called as the *cold-start problem* and is particularly challenging for LDQA, since the tasks may require domain knowledge from workers (e.g. knowledge of a language for detecting incorrectly labeled language tags).

Existing literature on crowdsourcing that addresses the cold-start problem is not applicable to LDQA due to several reasons [2,8]. Firstly, the manual creation of (GSTs) with known correct responses is expensive and difficult to scale [15]. Secondly, the effects of domain specific knowledge on the reliability of workers is not considered in existing literature [15]. Moreover, assignments using social network profiles of workers require significant information about workers and their friends, which poses a privacy problem [2].

We introduce the *Adapative Crowdsourcing for Linked Data Quality Assessment* (ACRyLIQ), a novel approach that addresses the cold-start problem by exploiting a generalized knowledge base. ACRyLIQ estimates the reliability of a worker using test questions generated from the knowledge base. Subsequently, the estimated reliability is used for adaptive task assignment to the best workers. Indeed, with the generality of the DBpedia [12] (the Linked Data version of the Wikipedia), it is not difficult to find facts related to most topics or domains. As a consequence, a large quantity of domain specific and mostly correct facts can be obtained to test the knowledge of workers. Thus, the fundamental research question addressed in the paper is: *How can we estimate the reliability of crowd workers using facts from DBpedia to achieve high accuracy of LDQA tasks though adaptive task assignment*? The core contributions of this paper are:

- A novel approach, ACRyLIQ, to generate test questions from DBpedia. The test questions are used for estimating the reliability of workers while considering the domain-specific topics associated with LDQA tasks.
- A comparative study of the proposed approach against baseline approaches on LDQA tasks using two real datasets. The first dataset considers language verification tasks for five different languages. The second dataset considers entity matching tasks for five topics: (i) Books, (ii) Nature, (iii) Anatomy, (iv) Places, and (v) Economics.
- Evaluation of the proposed and baseline approaches by employing workers from Amazon Mechanical Turk. The results demonstrate that our proposed approach achieves high accuracy without the need for gold-standard task.

2 Preliminaries

In this section, we introduce the core concepts used throughout this paper. Furthermore, we highlight the key assumptions associated with those concepts.

Definition 1 (Topics). *Given a Linked Data dataset, let S be a set of topics associated with the dataset. Each topic $s \in S$ specifies an area of knowledge that is differentiated from other topics.*

For instance, consider a dataset consisting of review articles about books. An article might refer to various topics such as "Books", "Political-biographies", "1960-novels", etc. We assume that similar topics are grouped together and there is minimum overlap between the topics (or topic groups) in the set S.

Definition 2 (Tasks). *Let $T = \{t_1, t_2, ..., t_n\}$ be the set of LDQA tasks for the dataset. Each task $t_i \in T$ is a multiple-choice question with an unknown correct response r_i^* that must be generated through crowdsourcing.*

For instance, a task might ask workers to judge whether two review articles are referring to the same book. We assume that the set of tasks T is partitioned according to topics associated with the dataset; hence, each task is associated with a topic. In practice, it is possible that a task might be associated with more than one topic. In such a case, the primary topic of each task can be chosen using a relevance ranking. If there is no obvious ranking, then the primary topic of a task can be chosen arbitrarily.

Definition 3 (Workers). *Let $W = \{w_1, w_2, ..., w_m\}$ be the set of workers that are willing to perform tasks. Workers arrive in an online manner and request tasks; in addition, each worker $w_j \in W$ has a latent reliability $p_{i,j}$ on task $t_i \in T$.*

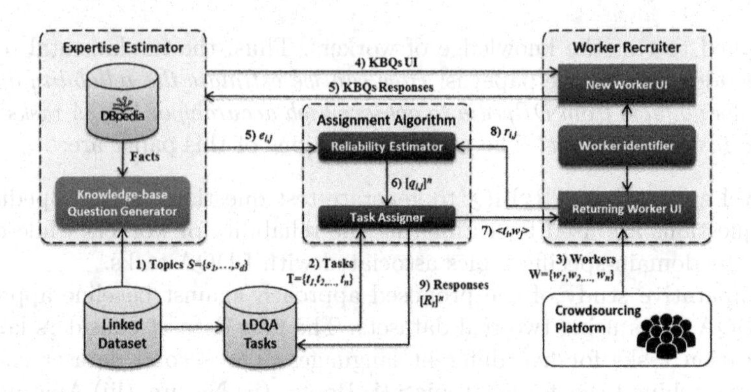

Fig. 1. An overview of the proposed ACRyLIQ approach.

If worker w_j performs task t_i and provides the response $r_{i,j}$, then the probability of $r_i = r_{i,j}$ is $p_{i,j}$ and the probability of $r_i \neq r_{i,j}$ is $1 - p_{i,j}$. Without the loss of generality we assume that $p_{i,j} \in [0, 1]$. For instance, a worker who is well versed on the topic "1960-novels" has higher likelihood of providing correct responses to tasks associated with the same topic.

Let $R_i = \{r_{i,j} \mid w_j \in W_i\}$ be the set of responses collected from the workers $W_i \subset W$ assigned to the task t_i. We use majority-vote on R_i to generate the response $\hat{r}_i$. The goal of an assignment algorithm is to find the best workers for a task such that the estimated response $\hat{r}_i$ is accurate. Therefore, the assignment algorithm must estimate the reliabilities of workers. The set of workers W_i for a task t_i can be chosen such that $\sum_{w_j \in W_i} p_{i,j}$ is maximized. In case of arbitrary reliabilities, the assignment algorithm may not be able to find good worker. We assume that the reliability of a worker on tasks associated with the same topic remain approximately the same. That is $p_{i,j} \sim p_{i',j}$ for all w_j when t_i and $t_{i'}$ are associated with the same topic.

Definition 4 (Gold-Standard Tasks). *The subset of tasks $T_G \subset T$ with known correct responses (most often from the same dataset).*

Existing approaches for adaptive task assignment use gold-standard task to estimate reliabilities of workers which imposes additional costs for the collection of correct responses from domain experts [3,8]. Furthermore, it is difficult to generate gold-standard task for complex or knowledge-intensive tasks [15].

Definition 5 (Knowledge Base). *A set of facts F related to the set of topics S where each fact belongs to exactly one topic.*

We assume access to a knowledge base that contains facts F related to the dataset of LDQA tasks. Similar to the partitioning of tasks, the facts in the knowledge base are divided in to $|S|$ partitions. Next, we describe a novel approach for estimating the reliabilities of workers by exploiting a knowledge base.

3 Reliability Estimation Using Knowledge Base

Figure 1 illustrates the proposed approach, for reliability estimation and adaptive task assignment, that uses DBpedia as a generalized knowledge base of facts[2]. First, the topics S are used for selecting facts from DBpedia and generating test questions, referred to as knowledge base question (KBQs). Then, each new worker w_j is given the KBQs and worker's responses to KBQs are used to generated estimated reliabilities $q_{i,j}$. Finally, the estimated reliabilities are used for assigning LDQA tasks to workers and estimation of task responses. Note that a similar approach can be applied to knowledge bases other than DBpedia. The number of facts in a knowledge base can be in millions which raises the KBQ selection challenge: *How to choose a set of facts from the knowledge base such that the reliability of a worker on the KBQs highly correlates with their reliability on LDQA tasks*? The goal is to minimize the difference between the estimated reliability $q_{i,j}$ and the true reliability $p_{i,j}$.

3.1 KBQs Selection Problem

Recent research has shown that the reliability of a worker tends to be comparable on similar topics [3]. Therefore, we propose to use the similarity between tasks and facts to address the KBQs selection problem. The intuition is that the similarity quantifies the influence of workers' response to both facts and tasks. Since the similarity can be defined in terms of textual comparisons or detailed semantics, we detail the similarity measure used in this paper in Sect. 5. Given the similarity measure $sim(t, f)$, next we formalize the KBQs selection problem.

Let Q be the set of KBQs generated from facts F in the knowledge base. The number of KBQs is fixed at Φ to control the overhead costs of reliability estimation. Based on their associated facts, the KBQs in Q are also divided into $|S|$ partitions. Let Q_s be the set of KBQs selected for topic s. The probability that a KBQ is selected for topic s is $\mathbb{P}(Q_s) = |Q_s|/\Phi$. We define the entropy of the set Q according to a measure based on Shannon entropy [17], that is $\mathbb{H}(Q) = -\sum_{s \in S} P(Q_s) \cdot \ln P(Q_s)$. The intuition is to generate a diverse set of KBQs. A higher value of entropy means that more topics are covered with equal number of KBQs; hence, it is desirable to maximize the entropy of Q.. Besides entropy, the objective is to generate KBQs that have high influence on the tasks. Influence is the positive correlation between the accuracy of worker responses to KBQ and the accuracy of the same worker on tasks. The next section details a parametric algorithm that addresses the KBQs selection problem.

3.2 KBQs Selection Algorithm

We devise a greedy algorithm for the KBQs selection problem, as shown in Algorithm 1. The algorithm assumes availability of a similarity measure $sim(t, f)$ between LDQA tasks and facts in the knowledge base. The algorithm starts with

[2] A DBpedia triple is considered a fact.

Algorithm 1. KBQ Selection Algorithm

Require: T, S, F, Φ, β
1: $Q \leftarrow \emptyset$
2: **for** $i = 1, ..., \Phi$ **do**
3: $F \leftarrow F - Q$
4: **for** $f_k \in F$ **do**
5: $\Delta_k = \mathbb{H}(Q \cup f_k) - \mathbb{H}(Q)$
6: **end for**
7: $\hat{f} = \text{argmax}_{f_k \in F} \beta \cdot \Delta_k + (1 - \beta) \sum_{t_i \in T} sim(t_i, f_k)$
8: $Q \leftarrow Q \cup \hat{f}$
9: **end for**
10: **return** Q

an empty set of facts Q and then iteratively selects Φ facts from the knowledge base F. For each new fact f_k, the algorithm calculates the difference between the entropy of Q and the entropy of $Q \cup f_k$ (Line 5). Then it selects the fact $\hat{f}$ that maximizes the entropy difference and the similarity with the tasks, and β is used as β (Line 7). The computational complexity of Algorithm 1 is $\mathcal{O}(\Phi|F||T|)$. Understandably, a performance bottleneck can be the number of facts in the knowledge base. This necessitates an effective pruning strategy to exclude facts that are very different from tasks or have little benefit for reliability estimation. Section 5.2 discusses one such strategy that is employed to reduce the search space when selecting facts from DBpedia.

4 Adaptive Task Assignment

Given the set of KBQs, we extend an existing adaptive task assignment algorithm that uses gold-standard task to estimate worker reliabilities [3]. This algorithm also serves as the baseline during the evaluation of our proposed approach. Algorithm 2 lists our algorithm for adaptive task assignment that uses KBQs for reliability estimates. The algorithm expects a set of tasks T, a set of KBQs Q, and three control parameters (i.e. λ, α, γ). The parameter λ specifies the number of unique workers $|W_i|$ to be assigned to each task. The similarity-accuracy trade-off parameter is α and the number of iterations is γ. The algorithm consists of two distinct phases: (i) offline initialization and (ii) online task assignment.

The offline initialization phase (Lines 1–18) consists of following steps. The algorithm starts by combining LDQA tasks and KBQs (Lines 2–3) and calculates their similarity based on the topics shared between them (Line 4). For instance, LDQA tasks or KBQs belonging to the same topic are assigned a similarity value between 0 and 1. Next, the algorithm normalizes the similarity scores (Lines 5–9). The similarity scores in matrix $\hat{\boldsymbol{Z}}$ are further weighted using the parameter α to control the effect of similarity on reliability estimation (Lines 10–17). Parameter γ controls the number of iterations used for the adjustment of similarity scores.

Algorithm 2. Adaptive Assignment Algorithm

Require: $T, Q, \lambda, \alpha, \gamma$
1: $T \leftarrow T \cup Q$ {Combine tasks and KBQs}
2: $n \leftarrow |T|$
3: $\boldsymbol{Z} \leftarrow TopicSimilarityMatrix(T)$ {Topic similarity matrix}
4: $\boldsymbol{D} \leftarrow [0]^{n \times n}$
5: **for** $i = 1, ..., n$ **do**
6: $\quad D_{i,i} = \sum_{j=1}^{n} Z_{i,j}$
7: **end for**
8: $\hat{\boldsymbol{Z}} \leftarrow \boldsymbol{D}^{-1/2}\boldsymbol{Z}\boldsymbol{D}^{-1/2}$
9: **for** $t_i \in T$ **do**
10: $\quad \boldsymbol{p_i} \leftarrow [0]^n$
11: $\quad p_{i,i} \leftarrow 1$
12: $\quad \boldsymbol{q_i} \leftarrow \boldsymbol{p_i}$
13: $\quad$ **for** $g = 1, ..., \gamma$ **do**
14: $\quad\quad \boldsymbol{p_i} \leftarrow \frac{1}{1+\alpha}\boldsymbol{p_i}\hat{\boldsymbol{Z}} + \frac{\alpha}{1+\alpha}\boldsymbol{q_i}$
15: $\quad$ **end for**
16: **end for**
17: $c \leftarrow 0$ {Initialize assignments counter}
18: $R \leftarrow \emptyset$ {Initialize response set}
19: **for** $c < n\lambda$ **do**
20: $\quad (w_j, C_j) \leftarrow getNextWorker()$ {Worker requests C_j tasks}
21: $\quad$ **if** w_j is a new worker **then**
22: $\quad\quad$ Assign Q KBQs to worker
23: $\quad\quad \boldsymbol{q_j} \leftarrow ObservedAccuracy(Q)$
24: $\quad\quad \boldsymbol{p_j} \leftarrow \sum_{q_{i,j}} q_{i,j} \cdot \boldsymbol{p_i}$
25: $\quad$ **end if**
26: $\quad \mathcal{T} = \{\tau | \tau \subset T, |\tau| = C_j\}$
27: $\quad T_j^* = \text{argmax}_{\tau \in \mathcal{T}} \sum_{t_i \in \tau} p_{i,j}$
28: $\quad$ Assign T_j^* to worker w_j to get R_j responses.
29: $\quad c \leftarrow c + C_j$
30: $\quad R \leftarrow R \cup R_j$
31: **end for**
32: **return** R

The online task assignment phase (Lines 19–31) proceeds in iterations as workers arrive dynamically and request tasks. The task assignment process stops when all tasks have received λ responses. Each dynamically arriving worker w_j requests C_j tasks (Line 22). If the worker w_j is requesting tasks for the first time, then the set of KBQs is assigned to the worker (Line 24). Based on the responses to the KBQs, an estimated reliability vector $\boldsymbol{p_j}$ is generated for the worker w_j (Lines 25–26). The set of tasks, for which the w_j has highest reliabilities, is assigned to the worker (Lines 28–30). At the end of an iteration, the assignment counter and response sets are updated. The computational complexity of the offline initialization phase is $\mathcal{O}(n^2)$ and the online assignment phase is $\mathcal{O}(n)$.

5 Evaluation Methodology

For the purpose evaluation, we collected responses to KBQs and LDQA from real workers on Amazon Mechanical Turk. Since repeated deployment of the assignment algorithm with actual workers is known to be difficult and expensive [3,18], we employed a simulation-based evaluation methodology to compare the algorithms using collected responses. Each run of the algorithm is initialized with specific tasks and worker conditions.

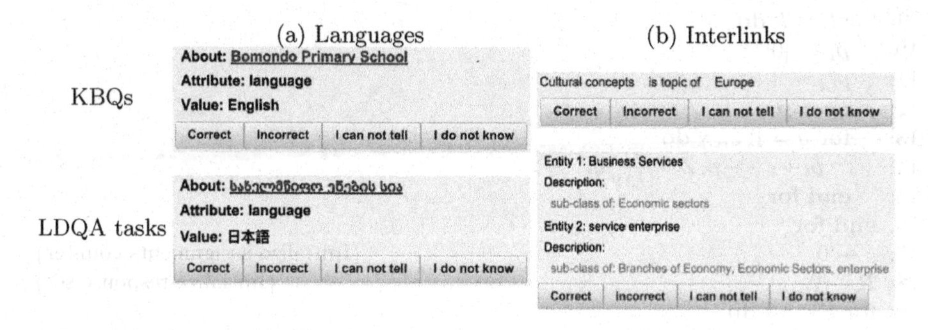

Fig. 2. Examples of KBQs and LDQA tasks for the Languages and Interlinks datasets.

5.1 LDQA Tasks

The LDQA tasks were based on two real-world datasets, as shown in Fig. 2. The datasets are summarized below:

- **Languages Dataset:** These tasks represent the syntactic validity of datasets [20], that is, the use of correct datatypes (in this case correct language tags) for literals. The tasks are based on the LinkedSpending[3] dataset, which is a Linked Data version of the OpenSpending project[4]. The dataset contains financial budget and spending data from governments from all over the world. As OpenSpending does not contain language tags for its entities, the 981 values of LinkedSpending entities only contain plain literals. In an effort to accurately identify the missing language tags, we first applied an automated language detection library[5] to generate a dataset containing the entities and the corresponding language. Out of the 40 distinct languages detected, 25 entity language pairs were randomly chosen to generate tasks. The correct responses for tasks were created with the help of language translation tools.
- **Interlinks Dataset:** These tasks represent the interlinking quality of a dataset [20], specifically about the presence of correct interlinks between

[3] http://linkedspending.aksw.org/.
[4] https://openspending.org/.
[5] https://github.com/optimaize/language-detector.

datasets. The tasks were generated from the Ontology Alignment Evaluation Initiative[6] (OAEI) datasets that cover five topics: (i) Books, (ii) Geography, (iii) Nature, (iv) Economics and (v) Anatomy. Each task required workers to examine two entities along with their corresponding information and evaluate whether they are related to each other (e.g. if they are the same). The correct responses for these tasks are available along with the OAEI datasets.

5.2 KBQs Selection with Pre-pruning

Since DBpedia contains more than three billion facts[7], it was essential to devise a pruning strategy to assist the KBQs selection process. In order to reduce the search space, we employed pre-pruning with the help of a string similarity tool. We used the LIMES tool [13], which employs time-efficient approaches for large-scale link discovery based on the characteristics of metric spaces. In particular, we used LIMES to find resources from the DBpedia dataset similar to the entities mentioned in the tasks. The triples associated with similar resources were used as facts for KBQ generation. As a pruning strategy, we could restrict the resources for each domain by specifying the particular class from DBpedia. For example, the resources for the topic "Books" were restricted to instances of the DBpedia class "Book". For the Languages dataset, the specification of the language in the LIMES configuration file assisted in selecting resources that were in that particular language. LIMES also supports the specification of using a specific string similarity metric, as well as the corresponding thresholds. In our case we used the "Jaro" similarity metric [19] and retrieved all the resources above threshold of 0.5. Figure 2 shows the examples of knowledge base questions generated for both datasets, as summarized below:

- **Language Dataset:** A set of 10 KBQs for the Languages dataset was generated from DBpedia. Each test question asks the worker to identify whether a value has a correct language tag.
- **Interlinks Dataset:** The KBQs were focused towards estimating the expertise of a worker on the topics associated with the Interlinks dataset. Triples with DBpedia property "is subject of" were used to generate the 10 KBQs. Each question required workers to identify whether one entity is related to (i.e. is subject of) another entity.

5.3 Compared Approaches and Metrics

We evaluated the performance of three reliability estimation approaches: (i) the proposed KBQ approach, (ii) the existing GST approach and the baseline *randomly generated estimates* (RND) approach. The following metrics were used to report the performance of algorithms:

[6] http://oaei.ontologymatching.org/.
[7] As of 2014 http://wiki.dbpedia.org/about.

- *Average Accuracy:* The primary metric for the performance is the average accuracy of the final aggregated responses over all tasks.
- *Overhead Costs:* The total overhead costs paid to workers due to the KBQs or GSTs used for estimating the worker reliabilities.

6 Experimental Results

Revisiting our research question, we aim to estimate the reliability of a worker using KBQs in an effort to assign the best workers for LDQA tasks. In the following, we first report the results of data collected from real workers and then present the results of simulation experiments performed for the evaluation of the proposed approach. Table 1 shows the experimental parameters and their default values.

Table 1. Summary of experiment parameters and their default values (in bold font).

Parameter	Description	Values
λ	Assignment size per task	**3**, 5, 7
α	The similarity-accuracy trade-off parameter	0, **0.5**, 1
β	The similarity-entropy trade-off parameter	0.25, **0.5**, 0.75
γ	The number of iterations	0.25, **0.5**, 0.75
n	Number of LDQA tasks i.e. $\|T\|$	**15**
m	Number of crowd workers i.e. $\|W\|$	**60**
Φ	The number for KBQs or GSTs per worker	5, **10**, 15

6.1 Diverse Reliability of Crowd Workers

The KBQs and LDQA tasks were posted on the a dedicate web server[8]. We used Amazon Mechanical Turk to hire 60 Master workers. The workers were paid a wage at the rate of $1.5 for 30 min spent on the tasks. Worker was first asked to provide background information such as: the region that they belong to, their self-assessed knowledge about the five topics (of the Interlinking dataset), the amount of years they have spoken each language (of the Languages dataset) and their native language. Then the worker was asked to answer the 10 KBQs for each dataset. The sequence of KBQs was randomized for each worker. Finally, the worker was asked to respond to the set of 25 LDQA tasks for each dataset. Workers took nine minutes, on the average, to complete the background information, the KBQs, and the LDQA tasks.

We used the first 10 tasks in both datasets as the gold-standard task. Figure 3 shows the average reliability of workers in terms of the languages in the Languages dataset and topics in the Interlinks dataset. Note that the workers are less reliable on Asian languages and their standard deviation of reliability is

[8] http://dataevaluation.aksw.org.

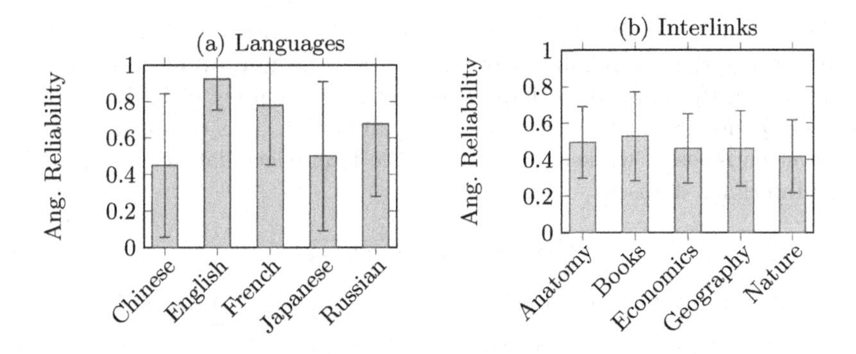

Fig. 3. Average reliability of workers on all 25 tasks for both datasets.

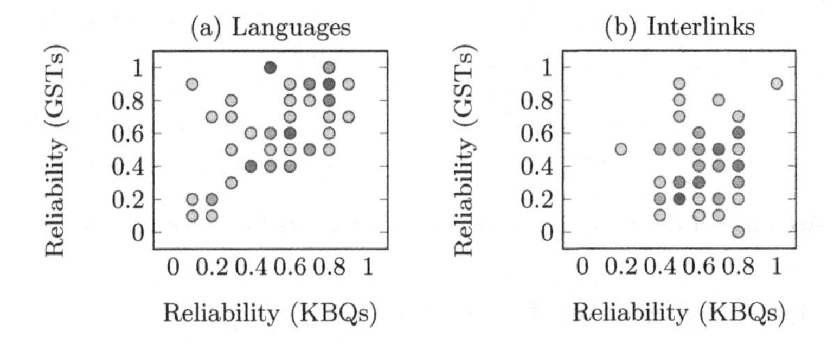

Fig. 4. Reliability of the 60 real workers on KBQs and GSTs.

high. Workers exhibit high reliability on European languages with low standard deviation. The average reliability is low across the topics in the Interlinks datasets. Figure 4 shows the relationship between the reliability of workers on KBQs and GSTs. The Pearson correlation between the two reliabilities is 0.545 and 0.226 for the Languages and Interlinks datasets, respectively.

6.2 Accuracy of Compared Approaches

We compared the average accuracy of the proposed approach against two baseline approaches: RNDs and GSTs. We also varied the λ parameters to study its effects on the performance of each approach. Figure 5 shows the accuracy on the Language and Interlinking tasks, based on 30 runs of each approach under the same settings. In general, the accuracy for both the KBQ and the GST approach is better than the baseline RND approach. This underlines the effectiveness of the adaptive task assignment algorithm in finding reliable workers for LDQA tasks. We compared the accuracy of the KBQ approach against the RND and GST approaches using the t-test, on the Languages dataset. The difference between the KBQ approach and the RND approach is significant with $t(178) = 13.745$ and $p < 0.05$. The difference between the KBQ approach and the GST approach is also significant with $t(178) = 3.719$ and $p < 0.05$.

These results establish the effectiveness of adaptive task assignment in exploiting the diverse reliability of workers for the improvement of accuracy. In the case of the Languages dataset, the average accuracy of RND is closer to both KBQ and GST, although still statistically lower. This can be attributed to the lower variance of worker reliability of the Languages dataset in comparison to the Interlinks dataset.

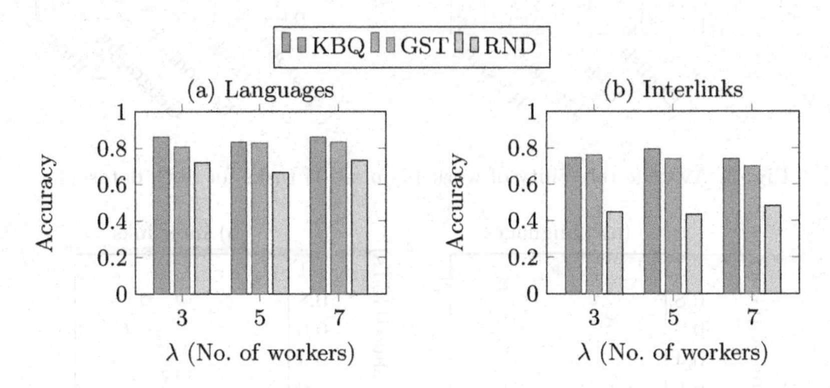

Fig. 5. Accuracy of the reliability estimation approaches for Languages and Interlinks.

6.3 Effects of Algorithm Parameters

We also studied the effects of the three algorithm parameters (i.e. Φ, α, β) on the average accuracy. For this purpose, we used a subset of the DBpedia resources using the pre-pruning strategies discussed earlier (c.f. Sect. 5.2). These resources were utilized to generate 74,993 KBQs that were used for the experiments. The similarity values between the KBQs and LDQA tasks were calculated using the Jaro-Winkler similarity measure. We simulated the answers of the workers on the 74,993 KBQs by training a logistic regression model from their answers to the 10 KBQs presented to them earlier. The model accuracy was more than 72% on test instances. We used this model to analyze the effects of different parameters on the performance the proposed algorithm.

The parameter Φ defines the budget for the overhead costs due to KBQs. Table 2 shows that the accuracy increases with increase in Φ; however, the relative increase is marginal. This indicates that even at the small cost Φ of estimating reliabilities through KBQs, the assignment algorithm achieves high accuracy.

Table 2. Effects of parameters Φ and β on the accuracy for the Interlinks dataset.

	Overhead costs budget			Similarity-entropy trade-off		
λ	$\Phi = 5$	$\Phi = 10$	$\Phi = 15$	$\beta = 0.25$	$\beta = 0.5$	$\beta = 0.75$
3	0.709 ± 0.029	0.716 ± 0.033	0.718 ± 0.034	0.684 ± 0.028	0.780 ± 0.025	0.749 ± 0.032
5	0.716 ± 0.033	0.758 ± 0.030	0.758 ± 0.028	0.760 ± 0.030	0.760 ± 0.027	0.740 ± 0.027
7	0.744 ± 0.032	0.727 ± 0.026	0.742 ± 0.025	0.733 ± 0.025	0.733 ± 0.030	0.736 ± 0.034

The parameter β controls the similarity-entropy trade-off. As shown in Table 2, the highest accuracy is achieved for $\beta = 0.5$. Meaning that any non-extreme value for the similarity-entropy trade-off parameter is sufficient. Similar results were observed for similarity-accuracy trade-off parameter α. In general, the conservative values of these parameters do not have significant effects on performance. However, this might change with a larger number of tasks with multiple topics.

7 Discussion and Limitations

The majority of existing literature on on adaptive task assignment in crowdsourcing considers GSTs for the cold-start problem [3,8,22]. Generating GSTs in itself is a difficult and expensive process [15]. Especially when the accuracy of task responses is not measurable. A key strength of our proposed approach is the applicability to such scenarios. It provides a quick and inexpensive method of estimating the reliability and expertise of workers. This approach is particularly suited for complex or knowledge-intensive tasks.

Our approach has three main limitations. First, the assumption that both facts and tasks are partitioned according to the same set of topics. In practice, this assumption can be relaxed by using a mapping between topics of facts and topics of tasks. A similar approach was employed for alignment of topics for the Interlinks dataset. Second, the approach assumes that the majority of the facts, that are used for the generation of KBQs, are correct. If a high percentage of incorrect facts are used for generating KBQs then our approach can misjudge the reliability of workers on tasks. Third, it assumes that the domain topics are mutually exclusive. This underlines that need for reconsideration of the entropy measure when the domain topics are overlapping.

The experiments presented in this paper is also limited in terms of scalability. In the case of DBpedia, pre-pruning can be utilized to limit the facts to the core DBpedia ontology and SKOS concepts. The facts can also be filtered according to the ratings of their associated articles in Wikipedia. The evaluation is also limited in terms of the overhead costs of the KBQs selection algorithm. The Languages and Interlinks dataset represent two types of LDQA tasks which can also be seen as a limitation of the experimental evaluation. However, an extension of the proposed approach to other types of LDQA tasks should be straight forward.

8 Related Work

At a technical level, specific Linked Data management tasks have been subject to crowdsourcing, including entity linking [6], ontology alignment [16], and quality assurance [1,21]. However, none of these proposals consider adaptive task assignment or the cold-start problem. Noy et al. performed a comparative study of crowd workers against student and domain experts on ontology engineering tasks [4,14]. Their study highlighted the need for improved filtering methods for workers; however, they did not propose algorithms for KBQs generation or adaptive task assignment.

Within the literature on crowdsourcing, several approaches have been proposed for adaptive task assignment. Ho et al. proposed primal-dual techniques for adaptive task assignment of classification tasks using GSTs [8]. Their approach estimates reliability of a worker against different types of tasks instead of topics. Zhou et al. proposed a multi-armed bandit approach for assigning top-K workers to a task, and their approach also uses GSTs [22]. Another approach focused on dynamic estimation of worker expertise based on conformity of workers with the majority responses [18]. Ipeirotis et al. proposed an approach for separating worker bias from reliability estimation [11]. Such an approach is complimentary to our algorithm for reducing the influence of spammers on task responses. Oleson et al. proposed a manual audit approach to quality control in crowdsourcing by generating gold-standard task with different types of errors previously observed in different tasks [15]. By comparison, our approach focuses on automated selection of knowledge base questions for quality control in crowdsourcing. Hassan et al. used a hybrid approach of self-rating and gold-standard task for estimating the expertise of workers [5]. By comparison, our approach uses DBpedia facts for estimation of worker expertise.

9 Conclusion and Future Work

In this paper, we presented ACRyLIQ, a novel approach to estimate the reliability of crowd workers. ACRyLIQ supports the adaptive task assignment process for achieving high accuracy of Linked Data Quality Assessment tasks using crowdsourcing. The proposed approach leverages a generalized knowledge base, in this case DBpedia, to generate test questions for new workers. These test questions are used to estimate the reliability of workers on diverse tasks. ACRyLIQ employs a similarity measure to find good candidate questions, and it uses an entropy measure to maximize the diversity of the selected questions. The adaptive task assignment algorithm exploits the test questions for estimating the reliability. We evaluated the proposed approach using crowdsourced data collected from real workers on Amazon Mechanical Turk. The results suggest that ACRyLIQ is able to achieve high accuracy without using gold-standard task.

As part of the future work, we plan to apply our approach to larger datasets within multiple domains. We also plan to further investigate the relationship between the reliability of workers and the semantic similarity of facts and tasks. A detailed study is needed to understand the relationship between the reliability of workers and the semantic similarity of DBpedia facts and crowdsourcing tasks.

Acknowledgement. This work has been supported in part by the Science Foundation Ireland (SFI) under grant No. SFI/12/RC/2289 and the Seventh EU Framework Programme (FP7) from ICT grant agreement No. 619660 (WATERNOMICS).

References

1. Acosta, M., Zaveri, A., Simperl, E., Kontokostas, D., Auer, S., Lehmann, J.: Crowdsourcing linked data quality assessment. In: Alani, H., et al. (eds.) ISWC 2013. LNCS, vol. 8219, pp. 260–276. Springer, Heidelberg (2013). doi:10.1007/978-3-642-41338-4_17
2. Difallah, D.E., Demartini, G., Cudrè-Mauroux, P.: Pick-a-crowd: tell me what you like, and i'll tell you what to do. In: Proceedings of the 22nd International Conference on World Wide Web, pp. 367–374 (2013)
3. Fan, J., et al.: iCrowd: an adaptive crowdsourcing framework. In: Proceedings of the 2015 ACM SIGMOD International Conference on Management of Data, pp. 1015–1030. ACM (2015)
4. Ghazvinian, A., Noy, N.F., Musen, M.A., et al.: Creating mappings for ontologies in biomedicine: simple methods work. In: AMIA (2009)
5. Ul Hassan, U., O'Riain, S., Curry, E.: Effects of expertise assessment on the quality of task routing in human computation. In: Proceedings of the 2nd International Workshop on Social Media for Crowdsourcing and Human Computation, Paris, France (2013)
6. Ul Hassan, U., O'Riain, S., Curry, E.: Leveraging matching dependencies for guided user feedback in linked data applications. In: Proceedings of the 9th International Workshop on Information Integration on the Web, pp. 1–6. ACM Press (2012)
7. Heath, T., Bizer, C.: Linked Data: Evolving the Web Into a Global Data Space, vol. 1. Morgan & Claypool Publishers, San Rafael (2011)
8. Ho, C.-J., Jabbari, S., Vaughan, J.W.: Adaptive task assignment for crowdsourced classification. In: Proceedings of the 30th International Conference on Machine Learning (ICML-13), pp. 534–542 (2013)
9. Howe, J.: The rise of crowdsourcing. Wired Mag. **14**(6), 1–4 (2006)
10. Ipeirotis, P.G.: Analyzing the amazon mechanical turk marketplace. XRDS: Crossroads ACM Mag. Students **17**(2), 16–21 (2010)
11. Ipeirotis, P.G., Provost, F., Wang, J.: Quality management on amazon mechanical turk. In: Proceedings of the ACM SIGKDD Workshop on Human Computation, pp. 64–67. ACM (2010)
12. Lehmann, J., et al.: DBpedia - a large-scale, multilingual knowledge base extracted from wikipedia. Semant. Web J. **6**(2), 167–195 (2015)
13. Ngonga Ngomo, A.-C., Auer, S.: LIMES - a time-efficient approach for large-scale link discovery on the web of data. In: Proceedings of IJCAI (2011)
14. Noy, N.F., et al.: Mechanical turk as an ontology engineer?: using microtasks as a component of an ontology-engineering workflow. In: Proceedings of the 5th Annual ACM Web Science Conference, pp. 262–271 (2013)
15. Oleson, D., et al.: Programmatic gold: targeted and scalable quality assurance in crowdsourcing. In: Human Computation 11.11 (2011)
16. Sarasua, C., Simperl, E., Noy, N.F.: CrowdMap: crowdsourcing ontology alignment with microtasks. In: Cudré-Mauroux, P., et al. (eds.) ISWC 2012. LNCS, vol. 7649, pp. 525–541. Springer, Heidelberg (2012). doi:10.1007/978-3-642-35176-1_33
17. Shannon, C.E.: A mathematical theory of communication. ACM SIGMOBILE Mob. Comput. Commun. Rev. **5**(1), 3–55 (2001)
18. Tarasov, A., Delany, S.J., Namee, B.M.: Dynamic estimation of worker reliability in crowdsourcing for regression tasks: making it work. In: Expert Systems with Applications 41.14, pp. 6190–6210 (2014)

19. Winkler, W.: String comparator metrics and enhanced decision rules in the Fellegi-Sunter model of record linkage. In: Proceedings of the Section on Survey Research Methods (American Statistical Association), pp. 354–359 (1990)
20. Zaveri, A., et al.: Quality assessment for linked data: a survey. Semant. Web J. 7(1), 63–93 (2016)
21. Zaveri, A., et al.: User-driven quality evaluation of DBpedia. In: Proceedings of the 9th International Conference on Semantic Systems, pp. 97–104. ACM (2013)
22. Zhou, Y., Chen, X., Li, J.: Optimal PAC multiple arm identification with applications to crowdsourcing. In: Proceedings of the 31st International Conference on Machine Learning (ICML-14), pp. 217–225 (2014)

The Semantic Web in an SMS

Onno Valkering, Victor de Boer(✉), Gossa Lô, Romy Blankendaal, and Stefan Schlobach

Vrije Universiteit Amsterdam, Amsterdam, The Netherlands
{o.a.b.valkering,r.a.m.blankendaal}@student.vu.nl,
{v.de.boer,a.g.lo,k.s.schlobach}@vu.nl

Abstract. Many ICT applications and services, including those from the Semantic Web, rely on the Web for the exchange of data. This includes expensive server and network infrastructures. Most rural areas of developing countries are not reached by the Web and its possibilities, while at the same time the ability to share knowledge has been identified as a key enabler for development. To make widespread knowledge sharing possible in these rural areas, the notion of the Web has to be downscaled based on the specific low-resource infrastructure in place. In this paper, we introduce SPARQL over SMS, a solution for Web-like exchange of RDF data over cellular networks in which HTTP is substituted by SMS. We motivate and validate this through two use cases in West Africa. We present the design and implementation of the solution, along with a data compression method that combines generic compression strategies and strategies that use Semantic Web specific features to reduce the size of RDF before it is transferred over the low-bandwidth cellular network.

1 Introduction

The Semantic Web by design builds on, and relies on, the Web infrastructure for data exchange. This includes sophisticated server and network infrastructures which are unavailable in many rural areas of developing countries. These areas are not reached by the web and its possibilities while at the same time the ability to share knowledge has been identified as a key enabler for development. To make knowledge sharing possible in rural developing areas, the notion of the Web has to be *downscaled* based on the specific low-resource infrastructure in place [6].

Data sharing solutions, such as those based on Semantic Web and Linked Data technologies, should not only be accessible to those with abundant resources and reliable infrastructures, but also in low-resource environments. The flexible graph models of the Semantic Web and its language-agnostic nature make it especially useful for data sharing in these location, because of the many different spoken languages and customs. In [2] we show that locally produced market data, stored as RDF, is produced through, and used in, a voice-interface accessible for low-literate users in their preferred language. We also identified opportunities for data sharing and integration. More recently, we developed the Kasadaka[1],

[1] http://www.kasadaka.com. "Kasadaka" roughly translates to "Talking Box" in a number of Ghanaian languages.

E. Blomqvist et al. (Eds.): EKAW 2016, LNAI 10024, pp. 697–712, 2016.
DOI: 10.1007/978-3-319-49004-5_45

a low-resource prototyping and computing platform which uses semantic technologies specialized at developing multi-modal user interfaces, e.g. touchscreens or voice, for low literacy in rural areas of development countries. While this widens the applicability and possible use-cases, a core problem remains; the lack of infrastructure for Web-like sharing of information in the targeted rural areas.

The main challenge lies in the unavailability of network connections. Especially in many rural areas of developing countries, internet connections are either missing or extremely unreliable. Internet penetration is estimated to be 28.6 % of the population in Africa as a whole (compared to 52.8 % in the rest of the world), with some countries reaching considerably less of their population: for example 7.0 % in Mali[2]. These numbers include both urban and rural areas and in the latter, internet penetration is virtually non-existent. The Semantic Web is built on top of the Internet (TCP/IP) and Web infrastructure (including HTTP) and as such when no Internet is available, it is unusable. However, we can design solutions to implement Web-like data sharing using alternative networking capabilities available in low-resource environments.

We present a specific downscaling solution for exchanging (RDF) data in which HTTP is substituted by SMS to enable Web-like exchange of data over cellular networks. We show the viability of this solution in two different ways:

1. Technological: we identify three main technological problems when using an SMS protocol as semantic data transfer protocol, message size, the asynchronous nature of the protocol and how to deal with pagination issues. Our solution is validated w.r.t. each of those problems with a variety of methods, which includes a large-scale empirical comparison of compression size.
2. Societal: Using two use-cases from Sub-Saharan Africa (one from Ghana, one from Mali) we will show how SMS-based Semantic Web can practically address the knowledge sharing needs of rural communities. We introduce these in Sect. 3 and validate our solution against these cases in Sect. 6.

While in this paper, we present a practical solution to low-bandwidth knowledge sharing, our investigation will also be more generally useful to understand how Semantic Web principles and practices can be separated from the infrastructure layers that often are assumed to be prerequisites.

2 Related Work

SMS as a data channel has been proposed in other ICT for Development (ICT4D) cases, for example in [8]. Mobile banking -including through SMS- has been well-established in many developing economies (cf. [10]). A number of Social Network Services such as Twitter, Facebook as well as the Google search engine allow for accessing those services through SMS[3]. Mostly, this deals with machine-to-human interaction and not, as in our case for machine-to-machine (M2M)

[2] As of November 2015 http://www.internetworldstats.com.

[3] http://www.digitaltrends.com/mobile/sms-your-way-back-to-the-web.

interaction. Related work in semantic data exchange in low-resource network environments includes the Entity Registry System (ERS) [3], an open-source entity registry specifically designed for environments with ad-hoc and/or unreliable network connectivity, as is often the case in rural areas. It allows for Linked Data without using the centralised components that make up the Web infrastructure. ERS has mechanisms to deal with interval-based network connectivity (e.g. a mobile truck that functions as an access point) and is resistant against packet loss. DakNet provides similar solutions where ad-hoc wireless networks are combined with asynchronous networking, also including mobile access points [11]. Whereas these solutions also implement Web-like data exchange without Web infrastructure, they focus mainly on local networks and rely on the availability of partial Internet connectivity.

Another way of transferring data without Internet is through so-called Sneakernets, where data is exchanged by physically moving removable media or hard disks. For large-scale non-immediate data transfer, this is a viable solution [5] which can be combined with solutions such as the one presented in this paper.

In this paper, we focus on semantic data exchange using the SPARQL protocol. There are other opportunities for accessing RDF data over a network. Two examples are simple URI dereferencing and the use of Linked Data Fragments [13]. Compared to these methods, accessing RDF using SPARQL typically takes more computing resources on the client and server devices itself, but allows for more fine-grained querying by which bandwidth can be limited. For our specific cases, saving bandwidth is a key issue, which is why we use SPARQL. It is interesting to further investigate the trade-off between computational and networking resources in these specific ICT4D cases.

3 Information Sharing in the Absence of the Web

As early as 2011 we pointed to some negative effects of the Semantic Web's reliance on Web infrastructure [6], which effectively made this technology inaccessible for a majority of the world population. Through a number of research projects in Sub-Saharan Africa we have since then identified numerous use cases that rely on knowledge sharing. The recent Kasadaka project builds on information acquired in Burkina Faso, Mali, Ghana and Niger and aims at providing information to people living in rural communities for several different use-cases. It provides a generic platform, which enables voice- and SMS-based communication over GSM and can be deployed in communities and owned and maintained by local stakeholders. The platform can host different information sharing services, accessible through simple icon-based visual interfaces or voice interfaces callable from any mobile phone as users especially those in remote rural villages are often low-literate and speak local languages. We regularly ran into conceptual and technical problems for which the Knowledge Engineering community has already provided robust methods, as most real use cases require data and knowledge sharing across communities and devices. We here describe two cases that have been co-developed with local stakeholders in rural West Africa.

3.1 The DigiVet Case

One of the information needs identified for and by rural farmers is on animal health, in particular on diagnosing animals. DigiVet is a voice-based veterinary information service that support subsistence farmers in making the decision whether or not to visit a veterinarian. Animal diseases spread within and between villages and can often be cured merely with the intervention of a veterinarian. The problem that arises in these rural areas is that local expertise is often lacking and poor infrastructures (poor roads, lack of electricity) prevent access to information and sharing of knowledge. Farmers need information on animal diseases, disease patterns, diagnosis and symptoms to take preventive action and preclude cattle loss, but this cannot be shared easily over large distances.

DigiVet includes a simple interface which presents farmers with a set of symptom related questions on a touchscreen. It is based around a knowledge base[4] developed by interviewing veterinarians working in rural Northern Ghana. The system provides a diagnosis whether or not a farmer should contact a veterinarian. DigiVet relies on semantic data exchange between farmers and veterinarians at large distances. While the used Kasadaka platform is suitable for the creating the interface for diagnosing, there is currently no technology to cater the necessary semantic data exchange.

3.2 The RadioMarché Case

The RadioMarché case, introduced in [1], is a market information system designed to gather and distribute information about offerings of specific produce on local markets in the Tominian region of Mali. To allow low-literate stakeholders to retrieve market information in the absence of internet connection, a voice-accessible service was built that can be called from any mobile phone. The service can be called by local farmers in their own language to retrieve this information. Community radio hosts retrieve and broadcast local offerings on the radio. The system was developed and deployed in 2012 [7]. The gathered product offering data was ported to the RDF data model and a Semantic Web compliant version was developed. The benefits of linking market data to external data sources and using this for visualization and improved data analyses, in particular for stakeholders such as NGOs or bulk buyers is described in [2].

4 A Platform for Semantic Web in an SMS

Our goal is to make the use of Semantic Web applications possible in areas lacking an infrastructure to support Web-like exchange of data. The intent is not to create an isolated network that mimics Semantic Web practices, but rather to develop a mechanism that supports the retrieval and manipulation of RDF data across different kind of networks infrastructures. We want to achieve this without imposing additional network-specific operations for application developers. This

[4] https://github.com/biktorrr/digivetkb.

means that applications can still be developed for HTTP, the mechanism sits in between, converting messages to be able to cross the specific low-bandwidth network in place, without requiring the application to be adjusted.

4.1 SPARQL over SMS

The rural areas in West Africa targeted by our use cases only have cellular networks available for digital communication. Applications deployed in these areas can make use of SMS for M2M transfer of data, instead of HTTP as part of a Web-based network. Noteworthy practical differences between SMS-based networks and Web-based networks are:

- SMS-based network agents are identified by phone numbers instead of URLs;
- The size of an SMS is limited up to 160 bytes[5];
- SMS implements a one-way messaging pattern, whereas HTTP implements a request-response messaging pattern.

To transfer HTTP messages, produced by Semantic Web applications, over a SMS-based networks, a conversion mechanism is required. This mechanism, in addition to the above-mentioned differences, must take into account these case-specific requirements:

- the number of messages sent should be as low as possible, in view of costs;
- the mechanism should be possible to run on affordable hardware[6].

Although the costs per transferred byte are relatively high for SMSes, it builds on existing infrastructure which has a global reach including many rural areas of development countries. Also, the required hardware to be able to send SMSes is affordable and widely available.

Our implementation of the described mechanism is called *SPARQL over SMS*. By supporting the CONSTRUCT and INSERT/DELETE DATA query forms a basic usage of Semantic Web for M2M communication is realized. We select this subset of SPARQL as it involves simple data transfer using RDF triples. SELECT query responses (where they are not part of a CONSTRUCT query) take the form of result tables of arbitrary sizes and are harder to optimize.

Figure 1 gives an overview of SPARQL over SMS. In the context of SPARQL over SMS, application that can both send and receive SPARQL queries are called *agents*. The *converter* is a key component responsible for the conversion between an HTTP SPARQL request and an SMS-optimized equivalent. Different options for sending and receiving SMSes are supported by the converter, such as a GSM dongle or an online SMS service. This allows the converter to be deployed in various scenarios. A converter deployed in a data-center could be used to share data with a triple store running on low-resource hardware deployed in the field. This can be useful when aggregation of data from multiple devices is desirable.

[5] Based on the encoding used: 8-bit supports 140 characters, 7-bit up to 160 characters.
[6] Such as a Raspberry Pi computer: https://www.raspberrypi.org.

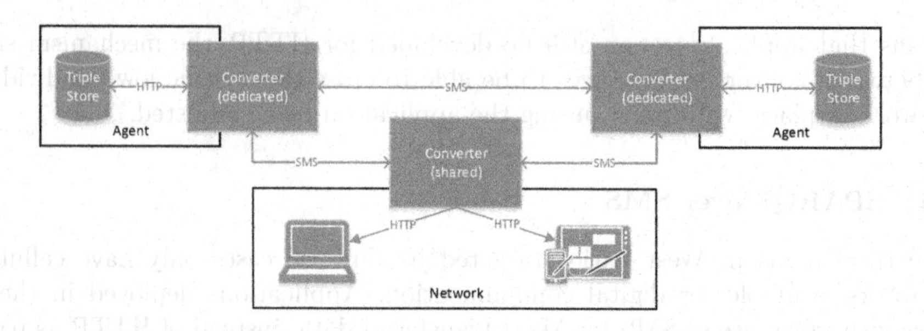

Fig. 1. SPARQL over SMS Overview

A converter instance supports two modes: shared and dedicated. A dedicated converter maps a phone number directly to a particular agent, making it possible for the agent to both send and receive SPARQL queries. Shared converters can serve multiple agents, so that a phone number (assigned to a single GSM dongle) cannot be mapped to a single agent. In this case, the agents can only perform outgoing SPARQL queries but cannot be the target of incoming queries. Sending SPARQL queries requires an endpoint URL identifying the target receiver. To allow the targeting of an agent in an SMS-network, the converter can provide a URL representation of an arbitrary phone number. For example, the format of a SPARQL endpoint URL is: *http://{converter hostname}/agent/{phone number}/sparql*. SPARQL requests sent to such an endpoint are captured by the converter and sent to the phone number. The converter then receives the query and runs it on the triple store of the associated agent. The result is then send back to the initial converter which returns is as the response to the SPARQL request in the specified format.

4.2 SMS Message Structure and Conversion

SMSes sent between converters follow a specific structure. Five characters of each SMS are reserved for metadata for which the basic 7-bit character set, as specific by the GSM 03.38 character set[7], is used. This includes the message type, message identifier, and position for multi-part messages. Excluding the non-print characters there are 125 different characters left that can be used. A single character can thus express a numerical value of 1 to and including 125.

HTTP to SMS. The converter creates optimized representations of HTTP SPARQL queries and results. In the case of a SPARQL query the encoding routine is based on the SPARQL query form to perform fine-grained optimizations. The compact representation is optionally split into multiple SMSes if it exceeds the character limitation of a single SMS. The position of each part will be indicated by the multi-part position character in the metadata. After conversion, the compact representation is send over SMS to the phone number extracted from the endpoint URL used to send the SPARQL query to the converter.

[7] http://www.3gpp.org/DynaReport/23038.htm.

SMS to HTTP. When a converter receives an SMS, HTTP representations are reconstructed. As different encoding routines might be used, the appropriate routine is based on the message type defined in the metadata. Decoding cannot guarantee a result exactly identical to the original message. The resulting message might thus not be syntactically equivalent, but it will be semantically equivalent. Multi-part messages are concatenated based on the multi-part position in the SMS metadata.

5 Research Challenges

In the design of our solution, converters are used to transfer SPARQL queries and results between Web- and SMS-based networks. However, to develop a workable solution, a number of challenges need to be addressed. In this section we outline the different challenges, namely: how to reduce the size of RDF data to allow for efficient transfer over SMS, issues around asynchronicity of communication and how to deal with unpredictable query result sizes.

5.1 Small RDF Data Compression

The serialization format has a great effect on the size of an RDF file, and thereby on the amount of SMSes needed to transfer the data. The costs associated with SMSes restricts us to cases with small amounts of triples which. Still, to save costs associated with the sending SMSes, we want to use the combination of RDF serialization and compression that is most efficient, in terms of transfer size, for such small RDF data sets.

Experimental Setup. To identify the best serialization and compression combination we run experiments on RDF data sets provided by the LOD Laundromat [12]. These RDF files are crawled from multiple Linked Data sources, making it a realistic representation of real-world RDF data sets. Our benchmark consists of 232,822 RDF files with size between 1 and 1000 triples. This large-scale experiment ensures that we test across many types of data sets and various characteristics which might influence serialization and compression.

The files were converted to various serializations (RDF/XML, Turtle, HDT and EXI). RDF/XML and Turtle are plain text serialization formats specifically designed for RDF data. The binary format "Header, Dictionary, Triples" (HDT) is a data structure developed to compactly store and exchange RDF data without sacrificing the ability to query the data [4]. Efficient XML Interchange (EXI) is a binary format designed to create compact representations of XML and has been proposed for efficient RDF exchange in constrained embedded networks [9]. For each format, including the original N-Triples format, the file size is recorded before and after applying gzip compression. The default implementations of RDFLib[8], HDT[9] and EXIficient[10] have been used.

[8] https://github.com/RDFLib/rdflib.
[9] https://github.com/rdfhdt/hdt-java.
[10] https://github.com/EXIficient/exificient.

Results. Table 1 lists per format the average file size w.r.t to the original N-Triples format. The best compression per bin is marked bold. The binary gzip, HDT and EXI formats include Base64 encoding overhead. We first look at the results from the RDF files in the range of 1 to 100 triples (81,492 in total) in bins of 10. As expected, the size reduction compared to the original format increases with the number of triples, due to more syntactic redundancy. HDTs are bigger than the original due to the HDT's metadata. With files up to 30 gzip compressed N-Triples outperforms the other formats. Above 50 triples, the compressed Turtle format outperforms compressed N-Triples. For sets between 30–60 triples, the uncompressed EXI format performs similar to compressed N-Triples and compressed Turtle. Applying gzip compression to EXI hardly has any effect and even increases the file size in most cases.

When considering RDF files between 100–1000 triples we note that gzip compressed Turtle results in the best compression. In addition, RDF/XML stagnates around 51 % and Turtle around 36 % of size compared to N-Triples. For files with 600+ triples gzip compressed HDT drops below the size of EXI, at the cost of losing the ability to directly query the HDT files due to an additional layer of gzip compression. We conclude that for the smallest data sets ($\leq$40 triples), gzip compressed N-Triples is preferable. For data sets between 40–1000 triples, gzip compressed Turtle serialization scores best. The reason N-Triples performs better than Turtle for the smallest data sets can be the added overhead of prefixes in Turtle[11]. In our implementation, we decided to dynamically select the appropriate serialization (N-Triples or Turtle) based on the number of triples in the SPARQL result.

5.2 Shared Vocabulary/Semantic RDF Data Compression

Section 5.1 focused on serialization and gzip compression. These generic strategies consider only the syntactical representation of RDF. In order to reduce the size of RDF data even more, we also tested two compression strategies focused on RDF content-specific aspects. We do this on the basis of RDF vocabularies that define reusable definitions for common properties and/or types.

Experiment Setup. We experimented with 30 popular vocabularies[12]. To 3,577 RDF data sets from the previous experiment dictionary-encoding and reasoning, based on the RDF vocabularies, was applied. The experiment had three rounds with an increasing number of vocabularies (most popular 10, 20 and 30). A single HDT file, containing combinations of vocabularies, is generated in each round, for which the dictionary-component is used as dictionary encoding. During the encoding all the URIs that occur in one of the vocabularies are replaced with a placeholder containing the identifier generated by HDT for an URI.

We use reasoning to find semantic redundancies in RDF data sets, based on the vocabularies. The implemented reasoner searches for redundancies based on

[11] The used Turtle serializer adds RDF, RDFS, XSD and XML prefixes by default.

[12] Including YAGO, FOAF, and SKOS. Based on http://prefix.cc/popular/all.

Table 1. Results of the LOD lab compression experiment (N-Triples = 100 %)

No. Triples	N-Triples+ Gzip	RDF/ XML	RDF/ XML+Gzip	Turtle	Turtle+ Gzip	HDT	HDT+ Gzip	EXI	EXI+ Gzip	Comb. method
1–10	**50.7**	103.8	77.0	102.0	70.3	495.5	180.1	57.5	65.9	44.2
11–20	**22.5**	62.0	27.1	50.5	24.2	122.2	47.0	23.3	24.9	18.9
21–30	**16.2**	58.2	18.5	48.7	16.3	79.5	31.1	16.5	17.5	13.6
31–40	28.3	69.1	30.9	62.1	28.6	86.5	40.7	**28.2**	29.1	23.5
41–50	9.8	51.2	10.2	42.3	**8.6**	38.1	14.8	9.3	9.7	8.0
51–60	17.2	59.2	17.5	50.1	15.9	50.5	22.8	**15.8**	16.3	8.7
61–70	11.8	58.5	12.4	42.4	**10.0**	43.0	17.7	11.1	11.6	6.0
71–80	8.8	54.8	8.5	40.9	**7.0**	31.6	11.2	7.5	7.8	6.4
81–90	6.7	52.0	6.3	40.6	**5.1**	25.4	9.1	5.8	6.0	4.4
91–100	8.1	54.9	7.6	40.4	**6.2**	26.9	9.7	6.8	7.0	5.7
101–200	8.8	62.0	8.3	39.2	**6.7**	24.7	10.1	7.6	7.9	5.7
201–300	4.8	50.8	3.6	39.0	**2.8**	13.4	4.0	3.6	3.6	2.7
301–400	4.8	51.5	3.3	37.7	**2.5**	11.4	3.3	3.0	3.1	2.5
401–500	4.4	51.5	2.9	37.4	**2.2**	10.4	2.7	2.6	2.7	2.2
501–600	5.0	53.8	3.4	38.7	**2.5**	8.9	3.0	2.9	3.0	2.4
601–700	4.1	51.0	2.5	35.9	**1.7**	8.5	2.2	2.3	2.4	1.7
701–800	4.5	51.1	2.7	36.2	**1.9**	8.1	2.1	2.4	2.4	1.9
801–900	4.4	51.1	2.6	36.4	**1.8**	7.9	1.9	2.3	2.3	1.8
901–1000	4.1	50.9	2.4	36.5	**1.7**	7.7	**1.7**	2.1	2.1	1.7

twelve RDFS entailment patterns[13], as well as rules for two OWL properties: *owl:SymmetricProperty* and *owl:inverseOf*. Semantically redundant triples are removed from the data set. Only explicitly defined triples from the RDF data set and vocabularies are considered. Therefore, it is not guaranteed that the final result is always the smallest set of triples possible.

Results. The two compression strategies are measured independently during the experiment. Precondition for these compression strategies is that the subjected RDF data set must use one of the considered vocabularies, which makes that these compression strategies do not always have a size reducing effect.

Size reduction averages have been calculated only for results that led to an actual size reduction, grouped by number of triples in bins of 100. Based on these averages we have found that reasoning based on the top 10 vocabularies has minimal effect, up to 3 % average size reduction. Using an additional 10 vocabularies increases the average size reduction across all bins, resulting in average size reduction ranging from 8.7 % to 13.4 %. Using the top 30 vocabularies has no advantage over the top 20 vocabularies when using reasoning. The dictionary-encoding achieves around 6.5 % average size reduction based on the top 10 vocabularies. This is slightly improved to around 8 % when using the top 20 vocabularies. An additional, but minimal improvement, can be obtained when using all the 30 vocabularies for dictionary-encoding. Furthermore, it stands out

[13] https://www.w3.org/TR/rdf11-mt/#rdfs-entailment.

that the dictionary-encoding could be used more consistently, 96 % of the files could be reduced by using dictionary-encoding against 31 % for reasoning. Based on these results we conclude that it is best to use the top 20 popular vocabularies when using the reasoning and dictionary-encoding compression strategies. The minimal improvement of using the top 30 for dictionary-encoding is not commensurate to the increase of processing duration and maintenance efforts introduced by the additional 10 vocabularies.

We combined vocabulary based compression strategies with the generic compression strategies from Sect. 5.1 to form a RDF compression method for SPARQL over SMS. The LOD Lab data sets, as described in Sect. 5.1, have been subjected to this combined method to measure the performance. The results are listed in Table 1. It shows that the added reasoning and dictionary-encoding strategies are especially effective when compressing the smallest RDF data sets (1–200 triples). As the number of triples increases, the syntactical compression strategies become more efficient and gradually make the vocabulary based compression strategies less beneficial. This is also follows from Table 2, the added vocabulary based compression provides a head-start in terms of the average number of triples that can be sent per SMS. This holds up to 10 SMSes.

Table 2. Average number of triples that can be send based on the number of SMSes

Nr. of SMSes	Only serialization and compression	Added shared vocabulary compression
1	0	0
2	3	3
3	6	8
4	9	16
5	21	24
6	66	84
7	84	98
8	116	126
9	175	189
10	301	301

5.3 Blending Synchronous and Asynchronous Messaging

SPARQL exchanges follow a request-response messaging pattern. When a query is sent as a request over the network, the receiver processes the request and composes a response with the query result. A single connection is used and kept open during the transfer, making it a synchronous operation. Sending SMSes follows a one-way messaging pattern. Each message is a standalone message that not enforces a follow-up response. As the connection is terminated after a message has been delivered sending of SMS is an asynchronous operation.

With SPARQL over SMS we want to seamlessly transfer messages from Web-based networks to SMS-based networks and vice-versa. For this purpose, we need to harmonize the two different messaging patterns. The initial implementation of SPARQL over SMS keeps HTTP connections open during the data transfer. After sending a SPARQL query over SMS, the converter waits until the corresponding result response comes in. The query result response is of a different message type, but it can be correlated to the original request by using the message identifier for correlation.

This implementation is functional, but might not be a optimal considering the deployment in rural areas described in our use cases. For example, due to temporary loss of connectivity the response message might be available after hours or even days. It is questionable if the low-resource hardware can hold open multiple connections for that period. Even if it is capable of doing so, there is a genuine risk of a sudden power outage that will result in a loss of all open connections requiring retries. In our specific use-cases this would result in additional, unnecessary, costs. Additional efforts are required to create an asynchronous-supporting version of SPARQL for situations when a response is not expected within a seconds- or minutes-long time span.

5.4 Unpredictable Query Result Sizes

A simple looking SPARQL query might yield an unexpectedly large result. To be thrifty with sending SMSes, we want to restrict the amount of SMSes that will be send. We have considered two options to achieve this, SPARQL pagination and pagination on a SMS level. SPARQL provides the LIMIT, OFFSET and ORDER BY keywords that can be used to implement pagination, but the SPARQL query result will not include pagination information, e.g. the total number of results available. This means it is not possible to tell if all results have been retrieved yet. Another issue is the possibility for a triple to have a very long literal object that can span multiple SMSes. Relying only on SPARQL pagination for regulating the result size is not sufficient to regulate the number of SMSes send. Pagination on the SMS level would only introduce the option to decide whether or not to continue receiving the SMSes. Since a partial result from a complete SPARQL result cannot be created: it is the whole SPARQL result or nothing. This would also alter the way SPARQL over SMS must be used compared to the SPARQL standard, due to the addition of pagination operations which can not be ignored. With our implementation the SPARQL result is directly, after compression, send through SMS. If the result does not fit in a single SMS it will split up the message, on arbitrary points, into multiple SMSes. If a hard set restriction is reached and stops sending SMSes the receiver cannot read the message properly due to missing parts: partial results are not supported. Therefore, we consider above-limit SPARQL results as an error.

6 Practical Validation

6.1 Implementation and Integration

SPARQL over SMS[14] is developed with integration with other services in mind and can be deployed on various operating systems and devices. For validation, we used SPARQL over SMS in combination with Kasadaka[15]. This combination runs on widely available and affordable hardware.

6.2 Evaluation in Four Scenarios

For the two use case in Sect. 3, we are developing services using Kasadaka running on low-resource devices. The communication between two remotely deployed devices is key. Our setup consists of two Raspberry Pi 2 computers with both the DigiVet and RadioMarché[16] data sets loaded in a ClioPatria[17] triple store. Four SPARQL queries that correspond to two scenarios per use case are tested to determine the amount of SMSes required both with our RDF compression method and without (plain RDF/XML). To improve the shared vocabulary compression strategies, the use case vocabularies are added to vocabularies used.

Extending the DigiVet Application. Combining DigiVet with SPARQL over SMS adds new data sharing options. With two new scenarios, from the perspective of the veterinarian, we demonstrate how the DigiVet application can be extended using the new features. First we consider a veterinarian interested in types and frequencies of animal disease symptoms occurring near Walewale, Ghana. The SPARQL query in Listing 1.1 answers this question. Sending the query requires 3 SMSes and yields a result of 7 triples. Returning the result takes 3 SMSes with our solution (without compression it takes 14 SMSes). As a second scenario for the DigiVet use case we consider the need of a veterinarian

Listing 1.1. Digivet SPARQL query for 1st scenario

```
PREFIX foaf: <http://xmlns.com/foaf/0.1/>
PREFIX dv: <https://w3id.org/w4ra/digivet/>

CONSTRUCT {
  ?sym dv:occurance_count ?count
}
WHERE {
  SELECT ?sym (COUNT(?sym) as ?count) WHERE {
    ?person foaf:based_near <http://sws.geonames.org/2294174/> .
    ?person dv:has_case ?case .
    ?case dv:has_symptom ?sym
  }
  GROUP BY ?sym }
```

[14] https://github.com/onnovalkering/sparql-over-sms, available as open source.

[15] https://github.com/abaart/KasaDaka.

[16] A clone of the store is available at http://semanticweb.cs.vu.nl/radiomarche.

[17] http://cliopatria.swi-prolog.org.

to update the animal diseases knowledge base present on a DigiVet deployment. As an example, the SPARQL query in Listing 1.2 can be used to add a new disease ("Black Leg") with associated symptoms to the triple store. Transferring this query requires 3 SMS messages (5 without using compression).

Listing 1.2. DigiVet SPARQL query for 2nd scenario

```
PREFIX rdfs: <http://www.w3.org/2000/01/rdf-schema#>
PREFIX foaf: <http://xmlns.com/foaf/0.1/>
PREFIX dv: <https://w3id.org/w4ra/digivet/>

INSERT DATA {
    dv:black_leg a dv:Disease .
    dv:black_leg rdfs:label"Black_leg"@en .
    dv:Cow dv:canCarryDisease dv:black_leg .
    dv:Sheep dv:canCarryDisease dv:black_leg .
    dv:unwillingnessToMove dv:symptom_for_disease dv:black_leg .
    dv:rapidBreathing dv:symptom_for_disease dv:black_leg .
    dv:lameness dv:symptom_for_disease dv:black_leg .
    dv:appetiteLoss dv:symptom_for_disease dv:black_leg .
    dv:fever dv:symptom_for_disease dv:black_leg .
    dv:swellingThigh dv:symptom_for_disease dv:black_leg .}
```

Extending the RadioMarche Application. For the RadioMarché service, we consider two scenarios. The first involves the retrieval of the current offerings, including the phone number of the advertisers, in the Mafoune and Mandiakuy regions of Mali. Using a CONSTRUCT query (Listing 1.3), this information is retrieved as an RDF graph from a RadioMarché installation. Sending the query through our solution requires 3 SMS messages (4 without using compression). The query result consists of 152 triples in total and could be transferred using 8 SMS messages (121 SMS messages would have been required without compression). This shows the economic impact of the compression step. As a second scenario, we perform an INSERT DATA query (Listing 1.4) to add product labels in more languages. The query creates ten new triples. Our solution requires only 3 SMS messages, half of the uncompressed number.

Listing 1.3. RadioMarché SPARQL query for 1st scenario

```
PREFIX rdfs: <http://www.w3.org/2000/01/rdf-schema#>
PREFIX rm: <http://purl.org/collections/w4ra/radiomarche/>

CONSTRUCT {
    ?contact rm:contact_tel ?tel .
    ?contact rm:has_offering ?offering .
    ?offering rdfs:label ?prod_name
} WHERE {
    ?offering a rm:Offering .
    ?offering rm:has_contact ?contact .
    ?offering rm:prod_name ?prod .
    ?prod rdfs:label ?prod_name .
    ?contact rm:contact_tel ?tel .
    ?contact rm:zone ?zone .
    FILTER (?zone IN (rm:zone_Mafoune, rm:zone_Mandiakuy)) }
```

Listing 1.4. RadioMarché SPARQL query for 2nd scenario

```
PREFIX rdfs: <http://www.w3.org/2000/01/rdf-schema#>
PREFIX rm: <http://purl.org/collections/w4ra/radiomarche/>
INSERT DATA {
    rm:product-Beurre_de_karite rdfs:label "Shea butter"@en .
    rm:product-Beurre_de_karite rdfs:label "La manteca de karit"@es .
    rm:product-Miel_liquide rdfs:label "Honey"@en .
    rm:product-Miel_liquide rdfs:label "Miel"@es .
    rm:product-Amande_de_karite rdfs:label "Shea nuts"@en .
    rm:product-Amande_de_karite rdfs:label "Nueces de karit"@es .
    rm:product-Tamarin rdfs:label "Tamarind"@en .
    rm:product-Tamarin rdfs:label "Tamarindo"@es .
    rm:product-Graine_de_nere rdfs:label "Nere seeds"@en .
    rm:product-Graine_de_nere rdfs:label "Semillas Nere"@es . }
```

Discussion. Table 3 summarizes the results for all scenarios. It shows the number of SMSes needed to transfer the query as well as the query response. For the realistic use cases, the amount of SMS per query is limited. We also list the total costs per query by converting current local SMS rates from two providers to US Dollars[18]. This suggests that, although expensive, the use case could potentially be made economically viable. The number of SMSes required to transfer the SPARQL results confirm to the estimations in Table 2.

Table 3. Summary of the four validation scenarios

Scenario	Location	Query type	Request size in nr. of SMS	Request est. cost (USD)	Response size in nr. of SMS	Response est. cost (USD)
Digivet Sc.1	Ghana	CONSTRUCT	3	0.042	3	0.042
Digivet Sc.2	Ghana	INSERT	3	0.042	n.a	
RadioMarché Sc.1	Mali	CONSTRUCT	3	0.105	8	0.280
RadioMarché Sc.2	Mali	INSERT	3	0.105	n.a	

7 Conclusions

We show that using the Semantic Web for data sharing is possible in areas without a Web infrastructure. We developed a conversion module that translates SPARQL over HTTP requests to SMSes and decodes these messages at the other end. SPARQL over SMS is an example of downscaling the Semantic Web to the infrastructure in place, in our case SMS. Extending the Kasadaka platform with this M2M communication functionality adds new possibilities for Semantic Web applications. Our solution integrates easily with other data sharing solutions since it does not create an isolated SMS-network but presents a conversion mechanism.

We investigated a number of challenges around porting SPARQL data exchange using SMS. Several RDF compression strategies are evaluated based

[18] For Mali, we assume an average cost of 20CFA = 0.035USD per SMS http://www.orangemali.com/2/particuliers/28/34/les-prepayes-113.html (accessed April 2016). For Ghana, we assume 0.055GH = 0.014USD per SMS http://support.vodafone.com.gh/customer/portal/articles/1823814-sms (accessed April 2016).

on real-world small data sets, leading us to a dynamic compression method that combines the generic serialization and text compression strategies with strategies using shared vocabulary. We show the viability of sending small RDF data sets using SPARQL over SMS and elaborate this in four scenarios from two realistic use cases. Future work consist of further development and deployment of solutions which include SPARQL over SMS in the field and designing longer term evaluations for these and new ICT4D use cases.

The current SPARQL over SMS has several limitations and opportunities for improvement. First, the reasoning that is used to eliminate semantic redundancies is based on a limited number of RDFS and OWL patterns and is restricted in terms of the search depth. Second, the SMS transfer mechanism is not yet fitted to properly deal with unexpected faults or partial transfers. We are looking at methods from systems such as the aforementioned ERS. Furthermore, not yet all SPARQL operations are supported. To achieve full compatibility, these will have to be implemented. Lastly, the implementation used to send and receive SMSes only supports 8-bit SMSes (140 characters). Using 7-bit SMSes (160 character) can further increase efficiency. The intent is to conduct further tests, by deploying SPARQL over SMS in the field, to identify the effects of these limitations and to validate the solution in real-world conditions. These field tests will include research into the economic viability of these solutions as discussed in [7] and look, for example, at integrating mobile-based payment plans.

Although SPARQL over SMS is developed based on ICT4D cases, it is applicable to other low-bandwidth cases. For example in the context of disaster-management or Internet of Things. The technologies of SPARQL over SMS are platform independent and it can be ported to other cases and platforms.

Finally, in this paper, we presented a specific approach for decoupling the principles and practices of the Semantic Web from the underlying implementation. This shows that these principles are still valid and valuable without the availability of Internet and a Web infrastructure. The more non-Web-Based networks are supported, the greater the reach of Semantic Web will be, as knowledge can be send across multiple types of networks in a standardized fashion.

References

1. de Boer, V., De Leenheer, P., Bon, A., Gyan, N.B., van Aart, C., Guéret, C., Tuyp, W., Boyera, S., Allen, M., Akkermans, H.: RadioMarché: distributed voice- and web-interfaced market information systems under rural conditions. In: Ralyté, J., Franch, X., Brinkkemper, S., Wrycza, S. (eds.) CAiSE 2012. LNCS, vol. 7328, pp. 518–532. Springer, Heidelberg (2012)
2. de Boer, V., Gyan, N.B., Bon, A., Tuyp, W., van Aart, C., Akkermans, H.: A dialogue with linked data: voice-based access to market data in the Sahel. In: Semantic Web (2013)
3. Charlaganov, M., Cudré-Mauroux, P., Dinu, C., Guéret, C., Grund, M., Macicas, T.: The Entity Registry System: Implementing 5-Star Linked Data Without the Web. arXiv preprint arXiv:1308.3357 (2013)

4. Fernández, J.D., Martínez-Prieto, M.A., Gutiérrez, C., Polleres, A., Arias, M.: Binary RDF representation for publication and exchange (HDT). Web Semant.: Sci. Serv. Agents World Wide Web **19**, 22–41 (2013)
5. Gray, J., Chong, W., Barclay, T., Szalay, A., Vandenberg, J.: TeraScale SneakerNet: Using Inexpensive Disks for Backup, Archiving, and Data Exchange. arXiv preprint cs/0208011 (2002)
6. Guéret, C., Schlobach, S., De Boer, V., Bon, A., Akkermans, H.: Is data sharing the privilege of a few? Bringing linked data to those without the Web. In: ISWC 2011 Outrageous Ideas Track, pp. 1–4 (2011). Best Paper Award
7. Gyan, N.B.: The web, speech technologies and rural development in West Africa: an ICT4D approach. Ph.D. thesis, Vrije Universiteit Amsterdam (2016)
8. Heeks, R.: ICT4D 2.0: the next phase of applying ICT for international development. Computer **41**(6), 26–33 (2008)
9. Käbisch, S., Peintner, D., Anicic, D.: Standardized and efficient RDF encoding for constrained embedded networks. In: Gandon, F., Sabou, M., Sack, H., d'Amato, C., Cudré-Mauroux, P., Zimmermann, A. (eds.) ESWC 2015. LNCS, vol. 9088, pp. 437–452. Springer, Heidelberg (2015). doi:10.1007/978-3-319-18818-8_27
10. Medhi, I., Ratan, A., Toyama, K.: Mobile-banking adoption and usage by low-literate, low-income users in the developing world. In: Aykin, N. (ed.) IDGD 2009. LNCS, vol. 5623, pp. 485–494. Springer, Heidelberg (2009). doi:10.1007/978-3-642-02767-3_54
11. Pentland, A., Fletcher, R., Hasson, A.: DakNet: rethinking connectivity in developing nations. Computer **37**(1), 78–83 (2004). http://dx.doi.org/10.1109/MC.2004.1260729
12. Rietveld, L., Beek, W., Schlobach, S.: LOD lab: experiments at LOD scale. In: Arenas, M., et al. (eds.) ISWC 2015. LNCS, vol. 9367, pp. 339–355. Springer, Heidelberg (2015). doi:10.1007/978-3-319-25010-6_23
13. Verborgh, R., Vander Sande, M., Colpaert, P., Coppens, S., Mannens, E., Van de Walle, R.: Web-scale querying through linked data fragments. In: LDOW (2014)

Extraction and Visualization of TBox Information from SPARQL Endpoints

Marc Weise[1], Steffen Lohmann[2(✉)], and Florian Haag[1]

[1] Institute for Visualization and Interactive Systems (VIS), University of Stuttgart, Universitätsstraße 38, 70569 Stuttgart, Germany
weisemc@studi.informatik.uni-stuttgart.de,
florian.haag@vis.uni-stuttgart.de

[2] Fraunhofer Institute for Intelligent Analysis and Information Systems (IAIS), Schloss Birlinghoven, 53757 Sankt Augustin, Germany
steffen.lohmann@iais.fraunhofer.de

Abstract. The growing amount of data being published as Linked Data has a huge potential, but the usage of this data is still cumbersome, especially for non-technical users. Visualizations can help to get a better idea of the type and structure of the data available in some SPARQL endpoint, and can provide a useful starting point for querying and analysis. We present an approach for the extraction and visualization of TBox information from Linked Data. SPARQL queries are used to infer concept information from the ABox of a given endpoint, which is then gradually added to an interactive VOWL graph visualization. We implemented the approach in a web application, which was tested on several SPARQL endpoints and evaluated in a qualitative user study with promising results.

Keywords: Linked data · Concept extraction · Visualization · Ontology · SPARQL · RDF · OWL · TBox

1 Introduction

A huge amount of Linked Data has been published in recent years and is ready for consumption [7]. A large portion of this data is available in RDF format and can be queried using the standardized query language SPARQL. The data often does not adhere to a strict schema, but typically different ontologies and vocabularies are used to describe it in a flexible way. On the one hand, this flexibility is an important characteristic and benefit of Linked Data; on the other hand, it can render it difficult to get an idea of what data is actually provided by a SPARQL endpoint. Visualizations help to get a better overview of the type and structure of the available data and can provide a useful starting point for further querying and analysis.

Sometimes, information on the used schema—the so-called TBox—is supplied directly by the SPARQL endpoint. This is, however, not always the case: The TBox information may be incomplete, much more generic than the data actually

E. Blomqvist et al. (Eds.): EKAW 2016, LNAI 10024, pp. 713–728, 2016.
DOI: 10.1007/978-3-319-49004-5_46

found on the server, or simply unavailable. Furthermore, due to the inherent capability of Linked Data to make use of parts of different ontologies, it is not necessarily obvious which definitions from which ontologies are instantiated in a given dataset.

This information is, however, important for various kinds of users of Linked Data: A developer looking for test cases to verify the behavior of a Linked Data application with respect to different SPARQL endpoints; a data curator checking whether a new dataset with an underspecified TBox can be easily integrated or aligned with the existing ontologies; an analyst trying to determine whether a Linked Data source contains information about a specific kind of connection between certain entities—all of them may need to know TBox information, such as relationships between classes in a given dataset, even when that dataset does not contain any explicit TBox information.

In this paper, we present an approach to extract and visualize TBox information from SPARQL endpoints. Rather than relying on given TBox information, we infer what a TBox for the available ABox data could reasonably look like based on several SPARQL queries.

This TBox information is then incrementally added to an interactive graph visualization based upon the Visual Notation for OWL Ontologies (VOWL) [11, 12]. We chose a node-link-based graph visualization, as it allows users to grasp certain structural criteria at a single glance, such as the presence of highly linked central classes or largely disjoint clusters of classes, before examining subgraphs in depth to analyze the details. In doing so, we had to slightly adapt VOWL to cope with the challenges of visualizing information extracted from Linked Data. We implemented the approach in a web application and tested it on several SPARQL endpoints. These tests and the results of a user evaluation confirmed that the approach is usable and helpful to get a better understanding of the type and structure of the data provided by a SPARQL endpoint. We also run performance tests revealing that the extraction can be done in reasonable time on a set of public SPARQL endpoints.

The rest of the paper is structured as follows: In Sect. 2, we summarize related work on the extraction and visualization of concept information from Linked Data. In Sect. 3, we describe our approach of inferring TBox information via SPARQL. Section 4 summarizes the slightly modified VOWL notation being used to visualize the extracted TBox information. The implementation of the approach is presented in Sect. 5 and evaluated in Sect. 6. Section 7 concludes the paper and presents ideas for future work.

2 Related Work

There are surprisingly few works concerning the extraction and visualization of concept information from Linked Data. Presutti et al. describe an approach of extracting *core knowledge* [14] from Linked Data by identifying knowledge patterns. They create a dataset knowledge architecture using extracted *type-property paths* and statistics about the property usage. Central types and properties are identified by their betweenness and number of instances. In contrast to our approach, they focus on the recognition of patterns in the data but not on the extraction and visualization of concept information.

Peroni et al. developed an approach for the automatic identification of *key concepts* [13]. Different from our work, the approach runs on ontologies and not on Linked Data. They use a couple of metrics, such as the length of concept names and their centrality in the graph structure, to find natural categories in the dataset. The basic idea is that key concepts should have both a high information density and concept coverage throughout the ontology. The concepts are also weighted by their popularity, which is defined as the number of results found by a search engine.

Visual query languages—such as QueryVOWL [5], which is based on the VOWL notation like our approach—use visual elements to represent Linked Data, as well. However, instead of being automatically generated, the visualization is assembled by the user in order to express a subgraph that is presumed to exist in the dataset. To do so, the user needs to already have an idea about the structure of the data before starting the querying process.

In contrast, RelFinder [8] extracts relations between two or more entities and visualizes them in a force-directed graph. It thereby assists users to discover unknown connections between individuals, but does not help to gain any overview of overall structures or larger subsets of individuals in a dataset. As in our approach, the information is extracted by generated SPARQL queries and dynamically added to the visualization. Similarly, LodLive [2] allows to browse Linked Data using a dynamic graph visualization. Starting with one resource, the user can explore a given dataset by expanding properties and navigating from one resource to the next. However, both RelFinder and LodLive focus on the ABox of a SPARQL endpoint and visualize only parts of the TBox but do not provide any overview visualization.

Other works are concerned with the recommendation of concepts based on Linked Data [4,16], the extraction of concepts from text using Linked Data [3], or follow general approaches of applying *formal concept analysis* to the Semantic Web [9].

3 Extraction of TBox Information

Our extraction approach uses a *class-centric perspective*, i.e., classes are extracted first and define the view on the Linked Data source. These classes are then connected by object properties and enriched by datatypes. We have chosen this approach, as a *class-centric perspective* is very common in ontology engineering and fits well with the node-link paradigm of the graph visualization that the VOWL notation is based upon [11].

The extraction is realized by dynamically generated SPARQL queries which reveal the TBox information from the ABox of a given dataset based on a couple of assumptions. For these queries, we had to find a trade-off between the number of required requests and the complexity of the queries. Since the SPARQL endpoints of Linked Data sources can have strict limits in terms of execution time, the queries must not be too complex. At the same time, we were aiming for displaying parts of the retrieved TBox information as soon as possible, hence short response times were important, as well. Therefore, our priority was on using simple SPARQL queries, while we were also interested in limiting the total number of requests.

The SPARQL queries are sent in a stepwise approach based on a couple of assumptions that are detailed in the following:

Step 1: Extract Classes with Most Instances

A generic SPARQL query asking for the n classes with the most instances is sent to the endpoint (where n is a user-defined upper limit). Listing 1.1 shows this query for the default limit of $n = 10$. The results of this query serve as a starting point for further extractions.

This approach is based on the assumption that a dataset is well represented by the classes having the most instances. On the other hand, these classes are often also the more generic ones. Therefore, we integrated three strategies to avoid a too generic visualization:

1. All built-in classes and properties of RDF, RDFS, OWL and optionally SKOS are contained in a blacklist that is filtered by default.
2. Users can customize this blacklist by adding or removing classes according to their needs. For instance, they can remove `owl:Thing` from the list to include it in the visualization or add `foaf:Agent` to filter it too.
3. Users can increase the number n of retrieved classes if the n initially retrieved classes are too generic, by changing the upper limit of retrieved classes accordingly.

```
SELECT DISTINCT ?class (COUNT(?sub) AS ?instanceCount)
WHERE {
  ?sub a ?class .
}
GROUP BY ?class
ORDER BY DESC(?instanceCount)
LIMIT 10 OFFSET 0
```

Listing 1.1. SPARQL query retrieving the $n = 10$ classes with the most instances

Step 2: Detect Subclasses, Equivalent and Disjoint Classes

Based on the n extracted classes with the most instances, further SPARQL queries are sent to the endpoint in order to detect classes that can be considered equivalent, subclasses, or disjoint classes. This is done by a pairwise comparison of the numbers of shared instances for all n classes, using the following assumptions:

1. If the number of shared instances of two classes is equal to the number of instances of each individual class, the classes are assumed to be extensionally equivalent.
2. If the number of shared instances of two classes is equal to the number of instances of the class having fewer instances, the class with fewer instances is a proper subset of the other class, which indicates a subclass relation between the two classes.

3. If there are no common instances at all, the two classes are considered to be disjoint.

All three assumptions are based entirely on the ABox information. For instance, two classes might not be explicitly defined as disjoint in their ontologies; however, if they do not share any instances in a given dataset, a disjoint relationship will be inferred following the above assumption. This provides users with the information that any search for individuals in that dataset which belong to both classes will be in vain.

Similarly, two classes do not necessarily need to be equivalent if the number of shared instances of the classes is equal to the number of instances of each individual class. However, we assume at least an extensional equivalence between the classes for the TBox visualization, even though the intensional meaning of the classes may be completely different.

Step 3: Retrieve Object Properties

In the third step, properties between the instances of the classes are retrieved. As with the classes, we retrieve the most frequently used properties first, i.e., properties with the greatest number of subject individuals (see example in Listing 1.2). This also includes property loops, i.e., properties where the subject and object individuals are from the same class.

As there can be a huge amount of different properties between the instances of two classes, we retrieve the properties in an incremental manner. When using a single SPARQL query, the execution of the query could take a very long time, possibly too long for SPARQL endpoints that have a strict limit for the execution time. Therefore, we choose the following approach in our implementation: Starting with a limit of l properties, this limit is doubled with each SPARQL query sent until all properties are retrieved.

It must also be considered that due to the pairwise retrieval in both step two and three of the extraction process, the number of SPARQL requests that need to be sent grows quadratically with the number of classes n retrieved in the first step (i.e., $N_{requests} \in \mathcal{O}(n^2)$). Thus, we recommend to select the number of classes n that are initially retrieved with care and in accordance to the endpoint performance (we currently use $n = 10$ as default, cf. Listing 1.1).

```
SELECT (COUNT(?originInstance) as ?count) ?prop
WHERE {
  ?originInstance a <http://dbpedia.org/ontology/Agent> .
  ?targetInstance a <http://xmlns.com/foaf/0.1/Document> .
  ?originInstance ?prop ?targetInstance .
}
GROUP BY ?prop
ORDER BY DESC(?count)
LIMIT 10 OFFSET 0
```

Listing 1.2. SPARQL query retrieving the $l = 10$ most often used object properties linking instances of the DBpedia classes `Agent` and `Document`

Step 4: Retrieve Datatypes and Datatype Properties

In the fourth step, we retrieve datatypes linked with the instances of the extracted classes. This step can be performed either after the third step or in parallel to it. We recommend a parallel execution to avoid the impression that there are no datatypes defined for the retrieved classes due to the delayed retrieval and visualization (remember that we visualize the information in a stepwise manner as soon as it is extracted).

For each class, we send queries that retrieve up to m datatypes that are most often used with the instances of that class (see Listing 1.3). After the datatypes are retrieved, the properties that connect the instances of the classes with these datatypes are queried in a second step (see Listing 1.4).

The reason for this two-step approach is again the restricted execution time of many SPARQL endpoints. In addition, it supports our goal of displaying the extracted information as quickly as possible in the visualization, even if it is still incomplete. This requires that we use placeholders as labels for the datatype properties in the visualization as long as the actual properties are unknown.

```
SELECT (COUNT(?val) AS ?valCount) ?valType
WHERE {
  ?instance a <http://dbpedia.org/ontology/Agent> .
  ?instance ?prop ?val .
  BIND(DATATYPE(?val) AS ?valType) .
}
GROUP BY ?valType
ORDER BY DESC(?valCount)
LIMIT 10
```

Listing 1.3. SPARQL query retrieving the $m = 10$ datatypes most often linked to the DBpedia class `Agent`

```
SELECT DISTINCT ?prop
WHERE {
  ?instance a <http://dbpedia.org/ontology/Agent> .
  ?instance ?prop ?val .
  FILTER (
      DATATYPE(?val) = <http://www.w3.org/2001/
          XMLSchema#string>
  )
}
LIMIT 10 OFFSET 0
```

Listing 1.4. SPARQL query retrieving properties between instances of the DBpedia class `Agent` and the linked datatype `string`

4 Visualization

The visualization of the extracted TBox Information is based on version 2 of the Visual Notation for OWL Ontologies (VOWL 2) [11,12]. We had to make some minor modifications to VOWL in order to address the peculiarities arising when visualizing information extracted from Linked Data.

4.1 Classes

In accordance with VOWL, extracted classes are represented as circle nodes in a force-directed layout (see Fig. 1a). Each class is displayed at most once in the visualization. The radii of the circles refer to the number of instances of the classes. Starting with a default size, the radii are scaled logarithmically, which usually results in a more homogeneous and aesthetically pleasing visualization. The circles are labeled with the linked `rdfs:label`, or, if no label is available, the last part of the URI of the class. If two or more classes are identified as equal, they are merged into a single circle drawn with a double border. Disjointedness and subclass relations are displayed by a dashed line between the circles. Subclass relations additionally have an empty arrowhead which points at the superclass.

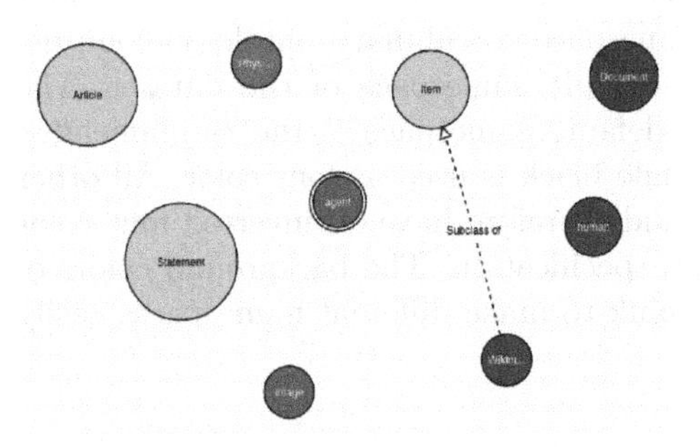

(a) Visualization resulting after the detection of subclasses, equivalent and disjoint classes, i.e., after the second step of the extraction process

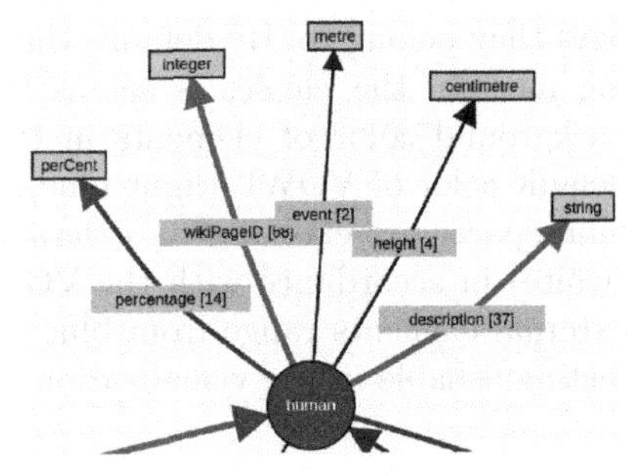

(b) Visualization of datatypes often used with instances of human and corresponding properties after the complete extraction

Fig. 1. Visualizations of parts of the DBpedia dataset at different stages of the extraction

4.2 Properties

Extracted properties are shown as directed and labeled lines (edges) linking the nodes, like it is specified by VOWL. In contrast to classes, properties can occur multiple times in the visualization. Different from VOWL, multiple properties between instances of the same pair of classes are merged into one line to avoid potentially large numbers of edges between the same two class nodes, which would clutter the visualization. The more different properties exist between the instances of two classes, the broader the referring line is drawn. This mapping is scaled with the square root, and the line width is restricted to a reasonable maximum to preserve the readability. If different properties are merged into one line, the property which occurs most often is considered most important—analogous to the class extraction principle. Therefore, the label of this property is shown on the line together with the number of properties that have been merged given in brackets. The arrowhead of a line indicates the direction of a set of properties.

4.3 Datatypes

Datatypes, such as strings, dates, or numbers, are displayed as yellow rectangles with a black border, as specified by VOWL. Similar to properties, a specific datatype can occur multiple times in the visualization but at most once per class. This multiplication prevents datatypes which are used by instances of different classes to get into the focus of the visual representation. Datatype properties linking the class instances with the datatypes are shown as edges with a green label (see Fig. 1b).

4.4 Namespaces

The background color of elements refers to the namespace (i.e., vocabulary, ontology) they belong to. By default, the namespace (vocabulary, ontology) comprising most of the classes is set as the default namespace of the dataset. The background color of elements in this default namespace is the recommended default color of VOWL (light blue), while black is used as font color. All other namespaces are classified as *external* and therefore have an inverted font color (white) in accordance with the VOWL specification. The background colors of external elements range from blue to pink to make different namespaces easily distinguishable in the visualization.

5 Implementation

We developed a web application that implements the presented TBox extraction and visualization approach. It is called LD-VOWL and uses JavaScript to generate and send the SPARQL queries via HTTP-GET requests in order to extract the TBox information. The extracted information is visualized using SVG and CSS. Furthermore, the application makes use of the visualization toolkit D3 [1] for computing and displaying the force-directed graph.[1]

User Interface. The user interface of LD-VOWL is inspired by WebVOWL [10] and consists of three views:

1. The *start view* where the user can input the SPARQL endpoint to be analyzed by either entering a URL or selecting one from a list of predefined endpoints.
2. The main view (see Fig. 2) shows the visualization of the extracted TBox information. It is complemented by a sidebar with controls, filters, and details.
3. The *settings view* where the user can adjust the extraction by editing the blacklist or the language of labels, among others.

Interaction. There are many possible ways for the user to interact with the visualization. Following the popular information seeking mantra of "overview first, zoom and filter, then details-on-demand" [15], the user starts with an *overview*

[1] A live demo of LD-VOWL is available at: http://ldvowl.visualdataweb.org.

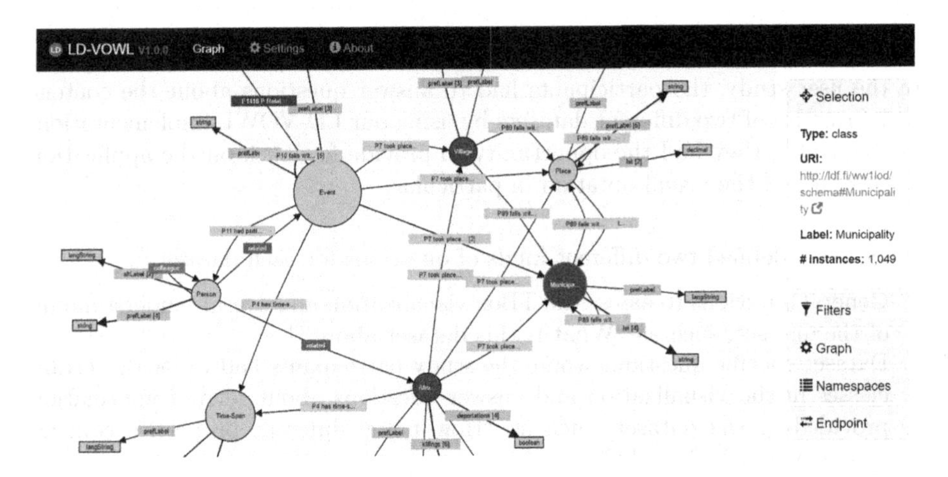

Fig. 2. LD-VOWL applied to a SPARQL endpoint with data about World War I (Linked Data Finland – World War I as Linked Open Data: http://ldf.fi/ww1lod)

and can use *zooming* and panning to adjust the visible area and position of the graph that is shown in the main view.

Furthermore, LD-VOWL provides options to *filter* different types of elements in the graph: The user can choose whether datatypes, property loops, subclass relations or disjoint classes are displayed in the graph by selecting corresponding checkboxes in the filter menu of the sidebar.

All nodes and edges in the graph can be selected to get *details on demand.* Depending on the selected element, this information can, for example, include the exact number of class instances, a list of all properties visually represented by a line, or comments describing the element. All URIs are displayed as hyperlinks, and users can click on them to view further information (if available).

As the layout of the graph visualization is force-directed, positions of nodes are random in the beginning and may not be optimal when the graph dynamically unfolds during the extraction. Therefore, the user can use drag-and-drop to move the nodes around in order to reduce edge crossings and overlappings. To further adjust the graph layout, users can change the length of the edges representing object and datatype properties.

Finally, users can control the namespace classification by flagging namespaces as beloning to the main vocabulary or being marked as external. Users can also decide whether different colors should be used for the external namespaces or not.

6 Evaluation

We evaluated the presented approach and implementation in two different ways: First, we conducted a qualitative user study to assess the usability and usefulness of the TBox visualization extracted from SPARQL endpoints. Second, we tested our approach on different endpoints to check whether it can be executed in reasonable time.

6.1 Qualitative User Study

In the user study, the participants had to answer questions about the content and structure of two different datasets by using our LD-VOWL implementation. Subsequently, they had the opportunity to provide feedback on the application in general and the visual notation in particular.

Tasks. We defined two different kinds of questions for each dataset:[2]

1. General questions to assess the TBox visualization and overall understanding of the dataset, such as "What is this dataset about?"
2. Dataset-specific questions where the study participants had to locate certain classes in the visualization and answer questions about related information provided in the dataset, such as "How many different properties connect classes X and Y directly?"

Datasets. Two SPARQL endpoints with a rather specific topic were chosen to be examined by the participants of the study. The first endpoint contains a dataset about Nobel prize laureates[3] and creates a rather sparse graph (see Fig. 3); the second one is about Austrian skiers[4] and the graph is much more dense (see Fig. 4).

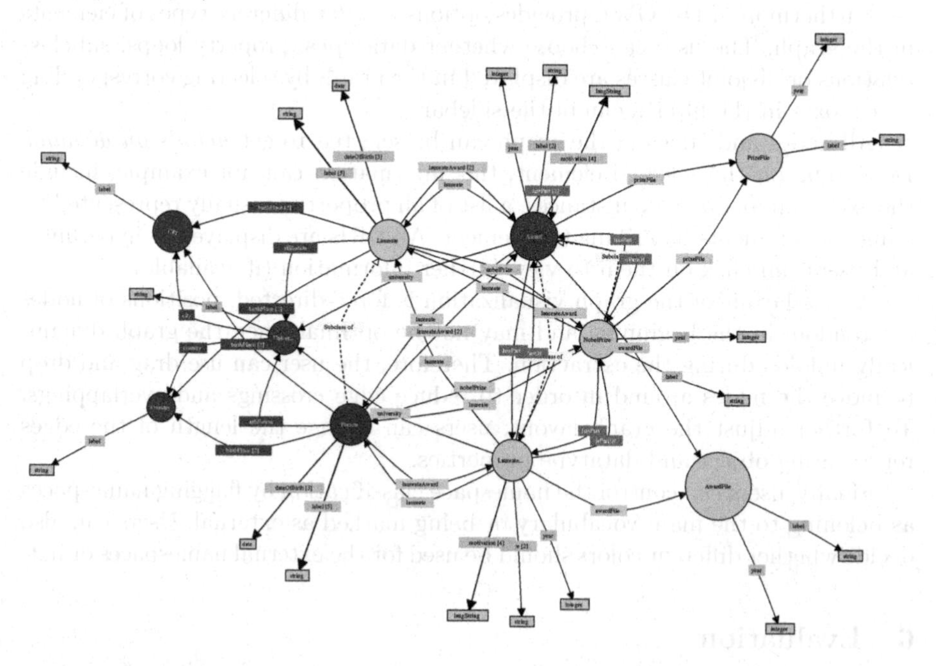

Fig. 3. Visualization of TBox information extracted from the Nobel prize dataset

[2] The complete list of questions is available at http://ldvowl.visualdataweb.org.

[3] http://data.nobelprize.org/sparql.

[4] http://vocabulary.semantic-web.at/PoolParty/sparql/AustrianSkiTeam.

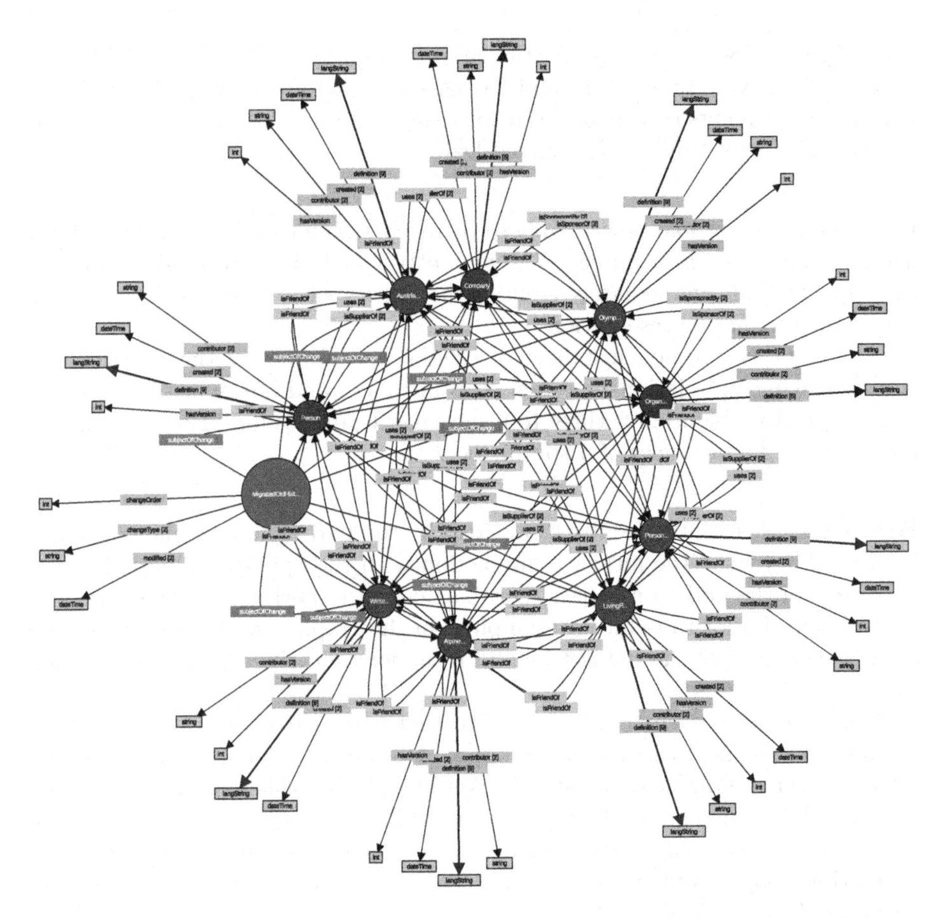

Fig. 4. Visualization of TBox information extracted from the Austrian ski team dataset

When comparing the LD-VOWL visualizations with the ontologies used by the endpoints, some information that is only found by LD-VOWL can be spotted. For instance, while the data on Nobel prizes is centered around its own ontology[5], the LD-VOWL output reveals that for several hundred Nobel laureates, birth and death places are supplied based upon the DBpedia ontology[6]. At the time of writing, this additional information is not hinted at in the Nobel prize ontology. The data about the Austrian skiers is not directly based upon any ontology in particular, but it uses a mixture of definitions from several ontologies, such as classes from DBpedia[7]. The connections between these definitions become visible in the LD-VOWL visualization.

[5] http://data.nobelprize.org/terms.
[6] http://dbpedia.org/ontology.
[7] http://dbpedia.org/class.

Participants. Seven subjects participated in the user study (six male and one female). They were between 17 and 25 years old (mean age of 21, 4 years) and were mainly students recruited at the university campus. Only one of them stated to have prior knowledge about Linked Data.

Results. All participants were able to correctly summarize the content of both datasets. Most of the summaries were pretty exact. In almost all cases, the participants could immediately identify the classes with most instances by the size of their visual representation.

All relations between instances of classes were found by the participants, and by looking up relations between class instances and datatypes, the participants were also able to give three to five correct examples for TBox information provided for the instances of a specific class.

Feedback. Most of the feedback provided by the participants of the study was positive. They liked the intuitive interaction with the application and the fluid animation of the nodes and edges. One participant remarked that moving nodes around helped a lot to reduce overlappings in the graph. Another one called the class-centric view of the dataset helpful to understand the ontology underlying the dataset.

Some of the participants did not like that a moved node was pulled back in its initial position when it was dropped. However, by adding a *pick-and-pin* mode to the force-directed layout as in WebVOWL [10], this issue could be easily resolved. Two participants found it rather difficult to trace edges in the dense graph. One of them suggested a mode in the application enabling the selection of multiple elements in the graph and highlighting the relations between these elements. This would make it much easier to follow relations especially in dense graphs.

6.2 Extraction Performance

Beside the user study, which focused on the usability and usefulness of the extracted overview visualization, we also evaluated the performance of the extraction approach. The goal was to determine whether the extraction process of LD-VOWL is fast enough to be used in practice. We consider short waiting times comparable to those when copying files acceptable in those contexts. For this purpose, we measured the number of SPARQL queries sent, and the total time needed to extract the TBox information from a set of ten public SPARQL endpoints with datasets of different size.

Procedure. The extraction was performed with the default limit of $n = 10$ classes (cf. Sect. 3). Before it was started, the browser cache and local storage was cleared to avoid client-side caching. Next to the total execution time and the number of successful and failed queries, the total number of triples was

Table 1. List of SPARQL endpoints queried in the performance evaluation

#	Name	URL
1	Transparency International	http://transparency.270a.info/sparql
2	Archiveshub	http://data.archiveshub.ac.uk/sparql
3	DBCLS	http://data.allie.dbcls.jp/sparql
4	Ecuador Research	http://data.utpl.edu.ec/ecuadorresearch/lod/sparql
5	Nobelprize	http://data.nobelprize.org/sparql
6	Springer LOD	http://lod.springer.com/sparql
7	DBpedia	http://dbpedia.org/sparql
8	Scientific Ocean Drilling	http://data.oceandrilling.org/sparql
9	Linkedspending	http://linkedspending.aksw.org/sparql
10	Isidore	http://www.rechercheisidore.fr/sparql

determined for each endpoint by using a generic SPARQL query matching any triple. Furthermore, the software being used at the endpoint was noted by looking at the HTTP headers and any endpoint-related websites. As the hardware on which the software is running cannot be determined remotely, it could not be considered in the performance evaluation.

Endpoints. The SPARQL endpoints for the performance evaluation were selected out of a list of publicly available endpoints[8]. Criteria for the selection were an enabled Cross-Origin-Resource-Sharing (CORS) and the reliably of the endpoint in previous tests during the development of the LD-VOWL application. A list of all selected endpoints and their URLs is shown in Table 1.

Results. The results of the performance evaluation are shown in the two scatter plots in Fig. 5. The total amount of time needed for the complete extraction differs a lot across the different SPARQL endpoints (see Fig. 5a). The time spans we measured ranged from less than one minute (e.g., Transparency International) to more than fourteen minutes (e.g., DBpedia).

On some endpoints, the extraction needed significantly less time (only a few seconds). These endpoints were not included, as it can be assumed that they perform some server-side caching of results.

Furthermore, the throughput varied much across endpoints: Although some endpoints answered between one and two hundred queries per minute, other endpoints were only capable to respond to twenty or thirty queries (see Fig. 5b).

With an increasing size of the dataset, the time needed for the TBox extraction seems to increase, whereas the query throughput seems to decline. Aside from that, there are also performance differences between endpoints with datasets of similar size (especially for the smaller datasets), which might be caused by different hardware and/or software.

[8] SPARQL Endpoints Status: http://sparqles.okfn.org.

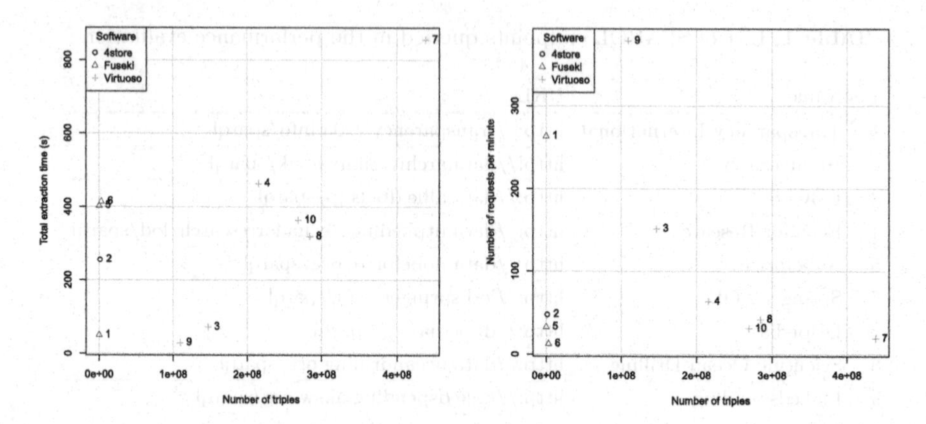

(a) Total time in seconds needed for the TBox extraction (b) Number of answered requests per minute by endpoint

Fig. 5. LD-VOWL endpoint performance

The evaluation of the extraction performance also shows the problem that some SPARQL queries need much more time than others. These stragglers can block the overall progress of the extraction. However, the extraction performance could be improved by adding an intelligent query scheduler that estimates the execution time of each SPARQL query using statistics or machine learning, as proposed in [6].

Scalability. Summing up the results, we can draw the following conclusion about the scalability of the presented approach: It is possible to extract TBox information from Linked Data in an acceptable time; however, the scalability is limited for the following reasons:

1. The number of SPARQL queries required to retrieve the relations increases quadratically with the number of extracted classes (cf. Sect. 3).
2. The throughput of the requests seems to decrease with an increasing size of the dataset.
3. The readability of the graph and the traceability of the edges decreases with the density of the graph due to visual clutter.
4. The response speed of the servers can vary a lot. We hope that further development of SPARQL processors as well as caching solutions can help alleviate this issue in the future.

7 Conclusion and Future Work

To unleash the full potential of Linked Data, it is important that users can get a quick overview of the type and structure of the data provided by a SPARQL endpoint. In this paper, we presented an approach to extract and visualize TBox information from SPARQL endpoints. It uses a number of SPARQL queries that help to structure the data and reveal how it is described by ontologies, based on a set of assumptions. This TBox information is then incrementally added to an interactive graph visualization using a slightly adapted version of the VOWL notation.

We implemented the approach in the web application LD-VOWL and tested it on several SPARQL endpoints. The results of these tests and a user study showed that the approach can create a comprehensible overview of the content and structure of a dataset within a few seconds to minutes, depending on the size of the dataset and the performance of the endpoint. However, the results also revealed that the scalability of the approach is limited due to an increasing execution time of the SPARQL queries on huge datasets and the deterioration of the readability in increasingly dense graphs.

There are several possibilities to improve the approach and implementation in the future. One direction of research would be to improve the performance of the extraction by implementing the aforementioned request scheduler. Other research could look into the extraction of further TBox information, such as inverse properties or set operators, by developing corresponding assumptions and extraction patterns. The LD-VOWL application would benefit from additional interactive features, such as a pick-and-pin mode or an advanced highlighting that enables users to select multiple nodes in the graph and indicates the relationships between them.

Finally, the visual scalability is an important issue for future research by investigating possibilities of how to show larger graphs and comprehensive overview visualizations in a more compact way.

References

1. Bostock, M., Ogievetsky, V., Heer, J.: D^3 data-driven documents. IEEE Trans. Vis. Comput. Graph. **17**(12), 2301–2309 (2011)
2. Camarda, D.V., Mazzini, S., Antonuccio, A.: LodLive, exploring the web of data. In: 8th International Conference on Semantic Systems (I-SEMANTICS 2012), pp. 197–200. ACM (2012)
3. Cortis, K.: ACE: a concept extraction approach using linked open data. In: Concept Extraction Challenge at the Workshop on 'Making Sense of Microposts'. In: CEUR Workshop Proceedings, vol. 1019, pp. 31–35. CEUR-WS.org (2013)
4. Damljanovic, D., Stankovic, M., Laublet, P.: Linked data-based concept recommendation: comparison of different methods in open innovation scenario. In: Simperl, E., Cimiano, P., Polleres, A., Corcho, O., Presutti, V. (eds.) ESWC 2012. LNCS, vol. 7295, pp. 24–38. Springer, Heidelberg (2012)
5. Haag, F., Lohmann, S., Siek, S., Ertl, T.: QueryVOWL: a visual query notation for linked data. In: Gandon, F., Guéret, C., Villata, S., Breslin, J., Faron-Zucker, C., Zimmermann, A. (eds.) ESWC 2015. LNCS, vol. 9341, pp. 387–402. Springer, Heidelberg (2015). doi:10.1007/978-3-319-25639-9_51
6. Hasan, R., Gandon, F.L.: A machine learning approach to SPARQL query performance prediction. In: WI-IAT (2), pp. 266–273. IEEE Computer Society (2014)

7. Heath, T., Bizer, C.: Linked Data: Evolving the Web into a Global Data Space. Synthesis Lectures on the Semantic Web. Morgan & Claypool Publishers, San Rafael (2011)
8. Heim, P., Hellmann, S., Lehmann, J., Lohmann, S., Stegemann, T.: RelFinder: revealing relationships in RDF knowledge bases. In: Chua, T.-S., Kompatsiaris, Y., Mérialdo, B., Haas, W., Thallinger, G., Bailer, W. (eds.) SAMT 2009. LNCS, vol. 5887, pp. 182–187. Springer, Heidelberg (2009)
9. Kirchberg, M., Leonardi, E., Tan, Y.S., Link, S., Ko, R.K.L., Lee, B.S.: Formal concept discovery in semantic web data. In: Domenach, F., Ignatov, D.I., Poelmans, J. (eds.) ICFCA 2012. LNCS, vol. 7278, pp. 164–179. Springer, Heidelberg (2012)
10. Lohmann, S., Link, V., Marbach, E., Negru, S.: WebVOWL: web-based visualization of ontologies. In: Lambrix, P., Hyvönen, E., Blomqvist, E., Presutti, V., Qi, G., Sattler, U., Ding, Y., Ghidini, C. (eds.) EKWA 2014 Satellite Events. LNCS, vol. 8982, pp. 154–158. Springer, Heidelberg (2015)
11. Lohmann, S., Negru, S., Haag, F., Ertl, T.: Visualizing ontologies with VOWL. Semant. Web **7**(4), 399–419 (2016)
12. Negru, S., Lohmann, S., Haag, F.: VOWL: visual notation for OWL ontologies (2014). http://purl.org/vowl/
13. Peroni, S., Motta, E., d'Aquin, M.: Identifying key concepts in an ontology, through the integration of cognitive principles with statistical and topological measures. In: Domingue, J., Anutariya, C. (eds.) ASWC 2008. LNCS, vol. 5367, pp. 242–256. Springer, Heidelberg (2008)
14. Presutti, V., Aroyo, L., Adamou, A., Schopman, B.A.C., Gangemi, A., Schreiber, G.: Extracting core knowledge from linked data. In: 2nd International Workshop on Consuming Linked Data (COLD 2011), CEUR Workshop Proceedings, vol. 782. CEUR-WS.org (2011)
15. Shneiderman, B.: The eyes have it: a task by data type taxonomy for information visualizations. In: 1996 IEEE Symposium on Visual Languages (VL 1996), pp. 336–343. IEEE (1996)
16. Stankovic, M., Breitfuss, W., Laublet, P.: Linked-data based suggestion of relevant topics. In: 7th International Conference on Semantic Systems (I-SEMANTICS 2011), pp. 49–55. ACM (2011)

In-Use Papers

Learning Domain Labels Using Conceptual Fingerprints: An In-Use Case Study in the Neurology Domain

Zubair Afzal, George Tsatsaronis(✉), Marius Doornenbal, Pascal Coupet, and Michelle Gregory

Content and Innovation Group, Operations Division, Elsevier B.V., Radarweg 29, 1043 NX Amsterdam, The Netherlands
{m.afzal.1,g.tsatsaronis,M.Doornenbal,p.coupet,m.gregory}@elsevier.com

Abstract. Modelling a science domain for the purposes of thematically categorizing the research work and enabling better browsing and search can be a daunting task, especially if a specialized taxonomy or ontology does not exist for this domain. *Elsevier*, the largest academic publisher, faces this challenge often, for the needs of supporting the journals submission system, but also for supplying *ScienceDirect* and *Scopus*, two flagship platforms of the company, with sufficient metadata, such as conceptual labels that characterize the research works, which can improve the user experience in browsing and searching the literature. In this paper we describe an *Elsevier* in-use case study of learning appropriate domain labels from a collection of 6,357 full text articles in the neurology domain, exploring different document representations and clustering mechanisms. Besides the baseline approaches for document representation (e.g., bag-of-words) and their variations (e.g., n-grams), we employ a novel in-house methodology which produces conceptual fingerprints of the research articles, starting from a general domain taxonomy, such as the Medical Subject Headings (*MeSH*). A thorough empirical evaluation is presented, using a variety of clustering mechanisms and several validity indices to evaluate the resulting clusters. Our results summarize the best practices in modelling this specific domain and we report on the advantages and disadvantages of using the different clustering mechanisms and document representations that were examined, with the aim to learn appropriate conceptual labels for this domain.

Keywords: Document labeling · Document clustering · Conceptual fingerprints · Domain taxonomy · Neurology domain · Clustering evaluation · Best practices

1 Introduction

Reasoning with the content of text documents constitutes a key challenge to every intelligent document management system, and enables a wide variety of applications, such as knowledge discovery, question answering, and thematic

E. Blomqvist et al. (Eds.): EKAW 2016, LNAI 10024, pp. 731–745, 2016.
DOI: 10.1007/978-3-319-49004-5_47

browsing of document collections. A fundamental step for the creation of such an infrastructure is the application of methods that enable the automated annotation of unrestricted text with ontology concepts. These techniques can add important metadata to the indexed documents, such as concepts that describe at a semantic level the documents' contents, but they can also be used to populate ontologies from text. The end effect is the creation of an expanded, semantic index based on which a next generation of search engines is possible, e.g., [5], or other important applications are feasible, such as natural language question answering [13].

A key challenge is the identification of an appropriate conceptual label set that describes the underlying document collection, and covers the domain sufficiently. Though for many domains, and especially in life sciences, there exist general taxonomy or ontologies which can act as a basis for this label set, e.g., the *NLM*'s *Medical Subject Headings* (*MeSH*), it is still a great challenge to restrict the labels only to those that are important to describe the specific document collection and their domain. Such a restriction is especially meaningful for the aforementioned applications which are focused on precision, and where the usage of a wider label set may cause drifting and add further ambiguity problems to the annotation process.

Motivated by this challenge, and given that in *Elsevier* the process of annotating scientific articles with semantic metadata is a very important component that affects a large number of products (e.g., *Scopus, ScienceDirect*) and processes, (e.g., search and browsing, finding appropriate reviewers for submitted articles), in this paper we present the results of an in-use case study with the objective to learn an appropriate concept label set for scientific articles on the neurology domain. The methodology uses clustering of the document set and cluster labelling techniques to characterize the clusters. The contributions of this work can be summarized in the following: (a) presentation of the applied methodology that uses existing tools and techniques in order to process full text articles from the neurology domain, cluster them and extract conceptual labels, (b) thorough empirical evaluation of the resulting clusters and labels using well known clustering validity indices and, (c) discussion on the best practices and the settings of the methodology that worked best. The rest of the paper is organized as follows: Sect. 2 discusses the related work. Section 3 presents the details of the applied methodology, as well as the evaluation metrics that are used. Section 4 presents the experimental setup and the results. Section 5 provides an analytical discussion on the best practices and the technical settings that worked best for this specific use case. Finally, Sect. 6 concludes and gives pointers to future research directions.

2 Background and Related Work

The current case study aims at the extraction of an appropriate conceptual label set from scientific articles of the neurology domain. The methodology that we apply, presented in Sect. 3, involves document clustering, application of clustering validity criteria, and clusters' labelling techniques. Therefore, the related work stretches across these three areas.

Text document clustering has a long history of several decades of research [1]. Some of the early works on text document clustering date back at the beginning of the 1980′s (e.g., [15]). More recent works on text document clustering have focused on the benefits that semantic metadata or measures of semantic similarity and relatedness may add at the document clustering process. For example, Fodeh and Punch discuss how the incorporation of semantic features from an ontology can reduce the number of features required to perform the document clustering by 90 % [6]. Similarly, Staab and Hotho had explored earlier the embedding of ontology-based heuristics for feature selection and feature aggregation during the pre-processing phase of the documents to be clustered, with positive findings [12]. Batet et al. [2] also conclude that adding semantic metadata information from a set of ontologies for the purposes of interpreting the document content and the resulting clusters is beneficial. Dagher and Fung also explored a similar direction [3], introducing the *Subject Vector Space Model*, which represents the documents to be clustered in a vectorial representation where dimensions are the domain subjects taken from *WordNet*. Zhao and Karypis [16] are driven to similar conclusions regarding the benefits of a topic-driven document clustering approach, where a given set of topics is assumed as input. Such a topic set can be inferred from the usage of an underlying ontology with which the documents may be annotated, and the most frequent concepts can be retained as a starting list. Finally, in our previous work [10], we have also explored the embedding of semantic information in the clustering process, by designing semantic smoothing kernels in order to compute the document similarity during the clustering process. The kernels utilize *WordNet* and *Wikipedia* to measure the semantic similarity of the documents' terms, and the results showed a great improvement in performance over baseline approaches in text clustering benchmarks.

In this work we build on the experience of the aforementioned studies for the text document clustering process, and we utilize *Elsevier*'s in-house tool for document annotation with ontological concepts, in order to enhance the documents' representation with semantic metadata. The tool, named the *Fingerprint Engine* (*FPE*) [14] will be described in more detail in Sect. 3.2.

With regards to the clustering evaluation, in the bibliography there are well established clustering validity indices [8]. The most important criterion on which clustering validity measures are applicable in a case study pertains to whether there is a *gold standard set* on which the documents have been manually clustered and the clusters have been manually curated. In our case study there is no such benchmark set, as we aim at inferring an appropriate, compact clustering of the documents from which we can extract the conceptual labels for the domain. Therefore, in our case only indices such as the *Silhouette* [11], *Inertia* and *Davies-Bouldin* (*DB*) [4] are applicable. These are discussed in Sect. 3.4.

Finally, the cluster labelling process is a topic that is also widely explored and discussed in the literature [9], with the main approaches utilizing statistical measures of co-occurrence around the clusters terms (e.g., *Pointwise Mutual Information*, X^2), or taking into account the centroid vectorial representations of the clusters and retaining the terms with the highest weight (e.g., highest *TF-IDF* score). In the following section we discuss in detail the approaches we used.

3 Learning Domain Labels from Scientific Articles

In this section we present the applied methodology for learning appropriate domain labels from scientific full text articles. The methodology, an overview of which is illustrated in Fig. 1, comprises a pre-processing step, a document representation module responsible for extracting the different representations that will be used for the clustering, a clustering module, a clustering evaluation module, a representation selection step, a clusters' labelling module, and the final step of extracting the actual labels. In the following, we are presenting the details of each module, and the alternative settings we tried for each.

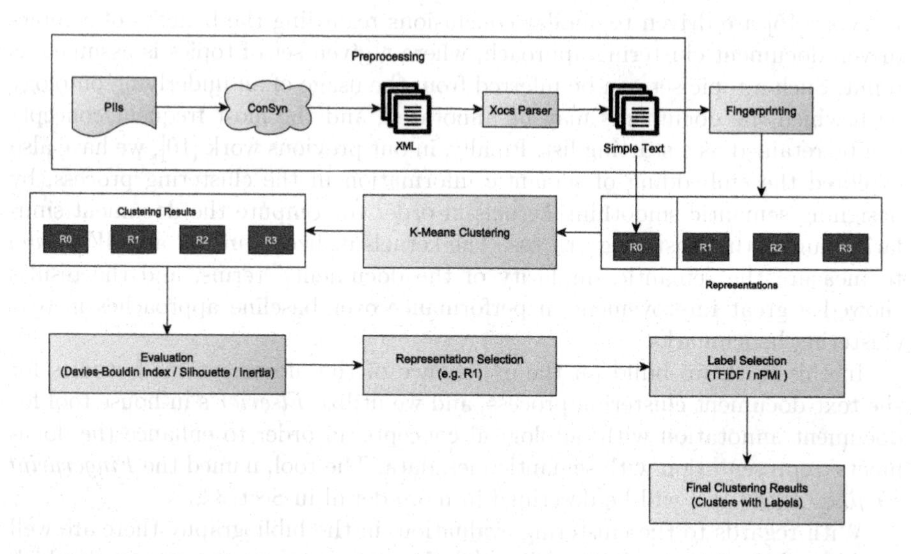

Fig. 1. Overview of *Elsevier*'s methodology applied for learning appropriate labels from a document collection.

3.1 Pre-processing Module

The first step involves the selection of document *Publisher Item Identifiers* (*PII*s), which are the *Elsevier*'s internal unique article identifiers, and the articles' parsing from the original sources maintained in *XML* format in the document repository. The documents' *PII*s are selected on the basis of which domain needs to be covered. In our case study we focus in the neurology domain, and, therefore, we identified as a first step the top books in the field (e.g., the *Fundamental Neuroscience*[1]). As a second step, we collected the *PIIs* of all of the books' chapters, and then fetched them from our content store API (*ConSyn*) in their original *XML* sources. As a third and final pre-processing step, we parsed the original sources, and retained from the chapters their title, section structure and titles, and full text. Therefore, as a result of this pre-processing step, we have a flat document format of all of the full text articles (book chapters) that we need to process, and this can in turn be used directly as input to several popular tools for further processing, e.g., the *Python*'s *scikit-learn* package[2] for performing machine learning.

3.2 Document Representation and Conceptual Fingerprinting Module

Results from the previous step (flat text documents) were used to generate 4 different document representations for the document clustering. We define these representations as $R0$, $R1$, $R2$, and $R3$, and we explain those, as well as some variations of those, in detail in the following. For the $R1$, $R2$ and the $R3$ representations, the output of the *Elsevier*'s *Fingerprint Engine* (*FPE*) was used. Therefore, we start the description by providing some detailed information on how the fingerprinting mechanism works.

Elsevier's Fingerprint Engine: The *Elsevier*'s *Fingerprint Engine* (*FPE*) is an in-house developed technology for annotating unstructured text with ontological concepts [14]. In fact, the *FPE* is a concept annotation system that might be compared to several biomedical concept annotators as evaluated in [7]. The key difference is that the *FPE* is applicable to all domains, ranging from engineering to medicine. The actual fingerprinting process is a concept annotation process that can be applied to texts of any size, from single lines to abstracts, or, conceivably, to full text articles. The process involves the consecutive execution of a number of *NLP* steps, where each step builds on the results of previous steps. A modular design of the *FPE*, comparable to that of similar frameworks as *UIMA* or *GATE* was chosen to allow for tailor-cut text-processing pipelines to meet the specific requirements of each situation and science domain. For all domains the processing pipeline includes steps of tokenization, input analysis and normalization, expansion of abbreviations and coordinations, a number of entity annotations (names, institutions, and citations) and part-of-speech tagging. These steps are preparatory relative to the term annotation step, in which the text is scanned for the occurrence of terms as defined in the target thesaurus or vocabulary that is designated as the relevant set of concepts for the domain. In this case study, the target vocabulary are all the terms of the *Medical Subject Headings* (*MeSH*) taxonomy. During the term annotation step, textual variations such as normalization and spelling differences, punctuation

[1] http://store.elsevier.com/product.jsp?isbn=9780123858702.
[2] http://scikit-learn.org/stable/.

and word order variations, and part-of-speech tags are either ignored or taken into account, depending on the nature of the terms sought for. After the term annotation task has been performed, annotated terms are evaluated in a number of disambiguation steps, which establish certainty on the question whether an annotated term candidate really designates the concept as it is defined in the target vocabulary. Word sense disambiguation (*WSD*) is an indispensable component of *NLP* solutions that claim to provide semantic annotation. The *FPE* employs disambiguation techniques based on pattern-based rules, as well as statistical co-occurrence methods, unification methods and thesaurus-based co-occurrence methods. The end effect of applying the *FPE* in a textual corpus, using *MeSH* as a target thesaurus in our case, is a conceptual fingerprint that can be produced for each article, or even section, i.e., a list of ranked of concepts from *MeSH*, that characterize the article or the section. The ranking of the identified concepts is produced from the *FPE* internally, and considers a large variety of scores, such as matching context and annotated term overlap. Therefore, *FPE* can be thought of as a module that maps texts onto a domain vocabulary, in our case *MeSH* headings.

***R0*, or *Full*, Document Representation:** This is our baseline representation used commonly for text processing tasks. It is the vectorial representation of documents, with the dimensions being the terms in the document, and their associated *Term Frequency-Inverse Document Frequency* (*TF-IDF*) scores. All parts of the documents were considered, and a typical stopword removal and lemmatization was performed on the documents' content. For the creation of the *TF-IDF* vectors of the documents we used the *Python*'s *scikit-learn* package.

***R1* Document Representation:** This representation contains the top 10 *FPE* fingerprint concepts identified per chapter section. The section titles were not included in this representation. The top 10 concepts were selected using the ranks given by the *FPE*. Therefore, *R*1 is the native conceptual representation of documents, with the concepts coming from the *MeSH* headings, via the application of the *FPE*. The representation is again vectorial for each document, with the dimensions being the *FPE* concepts from *MeSH*, and their associated *TF-IDF* values.

***R2* Document Representation:** This representation is an expansion of the *R*1 vectorial representation, enhanced by the terms contained in the chapters' section titles. Therefore, this representation can be considered as a mixture of *FPE* produced concepts from the whole chapter, and chapter's terms, with the terms coming only from the chapters' section titles. As a result, the documents vectors have as dimensions *FPE* concepts from *MeSH*, and section titles' terms, all of them with their associated *TF-IDF* values.

***R3* Document Representation:** This representation is an expansion of the *R*2 vectorial representation, enhanced further by the keywords of the chapter. Therefore, this representation is a wider mixture of *FPE* produced concepts from the whole chapter, and chapter's terms, with the terms coming only from chapters' section titles and keywords. As a result, there are even more dimensions in the documents' *TF-IDF* vectors, which now contain *FPE* concepts from *MeSH*, and terms coming from the documents' section titles and chapters' keywords.

Normalized vs. Non-normalized *TF-IDF* Vector Variations: For each of the $R0$-$R3$ representations, we used in our experiments both the normalized to the vectors' length and the non-normalized *TF-IDF* vector representations of the documents.

n-gram Variations: For each of the representations we also used their term n-gram variations. In these variations, n-grams of terms were extracted (with $n \leq 3$) considering their occurrence on text. For the $R0$ representation, and the term parts of the $R2$ and $R3$ (section title terms and keywords), this is straightforward. For the *FPE* n-grams, once the top-10 per section were extracted, we went back to the originally produced annotations from the *FPE*, isolated the *MeSH* concepts' occurrence, and extracted the n-grams from this occurrence.

3.3 Documents' Clustering Module

For the purpose of the documents' clustering with all of the aforementioned representations and variations we used the k-means clustering algorithm, with a varying number of input clusters (k) ranging from 2 to 100, and a step of 2. For each number of clusters tried, the algorithm was executed 30 times, as k-means is known to be unstable when starting from random seeds. We also tried a variation of k-means, called *Minibatch* k-means, with the same setup. This variation is computationally more efficient when clustering a large number of data points, and especially points with many dimensions, such as textual data. Finally, we tried one clustering algorithm from the category of the density-based clustering schemes, namely the *DBSCAN* algorithm, and an algorithm from the category of the kernel-based clustering schemes, namely the *MeanShift* algorithm. For the implementations of the 4 used clustering algorithms we utilized the *Python*'s *scikit-learn* package.

3.4 Clustering Validation Module

This module implements three measures that can be used as clustering validity criteria, namely the *Silhouette* score [11], the *Davies-Bouldin* index [4], and *Inertia*. All three of these criteria can be computed on a resulting clustering scheme without the need to have a *golden solution* of what the output is supposed to be, in order to compare the results with the actual clustering output. Having the values of these criteria for all the clustering setups, we are able to draw a conclusion on which setup worked best, and, therefore, which should be the parameters and settings for the final clustering, from which the conceptual labels are extracted. Below follow the details of the three measures.

***Silhouette* Score:** Assume that the data have been clustered via any technique, such as k-means, into k clusters. For each datum i, let $a(i)$ be the average dissimilarity of i with all other data within the same cluster. We can interpret

$a(i)$ as how well i is assigned to its cluster; the smaller the value, the better the assignment. We then define the average dissimilarity of point i to a cluster c as the average of the distance from i to all points in c. Let $b(i)$ be the lowest average dissimilarity of i to any other cluster, of which i is not a member.

The cluster with this lowest average dissimilarity is said to be the *"neighbouring cluster"* of i because it is the next best fit cluster for datum i. *Silhouette* ($s(i)$) for datum i can now be defined as shown in the following equation [11], and can be measured on the whole clustering output by averaging the $s(i)$ values over all data points i. From Eq. 1 it follows that if the $s(i)$ value is close to 1, datum i is clustered appropriately, while if it is -1, it should have been assigned to the neighbouring cluster. A value close to 0 indicates that i lies at the border of the two clusters.

$$s(i) = \frac{b(i) - a(i)}{\max a(i), b(i)} \tag{1}$$

***Davies-Bouldin* Index (DB):** Assume two measures: one measure that captures the *"scatter"* within a cluster C_i, and one measure that captures the *"separation"* between two clusters C_i and C_j. Let's call them S_i and $M_{i,j}$ respectively. The higher the S_i the bigger the *"scatter"* within cluster C_i. Similarly, higher $M_{i,j}$ means higher degree of *"separation"* between clusters C_i and C_j (i.e., clusters are very well separated and there aren't many overlaps). These two measures can be defined as shown in the following equations:

$$S_i = \frac{1}{T_i} \sum_{j=1}^{T_i} ||X_j - A_i||_p \tag{2}$$

where A_i is the centroid of cluster C_i, T_i its size, X_j the vector of datum j, and p is the p-norm, e.g., with $p = 1$ we have the Manhattan distance, with $p = 2$ the Euclidean. We are using $p = 2$ in all of our experiments (Euclidean distance), as this is more intuitive for measuring distances between document vectors.

$$M_{i,j} = ||A_i - A_j||_p \tag{3}$$

where A_i and A_j the centroids of clusters C_i and C_j respectively.
Using Eqs. 2 and 3 we can define a measure of how good a clustering is. Focusing only on the two clusters C_i and C_j this can be as follows:

$$R_{i,j} = \frac{S_i + S_j}{M_{i,j}} \tag{4}$$

Essentially we want to minimize $R_{i,j}$ for all pairs of clusters C_i and c_j. Therefore, if we compute the measure shown in the following equation for cluster C_i, we measure how good the clustering is from the point of view of compactness; the lower the value of D_i, the better the clustering for cluster C_i.

$$D_i = \max_{j \neq i} R_{i,j} \tag{5}$$

Generalizing this measure for all clusters, gives us the DB index showing in the following equation, where N is the number of all clusters.

$$DB = \frac{1}{N}\sum_{i=1}^{N} D_i \tag{6}$$

Inertia. Inertia is defined as the within-cluster sum of squares of the distances of the cluster's points to the centroid. It is, therefore, recognized as a measure of internal coherence of the clusters, suffering, however, in the cases that clusters are non-convex, or non-isotropic. This means that it performs poorly in elongated clusters or manifolds with irregular shapes. For a cluster C_i inertia can be computed as follows, and can be averaged over all clusters in order to compute the inertia of the complete clustering result.

$$\sum_{j=1}^{T_i} ||X_j - A_i||_p \tag{7}$$

3.5 Selection of Final Clustering Parameters and Document Representation:

Given the application of the aforementioned clustering schemes with a variety of parameters, and the values of the clustering validity criteria, we can select in this step the clustering set-up (e.g., k for the number of clusters), and the document representation, or variation, (e.g., $R1$) that seems to provide the most compact and stable clustering according to the presented criteria.

3.6 Clusters' Labelling and Final Labels Extraction

With the clustering parameters and document representations selected, in this module we label the clusters, which will be used in turn as the domain labels. In order to label the clusters we have used two methods, both of them applied to the terms, and to the *FPE* concepts (*MeSH* headings) of the documents in the clusters. Therefore, as a result we have four sets of clusters labels. The two methods involve the selection of the top-n terms (or concepts) from each cluster either based on their *TF-IDF* values, or based on their *Normalized Pointwise Mutual Information* (*nPMI*) scores. To compute the terms' (or concepts') scores for each of the measures, we handle each cluster as a huge document, which is created by collating all the clusters' documents together. We will not explain how the well known *TF-IDF* scoring works [9], but we will show how the *nPMI* is computed in this case.

Given a cluster C and a term T, the *nPMI* score of T for cluster C is shown in the following equation:

$$nPMI(C;T) = \frac{PMI(C;T)}{-\log[P(C,T)]} \tag{8}$$

where $PMI(C;T)$ is defined as:

$$PMI(C;T) = \log \frac{P(C,T)}{P(C)P(T)} = \log \frac{P(T|C)}{P(T)} \tag{9}$$

The variable C is associated with membership in a cluster, and the variable T is associated with the presence of a term. Either variable can have a value of 0 or 1. Therefore, $p(C = 1)$ represents the probability that a randomly selected document is a member of a particular cluster (in this case C), and $p(C = 0)$ represents the probability that it isn't. Similarly, $p(T = 1)$ represents the probability that a randomly selected document contains term T, and $p(T = 0)$ represents the probability that it doesn't. Therefore, we only need to measure $p(C = 1; T = 1)$, $p(C = 1)$ and $p(T = 1)$. With these we can calculate $PMI(C = 1; T = 1)$, meaning how much information the events $C = 1$ and $T = 1$ share, given of course as input the cluster C and the term T. Dividing the $PMI(C = 1; T = 1)$ with the $-\log(p(C = 1, T = 1))$ we get the $nPMI(C = 1, T = 1)$. Finally, for each cluster C, we can order the terms T (or concepts) that describe it better, using these $nPMI$ scores. The closer to 1 the scores are, the better for this cluster the terms are.

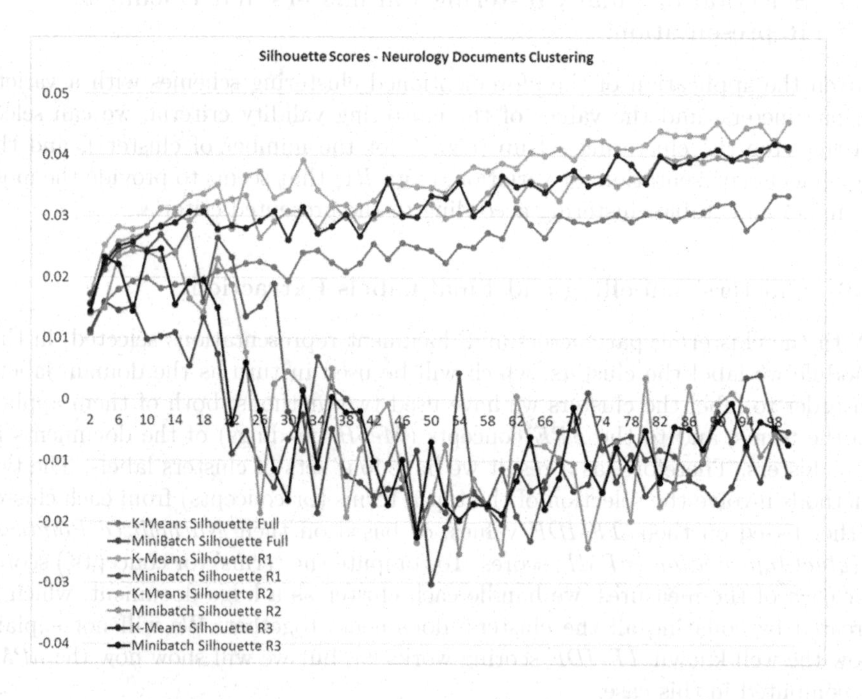

Fig. 2. Silhouette scores of k-means and *Minibatch* k-means using the 4 document representations, normalized *TF-IDF* vectors and $2 \leq k \leq 100$ and a step of 2.

4 Experimental Setup and Results

For our case study we collected 6,357 full texts of book chapters from fundamental *Elsevier*'s books in the field of *neurology* and *neuroscience*. We then applied the methodology presented in the previous section, in order to find a balanced and compact clustering of the documents from which we can label the clusters and extract the domain terminology. In the following we summarize our findings.

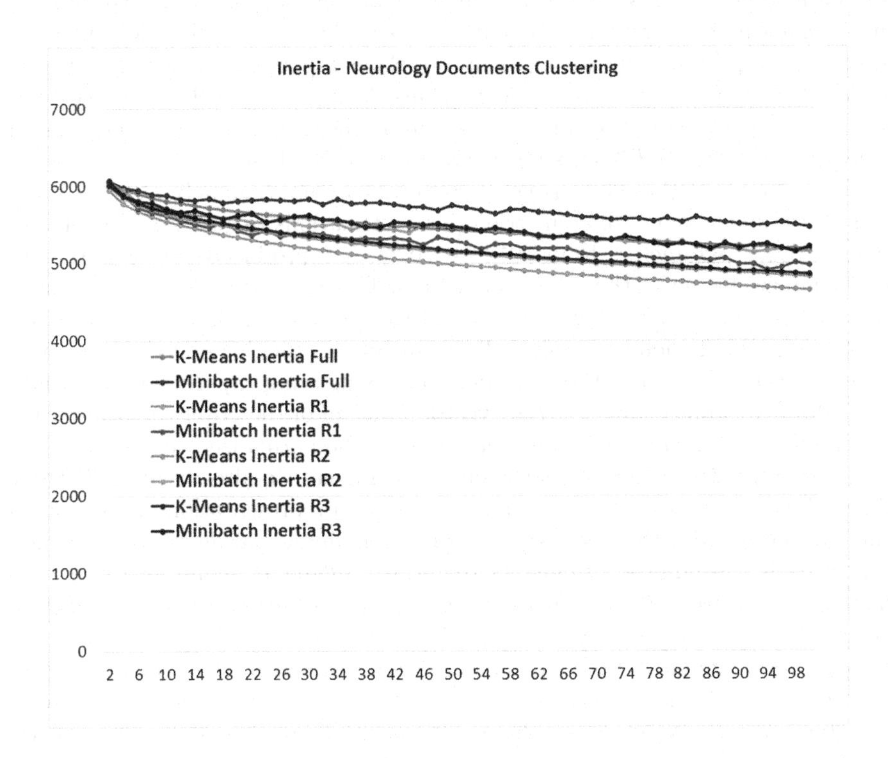

Fig. 3. Inertia scores of k-means and *Minibatch* k-means using the 4 document representations, normalized *TF-IDF* vectors and $2 \leq k \leq 100$ and a step of 2.

4.1 Clustering Performance and Documents' Best Representation

In Fig. 2 we present the *Silhouette* scores for the k-means and *Minibatch* k-means algorithms, introduced in Sect. 3.3, with k ranging from 2 to 100 and a step of 2, for all of the 4 representations, using normalization of the *TF-IDF* vectors on the vector length. Inertia scores for the same setups are shown in Fig. 3. Given that higher *Silhouette* scores indicate better cluster separation, and lower *inertia* scores more compact clusters, we can observe the following from these results: (1) the k-means algorithm provides much better clusters than the *Minibatch* k-means, (2) the *R1* representation, i.e., using the *MeSH* concept fingerprints from the *FPE*, seems to enable a better clustering of the documents overall, and, (3) the incorporation of terms, in addition to the concepts (i.e., representations $R2$ and $R3$) from section titles and keywords, almost always improves the clustering results compared to the baseline representation ($R0$ or *Full*).

Similar conclusions can be drawn by looking at the top plot of Fig. 4 which shows the *DB* index of the k-means algorithm for the same setup. The $R1$-$R3$ representations are performing consistently better than the $R0$ baseline, therefore indicating that clustering with the usage of the conceptual fingerprints produces much more stable and compact clusters. The rest of the plots in Fig. 4 are evaluating the n-gram variation on $R1$-$R3$, compared to their unigram variations. It is quite evident that the n-grams do not add much to the stability and the compactness of the clusters. Therefore, the simpler and smaller unigram representations of $R1$-$R3$ are sufficiently good for the task.

Regarding the other clustering schemes that we used, namely *DBSCAN* and *Meanshift*, under many different tried parametrizations (e.g., for $DBSCAN$ we examined eps parameters from 10^{-10} to 10^5) they produced one or maximum two different clusters of the 6, 357 documents. This means that the density-based approaches, in such a narrow domain, fail to separate the document space and result to one big region containing all documents, as the document space is very tight. We also tried all of the aforementioned clustering experiments with non-normalized versions of the *TF-IDF* vectors, but all of the criteria in these cases indicated worse clustering output compared to the normalized representations.

Given that with the aforementioned experiments we already identified that simple k-means works better with a concept-based representation of documents, and normalized *TF-IDF* vectors, the only remaining parameter for the final clustering is the number of clusters to retain. *Silhouette* scores in Fig. 2 show that for k between 20 and 40 we have very good cluster separation. *Inertia* inevitably always drops as the number of clusters increases. *DB* also shows great stability for a similar k, with values more than 0.5, which means there is a very good clustering for these setups. An alternative option according to *DB* would have been $k > 70$, which would greatly increase the number of clusters and the number of labels to be extracted. Therefore, for the final clustering we retained $k = 34$, after also manually exploring what the clusters look like for the values of k between 20 and 40.

4.2 Extraction of Clusters' Labels

For the quality of the extracted clusters' labels, which can be used as domain labels, we consulted the opinion of *Elsevier*'s editors of the most prominent journals in the *neurology* and *neuroscience* domains. The editors received a column formatted file with four sets of 10 labels per cluster: two providing terms as labels using the top *TF-IDF* and the top *nPMI* terms, and two providing concepts as labels, using the top *TF-IDF* and the top *nPMI* concepts. All of the editors preferred the labels that were produced from the top-*nPMI* selection, from the *MeSH* concepts. As an example, here are the 10 *nPMI* labels produced from

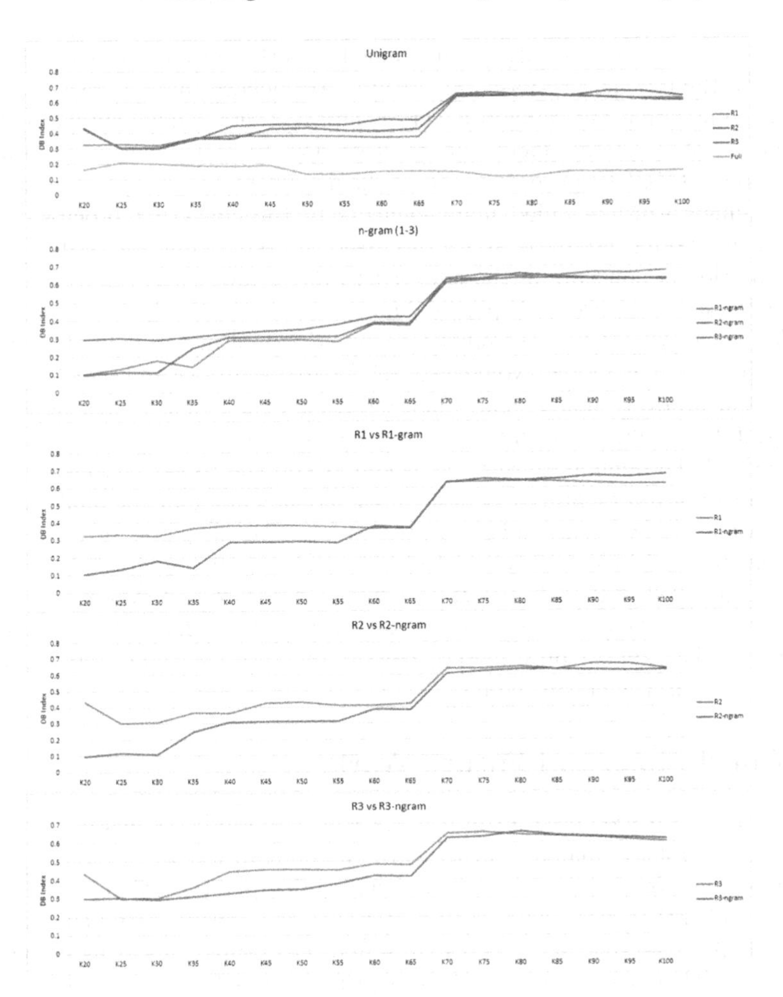

Fig. 4. *DB* index for k-means using the 4 document representations, normalized *TF-IDF* vectors and $20 \leq k \leq 100$ with a step of 2.

a cluster with 233 documents: *transference (psychology)*, *imagery (psychotherapy)*, *psychoanalysis*, *countertransference (psychology)*, *generalization (psychology)*, *free association*, *conscience*, *marital therapy*, *cathexis*, and *reinforcement*. The corresponding term labels were: *therapist*, *treatment*, *patient*, *therapy*, *behavior*, *client*, *patients*, *anxiety*, *study*, and *psychotherapy*. It is quite obvious in this case that the first set is much more informative and provides better quality labels.

5 Summarizing the Best Practices

We employed several techniques to cluster very similar articles within the neurology domain and also to learn appropriate concept labels automatically from each cluster. The k-means algorithm seemed to perform well, compared to *Mini-batch* and the two density-based algorithms. We believe that the reason for this lies in the fact that the domain is very compact, there is huge overlap between the documents, and the documents lie very close to each other, even in a high dimensional *TF-IDF* space. In terms of the document's representation, the conceptual fingerprinting added many benefits to the clustering process. This is not a new finding, however, as we discussed in Sect. 2, but we verify that the incorporation of concepts in the clustering process aids a lot, also when we are addressing a very dense and specific domain.

We would, therefore, recommend to practitioners in the field that in such dense domains they should avoid the density-based solutions and they should find ways to incorporate background knowledge, e.g., concepts, for the documents' representation, perhaps using a general domain taxonomy as a basis. For the diagnosis of the domain density, we would advise the usage of a measure such as cosine, on a representative sample of the documents, and the analysis of the distribution of the cosine values measured between all pairs. In terms of the labelling process, selecting labels from the conceptual fingerprints seemed to have worked well, using the *nPMI* as selection measure and ranking score. The reason for this is that the *nPMI* tends to select exclusive labels for each cluster, i.e., labels that appear only in that cluster, whereas using *TF-IDF* results in great overlap in the cluster labels.

6 Conclusions and Future Work

In this paper we presented an in-use *Elsevier* case study to cluster documents from the neurology domain and to learn the domain labels. Several clustering schemes were compared and evaluated using different textual representations. We concluded that k-means worked better than the density-based clustering for this task. A large range of k was tested to find the optimal clustering boundaries. We showed that using conceptual fingerprints in the document representation works best, and the concept labels are more compact. A great advantage of the concept labels is that they can directly be mapped to different taxonomies and a structure can be extracted or derived easily. It also became apparent that clustering documents from such a dense domain creates the challenge of highly overlapping clustering regions. This motivates us to experiment in the future with clustering techniques that are based on the *Latent Dirichlet Allocation* (*LDA*), which can assign a document to more than one cluster, and also in other domains.

References

1. Aggarwal, C.C., Zhai, C.: A survey of text clustering algorithms. In: Aggarwal, C.C., Zhai, C. (eds.) Mining Text Data, pp. 77–128. Springer, Heidelberg (2012)
2. Batet, M., Valls, A., Gibert, K., Sánchez, D.: Semantic clustering using multiple ontologies. In: Proceedings of the 13th International Conference of the Catalan Association for Artificial Intelligence, pp. 207–216 (2010)

3. Dagher, G.G., Fung, B.C.: Subject-based semantic document clustering for digital forensic investigations. Data Knowl. Eng. **86**, 224–241 (2013)
4. Davies, D.L., Bouldin, D.W.: A cluster separation measure. IEEE Trans. Pattern Anal. Mach. Intell. **1**(2), 224–227 (1979)
5. Dietze, H., Schroeder, M.: GoWeb: a semantic search engine for the life science web. BMC Bioinform. **10**(S–10), 7 (2009)
6. Fodeh, S.J., Punch, W.F., Tan, P.: On ontology-driven document clustering using core semantic features. Knowl. Inf. Syst. **28**(2), 395–421 (2011)
7. Funk, C., Baumgartner, W., Garcia, B., Roeder, C., Bada, M., Cohen, K.B., Hunter, L.E., Verspoor, K.: Large-scale biomedical concept recognition: an evaluation of current automatic annotators and their parameters. BMC Bioinform. **15**(1), 1–29 (2014)
8. Halkidi, M., Batistakis, Y., Vazirgiannis, M.: On clustering validation techniques. J. Intell. Inf. Syst. **17**(2–3), 107–145 (2001)
9. Manning, C.D., Raghavan, P., Schütze, H.: Introduction to Information Retrieval. Cambridge University Press, Cambridge (2008)
10. Nasir, J.A., Varlamis, I., Karim, A., Tsatsaronis, G.: Semantic smoothing for text clustering. Knowl.-Based Syst. **54**, 216–229 (2013)
11. Rousseeuw, P.J.: Silhouettes: a graphical aid to the interpretation and validation of cluster analysis. J. Comput. Appl. Math. **20**, 53–65 (1987)
12. Staab, S., Hotho, A.: Ontology-based text document clustering. In: Kłopotek, M.A., Wierzchoń, S.T., Trojanowski, K. (eds.) IIPWM 2003, pp. 451–452. Springer, Heidelberg (2003)
13. Tsatsaronis, G., Balikas, G., Malakasiotis, P., Partalas, I., Zschunke, M., Alvers, M.R., Weissenborn, D., Krithara, A., Petridis, S., Polychronopoulos, D., Almirantis, Y., Pavlopoulos, J., Baskiotis, N., Gallinari, P., Artières, T., Ngonga, A., Heino, N., Gaussier, É., Barrio-Alvers, L., Schroeder, M., Androutsopoulos, I., Paliouras, G.: An overview of the BIOASQ large-scale biomedical semantic indexing and question answering competition. BMC Bioinform. **16**, 138 (2015)
14. Vestdam, T., Rasmussen, H., Doornenbal, M.: Black magic meta data - get a glimpse behind the scene. Procedia Comput. Sci. **33**, 239–244 (2014)
15. Willet, P.: Document clustering using an inverted file approach. J. Inf. Sci. **2**, 223–231 (1980)
16. Zhao,Y., Karypis, G.: Topic-driven clustering for document datasets. In: Proceedings of the SDM, pp. 358–369 (2005)

Semantic Business Process Regulatory Compliance Checking Using LegalRuleML

Guido Governatori[1(✉)], Mustafa Hashmi[1], Ho-Pun Lam[1], Serena Villata[2], and Monica Palmirani[3]

[1] Data61, CSIRO, Spring Hill, QLD 4000, Australia
{guido.governatori,mustafa.hashmi,ho-pun.lam}@data61.csiro.au
[2] Université Côte d'Azur, CNRS, Inria, I3S, Rocquencourt, France
villata@i3s.unice.fr
[3] CIRSFID, University of Bologna, Bologna, Italy
monica.palmirani@unibo.it

Abstract. Legal documents are the source of norms, guidelines, and rules that often feed into different applications. In this perspective, to foster the need of development and deployment of different applications, it is important to have a sufficiently expressive conceptual framework such that various heterogeneous aspects of norms can be modeled and reasoned with. In this paper, we investigate how to exploit Semantic Web technologies and languages, such as LegalRuleML, to model a legal document. We show how the semantic annotations can be used to empower a business process (regulatory) compliance system and discuss the challenges of adapting a semantic approach to legal domain.

1 Introduction

Business Process Management (BPM) is a set of methodologies to capture, model and control in an integrated way all those activities that take place in an environment defining an enterprise [6]. Companies are subject to regulations. Non-compliance to such regulations would not only affect the added-value of the business processes, but may also result in judiciary pursuits. The scope of norms is to regulate the behaviour of their subjects and to define what is legal and what is not [18]. In BPM, checking the *compliance* of a business process with respect to a set of relevant regulations means to identify whether a process violates or not a set of norms. Consequently, to ensure business processes are compliant we need two components: (i) a conceptually sound formal representation of a business process, and (ii) a conceptually sound formalism to model and reason with the norms derived by the regulations. The task of modelling legal norms requires substantial human effort and powerful languages to capture the semantics of the normative systems and their dynamics. This is one of the reasons why existing compliance frameworks [7,21] are not fully satisfiable for companies.

Supported by EU H2020 research and innovation programme under the Marie Skłodowska-Curie grant agreement No. 690974 for the project *MIREL: MIning and REasoning with Legal texts.*

E. Blomqvist et al. (Eds.): EKAW 2016, LNAI 10024, pp. 746–761, 2016.
DOI: 10.1007/978-3-319-49004-5_48

We present an application of the semantic business process regulatory compliance checking where we rely on the semantics of LegalRuleML [1,2] for the representation of the norms and their dynamics. We discuss and analyse different but comparable ways to model the semantics of norms as well as their dynamics (e.g., new versions of certain regulations are proposed). Moreover, we show how this semantic modelling phase, with tasks coupled with semantic annotations, can be exploited to address and improve the regulatory compliance checking process, and answer companies' needs about compliance checking.

We experiment our approach on two versions of the Australian Telecommunications Consumer Protections Code[1] (hereafter, the Code). Our evaluation, in collaboration with an industry partner whose details cannot be disclosed for commercial reasons, shows that the proposed approach overcomes some of the drawbacks of standard non-semantic approaches add example and references to compliance checking.

The paper is organised as follows. In Sect. 2, we describe the context in which our approach has been conceived, answering the needs expressed by an industry partner. Section 3 describes how we model norms using LegalRuleML, and how we exploit the semantics of LegalRuleML to perform semantic regulatory compliance checking. In Sect. 4, we report on the results of the evaluation of the system, and discuss the insights inferred from this experience Sect. 4.2. Finally, we compare with the related literature in Sect. 5 and draw some conclusions.

2 Business Process Compliance

Regulatory compliance is a set of activities that aims to ensure that organisations' core business processes do not violate relevant regulations, in the jurisdiction in which the business is situated, governing the (industry) sectors where the organisation operates. Essentially, compliance connects two distinct domains: the legal domain and the business process domain.

Legal domain describes the legal boundaries for organisations by imposing conditions that detail which actions can be considered legal and which actions must be avoided during the execution of business process to stay compliant. Such legal boundaries can stem from normative documents (e.g., a code, bill, or an act) or organisation's internal policies (e.g., strategy documents or internal controls).

Business process domain, on the other hand, details how business activities should be carried out. A business process is a self-contained, temporally ordered set of activities describing how a process should be executed to achieve a business goal. Typically, it describes what needs to be done and when (control-flow), what resources is needed, who is/are involved (data and time), etc [17]. Many different formalisms (e.g., Petri-Nets. Process Algebra, ...) and notations (e.g., BPMN,YAWL, EPC, ...) have been proposed to present business processes. Apart from the differences in notations, typically a business process language is composed of the following minimal set of elements, namely: *tasks* (representing

[1] http://www.commsalliance.com.au/Documents/all/codes/c628.

complex business activities), *connectors* (defining the relationships among the tasks of the process, i.e., sequence, AND-Split, AND-Join, XOR-Split, XOR-Join). The combination of tasks and connectors defines the possible ways in which a process can be executed–where a possible way, called *trace*, is a sequence of tasks executed by respecting the order given by the connectors.

However, compliance is not only about the tasks that an organisation has to perform to achieve its business objectives. It concerns also on their effects (i.e., how the activities in the tasks changes the environment in which they operate), and the artefacts produced by the tasks (e.g., the data resulting from executing a task or modified by the task) [17]. Hence, to check whether a business process complies with the relevant regulations, an annotated business process model and the formal representation of regulations is needed. Accordingly, Governatori and Sadiq [23] introduced the idea of *compliance-by-design* in which business processes are supplemented with additional information (by means of annotations representing the formalised regulations) to ensure that a business process is compliant with relevant normative frameworks before its actual deployment[2].

We report the results of a project in cooperation with an industry partner (a small-to-medium Australian Telco with 50K–100K customers) subject to the Code. The main objective is to exploit semantic technologies to empower the compliance-by-design methodology mentioned above. More precisely, our objectives are as follows:

- Model the regulatory code as well as its dynamics using a (machine-readable) semantic framework such that differences and connections between two versions of the code (namely, 2012 and 2016) can be automatically identified;
- Capturing the tasks, their effects, and the artefacts resulting from them by means of *semantic annotations*; and
- Extend the architecture of the *Regorous Process Designer*, a business process compliance checker based on the compliance-by-design approach proposed in [15], to account for the semantic annotations.

3 The Framework

We first present an overview of LegalRuleML (Sect. 3.1), and explain how it can be exploited to model the Code (Sect. 3.2). Finally, we describe the semantic annotations in LegalRuleML we associated to the tasks of the processes, and how they are used to address semantic regulatory compliance checking (Sect. 3.3).

3.1 LegalRuleML: An Overview

LegalRuleML[3] is an effort to create a standard for the representation of norms[4]. It builds on the experience of RuleML to provide a rule representation language

[2] There are other approaches to compliance checking, namely: run-time and post-execution approaches, see [18] for details.

[3] https://tools.oasis-open.org/version-control/browse/wsvn/legalruleml/trunk/schemas/rdfs/modules/#_trunk_schemas_rdfs_modules.

[4] At the time of writing LegalRuleML is just about to enter in its public review phase.

and, at the same time, extends RuleML following the principles and guidelines proposed in [9] for rule language and legal reasoning. In particular, LegalRuleML offers facilities to model different types of norms (described below), deontic effects (e.g., obligations, prohibitions, permissions), and can specify preferences among them. In addition, it has features to capture the metadata of norms and other normative elements (such as jurisdiction, authorities, validity times, etc.), and has mechanisms to implement the so called *legal isomorphism* [3] principle that establishes the connection between a legal source or a norm, and the corresponding formal representation.

Accordingly, a LegalRuleML document consists ot (Fig. 1): *metadata*, *statements* and *contexts*. The metadata part is meant to contain the legal sources of the norms modelled by the document, and information about the (legal) temporal properties of the sources and the document itself, the jurisdiction where the norms are valid, and eventually details describing the authorities, authors, ... for the legal sources and the document.

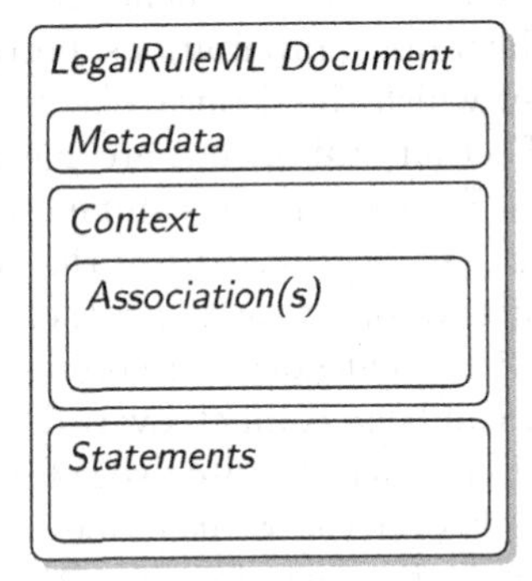

Fig. 1. LegalRuleML document structure.

The statements part contains the formal representation of the norms in form of rules or other expressions supported by the language, as depicted in Fig. 2.

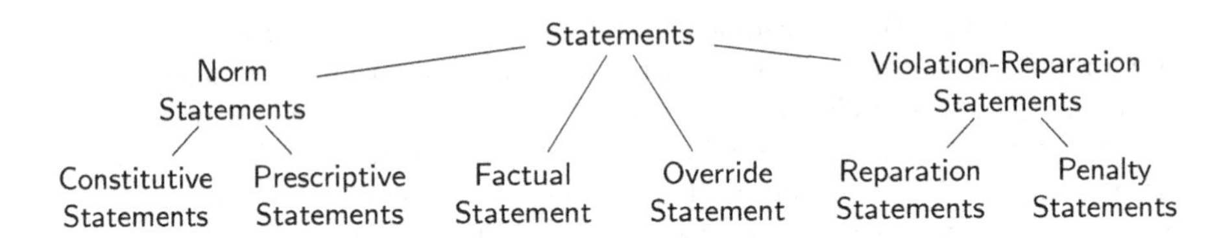

Fig. 2. Types of statements in LegalRuleML (adopted from [20]).

Normative statements follow the well known distinction of *constitutive* statements (rules) and *prescriptive* statements (rules) [24]. Constitutive rules are used to provide the definitions of the terms used in the document. For example, Chap. 2 of the Code provides the definitions of the terms used in the rest of Code. Often in legal documents, terms are defined defeasibly, thus the definition gives the base conditions that can be further extended or are subject to exceptions, e.g., "Complaint" is defined as:

> An expression of dissatisfaction made to a Supplier in relation to its Telecommunications Products or the complaints handling process itself, where a response or Resolution is explicitly or implicitly expected by the Consumer.
> An initial call to a provider to request a service or information or to request support is not necessarily a Complaint. An initial call to report a fault or service

> difficulty is not a Complaint. However, if a Customer advises that they want this initial call treated as a Complaint, the Supplier will also treat this initial call as a Complaint. If a Supplier is uncertain, a Supplier must ask a Customer if they wish to make a Complaint and must rely on the Customer's response.

Given the nature of the definition of the terms in legal reasoning, the definition of the terms can be captured by defeasible rules. In other terms, the constitutive rules provide the internal (defeasible) ontology used by the LegalRuleML document.

Prescriptive statements are rules that determine the deontic behaviour (such as obligations, prohibitions, permission) of the system and provide the conditions under which the deontic effects are in force.

Factual statements are meant to capture facts that are relevant in given cases. For instance, it can be used to specify that a particular manifestation of a norm (i.e., Section 8.4.1 of the 2012 version of the Code) is the same as another manifestation of the norm (i.e., Section 8.3.1 of the 2016 version of the Code).

Given that norms are represented by defeasible rules, and that two defeasible rules can be in conflict, override statements can be used to resolve the conflicts by specifying that, in case two rules in conflict fire at the same time, the stronger rule prevails over the weaker rule. Finally, violation and reparation statements offer convenient ways to formalise the penalties that can potentially apply for breaches of norms and information about how the violated norms can be compensated.

3.2 Modelling the Code and Its Dynamics

The Telecommunication Consumer Protections Code is the Australian industry code for the telecommunication industry and mandates that every operator has to provide annual compliance statements with the Code. The Code was enacted in September 2012, and it entered in force in April 2013. In 2015, the Code was revised and a new version enacted with some amendments entered in force in 2016. In this paper, we consider the 2012 and 2016 versions of the Code, and model them using LegalRuleML[5].

The first (and simplest) option is to model two versions of the Code using two separate LegalRuleML documents. However, after we compare the two versions of the Code, we realise that while there are differences, the vast majority of the definitions and prescriptions are just the same. Thus, modelling with this option will result in a large among of duplicated statements (rules) with exactly the same structure and meaning.

The second option is to utilise LegalRuleML's features to link statements with their legal sources. To be able to do this, we have first create a set of statements covering all the rules that can be inherited from the Code, irrespective to which version of the Code the rules has been mentioned, as shown below[6].

[5] The data modelled in LegalRuleML is available at: https://dl.dropboxusercontent.com/u/15116330/EKAW-dataset-legalruleml.zip.

[6] See Sect. 3.3 for how to model norms in LegalRuleML.

```
<lrml:Statements>
  <lrml:PrescriptiveStatement key="tcpc_ps1">
    ...
  </lrml:PrescriptiveStatement>
  <lrml:PrescriptiveStatement key="tcpc_ps2">
    ...
  </lrml:PrescriptiveStatement>
  <lrml:PrescriptiveStatement key="tcpc_ps3">
    ...
  </lrml:PrescriptiveStatement>
  ...
</lrml:Statements>
```

Next, we need to include the legal sources information about the two versions of the Code in the metadata section of the LegalRuleML document.

```
<lrml:LegalSources key="ls1">
  <lrml:LegalSource key="tcpc2012"
    sameAs="http://www.commsalliance.com.au/Documents/all/codes/c628#2012"/>
  <lrml:LegalSource key="tcpc2016"
    sameAs="http://www.commsalliance.com.au/Documents/all/codes/c628#2016"/>
</lrml:LegalSources>
```

Finally, we can make use of the `<lrml:Context>` element to create the associations between the legal sources and the statements, one for each version of the Code.

```
<lrml:Context key="tcpc2012-as">
  <lrml:Associations>
    <lrml:Association>
      <lrml:appliesSouce keyref="#tcpc2012/section\,8.2.1">
      <lrml:toTarget keyref="#tcpc_ps1">
    </lrml:Association>
    <lrml:Association>
      <lrml:appliesSouce keyref="#tcpc2012/section\,8.4.1">
      <lrml:toTarget keyref="tcps_ps2">
    </lrml:Association>
    ...
  </lrml:Associations>
<lrml:Context>

<lrml:Context key="tcpc2016-as">
  <lrml:Associations>
    <lrml:Association>
      <lrml:appliesSouce keyref="#tcpc2016/section\,8.2.1">
      <lrml:toTarget keyref="#tcpc_ps1">
    </lrml:Association>
    <lrml:Association>
      <lrml:appliesSouce keyref="#tcpc2016/section\,8.4.1">
      <lrml:toTarget keyref="#tcpc_ps3">
    </lrml:Association>
    ...
  </lrml:Associations>
</lrml:Context>
```

As can be seen from the snippet above, Section 8.2.1 is the same in both versions of the Code and is modelled by the same rule `tcpc_ps1`. On the other hand, Section 8.4.1 is different in the two versions of the Code. Thus the 2012 version is represented by the rule `tcpc_ps2`, while the 2016 by rule `tcpc_ps3`.

This option does not require us to duplicate the set of rules that are common among different versions of the legal document. The trade off is that we have to list all the associations for all the provisions in the Code and the corresponding rules. A more compact alternative would be to have a single association for each context with a single source (the entire version of the Code) and multiple targets (the rules corresponding to that version). However, this alternative has the drawback to loose the semantic information about the relationship between the sections of the Code and corresponding rules (syntactically, the correspondence could be regained by establishing a schema for labelling the key of the rules).

3.3 Business Process Regulatory Compliance

The set of traces T of a given business process describes the behaviour of the process insofar as it provides a description of all possible ways in which the process can be correctly executed. To check the semantic regulatory compliance of a process, we consider it as the set of its traces. The set of norms could vary from a particular regulation, to a specific statutory act, to a set of best practices, a standard, or simply a policy internal to an organisation or a combination of these types of prescriptive documents.

In this section, we provide an overview of the Regorous Process Designer, a business process regulatory compliance checker [11,16], and how we extended it by enriching both the representation of norms and the business process tasks with the LegalRuleML semantic model. Starting from the norms modelled in LegalRuleML presented in Sect. 3.2, we need now to add such semantic annotations in LegalRuleML to the tasks of the process, using them to record the data, resources and other information related to the single tasks in a process.

For the formal representation of the regulations, Regorous uses PCL (Process Compliance Logic) [10,14]. PCL is a simple, efficient, flexible rule-based logic, obtained from the combination of defeasible logic (for the efficient treatment of exceptions which are quite common in normative reasoning), and a deontic logic of violations. In PCL, a norm is represented by a rule of the kind $a_1, \ldots, a_n \Rightarrow c$ where $a_1, \ldots, a_n$ are the conditions of applicability of the rule and c is the *normative effect* of the rule. PCL distinguishes two normative effects: the first is that of introducing a definition for a new term, e.g., the following rule from the Code (2012) specifies that if a Customer requests information about a Complaint, then it is deemed a consumer complaint activity.

$$complaint, requestInformation \Rightarrow consumerComplaintActivity$$

The second normative effect is that of triggering obligations and other deontic notions. For obligations and permissions, we use the following notations:

- [P]p: p is permitted;
- [OP]p: p is a punctual obligation;
- [OM]p: p is a maintenance obligation;
- [OAPP]p: p an achievement preemptive perdurant obligation;
- [OAPNP]p: p is an achievement preemptive non-perdurant obligation;
- [OANPP]p: p an achievement non preemptive perdurant obligation;
- [OANPNP]p: p is an achievement non preemptive non-perdurant obligation;
- [OM]$\neg p$: p is prohibited.

Rules involving obligations and permissions are a bit more complex. Let us consider the following example from the Code (Section 8.1.1.a.x.E): "The supplier must implement, operate and comply with a Complaint handling process that is transparent, including prohibiting a Supplier from cancelling a Consumer's Telecommunications Service only because, being unable to Resolve a Complaint with their Supplier, that Consumer pursued their options for external dispute resolution". This provision is translated into PCL in the following rule:

$$\neg resolution, complaint, externalDisputeResolution \Rightarrow [\mathrm{OM}]\neg terminateService.$$

This rule establishes that in case there is a complaint (*complaint*) that has not been resolved to the satisfaction to the consumer ($\neg resolution$) and the consumer opted for the external dispute resolution option (*externalDisputeResolution*), then the provider has the prohibition to terminate the service (*terminateService*). For full description of PCL and its features, see [10, 14].

The above rule is translated using the LegalRuleML semantic model as follows:

```
<lrml:PrescriptiveStatement key="ps_tcpc_8_1_1_a_x_E">
  <ruleml:Rule key="tcpc_8_1_1_a_x_E">
    <lrml:Paraphrase>The supplier must implement, operate and comply with a
      Complaint handling process that is transparent, including E. prohibiting
      a Supplier from canceling a Consumer's Telecommunications Service only
      because, being unable to Resolve a Complaint with their Supplier, that
      Consumer pursued their options for external dispute resolution.
    </lrml:Paraphrase>
    <lrml:hasStrength>
      <lrml:DefeasibleStrength
        iri="http://spin.nicta.com.au/spindle/ruleStrength#defeasible"/>
    </lrml:hasStrength>
    <ruleml:if>
      <ruleml:And>
        <ruleml:Neg>
          <ruleml:Atom>
            <ruleml:Rel>resolution</ruleml:Rel>
          </ruleml:Atom>
        </ruleml:Neg>
        <ruleml:Atom>
          <ruleml:Rel>complaint</ruleml:Rel>
        </ruleml:Atom>
        <ruleml:Atom>
          <ruleml:Rel>external dispute resolution</ruleml:Rel>
        </ruleml:Atom>
      </ruleml:And>
    </ruleml:if>
    <ruleml:then>
      <lrml:Obligation iri="http://test.org/deontic#OM">
        <ruleml:Neg>
          <ruleml:Atom>
            <ruleml:Rel>terminate Service</ruleml:Rel>
          </ruleml:Atom>
        </ruleml:Neg>
      </lrml:Obligation>
    </ruleml:then>
  </ruleml:Rule>
</lrml:PrescriptiveStatement>
```

Enriching the regulatory compliance system with a semantic representation of the regulations the processes have to be checked against presents many advantages, i.e., a more insightful and precise representation of the semantics of the norms, and the possibility to keep track of the regulations' dynamics. However, compliance is not just about the tasks to be executed but also on what the tasks do, the way they change the data and the state of the artefacts related to the process, and the resources linked to it. Accordingly, process models must be enriched with such information [23]. For this reason, we decided to enrich process models with *semantic* annotations using the LegalRuleML model. Each task in a process model is then associated with a set of semantic annotations in LegalRuleML, representing the effects of the task. The approach can be used to model business process data compliance [17]. The set of effects in PCL and in LegalRuleML is just a set of literals in the underlying language. PCL and LegalRuleML are agnostic about the nature of the literals they uses. They can represent tasks, i.e., activities executed in a process, or propositions representing state variables.

An example of task annotated using LegalRuleML is the following: suppose that the complaint handling process of a telco contains a task called "Record Complaint". The Code (Section 8.5 of the 2012 version, and Section 8.4 of the 2016 version) specifies what information should be recoded for a complaint. Thus, the task "Record Compliant" indicates that such an activity is to be performed once a compliant as been verified as such, but, the process alone does not specify what data is recorded. Thus, such process model must be extended with the appropriate information. Note that it is beyond the scope of this paper to study how the annotations are generated, i.e., manually based on domain experts knowledge of the process or by examining database schemas associated to the task or programming script executed by the task [17]. Specifically, for the task "Record Complaint" the following literals (from the literals defined for the LegalRuleML document) are recorded as annotation for the task:

```
<taskEffects elementId="usertask15">
  <ruleml:Atom>
    <ruleml:Rel>record special circumstances</ruleml:Rel>
  </ruleml:Atom>
  <ruleml:Atom>
    <ruleml:Rel>record complaint issue</ruleml:Rel>
  </ruleml:Atom>
  <ruleml:Atom>
    <ruleml:Rel>record resolution sought</ruleml:Rel>
  </ruleml:Atom>
  <ruleml:Atom>
    <ruleml:Rel>record due date</ruleml:Rel>
  </ruleml:Atom>
  <ruleml:Atom>
    <ruleml:Rel>record complaint cause</ruleml:Rel>
  </ruleml:Atom>
</taskEffects>
```

Given an annotated process and the formalisation of the relevant regulation in LegalRuleML as we shown above, we can use the algorithm proposed in [14] to determine whether the annotated process model is compliant. Shortly, the procedure runs as follows:

- Generate an execution trace of the process.
- Traverse the trace:

  - for each task in the trace, cumulate the effects of the task using an update semantics (i.e., if an effect in the current task conflicts with previous annotation, update using the effects of the current tasks).
  - use the set of cumulated effects to determine which obligations enter into force at the current tasks, by calling a reasoner.
  - add the obligations obtained from the previous step to the set of obligations carried over from the previous task.
  - determine which obligations have been fulfilled, violated, or are pending; and if there are violated obligation check whether they have been compensated.
- repeat for all traces.

A process is evaluated as compliant if and only if all traces are compliant (all obligations have been fulfilled or if violated they have been compensated), or it is evaluated as weakly compliant if there is at least one trace that is compliant.

Soundness and completeness of the proposed methodology depend on the data (rules and semantic annotations) associated with a business process. The methodology is sound and complete provided that the rules are an appropriate interpretation of the norms, and the semantic annotations are complete. If this is the case the computational model supported by PCL properly simulates legal reasoning. Otherwise, there are two possible issues. The process is not compliant because some semantic annotation is missing. This is the case of unfulfilled obligation, that is, there is some obligation $[\mathrm{OANPP}]p$ that is force in some tasks (trace) but we do not have evidence for p in the tasks (trace). In this case Regorous report such issue and the user can add the information to some tasks, if appropriate. For the other case, it is possible to avoid some obligations by failing to trigger some rules. For example, given the rule $p \Rightarrow [\mathrm{OANPP}]q$, we can avoid the obligation of q if p is not an effect of some task. To handle this situation, Regorous asks for justification for the rules that are not used in the process (and it is, again, up to the user to provide the information if appropriate, p could be facultative, and there is no need to have it).

4 Evaluation

In this section, we first present the evaluation of our approach (Sect. 4.1), and then we discuss the lessons learned (Sect. 4.2).

4.1 Results

The approach proposed in this paper has been evaluated in a six week pilot project in collaboration with an industry partner (a small to medium Australian telecommunication service provider, about 70,000 customers at the time of the evaluation), and the regulator. For the evaluation, Chap. 8 of the Code on complaint handling was selected. A legal knowledge engineer from our group manually mapped Chap. 8 of the 2012 version of code in LegalRuleML, and XSLT transformations are used to translate the LegalRuleML representation in PCL as used by Regorous. The mapping of Chap. 8 took approximately 2 weeks. The chapter contains approximately 100 paragraphs, in addition to approximately 120 terms given in the Definitions and Interpretation section of the Code (Chap. 2). The mapping resulted in 176 LegalRuleML normative statements, containing 223 distinct RuleML atoms (`<ruleml:Atom>`), and 7 LegalRuleML overrides statements (`<lrml:overrides>`). Of the 176 normative statements, 33 were constitutive statements (`<lrml:ConstitutiveStatment>`) used to capture definitions of terms used in the remaining rules. Mapping the section required all features of PCL. The regulator examined the mapping, and they deemed it to be a suitable interpretation of the Code.

For the second phase of the evaluation, we had a series of 1-day workshops with the industry partner. The industry partner did not have formalized business processes. Thus, we worked with domain experts from the industry partner (who had not been previously exposed to BPM technology, but who were familiar with the industry code) to draw process models for the activities covered by the Code. The evaluation was carried out in two steps. In the first part, we modelled the processes as they were. It took two workshops to teach them how to model business processes, and to jointly model their existing processes related to complaint handling and managements of complaints and complaints procedures. The third 1-day workshop was dedicated to add the semantic annotation to the business processes. The domain experts were able to complete the task in one afternoon after they were instructed on how to do in the morning.

Regorous was able to identify several areas where the existing processes were not compliant with the new code. In some cases the industry partner was already aware of some of the areas requiring modifications of the existing processes given that the Code introduced totally new requirements (for example, the need to address in person or by phone complaint immediately). However, some of the compliance issues discovered by the tool were novel to the business analysts and were identified as genuine non-compliance issues to be resolved. Some of these issues where due to subtle changes in the Code while others were discovered by the deep analysis forced by the methodology implemented by Regorous which would be hard to detect with manual analysis. In the final part of the experiment, the existing processes were modified to comply with the Code based on the issues identified in the first phase. In addition, a few new business process models required by the new Code were designed. As result, we generated and annotated 6 process models. 5 of the 6 models are limited in size and they can be checked for compliance in seconds. The largest process contains 41 tasks, 12 decision points, XOR-Splits, (11 binary, 1 ternary). The shortest path in the model has 6 tasks, while the longest path consists of 33 tasks (with 2 loops), and the longest path without loop is 22 task long. It takes approximately 40 s to verify compliance for this process on a MacBook Pro 2.2 GHz Intel Core i7 processor with 8 GB of RAM (limited to 4 GB in Eclipse).

Due to a confidentiality agreement with the industry partner it is not possible to release the process models used in the evaluation. However, the Regorous compliance checker is available under a free evaluation license at http://www.regorous.com. The distribution includes some simple but realistic scenarios consisting of business process models and fragments of relevant regulations. The scenarios can evaluate the compliance of simple processes (10–15 tasks, a few decision nodes, and about 30–40 rules) in a matter of seconds running in a standard laptop with the same specifications as the computer used for the evaluation. Note that the complexity of checking the compliance of a business process is in function of the complexity of the underlying business process model (and linear in the number of rules and propositions). The problem has been shown to be NP-complete even when the correctness of processes is in PTIME, and the legal reasoning in the single tasks of a process is in PTIME as well [4]. Given that the complexity of the reasoning tasks in PCL is in PTIME [13], the complexity of a business process depends on the number of states traversed by the traces of a business process, which is potentially exponential in the number of tasks appearing in a process in function of the control flows nodes (connectors). The most complex process in the pilot study consists of approximately 24,000 states and processes included in the Regorous distribution are between 50 and 250 states. Thus, while the rate states/response time for the samples scenario is one order of magnitude larger than that of the pilot case, the response time for the process in the pilot case can be also extrapolated from the response time from the other processes based on the theoretical results on complexity when one accounts for initialisation time, and some optimisations in the implementation that allows us to avoid the computation in states that are already computed. Specifically, the set of traces corresponding to a process is represented as a tree, thus the states that are common to multiple traces are computed only once. Thus, for example, in a process of 10 tasks in sequence followed by an XOR-Split with two branches, the 10 initial states are common to the two branches, and it is pointless to compute them twice.

4.2 Lessons Learned and Future Work

The results presented in the previous section demonstrate the effectiveness of the semantically enriched business process regulatory compliance checking mechanism we proposed. Besides these considerations, some further positive and negative insight emerged during the evaluation we conducted with the industry partner:

Positive Feedback: First, we discovered, together with the domain experts of the industry partner, that exploiting a semantic model such as LegalRuleML allows us to embed much more information in the rules representing the regulatory code. This enhancement in the precision of the legal provisions leads to an enhancement of the regulatory checking phase, as we compared two more fine-grained representations of the Code and of the tasks, respectively. Second, the semantics of LegalRuleML allows for the evolution over time of the regulations to be compliant with. This has the advantage of tracking when a change occurred, and what is the context to be used depending if the compliance is verified against the new or the old version of the Code. This is also a valuable benefit of the adoption of this semantic model with respect to simple rule-based formats.

Negative Feedback: Even if the semantic model allows to provide a more faithful representation of the legal document, it is not straightforward to understand the semantic model of LegalRuleML and how it works. It required some time to the domain experts of the industry partner to understand how to match the rules present in the Code they were aware of, and the LegalRuleML semantics. As future work, we need to develop a graphical interface to interact with such a complex model so that examples of rules in natural language from existing regulations are the industry partner showed us that users are much better in using semantically enriched documents rather than in creating them. Finally, the showstopper is that the extraction of the rules from the legal documents is time consuming and there is a huge need to support the translation of such legal documents from natural language to their `xml` counterpart. This is another line we will address as future work.

5 Related Work

The problem of providing a machine-readable semantic representation of legal knowledge has been addressed in different domains, leading to the definition of various ontologies targeting different legal contexts. Among others, there are the Open Digital Rights Language (ODRL) for rights expressions[7], the Functional Ontology for Law [25] about normative knowledge, world knowledge, and responsibility knowledge, the Frame-Based Ontology of Law [19] about norms, acts and concepts descriptions, the IKF-IF-LEX Ontology for Norm Comparison [8] about agents, institutive and regulative norms, and norm dynamics, the LKIF-Core Ontology[8] including the OWL ontology of fundamental legal concepts [22]. However, all these works differ from LegalRuleML for what concerns *(i)* the use of rules to account for the specifics of the legal domain, and *(ii)* the use of a legal reasoning level on top of the ontological layer of the Semantic Web stack. LegalRuleML allows users to specify in different ways how legal documents evolve, and to keep track of these evolutions and connect them to each other, as we exploited in this paper.

To our knowledge, there is no other approach addressing the problem of a semantic business process regulatory compliance: in this work, we not only exploited Semantic Web technologies and languages to propose different modelling techniques to represent the legal information contained in the legal documents and their dynamics, i.e., the Code (2012 and 2016), but we empowered a business process compliance system with a semantic annotation of the rules and the processes. An approach to semantic business process compliance management has been proposed by El Kharbili and colleagues [5]. We share the idea of

[7] https://www.w3.org/ns/odrl/2/ODRL21.
[8] http://www.estrellaproject.org/lkif-core/.

making use of the advantages of semantic technologies for compliance management. However, the two approaches are different. First of all, we are interested in regulatory compliance, and not in business process management in general, which means that we need to exploit the powerful semantics of the LegalRuleML framework to convey the semantics of the rules extracted from the legal documents. We do not need to design a business policy and business rule ontology, as in [5]. Second, the proposed architecture considers also the dynamics of the legal documents to be checked the compliance with, by proposing alternative modeling solutions. Finally, we annotate the tasks included in the processes with the semantic of LegalRuleML, so that automated compliance checking is done at semantic level. Besides Regorous a few other compliance prototypes have been proposed. Here we consider some representative ones: Compass [7] and SeaFlows [21]. However, none of them exploits Semantic Web languages and technologies, and they are not compliant with the guidelines set up in [9] for rule languages for the representation of legal knowledge and legal reasoning. In addition, such approaches have severe limitations in modelling legal reasoning, since they do not provide a conceptually sound model of legal reasoning [12].

6 Conclusions

In this paper, we have presented a semantic approach to business process regulatory compliance checking. Regulatory compliance checking is a major challenge for companies and institutes, and being supported by automated techniques in such a verification phase results in a valuable gain of time and money. We have reported here our experience in applying Semantic Web technologies and languages to this challenging task in the context of a project with an industry partner. Accounting for the complexity and the required precision in modelling norms and regulations, and in checking whether a certain process and its related tasks are actually compliant with the normative system they are subject to, we propose a semantic approach based on the LegalRuleML semantic model. Our evaluation shows that our system is able to capture the semantics of the Code and to model its dynamics in a satisfactory way, and to efficiently check the compliance of processes with respect to this reference Code. The lessons learned during this project will guide our future work, that includes also the evaluation of the Regorous semantic system with larger processes, and applying this methodology with other kinds of regulations in order to make Regorous more flexible to the needs of the companies adopting it.

References

1. Athan, T., Boley, H., Governatori, G., Palmirani, M., Paschke, A., Wyner, A.: OASIS LegalRuleML. In: Francesconi, E., Verheij, B. (eds.) ICAIL 2013, pp. 3–12. ACM, New York (2013)
2. Athan, T., Governatori, G., Palmirani, M., Paschke, A., Wyner, A.: LegalRuleML: design principles and foundations. In: Faber, W., Paschke, A. (eds.) Reasoning Web 2015. LNCS, vol. 9203, pp. 151–188. Springer, Heidelberg (2015)

3. Bench-Capon, T., Coenen, F.P.: Isomorphism and legal knowledge based systems. Artif. Intell. Law **1**(1), 65–86 (1992)
4. Colombo Tosatto, S., Governatori, G., Kelsen, P.: Business process regulatory compliance is hard. IEEE Trans. Serv. Comput. **8**(6), 958–970 (2015)
5. Decreus, K., Poels, G., El Kharbili, M., Pulvermüller, E.: Policy-enabled goal-oriented requirements engineering for semantic business process management. Int. J. Intell. Syst. **25**(8), 784–812 (2010)
6. Dumas, M., La Rosa, M., Mendling, J., Reijers, H.A.: Fundamentals of Business Process Management. Springer, Heidelberg (2013)
7. Elgammal, A., Türetken, O., van den Heuvel, W.J.: Using patterns for the analysis and resolution of compliance violations. Int. J. Coop. Inf. Syst. **21**(1), 31–54 (2012)
8. Gangemi, A., Breuker, J.: Harmonizing legal ontologies. In: Deliverable 3,4 IST Project-2000-29243. Ontoweb (2002)
9. Gordon, T.F., Governatori, G., Rotolo, A.: Rules and norms: requirements for rule interchange languages in the legal domain. In: Governatori, G., Hall, J., Paschke, A. (eds.) RuleML 2009. LNCS, vol. 5858, pp. 282–296. Springer, Heidelberg (2009)
10. Governatori, G.: Representing business contracts in RuleML. Int. J. Coop. Inf. Syst. **14**(2–3), 181–216 (2005)
11. Governatori, G.: The Regorous approach to process compliance. In: EDOC 2015 Workshop, pp. 33–40. IEEE (2015)
12. Governatori, G., Hashmi, M.: No time for compliance. In: Hallé, S., Mayer, W. (eds.) EDOC 2015, pp. 9–18. IEEE, Piscataway (2015)
13. Governatori, G., Rotolo, A.: BIO logical agents: norms, beliefs, intentions in defeasible logic. J. Auton. Agents Multi Agent Syst. **17**(1), 36–69 (2008)
14. Governatori, G., Rotolo, A.: A conceptually rich model of business process compliance. In: Link, S., Ghose, A. (eds.) APCCM 2010, pp. 3–12. ACS, Washington, D.C. (2010)
15. Governatori, G., Sadiq, S.: The journey to business process compliance. In: Cardoso, J., van der Aalst, W. (eds.) Handbook of Research on BPM, pp. 426–454. IGI Global, Hershey (2009)
16. Governatori, G., Shek, S.: Regorous: a business process compliance checker. In: ICAIL 2013, pp. 245–246 (2013)
17. Hashmi, M., Governatori, G., Wynn, M.T.: Business process data compliance. In: Bikakis, A., Giurca, A. (eds.) RuleML 2012. LNCS, vol. 7438, pp. 32–46. Springer, Heidelberg (2012)
18. Hashmi, M., Governatori, G., Wynn, M.T.: Normative requirements for regulatory compliance: an abstract formal framework. Inf. Syst. Front. **18**(3), 429–455 (2016)
19. van Kralingen, R.: A conceptual frame-based ontology for the law. In: Proceedings of the 1st International Workshop on Legal Ontologies, pp. 6–17 (1997)
20. Lam, H.-P., Hashmi, M., Scofield, B.: Enabling reasoning with LegalRuleML. In: Alferes, J.J., Bertossi, L., Governatori, G., Fodor, P., Roman, D. (eds.) RuleML 2016. LNCS, vol. 9718, pp. 241–257. Springer, Heidelberg (2016). doi:10.1007/978-3-319-42019-6_16
21. Ly, L.T., Rinderle-Ma, S., Göser, K., Dadam, P.: On enabling integrated process compliance with semantic constraints in process management systems - requirements, challenges, solutions. Inf. Syst. Front. **14**(2), 195–219 (2012)
22. Rubino, R., Rotolo, A., Sartor, G.: An OWL ontology of fundamental legal concepts. In: van Engers, T.M. (ed.) JURIX 2006, pp. 101–110. IOS Press, Amsterdam (2006)

23. Sadiq, W., Governatori, G., Namiri, K.: Modeling control objectives for business process compliance. In: Alonso, G., Dadam, P., Rosemann, M. (eds.) BPM 2007. LNCS, vol. 4714, pp. 149–164. Springer, Heidelberg (2007)
24. Searle, J.: The Construction of Social Reality. The Free Press, New York (1996)
25. Valente, A., Breuker, J.: A functional ontology of law. Artif. Intell. Law **7**, 341–361 (1994)

An Open Repository Model for Acquiring Knowledge About Scientific Experiments

Martin J. O'Connor(✉), Marcos Martínez-Romero, Attila L. Egyedi, Debra Willrett, John Graybeal, and Mark A. Musen

Stanford Center for Biomedical Informatics Research, Stanford, CA 94305, USA
martin.oconnor@stanford.edu

Abstract. The availability of high-quality metadata is key to facilitating discovery in the large variety of scientific datasets that are increasingly becoming publicly available. However, despite the recent focus on metadata, the diversity of metadata representation formats and the poor support for semantic markup typically result in metadata that are of poor quality. There is a pressing need for a metadata representation format that provides strong interoperation capabilities together with robust semantic underpinnings. In this paper, we describe such a format, together with open-source Web-based tools that support the acquisition, search, and management of metadata. We outline an initial evaluation using metadata from a variety of biomedical repositories.

1 Introduction

To help tackle the reproducibility challenge in biomedical sciences, many funding agencies and journals are now demanding that experimental data are made publicly available [1]. These requirements have led to a dramatic increase in the availability of data sets derived from scientific experiments. High-quality metadata are seen as crucial to facilitate knowledge discovery with these data sets. The biomedical community has a strong history of tackling this metadata challenge by driving the development of *metadata templates* [2]. These templates focus on addressing the reproducibility challenge by providing detailed checklists of the metadata needed to describe particular types of experimental data sources. The key goal is to provide sufficient metadata to enable the source studies to be reproduced. A large number of public repositories have been built around these templates, greatly enhancing the ability of scientists to discover and share scientific knowledge [3–5].

While individual metadata templates can provide a standard format for a particular data source, they rarely share common structure or semantics. There is also a disconnect between the high-level checklist-based template definitions developed by scientific communities and the submission formats required by metadata repositories. Submission formats for biomedical repositories, for example, are typically spreadsheet-based, with a variety of *ad hoc* formats that require significant user effort to describe even very simple metadata content. These formats typically lack any standard way of semantically annotating templates. Despite the availability of a large number of

E. Blomqvist et al. (Eds.): EKAW 2016, LNAI 10024, pp. 762–777, 2016.
DOI: 10.1007/978-3-319-49004-5_49

controlled terminologies in biomedicine, submission templates have weak or nonexistent mechanisms for linking terms from these terminologies to metadata submissions. These difficulties combine to ensure that typical metadata submissions are poorly described and thus require significant post-processing to extract semantically useful content. There is a pressing need for a standardized approach for representing templates and semantically describing their content.

In this paper, we outline such an approach under development by the Center for Expanded Data Annotation and Retrieval (CEDAR) [6]. We define a lightweight standards-based template model that provides principled interoperation with Linked Open Data. We describe associated tools that use this model to provide an end-to-end workflow for metadata acquisition and management. The ultimate goal is to address the fragmented landscape of metadata submission tools and template formats.

2 Related Work

There are a number of ongoing research efforts to address the metadata challenge in the biomedical domain. In the United States, the National Institutes of Health's Big Data to Knowledge (BD2K) initiative [7] is funding an array of projects to tackle different dimensions of this challenge. The BD2K-funded BioCADDIE [12] project is tasked with building a search engine for metadata that describe a variety of experimental types. Our project, CEDAR [6], also supported by BD2K, focuses primarily on the metadata authoring process, and is building tools to enable users to easily create and submit high quality metadata.

These initiatives build on decades of work creating reusable templates. Early work involved the creation of *minimum information models* [2], primarily centered on describing metadata for laboratory experiments. A minimum information model specifies the core set of required and optional metadata for particular experiment types. These models serve as template building blocks and have spawned a cottage industry for defining community-based templates for a large variety of biological sources. These templates are used by a host of public biomedical repositories [3, 4]. Minimum information models by themselves, however, do not solve the metadata challenge. They often specify only a small core set of information and, typically, their individual fields do not require values from specific ontologies. As a result, the collected metadata can be sparsely populated and have low semantic value.

Several initiatives have concentrated on building tools to increase metadata quality. Foremost among these is the ISA Tools [8], which is a desktop application to allow curators to create spreadsheet-based submissions for metadata repositories. The Linked ISA [9] evolution of this tool provides a means to interoperate with Linked Open Data, effectively adding controlled term linkage to templates.

A key shortcoming is the absence of an open, interoperable format for metadata exchange. While the ISA Tools support the standard MAGE-TAB format [10], it is a spreadsheet-based representation with limited expressivity and extensibility. There is a need for an open format built on Web-based standards. The recently developed FAIR

Principles [11], which specify desirable properties of metadata, provide a set of important targets for this format to meet. Standard mechanisms of interoperating with Linked Open Data vocabularies are central to such a format.

3 Model and Implementation

We first developed a lightweight, abstract template model to specify the key aspects of template construction. This model represents the core structural characteristics of templates—the common entities and compositional patterns that define a template.

We then produced a concrete representation of the template model, emphasizing the addition of semantic markup and constraints. The concrete template model provides a consistent, interoperable information framework for defining templates and for creating and filling out metadata instances that correspond to those templates.

Finally, we developed a set of tools for creating metadata templates and for acquiring metadata to generate metadata instances.

3.1 Abstract Template Model

Our system needs to recursively compose templates from existing, more granular templates. In our model, we term these sub-templates *template elements*. Template elements constitute the building blocks of metadata templates. Template elements may contain one or more atomic pieces of information, such as a text or date field, or may be recursively composed from other template elements. *Template fields* are used to represent these atomic pieces of metadata. For example, a template field could be used to indicate the date at which a measurement was made for a particular scientific experiment. Template elements are used to recursively combine template fields or template elements to create more complex descriptions. For example, template fields Phone and Email could be contained in a template element called Contact Information, which could itself be contained in a template element called Person.

Figure 1 presents a basic overview of this model. The `Template`, `TemplateElement`, and `TemplateField` entities represent their namesake concepts. All entities have an `@type`[1] field that indicates the entity type, and are uniquely identifiable via an `@id` field. They also contain `title` and `description` fields. Template fields contain an `@value` field, which stores the field's value. A variety of built-in template field types are provided. These include a `TextField`, which represents a free text field, and a `ListField`, which represents a multiple-choice field. This set can be extended to incorporate additional field types. Both templates and template elements can optionally have fields or elements nested inside them. Template elements and fields can be grouped together in a `Template` to provide an overall description of a collection of metadata.

A *template instance* is created from a template. A template effectively serves as a structural specification of metadata instances conforming to that template.

[1] The @-prefixed notation follows JSON-LD; see Sect. 3.2.

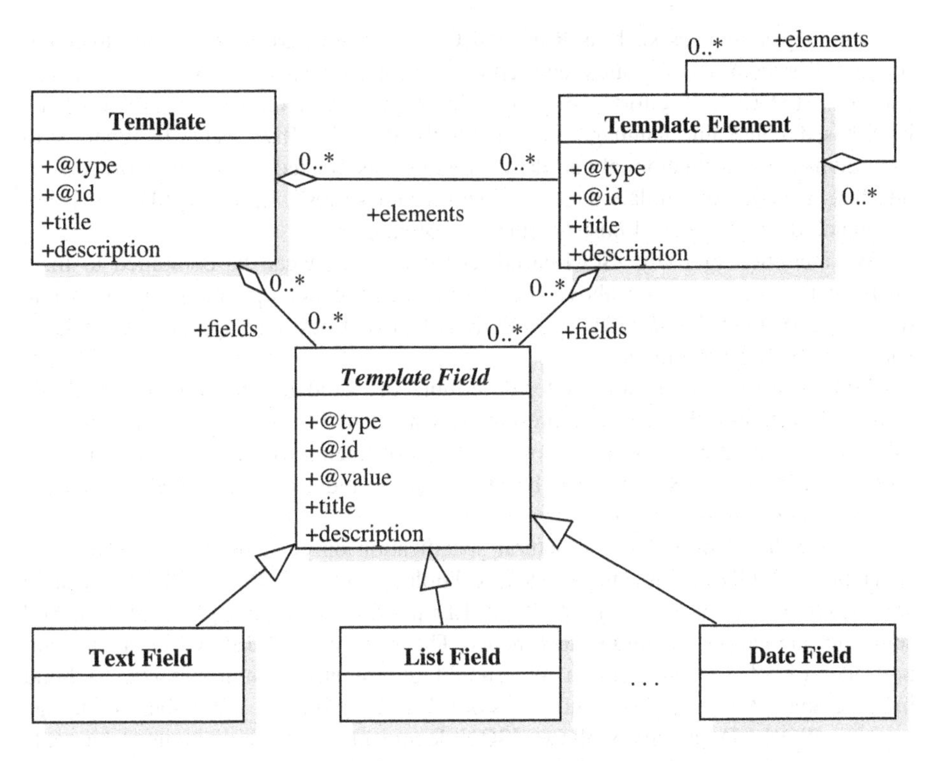

Fig. 1. Abstract template model showing main model components

This abstract model provides a structural specification of templates, elements and fields. In the next section we will outline the development of a concrete representation of this model. This representation extends beyond structural constraints and specifies how elements in the model are linked to controlled terms and how controlled-term–based value constraints can be specified for template instances.

3.2 Template Model Concrete Representation

The template model requires a machine-interpretable representation for software systems to work with the model programmatically. This representation must meet a variety of goals. Primarily, it must describe the structure of templates and the instances generated from these templates. It must also describe and constrain the various relationships between the entities in the model. Template representations must be conveniently serializable so that they can be provided via REST APIs and persisted to storage media. Ideally, the representation should be based on standard formats so that existing tools can be used to manage model entities. The representation should also permit easy validation, and easy indexing to support search. To enable interoperation with controlled terms, a standardized means to annotate templates with controlled terms is key. Finally, the template format must interoperate with Linked Open Data technologies such as RDF and OWL, and allow metadata to be represented as RDF graphs.

Ontology languages such as RDF and OWL are not particularly suited to representing the structural constraints required for this sort of data-centric problem [19]. The current SHACL [20] effort aims to address this shortcoming by providing an RDF-based constraint language to structurally describe RDF data. Its goal is to "communicate information about data structures associated with some process or interface, generate or validate data, or drive user interfaces". However, SHACL is not yet standardized and has almost no current tool support.

We identified two key JSON-based technologies that can be combined to meet many of the goals outlined above, while retaining full interoperation with semantic resources: JSON Schema [13], and JSON-LD [14]. Both are supported by a large variety of Web-centric tools.

JSON Schema is a technology for describing and validating the structure of JSON data. Its directives—themselves represented as standard JSON elements—can be used to provide a structural description of a JSON document. JSON documents that are specified with JSON Schema can be structurally validated against their associated schemas via off-the-shelf tools.

JSON Schema provides a structural specification only—it does not describe the semantics of JSON documents. A recent technology called JSON-LD ("Linked Data") was developed to meet this goal. JSON-LD provides a lightweight syntax to add semantic annotations to JSON documents. The key goals of JSON-LD are to support the use of Linked Data in Web-based programming environments, to build interoperable Web services, and to store Linked Data in JSON-based storage engines. JSON-LD effectively allows JSON documents and their contents to be made available as Linked Data, offering the potential for machine-interpretable RDF semantics [21].

We first outline how we use JSON Schema to describe the structure of templates and to constrain and validate the template instances generated from those templates. We then describe how we use JSON-LD to mark up these structural specifications, adding semantic content to these templates and instances. We show how this combination of JSON Schema and JSON-LD provides the capabilities to fully represent the template model and provide a strong bridge to semantic technologies[2].

3.2.1 Representing Template Structure Using JSON Schema

With JSON Schema we define the structure of the primary entities in the CEDAR template model. We first outline its use to define the three core entities in the model: template fields, template elements, and templates.

Representing Template Fields Using JSON Schema. Template fields are used to describe an atomic piece of metadata. Informally, they correspond to a single field in a form, which when filled out contains a single value. In principle, a template field could be stored as a simple JSON property value. However, in many cases we would like the option to add additional metadata to describe template fields. At a minimum, we want

[2] A complete template model specification can be found at http://metadatacenter.org/cedar-template-model.

users to record a name and description of each field. Hence, we use a JSON object to describe template fields.

The template field representation includes a value field, in addition to the other descriptive information. We first define a simple `value` field of type string to hold the raw value. We can also use the standard JSON Schema `title` and `description` fields to hold a name and description for the field.

For example, here is the definition of a Study Title template field, which contains the full name of a study represented as a single string[3]:

```
{
  "$schema": "http://json-schema.org/draft-04/schema#",
  "type": "object",
  "title": "Study Title", "description": "Study title template field",
  "properties": { "value": { "type": "string" } },
  "required": [ "value" ], "additionalProperties": false
}
```

A conforming instance of this template field could look as follows:

```
{ "value": "Immune Biomarkers" }
```

Representing Template Elements Using JSON Schema. Template elements support composition and can include a combination of multiple template fields and elements. They are represented using an approach equivalent to the one used to represent template fields. Again, we specify that a template element must be represented as a JSON object. We can then restrict each nested template field or template element using nested JSON Schema specifications.

For example, the definition of an Investigator template element is shown below. It contains one nested template field called `fullName`.

```
{
 "$schema": "http://json-schema.org/draft-04/schema#",
 "type": "object",
 "title": "Investigator",
 "description": "Investigator element",
 "properties": {
  "fullName": {
   "type": "object",
   "title": "Full Name",
   "description": "Full name template field",
   "properties": { "value": { "type": "string" } },
   "required": [ "value" ], "additionalProperties": false
   }
  },
  "required": [ "fullName" ], "additionalProperties": false
}
```

[3] A JSON Schema validator can be found at http://www.jsonschemavalidator.net.

A conforming *template element instance* could look like the following:

```
{ "fullName": { "value": "Dr. P.I." } }
```

Representing Templates Using JSON Schema. The representation of templates follows the same principles as template elements. Like template elements, templates can have nested elements and fields. A JSON Schema-encoded CEDAR template specification effectively defines the complete structure of a template instance.

Here is an example of a template containing a template field called `studyTitle` and a nested template element `pi` describing a principal investigator:

```
{
 "$schema": "http://json-schema.org/draft-04/schema#",
 "type": "object",
 "title": "Study", "description": "Study template",
 "properties": {
  "studyTitle": {
   "type": "object",
   "title": "Study Title", "description": "Study title template field",
   "properties": { "value": { "type": "string" } },
   "required": [ "value" ], "additionalProperties": false
  },
  "pi": {
   "type": "object",
   "title": "Investigator", "description": "Investigator element",
   "properties": {
     "fullName": {
      "type": "object",
      "title": "Full Name", "description": "Full name template field",
      "properties": { "value": { "type": "string" } },
      "required": [ "value" ], "additionalProperties": false
     }
   },
   "required": ["fullName" ], "additionalProperties": false
  }
 },
 "required": [ "studyTitle", "pi" ],
 "additionalProperties": false
}
```

A corresponding *template instance* could look like the following:

```
{
  "studyTitle": { "value": "Immune Biomarkers" },
  "pi": { "fullName": { "value": "Dr. P.I." } }
}
```

3.2.2 Representing Template Semantics Using JSON-LD

JSON Schema is useful for defining structural restrictions on JSON documents. It can also be used to specify basic type restrictions on field values. However, it provides a very basic set of built-in type restrictions. It also does not provide a way to add additional types or to interoperate with types defined in external sources.

As mentioned, JSON-LD [14] was developed to meet this goal. JSON-LD provides a lightweight syntax to add semantic annotations to JSON documents that can restrict the types and values of fields using terms from external vocabularies. Like JSON Schema, it adds some custom fields with well-known names to a JSON document to provide additional markup information.

JSON-LD provides four core fields to add semantic markup to JSON documents: `@context`, `@type`, `@id`, and `@value`. The `@context` field is used to define prefixes for controlled vocabularies and to map JSON properties to controlled vocabularies; the `@type` field indicates the semantic type of a JSON object; the `@id` field gives a unique identifier to a JSON object; finally, the `@value` field holds literal values, and can optionally be given a data type.

Here, for example, is a JSON-LD–enhanced template instance representing a study (with JSON-LD clauses in bold):

```
{
  "@type": "http://semantic-dicom.org/dcm#Study",
  "@id": "https://repo.metadatacenter.org/template_instances/55417",
  "@context": {
    "studyTitle": "https://schema.org/title",
    "pi": "https://mschema.org/property/hasPI"
  },
  "studyTitle": { "@value": "Immune biomarkers study" },
  "pi": {
    "@type": "https://schema.org/Person",
    "@context": {"fullName": "https://schema.org/name" },
    "fullName": { "@value": "Dr. P.I" }
  }
}
```

Note that we have added JSON-LD `@context`, `@type`, `@id`, and `@value` fields to provide semantic markup. The `@context` field maps JSON properties to properties in controlled vocabularies; the `@type` field indicates the semantic type of the instance, which in the case above is the Study class in the Radiation Oncology Ontology; the `@id` field gives a unique identifier to the template instance; finally, the `@value` field is used to store literal values. We use this field in our JSON-LD–enhanced content to replace the previous *ad hoc* `value` field we introduced earlier.

The JSON Schema specification can ensure that conforming instances are marked up with JSON-LD, both by demanding that specific fields are present and by restricting the content of those fields. For example, here is a JSON Schema template specification for the above study instance with clauses (marked in bold) ensuring that conforming instances carry appropriate JSON-LD markup:

```
{
 "$schema": "http://json-schema.org/draft-04/schema#",
 "@type": "https://repo.metadatacenter.org/core/Template",
 "@id": "https://repo.metadatacenter.org/templates/4353",
 "title": "Study", "description": "Study template",
 "properties": {
  "@context": {
    "properties": {
      "studyTitle": { "enum": [ "https://schema.org/title" ] },
      "pi": { "enum": [ "https://mschema.org/property/hasPI" ] }
    },
    "required": [ "studyTitle", "pi" ], "additionalProperties": false
  },
  "@type": { "enum": [ "http://semantic-dicom.org/dcm#Study" ] },
  "@id": { "type": "string", "format": "uri" },
  "studyTitle": { ... }, "pi": { ... }
 },
 "required": [ "@context", "@type", "@id", "studyTitle", "pi" ],
 "additionalProperties": false
}
```

As can be seen in this example, the JSON Schema template specification can ensure that template instances contain a significant amount of JSON-LD–encoded type information. Here, we are forcing the `@context`, `@type`, and `@id` fields in an instance to carry specific controlled terms. These instances can be automatically checked for conformance against the template specification. This use of JSON Schema is completely standard and instance validation can be performed with off-the-shelf tools. We also developed a JSON Schema-based validation schema that can be used to validate template, elements, and fields.

Generation of RDF from Template Instances. Since JSON-LD is effectively an RDF serialization, we can automatically produce RDF from JSON-LD marked up documents. Here, for example, is the RDF generated from the earlier study template instance (in Turtle syntax, with prefixes defined for clarity):[4]

```
@prefix rdf:    <http://www.w3.org/1999/02/22-rdf-syntax-ns#> .
@prefix dcm:    <http://semantic-dicom.org/dcm#> .
@prefix schema: <https://schema.org/> .
@prefix mscp:   <https://mschema.org/properties/> .
@prefix tinst:  <https://repo.metadatacenter.org/template_instances/> .
@prefix teinst: <https://repo.metadatacenter.org/element_instances/> .

tinst:55417 rdf:type      dcm:Study ;
            schema:title  "Immune biomarkers study" ;
            mscp:hasPI    teinst:5423 .
teinst:5423 rdf:type      schema:Person ;
            schema:name   "Dr. P.I" .
```

The ability to automatically generate RDF graphs from template instances provides a seamless path to interoperate with RDF-based Linked Data resources.

---

[4] A useful online JSON-LD tool can be found at http://json-ld.org/playground.

3.2.3 Value Constraints

JSON Schema allows us to express a very limited set of value constraints. It can, for example, state that the value of a field should be a particular value, or selected from a set of values. It can also restrict a field value to be of a particular type or format. We require more advanced constraints on field values, to specify that values must come from controlled terminologies. For example, we may specify that the value of a field should be selected from a set of URIs identifying classes from a particular ontology.

We provide five main constraint types for controlled terminology fields. We can constrain the possible values for a particular field to (1) classes from specific ontologies, (2) a set of classes, (3) ontology branches, (4) value sets, and (5) literals. Here, value sets are non-hierarchical collections of controlled terms. The possible values of a field can also be composed of some combination of the above five constraint types; the union of all five represents the set of possible field values.

To represent these value constraints, we use a `_valueConstraints` field that is placed inside a template field. The `_valueConstraints` field has five possible subfields for the five value constraint source types. We also include the ability to specify that the user may supply multiple entries when filling out template fields.

The top level JSON format adopted for our metadata instances is as follows:

```
{
  "_valueConstraints": {
    "ontologies": [ ... ],
    "classes": [ ... ],
    "branches": [ ... ],
    "valueSets": [ ... ],
    "literals": [ ... ],
    "multipleChoice": true
  }
}
```

The `ontologies` field specifies ontologies as controlled term sources for field values. Similarly, the `valueSets` field specifies values sets as term sources. The `branches` field is analogous to the `ontologies` field, but restricts values to branches within ontologies. The `classes` field indicates a set of classes as acceptable values. Finally, the `literals` field specifies a set of literals.

Here is an example showing a branches constraint for a field:

```
{
  "_valueConstraints": {
    "branches": [
      {
         "uri": "http://purl.obolibrary.org/obo/OBI_0000070",
         "includesRoot": false
      },
      {
         "uri": "http://www.co-ode.org/ontologies/galen#Assay",
         "includesRoot": false
      }
    ],
    "multipleChoice": true
  }
}
```

It restricts the possible values to classes in branches rooted in assay classes in the Ontology for Biomedical Investigations and in the GALEN ontology.

The four other types use a similar approach to constrain possible field values.

3.3 Template Design and Metadata Acquisition Tools

Using this model as a foundation, we built a set of tools for the acquisition, storage, search, and reuse of machine-readable metadata templates and instances [6]. We collectively refer to these tools as the CEDAR Workbench.[5]

Template Designer, Metadata Editor, and Metadata Explorer. We developed three highly interactive Web-based tools to simplify the process of managing metadata templates and instances (Fig. 2). The Template Designer allows users to create metadata templates. These templates are stored in CEDAR's template repository. The Metadata Editor tool uses these templates to automatically generate a forms-based acquisition interface for entering metadata. Entered metadata are stored as template instances in CEDAR's metadata repository. The Metadata Explorer tool can be used to perform faceted queries on the metadata stored in this repository.

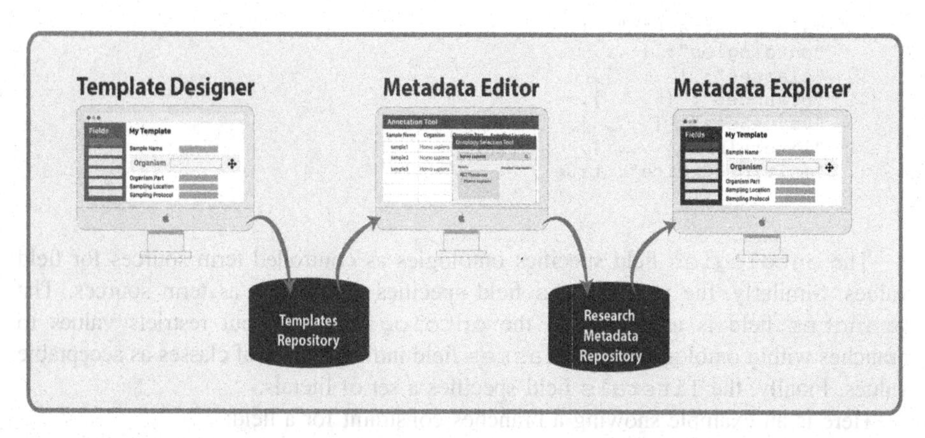

Fig. 2. Template and Metadata repositories, and Template Designer, Metadata Editor, and Metadata Explorer tools

A key focus is on interoperation with ontologies. Using interactive look-up services linked to the BioPortal ontology repository [15], the Template Designer allows template authors to find terms in ontologies to annotate their templates and to restrict the values of template fields. Users entering metadata using the Metadata Editor are prompted in real time with drop-down lists, auto-completion suggestions, and verification hints, significantly reducing their errors while speeding metadata entry and repair. This lookup is driven by the value constraints specified in templates.

[5] The CEDAR Workbench is available at https://cedar.metadatacenter.net.

Intelligent Authoring. To ease the burden of authoring high quality metadata, we developed a recommender framework that learns associations among data elements to suggest context-sensitive metadata values [18]. In the Metadata Editor the system can recommend possible values for metadata fields during submission. These value suggestions are based on analyses of instances in our metadata repository.

Web Services. We developed an array of REST-based Web services to manage model artifacts. These services provide functionality to store and query templates and instances. We also provide authentication and authorization services to restrict access to artifacts and to allow users to group their artifacts into collections.

4 Initial Evaluation

CEDAR is working with several major community-based groups to perform initial evaluations of our metadata model and tools. Core collaborating groups include (1) ISA [8], which develops tools for submitting metadata for biomedical experiments, (2) ImmPort [5], a data warehouse of immunology-related datasets, and (3) the Human Immunology Project Consortium (HIPC) [16], which designs new metadata templates and supplies experimental datasets to the ImmPort repository.

Our first evaluation assessed the ability of our model to represent real-world metadata models and instances from these collaborating groups. We also evaluated the model's ability to comprehensively link metadata elements and value constraints to controlled vocabularies and ontologies. We evaluated several other essential criteria: the use of the model-driven tools to support the development of model-specified templates; the instantiation of detailed metadata using the model; and the final submission of metadata to external repositories.

Modeling and Template Creation. In collaboration with the ISA, ImmPort, and HIPC teams, we jointly developed a template model (the 'CEDAR Study Model') to serve as a common representation for metadata in the ImmPort system and in a selection of ISA-populated repositories. The structure of this common model was heavily based on the ISA Model [8], which provides a rich description of metadata typically collected when performing experimental studies. The resulting model provides a comprehensive description of metadata for all studies in the ImmPort system, and of metadata in public repositories populated by the ISA tool chain.

We successfully used the Template Designer tool to create a representation of this common CEDAR Study Model. We also represented models from the LINCS Consortium [17] and from the Gene Expression Omnibus (GEO) [3] data repository using our template model. We further generated metadata instances following the corresponding instance model, as described below.

Metadata Acquisition. Using these templates we tested the ability to acquire metadata. The ImmPort team generated metadata instances conforming to our CEDAR Study Model for all 146 public studies in their system, and the ISA team performed a similar process for 300 publicly available studies. After correcting the model to address minor representation issues, we found all these target metadata instances could be represented in the core model.

Using the GEO template, we ingested all public GEO metadata as instances of this template. We indexed these instances using our Elasticsearch-based recommender system, and generated distribution profiles for all metadata fields. We then manually created experimental GEO submissions via this tool chain, and confirmed the relevance of the field-based recommendations that resulted [18].

Controlled Term Linkage. In addition to representing the structural constraints of the common studies model, we performed an analysis of all elements, fields, and value constrains in the model to identify appropriate controlled-term linkages for ImmPort metadata. In collaboration with members of the HIPC and ImmPort teams, we comprehensively annotated the model with appropriate controlled terms from a variety of biomedical ontologies. We also specified value constraints for controlled fields to ensure that the generated acquisition interfaces restricted acquired metadata to appropriate vocabularies. In the cases where custom value sets were required for fields, we used recently developed BioPortal functionality to create user-defined value sets. The final model (and tools that used it) successfully represented all type and value constraint information for the ImmPort metadata.

Metadata Submission. The final stage in the evaluation process involved assessing metadata submission to public repositories. We first targeted the NCBI BioSample [4] repository, which captures metadata about biological samples. In addition to a spreadsheet-based submission process, BioSample has an experimental XML-based submission portal. The structure of this submission is described in an XML Schema document, and an online validator is available to validate submissions.

For this test, we used the Template Designer to develop a template for the metadata described in the BioSample XML Schema. Starting with this template, we used our Metadata Editor to create sample metadata. We then transformed this metadata to the BioSample XML-based format and submitted it to the validator to ensure that it validated. In collaboration with NCBI, we have begun the process of developing a BioSample submission portal that can optionally be used by metadata submitters.

We are also working with the ImmPort representatives on our team to develop a similar submission pipeline for the ImmPort repository. Instead of manually generating a template using the Template Designer, the ImmPort team programmatically generated templates that conform to our template model. These templates form the basis of automatically generated user acquisition interfaces. With the ImmPort team, we are pursuing incorporation of our submission interface into their existing pipeline, toward acquiring real user submissions in the future.

Once these end-to-end submission pipelines are in place in production workflows, we will rigorously evaluate the quality of metadata collected compared to existing spreadsheet-based approaches.

5 Conclusion

We have described a standards-based template representation model that provides a common format for describing metadata templates and instances. The representation focuses on meeting the goals of the FAIR principles [11] and on interoperating with Linked Open Data. It also provides mechanisms to support template composition, with the aim of increasing reuse of descriptive metadata fragments across templates.

A key focus throughout is strongly linking metadata templates with controlled vocabularies and ontologies. In addition to semantically marking up the templates themselves, thereby specifying a rigorous link between template fields and well-specified semantic concepts, designers can specify ontology-based value constraints to ensure that metadata conforming to these templates are linked to controlled terminologies.

We developed a set of tools that leverages the template model to support the end-to-end management of metadata templates and instances, from initial template creation to final metadata submission. Two core tools focus on simplifying the process of managing templates: (1) the Template Designer allows users to create, search, and author metadata templates; this tool automatically produces a user interface specification from a template. (2) The Metadata Editor tool uses this interface specification to generate a forms-based acquisition interface for acquiring metadata conforming to the template. The system also provides an array of REST APIs, enabling access to all templates and metadata collected using those templates.

We performed a variety of internal evaluations with collaborators in the ISA [8], ImmPort [5], and HIPC [16] groups. These analyses focused on different stages of the metadata management process, from initial template creation, to final submission and validation. The evaluations demonstrated the fundamental viability of the template model. The next evaluation steps for the template model include supporting and evaluating user submissions to existing public metadata repositories, with the BioSample [4], ImmPort [5], and LINCS [17] repositories as the first targets. Soon, we expect to establish additional markup leveraging the JSON-LD constructs to support more advanced interoperability scenarios.

The use of JSON Schema provided a standard technology to represent all structural aspects of CEDAR's template model. No invention of new constructs was required—all structural constraints were captured directly by core JSON Schema clauses. Off-the-shelf tools proved usable for model validation. The use of JSON-LD provided a robust bridge between our model and semantic technologies. With JSON-LD we retained the full expressivity and power of semantic technologies, while gaining the practical advantage of the massive availability of Web-centric tooling. Since JSON-LD is effectively an RDF serialization, we can go back and forth between the two representations using standard tools. Again, no new JSON-LD constructs were required to represent the semantic markup required by CEDAR's model.

The JSON-based approach eased adoption by groups unfamiliar with semantic web technologies, and provided a low-overhead pathway for users to incrementally add semantic concepts. We have observed how JSON-LD facilitates incremental semantic enhancement of metadata, as our partner teams created simple structural metadata and then gradually added comprehensive semantic markup. We think such an incremental approach is key to bringing semantic web technologies into the wider Web community. We also note that several biomedical initiatives focused on metadata, such as Bio-CADDIE [12], have begun using JSON-LD to describe their metadata.

All software and schemas described in this paper are open source and available on GitHub [22]. We released a public alpha version of CEDAR in September, 2016.

Acknowledgments. CEDAR is supported by the National Institutes of Health through an NIH Big Data to Knowledge program under grant 1U54AI117925. NCBO is supported by the NIH Common Fund under grant U54HG004028. We appreciate the collaborations offered by the ImmPort, BioSharing, HIPC, and LINCS communities.

References

1. Borgman, C.L.: The conundrum of sharing research data. J. Am. Soc. Inform. Sci. Technol. **63**(6), 1059–1078 (2012)
2. Tenenbaum, J.D., Sansone, S.-A., Haendel, M.A.: A sea of standards for omics data: sink or swim? JAMIA **21**(2), 200–203 (2014)
3. Edgar, R., Domrachev, M., Lash, A.E.: Gene expression omnibus: NCBI gene expression and hybridization array data repository. Nucleic Acids Res. **30**(1), 207–210 (2002)
4. BioSample. http://www.ncbi.nlm.nih.gov/biosample. Accessed 15 Sept 2016
5. Bhattacharya, S., et al.: ImmPort: disseminating data to the public for the future of immunology. Immunol. Res. **58**(2–3), 234–239 (2014)
6. Musen, M.A., et al.: The center for expanded data annotation and retrieval. J. Am. Med. Inform. Assoc. **22**(6), 1148–1152 (2015)
7. BD2K. https://datascience.nih.gov/bd2k. Accessed 15 Sept 2016
8. Sansone, S.-A., Rocca-Serra, P., Field, D., et al.: Toward interoperable bioscience data. Nat. Genet. **44**(2), 121–126 (2012)
9. Rocca-Serra, P., Brandizi, M., Maquire, E., et al.: ISA software suite: supporting standards-compliant experimental annotation and enabling curation at the community level. Bioinformatics **26**(18), 2354–2356 (2010)
10. Rayner, T.D., et al.: A simple spreadsheet-based, MIAME-supportive format for microarray data: MAGE-TAB. BMC Bioinform. **7**(1), 489 (2006)
11. Wilkinson, M.D., et al.: The FAIR guiding principles for scientific data management and stewardship. Sci. Data **3**(1), 160018 (2016)
12. Nosek, B.A., et al.: Promoting an open research culture. Science **6242**(348), 1422–1425 (2015)
13. JSON Schema. http://json-schema.org. Accessed 15 Sept 2016
14. JSON-LD. http://json-ld.org. Accessed 15 Sept 2016
15. Musen, M.A., Noy, N.F., Shah, N.H., et al.: The national center for biomedical ontology. JAMIA **19**(2), 190–195 (2012)
16. Maecker, H., et al.: Standardizing immunophenotyping for the human immunology project. Nat. Rev. Immunol. **12**(3), 191–200 (2012)
17. LINCS. http://www.lincsproject.org. Accessed 15 Sept 2016
18. Panahiazar, M., et al.: Context aware recommendation engine for metadata submission. In: Workshop on Capturing Scientific Knowledge (2015)

19. Motik, B., Horrocks, I., Sattler, U.: Adding integrity constraints to OWL. In: OWLED, vol. 258 (2007)
20. SHACL. https://www.w3.org/TR/shacl/. Accessed 15 Sept 2016
21. JSON-LD Use Cases. https://www.w3.org/2013/dwbp/wiki/RDF_AND_JSON-LD_Use-Cases. Accessed 15 Sept 2016
22. CEDAR GitHub Organization. https://github.com/metadatacenter. Accessed 15 Sept 2016

OpenResearch: Collaborative Management of Scholarly Communication Metadata

Sahar Vahdati[1(✉)], Natanael Arndt[2], Sören Auer[1,3], and Christoph Lange[1,3]

[1] Enterprise Information Systems (EIS), University of Bonn, Bonn, Germany
{vahdati,auer}@cs.uni-bonn.de
[2] AKSW Group, Leipzig University, Leipzig, Germany
arndt@informatik.uni-leipzig.de
[3] Fraunhofer, Intelligent Analysis and Information Systems (IAIS), Sankt Augustin, Germany
math.semantic.web@gmail.com

Abstract. Scholars often need to search for matching, high-profile scientific events to publish their research results. Information about topical focus and quality of events is not made sufficiently explicit in the existing communication channels where events are announced. Therefore, scholars have to spend a lot of time on reading and assessing calls for papers but might still not find the right event. Additionally, events might be overlooked because of the large number of events announced every day. We introduce OpenResearch, a crowd sourcing platform that supports researchers in collecting, organizing, sharing and disseminating information about scientific events in a structured way. It enables quality-related queries over a multidisciplinary collection of events according to a broad range of criteria such as acceptance rate, sustainability of event series, and reputation of people and organizations. Events are represented in different views using map extensions, calendar and time-line visualizations. We have systematically evaluated the timeliness, usability and performance of OpenResearch.

Keywords: Scientific events · Collaborative knowledge acquisition · Semantic publishing · Semantic wikis · Linked data

1 Introduction

There is currently an era of departure to investigating how scholarly work and communication can be taken to the digital world. Much attention is devoted to new forms of publishing (e.g. semantic papers, micro-publications), open access, and free availability of publication metadata. Still, a large number of scholarly communication processes and artifacts (other than publications) are not currently well supported. This includes in particular information about events (conferences, workshops), projects, tools, funding calls etc. In particular for young researchers and interdisciplinary work it is of paramount importance to be able to easily identify venues, actors and organizations in a certain field and to assess their quality.

E. Blomqvist et al. (Eds.): EKAW 2016, LNAI 10024, pp. 778–793, 2016.
DOI: 10.1007/978-3-319-49004-5_50

Research results are published as scientific papers in journals and events such as conferences, workshops etc. Each component of this communication needs to be open and easily accessible. Besides conducting their actual research, scholars often need to search for scientific events to submit their research results to, for projects relevant to their research, for potential project partners and related research schools, for funding possibilities that support their particular research agenda, or for available tools supporting their research methodology. For lack of better support, scholars rely a lot on individual experience, recommendations from colleagues and informal community wisdom, they do simple Web searches or subscribe to mailing lists and are stuck with simplistic rankings such as calls for papers (CfPs) sorted by deadline. Domain specific mailing lists are a medium often used by conference and workshop organizers for posting initial, second, final calls for papers, as well as deadline extensions. But this situation leads to discussions on whether to allow calls for papers on the lists or threat them as spam[1] It is especially hard for subscribers to filter those calls according to their individual interests, or maybe explicitly subscribe to important information, such as deadline extensions or subsequent calls, on a specific event or an event series.

On the other hand, the quality of scientific events is directly connected to the research impact and the rankings of the scientific *papers* published by them. For example, the *Research Excellence Framework* (REF) for assessing the quality of research in UK higher education institutions, classifies publications by the venues they are published in. This facilitates assessing every researcher's impact based on the number of publications in conferences and journals. Providing such information to researchers supports them with a broader range of options and a comprehensive list of criteria while they are searching for events to submit their research contributions. To provide comprehensive information about scientific venues, projects, results etc., we present OpenResearch.org. OpenResearch is a platform for automating and crowd-sourcing the collection and integration of semantically structured metadata about scholarly communication. In particular, with regard to events, OpenResearch . . .

1. reduces the effort for researchers to find 'suitable' events (according to different metrics) to present their research results,
2. supports event organizers in visibly promoting their event,
3. establishes a comprehensive ranking of events by quality,
4. provides a cross-domain service recommending suitable submission targets to authors, and
5. supports easy and flexible data exploration using Linked Data technology: a structured dataset of conferences facilitates selection regarding fields of interest or quality of events.

OpenResearch empowers researchers of any field to collect, organize, share and disseminate information about scientific events, projects, organizations, funding sources and available tools. It enables the community to define views as queries

[1] Note a recent survey on calls for papers on the W3C mailing lists: https://lists.w3.org/Archives/Public/semantic-web/2016Mar/0108.html.

over the collected data; assuming sufficient data, such queries can enable rankings by relevance or quality. Driven by Semantic MediaWiki (SMW), OpenResearch provides a user interface for creating and editing semantically structured event profiles, tool and project descriptions, etc. in a collaborative wiki way. OpenResearch is part of a greater research and development agenda for enabling true open access to all types of scholarly communication metadata (beyond bibliographic ones) not just from a legal but also from a technical perspective. The work on OpenResearch is aligned with *OpenAIRE*, the Open Access Infrastructure for Research in Europe.

The remainder of this paper is organized as follows: Sect. 2 states the problem that OpenResearch intends to address. Section 3 presents the state of the art of existing services addressing the same problem. Section 4 establishes requirements for a system that can address the problem in a comprehensive way. Section 5 explains the approach and architecture of the OpenResearch platform. Section 6 presents the services that OpenResearch provides to its end users today. Section 7 discusses how we have assessed the time-lines, usability and performance of OpenResearch. Section 8 concludes and outlines future work.

2 Problem Statement

Challenge 1: Communication. Research communities use different communication channels to distribute event announcements and CfPs. Announcing CfPs through different mailing lists is the traditional but still most popular way of disseminating information about an event. Exploring the calls for papers posted on mailing lists of the Semantic Web community shows that 500 to 700 event announcements have been posted every year between 2006 and 2016 (approx. 15–30 % of the overall traffic). This shows that a large and widely spread amount of unstructured data about scientific events is increasingly being published via communication channels not specifically designed for this purpose. Due to the interdisciplinary nature of research, event organizers easily overlook relevant channels to announce their event. In addition, browsing through the CfPs in several channels to identify events that might be of interest is a time and effort consuming task.

Challenge 2: Structure. There are structural differences across events, for example, events with many co-located events or sub-events, or new events emerged from multiple smaller ones. One example for the latter is the Conference on Intelligent Computer Mathematics (CICM), which results from the convergence of four conferences that used to be separate but now are tracks of a single conference.[2] Scholars who want to find out whether an event matches their research interests therefore have to understand its structure; if they cannot find the desired information for the super-event, they will have to study the sub-events.

Challenge 3: Series. Most scientific events occur in series, whose individual editions take place in different locations with narrow topical changes. Researchers

[2] http://www.cicm-conference.org/.

often need to explore several resources to obtain an overview of the previous editions of an event series to be able to estimate the quality of the next upcoming event in this series.

Challenge 4: Addressing Different Stakeholders. Event organizers aim to attract as many submitters as possible to their events. Publishers want to know whether they should accept a particular event's proceedings in their renowned proceedings series. Potential PC members want to decide whether it is worth spending time in the reviewing process of an event. Similarly, sponsors and invited speakers need to decide whether a certain event is worth sponsoring or attending. Researchers receiving CfP emails have to distinguish whether the event is appropriate for presenting their work. Researchers searching for events through various communication channels assess events based on criteria such as thematic relevance, feasibility of the deadline, close location, low registration fee etc. The organizers of smaller events who plan to organize their event as a sub-event of a bigger event have to decide whether this is the *right* venue to co-locate with. These examples prove the importance of filtering events by topic and quality from the point of view of different stakeholders. Currently, the space of information around scientific events is organized in a cumbersome way, thus preventing events' stakeholders from making informed decisions, and preventing a competition of events around quality, economy and efficiency.

Strategies. Event organizers employ a number of strategies to cope with the challenges of advertising their event and engaging with the potential audience. They use multiple channels (mailing lists, social networks, homepages) to distribute CfPs. Some organizers plan deadline extensions in advance, as a strategy to attract more submissions. Some communities employ databases on top of mailing lists for announcing scientific events e.g., researchers in information systems and databases use the DBWorld database (cf. Sect. 3). The strategies mentioned so far target authors of submissions, whereas event organizers also have to find sponsors, high-profile program committee members and keynote speakers. This is currently done by contacting researchers or companies that the organizers know already. An approach for a centralized and holistic infrastructure for managing the information about scientific events was missing so far.

3 Related Work

CfP Classification and Annotation: *CFP Manager* [4] is an information extraction tool specific to the domain of computer science; it extracts metadata of events from an unstructured text representation of CfPs. Because of the different representations and terminologies of CfPs across research communities, this approach requires domain specific implementations. The extracted data is limited to the keywords used in the content of CfPs. In addition, CFP Manager does not support data curation workflows involving multiple stakeholders. Hurtado Martin et al. proposed an approach based on user profiles, which takes a scholar's recent publication list and recommends related CfPs using content analysis [3].

Xia et al. presented a classification method to filter CfPs by social tagging [10]. Wang et al. proposed another approach to classify CfPs by implementing three different methods but focus on comparing the classification methods rather than services to improve scientific communication [9].

Websites: *Google Scholar Metrics (GSM)*[3] provides ranked lists of conferences and journals by scientific field based on a 5-year impact analysis over the Google Scholar citation data. 20 top-ranked conferences and journals are shown for each (sub-)field. The ranking is based on the two metrics h5-index[4] and h5-median[5]. GSM's ranking method only considers the number of citations, whereas we intend to offer a multi-disciplinary service with a flexible search mechanism based on several quality metrics. *DBLP*[6], one of the most widely known bibliographic databases in computer science, provides information mainly about publications but also considers related entities such as authors, editors, conference proceedings and journals. Events, deadlines and subjects are out of DBLP's scope. DBLP allows event organizers to upload XML data with bibliographic data for ingestion. The dataset of DBLP is available as an RDF dump [7] *DBWorld*[8] collects data about upcoming events and other announcements in the field of databases and information systems. Each record comprises event title, deadline, event homepage and the full-text description. *WikiCFP*[9] is a popular service for publishing CfPs. Like DBWorld, WikiCFP only supports a limited set of structured event metadata (title, dates, deadlines), which results in limited search and exploration functionality. WikiCFP employs crawlers to track high-profile conferences. Although WikiCFP claims to be a semantic wiki, there is no collaborative authoring, versioning, minimal structure and the data is not downloadable as RDF or accessible via a SPARQL endpoint. *Cfplist*[10] works similar to WikiCFP but focuses on social science related subjects. Data is contributed by the community using an online form. *SemanticScholar*[11] offers a keyword-based search facility that shows metadata about publications and authors. It uses artificial intelligence methods in the back-end and retrieves results based on highly relevant hits with possibility of filtering.

Datasets: *ScholarlyData*[12] provides RDF dumps for scientific events. Conference-Ontology, a new data model developed for ScholarlyData, improves over already existing ontologies about scientific events such as the Semantic Web

[3] https://scholar.google.com/intl/en/scholar/metrics.html.
[4] h5-index is the h-index for articles published in the last 5 complete years.
[5] 5-median is the median number of citations for those articles in the h5-index.
[6] http://dblp.uni-trier.de/.
[7] http://dblp.l3s.de/d2r/.
[8] https://research.cs.wisc.edu/dbworld/.
[9] http://www.wikicfp.com/.
[10] https://www.cfplist.com/.
[11] https://www.semanticscholar.org.
[12] http://www.scholarlydata.org/dumps/.

Dog Food (SWDF) ontology. *Springer LOD*[13] is a portal publishing conference metadata collected from the traditional publishing process of Springer as Linked Open Data. All these conferences are related to Computer Science. The data is available through a SPARQL endpoint, which makes it possible to search or browse the data. A graph visualization of the results is also available. For each conference, there is information about its acronym, location and time, and a link to the conferences series. The aim of this service is to enrich Springer's own metadata and link them to related datasets in the LOD Cloud.

Other Services: *Conference.city*[14] is a new service initialized in 2016 that lists upcoming conferences by location. For each conference, title, date, deadline, location and number of views of its conference.city page are shown. Based on the location of the conference, Google plug-ins are used to recommend flights, accommodation and restaurants. The service collects data mainly from event homepages and from mailing lists. In addition, it allows users to add a conference using a form. *PapersInvited*[15] focuses on collecting CfPs from event organizers and attracting potential participants who already have access to the ProQuest service[16]. ProQuest acts as a hub between repositories holding rich and diverse scholarly data. The collected data is not made available to the public.

Conclusion: The comparison of currently available services in Table 1 shows that collaborative management of scholarly communication metadata in particular for events is not yet sufficiently supported.

4 Requirements

A collaborative and partially decentralized environment is required to enable community-based scientific data curation and extension, and to tap into the 'wisdom of the crowd' for elicitation and representation of metadata associated to scholarly communication. In particular, such a system is aimed to address the following requirements as services, which we have derived from the challenges C1–C4 pointed out in the problem statement and from the review of related work (R):

R1 It should be easily possible to create various views on the resulting data (addressing various communities), also in a collaborative way. (C1)
R2 Fine-grained and user extensible semantic representation of the (meta)data should be supported. (C1)
R3 The resulting ontological model should capture the relationships between various types of entities (e.g. event series, sub/super events, roles in event organization, etc.). (C2, C3)

[13] http://lod.springer.com/.
[14] http://conference.city/.
[15] http://www.papersinvited.com/.
[16] http://www.proquest.com/.

Table 1. Comparison of existing services.

Service	Entities	Event series	Sub-events	Quality criteria	Community contribution	Advanced search	LOD
CFP Manager	Events	✗	✗	✗	✗	✗	✗
GSM	Conferences, Journals	✗	✗	✗	✗	✗	✗
DBLP	Publications, Person	✗	✗	✗	✗	✗	✗
DBWorld	Events	✓	✓	✗	✗	✗	✗
WikiCFP	Events	✗	✗	✗	✓	✗	✗
Cfplist	Events	✗	✗	✗	✓	✓	✗
CiteSeer	Publications, Events	✗	✗	✗	✗	✗	✓
ScholarlyData	Events	✗	✗	✗	✗	✗	✓
Springer LOD	Events	✗	✗	✗	✗	✗	✓
Conference.city	Events	✗	✗	✗	✗	✗	✗
PapersInvited	Events	✗	✗	✗	✗	✗	✗
SemanticScholar	Publication, Person	✗	✗	✗	✗	✗	✗

R4 Different stakeholders of scholarly communication (event organizers, PC members, developers, etc.) have to be supported adequately. (C4)

R5 The data representation and view generation mechanisms should support fine-grained analyses (e.g. about the quality of events according to various indicators). (C4)

R6 The collaborative authoring and curation interfaces should be user friendly and enable novices to participate in the data gathering and curation processes.(C4)

R7 The system architecture should support automatic as well as manual/crowd-sourced data gathering from a variety of information sources. (R)

R8 All changes should be versioned to support tracking particular users' contributions and their review by the community. (R)

R9 The collected data should be easily reusable by application and service developers. (R)

5 Approach

The core of the OpenResearch approach is to balance manual/crowd-sourced contributions and automated methods. OpenResearch uses semantic descriptions of scientific events based on a comprehensive ontology; this enables distributed data collection by embedding markup in conference websites aligned with schema.org, and links to other portals and services. Semantic MediaWiki (SMW) serves as data curation interface employing semantic forms, templates various extensions and semantic annotations in the wiki markup. In the remainder, we describe the data model of OpenResearch and its architecture.

5.1 Data Model

The vocabulary used in OpenResearch reuses existing vocabularies from related domains, since reuse increases the value of semantic data. Existing related vocabularies are the *Semantic Web Conference Ontology* (SWC)[17], the *Semantic Web Portal Ontology* (SWPO)[18], and the *Funding, Research Administration and Projects Ontology* (FRAPO)[19], as well as schema.org. The SWC, SWPO and schema.org vocabularies provide means for modeling general events and SWC and SWPO also conferences. FRAPO provides terms to express scientific projects and their relations. *Conference Linked Data* (COLINDA)[20] contains information about scientific events collected from other systems such as WikiCFP and EventSeer and published as Linked Data, and the CfP ontology[21] provides means for modeling calls. A specific ontology for CfPs has been proposed in [8].

The property alignment is implemented using the SMW mechanism for importing vocabularies[22]. This includes definitions of the reused vocabularies in special vocabulary pages e.g. for SWC[23], which lists all imported properties and annotates them with SMW data types for the values. Wiki categories and properties are then aligned with the vocabulary terms using special *imported from* links. For instance *Category:Conference* is aligned to *swc:ConferenceEvent* with `[[imported from::swc:ConferenceEvent]]`. For modeling the calls and roles for a conference we defined new properties in our own vocabulary[24]. Figure 1[25] provides an example for using the data model. In contrast to the existing data model for calls and roles in the SWC ontology we are following a flat structure, which allows users, e.g., to directly attach a deadline to an event rather than creating a new instance for a call in addition to the actual event.

5.2 Architecture

Figure 2 depicts the three layers of OpenResearch's architecture: Data gathering, Data processing and Data representation.

Data Gathering and Scrapers. This layer supports ingestion, semantic lifting and integration of relevant information from various sources. To populate the OpenResearch knowledge base in addition to crowd-sourcing, we gather information from different sources. Sources can be available as Linked Data already, or structured, semi-structured and unstructured. SMW itself provides two options

[17] http://data.semanticweb.org/ns/swc/swc_2009-05-09.html.
[18] http://sw-portal.deri.org/ontologies/swportal.
[19] http://www.sparontologies.net/ontologies/frapo/source.html.
[20] http://colinda.org, offline at the time of this writing.
[21] http://sw.deri.org/2005/08/conf/cfp referenced by SWC, offline at time of writing.
[22] https://www.semantic-mediawiki.org/wiki/Help:Import_vocabulary.
[23] http://openresearch.org/MediaWiki:Smw_import_swc.
[24] http://openresearch.org/vocab/.
[25] Besides the usual prefix mappings that are available at http://prefix.cc/, we also use `wiki:` http://openresearch.org/Special:URIResolver/ and `export:` http://openresearch.org/Special:ExportRDF/.

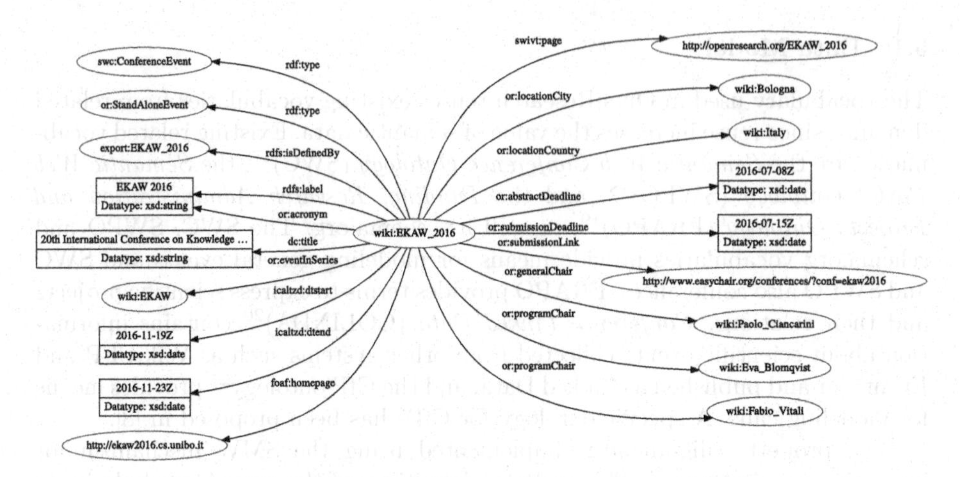

Fig. 1. An exemplary usage of the OpenResearch data model showing the EKAW 2016 resource.

for importing data: creation of individual pages/resources and bulk import[26] using the MediaWiki export format. Structured and semi-structured information can be imported as CSV and RDF: CSV files, prepared manually or obtained from WikiCFP via a crawler that we have implemented[27], can be transformed to the MediaWiki export format using the MediaWiki CSV Import[28] and then imported using the bulk importer; RDF datasets can be imported using the RDFIO MediaWiki extension[29].

Data Processing. This layer enables the storing and management of unstructured (text markup), semi-structured (annotations and infoboxes), structured data (RDF data adhering to an ontology) and schema data (the underlying ontology) Two database management systems are used in the OpenResearch architecture: one to store the schema-level information, the other to store the generated semantic triples. SMW supports multiple triple stores for storing the RDF graph, e.g., Blazegraph or Virtuoso. We use Blazegraph as it has been selected Wikimedia Foundation based on a performance and quality.[30] A MySQL relational database is used to store the templates, properties and, form names.

Data Exploring. This layer comprises various means for human and machine-readable consumption of the data. Several types of data representation are made possible by data exploration. CfPs are represented as individual wiki pages for each event instance, including a semantic representation of their metadata. SMW

26 https://www.mediawiki.org/wiki/Help:Export.
27 https://github.com/EIS-Bonn/OpenResearch/tree/master/wikiCFP.
28 http://mwcsvimport.pronique.com/; usage described at http://openresearch.org/OpenResearch:HowTo.
29 https://www.mediawiki.org/wiki/Extension:RDFIO.
30 https://goo.gl/NNm407.

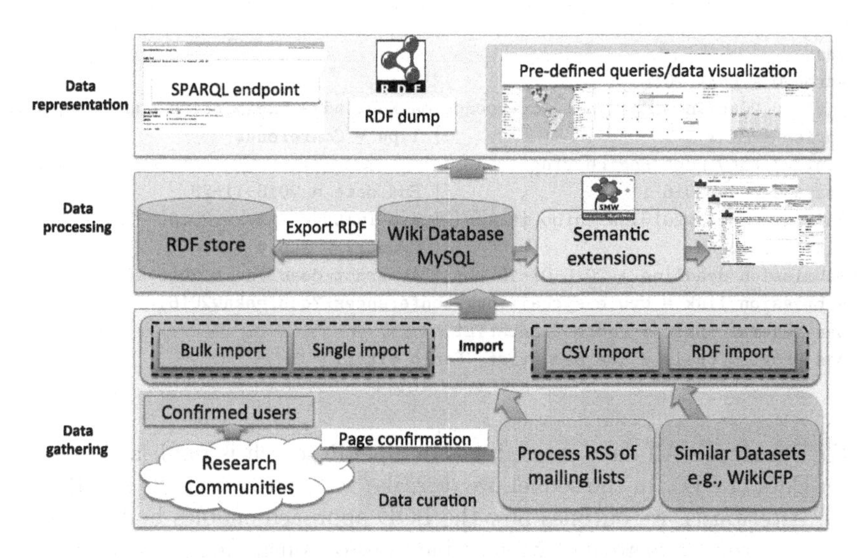

Fig. 2. OpenResearch Architecture

provides a full-text search facility and supports semantic queries. Queries and the visualization of their results are detailed in Sect. 6. Furthermore, the RDF triple store can be accessed using a SPARQL endpoint or downloadable RDF dump.

6 OpenResearch Services

On top of the basic architectural layers, OpenResearch offers services for different stakeholders of scientific communication. As a semantic wiki, it offers initial LOD services and semantic representation of metadata about events. We address the issues discussed in Sect. 2 by establishing a set of quality metrics for scientific events and implementing them as properties. We adopt the definition of quality as *fitness for use*, which, here, means the extent to which the specification of an event satisfies its stakeholders [5,6]. In the remainder of this section, the current services are explained in three categories: wiki pages, LOD services and queries.

Semantic Wiki Pages: SMW powers OpenResearch to provide semantic representation of CfPs as one wiki page per event. In OpenResearch, specific semantic forms have been designed for each type of entities to make content creation and revision as easy as possible for users. Properties of each semantic object are populated via fields in these semantic forms. The following example shows the generated SMW wiki markup containing general information about an event. Further information about committee members, extensions and other important dates can also be provided in other parts of the form. The complete textual representation of the CfPs can also be added as content of the wiki page with embedded semantic annotations.

```
{{Event
 | Acronym = EKAW 2016
 | Title = 20th International Conference on Knowledge Engineering and Management
 | Series = EKAW                          | Type = Conference
 | Field = Knowledge Engineering
 | Start date = 2016-11-19                | End date = 2016-11-23
 | Homepage = ekaw2016.cs.unibo.it        | has Twitter = @ekaw2016
 | City = Bologna                         | Country = Italy
 | Submission deadline = 2016-07-15       | Abstract deadline = 2016-07-08
 | submission link = www.easychair.org/conferences/?conf=ekaw2016,
 | has general chair = Paolo Ciancarini,
 | has program chair = Eva Blomqvist, Fabio Vitali
}}
```

LOD Services: All data created within OpenResearch is published as Linked Open Data (LOD). In the sequel, we describe ways for accessing OpenResearch LOD. Afterwards, we outlines how the LOD approach enables building further services on top by sketching two possible ways of consuming the OpenResearch LOD: interlinking with relevant datasets, and using OpenResearch LOD as external plug-in for the Fidus Writer scientific authoring platform[31].

Accessing OpenResearch LOD. An updated version of the OpenResearch dataset is produced daily and available for download and query[32]. The data is also queryable via a SPARQL endpoint[33]. In addition, the semantic representation of the metadata for each event is represented as an RDF feed in each page. The RDF feed for the EKAW 2016 resource is available at http://openresearch.org/Special:ExportRDF/EKAW_2016. To expose dereferenceable resources conforming with Linked Data best practices, the URI resolver provides URIs with content negotiation; e.g., for the EKAW 2016 resource the URI is http://openresearch.org/Special:URIResolver/EKAW_2016.

Interlinking. To increase the coherence of the data, we interlink the OpenResearch LOD with other relevant datasets. We are applying the same technical framework that we are using for OpenAIRE[34] Interlinking [1]. The following use cases enabled by interlinking show how the results of connecting the linked dataset of OpenResearch with other relevant datasets enhance the services:

1. *PC members recommendation:* one of the difficult and time-consuming tasks for event organizers is to collect a group of high-profile researchers as PC members. Interlinking OpenResearch LOD with datasets including author and person information such as ORCID[35] helps in this regard.

[31] https://www.fiduswriter.org/.
[32] https://zenodo.org/record/57899.
[33] http://openresearch.org/sparql.
[34] https://www.openaire.eu/.
[35] http://orcid.org/.

2. *Sponsoring recommendation:* it is often a challenge especially for smaller events to find local and international sponsors. On the other hand organizations and companies who want to gain visibility and decide whether or not to sponsor an event can use OpenResearch.

Integration with an Authoring Platforms. In this section we introduce our approach to improve the workflow of authoring processing [7]. The OpenResearch LOD will be plugged into the Fidus Writer authoring platform to improve the workflow in the following use cases:

1. *Venue recommendation:* One of the critical aspects in the process of writing and publishing is to find a suitable event to submit the scientific results. The OpenResearch dataset contains data about events annotated with corresponding scientific field as *:category* and keywords. We also annotate keywords from the content of the under-production scholarly document in the OSCOSS project that could be imported to the OpenResearch search services. For example, *Find all events in the computer science field that focus on data analysis, big data, knowledge engineering, linked data.* The result of queries can be shown to the authors with a user-friendly interface and filtering metrics such as deadline and location distance.
2. *Direct link to submission pages:* The OpenResearch data contains a property named *submission link* that provides a direct link to paper submission pages of events. The submission page of the targeted event can be made accessible easily from the authoring platform.
3. *Notification services:* there are different deadlines attached to the events that should be considered by authors such as abstract deadline, submission deadline or registration deadline as well as deadline extensions. Enabling notification services in the authoring platform will support both organizers and researchers.

Queries and Visualization of Results: To support the creation of various views, recommendations and ranked lists (by quality indicators), queries can be defined and executed using all defined properties and classes and the results can be embedded in wiki pages. For example, events can be ranked by acceptance rate using the corresponding properties in queries:

```
{{#ask:[[Category:Event]]
 | ?title = Name | ?Event in series = Series | ?Category | ?Acceptance rate
 | format = table | limit=10 | sort=Acceptance rate | order=desc
}}
```

It is also possible to capture the relationships between various types of entities (e.g. event series, sub/super events, roles of a person in event organization, etc.). Many popular views have been implemented in OpenResearch as pre-defined queries. Various display formats provided by SMW extensions are used to visualize the query results. Figure 3 shows a map view of the upcoming events using location-based filtering. Similarly, calendar and timeline views show

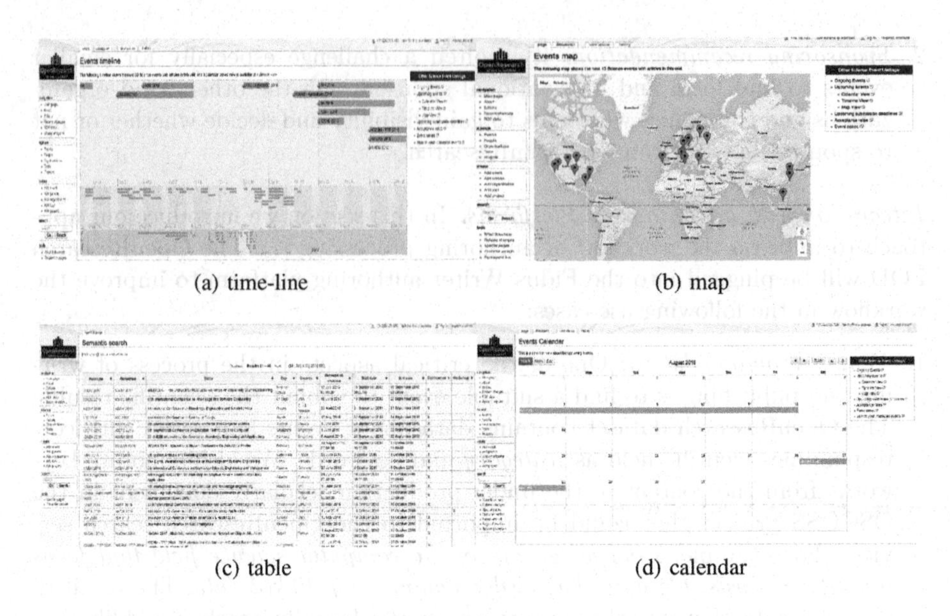

(a) time-line (b) map

(c) table (d) calendar

Fig. 3. Upcoming events in different visualizations

upcoming submission and notification deadlines as well as the events themselves. In addition, taking, for example, participation figures into account enables new indicators for measuring the quality and relevance of research that are not just based on citation counts [2]. Based on semantically enriched indicators, predefined SPARQL queries as well as form-based search facilities will be implemented for recommendation services.

7 Evaluation

The main objective of this work is to introduce a comprehensive approach for collaborative management of scholarly communication metadata with a special focus on events. We are for now mainly interested in collecting data, as this allows to provide more interesting analysis services. Nevertheless, we evaluated three aspects of OpenResearch including two surveys, performance measurements of the system as well as a usability analysis.

Timeliness Questionnaire: In a survey, we asked 40 researchers from different fields including Computer Science, Social Science to explain how they explore scientific events[36]. Over 75 % of the participants agree that having an event recommendation service is very relevant for them. For selecting an event to participate, all participants confirmed that they consider information that is not served directly by the current communication channels. Some of these criteria are networking possibilities, review quality, high-profile organizers, keynote speakers

[36] https://goo.gl/L02UU5.

and sponsors, low acceptance rate, having high quality co-located events, close location, citations counts for accepted papers of previous years. Participants indicated that they explore scientific events using: search engines, mailing lists, social media and personal contacts. Then, they assess the CfPs to find out whether that event satisfies their criteria. Over 85 % of the participants supported the idea of using a knowledge base for this purpose.

Usability Survey: We asked users to tell us about their experience wrt. the ease and usability of the system[37]. Overall 12 users participated in the survey; they have had several roles in scientific events (participant, PC member, event organizer and keynote speaker). 75 % of the users replied they had basic knowledge about wikis in general, however, half of them did not know about SMW. 66 % got familiarized easily with OpenResearch which shows its suitability for researchers of different fields. Again 66 % answered that they needed less than 5 min to add a single event which is relatively low time wrt. the time organizers need to announce their event in several channels. The average number of single events created by individual users is 10. More than half of the participants needed less than 5 min for a bulk upload. The participants largely agreed that these times are reasonable.

Objective comparison metrics	Data import	Complex queries
Time (s)	32.6	0.31
Memory (MB)	24.44	2.89
Number of pages	100	n/a
Number of queries	n/a	10

Performance Measurement: Currently, OpenResearch is running on a Debian server at the University of Bonn with 8 GB of RAM allocated. By private invitation (OpenResearch has not yet been publicly announced at a large scale), 70 users have been added during the last two months. Above 300 events have been added by the users during last two months and several bulk uploads of data are performed every week by the admins; each time 100 pages were created. The measured time for bulk import varies with the content of CfPs and reduces when events exist already in the system. The table below shows a performance measurements of OR w.r.t. the average time and memory usage for several bulk imports and complex queries running over the event query form.

8 Conclusion and Future Work

With regard to scholarly communication we are currently at a crossroad: On the one hand, there are commercial publishers and new incumbents such as social networks for researchers (e.g. ResearchGate, Academia.edu), which provide commercial services to the research community. Researchers either pay directly for

[37] https://goo.gl/HIIeEh.

these services by means of publication and access fees or indirectly (such as in the case of social networks) with their data. Either way, these commercial services strive to create a lock-in effect, which forces researchers to continue using these services without being able to migrate and choose competing services. On the other hand, there is an increasing push towards more open-access and open platforms for scholarly communication. Examples are open-access repositories such as arXiv, Zenodo, bibliographic metadata services such as DBLP and OpenAIRE, journal and conference management software and services such as Open Journal Systems and EasyChair or OpenCourseWare platforms such as SlideWiki.org. We see the work on OpenResearch presented in this article as a first step towards tighter interlinking and integrating of open services for scholarly communication.

In future, we envision to intensify data flows and service integration between OpenResearch and other open scholarly services. In particular, we are planning to import information from events' web pages, mailing lists and proceedings catalogs. Crawling event's web pages and extracting, e.g., embedded structured information such as schema.org RDFa or microdata, including the *Event* class and properties such as *name*, *organizer*, *location*, *startDate*, *endDate*, *subEvent*, or *superEvent*, keeps us up to date with the organizers. Extracting information from unstructured emails is challenging, but some emails have iCalendar attachments. Further information about events and their proceedings could be scraped from semi-structured listings such as the index page of the CEUR-WS.org open access workshop proceedings. Furthermore, we plan to relate events with other entities e.g., publications, projects, datasets.

Acknowledgments. This work has been partially funded by the European Commission with a grant for the project OpenAIRE (GA no. 643410) and by a grant for the project LEDS (GA no. 03WKCG11C) from the German Federal Ministry of Education and Research (BMBF). We thank Yakun Li for his technical work on openresearch.org.

References

1. Alexiou, G. OpenAIRE LOD services: scholarly communication data as linked data. In: 2nd Workshop, SAVE-SD. LNCS. Springer (2016)
2. Iorio, A.D., Lange, C., Dimou, A., Vahdati, S.: Semantic publishing challenge – assessing the quality of scientific output by information extraction and interlinking. In: Gandon, F., Cabrio, E., Stankovic, M., Zimmermann, A. (eds.) SemWebEval 2015. CCIS, vol. 548, pp. 65–80. Springer, Heidelberg (2015). doi:10.1007/978-3-319-25518-7_6
3. Hurtado Martín, G., Schockaert, S., Cornelis, C., Naessens, H.: An exploratory study on content-based filtering of call for papers. In: Lupu, M., Kanoulas, E., Loizides, F. (eds.) IRFC 2013. LNCS, vol. 8201, pp. 58–69. Springer, Heidelberg (2013). doi:10.1007/978-3-642-41057-4_7
4. Issertial, L., Tsuji, H.: Information extraction for call for paper. Int. J. Knowl. Syst. Sci. (IJKSS) **6**(4), 35–49 (2015)
5. Juran, J.M.: Juran's Quality Control Handbook, 4th edn. McGraw-Hill (Tx), New York (1974)

6. Knight, S.-A., Burn, J.M.: Developing a framework for assessing information quality on the World Wide Web. Informing Sci.: Int. J. Emerg. Transdiscipline **8**(5), 159–172 (2005)
7. Mayr, P., Momeni, F., Lange, C.: Opening communication social sciences: supporting open peer review with fidus writer. In: EA Conference (2016)
8. Tomberg, V., et al.: Towards, a comprehensive call ontology for research 2.0. In: i-KNOW. ACM (2011)
9. Wang, H.-D., Wu, J.: Collaborative filtering of call for papers. In: (7th Dec.). IEEE, pp. 963–970 (2015)
10. Xia, J., et al.: Optimizing academic conference classification using social tags. In: CSE (Hong Kong, China). IEEE, pp. 289–294 (2010)

Position Paper

Data-Driven RDF Property Semantic-Equivalence Detection Using NLP Techniques

Mariano Rico([✉]), Nandana Mihindukulasooriya, and Asunción Gómez-Pérez

Ontology Engineering Group, Universidad Politécnica de Madrid, Madrid, Spain
{mariano.rico,nmhinidu,asun}@fi.upm.es

Abstract. DBpedia extracts most of its data from Wikipedia's infoboxes. Manually-created "mappings" link infobox attributes to DBpedia ontology properties (dbo properties) producing most used DBpedia triples. However, infoxbox attributes without a mapping produce triples with properties in a different namespace (dbp properties). In this position paper we point out that (a) the number of triples containing dbp properties is significant compared to triples containing dbo properties for the DBpedia instances analyzed, (b) the SPARQL queries made by users barely use both dbp and dbo properties simultaneously, (c) as an exploitation example we show a method to automatically enhance SPARQL queries by using syntactic and semantic similarities between dbo properties and dbp properties.

Keywords: SPARQL query · Query enhancement · DBpedia · Spanish DBpedia · Property mapping

1 Introduction

DBpedia [1] is the central hub of the Linked Open Data (LOD) cloud because it provides a vast amount of information and most of the datasets in the LOD cloud link to DBpedia. The extraction process [2] in DBpedia generates properties of two types: (1) properties in the DBpedia ontology (we name these dbo properties), and (2) properties not in the DBpedia ontology (let us name them dbp properties). The dbp properties come from the attribute-value pairs found in Wikipedia infoboxes that has no manually-created mappings[1]. The analysis of the Spanish DBpedia (esDBpedia) found [3] that, despite the high number of mappings (100+ classes), for each 4 triples containing a dbo property there is 1 triple containing a dbp property. In this work, we extend this analysis to English and German DBpedia instances, with similar results. For instance, in the English DBpedia this ratio goes to almost one to one.

In this position paper we hypothesize that triples can not be accessed because most queries are comprised of dbo properties. DBpedia defines around 2500

[1] See DBpedia multilingual mappings at http://mappings.dbpedia.org.

E. Blomqvist et al. (Eds.): EKAW 2016, LNAI 10024, pp. 797–804, 2016.
DOI: 10.1007/978-3-319-49004-5_51

properties, but only 2 % infoboxes fields are mapped to the DBpedia ontology. Thus, there are many dbp properties in DBpedia: 58,239 for the English version, 17,111 for the Spanish and 12,167 for the German. Therefore, users that query the DBpedia endpoint by using SPARQL queries containing only dbo properties have no access to a significant amount of triples and could lead to null or incomplete results even if the relevant data is available in DBpedia.

In this work, we start by checking the assumption that users barely mix dbp and dbo properties in SPARQL queries. Later we provide a method to automatically identify the most similar dbp properties for a given dbo property. This method takes advantage of techniques from Natural Language Processing and Statistical Methods. The goal of the proposed method is to generate "automatic mappings" with a certain confidence level. These mappings can be manually approved by a specialist or through crowd-sourcing in a semi-automatic manner. Some examples point out that these mappings can enhance the SPARQL queries to generate better results by accessing more information in different DBpedia instances.

2 Background

In this section, we explore two hypotheses addressed in this paper. On the one hand, we analyze the amount of information described by using dbp properties in 3 DBpedia instances. On the other hand, we analyze how dbo and dbp properties are used in SPARQL queries made to the English DBpedia.

Firstly, Table 1 shows for three DBpedia instances (English, Spanish and German) the following data: the number of dbo and dbp properties, the number of triples containing those properties, and the top-10 dbp properties ordered by the number of triples containing those properties. The ratio dbp/dbo (number of triples with dbp properties per number of triples with dbo properties) goes to 0.95, 0.32, and 0.20 respectively. That is, the English DBpedia has the highest ratio, with almost as much triples containing dbo properties as triples containing dbp properties.

Secondly, we analyzed a SPARQL query log to evaluate the assumption that users do not frequently use dbp properties in their SPARQL queries. We used the Linked SPARQL Queries Dataset [4], which provides a RDF model to know details about SPARQL queries made to several endpoints. We explored the data from the English DBpedia to see how many queries use both dbp and dbo properties. Out of 1,208,762 distinct queries only 2,328 queries use both dbo and dbp properties in the same query. We made a similar analysis for agents (IPs): out of 3,041 distinct agents (IPs), only 473 use both dbp and dbo properties in the same SPARQL query. This illustrates that the majority of the SPARQL queries miss some portion of the data. We argue that this information can be reached by enhancing the SPARQL queries by using our proposed mappings between dbo and dbp properties.

Table 1. Top-10 dbp properties for the English, Spanish and German DBpedia instances (2015-04 version).

English DBpedia		Spanish DBpedia		German DBpedia	
dbp: http://dbpedia.org/property/					
URI	Triples	URI	Triples	URI	Triples
dbp:hasPhotoCollection	4,041,585	dbp:wikiPage UsesTemplate	3,402,499	dbp:name	494,852
dbp:name	4,021,368	dbp:nombre	558,837	dbp:geburtsort	305,063
dbp:title	1,452,504	dbp:título	327,498	dbp:kurzbeschreibung	283,695
dbp:subdivisionType	1,257,766	dbp:name	230,763	dbp:geburtsdatum	283,405
dbp:shortDescription	1,194,274	dbp:tipoSuperior	225,868	dbp:typ	232,702
dbp:dateOfBirth	1,023,951	dbp:horario	203,890	dbp:viaf	169,145
dbp:subdivisionName	1,004,294	dbp:imagen	183,887	dbp:gnd	165,362
dbp:goals	969,216	dbp:familia	152,430	dbp:jahre	156,498
dbp:placeOfBirth	908,819	dbp:title	144,724	dbp:sterbedatum	144,209
dbp:birthPlace	903,529	dbp:ordo	142,196	dbp:alternativnamen	143,893
#props dbp	58,239	#props dbp	17,111	#props dbp	12,167
#props dbo	1,338	#props dbo	559	#props dbo	534
#triples dbp	78,125,087	#triples dbp	28,234,292	#triples dbp	10,483,987
#triples dbo	82,369,408	#triples dbo	90,389,560	#triples dbo	50,750,486

3 An Approach for Automatically Enhancing SPARQL Queries

Figure 1 shows, from top to down, the process for finding 'similar' dbp properties for a given dbo property. The first step (figure top side) is to aggregate properties into groups according to their domain and range. The objective of this grouping is to work with smaller groups of properties with potentially similar semantics. For dbo properties, domain and range are specified by the DBpedia ontology, but dbp properties have no explicit domain or range. However, we can estimate domain and range by using tools such as LOUPE [5] (http://loupe.linkeddata.es) which provides domain and range for dbp properties analyzing the subject and the object of all triples containing a given dbp property. Following the figure, after this aggregation, properties that have *dbo:Person* as domain and *dbo:Place* as range are located in a smaller group which includes, among others, dbp properties like *dbp:birthPlace*, *dbp:birzPlace* and *dbp:deathPlace*, as well as dbo properties like *dbo:birthPlace* or *dbo:birthLocation*.

The second step involves processing each small group by using Natural Language pre-Processing which includes tokenization and stemming/lemmatization. Many dbp properties are compound words (e.g. birthPlace → (birth, place)). It is necessary to do some pre-processing for tokenizing those properties before applying linguistic techniques to find syntactic and semantic similarity. For dbp properties that use the camel case convention, this tokenization can be done easily by breaking the words using the camel case convention. For the rest, for instance the dbp properties that use all simple letters (e.g. *oldcode* or *testaverage*) or all capitals, dictionary tools that break the compound words into separate tokens of known words can be used. We also used other punctuation

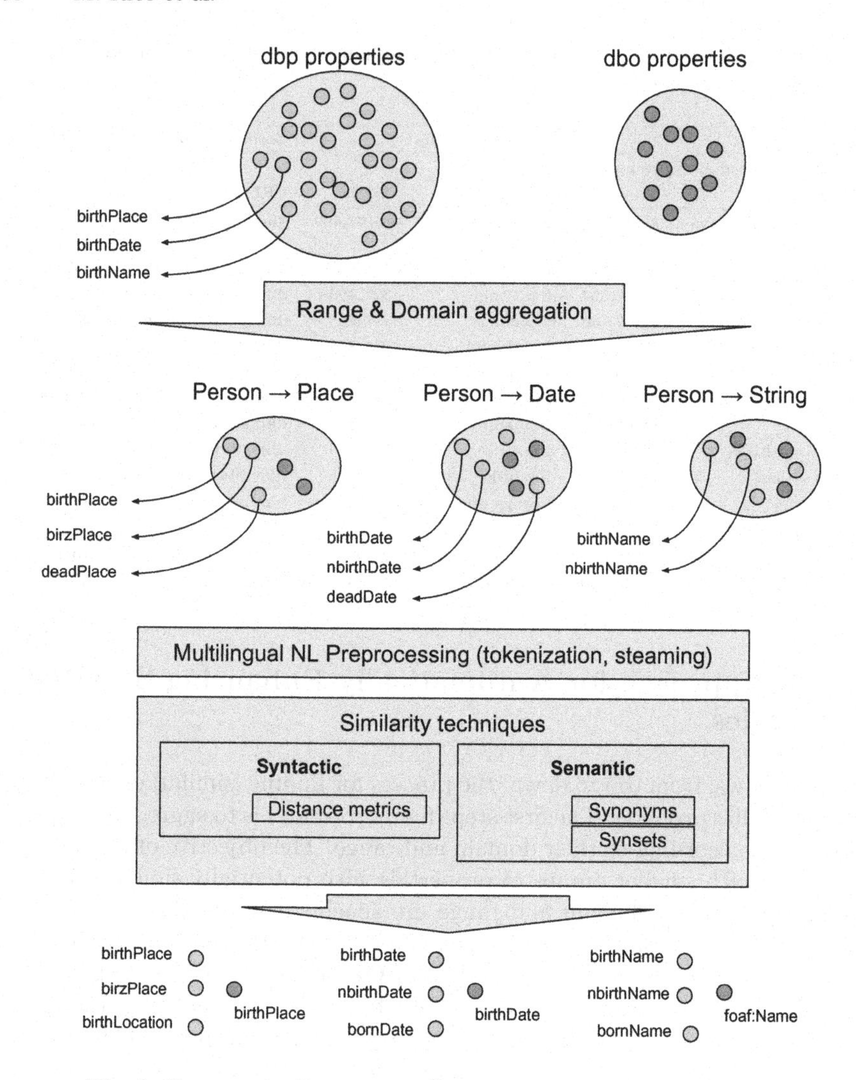

Fig. 1. Process pipeline to map dbp properties to dbo properties

marks such as brackets (e.g. *numEmployees(globally)*) for tokenization when they were applicable. In addition, lemmatization can be used for finding more results by normalizing the different variations such the inconsistent use of singular and plural words (e.g. coachTeams → (coach, team)).

As the majority of the dbo properties only have labels in English, when non-English dbp properties are detected in DBpedia instances such as the Spanish one or the German, translation tools are used to convert the property into English for mapping with the dbo property (e.g., geburtsort → birthPlace).

The third step comprises similarity techniques. The simplest is the syntactic distance, which includes classical string distance metrics (e.g. Jaro-Winkler distance, Damerau-Levenshtein distance), and token-based techniques (e.g. Jaccard similarity, Cosine Similarity). Several techniques can be used to identify different types of variations in dbp properties, for instance, edit distance-based measures such as Damerau-Levenshtein perform better for identifying typos but they are sensitive to substring locality. Using syntactic techniques such as string similarity we can identify that *dbp:birzPlace* means *dbo:birthPlace*. Semantics techniques go a step forward, and we have tested two 'semantic similarity' measures: (1) a dictionary-based method for synonyms and (2) a synsets-based method using WordNet. Semantic similarity allows us to identify that dbp properties like *dbp:birthLocation* or *dbp:cityOfBirth* are similar to the dbo property *dbo:birthPlace*. Further studies will be focused on finding the most accurate semantic-similarity methods for these tasks.

3.1 Enhancing SPARQL Queries by Using Dbp Properties

Knowing the dbp properties with the same meaning that a given dbo property, we can use them like in the example shown in Listing 1.1. Here we show a simple SPARQL query containing the property *dbo:birthPlace*. Listing 1.2 shows a query enhancement based only in dbp properties syntactically similar to *dbo:birthPlace*. We use VALUES, a SPARQL 1.1 feature equivalent to a set of UNION, which allow us a more compact representation. Notice that this query uses real properties available in the English DBpedia SPARQL endpoint.

```
1 PREFIX dbo: <http://dbpedia.org/ontology/>
2 select ?s ?bp {
3     ?s dbo:birthPlace ?bp .
4 }
```

Listing 1.1. Original SPARQL query

```
1  PREFIX dbo: <http://dbpedia.org/ontology/>
2  PREFIX dbp: <http://dbpedia.org/property/>
3
4  select ?s ?bp  where {
5   ?s ?p ?bp .
6   VALUES ?p {
7    dbo:birthPlace #typical dbo property
8    #Alternative dbp properties
9    dbp:birthPlcace dbp:birthplace
10   dbp:birhPlace   dbp:bithPlace
11   dbp:birtPlace   dbp:biRthPlace
12  }
13 }
```

Listing 1.2. Enhanced SPARQL query

Table 2. Example of dbp → dbo property mappings. Δ_1 is the enhancement for the example query in Listing 1.1.

DBpedia	dbo prop	dbp prop		Δ_1
		Syntactic	Semantic	
English	birthPlace	birthPlace birthplace placeofbirth birthPlac birthdplace birthPalce birthPlace PlaceOfBirth laceOfBirth oplaceOfBirth birthPlace. birthPlacE birthPalce birthPlae birthPace birthPlaxe birtPlace birthPlcace bithPlace brithPlace nbirthPlace birthplace birghPlace birthdplace biRthPlace birth placebirth placeOfBirth placOfBirth birthPlaceOf birthPlae	cityofbirth cityofbirthPlace cityOfBirth birthLocation	350%
Spanish	birthPlace	lugarDeNacimiento lugarNacimiento lugarNacimiento lugarnacimiento lugardenacimiento lugarNacimento lugarNaciento	ciudaddenacimiento ciudadDenacimiento paisdenacimiento paisNacimiento birthPlace birthplace placeOfBirth	221%
German	birthPlace	geburtsort birthplace birthPlace placeOfBirth placeofbirth	geburtsland countryofbirth	134%

4 Evaluation Example

As a complete evaluation would require more space, we only show an evaluation example to check our hypothesis that SPARQL query results can be improved by using dbp properties with the same semantics that the dbo properties used in a SPARQL query. Following the proposed method described in Sect. 3, we use the *dbo:birthPlace* property for the analysis. Table 2 shows the possible dbp properties mapping the *dbo:birthPlace* property for the three DBpedia instances analyzed, distinguishing between syntactic and semantic techniques as described in Sect. 3. Then, a simple query is used to analyze the number of results returned when only *dbo:birthPlace* is used (similar to Listing 1.1) and when an enhanced query is used (similar to Listing 1.2). This enhancement, denoted Δ_1 in the table leads to 350 % improvement in the case of English DBpedia (3,940,073 results instead of 1,211,868), 221 % improvement in the case of Spanish DBpedia (765,633 results instead of 346,515), and 132 % improvement in the case of German DBpedia (1,319,892 results instead of 986,323). These results illustrate that enhancing the queries using the approach proposed in this paper leads to better answers to the queries regarding the number of results. In the future, we plan to evaluate the correctness of the answers of the enhanced queries to assess if there is an impact on the quality of the results.

The queries used in the paper and the intermediate results are found in this supplementary material page[2].

[2] See http://tinyurl.com/EKAW2016paper129extras.

5 Related Work

Both Rahm and Bernstein [6], and Shvaiko and Euzenat [7] provide surveys of schema matching approaches and classify schema matching approaches into categories. The approach proposed in this paper combines several linguistic techniques that are mentioned in the survey including both syntactic and semantic techniques. Rinser et al. [8] propose a three-stage instance-based schema matching approach for mapping infoboxes from Wikipedias of different languages. The presented approach is only about Wikipedia, however it can be used to complement the property mappings proposed in this paper. Zhang et al. [9] propose Statistical Knowledge Patterns for identifying synonymous relations in large linked datasets. The method presented in this paper uses a similar technique for property clustering, but also compliment it with the NLP techniques. Palmero Aprosio et al. [10] emphasize the problem of non-mapped infoboxes in DBpedia and proposes an approach for automatic mapping generation applied to the Italian chapter of DBpedia.

6 Conclusions and Future Work

Our work starts by realizing that DBpedia triples are comprised not only by properties defined in the DBpedia ontology (dbo properties) but, to a big extent, by other properties (dbp properties). The DBpedia extraction process generates triples containing dbo properties when there is a mapping between a field in a Wikipedia infobox and a dbo property. But the extraction process also generates dbp properties for the fields in Wikipedia infoboxes that do not have such mapping. In the case of the English DBpedia, almost 50 % of all triples contain dbp properties in its predicate. Therefore, queries containing only dbo properties cannot access big parts of the DBpedia dataset.

In order to check the infra-utilization of dbp properties, we have analyzed a SPARQL query log repository containing SPARQL queries form several datasets, concluding that our hypothesis is correct at least for the English DBpedia.

As an initial application of this work, we have sketched a method to find the most similar dbp properties for a given dbo property. This could be used to automatically enhance SPARQL queries in order to get more results and we have shown some simple usage examples.

The proposed method depends on many parameters and we have applied them to three DBpedia instances (English, Spanish and German). Future work will explore the most adequate parameters for a wider set of local DBpedia instances. For instance, we should identify the most appropriated method and parameters for syntactic similarity. A too restrictive similarity parameters would not provide much more data, and too relaxed parameters could produce wrong results. Concerning semantic similarity we have to find a similar balance. In both cases we have to test the results with real users by means of a testing tool. This tool will allow us to get the best parameters, for a given language, in order to provide the most similar dbp properties for a given dbo.

But this method is only an example of the utility of dbp properties. We claim dbp properties as first-class citizens, and linked data tools should allow users to exploit them. We show LOUPE as an exploring tool, which allows 'property exploration' for both, dbo and dbp, properties.

In summary, dbp properties are a good complement for dbo properties in SPARQL queries because they give us access to a richer DBpedia.

Acknowledgments. This work was funded by the JCI-2012-12719 contract, the BES-2014-068449 grant under the 4V project (TIN2013-46238-C4-2-R), JC2015-00028 and UNPM13-4E-1814.

References

1. Auer, S., Bizer, C., Kobilarov, G., Lehmann, J., Cyganiak, R., Ives, Z.G.: DBpedia: a nucleus for a Web of open data. In: Aberer, K., et al. (eds.) ASWC 2007 and ISWC 2007. LNCS, vol. 4825, pp. 722–735. Springer, Heidelberg (2007)
2. Lehmann, J., Isele, R., Jakob, M., Jentzsch, A., Kontokostas, D., Mendes, P.N., Hellmann, S., Morsey, M., van Kleef, P., Auer, S., et al.: DBpedia-a large-scale, multilingual knowledge base extracted from Wikipedia. Semant. Web **6**(2), 167–195 (2015)
3. Mihindukulasooriya, N., Rico, M., García-Castro, R., Gómez-Pérez, A.: An analysis of the quality issues of the properties available in the Spanish DBpedia. In: Puerta, J.M., Gámez, J.A., Dorronsoro, B., Barrenechea, E., Troncoso, A., Baruque, B., Galar, M. (eds.) CAEPIA 2015. LNCS, vol. 9422, pp. 198–209. Springer International Publishing, Cham (2015)
4. Saleem, M., Ali, M.I., Hogan, A., Mehmood, Q., Ngomo, A.-C.N.: LSQ: the linked SPARQL queries dataset. In: Arenas, M., et al. (eds.) ISWC 2015. LNCS, vol. 9367, pp. 261–269. Springer, Heidelberg (2015). doi:10.1007/978-3-319-25010-6_15
5. Mihindukulasooriya, N., Villalon, M.P., García-Castro, R., Gómez-Pérez, A.: Loupe-An Online Tool for Inspecting Datasets in the Linked Data Cloud (2015)
6. Rahm, E., Bernstein, P.A.: A survey of approaches to automatic schema matching. VLDB J. **10**(4), 334–350 (2001)
7. Shvaiko, P., Euzenat, J.: A survey of schema-based matching approaches. In: Spaccapietra, S. (ed.) Journal on Data Semantics IV. LNCS, vol. 3730, pp. 146–171. Springer, Heidelberg (2005). doi:10.1007/11603412_5
8. Rinser, D., Lange, D., Naumann, F.: Cross-lingual entity matching and infobox alignment in Wikipedia. Inf. Syst. **38**(6), 887–907 (2013)
9. Zhang, Z., Gentile, A.L., Blomqvist, E., Augenstein, I., Ciravegna, F.: Statistical knowledge patterns: identifying synonymous relations in large linked datasets. In: Alani, H., et al. (eds.) ISWC 2013, Part I. LNCS, vol. 8218, pp. 703–719. Springer, Heidelberg (2013)
10. Palmero Aprosio, A., Giuliano, C., Lavelli, A.: Towards an automatic creation of localized versions of DBpedia. In: Alani, H., et al. (eds.) ISWC 2013, Part I. LNCS, vol. 8218, pp. 494–509. Springer, Heidelberg (2013)

Author Index